FUNDAMENTAL IDENTITIES

$$\sec x = \frac{1}{\cos x} \qquad\qquad \csc x = \frac{1}{\sin x}$$

$$\tan x = \frac{\sin x}{\cos x} \qquad\qquad \cot x = \frac{1}{\tan x}$$

$$\sin^2 x + \cos^2 x = 1 \qquad 1 + \tan^2 x = \sec^2 x \qquad 1 + \cot^2 x = \csc^2 x$$

$$\sin(-x) = -\sin x \qquad \cos(-x) = \cos x \qquad \tan(-x) = -\tan x$$

COFUNCTION IDENTITIES

$$\sin\left(\frac{\pi}{2} - x\right) = \cos x \qquad\qquad \cos\left(\frac{\pi}{2} - x\right) = \sin x$$

$$\tan\left(\frac{\pi}{2} - x\right) = \cot x \qquad\qquad \cot\left(\frac{\pi}{2} - x\right) = \tan x$$

$$\sec\left(\frac{\pi}{2} - x\right) = \csc x \qquad\qquad \csc\left(\frac{\pi}{2} - x\right) = \sec x$$

REDUCTION IDENTITIES

$$\sin(x + \pi) = -\sin x \qquad\qquad \sin\left(x + \frac{\pi}{2}\right) = \cos x$$

$$\cos(x + \pi) = -\cos x \qquad\qquad \cos\left(x + \frac{\pi}{2}\right) = -\sin x$$

$$\tan(x + \pi) = \tan x \qquad\qquad \tan\left(x + \frac{\pi}{2}\right) = -\cot x$$

ADDITION AND SUBTRACTION FORMULAS

$$\sin(x + y) = \sin x \cos y + \cos x \sin y$$

$$\sin(x - y) = \sin x \cos y - \cos x \sin y$$

$$\cos(x + y) = \cos x \cos y - \sin x \sin y$$

$$\cos(x - y) = \cos x \cos y + \sin x \sin y$$

$$\tan(x + y) = \frac{\tan x + \tan y}{1 - \tan x \tan y} \qquad \tan(x - y) = \frac{\tan x - \tan y}{1 + \tan x \tan y}$$

DOUBLE-ANGLE FORMULAS

$$\sin 2x = 2 \sin x \cos x \qquad \cos 2x = \cos^2 x - \sin^2 x$$

$$= 2\cos^2 x - 1$$

$$\tan 2x = \frac{2 \tan x}{1 - \tan^2 x} \qquad\qquad = 1 - 2\sin^2 x$$

FORMULAS FOR REDUCING POWERS

$$\sin^2 x = \frac{1 - \cos 2x}{2} \qquad\qquad \cos^2 x = \frac{1 + \cos 2x}{2}$$

$$\tan^2 x = \frac{1 - \cos 2x}{1 + \cos 2x}$$

HALF-ANGLE FORMULAS

$$\sin\frac{u}{2} = \pm\sqrt{\frac{1 - \cos u}{2}} \qquad\qquad \cos\frac{u}{2} = \pm\sqrt{\frac{1 + \cos u}{2}}$$

$$\tan\frac{u}{2} = \frac{1 - \cos u}{\sin u} = \frac{\sin u}{1 + \cos u}$$

PRODUCT-TO-SUM AND SUM-TO-PRODUCT IDENTITIES

$$\sin u \cos v = \frac{1}{2}[\sin(u + v) + \sin(u - v)]$$

$$\cos u \sin v = \frac{1}{2}[\sin(u + v) - \sin(u - v)]$$

$$\cos u \cos v = \frac{1}{2}[\cos(u + v) + \cos(u - v)]$$

$$\sin u \sin v = \frac{1}{2}[\cos(u - v) - \cos(u + v)]$$

$$\sin x + \sin y = 2 \sin \frac{x + y}{2} \cos \frac{x - y}{2}$$

$$\sin x - \sin y = 2 \cos \frac{x + y}{2} \sin \frac{x - y}{2}$$

$$\cos x + \cos y = 2 \cos \frac{x + y}{2} \cos \frac{x - y}{2}$$

$$\cos x - \cos y = -2 \sin \frac{x + y}{2} \sin \frac{x - y}{2}$$

THE LAWS OF SINES AND COSINES

The Law of Sines

$$\frac{\sin A}{a} = \frac{\sin B}{b} = \frac{\sin C}{c}$$

The Law of Cosines

$$a^2 = b^2 + c^2 - 2bc \cos A$$

$$b^2 = a^2 + c^2 - 2ac \cos B$$

$$c^2 = a^2 + b^2 - 2ab \cos C$$

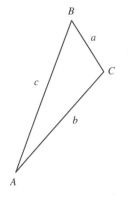

COMPLEX NUMBERS

For the complex number $z = a + bi$

the **conjugate** is $\bar{z} = a - bi$

the **modulus** is $|z| = \sqrt{a^2 + b^2}$

the **argument** is θ, where $\tan \theta = b/a$

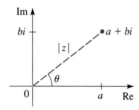

Polar form of a complex number

For $z = a + bi$, the **polar form** is

$$z = r(\cos \theta + i \sin \theta)$$

where $r = |z|$ is the modulus of z and θ is the argument of z

De Moivre's Theorem

$$z^n = [r(\cos \theta + i \sin \theta)]^n = r^n(\cos n\theta + i \sin n\theta)$$

$$\sqrt[n]{z} = [r(\cos \theta + i \sin \theta)]^{1/n}$$

$$= r^{1/n}\left(\cos \frac{\theta + 2k\pi}{n} + i \sin \frac{\theta + 2k\pi}{n}\right)$$

where $k = 0, 1, 2, \ldots, n - 1$

ROTATION OF AXES

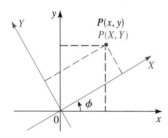

Rotation of axes formulas

$$x = X \cos \phi - Y \sin \phi$$
$$y = X \sin \phi + Y \cos \phi$$

Angle-of-rotation formula for conic sections

To eliminate the xy-term in the equation

$$Ax^2 + Bxy + Cy^2 + Dx + Ey + F = 0$$

rotate the axis by the angle ϕ that satisfies

$$\cot 2\phi = \frac{A - C}{B}$$

POLAR COORDINATES

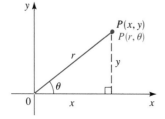

$$x = r \cos \theta$$
$$y = r \sin \theta$$
$$r^2 = x^2 + y^2$$
$$\tan \theta = \frac{y}{x}$$

POLAR EQUATIONS OF CONICS

The graph of a polar equation of the form

$$r = \frac{ed}{1 \pm e \cos \theta} \qquad \text{or} \qquad r = \frac{ed}{1 \pm e \sin \theta}$$

is a conic with eccentricity e and with one focus at the origin. The conic is

1. a parabola if $e = 1$.
2. an ellipse if $0 < e < 1$.
3. a hyperbola if $e > 1$.

HARMONIC MOTION

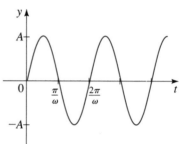

$$y = A \sin \omega t$$

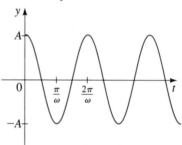

$$y = A \cos \omega t$$

amplitude: A

period: $p = \frac{2\pi}{\omega}$

frequency: $f = \frac{1}{p} = \frac{\omega}{2\pi}$

DAMPED HARMONIC MOTION

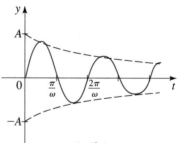

$$y = Ae^{-ct} \sin \omega t$$

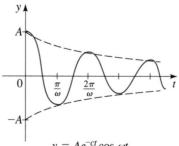

$$y = Ae^{-ct} \cos \omega t$$

damping constant: c

TRIGONOMETRY

ABOUT THE AUTHORS

JAMES STEWART received his MS from Stanford University and his PhD from the University of Toronto. He did research at the University of London and was influenced by the famous mathematician George Polya at Stanford University. Stewart is Professor Emeritus at McMaster University and is currently Professor of Mathematics at the University of Toronto. His research field is harmonic analysis and the connections between mathematics and music. James Stewart is the author of a best-selling calculus textbook series published by Brooks/Cole, Cengage Learning, including *Calculus, Calculus: Early Transcendentals*, and *Calculus: Concepts and Contexts*; a series of precalculus texts; and a series of high-school mathematics textbooks.

LOTHAR REDLIN grew up on Vancouver Island, received a Bachelor of Science degree from the University of Victoria, and received a PhD from McMaster University in 1978. He subsequently did research and taught at the University of Washington, the University of Waterloo, and California State University, Long Beach. He is currently Professor of Mathematics at The Pennsylvania State University, Abington Campus. His research field is topology.

SALEEM WATSON received his Bachelor of Science degree from Andrews University in Michigan. He did graduate studies at Dalhousie University and McMaster University, where he received his PhD in 1978. He subsequently did research at the Mathematics Institute of the University of Warsaw in Poland. He also taught at The Pennsylvania State University. He is currently Professor of Mathematics at California State University, Long Beach. His research field is functional analysis.

Stewart, Redlin, and Watson have also published *Precalculus: Mathematics for Calculus, Algebra and Trigonometry; College Algebra;* and (with Phyllis Panman) *College Algebra: Concepts and Contexts*.

ABOUT THE COVER

The cover photograph shows the La Rioja Bodegas Ysios winery complex in Laguardia, Spain. Its undulating roof in the shape of a sine curve creates a strong wave-like profile. Built from 1998 to 2001, it was designed by Santiago Calatrava, a Spanish architect. Calatrava has always been very interested in how mathematics can help him realize the buildings he imagines. As a young student, he taught himself descriptive geometry from books to represent three-dimensional objects in two dimensions. Trained as both an engineer and an architect, he wrote a doctoral thesis in 1981 entitled "On the Foldability of Space Frames," which is filled with mathematics, especially geometric transformations. His strength as an engineer enables him to be daring in his architecture.

SECOND EDITION

TRIGONOMETRY

JAMES STEWART
McMASTER UNIVERSITY AND UNIVERSITY OF TORONTO

LOTHAR REDLIN
THE PENNSYLVANIA STATE UNIVERSITY

SALEEM WATSON
CALIFORNIA STATE UNIVERSITY, LONG BEACH

With the assistance of Phyllis Panman

BROOKS/COLE
CENGAGE Learning

Australia • Brazil • Japan • Korea • Mexico • Singapore • Spain • United Kingdom • United States

BROOKS/COLE
CENGAGE Learning·

***Trigonometry,* Second Edition**
James Stewart, Lothar Redlin, Saleem Watson

Acquisitions Editor: Gary Whalen

Developmental Editor: Stacy Green

Assistant Editor: Cynthia Ashton

Editorial Assistant: Sabrina Black

Media Editor: Lynh Pham

Marketing Manager: Mandee Eckersley

Marketing Assistant: Shannon Maier

Marketing Communications Manager:
Darlene Macanan

Content Project Manager: Jennifer Risden

Design Director: Rob Hugel

Art Director: Vernon Boes

Print Buyer: Karen Hunt

Rights Acquisitions Specialist:
Roberta Broyer

Production Service: Graphic World Inc.

Text Designer: Lisa Henry

Photo Researcher: Bill Smith Group

Copy Editor: Barbara Willette

Illustrators: Matrix Art Services,
Precision Graphics

Cover Designer: Lisa Henry

Cover Image: Cephas Picture Library / Alamy

Compositor: Graphic World Inc.

For product information and technology assistance, contact us at
Cengage Learning Customer & Sales Support, 1-800-354-9706
For permission to use material from this text or product,
submit all requests online at **www.cengage.com/permissions**
Further permissions questions can be e-mailed to
permissionrequest@cengage.com

Library of Congress Control Number: 2011932671

ISBN-13: 978-1-111-57448-2

ISBN-10: 1-111-57448-0

Brooks/Cole
20 Davis Drive
Belmont, CA 94002-3098
USA

Cengage Learning is a leading provider of customized learning solutions with office locations around the globe, including Singapore, the United Kingdom, Australia, Mexico, Brazil, and Japan. Locate your local office at **www.cengage.com/global**

Cengage Learning products are represented in Canada by Nelson Education, Ltd.

To learn more about Brooks/Cole, visit **www.cengage.com/brookscole**

Purchase any of our products at your local college store or at our preferred online store **www.cengagebrain.com**

Printed in the United States of America
1 2 3 4 5 6 7 15 14 13 12 11

CONTENTS

PREFACE

In this second edition of *Trigonometry* we have made several changes designed to enhance the utility of the book as an instructional tool for teachers and as a learning tool for students. But we have retained the overall structure and the main features that have contributed to the success of this book. In particular, in this edition, as in the first, the two approaches to defining the trigonometric functions—the unit circle approach and the right triangle approach—may be taught in either order. The parallel presentation of these approaches, each with its own chapter and with its relevant applications, underscores the purpose of defining the trigonometric functions in these different ways. Another way to teach trigonometry is to intertwine the two approaches. Our organization makes it easy to do this without obscuring the fact that the different approaches involve distinct representations of the same functions.

The changes in this edition include a restructuring of the exercise sets to better align the exercises with the examples of each section. Each exercise set now begins with *Concepts* exercises, which encourage students to work with basic concepts and to use mathematical vocabulary appropriately. Several chapters have been reorganized and rewritten (as described in the following) and we have added a new chapter on vectors in two and three dimensions.

Trigonometry has many interesting and powerful applications. We have integrated many applications in the examples and exercise sets. In the *Focus on Modeling* sections following each chapter we highlight applications of trigonometry in modeling real-world situations. In addition, the *Mathematics in the Modern World* vignettes describe several modern applications of mathematics.

Many of the changes in this edition have been drawn from our own experience in teaching, but, more importantly, we have listened carefully to the users of the current edition. We are grateful for the many letters and e-mails we have received from users of this book, instructors as well as students, recommending changes and suggesting additions. Many of these have helped greatly in making this edition even more user-friendly.

New to the Second Edition

- **Exercise Sets** More than 20% of the exercises are new. This includes new *Concepts* Exercises for each section. Key exercises are now linked to examples in the text.

- **Book Companion Website** A new website, **www.stewartmath.com,** contains *Discovery Projects* for each chapter and *Focus on Problem Solving* sections that highlight different problem-solving principles outlined in the Prologue.

- **Chapter 1 Fundamentals** This chapter has been rewritten to focus more sharply on the fundamental and crucial concept of *function*. A new section entitled "Getting

Information from the Graph of a Function" encourages students to think of the graph of a function as a tool for understanding the behavior of the function.

- **Chapter 2 Trigonometric Functions: Unit Circle Approach** This chapter includes a new section on inverse trigonometric functions and their graphs (Section 2.5). Introducing this topic here reinforces the function concept in the context of trigonometry.

- **Chapter 3 Trigonometric Functions: Right Triangle Approach** This chapter includes a new section on inverse trigonometric functions and right triangles (Section 3.4), which is needed in applying the Laws of Sines and Cosines in the following section, as well as for solving trigonometric equations in Chapter 4.

- **Chapter 4 Analytic Trigonometry** This chapter has been completely revised. There are two new sections on trigonometric equations (Sections 4.4 and 4.5). The material on this topic has been expanded and revised.

- **Chapter 5 Polar Coordinates and Parametric Equations** This chapter is now more sharply focused on the concept of a coordinate system. The section on plane curves and parametric equations (Section 5.4) is new to this chapter. The material on vectors is now in its own chapter.

- **Chapter 6 Vectors in Two and Three Dimensions** This is a new chapter with a new *Focus on Modeling* section.

- **Chapter 7 Conic Sections** This chapter is now more closely devoted to the topic of analytic geometry, especially the conic sections; the section on parametric equations has been moved to Chapter 5.

- **Chapter 8 Exponential and Logarithmic Functions** The material on the natural exponential function is now in a separate section.

Teaching with the Help of This Book

We are keenly aware that good teaching comes in many forms and that there are many different approaches to teaching the concepts and skills of trigonometry. The organization of the topics in this book is designed to accommodate different teaching styles. The trigonometry chapters have been organized so that either the unit circle approach or the right triangle approach can be taught first. Here are other special features that can be used to complement different teaching styles.

EXERCISE SETS The most important way to foster conceptual understanding and hone technical skill is through the problems that the instructor assigns. To that end we have provided a wide selection of exercises.

- **Concepts Exercises** These exercises ask students to use mathematical language to state fundamental facts about the topics of each section.

- **Skills Exercises** Each exercise set is carefully graded, progressing from basic skill-development exercises to more challenging problems requiring synthesis of previously learned material with new concepts.

- **Applications Exercises** We have included substantial applied problems that we believe will capture the interest of students.

- **Discovery, Writing, and Group Learning** Each exercise set ends with a block of exercises labeled *Discovery* ■ *Discussion* ■ *Writing*. These exercises are designed to encourage students to experiment, preferably in groups, with the concepts developed in the section, and then to write about what they have learned, rather than simply look for "the answer."

- **Now Try Exercise ...** At the end of each example in the text the student is directed to a similar exercise in the section that helps reinforce the concepts and skills developed in that example (see, for instance, page 5).

- **Check Your Answer** Students are encouraged to check whether an answer they obtained is reasonable. This is emphasized in *Check Your Answer* sidebars that accompany the examples. (See, for instance, pages 80 and 448.)
- **Answers** Brief answers to odd-numbered exercises in each section (including the review exercises), and to all questions in the Concepts Exercises and Chapter Tests, are given in the back of the book.

REVIEW SECTIONS AND CHAPTER TESTS Each chapter ends with an extensive review section including the following:

- **Concept Check** The *Concept Check* at the end of each chapter is designed to get the students to think about and explain in their own words the ideas presented in the chapter. These can be used as writing exercises, in a classroom discussion setting, or for personal study.
- **Review Exercises** The *Review Exercises* at the end of each chapter recapitulate the basic concepts and skills of the chapter and include exercises that combine the different ideas learned in the chapter.
- **Chapter Test** The review sections conclude with a *Chapter Test* designed to help students gauge their progress.

FLEXIBLE APPROACH TO TRIGONOMETRY The chapters of this text have been written so that either the right triangle approach or the unit circle approach may be taught first. Putting these two approaches in different chapters, each with its relevant applications, helps to clarify the purpose of each approach. The chapters introducing trigonometry are as follows:

- **Chapter 2 Trigonometric Functions: Unit Circle Approach** This chapter introduces trigonometry through the unit circle approach. This approach emphasizes that the trigonometric functions are functions of real numbers, just like the polynomial and exponential functions with which students are already familiar.
- **Chapter 3 Trigonometric Functions: Right Triangle Approach** This chapter introduces trigonometry through the right triangle approach. This approach builds on the foundation of a conventional high-school course in trigonometry.

Another way to teach trigonometry is to intertwine the two approaches. Some instructors teach this material in the following order: Sections 2.1, 2.2, 3.1, 3.2, 3.3, 2.3, 2.4, 2.5, 2.6, 3.4, 3.5, and 3.6. Our organization makes it easy to do this without obscuring the fact that the two approaches involve distinct representations of the same functions.

GRAPHING CALCULATORS We make use of graphing calculators and computers in examples and exercises throughout the book. Our calculator-oriented examples are always preceded by examples in which students must graph or calculate by hand, so that they can understand precisely what the calculator is doing when they later use it to simplify the routine, mechanical part of their work. The graphing calculator sections, subsections, examples, and exercises, all marked with the special symbol 🖩, are optional and may be omitted without loss of continuity.

FOCUS ON MODELING The "modeling" theme has been used throughout to unify and clarify the many applications of trigonometry.

- **Constructing Models** There are numerous applied problems throughout the book where students are given a model to analyze (see, for instance, pages 148, 200, and 427). But the material on modeling, in which students are required to *construct* mathematical models, has been organized into clearly defined sections and subsections (see, for example, pages 146 and 456).
- **Focus on Modeling** Each chapter concludes with a *Focus on Modeling* section. The first such section, after Chapter 1, introduces the basic idea of modeling a real-

life situation by fitting lines to data (linear regression). Other sections present ways in which trigonometric functions can be used to model familiar phenomena from the sciences and from everyday life (see, for example, pages 159, 305, and 412).

MATHEMATICAL VIGNETTES Throughout the book we make use of the margins to provide historical notes, key insights, or applications of mathematics in the modern world. These serve to enliven the material and show that mathematics is an important, vital activity, and that even at this elementary level it is fundamental to everyday life.

- **Mathematical Vignettes** These vignettes include biographies of interesting mathematicians and often include a key insight that the mathematician discovered and which is relevant to trigonometry. (See, for instance, the vignettes on Fourier, page 233; Agnesi, page 295; and Paths of Comets, page 381).
- **Mathematics in the Modern World** This is a series of vignettes that emphasizes the central role of mathematics in current advances in technology and the sciences (see pages 33, 134, and 331, for example).

BOOK COMPANION WEBSITE A website that accompanies this book is located at **www.stewartmath.com**. The site includes many useful resources for teaching trigonometry, including the following:

- **Discovery Projects** Discovery Projects for each chapter are available on the website. Each project provides a challenging but accessible set of activities that enable students (perhaps working in groups) to explore in greater depth an interesting aspect of the topic they have just learned. (See, for instance, the Discovery Projects *Predator/Prey Models, Similarity,* or *Where to Sit at the Movies, and Fractals.*)
- **Focus on Problem Solving** Several *Focus on Problem Solving* sections are available on the website. Each such section highlights one of the problem-solving principles introduced in the Prologue and includes several challenging problems. (See, for instance, *Recognizing Patterns, Using Analogy, Introducing Something Extra, Taking Cases,* and *Working Backward.*)

PREREQUISITE MATERIAL We have included several review sections as appendices, primarily as a handy reference for the basic concepts that are preliminary to this course. As much or as little of these appendices can be used, depending on the background of the students.

- **Appendix A Algebra Review** The appendix contains the fundamental concepts from algebra that a student needs to begin a trigonometry course.
- **Appendix B Geometry Review** This appendix contains fundamental results from geometry that are used in the text.
- **Appendix C Graphing Calculators** This appendix gives general guidelines for using a graphing calculator.
- **Appendix D Complex Numbers** This appendix reviews the basic algebraic operations involving complex numbers.

Acknowledgments

We thank the following reviewers for their thoughtful and constructive comments.

REVIEWERS FOR THE FIRST EDITION Michelle Benedict, Augusta State University; Linda Crawford, Augusta State University; Edward Dixon, Tennessee Technological University; Richard Dodge, Jackson Community College; Floyd Downs, Arizona State University at Tempe; Vivian G. Kostyk, Inver Hills Community College; Marjorie Kreienbrink, University of Wisconsin–Waukesha; Donna Krichiver, Johnson Community College; Wayne Lewis, University of Hawaii and Honolulu Community College; Adam Lutoborski, Syracuse University; Heather C. McGilvray, Seattle University; Keith Oberlander, Pasadena City College; Christine Panoff, University of Michigan at Flint; Susan Piliero, Cornell

University; Gregory St. George, University of Montana; Gary Stoudt, Indiana University of Pennsylvania; Arnold Vobach, University of Houston; Tom Walsh, City College of San Francisco; Muserref Wiggins, University of Akron; Diane Williams, Northern Kentucky University; Suzette Wright, University of California at Davis; and Yison Yang, Polytech University.

REVIEWERS FOR THE PRECALCULUS SERIES Raji Baradwaj, UMBC; Chris Herman, Lorain County Community College; Irina Kloumova, Sacramento City College; Jim McCleery, Skagit Valley College, Whidbey Island Campus; Sally S. Shao, Cleveland State University; David Slutzky, Gainesville State College; Edward Stumpf, Central Carolina Community College; Ricardo Teixeira, University of Texas at Austin; Anna Wlodarczyk, Florida International University, and Taixi Xu, Southern Polytechnic State University.

We are grateful to our colleagues who continually share with us their insights into teaching mathematics. We especially thank Andrew Bulman-Fleming for writing the Solutions Manual and Doug Shaw at the University of Northern Iowa for writing the Instructor Guide and the Study Guide.

We thank Kate Mannix, our production service, for her steady and dependable guidance in bringing this book to completion. We thank Martha Emry for the energy, devotion, and experience she has provided to this series. We thank Barbara Willette, our copy editor, for her attention to every detail in the manuscript. We thank Jade Meyers and his staff at Matrix for their attractive and accurate graphs and Network Graphics for bringing many of our illustrations to life. We thank our designer Lisa Henry for the elegant and appropriate design for the interior of the book.

At Brooks/Cole we especially thank Stacy Green, developmental editor, for guiding and facilitating every aspect of the production of this book. Of the many Brooks/Cole staff involved in this project we particularly thank the following: Jennifer Risden, project content manager; Cynthia Ashton, assistant editor; Lynh Pham, media editor; and Vernon Boes, art director. They have all done an outstanding job.

Numerous other people were involved in the production of this book—including permissions editors, photo researchers, text designers, compositors, proof readers, printers, and many more. We thank them all.

Above all, we thank our editor Gary Whalen. His vast editorial experience, his extensive knowledge of current issues in the teaching of mathematics, and especially his deep interest in mathematics textbooks have been invaluable resources in the writing of this book.

INSTRUCTOR RESOURCES

Printed

Complete Solution Manual
ISBN-10: 1-133-11315-X; ISBN-13: 978-1-133-11315-7
The complete solutions manual provides worked-out solutions to all of the problems in the text.

Instructor's Guide   ISBN-10: 1-111-57476-6; ISBN-13: 978-1-111-57476-5
Doug Shaw, author of the Instructor Guides for the widely used Stewart calculus texts, wrote this helpful teaching companion. It contains points to stress, suggested time to allot, text discussion topics, core materials for lectures, workshop/discussion suggestions, group work exercises in a form suitable for handout, solutions to group work exercises, and suggested homework problems.

Media

Enhanced WebAssign   ISBN-10: 0-538-73810-3; ISBN-13: 978-0-538-73810-1
Exclusively from Cengage Learning, Enhanced WebAssign® offers an extensive online program for *Trigonometry* to encourage the practice that's so critical for concept mastery. The meticulously crafted pedagogy and exercises in this text become even more effective in Enhanced WebAssign, supplemented by multimedia tutorial support and immediate feedback as students complete their assignments. Algorithmic problems allow you to assign unique versions to each student. The Practice Another Version feature (activated at your discretion) allows students to attempt the questions with new sets of values until they feel confident enough to work the original problem. Students benefit from a new Premium eBook with highlighting and search features; Personal Study Plans (based on diagnostic quizzing) that identify chapter topics they still need to master; and links to video solutions, interactive tutorials, and even live online help.

ExamView Computerized Testing
ExamView® testing software allows instructors to quickly create, deliver, and customize tests for class in print and online formats, and features automatic grading. Includes a test bank with hundreds of questions customized directly to the text. ExamView is available within the PowerLecture CD-ROM.

Solution Builder   www.cengage.com/solutionbuilder
This online instructor database offers complete worked solutions to all exercises in the text, allowing you to create customized, secure solutions printouts (in PDF format) matched exactly to the problems you assign in class.

PowerLecture with ExamView
ISBN-10: 1-111-98995-8; ISBN-13: 978-1-111-98995-8
This CD-ROM provides the instructor with dynamic media tools for teaching. Create, deliver, and customize tests (both print and online) in minutes with ExamView® Computerized Testing Featuring Algorithmic Equations. Easily build solution sets for homework or exams using Solution Builder's online solutions manual. Microsoft® PowerPoint® lecture slides and figures from the book are also included on this CD-ROM.

STUDENT RESOURCES

Printed

Student Solution Manual
ISBN-10: 1-133-10352-9; ISBN-13: 978-1-133-10352-3
Contains fully worked-out solutions to all of the odd-numbered exercises in the text, giving students a way to check their answers and ensure that they took the correct steps to arrive at an answer.

Study Guide   ISBN-10: 1-111-98985-0; ISBN-13: 978-1-111-98985-9
This carefully crafted learning resource helps students develop their problem-solving skills while reinforcing their understanding with detailed explanations, worked-out examples, and practice problems. Students will also find listings of key ideas to master. Each section of the main text has a corresponding section in the Study Guide.

Media

Enhanced WebAssign   ISBN-10: 0-538-73810-3; ISBN-13: 978-0-538-73810-1
Exclusively from Cengage Learning, Enhanced WebAssign® offers an extensive online program for *Trigonometry* to encourage the practice that's so critical for concept mastery. You'll receive multimedia tutorial support as you complete your assignments. You'll also benefit from a new Premium eBook with highlighting and search features; Personal Study Plans (based on diagnostic quizzing) that identify chapter topics you still need to master; and links to video solutions, interactive tutorials, and even live online help.

Book Companion Website
A new website, **www.stewartmath.com,** contains *Discovery Projects* for each chapter and *Focus on Problem Solving* sections that highlight different problem-solving principles outlined in the Prologue.

CengageBrain.com
Visit **www.cengagebrain.com** to access additional course materials and companion resources. At the CengageBrain.com home page, search for the ISBN of your title (from the back cover of your book) using the search box at the top of the page. This will take you to the product page where free companion resources can be found.

Text-Specific DVDs   ISBN-10: 1-111-98984-2; ISBN-13: 978-1-111-98984-2
The text-specific DVDs include new learning objective based lecture videos. These DVDs provide comprehensive coverage of the course—along with additional explanations of concepts, sample problems, and applications—to help students review essential topics.

This textbook was written for you to use as a guide to mastering trigonometry. Here are some suggestions to help you get the most out of your course.

First of all, you should read the appropriate section of text *before* you attempt your homework problems. Reading a mathematics text is quite different from reading a novel, a newspaper, or even another textbook. You may find that you have to reread a passage several times before you understand it. Pay special attention to the examples, and work them out yourself with pencil and paper as you read. Then do the linked exercises referred to in *"Now Try Exercise . . ."* at the end of each example. With this kind of preparation you will be able to do your homework much more quickly and with more understanding.

Don't make the mistake of trying to memorize every single rule or fact you may come across. Mathematics doesn't consist simply of memorization. Mathematics is a *problem-solving art*, not just a collection of facts. To master the subject you must solve problems— lots of problems. Do as many of the exercises as you can. Be sure to write your solutions in a logical, step-by-step fashion. Don't give up on a problem if you can't solve it right away. Try to understand the problem more clearly—reread it thoughtfully and relate it to what you have learned from your teacher and from the examples in the text. Struggle with it until you solve it. Once you have done this a few times you will begin to understand what mathematics is really all about.

Answers to the odd-numbered exercises, as well as all the answers to each chapter test, appear at the back of the book. If your answer differs from the one given, don't immediately assume that you are wrong. There may be a calculation that connects the two answers and makes both correct. For example, if you get $1/(\sqrt{2} - 1)$ but the answer given is $1 + \sqrt{2}$, your answer *is* correct, because you can multiply both numerator and denominator of your answer by $\sqrt{2} + 1$ to change it to the given answer.

The symbol ⊘ is used to warn against committing an error. We have placed this symbol in the margin to point out situations where we have found that many of our students make the same mistake.

ABBREVIATIONS

cm	centimeter		**mg**	milligram
dB	decibel		**MHz**	megahertz
F	farad		**mi**	mile
ft	foot		**min**	minute
g	gram		**mL**	milliliter
gal	gallon		**mm**	millimeter
h	hour		**N**	Newton
H	henry		**qt**	quart
Hz	Hertz		**oz**	ounce
in.	inch		**s**	second
J	Joule		**Ω**	ohm
kcal	kilocalorie		**V**	volt
kg	kilogram		**W**	watt
km	kilometer		**yd**	yard
kPa	kilopascal		**yr**	year
L	liter		**°C**	degree Celsius
lb	pound		**°F**	degree Fahrenheit
lm	lumen		**K**	Kelvin
m	meter		$\Rightarrow$	implies
M	mole of solute per liter of solution		$\Leftrightarrow$	is equivalent to

MATHEMATICAL VIGNETTES

MATHEMATICS IN THE MODERN WORLD

Chuck Painter/Stanford News Service

GEORGE POLYA (1887–1985) is famous among mathematicians for his ideas on problem solving. His lectures on problem solving at Stanford University attracted overflow crowds whom he held on the edges of their seats, leading them to discover solutions for themselves. He was able to do this because of his deep insight into the psychology of problem solving. His well-known book *How To Solve It* has been translated into 15 languages. He said that Euler (see page 541) was unique among great mathematicians because he explained how he found his results. Polya often said to his students and colleagues, "Yes, I see that your proof is correct, but how did you discover it?" In the preface to *How To Solve It,* Polya writes, "A great discovery solves a great problem but there is a grain of discovery in the solution of any problem. Your problem may be modest; but if it challenges your curiosity and brings into play your inventive faculties, and if you solve it by your own means, you may experience the tension and enjoy the triumph of discovery."

The ability to solve problems is a highly prized skill in many aspects of our lives; it is certainly an important part of any mathematics course. There are no hard-and-fast rules that will ensure success in solving problems. However, in this Prologue we outline some general steps in the problem-solving process and we give principles that are useful in solving certain problems. These steps and principles are just common sense made explicit. They have been adapted from George Polya's insightful book *How To Solve It*.

1. Understand the Problem

The first step is to read the problem and make sure that you understand it. Ask yourself the following questions:

> *What is the unknown?*
> *What are the given quantities?*
> *What are the given conditions?*

For many problems it is useful to

> *draw a diagram*

and identify the given and required quantities on the diagram. Usually, it is necessary to

> *introduce suitable notation*

In choosing symbols for the unknown quantities, we often use letters such as a, b, c, m, n, x, and y, but in some cases it helps to use initials as suggestive symbols, for instance, V for volume or t for time.

2. Think of a Plan

Find a connection between the given information and the unknown that enables you to calculate the unknown. It often helps to ask yourself explicitly: "How can I relate the given to the unknown?" If you don't see a connection immediately, the following ideas may be helpful in devising a plan.

▶ Try to Recognize Something Familiar

Relate the given situation to previous knowledge. Look at the unknown and try to recall a more familiar problem that has a similar unknown.

▶ Try to Recognize Patterns

Certain problems are solved by recognizing that some kind of pattern is occurring. The pattern could be geometric, numerical, or algebraic. If you can see regularity or repetition in a problem, then you might be able to guess what the pattern is and then prove it.

▶ Use Analogy

Try to think of an analogous problem, that is, a similar or related problem but one that is easier than the original. If you can solve the similar, simpler problem, then it might give you the clues you need to solve the original, more difficult one. For instance, if a problem involves very large numbers, you could first try a similar problem with smaller numbers. Or if the problem is in three-dimensional geometry, you could look for something similar in two-dimensional geometry. Or if the problem you start with is a general one, you could first try a special case.

▶ Introduce Something Extra

You might sometimes need to introduce something new—an auxiliary aid—to make the connection between the given and the unknown. For instance, in a problem for which a diagram is useful, the auxiliary aid could be a new line drawn in the diagram. In a more algebraic problem the aid could be a new unknown that relates to the original unknown.

▶ Take Cases

You might sometimes have to split a problem into several cases and give a different argument for each case. For instance, we often have to use this strategy in dealing with absolute value.

▶ Work Backward

Sometimes it is useful to imagine that your problem is solved and work backward, step by step, until you arrive at the given data. Then you might be able to reverse your steps and thereby construct a solution to the original problem. This procedure is commonly used in solving equations. For instance, in solving the equation $3x - 5 = 7$, we suppose that x is a number that satisfies $3x - 5 = 7$ and work backward. We add 5 to each side of the equation and then divide each side by 3 to get $x = 4$. Since each of these steps can be reversed, we have solved the problem.

▶ Establish Subgoals

In a complex problem it is often useful to set subgoals (in which the desired situation is only partially fulfilled). If you can attain or accomplish these subgoals, then you might be able to build on them to reach your final goal.

▶ Indirect Reasoning

Sometimes it is appropriate to attack a problem indirectly. In using **proof by contradiction** to prove that P implies Q, we assume that P is true and Q is false and try to see why this cannot happen. Somehow we have to use this information and arrive at a contradiction to what we absolutely know is true.

▶ Mathematical Induction

In proving statements that involve a positive integer n, it is frequently helpful to use the Principle of Mathematical Induction. (See for example, *Precalculus 6e* by Stewart, Redlin, and Watson, Section 12.5.)

3. Carry Out the Plan

In Step 2, a plan was devised. In carrying out that plan, you must check each stage of the plan and write the details that prove that each stage is correct.

4. Look Back

Having completed your solution, it is wise to look back over it, partly to see whether any errors have been made and partly to see whether you can discover an easier way to solve the problem. Looking back also familiarizes you with the method of solution, which may be useful for solving a future problem. Descartes said, "Every problem that I solved became a rule which served afterwards to solve other problems."

We illustrate some of these principles of problem solving with an example.

PROBLEM | Average Speed

A driver sets out on a journey. For the first half of the distance, she drives at the leisurely pace of 30 mi/h; during the second half she drives 60 mi/h. What is her average speed on this trip?

THINKING ABOUT THE PROBLEM

It is tempting to take the average of the speeds and say that the average speed for the entire trip is

$$\frac{30 + 60}{2} = 45 \text{ mi/h}$$

But is this simple-minded approach really correct?

Try a special case ▶   Let's look at an easily calculated special case. Suppose that the total distance traveled is 120 mi. Since the first 60 mi is traveled at 30 mi/h, it takes 2 h. The second 60 mi is traveled at 60 mi/h, so it takes one hour. Thus, the total time is $2 + 1 = 3$ hours and the average speed is

$$\frac{120}{3} = 40 \text{ mi/h}$$

So our guess of 45 mi/h was wrong.

SOLUTION

Understand the problem ▶   We need to look more carefully at the meaning of average speed. It is defined as

$$\text{average speed} = \frac{\text{distance traveled}}{\text{time elapsed}}$$

Introduce notation ▶   Let d be the distance traveled on each half of the trip. Let t_1 and t_2 be the times taken for the first and second halves of the trip. Now we can write down the information we have **State what is given ▶** been given. For the first half of the trip we have

$$30 = \frac{d}{t_1}$$

and for the second half we have

$$60 = \frac{d}{t_2}$$

Identify the unknown ▶   Now we identify the quantity that we are asked to find:

$$\text{average speed for entire trip} = \frac{\text{total distance}}{\text{total time}} = \frac{2d}{t_1 + t_2}$$

Connect the given with the unknown ▶   To calculate this quantity, we need to know t_1 and t_2, so we solve the above equations for these times:

$$t_1 = \frac{d}{30} \qquad t_2 = \frac{d}{60}$$

Don't feel bad if you can't solve these problems right away. Problems 1 and 4 were sent to Albert Einstein by his friend Wertheimer. Einstein (and his friend Bucky) enjoyed the problems and wrote back to Wertheimer. Here is part of his reply:

Your letter gave us a lot of amusement. The first intelligence test fooled both of us (Bucky and me). Only on working it out did I notice that no time is available for the downhill run! Mr. Bucky was also taken in by the second example, but I was not. Such drolleries show us how stupid we are!

(See *Mathematical Intelligencer,* Spring 1990, page 41.)

Now we have the ingredients needed to calculate the desired quantity:

$$\text{average speed} = \frac{2d}{t_1 + t_1} = \frac{2d}{\dfrac{d}{30} + \dfrac{d}{60}}$$

$$= \frac{60(2d)}{60\left(\dfrac{d}{30} + \dfrac{d}{60}\right)} \qquad \text{Multiply numerator and denominator by 60}$$

$$= \frac{120d}{2d + d} = \frac{120d}{3d} = 40$$

So the average speed for the entire trip is 40 mi/h. ■

PROBLEMS

1. **Distance, Time, and Speed**   An old car has to travel a 2-mile route, uphill and down. Because it is so old, the car can climb the first mile—the ascent—no faster than an average speed of 15 mi/h. How fast does the car have to travel the second mile—on the descent it can go faster, of course—to achieve an average speed of 30 mi/h for the trip?

2. **Comparing Discounts**   Which price is better for the buyer, a 40% discount or two successive discounts of 20%?

3. **Cutting Up a Wire**   A piece of wire is bent as shown in the figure. You can see that one cut through the wire produces four pieces and two parallel cuts produce seven pieces. How many pieces will be produced by 142 parallel cuts? Write a formula for the number of pieces produced by n parallel cuts.

4. **Amoeba Propagation**   An amoeba propagates by simple division; each split takes 3 minutes to complete. When such an amoeba is put into a glass container with a nutrient fluid, the container is full of amoebas in one hour. How long would it take for the container to be filled if we start with not one amoeba, but two?

5. **Batting Averages**   Player A has a higher batting average than player B for the first half of the baseball season. Player A also has a higher batting average than player B for the second half of the season. Is it necessarily true that player A has a higher batting average than player B for the entire season?

6. **Coffee and Cream**   A spoonful of cream is taken from a pitcher of cream and put into a cup of coffee. The coffee is stirred. Then a spoonful of this mixture is put into the pitcher of cream. Is there now more cream in the coffee cup or more coffee in the pitcher of cream?

7. **Wrapping the World**   A ribbon is tied tightly around the earth at the equator. How much more ribbon would you need if you raised the ribbon 1 ft above the equator everywhere? (You don't need to know the radius of the earth to solve this problem.)

8. **Ending Up Where You Started**   A woman starts at a point P on the earth's surface and walks 1 mi south, then 1 mi east, then 1 mi north, and finds herself back at P, the starting point. Describe all points P for which this is possible.   [*Hint:* There are infinitely many such points, all but one of which lie in Antarctica.]

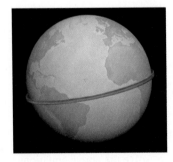

Many more problems and examples that highlight different problem-solving principles are available at the book companion website: **www.stewartmath.com**. You can try them as you progress through the book.

CHAPTER 1

FUNDAMENTALS

The main topics of this textbook are the trigonometric *functions* and their use in modeling real-world phenomena. So in this first chapter we review the concept of function. You are probably already familiar with functions and how to graph a function in a coordinate plane, but it's helpful to take another look at this all-important topic. We study transformations of functions, inverse functions, and how to obtain information from the graph of a function. In particular, in this chapter we investigate properties of linear functions—functions whose graphs are lines. By contrast, we'll see in subsequent chapters that the trigonometric functions have graphs that rise and fall indefinitely—like the motions of the tides or of a vibrating leaf.

1.1 COORDINATE GEOMETRY

The Coordinate Plane ▶ The Distance and Midpoint Formulas ▶ Graphs of Equations in Two Variables ▶ Intercepts ▶ Circles ▶ Symmetry

The *coordinate plane* is the link between algebra and geometry. In the coordinate plane we can draw graphs of algebraic equations. The graphs, in turn, allow us to "see" the relationship between the variables in the equation. In this section we study the coordinate plane.

▼ The Coordinate Plane

The Cartesian plane is named in honor of the French mathematician René Descartes (1596–1650), although another Frenchman, Pierre Fermat (1601–1665), also invented the principles of coordinate geometry at the same time. (See their biographies on pages 59 and 241.)

Just as points on a line can be identified with real numbers to form the coordinate line, points in a plane can be identified with ordered pairs of numbers to form the **coordinate plane** or **Cartesian plane**. To do this, we draw two perpendicular real lines that intersect at 0 on each line. Usually, one line is horizontal with positive direction to the right and is called the **x-axis**; the other line is vertical with positive direction upward and is called the **y-axis**. The point of intersection of the x-axis and the y-axis is the **origin O**, and the two axes divide the plane into four **quadrants**, labeled I, II, III, and IV in Figure 1. (The points *on* the coordinate axes are not assigned to any quadrant.)

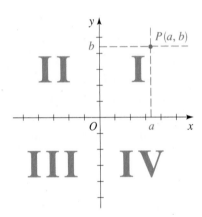

FIGURE 1   **FIGURE 2**

Although the notation for a point (a, b) is the same as the notation for an open interval (a, b), the context should make clear which meaning is intended.

Any point P in the coordinate plane can be located by a unique **ordered pair** of numbers (a, b), as shown in Figure 1. The first number a is called the **x-coordinate** of P; the second number b is called the **y-coordinate** of P. We can think of the coordinates of P as its "address," because they specify its location in the plane. Several points are labeled with their coordinates in Figure 2.

EXAMPLE 1 | Graphing Regions in the Coordinate Plane

Describe and sketch the regions given by each set.

(a) $\{(x, y) \mid x \geq 0\}$ **(b)** $\{(x, y) \mid y = 1\}$ **(c)** $\{(x, y) \mid |y| < 1\}$

SOLUTION

(a) The points whose x-coordinates are 0 or positive lie on the y-axis or to the right of it, as shown in Figure 3(a).

(b) The set of all points with y-coordinate 1 is a horizontal line one unit above the x-axis, as in Figure 3(b).

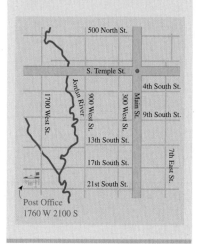
(c) Recall that

$$|y| < 1 \qquad \text{if and only if} \qquad -1 < y < 1$$

(see Appendix A.5). So the given region consists of those points in the plane whose y-coordinates lie between -1 and 1. Thus, the region consists of all points that lie between (but not on) the horizontal lines $y = 1$ and $y = -1$. These lines are shown as broken lines in Figure 3(c) to indicate that the points on these lines do not lie in the set.

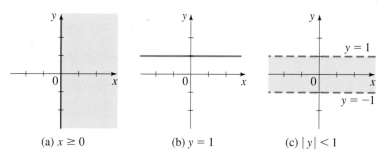

(a) $x \geq 0$ (b) $y = 1$ (c) $|y| < 1$

FIGURE 3

✎ NOW TRY EXERCISES **23**, **25**, AND **29**

▼ The Distance and Midpoint Formulas

We now find a formula for the distance $d(A, B)$ between two points $A(x_1, y_1)$ and $B(x_2, y_2)$ in the plane. The distance between points a and b on a number line is $d(a, b) = |b - a|$ (see Appendix A.1). So from Figure 4 we see that the distance between the points $A(x_1, y_1)$ and $C(x_2, y_1)$ on a horizontal line must be $|x_2 - x_1|$, and the distance between $B(x_2, y_2)$ and $C(x_2, y_1)$ on a vertical line must be $|y_2 - y_1|$.

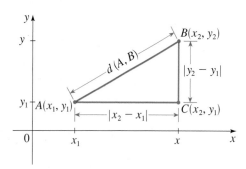

FIGURE 4

Since triangle ABC is a right triangle, the Pythagorean Theorem gives

$$d(A, B) = \sqrt{|x_2 - x_1|^2 + |y_2 - y_1|^2} = \sqrt{(x_2 - x_1)^2 + (y_2 - y_1)^2}$$

DISTANCE FORMULA

The distance between the points $A(x_1, y_1)$ and $B(x_2, y_2)$ in the plane is

$$d(A, B) = \sqrt{(x_2 - x_1)^2 + (y_2 - y_1)^2}$$

EXAMPLE 2 | Applying the Distance Formula

Which of the points $P(1, -2)$ or $Q(8, 9)$ is closer to the point $A(5, 3)$?

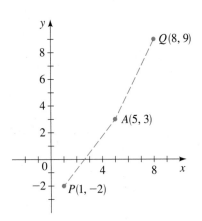

FIGURE 5

SOLUTION By the Distance Formula we have

$$d(P, A) = \sqrt{(5 - 1)^2 + [3 - (-2)]^2} = \sqrt{4^2 + 5^2} = \sqrt{41}$$

$$d(Q, A) = \sqrt{(5 - 8)^2 + (3 - 9)^2} = \sqrt{(-3)^2 + (-6)^2} = \sqrt{45}$$

This shows that $d(P, A) < d(Q, A)$, so P is closer to A (see Figure 5).

✎ NOW TRY EXERCISE **33**

Now let's find the coordinates (x, y) of the midpoint M of the line segment that joins the point $A(x_1, y_1)$ to the point $B(x_2, y_2)$. In Figure 6 notice that triangles APM and MQB are congruent because $d(A, M) = d(M, B)$ and the corresponding angles are equal.

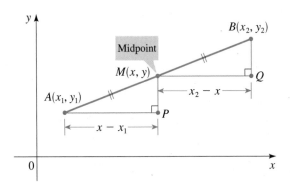

FIGURE 6

It follows that $d(A, P) = d(M, Q)$, so

$$x - x_1 = x_2 - x$$

Solving this equation for x, we get $2x = x_1 + x_2$, so $x = \dfrac{x_1 + x_2}{2}$. Similarly, $y = \dfrac{y_1 + y_2}{2}$.

MIDPOINT FORMULA

The midpoint of the line segment from $A(x_1, y_1)$ to $B(x_2, y_2)$ is

$$\left(\frac{x_1 + x_2}{2}, \frac{y_1 + y_2}{2} \right)$$

EXAMPLE 3 | Applying the Midpoint Formula

Show that the quadrilateral with vertices $P(1, 2)$, $Q(4, 4)$, $R(5, 9)$, and $S(2, 7)$ is a parallelogram by proving that its two diagonals bisect each other.

SOLUTION If the two diagonals have the same midpoint, then they must bisect each other. The midpoint of the diagonal PR is

$$\left(\frac{1 + 5}{2}, \frac{2 + 9}{2} \right) = \left(3, \frac{11}{2} \right)$$

and the midpoint of the diagonal QS is

$$\left(\frac{4 + 2}{2}, \frac{4 + 7}{2} \right) = \left(3, \frac{11}{2} \right)$$

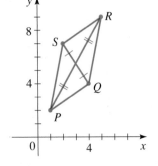

FIGURE 7

so each diagonal bisects the other, as shown in Figure 7. (A theorem from elementary geometry states that the quadrilateral is therefore a parallelogram.)

✎ NOW TRY EXERCISE **37**

▼ Graphs of Equations in Two Variables

An **equation in two variables**, such as $y = x^2 + 1$, expresses a relationship between two quantities. A point (x, y) **satisfies** the equation if it makes the equation true when the values for x and y are substituted into the equation. For example, the point $(3, 10)$ satisfies the equation $y = x^2 + 1$ because $10 = 3^2 + 1$, but the point $(1, 3)$ does not, because $3 \neq 1^2 + 1$.

Fundamental Principle of Analytic Geometry

A point (x, y) lies on the graph of an equation if and only if its coordinates satisfy the equation.

THE GRAPH OF AN EQUATION

The **graph** of an equation in x and y is the set of all points (x, y) in the coordinate plane that satisfy the equation.

The graph of an equation is a curve, so to graph an equation, we plot as many points as we can, then connect them by a smooth curve.

EXAMPLE 4 | Sketching a Graph by Plotting Points

Sketch the graph of the equation $2x - y = 3$.

SOLUTION We first solve the given equation for y to get

$$y = 2x - 3$$

This helps us calculate the y-coordinates in the following table.

x	$y = 2x - 3$	(x, y)
-1	-5	$(-1, -5)$
0	-3	$(0, -3)$
1	-1	$(1, -1)$
2	1	$(2, 1)$
3	3	$(3, 3)$
4	5	$(4, 5)$

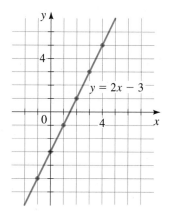

FIGURE 8

Of course, there are infinitely many points on the graph, and it is impossible to plot all of them. But the more points we plot, the better we can imagine what the graph represented by the equation looks like. We plot the points we found in Figure 8; they appear to lie on a line. So we complete the graph by joining the points by a line. (In Section 1.2 we verify that the graph of this equation is indeed a line.)

✎ NOW TRY EXERCISE **59**

EXAMPLE 5 | Sketching a Graph by Plotting Points

Sketch the graph of the equation $y = x^2 - 2$.

SOLUTION We find some of the points that satisfy the equation in the following table. In Figure 9 we plot these points and then connect them by a smooth curve. A curve with this shape is called a *parabola*.

x	$y = x^2 - 2$	(x, y)
-3	7	$(-3, 7)$
-2	2	$(-2, 2)$
-1	-1	$(-1, -1)$
0	-2	$(0, -2)$
1	-1	$(1, -1)$
2	2	$(2, 2)$
3	7	$(3, 7)$

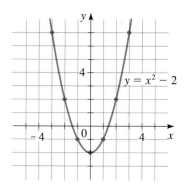

FIGURE 9

✎ NOW TRY EXERCISE **63**

EXAMPLE 6 │ Graphing an Absolute Value Equation

Sketch the graph of the equation $y = |x|$.

SOLUTION   We make a table of values:

x	$y = \lvert x \rvert$	(x, y)
-3	3	$(-3, 3)$
-2	2	$(-2, 2)$
-1	1	$(-1, 1)$
0	0	$(0, 0)$
1	1	$(1, 1)$
2	2	$(2, 2)$
3	3	$(3, 3)$

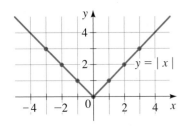

FIGURE 10

In Figure 10 we plot these points and use them to sketch the graph of the equation.

✎ NOW TRY EXERCISE **75**   ■

▼ Intercepts

The *x*-coordinates of the points where a graph intersects the *x*-axis are called the **x-intercepts** of the graph and are obtained by setting $y = 0$ in the equation of the graph. The *y*-coordinates of the points where a graph intersects the *y*-axis are called the **y-intercepts** of the graph and are obtained by setting $x = 0$ in the equation of the graph.

DEFINITION OF INTERCEPTS

Intercepts	**How to find them**	**Where they are on the graph**
x-intercepts: The *x*-coordinates of points where the graph of an equation intersects the *x*-axis	Set $y = 0$ and solve for x	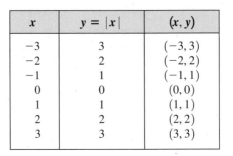
y-intercepts: The *y*-coordinates of points where the graph of an equation intersects the *y*-axis	Set $x = 0$ and solve for y	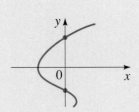

EXAMPLE 7 │ Finding Intercepts

Find the *x*- and *y*-intercepts of the graph of the equation $y = x^2 - 2$.

SOLUTION   To find the *x*-intercepts, we set $y = 0$ and solve for *x*. Thus

$$0 = x^2 - 2 \qquad \text{Set } y = 0$$

$$x^2 = 2 \qquad \text{Add 2 to each side}$$

$$x = \pm\sqrt{2} \qquad \text{Take the square root}$$

The *x*-intercepts are $\sqrt{2}$ and $-\sqrt{2}$.

To find the y-intercepts, we set $x = 0$ and solve for y. Thus

$$y = 0^2 - 2 \qquad \text{Set } x = 0$$
$$y = -2$$

The y-intercept is -2.

The graph of this equation was sketched in Example 5. It is repeated in Figure 11 with the x- and y-intercepts labeled.

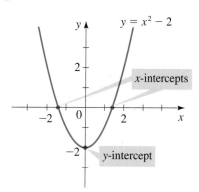

FIGURE 11

NOW TRY EXERCISE **65**

▼ Circles

So far, we have discussed how to find the graph of an equation in x and y. The converse problem is to find an equation of a graph, that is, an equation that represents a given curve in the xy-plane. Such an equation is satisfied by the coordinates of the points on the curve and by no other point. This is the other half of the fundamental principle of analytic geometry as formulated by Descartes and Fermat. The idea is that if a geometric curve can be represented by an algebraic equation, then the rules of algebra can be used to analyze the curve.

As an example of this type of problem, let's find the equation of a circle with radius r and center (h, k). By definition the circle is the set of all points $P(x, y)$ whose distance from the center $C(h, k)$ is r (see Figure 12). Thus P is on the circle if and only if $d(P, C) = r$. From the distance formula we have

$$\sqrt{(x - h)^2 + (y - k)^2} = r$$
$$(x - h)^2 + (y - k)^2 = r^2 \qquad \text{Square each side}$$

This is the desired equation.

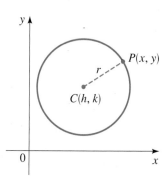

FIGURE 12

EQUATION OF A CIRCLE

An equation of the circle with center (h, k) and radius r is

$$(x - h)^2 + (y - k)^2 = r^2$$

This is called the **standard form** for the equation of the circle. If the center of the circle is the origin $(0, 0)$, then the equation is

$$x^2 + y^2 = r^2$$

EXAMPLE 8 | Graphing a Circle

Graph each equation.

(a) $x^2 + y^2 = 25$ **(b)** $(x - 2)^2 + (y + 1)^2 = 25$

SOLUTION

(a) Rewriting the equation as $x^2 + y^2 = 5^2$, we see that this is an equation of the circle of radius 5 centered at the origin. Its graph is shown in Figure 13.

(b) Rewriting the equation as $(x - 2)^2 + (y + 1)^2 = 5^2$, we see that this is an equation of the circle of radius 5 centered at $(2, -1)$. Its graph is shown in Figure 14.

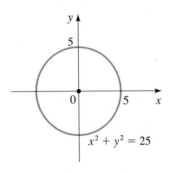

FIGURE 13

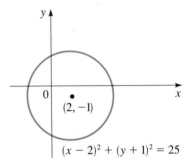

FIGURE 14

✎ NOW TRY EXERCISES **87** AND **89**

EXAMPLE 9 | Finding an Equation of a Circle

(a) Find an equation of the circle with radius 3 and center $(2, -5)$.

(b) Find an equation of the circle that has the points $P(1, 8)$ and $Q(5, -6)$ as the endpoints of a diameter.

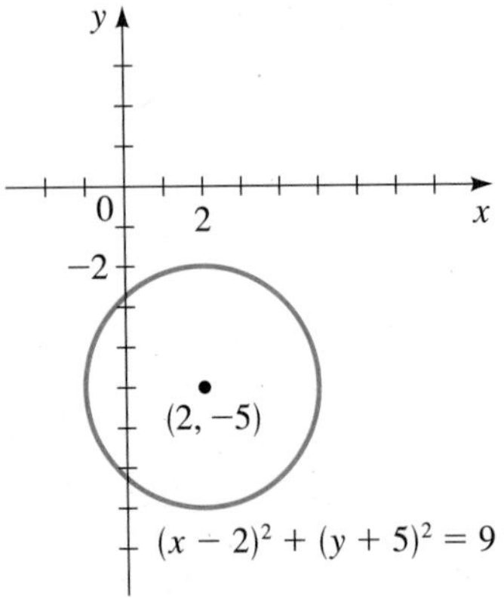

SOLUTION

(a) Using the equation of a circle with $r = 3$, $h = 2$, and $k = -5$, we obtain

$$(x - 2)^2 + (y + 5)^2 = 9$$

The graph is shown in Figure 15.

(b) We first observe that the center is the midpoint of the diameter PQ, so by the Midpoint Formula the center is

$$\left(\frac{1 + 5}{2}, \frac{8 - 6}{2} \right) = (3, 1)$$

The radius r is the distance from P to the center, so by the Distance Formula

$$r^2 = (3 - 1)^2 + (1 - 8)^2 = 2^2 + (-7)^2 = 53$$

Therefore, the equation of the circle is

$$(x - 3)^2 + (y - 1)^2 = 53$$

The graph is shown in Figure 16.

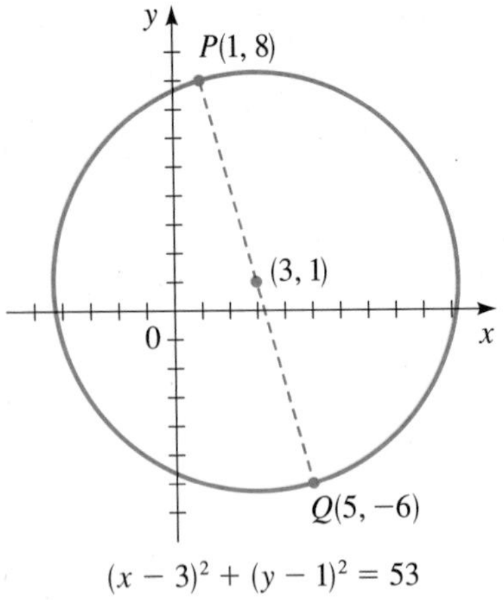

✎ NOW TRY EXERCISES **93** AND **97**

Let's expand the equation of the circle in the preceding example.

$(x - 3)^2 + (y - 1)^2 = 53$	Standard form
$x^2 - 6x + 9 + y^2 - 2y + 1 = 53$	Expand the squares
$x^2 - 6x + y^2 - 2y = 43$	Subtract 10 to get expanded form

Suppose we are given the equation of a circle in expanded form. Then to find its center and radius, we must put the equation back in standard form. That means that we must reverse the steps in the preceding calculation, and to do that we need to know what to add to an expression like $x^2 - 6x$ to make it a perfect square—that is, we need to complete the square, as in the next example.

In the left margin:

FIGURE 15

$(x - 2)^2 + (y + 5)^2 = 9$

$(2, -5)$

FIGURE 16

$P(1, 8)$

$(3, 1)$

$Q(5, -6)$

$(x - 3)^2 + (y - 1)^2 = 53$

Completing the square is used in many contexts in algebra. In Appendix A.4 we use completing the square to solve quadratic equations.

EXAMPLE 10 | Identifying an Equation of a Circle

Show that the equation $x^2 + y^2 + 2x - 6y + 7 = 0$ represents a circle, and find the center and radius of the circle.

SOLUTION   We first group the x-terms and y-terms. Then we complete the square within each grouping. That is, we complete the square for $x^2 + 2x$ by adding $\left(\frac{1}{2} \cdot 2\right)^2 = 1$, and we complete the square for $y^2 - 6y$ by adding $\left[\frac{1}{2} \cdot (-6)\right]^2 = 9$.

$$(x^2 + 2x \quad) + (y^2 - 6y \quad) = -7 \qquad \text{Group terms}$$

⊘ We must add the same numbers to *each side* to maintain equality.

$$(x^2 + 2x + 1) + (y^2 - 6y + 9) = -7 + 1 + 9 \qquad \begin{array}{l}\text{Complete the square by}\\ \text{adding 1 and 9 to each side}\end{array}$$

$$(x + 1)^2 + (y - 3)^2 = 3 \qquad \text{Factor and simplify}$$

Comparing this equation with the standard equation of a circle, we see that $h = -1$, $k = 3$, and $r = \sqrt{3}$, so the given equation represents a circle with center $(-1, 3)$ and radius $\sqrt{3}$.

✎ NOW TRY EXERCISE **103**   ■

▼ Symmetry

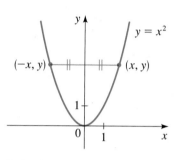

FIGURE 17

Figure 17 shows the graph of $y = x^2$. Notice that the part of the graph to the left of the y-axis is the mirror image of the part to the right of the y-axis. The reason is that if the point (x, y) is on the graph, then so is $(-x, y)$, and these points are reflections of each other about the y-axis. In this situation we say that the graph is **symmetric with respect to the y-axis**. Similarly, we say that a graph is **symmetric with respect to the x-axis** if whenever the point (x, y) is on the graph, then so is $(x, -y)$. A graph is **symmetric with respect to the origin** if whenever (x, y) is on the graph, so is $(-x, -y)$.

DEFINITION OF SYMMETRY

Type of symmetry	How to test for symmetry	What the graph looks like (figures in this section)	Geometric meaning
Symmetry with respect to the x-axis	The equation is unchanged when y is replaced by $-y$	*graph with points (x, y) and $(x, -y)$* (Figures 13, 18)	Graph is unchanged when reflected in the x-axis
Symmetry with respect to the y-axis	The equation is unchanged when x is replaced by $-x$	*graph with points $(-x, y)$ and (x, y)* (Figures 9, 10, 11, 13, 17)	Graph is unchanged when reflected in the y-axis
Symmetry with respect to the origin	The equation is unchanged when x is replaced by $-x$ and y by $-y$	*graph with points (x, y) and $(-x, -y)$* (Figures 13, 19)	Graph is unchanged when rotated 180° about the origin

The remaining examples in this section show how symmetry helps us sketch the graphs of equations.

EXAMPLE 11 | Using Symmetry to Sketch a Graph

Test the equation $x = y^2$ for symmetry and sketch the graph.

SOLUTION   If y is replaced by $-y$ in the equation $x = y^2$, we get

$$x = (-y)^2 \qquad \text{Replace } y \text{ by } -y$$
$$x = y^2 \qquad \text{Simplify}$$

and so the equation is unchanged. Therefore, the graph is symmetric about the x-axis. But changing x to $-x$ gives the equation $-x = y^2$, which is not the same as the original equation, so the graph is not symmetric about the y-axis.

We use the symmetry about the x-axis to sketch the graph by first plotting points just for $y > 0$ and then reflecting the graph in the x-axis, as shown in Figure 18.

y	$x = y^2$	(x, y)
0	0	$(0, 0)$
1	1	$(1, 1)$
2	4	$(4, 2)$
3	9	$(9, 3)$

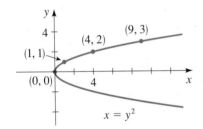

FIGURE 18

🔖 NOW TRY EXERCISE **77**

EXAMPLE 12 | Using Symmetry to Sketch a Graph

Test the equation $y = x^3 - 9x$ for symmetry and sketch its graph.

SOLUTION   If we replace x by $-x$ and y by $-y$ in the equation, we get

$$-y = (-x)^3 - 9(-x) \qquad \text{Replace } x \text{ by } -x \text{ and } y \text{ by } -y$$
$$-y = -x^3 + 9x \qquad \text{Simplify}$$
$$y = x^3 - 9x \qquad \text{Multiply by } -1$$

and so the equation is unchanged. This means that the graph is symmetric with respect to the origin. We sketch it by first plotting points for $x > 0$ and then using symmetry about the origin (see Figure 19).

x	$y = x^3 - 9x$	(x, y)
0	0	$(0, 0)$
1	-8	$(1, -8)$
1.5	-10.125	$(1.5, -10.125)$
2	-10	$(2, -10)$
2.5	-6.875	$(2.5, -6.875)$
3	0	$(3, 0)$
4	28	$(4, 28)$

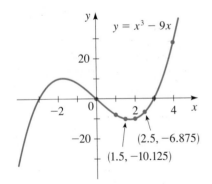

FIGURE 19

🔖 NOW TRY EXERCISE **79**

1.1 EXERCISES

CONCEPTS

1. The point that is 3 units to the right of the y-axis and 5 units below the x-axis has coordinates (____, ____).

2. The distance between the points (a, b) and (c, d) is _____. So the distance between $(1, 2)$ and $(7, 10)$ is _____.

3. The point midway between (a, b) and (c, d) is _____. So the point midway between $(1, 2)$ and $(7, 10)$ is _____.

4. If the point $(2, 3)$ is on the graph of an equation in x and y, then the equation is satisfied when we replace x by _____ and y by _____. Is the point $(2, 3)$ on the graph of the equation $2y = x + 1$?

5. (a) To find the x-intercept(s) of the graph of an equation, we set _____ equal to 0 and solve for _____. So the x-intercept of $2y = x + 1$ is _____.

(b) To find the y-intercept(s) of the graph of an equation, we set _____ equal to 0 and solve for _____. So the y-intercept of $2y = x + 1$ is _____.

6. The graph of the equation $(x - 1)^2 + (y - 2)^2 = 9$ is a circle with center (____, ____) and radius _____.

SKILLS

7. Plot the given points in a coordinate plane.
$(2, 3), (-2, 3), (4, 5), (4, -5), (-4, 5), (-4, -5)$

8. Find the coordinates of the points shown in the figure.

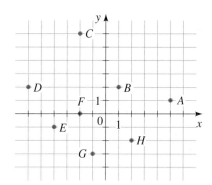

9–12 ■ A pair of points is graphed.
(a) Find the distance between them.
(b) Find the midpoint of the segment that joins them.

9.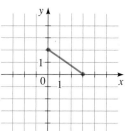

10.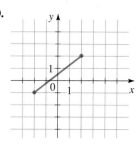

11.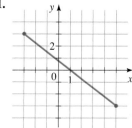

12.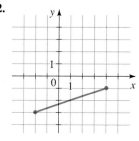

13–18 ■ A pair of points is given.
(a) Plot the points in a coordinate plane.
(b) Find the distance between them.
(c) Find the midpoint of the segment that joins them.

13. $(0, 8), (6, 16)$ **14.** $(-2, 5), (10, 0)$

15. $(-3, -6), (4, 18)$ **16.** $(-1, -1), (9, 9)$

17. $(6, -2), (-6, 2)$ **18.** $(0, -6), (5, 0)$

19. Draw the rectangle with vertices $A(1, 3), B(5, 3), C(1, -3)$, and $D(5, -3)$ on a coordinate plane. Find the area of the rectangle.

20. Draw the parallelogram with vertices $A(1, 2), B(5, 2), C(3, 6)$, and $D(7, 6)$ on a coordinate plane. Find the area of the parallelogram.

21. Plot the points $A(1, 0), B(5, 0), C(4, 3)$, and $D(2, 3)$ on a coordinate plane. Draw the segments AB, BC, CD, and DA. What kind of quadrilateral is $ABCD$, and what is its area?

22. Plot the points $P(5, 1), Q(0, 6)$, and $R(-5, 1)$ on a coordinate plane. Where must the point S be located so that the quadrilateral $PQRS$ is a square? Find the area of this square.

23–32 ■ Sketch the region given by the set.

23. $\{(x, y) \mid x \geq 3\}$ **24.** $\{(x, y) \mid y < 3\}$

25. $\{(x, y) \mid y = 2\}$ **26.** $\{(x, y) \mid x = -1\}$

27. $\{(x, y) \mid 1 < x < 2\}$ **28.** $\{(x, y) \mid 0 \leq y \leq 4\}$

29. $\{(x, y) \mid |x| > 4\}$ **30.** $\{(x, y) \mid |y| \leq 2\}$

31. $\{(x, y) \mid x \geq 1 \text{ and } y < 3\}$

32. $\{(x, y) \mid |x| \leq 2 \text{ and } |y| \leq 3\}$

33. Which of the points $A(6, 7)$ or $B(-5, 8)$ is closer to the origin?

34. Which of the points $C(-6, 3)$ or $D(3, 0)$ is closer to the point $E(-2, 1)$?

35. Which of the points $P(3, 1)$ or $Q(-1, 3)$ is closer to the point $R(-1, -1)$?

36. (a) Show that the points $(7, 3)$ and $(3, 7)$ are the same distance from the origin.

(b) Show that the points (a, b) and (b, a) are the same distance from the origin.

37. Show that the triangle with vertices $A(0, 2)$, $B(-3, -1)$, and $C(-4, 3)$ is isosceles.

38. Find the area of the triangle shown in the figure.

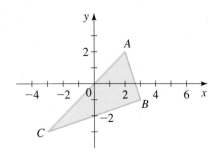

39. Refer to triangle ABC in the figure below.

(a) Show that triangle ABC is a right triangle by using the converse of the Pythagorean Theorem (see page 529).

(b) Find the area of triangle ABC.

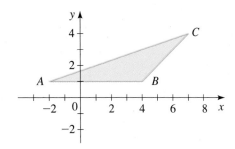

40. Show that the triangle with vertices $A(6, -7)$, $B(11, -3)$, and $C(2, -2)$ is a right triangle by using the converse of the Pythagorean Theorem. Find the area of the triangle.

41. Show that the points $A(-2, 9)$, $B(4, 6)$, $C(1, 0)$, and $D(-5, 3)$ are the vertices of a square.

42. Show that the points $A(-1, 3)$, $B(3, 11)$, and $C(5, 15)$ are collinear by showing that $d(A, B) + d(B, C) = d(A, C)$.

43. Find a point on the y-axis that is equidistant from the points $(5, -5)$ and $(1, 1)$.

44. Find the lengths of the medians of the triangle with vertices $A(1, 0)$, $B(3, 6)$, and $C(8, 2)$. (A *median* is a line segment from a vertex to the midpoint of the opposite side.)

45. Plot the points $P(-1, -4)$, $Q(1, 1)$, and $R(4, 2)$ on a coordinate plane. Where should the point S be located so that the figure $PQRS$ is a parallelogram?

46. If $M(6, 8)$ is the midpoint of the line segment AB and if A has coordinates $(2, 3)$, find the coordinates of B.

47. (a) Sketch the parallelogram with vertices $A(-2, -1)$, $B(4, 2)$, $C(7, 7)$, and $D(1, 4)$.

(b) Find the midpoints of the diagonals of this parallelogram.

(c) From part (b) show that the diagonals bisect each other.

48. The point M in the figure is the midpoint of the line segment AB. Show that M is equidistant from the vertices of triangle ABC.

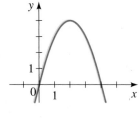

49–52 ■ Determine whether the given points are on the graph of the equation.

49. $x - 2y - 1 = 0$; $\quad (0, 0), (1, 0), (-1, -1)$

50. $y(x^2 + 1) = 1$; $\quad (1, 1), (1, \frac{1}{2}), (-1, \frac{1}{2})$

51. $x^2 + xy + y^2 = 4$; $\quad (0, -2), (1, -2), (2, -2)$

52. $x^2 + y^2 = 1$; $\quad (0, 1), \left(\dfrac{1}{\sqrt{2}}, \dfrac{1}{\sqrt{2}} \right), \left(\dfrac{\sqrt{3}}{2}, \dfrac{1}{2} \right)$

53–56 ■ An equation and its graph are given. Find the x- and y-intercepts.

53. $y = 4x - x^2$

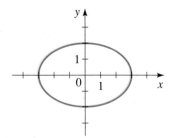

54. $\dfrac{x^2}{9} + \dfrac{y^2}{4} = 1$

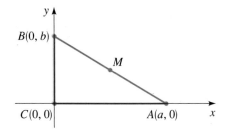

55. $x^4 + y^2 - xy = 16$

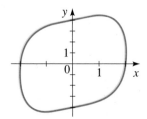

56. $x^2 + y^3 - x^2 y^2 = 64$

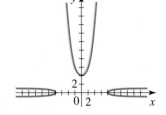

57–76 ■ Make a table of values and sketch the graph of the equation. Find the x- and y-intercepts and test for symmetry.

57. $y = -x + 4$ **58.** $y = 3x + 3$

59. $2x - y = 6$ **60.** $x + y = 3$

61. $y = 1 - x^2$ **62.** $y = x^2 + 2$

63. $4y = x^2$ **64.** $8y = x^3$

65. $y = x^2 - 9$ **66.** $y = 9 - x^2$

67. $xy = 2$ **68.** $y = \sqrt{x + 4}$

69. $y = \sqrt{4 - x^2}$ **70.** $y = -\sqrt{4 - x^2}$

71. $x + y^2 = 4$ **72.** $x = y^3$

73. $y = 16 - x^4$ **74.** $x = |y|$

75. $y = 4 - |x|$ **76.** $y = |4 - x|$

77–82 ■ Test the equation for symmetry.

77. $y = x^4 + x^2$ **78.** $x = y^4 - y^2$

79. $x^2y^2 + xy = 1$ **80.** $x^4y^4 + x^2y^2 = 1$

81. $y = x^3 + 10x$ **82.** $y = x^2 + |x|$

83–86 ■ Complete the graph using the given symmetry property.

83. Symmetric with respect to the y-axis

84. Symmetric with respect to the x-axis

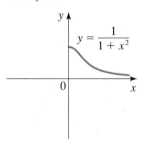

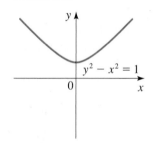

85. Symmetric with respect to the origin

86. Symmetric with respect to the origin

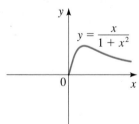

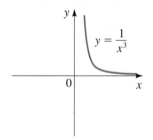

87–92 ■ Find the center and radius of the circle and sketch its graph

87. $x^2 + y^2 = 9$ **88.** $x^2 + y^2 = 5$

89. $(x - 3)^2 + y^2 = 16$ **90.** $x^2 + (y - 2)^2 = 4$

91. $(x + 3)^2 + (y - 4)^2 = 25$ **92.** $(x + 1)^2 + (y + 2)^2 = 36$

93–100 ■ Find an equation of the circle that satisfies the given conditions.

93. Center $(2, -1)$;   radius 3

94. Center $(-1, -4)$;   radius 8

95. Center at the origin;   passes through $(4, 7)$

96. Center $(-1, 5)$;   passes through $(-4, -6)$

97. Endpoints of a diameter are $P(-1, 1)$ and $Q(5, 9)$

98. Endpoints of a diameter are $P(-1, 3)$ and $Q(7, -5)$

99. Center $(7, -3)$;   tangent to the x-axis

100. Circle lies in the first quadrant, tangent to both x-and y-axes; radius 5

101–102 ■ Find the equation of the circle shown in the figure.

101.   **102.**

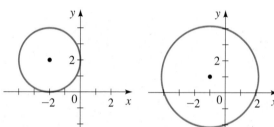

103–108 ■ Show that the equation represents a circle, and find the center and radius of the circle.

103. $x^2 + y^2 - 4x + 10y + 13 = 0$

104. $x^2 + y^2 + 6y + 2 = 0$

105. $x^2 + y^2 - \frac{1}{2}x + \frac{1}{2}y = \frac{1}{8}$

106. $x^2 + y^2 + \frac{1}{2}x + 2y + \frac{1}{16} = 0$

107. $2x^2 + 2y^2 - 3x = 0$

108. $3x^2 + 3y^2 + 6x - y = 0$

109–110 ■ Sketch the region given by the set.

109. $\{(x, y) \mid x^2 + y^2 \leq 1\}$

110. $\{(x, y) \mid x^2 + y^2 > 4\}$

111. Find the area of the region that lies outside the circle $x^2 + y^2 = 4$ but inside the circle

$$x^2 + y^2 - 4y - 12 = 0$$

112. Sketch the region in the coordinate plane that satisfies both the inequalities $x^2 + y^2 \leq 9$ and $y \geq |x|$. What is the area of this region?

APPLICATIONS

113. Distances in a City A city has streets that run north and south and avenues that run east and west, all equally spaced. Streets and avenues are numbered sequentially, as shown in the figure. The *walking* distance between points A and B is 7 blocks—that is, 3 blocks east and 4 blocks north. To find the *straight-line* distance d, we must use the Distance Formula.

    (a) Find the straight-line distance (in blocks) between A and B.

(b) Find the walking distance and the straight-line distance between the corner of 4th St. and 2nd Ave. and the corner of 11th St. and 26th Ave.

(c) What must be true about the points P and Q if the walking distance between P and Q equals the straight-line distance between P and Q?

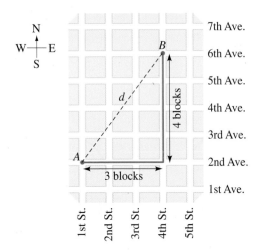

114. Halfway Point Two friends live in the city described in Exercise 113, one at the corner of 3rd St. and 7th Ave., the other at the corner of 27th St. and 17th Ave. They frequently meet at a coffee shop halfway between their homes.

(a) At what intersection is the coffee shop located?

(b) How far must each of them walk to get to the coffee shop?

115. Orbit of a Satellite A satellite is in orbit around the moon. A coordinate plane containing the orbit is set up with the center of the moon at the origin, as shown in the graph, with distances measured in megameters (Mm). The equation of the satellite's orbit is

$$\frac{(x-3)^2}{25} + \frac{y^2}{16} = 1$$

(a) From the graph, determine the closest and the farthest that the satellite gets to the center of the moon.

(b) There are two points in the orbit with y-coordinates 2. Find the x-coordinates of these points, and determine their distances to the center of the moon.

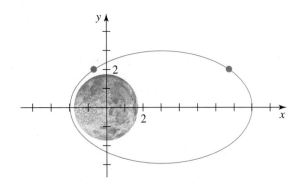

116. Shifting the Coordinate Plane Suppose that each point in the coordinate plane is shifted 3 units to the right and 2 units upward.

(a) The point $(5, 3)$ is shifted to what new point?

(b) The point (a, b) is shifted to what new point?

(c) What point is shifted to $(3, 4)$?

(d) Triangle ABC in the figure has been shifted to triangle $A'B'C'$. Find the coordinates of the points A', B', and C'.

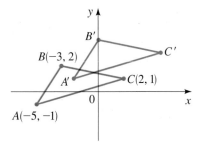

117. Reflecting in the Coordinate Plane Suppose that the y-axis acts as a mirror that reflects each point to the right of it into a point to the left of it.

(a) The point $(3, 7)$ is reflected to what point?

(b) The point (a, b) is reflected to what point?

(c) What point is reflected to $(-4, -1)$?

(d) Triangle ABC in the figure is reflected to triangle $A'B'C'$. Find the coordinates of the points A', B', and C'.

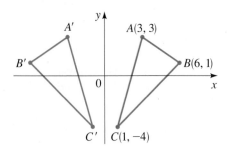

118. Completing a Line Segment Plot the points $M(6, 8)$ and $A(2, 3)$ on a coordinate plane. If M is the midpoint of the line segment AB, find the coordinates of B. Write a brief description of the steps you took to find B, and your reasons for taking them.

119. Completing a Parallelogram Plot the points $P(0, 3)$, $Q(2, 2)$, and $R(5, 3)$ on a coordinate plane. Where should the point S be located so that the figure $PQRS$ is a parallelogram? Write a brief description of the steps you took and your reasons for taking them.

120. Circle, Point, or Empty Set? Complete the squares in the general equation $x^2 + ax + y^2 + by + c = 0$ and simplify the result as much as possible. Under what conditions on the coefficients a, b, and c does this equation represent a circle? A single point? The empty set? In the case that the equation does represent a circle, find its center and radius.

121. Do the Circles Intersect?

(a) Find the radius of each circle in the pair and the distance between their centers; then use this information to determine whether the circles intersect.

(i) $(x - 2)^2 + (y - 1)^2 = 9$;
$(x - 6)^2 + (y - 4)^2 = 16$

(ii) $x^2 + (y - 2)^2 = 4$;
$(x - 5)^2 + (y - 14)^2 = 9$

(iii) $(x - 3)^2 + (y + 1)^2 = 1$;
$(x - 2)^2 + (y - 2)^2 = 25$

(b) How can you tell, just by knowing the radii of two circles and the distance between their centers, whether the circles intersect? Write a short paragraph describing how you would decide this and draw graphs to illustrate your answer.

122. Making a Graph Symmetric The graph shown in the figure is not symmetric about the x-axis, the y-axis, or the origin. Add more line segments to the graph so that it exhibits the indicated symmetry. In each case, add as little as possible.

(a) Symmetry about the x-axis

(b) Symmetry about the y-axis

(c) Symmetry about the origin

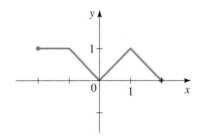

1.2 LINES

The Slope of a Line ▶ Point-Slope Form of the Equation of a Line ▶ Slope-Intercept Form of the Equation of a Line ▶ Vertical and Horizontal Lines ▶ General Equation of a Line ▶ Parallel and Perpendicular Lines ▶ Modeling with Linear Equations: Slope as Rate of Change

In this section we find equations for straight lines lying in a coordinate plane. The equations will depend on how the line is inclined, so we begin by discussing the concept of slope.

▼ The Slope of a Line

We first need a way to measure the "steepness" of a line, or how quickly it rises (or falls) as we move from left to right. We define *run* to be the distance we move to the right and *rise* to be the corresponding distance that the line rises (or falls). The *slope* of a line is the ratio of rise to run:

$$\text{slope} = \frac{\text{rise}}{\text{run}}$$

Figure 1 shows situations in which slope is important. Carpenters use the term *pitch* for the slope of a roof or a staircase; the term *grade* is used for the slope of a road.

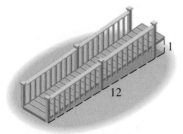

Slope of a ramp
Slope $= \frac{1}{12}$

Pitch of a roof
Slope $= \frac{1}{3}$

Grade of a road
Slope $= \frac{8}{100}$

FIGURE 1

If a line lies in a coordinate plane, then the **run** is the change in the *x*-coordinate and the **rise** is the corresponding change in the *y*-coordinate between any two points on the line (see Figure 2). This gives us the following definition of slope.

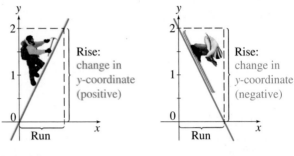

Rise: change in *y*-coordinate (positive)

Run

Rise: change in *y*-coordinate (negative)

Run

FIGURE 2

SLOPE OF A LINE

The **slope** m of a nonvertical line that passes through the points $A(x_1, y_1)$ and $B(x_2, y_2)$ is

$$m = \frac{\text{rise}}{\text{run}} = \frac{y_2 - y_1}{x_2 - x_1}$$

The slope of a vertical line is not defined.

The slope is independent of which two points are chosen on the line. We can see that this is true from the similar triangles in Figure 3:

$$\frac{y_2 - y_1}{x_2 - x_1} = \frac{y_2' - y_1'}{x_2' - x_1'}$$

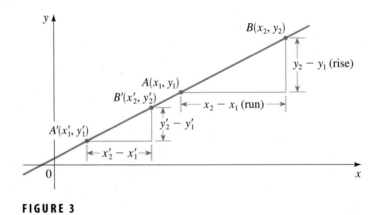

FIGURE 3

Figure 4 shows several lines labeled with their slopes. Notice that lines with positive slope slant upward to the right, whereas lines with negative slope slant downward to the right. The steepest lines are those for which the absolute value of the slope is the largest; a horizontal line has slope zero.

FIGURE 4 Lines with various slopes

EXAMPLE 1 | Finding the Slope of a Line Through Two Points

Find the slope of the line that passes through the points $P(2, 1)$ and $Q(8, 5)$.

SOLUTION Since any two different points determine a line, only one line passes through these two points. From the definition the slope is

$$m = \frac{y_2 - y_1}{x_2 - x_1} = \frac{5 - 1}{8 - 2} = \frac{4}{6} = \frac{2}{3}$$

This says that for every 3 units we move to the right, the line rises 2 units. The line is drawn in Figure 5.

✎ NOW TRY EXERCISE **5**   ■

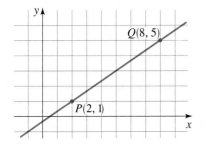

FIGURE 5

▼ Point-Slope Form of the Equation of a Line

Now let's find the equation of the line that passes through a given point $P(x_1, y_1)$ and has slope m. A point $P(x, y)$ with $x \neq x_1$ lies on this line if and only if the slope of the line through P_1 and P is equal to m (see Figure 6), that is,

$$\frac{y - y_1}{x - x_1} = m$$

This equation can be rewritten in the form $y - y_1 = m(x - x_1)$; note that the equation is also satisfied when $x = x_1$ and $y = y_1$. Therefore, it is an equation of the given line.

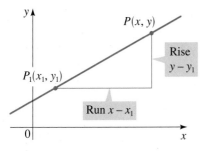

FIGURE 6

POINT-SLOPE FORM OF THE EQUATION OF A LINE

An equation of the line that passes through the point (x_1, y_1) and has slope m is

$$y - y_1 = m(x - x_1)$$

EXAMPLE 2 | Finding the Equation of a Line with Given Point and Slope

(a) Find an equation of the line through $(1, -3)$ with slope $-\frac{1}{2}$.

(b) Sketch the line.

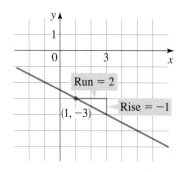

FIGURE 7

SOLUTION

(a) Using the point-slope form with $m = -\frac{1}{2}$, $x_1 = 1$, and $y_1 = -3$, we obtain an equation of the line as

$$y + 3 = -\tfrac{1}{2}(x - 1) \qquad \text{Slope } m = -\tfrac{1}{2}, \text{ point } (1, -3)$$

$$2y + 6 = -x + 1 \qquad \text{Multiply by 2}$$

$$x + 2y + 5 = 0 \qquad \text{Rearrange}$$

(b) The fact that the slope is $-\frac{1}{2}$ tells us that when we move to the right 2 units, the line drops 1 unit. This enables us to sketch the line in Figure 7.

✎ NOW TRY EXERCISE **19** ∎

EXAMPLE 3 | Finding the Equation of a Line Through Two Given Points

Find an equation of the line through the points $(-1, 2)$ and $(3, -4)$.

SOLUTION The slope of the line is

$$m = \frac{-4 - 2}{3 - (-1)} = -\frac{6}{4} = -\frac{3}{2}$$

We can use *either* point, $(-1, 2)$ *or* $(3, -4)$, in the point-slope equation. We will end up with the same final answer.

Using the point-slope form with $x_1 = -1$ and $y_1 = 2$, we obtain

$$y - 2 = -\tfrac{3}{2}(x + 1) \qquad \text{Slope } m = -\tfrac{3}{2}, \text{ point } (-1, 2)$$

$$2y - 4 = -3x - 3 \qquad \text{Multiply by 2}$$

$$3x + 2y - 1 = 0 \qquad \text{Rearrange}$$

✎ NOW TRY EXERCISE **23** ∎

▼ Slope-Intercept Form of the Equation of a Line

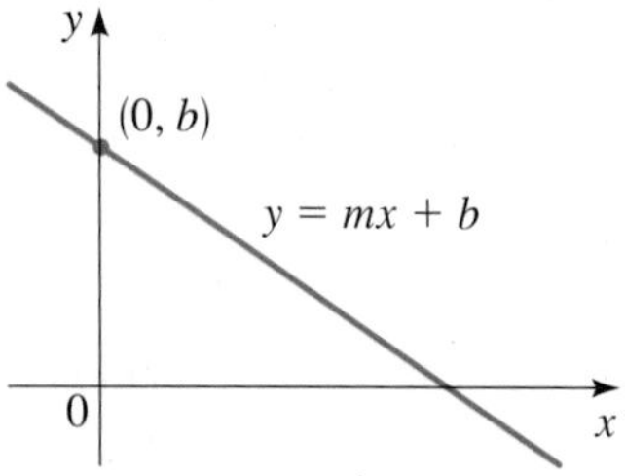

Suppose a nonvertical line has slope m and y-intercept b (see Figure 8). This means that the line intersects the y-axis at the point $(0, b)$, so the point-slope form of the equation of the line, with $x = 0$ and $y = b$, becomes

$$y - b = m(x - 0)$$

This simplifies to $y = mx + b$, which is called the **slope-intercept form** of the equation of a line.

FIGURE 8

> **SLOPE-INTERCEPT FORM OF THE EQUATION OF A LINE**
>
> An equation of the line that has slope m and y-intercept b is
>
> $$y = mx + b$$

EXAMPLE 4 | Lines in Slope-Intercept Form

(a) Find the equation of the line with slope 3 and y-intercept -2.

(b) Find the slope and y-intercept of the line $3y - 2x = 1$.

SOLUTION

(a) Since $m = 3$ and $b = -2$, from the slope-intercept form of the equation of a line we get

$$y = 3x - 2$$

(b) We first write the equation in the form $y = mx + b$:

$$3y - 2x = 1$$
$$3y = 2x + 1 \qquad \text{Add } 2x$$
$$y = \tfrac{2}{3}x + \tfrac{1}{3} \qquad \text{Divide by 3}$$

From the slope-intercept form of the equation of a line, we see that the slope is $m = \tfrac{2}{3}$ and the y-intercept is $b = \tfrac{1}{3}$.

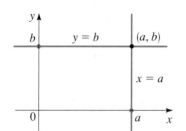

Slope y-intercept

$y = \tfrac{2}{3}x + \tfrac{1}{3}$

✎ NOW TRY EXERCISES **25** AND **47** ◼

▼ Vertical and Horizontal Lines

If a line is horizontal, its slope is $m = 0$, so its equation is $y = b$, where b is the y-intercept (see Figure 9). A vertical line does not have a slope, but we can write its equation as $x = a$, where a is the x-intercept, because the x-coordinate of every point on the line is a.

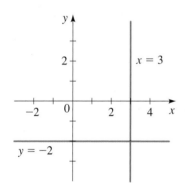

FIGURE 9

> **VERTICAL AND HORIZONTAL LINES**
>
> An equation of the vertical line through (a, b) is $x = a$.
> An equation of the horizontal line through (a, b) is $y = b$.

EXAMPLE 5 | Vertical and Horizontal Lines

(a) An equation for the vertical line through $(3, 5)$ is $x = 3$.

(b) The graph of the equation $x = 3$ is a vertical line with x-intercept 3.

(c) An equation for the horizontal line through $(8, -2)$ is $y = -2$.

(d) The graph of the equation $y = -2$ is a horizontal line with y-intercept -2.

The lines are graphed in Figure 10.

FIGURE 10

✎ NOW TRY EXERCISES **29** AND **33** ◼

▼ General Equation of a Line

A **linear equation** is an equation of the form

$$Ax + By + C = 0$$

where A, B, and C are constants and A and B are not both 0. The equation of a line is a linear equation:

- A nonvertical line has the equation $y = mx + b$ or $-mx + y - b = 0$, which is a linear equation with $A = -m$, $B = 1$, and $C = -b$.
- A vertical line has the equation $x = a$ or $x - a = 0$, which is a linear equation with $A = 1$, $B = 0$, and $C = -a$.

Conversely, the graph of a linear equation is a line.

- If $B \neq 0$, the equation becomes

$$y = -\frac{A}{B}x - \frac{C}{B} \qquad \text{Divide by } B$$

and this is the slope-intercept form of the equation of a line (with $m = -A/B$ and $b = -C/B$).

- If $B = 0$, the equation becomes

$$Ax + C = 0 \qquad \text{Set } B = 0$$

or $x = -C/A$, which represents a vertical line.

We have proved the following.

GENERAL EQUATION OF A LINE

The graph of every **linear equation**

$$Ax + By + C = 0 \qquad (A, B \text{ not both zero})$$

is a line. Conversely, every line is the graph of a linear equation.

EXAMPLE 6 | Graphing a Linear Equation

Sketch the graph of the equation $2x - 3y - 12 = 0$.

SOLUTION 1 Since the equation is linear, its graph is a line. To draw the graph, it is enough to find any two points on the line. The intercepts are the easiest points to find.

x-intercept: Substitute $y = 0$, to get $2x - 12 = 0$, so $x = 6$

y-intercept: Substitute $x = 0$, to get $-3y - 12 = 0$, so $y = -4$

With these points we can sketch the graph in Figure 11.

SOLUTION 2 We write the equation in slope-intercept form:

$$2x - 3y - 12 = 0$$
$$2x - 3y = 12 \qquad \text{Add 12}$$
$$-3y = -2x + 12 \qquad \text{Subtract } 2x$$
$$y = \tfrac{2}{3}x - 4 \qquad \text{Divide by } -3$$

This equation is in the form $y = mx + b$, so the slope is $m = \tfrac{2}{3}$ and the y-intercept is $b = -4$. To sketch the graph, we plot the y-intercept and then move 3 units to the right and 2 units up as shown in Figure 12.

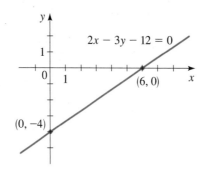

FIGURE 11

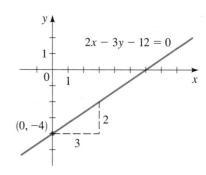

FIGURE 12

✎ NOW TRY EXERCISE **53** ∎

▼ Parallel and Perpendicular Lines

Since slope measures the steepness of a line, it seems reasonable that parallel lines should have the same slope. In fact, we can prove this.

PARALLEL LINES

Two nonvertical lines are parallel if and only if they have the same slope.

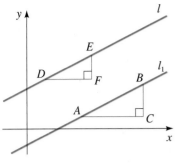

FIGURE 13

PROOF Let the lines l_1 and l_2 in Figure 13 have slopes m_1 and m_2. If the lines are parallel, then the right triangles ABC and DEF are similar, so

$$m_1 = \frac{d(B, C)}{d(A, C)} = \frac{d(E, F)}{d(D, F)} = m_2$$

Conversely, if the slopes are equal, then the triangles will be similar, so $\angle BAC = \angle EDF$ and the lines are parallel. ∎

EXAMPLE 7 | Finding the Equation of a Line Parallel to a Given Line

Find an equation of the line through the point $(5, 2)$ that is parallel to the line $4x + 6y + 5 = 0$.

SOLUTION First we write the equation of the given line in slope-intercept form.

$$4x + 6y + 5 = 0$$

$$6y = -4x - 5 \qquad \text{Subtract } 4x + 5$$

$$y = -\tfrac{2}{3}x - \tfrac{5}{6} \qquad \text{Divide by 6}$$

So the line has slope $m = -\tfrac{2}{3}$. Since the required line is parallel to the given line, it also has slope $m = -\tfrac{2}{3}$. From the point-slope form of the equation of a line, we get

$$y - 2 = -\tfrac{2}{3}(x - 5) \qquad \text{Slope } m = -\tfrac{2}{3}, \text{ point } (5, 2)$$

$$3y - 6 = -2x + 10 \qquad \text{Multiply by 3}$$

$$2x + 3y - 16 = 0 \qquad \text{Rearrange}$$

Thus, the equation of the required line is $2x + 3y - 16 = 0$.

✎ NOW TRY EXERCISE **31** ∎

The condition for perpendicular lines is not as obvious as that for parallel lines.

PERPENDICULAR LINES

Two lines with slopes m_1 and m_2 are perpendicular if and only if $m_1 m_2 = -1$, that is, their slopes are negative reciprocals:

$$m_2 = -\frac{1}{m_1}$$

Also, a horizontal line (slope 0) is perpendicular to a vertical line (no slope).

PROOF In Figure 14 we show two lines intersecting at the origin. (If the lines intersect at some other point, we consider lines parallel to these that intersect at the origin. These lines have the same slopes as the original lines.)

If the lines l_1 and l_2 have slopes m_1 and m_2, then their equations are $y = m_1 x$ and $y = m_2 x$. Notice that $A(1, m_1)$ lies on l_1 and $B(1, m_2)$ lies on l_2. By the Pythagorean Theorem and its converse (see Appendix B.2, page 529), $OA \perp OB$ if and only if

$$[d(O, A)]^2 + [d(O, B)]^2 = [d(A, B)]^2$$

By the Distance Formula this becomes

$$(1^2 + m_1^2) + (1^2 + m_2^2) = (1 - 1)^2 + (m_2 - m_1)^2$$

$$2 + m_1^2 + m_2^2 = m_2^2 - 2m_1 m_2 + m_1^2$$

$$2 = -2m_1 m_2$$

$$m_1 m_2 = -1 \qquad ∎$$

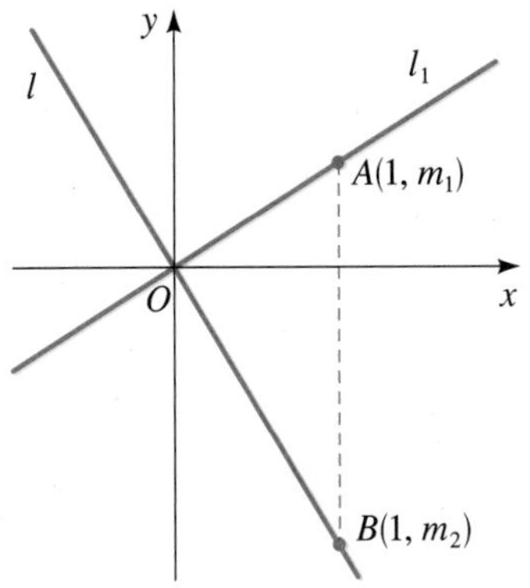

FIGURE 14

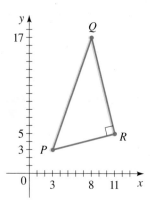

FIGURE 15

EXAMPLE 8 | Perpendicular Lines

Show that the points $P(3, 3)$, $Q(8, 17)$, and $R(11, 5)$ are the vertices of a right triangle.

SOLUTION The slopes of the lines containing PR and QR are, respectively,

$$m_1 = \frac{5 - 3}{11 - 3} = \frac{1}{4} \quad \text{and} \quad m_2 = \frac{5 - 17}{11 - 8} = -4$$

Since $m_1 m_2 = -1$, these lines are perpendicular, so PQR is a right triangle. It is sketched in Figure 15.

✎. NOW TRY EXERCISE **57** ■

EXAMPLE 9 | Finding an Equation of a Line Perpendicular to a Given Line

Find an equation of the line that is perpendicular to the line $4x + 6y + 5 = 0$ and passes through the origin.

SOLUTION In Example 7 we found that the slope of the line $4x + 6y + 5 = 0$ is $-\frac{2}{3}$. Thus, the slope of a perpendicular line is the negative reciprocal, that is, $\frac{3}{2}$. Since the required line passes through $(0, 0)$, the point-slope form gives

$$y - 0 = \tfrac{3}{2}(x - 0) \qquad \text{Slope } m = \tfrac{3}{2}, \text{ point } (0, 0)$$
$$y = \tfrac{3}{2}x \qquad\qquad \text{Simplify}$$

✎. NOW TRY EXERCISE **35** ■

EXAMPLE 10 | Graphing a Family of Lines

Use a graphing calculator to graph the family of lines

$$y = 0.5x + b$$

for $b = -2, -1, 0, 1, 2$. What property do the lines share?

SOLUTION The lines are graphed in Figure 16 in the viewing rectangle $[-6, 6]$ by $[-6, 6]$. The lines all have the same slope, so they are parallel.

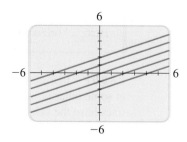

FIGURE 16 $y = 0.5x + b$

✎. NOW TRY EXERCISE **41** ■

▼ Modeling with Linear Equations: Slope as Rate of Change

When a line is used to model the relationship between two quantities, the slope of the line is the **rate of change** of one quantity with respect to the other. For example, the graph in

Figure 17(a) gives the amount of gas in a tank that is being filled. The slope between the indicated points is

$$m = \frac{6 \text{ gallons}}{3 \text{ minutes}} = 2 \text{ gal/min}$$

The slope is the *rate* at which the tank is being filled, 2 gallons per minute. In Figure 17(b) the tank is being drained at the *rate* of 0.03 gallon per minute, and the slope is -0.03.

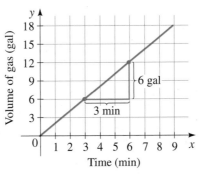

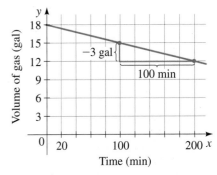

(a) Tank filled at 2 gal/min
Slope of line is 2

(b) Tank drained at 0.03 gal/min
Slope of line is -0.03

FIGURE 17

The next two examples give other situations in which the slope of a line is a rate of change.

EXAMPLE 11 | Slope as Rate of Change

A dam is built on a river to create a reservoir. The water level w in the reservoir is given by the equation

$$w = 4.5t + 28$$

where t is the number of years since the dam was constructed and w is measured in feet.

(a) Sketch a graph of this equation.

(b) What do the slope and w-intercept of this graph represent?

SOLUTION

(a) This equation is linear, so its graph is a line. Since two points determine a line, we plot two points that lie on the graph and draw a line through them.

> When $t = 0$, then $w = 4.5(0) + 28 = 28$, so $(0, 28)$ is on the line.
>
> When $t = 2$, then $w = 4.5(2) + 28 = 37$, so $(2, 37)$ is on the line.

The line that is determined by these points is shown in Figure 18.

(b) The slope is $m = 4.5$; it represents the rate of change of water level with respect to time. This means that the water level *increases* 4.5 ft per year. The w-intercept is 28 and occurs when $t = 0$, so it represents the water level when the dam was constructed.

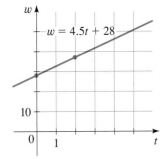

FIGURE 18

✎ NOW TRY EXERCISE **69**   ■

EXAMPLE 12 | Linear Relationship Between Temperature and Elevation

(a) As dry air moves upward, it expands and cools. If the ground temperature is 20 °C and the temperature at a height of 1 km is 10 °C, express the temperature T (in °C) in terms of the height h (in km). (Assume that the relationship between T and h is linear.)

(b) Draw the graph of the linear equation. What does its slope represent?

(c) What is the temperature at a height of 2.5 km?

SOLUTION

(a) Because we are assuming a linear relationship between T and h, the equation must be of the form

$$T = mh + b$$

where m and b are constants. When $h = 0$, we are given that $T = 20$, so

$$20 = m(0) + b$$
$$b = 20$$

Thus, we have

$$T = mh + 20$$

When $h = 1$, we have $T = 10$, so

$$10 = m(1) + 20$$
$$m = 10 - 20 = -10$$

The required expression is

$$T = -10h + 20$$

(b) The graph is sketched in Figure 19. The slope is $m = -10°C/km$, and this represents the rate of change of temperature with respect to distance above the ground. So the temperature *decreases* 10°C per kilometer of height.

(c) At a height of $h = 2.5$ km the temperature is

$$T = -10(2.5) + 20 = -25 + 20 = -5°C$$

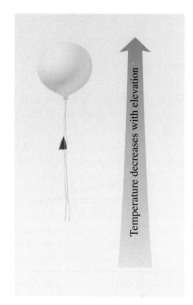

Temperature decreases with elevation

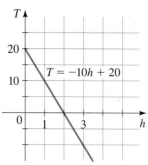

FIGURE 19

NOW TRY EXERCISE **73**

1.2 EXERCISES

CONCEPTS

1. We find the "steepness," or slope, of a line passing through two points by dividing the difference in the _____-coordinates of these points by the difference in the _____-coordinates. So the line passing through the points (0, 1) and (2, 5) has slope

_____.

2. A line has the equation $y = 3x + 2$.

   (a) This line has slope _____.

   (b) Any line parallel to this line has slope _____.

   (c) Any line perpendicular to this line has slope _____.

3. The point-slope form of the equation of the line with slope 3 passing through the point (1, 2) is _____.

4. (a) The slope of a horizontal line is _____. The equation of the horizontal line passing through (2, 3) is _____.

   (b) The slope of a vertical line is _____. The equation of the vertical line passing through (2, 3) is _____.

SKILLS

5–12 ■ Find the slope of the line through P and Q.

5. $P(0, 0), Q(4, 2)$ **6.** $P(0, 0), Q(2, -6)$

7. $P(2, 2), Q(-10, 0)$ **8.** $P(1, 2), Q(3, 3)$

9. $P(2, 4), Q(4, 3)$ **10.** $P(2, -5), Q(-4, 3)$

11. $P(1, -3), Q(-1, 6)$ **12.** $P(-1, -4), Q(6, 0)$

13. Find the slopes of the lines l_1, l_2, l_3, and l_4 in the figure below.

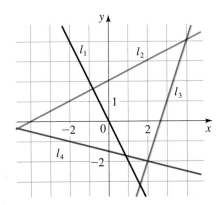

14. (a) Sketch lines through $(0, 0)$ with slopes $1, 0, \frac{1}{2}, 2$, and -1.

(b) Sketch lines through $(0, 0)$ with slopes $\frac{1}{3}, \frac{1}{2}, -\frac{1}{3}$, and 3.

15–18 ■ Find an equation for the line whose graph is sketched.

15.

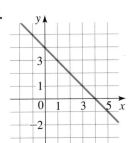

16.

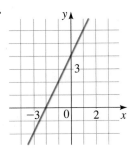

17.

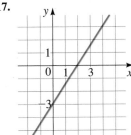

18.

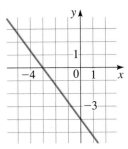

19–38 ■ Find an equation of the line that satisfies the given conditions.

19. Through $(2, 3)$; slope 5

20. Through $(-2, 4)$; slope -1

21. Through $(1, 7)$; slope $\frac{2}{3}$

22. Through $(-3, -5)$; slope $-\frac{7}{2}$

23. Through $(2, 1)$ and $(1, 6)$

24. Through $(-1, -2)$ and $(4, 3)$

25. Slope 3; y-intercept -2

26. Slope $\frac{2}{5}$; y-intercept 4

27. x-intercept 1; y-intercept -3

28. x-intercept -8; y-intercept 6

29. Through $(4, 5)$; parallel to the x-axis

30. Through $(4, 5)$; parallel to the y-axis

31. Through $(1, -6)$; parallel to the line $x + 2y = 6$

32. y-intercept 6; parallel to the line $2x + 3y + 4 = 0$

33. Through $(-1, 2)$; parallel to the line $x = 5$

34. Through $(2, 6)$; perpendicular to the line $y = 1$

35. Through $(-1, -2)$; perpendicular to the line $2x + 5y + 8 = 0$

36. Through $\left(\frac{1}{2}, -\frac{2}{3}\right)$; perpendicular to the line $4x - 8y = 1$

37. Through $(1, 7)$; parallel to the line passing through $(2, 5)$ and $(-2, 1)$

38. Through $(-2, -11)$; perpendicular to the line passing through $(1, 1)$ and $(5, -1)$

39. (a) Sketch the line with slope $\frac{3}{2}$ that passes through the point $(-2, 1)$.

(b) Find an equation for this line.

40. (a) Sketch the line with slope -2 that passes through the point $(4, -1)$.

(b) Find an equation for this line.

41–44 ■ Use a graphing device to graph the given family of lines in the same viewing rectangle. What do the lines have in common?

41. $y = -2x + b$ for $b = 0, \pm 1, \pm 3, \pm 6$

42. $y = mx - 3$ for $m = 0, \pm 0.25, \pm 0.75, \pm 1.5$

43. $y = m(x - 3)$ for $m = 0, \pm 0.25, \pm 0.75, \pm 1.5$

44. $y = 2 + m(x + 3)$ for $m = 0, \pm 0.5, \pm 1, \pm 2, \pm 6$

45–56 ■ Find the slope and y-intercept of the line and draw its graph.

45. $x + y = 3$

46. $3x - 2y = 12$

47. $x + 3y = 0$

48. $2x - 5y = 0$

49. $\frac{1}{2}x - \frac{1}{3}y + 1 = 0$

50. $-3x - 5y + 30 = 0$

51. $y = 4$

52. $x = -5$

53. $3x - 4y = 12$

54. $4y + 8 = 0$

55. $3x + 4y - 1 = 0$

56. $4x + 5y = 10$

57. Use slopes to show that $A(1, 1)$, $B(7, 4)$, $C(5, 10)$, and $D(-1, 7)$ are vertices of a parallelogram.

58. Use slopes to show that $A(-3, -1)$, $B(3, 3)$, and $C(-9, 8)$ are vertices of a right triangle.

59. Use slopes to show that $A(1, 1)$, $B(11, 3)$, $C(10, 8)$, and $D(0, 6)$ are vertices of a rectangle.

60. Use slopes to determine whether the given points are collinear (lie on a line).

(a) $(1, 1), (3, 9), (6, 21)$

(b) $(-1, 3), (1, 7), (4, 15)$

61. Find an equation of the perpendicular bisector of the line segment joining the points $A(1, 4)$ and $B(7, -2)$.

62. Find the area of the triangle formed by the coordinate axes and the line

$$2y + 3x - 6 = 0$$

63. **(a)** Show that if the x- and y-intercepts of a line are nonzero numbers a and b, then the equation of the line can be written in the form

$$\frac{x}{a} + \frac{y}{b} = 1$$

This is called the **two-intercept form** of the equation of a line.

(b) Use part (a) to find an equation of the line whose x-intercept is 6 and whose y-intercept is -8.

64. **(a)** Find an equation for the line tangent to the circle $x^2 + y^2 = 25$ at the point $(3, -4)$. (See the figure.)

(b) At what other point on the circle will a tangent line be parallel to the tangent line in part (a)?

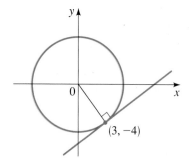

APPLICATIONS

65. **Grade of a Road** West of Albuquerque, New Mexico, Route 40 eastbound is straight and makes a steep descent toward the city. The highway has a 6% grade, which means that its slope is $-\frac{6}{100}$. Driving on this road, you notice from elevation signs that you have descended a distance of 1000 ft. What is the change in your horizontal distance?

6% grade

1000 ft

66. **Global Warming** Some scientists believe that the average surface temperature of the world has been rising steadily. The average surface temperature can be modeled by

$$T = 0.02t + 15.0$$

where T is temperature in °C and t is years since 1950.

(a) What do the slope and T-intercept represent?

(b) Use the equation to predict the average global surface temperature in 2050.

67. **Drug Dosages** If the recommended adult dosage for a drug is D (in mg), then to determine the appropriate dosage c for a child of age a, pharmacists use the equation

$$c = 0.0417D(a + 1)$$

Suppose the dosage for an adult is 200 mg.

(a) Find the slope. What does it represent?

(b) What is the dosage for a newborn?

68. **Flea Market** The manager of a weekend flea market knows from past experience that if she charges x dollars for a rental space at the flea market, then the number y of spaces she can rent is given by the equation $y = 200 - 4x$.

(a) Sketch a graph of this linear equation. (Remember that the rental charge per space and the number of spaces rented must both be nonnegative quantities.)

(b) What do the slope, the y-intercept, and the x-intercept of the graph represent?

69. **Production Cost** A small-appliance manufacturer finds that if he produces x toaster ovens in a month, his production cost is given by the equation

$$y = 6x + 3000$$

(where y is measured in dollars).

(a) Sketch a graph of this linear equation.

(b) What do the slope and y-intercept of the graph represent?

70. **Temperature Scales** The relationship between the Fahrenheit (F) and Celsius (C) temperature scales is given by the equation $F = \frac{9}{5}C + 32$.

(a) Complete the table to compare the two scales at the given values.

(b) Find the temperature at which the scales agree. [*Hint:* Suppose that a is the temperature at which the scales agree. Set $F = a$ and $C = a$. Then solve for a.]

C	F
$-30°$	
$-20°$	
$-10°$	
$0°$	
	$50°$
	$68°$
	$86°$

71. **Crickets and Temperature** Biologists have observed that the chirping rate of crickets of a certain species is related to temperature, and the relationship appears to be very nearly linear. A cricket produces 120 chirps per minute at 70°F and 168 chirps per minute at 80°F.

(a) Find the linear equation that relates the temperature t and the number of chirps per minute n.

(b) If the crickets are chirping at 150 chirps per minute, estimate the temperature.

72. **Depreciation** A small business buys a computer for $4000. After 4 years the value of the computer is expected to be $200. For accounting purposes the business uses *linear depreciation* to assess the value of the computer at a given time.

This means that if V is the value of the computer at time t, then a linear equation is used to relate V and t.

(a) Find a linear equation that relates V and t.

(b) Sketch a graph of this linear equation.

(c) What do the slope and V-intercept of the graph represent?

(d) Find the depreciated value of the computer 3 years from the date of purchase.

73. Pressure and Depth At the surface of the ocean the water pressure is the same as the air pressure above the water, 15 lb/in². Below the surface the water pressure increases by 4.34 lb/in² for every 10 ft of descent.

(a) Find an equation for the relationship between pressure and depth below the ocean surface.

(b) Sketch a graph of this linear equation.

(c) What do the slope and y-intercept of the graph represent?

(d) At what depth is the pressure 100 lb/in²?

Water pressure increases with depth

74. Distance, Speed, and Time Jason and Debbie leave Detroit at 2:00 P.M. and drive at a constant speed, traveling west on I-90. They pass Ann Arbor, 40 mi from Detroit, at 2:50 P.M.

(a) Express the distance traveled in terms of the time elapsed.

(b) Draw the graph of the equation in part (a).

(c) What is the slope of this line? What does it represent?

75. Cost of Driving The monthly cost of driving a car depends on the number of miles driven. Lynn found that in May her driving cost was $380 for 480 mi and in June her cost was $460 for 800 mi. Assume that there is a linear relationship between the monthly cost C of driving a car and the distance driven d.

(a) Find a linear equation that relates C and d.

(b) Use part (a) to predict the cost of driving 1500 mi per month.

(c) Draw the graph of the linear equation. What does the slope of the line represent?

(d) What does the y-intercept of the graph represent?

(e) Why is a linear relationship a suitable model for this situation?

76. Manufacturing Cost The manager of a furniture factory finds that it costs $2200 to manufacture 100 chairs in one day and $4800 to produce 300 chairs in one day.

(a) Assuming that the relationship between cost and the number of chairs produced is linear, find an equation that expresses this relationship. Then graph the equation.

(b) What is the slope of the line in part (a), and what does it represent?

(c) What is the y-intercept of this line, and what does it represent?

DISCOVERY ■ DISCUSSION ■ WRITING

77. What Does the Slope Mean? Suppose that the graph of the outdoor temperature over a certain period of time is a line. How is the weather changing if the slope of the line is positive? If it is negative? If it is zero?

78. Collinear Points Suppose that you are given the coordinates of three points in the plane and you want to see whether they lie on the same line. How can you do this using slopes? Using the Distance Formula? Can you think of another method?

1.3 WHAT IS A FUNCTION?

| Functions All Around Us ▶ Definition of Function ▶ Evaluating a Function ▶ Domain of a Function ▶ Four Ways to Represent a Function

In this section we explore the idea of a function and then give the mathematical definition of function.

▼ Functions All Around Us

In nearly every physical phenomenon we observe that one quantity depends on another. For example, your height depends on your age, the temperature depends on the date, the cost of mailing a package depends on its weight (see Figure 1). We use the term *function* to describe this dependence of one quantity on another. That is, we say the following:

- Height is a function of age.
- Temperature is a function of date.
- Cost of mailing a package is a function of weight.

The U.S. Post Office uses a simple rule to determine the cost of mailing a first-class parcel on the basis of its weight. But it's not so easy to describe the rule that relates height to age or the rule that relates temperature to date.

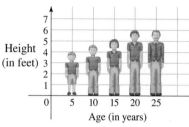

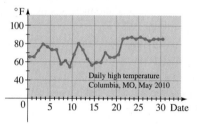

w (ounces)	Postage (dollars)
$0 < w \leq 1$	1.22
$1 < w \leq 2$	1.39
$2 < w \leq 3$	1.56
$3 < w \leq 4$	1.73
$4 < w \leq 5$	1.90
$5 < w \leq 6$	2.07

FIGURE 1

Height is a function of age.   Temperature is a function of date.   Postage is a function of weight.

Can you think of other functions? Here are some more examples:

- The area of a circle is a function of its radius.
- The number of bacteria in a culture is a function of time.
- The weight of an astronaut is a function of her elevation.
- The price of a commodity is a function of the demand for that commodity.

The rule that describes how the area A of a circle depends on its radius r is given by the formula $A = \pi r^2$. Even when a precise rule or formula describing a function is not available, we can still describe the function by a graph. For example, when you turn on a hot water faucet, the temperature of the water depends on how long the water has been running. So we can say:

- The temperature of water from the faucet is a function of time.

Figure 2 shows a rough graph of the temperature T of the water as a function of the time t that has elapsed since the faucet was turned on. The graph shows that the initial temperature of the water is close to room temperature. When the water from the hot water tank reaches the faucet, the water's temperature T increases quickly. In the next phase, T is constant at the temperature of the water in the tank. When the tank is drained, T decreases to the temperature of the cold water supply.

FIGURE 2 Graph of water temperature T as a function of time t

▼ Definition of Function

A function is a rule. To talk about a function, we need to give it a name. We will use letters such as $f, g, h, \ldots$ to represent functions. For example, we can use the letter f to represent a rule as follows:

"f"   is the rule   "square the number"

When we write $f(2)$, we mean "apply the rule f to the number 2." Applying the rule gives $f(2) = 2^2 = 4$. Similarly, $f(3) = 3^2 = 9$, $f(4) = 4^2 = 16$, and in general $f(x) = x^2$.

DEFINITION OF A FUNCTION

A **function** f is a rule that assigns to each element x in a set A exactly one element, called $f(x)$, in a set B.

The ☑ key on your calculator is a good example of a function as a machine. First you input x into the display. Then you press the key labeled ☑. (On most *graphing* calculators the order of these operations is reversed.) If $x < 0$, then x is not in the domain of this function; that is, x is not an acceptable input, and the calculator will indicate an error. If $x \geq 0$, then an approximation to $\sqrt{x}$ appears in the display, correct to a certain number of decimal places. (Thus, the ☑ key on your calculator is not quite the same as the exact mathematical function f defined by $f(x) = \sqrt{x}$.)

We usually consider functions for which the sets A and B are sets of real numbers. The symbol $f(x)$ is read "f of x" or "f at x" and is called the **value of f at x**, or the **image of x under f**. The set A is called the **domain** of the function. The **range** of f is the set of all possible values of $f(x)$ as x varies throughout the domain, that is,

$$\text{range of } f = \{f(x) \mid x \in A\}$$

The symbol that represents an arbitrary number in the domain of a function f is called an **independent variable**. The symbol that represents a number in the range of f is called a **dependent variable**. So if we write $y = f(x)$, then x is the independent variable and y is the dependent variable.

It is helpful to think of a function as a **machine** (see Figure 3). If x is in the domain of the function f, then when x enters the machine, it is accepted as an **input** and the machine produces an **output** $f(x)$ according to the rule of the function. Thus, we can think of the domain as the set of all possible inputs and the range as the set of all possible outputs.

FIGURE 3 Machine diagram of f

Another way to picture a function is by an **arrow diagram** as in Figure 4. Each arrow connects an element of A to an element of B. The arrow indicates that $f(x)$ is associated with x, $f(a)$ is associated with a, and so on.

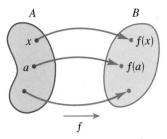

FIGURE 4 Arrow diagram of f

EXAMPLE 1 | Analyzing a Function

A function f is defined by the formula

$$f(x) = x^2 + 4$$

(a) Express in words how f acts on the input x to produce the output $f(x)$.

(b) Evaluate $f(3)$, $f(-2)$, and $f(\sqrt{5})$.

(c) Find the domain and range of f.

(d) Draw a machine diagram for f.

SOLUTION

(a) The formula tells us that f first squares the input x and then adds 4 to the result. So f is the function

"square, then add 4"

(b) The values of f are found by substituting for x in the formula $f(x) = x^2 + 4$.

$$f(3) = 3^2 + 4 = 13 \qquad \text{Replace } x \text{ by } 3$$

$$f(-2) = (-2)^2 + 4 = 8 \qquad \text{Replace } x \text{ by } -2$$

$$f(\sqrt{5}) = (\sqrt{5})^2 + 4 = 9 \qquad \text{Replace } x \text{ by } \sqrt{5}$$

(c) The domain of f consists of all possible inputs for f. Since we can evaluate the formula $f(x) = x^2 + 4$ for every real number x, the domain of f is the set $\mathbb{R}$ of all real numbers.

 The range of f consists of all possible outputs of f. Because $x^2 \geq 0$ for all real numbers x, we have $x^2 + 4 \geq 4$, so for every output of f we have $f(x) \geq 4$. Thus, the range of f is $\{y \mid y \geq 4\} = [4, \infty)$.

(d) A machine diagram for f is shown in Figure 5.

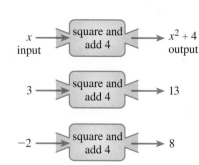

x input → square and add 4 → $x^2 + 4$ output

3 → square and add 4 → 13

−2 → square and add 4 → 8

FIGURE 5 Machine diagram

✎ NOW TRY EXERCISES **9, 13, 17,** AND **43**

▼ Evaluating a Function

In the definition of a function the independent variable x plays the role of a placeholder. For example, the function $f(x) = 3x^2 + x - 5$ can be thought of as

$$f(\,\blacksquare\,) = 3 \cdot \blacksquare^2 + \blacksquare - 5$$

To evaluate f at a number, we substitute the number for the placeholder.

EXAMPLE 2 | Evaluating a Function

Let $f(x) = 3x^2 + x - 5$. Evaluate each function value.

(a) $f(-2)$ **(b)** $f(0)$ **(c)** $f(4)$ **(d)** $f\left(\frac{1}{2}\right)$

SOLUTION To evaluate f at a number, we substitute the number for x in the definition of f.

(a) $f(-2) = 3 \cdot (-2)^2 + (-2) - 5 = 5$

(b) $f(0) = 3 \cdot 0^2 + 0 - 5 = -5$

(c) $f(4) = 3 \cdot (4)^2 + 4 - 5 = 47$

(d) $f\left(\frac{1}{2}\right) = 3 \cdot \left(\frac{1}{2}\right)^2 + \frac{1}{2} - 5 = -\frac{15}{4}$

✎ NOW TRY EXERCISE **19**

EXAMPLE 3 | A Piecewise Defined Function

A cell phone plan costs \$39 a month. The plan includes 400 free minutes and charges 20¢ for each additional minute of usage. The monthly charges are a function of the number of minutes used, given by

$$C(x) = \begin{cases} 39 & \text{if } 0 \leq x \leq 400 \\ 39 + 0.20(x - 400) & \text{if } x > 400 \end{cases}$$

Find $C(100)$, $C(400)$, and $C(480)$.

SOLUTION Remember that a function is a rule. Here is how we apply the rule for this function. First we look at the value of the input x. If $0 \leq x \leq 400$, then the value of $C(x)$ is 39. On the other hand, if $x > 400$, then the value of $C(x)$ is $39 + 0.20(x - 400)$.

Since $100 \leq 400$, we have $C(100) = 39$.

Since $400 \leq 400$, we have $C(400) = 39$.

Since $480 > 400$, we have $C(480) = 39 + 0.20(480 - 400) = 55$.

Thus, the plan charges $39 for 100 minutes, $39 for 400 minutes, and $55 for 480 minutes.

> A **piecewise defined function** is defined by different formulas on different parts of its domain. The function C of Example 3 is piecewise defined.

◥ NOW TRY EXERCISE **27** ■

EXAMPLE 4 | Evaluating a Function

If $f(x) = 2x^2 + 3x - 1$, evaluate the following.

(a) $f(a)$ **(b)** $f(-a)$

(c) $f(a + h)$ **(d)** $\dfrac{f(a + h) - f(a)}{h}, \quad h \neq 0$

> Expressions like the one in part (d) of Example 4 occur frequently in calculus; they are called *difference quotients*, and they represent the average change in the value of f between $x = a$ and $x = a + h$.

SOLUTION

(a) $f(a) = 2a^2 + 3a - 1$

(b) $f(-a) = 2(-a)^2 + 3(-a) - 1 = 2a^2 - 3a - 1$

(c) $f(a + h) = 2(a + h)^2 + 3(a + h) - 1$

$$= 2(a^2 + 2ah + h^2) + 3(a + h) - 1$$

$$= 2a^2 + 4ah + 2h^2 + 3a + 3h - 1$$

(d) Using the results from parts (c) and (a), we have

$$\frac{f(a + h) - f(a)}{h} = \frac{(2a^2 + 4ah + 2h^2 + 3a + 3h - 1) - (2a^2 + 3a - 1)}{h}$$

$$= \frac{4ah + 2h^2 + 3h}{h} = 4a + 2h + 3$$

◥ NOW TRY EXERCISE **35** ■

EXAMPLE 5 | The Weight of an Astronaut

If an astronaut weighs 130 pounds on the surface of the earth, then her weight when she is h miles above the earth is given by the function

$$w(h) = 130\left(\frac{3960}{3960 + h}\right)^2$$

(a) What is her weight when she is 100 mi above the earth?

(b) Construct a table of values for the function w that gives her weight at heights from 0 to 500 mi. What do you conclude from the table?

SOLUTION

(a) We want the value of the function w when $h = 100$; that is, we must calculate $w(100)$.

$$w(100) = 130\left(\frac{3960}{3960 + 100}\right)^2 \approx 123.67$$

So at a height of 100 mi she weighs about 124 lb.

The weight of an object on or near the earth is the gravitational force that the earth exerts on it. When in orbit around the earth, an astronaut experiences the sensation of "weightlessness" because the centripetal force that keeps her in orbit is exactly the same as the gravitational pull of the earth.

(b) The table gives the astronaut's weight, rounded to the nearest pound, at 100-mile increments. The values in the table are calculated as in part (a).

h	$w(h)$
0	130
100	124
200	118
300	112
400	107
500	102

The table indicates that the higher the astronaut travels, the less she weighs.

✎ NOW TRY EXERCISE **71** ■

▼ The Domain of a Function

Recall that the *domain* of a function is the set of all inputs for the function. The domain of a function may be stated explicitly. For example, if we write

$$f(x) = x^2 \qquad 0 \le x \le 5$$

then the domain is the set of all real numbers x for which $0 \le x \le 5$. If the function is given by an algebraic expression and the domain is not stated explicitly, then by convention *the domain of the function is the domain of the algebraic expression—that is, the set of all real numbers for which the expression is defined as a real number.* For example, consider the functions

$$f(x) = \frac{1}{x - 4} \qquad g(x) = \sqrt{x}$$

The function f is not defined at $x = 4$, so its domain is $\{x \mid x \ne 4\}$. The function g is not defined for negative x, so its domain is $\{x \mid x \ge 0\}$.

EXAMPLE 6 | Finding Domains of Functions

Find the domain of each function.

(a) $f(x) = \dfrac{1}{x^2 - x}$ **(b)** $g(x) = \sqrt{9 - x^2}$ **(c)** $h(t) = \dfrac{t}{\sqrt{t + 1}}$

SOLUTION

(a) A rational expression is not defined when the denominator is 0. Since

$$f(x) = \frac{1}{x^2 - x} = \frac{1}{x(x - 1)}$$

we see that $f(x)$ is not defined when $x = 0$ or $x = 1$. Thus, the domain of f is

$$\{x \mid x \ne 0, x \ne 1\}$$

The domain may also be written in interval notation as

$$(\infty, 0) \cup (0, 1) \cup (1, \infty)$$

(b) We can't take the square root of a negative number, so we must have $9 - x^2 \ge 0$. So, $x^2 \le 9$, and hence, we can solve this inequality to find that $-3 \le x \le 3$. Thus, the domain of g is

$$\{x \mid -3 \le x \le 3\} = [-3, 3]$$

(c) We can't take the square root of a negative number, and we can't divide by 0, so we must have $t + 1 > 0$, that is, $t > -1$. So the domain of h is

$$\{t \mid t > -1\} = (-1, \infty)$$

✎ NOW TRY EXERCISES **47** AND **51**

▼ Four Ways to Represent a Function

To help us understand what a function is, we have used machine and arrow diagrams. We can describe a specific function in the following four ways:

- verbally (by a description in words)
- algebraically (by an explicit formula)
- visually (by a graph)
- numerically (by a table of values)

A single function may be represented in all four ways, and it is often useful to go from one representation to another to gain insight into the function. However, certain functions are described more naturally by one method than by the others. An example of a verbal description is the following rule for converting between temperature scales:

> "To find the Fahrenheit equivalent of a Celsius temperature,
> multiply the Celsius temperature by $\frac{9}{5}$, then add 32."

In Example 7 we see how to describe this verbal rule or function algebraically, graphically, and numerically. A useful representation of the area of a circle as a function of its radius is the algebraic formula

$$A(r) = \pi r^2$$

The graph produced by a seismograph (see the box on the next page) is a visual representation of the vertical acceleration function $a(t)$ of the ground during an earthquake. As a final example, consider the function $C(w)$, which is described verbally as "the cost of mailing a first-class letter with weight w." The most convenient way of describing this function is numerically—that is, using a table of values.

MATHEMATICS IN THE MODERN WORLD

Courtesy of NASA

Error-Correcting Codes

The pictures sent back by the *Pathfinder* spacecraft from the surface of Mars on July 4, 1997, were astoundingly clear. But few viewing these pictures were aware of the complex mathematics used to accomplish that feat. The distance to Mars is enormous, and the background noise (or static) is many times stronger than the original signal emitted by the spacecraft. So, when scientists receive the signal, it is full of errors. To get a clear picture, the errors must be found and corrected. This same problem of errors is routinely encountered in transmitting bank records when you use an ATM machine or voice when you are talking on the telephone.

To understand how errors are found and corrected, we must first understand that to transmit pictures, sound, or text we transform them into bits (the digits 0 or 1; see page 116). To help the receiver recognize errors, the message is "coded" by inserting additional bits. For example, suppose you want to transmit the message "10100." A very simple-minded code is as follows: Send each digit a million times. The person receiving the message reads it in blocks of a million digits. If the first block is mostly 1's, he concludes that you are probably trying to transmit a 1, and so on. To say that this code is not efficient is a bit of an understatement; it requires sending a million times more data than the original message. Another method inserts "check digits." For example, for each block of eight digits insert a ninth digit; the inserted digit is 0 if there is an even number of 1's in the block and 1 if there is an odd number. So, if a single digit is wrong (a 0 changed to a 1, or vice versa), the check digits allow us to recognize that an error has occurred. This method does not tell us where the error is, so we can't correct it. Modern error-correcting codes use interesting mathematical algorithms that require inserting relatively few digits but that allow the receiver to not only recognize, but also correct, errors. The first error-correcting code was developed in the 1940s by Richard Hamming at MIT. It is interesting to note that the English language has a built-in error correcting mechanism; to test it, try reading this error-laden sentence: Gve mo libty ox giv ne deth.

We will be using all four representations of functions throughout this book. We summarize them in the following box.

FOUR WAYS TO REPRESENT A FUNCTION

Verbal

Using words:

"To convert from Celsius to Fahrenheit, multiply the Celsius temperature by $\frac{9}{5}$, then add 32."

Relation between Celsius and Fahrenheit temperature scales

Algebraic

Using a formula:

$$A(r) = \pi r^2$$

Area of a circle

Visual

Using a graph:

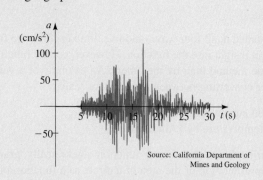

Source: California Department of Mines and Geology

Vertical acceleration during an earthquake

Numerical

Using a table of values:

w (ounces)	$C(w)$ (dollars)
$0 < w \le 1$	1.22
$1 < w \le 2$	1.39
$2 < w \le 3$	1.56
$3 < w \le 4$	1.73
$4 < w \le 5$	1.90
$\vdots$	$\vdots$

Cost of mailing a first-class parcel

EXAMPLE 7 | Representing a Function Verbally, Algebraically, Numerically, and Graphically

Let $F(C)$ be the Fahrenheit temperature corresponding to the Celsius temperature C. (Thus, F is the function that converts Celsius inputs to Fahrenheit outputs.) The box above gives a verbal description of this function. Find ways to represent this function

(a) Algebraically (using a formula)

(b) Numerically (using a table of values)

(c) Visually (using a graph)

SOLUTION

(a) The verbal description tells us that we should first multiply the input C by $\frac{9}{5}$ and then add 32 to the result. So we get

$$F(C) = \tfrac{9}{5}C + 32$$

(b) We use the algebraic formula for F that we found in part (a) to construct a table of values:

C (Celsius)	F (Fahrenheit)
−10	14
0	32
10	50
20	68
30	86
40	104

(c) We use the points tabulated in part (b) to help us draw the graph of this function in Figure 6.

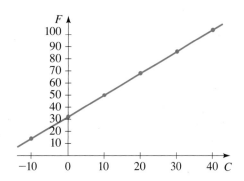

FIGURE 6 Celsius and Fahrenheit

✎. NOW TRY EXERCISE **65**

1.3 EXERCISES

CONCEPTS

1. If a function f is given by the formula $y = f(x)$, then $f(a)$ is the _____ of f at $x = a$.

2. For a function f, the set of all possible inputs is called the _____ of f, and the set of all possible outputs is called the _____ of f.

3. **(a)** Which of the following functions have 5 in their domain?
$$f(x) = x^2 - 3x \qquad g(x) = \frac{x - 5}{x} \qquad h(x) = \sqrt{x - 10}$$

(b) For the functions from part (a) that *do* have 5 in their domain, find the value of the function at 5.

4. A function is given algebraically by the formula $f(x) = (x - 4)^2 + 3$. Complete these other ways to represent f:

(a) *Verbal:* "Subtract 4, then _____ and _____.
(b) *Numerical:*

x	$f(x)$
0	19
2	
4	
6	

SKILLS

5–8 ■ Express the rule in function notation. (For example, the rule "square, then subtract 5" is expressed as the function $f(x) = x^2 - 5$.)

5. Add 3, then multiply by 2   **6.** Divide by 7, then subtract 4

7. Subtract 5, then square

8. Take the square root, add 8, then multiply by $\frac{1}{3}$

9–12 ■ Express the function (or rule) in words.

✎. **9.** $h(x) = x^2 + 2$ **10.** $k(x) = \sqrt{x + 2}$

11. $f(x) = \dfrac{x - 4}{3}$ **12.** $g(x) = \dfrac{x}{3} - 4$

13–14 ■ Draw a machine diagram for the function.

✎. **13.** $f(x) = \sqrt{x - 1}$ **14.** $f(x) = \dfrac{3}{x - 2}$

15–16 ■ Complete the table.

15. $f(x) = 2(x - 1)^2$ **16.** $g(x) = |2x + 3|$

x	$f(x)$
−1	
0	
1	
2	
3	

x	$g(x)$
−3	
−2	
0	
1	
3	

17–26 ■ Evaluate the function at the indicated values.

✎. **17.** $f(x) = x^2 - 6$; $f(-3), f(3), f(0), f(\frac{1}{2}), f(10)$

18. $f(x) = x^3 + 2x$; $f(-2), f(1), f(0), f(\frac{1}{3}), f(0.2)$

✎. **19.** $f(x) = 2x + 1$;

$f(1), f(-2), f(\frac{1}{2}), f(a), f(-a), f(a + b)$

20. $f(x) = x^2 + 2x$;

$f(0), f(3), f(-3), f(a), f(-x), f\left(\dfrac{1}{a}\right)$

21. $g(x) = \dfrac{1 - x}{1 + x}$;

$g(2), g(-2), g(\frac{1}{2}), g(a), g(a - 1), g(-1)$

22. $h(t) = t + \dfrac{1}{t}$;

$h(1), h(-1), h(2), h(\frac{1}{2}), h(x), h\left(\dfrac{1}{x}\right)$

23. $f(x) = 2x^2 + 3x - 4$;

$f(0), f(2), f(-2), f(\sqrt{2}), f(x + 1), f(-x)$

24. $f(x) = x^3 - 4x^2$;

$f(0), f(1), f(-1), f(\frac{3}{2}), f\left(\dfrac{x}{2}\right), f(x^2)$

25. $f(x) = 2|x - 1|$;

$f(-2), f(0), f(\frac{1}{2}), f(2), f(x + 1), f(x^2 + 2)$

26. $f(x) = \dfrac{|x|}{x}$;

$f(-2), f(-1), f(0), f(5), f(x^2), f\left(\dfrac{1}{x}\right)$

27–30 ■ Evaluate the piecewise defined function at the indicated values.

27. $f(x) = \begin{cases} x^2 & \text{if } x < 0 \\ x + 1 & \text{if } x \geq 0 \end{cases}$

$f(-2), f(-1), f(0), f(1), f(2)$

28. $f(x) = \begin{cases} 5 & \text{if } x \leq 2 \\ 2x - 3 & \text{if } x > 2 \end{cases}$

$f(-3), f(0), f(2), f(3), f(5)$

29. $f(x) = \begin{cases} x^2 + 2x & \text{if } x \leq -1 \\ x & \text{if } -1 < x \leq 1 \\ -1 & \text{if } x > 1 \end{cases}$

$f(-4), f(-\frac{3}{2}), f(-1), f(0), f(25)$

30. $f(x) = \begin{cases} 3x & \text{if } x < 0 \\ x + 1 & \text{if } 0 \leq x \leq 2 \\ (x - 2)^2 & \text{if } x > 2 \end{cases}$

$f(-5), f(0), f(1), f(2), f(5)$

31–34 ■ Use the function to evaluate the indicated expressions and simplify.

31. $f(x) = x^2 + 1$; $f(x + 2), f(x) + f(2)$

32. $f(x) = 3x - 1$; $f(2x), 2f(x)$

33. $f(x) = x + 4$; $f(x^2), (f(x))^2$

34. $f(x) = 6x - 18$; $f\left(\dfrac{x}{3}\right), \dfrac{f(x)}{3}$

35–42 ■ Find $f(a)$, $f(a + h)$, and the difference quotient $\dfrac{f(a + h) - f(a)}{h}$, where $h \neq 0$.

35. $f(x) = 3x + 2$

36. $f(x) = x^2 + 1$

37. $f(x) = 5$

38. $f(x) = \dfrac{1}{x + 1}$

39. $f(x) = \dfrac{x}{x + 1}$

40. $f(x) = \dfrac{2x}{x - 1}$

41. $f(x) = 3 - 5x + 4x^2$

42. $f(x) = x^3$

43–64 ■ Find the domain of the function.

43. $f(x) = 2x$

44. $f(x) = x^2 + 1$

45. $f(x) = 2x, \quad -1 \leq x \leq 5$

46. $f(x) = x^2 + 1, \quad 0 \leq x \leq 5$

47. $f(x) = \dfrac{1}{x - 3}$

48. $f(x) = \dfrac{1}{3x - 6}$

49. $f(x) = \dfrac{x + 2}{x^2 - 1}$

50. $f(x) = \dfrac{x^4}{x^2 + x - 6}$

51. $f(x) = \sqrt{x - 5}$

52. $f(x) = \sqrt[4]{x + 9}$

53. $f(t) = \sqrt[3]{t - 1}$

54. $g(x) = \sqrt{7 - 3x}$

55. $h(x) = \sqrt{2x - 5}$

56. $G(x) = \sqrt{x^2 - 9}$

57. $g(x) = \dfrac{\sqrt{2 + x}}{3 - x}$

58. $g(x) = \dfrac{\sqrt{x}}{2x^2 + x - 1}$

59. $g(x) = \sqrt[4]{x^2 - 6x}$

60. $g(x) = \sqrt{x^2 - 2x - 8}$

61. $f(x) = \dfrac{3}{\sqrt{x - 4}}$

62. $f(x) = \dfrac{x^2}{\sqrt{6 - x}}$

63. $f(x) = \dfrac{(x + 1)^2}{\sqrt{2x - 1}}$

64. $f(x) = \dfrac{x}{\sqrt[4]{9 - x^2}}$

65–68 ■ A verbal description of a function is given. Find **(a)** algebraic, **(b)** numerical, and **(c)** graphical representations for the function.

65. To evaluate $f(x)$, divide the input by 3 and add $\frac{2}{3}$ to the result.

66. To evaluate $g(x)$, subtract 4 from the input and multiply the result by $\frac{3}{4}$.

67. Let $T(x)$ be the amount of sales tax charged in Lemon County on a purchase of x dollars. To find the tax, take 8% of the purchase price.

68. Let $V(d)$ be the volume of a sphere of diameter d. To find the volume, take the cube of the diameter, then multiply by π and divide by 6.

APPLICATIONS

69. Production Cost   The cost C in dollars of producing x yards of a certain fabric is given by the function

$$C(x) = 1500 + 3x + 0.02x^2 + 0.0001x^3$$

(a) Find $C(10)$ and $C(100)$.
(b) What do your answers in part (a) represent?
(c) Find $C(0)$. (This number represents the *fixed costs*.)

70. Area of a Sphere The surface area S of a sphere is a function of its radius r given by

$$S(r) = 4\pi r^2$$

(a) Find $S(2)$ and $S(3)$.
(b) What do your answers in part (a) represent?

71. Torricelli's Law A tank holds 50 gallons of water, which drains from a leak at the bottom, causing the tank to empty in 20 minutes. The tank drains faster when it is nearly full because the pressure on the leak is greater. **Torricelli's Law** gives the volume of water remaining in the tank after t minutes as

$$V(t) = 50\left(1 - \frac{t}{20}\right)^2 \qquad 0 \le t \le 20$$

(a) Find $V(0)$ and $V(20)$.
(b) What do your answers to part (a) represent?
(c) Make a table of values of $V(t)$ for $t = 0, 5, 10, 15, 20$.

72. How Far Can You See? Because of the curvature of the earth, the maximum distance D that you can see from the top of a tall building or from an airplane at height h is given by the function

$$D(h) = \sqrt{2rh + h^2}$$

where $r = 3960$ mi is the radius of the earth and D and h are measured in miles.
(a) Find $D(0.1)$ and $D(0.2)$.
(b) How far can you see from the observation deck of Toronto's CN Tower, 1135 ft above the ground?
(c) Commercial aircraft fly at an altitude of about 7 mi. How far can the pilot see?

73. Blood Flow As blood moves through a vein or an artery, its velocity v is greatest along the central axis and decreases as the distance r from the central axis increases (see the figure). The formula that gives v as a function of r is called the **law of laminar flow**. For an artery with radius 0.5 cm, the relationship between v (in cm/s) and r (in cm) is given by the function

$$v(r) = 18{,}500(0.25 - r^2) \qquad 0 \le r \le 0.5$$

(a) Find $v(0.1)$ and $v(0.4)$.
(b) What do your answers to part (a) tell you about the flow of blood in this artery?
(c) Make a table of values of $v(r)$ for $r = 0, 0.1, 0.2, 0.3, 0.4, 0.5$.

74. Pupil Size When the brightness x of a light source is increased, the eye reacts by decreasing the radius R of the pupil. The dependence of R on x is given by the function

$$R(x) = \sqrt{\frac{13 + 7x^{0.4}}{1 + 4x^{0.4}}}$$

where R is measured in millimeters and x is measured in appropriate units of brightness.
(a) Find $R(1)$, $R(10)$, and $R(100)$.
(b) Make a table of values of $R(x)$.

75. Relativity According to the Theory of Relativity, the length L of an object is a function of its velocity v with respect to an observer. For an object whose length at rest is 10 m, the function is given by

$$L(v) = 10\sqrt{1 - \frac{v^2}{c^2}}$$

where c is the speed of light (300,000 km/s).
(a) Find $L(0.5c)$, $L(0.75c)$, and $L(0.9c)$.
(b) How does the length of an object change as its velocity increases?

76. Income Tax In a certain country, income tax T is assessed according to the following function of income x:

$$T(x) = \begin{cases} 0 & \text{if } 0 \le x \le 10{,}000 \\ 0.08x & \text{if } 10{,}000 < x \le 20{,}000 \\ 1600 + 0.15x & \text{if } 20{,}000 < x \end{cases}$$

(a) Find $T(5{,}000)$, $T(12{,}000)$, and $T(25{,}000)$.
(b) What do your answers in part (a) represent?

77. Internet Purchases An Internet bookstore charges $15 shipping for orders under $100 but provides free shipping for orders of $100 or more. The cost C of an order is a function of the total price x of the books purchased, given by

$$C(x) = \begin{cases} x + 15 & \text{if } x < 100 \\ x & \text{if } x \ge 100 \end{cases}$$

(a) Find $C(75)$, $C(90)$, $C(100)$, and $C(105)$.
(b) What do your answers in part (a) represent?

78. Cost of a Hotel Stay A hotel chain charges $75 each night for the first two nights and $50 for each additional night's stay. The total cost T is a function of the number of nights x that a guest stays.

(a) Complete the expressions in the following piecewise defined function.

$$T(x) = \begin{cases} \rule{1cm}{0.3mm} & \text{if } 0 \le x \le 2 \\ \rule{1cm}{0.3mm} & \text{if } x > 2 \end{cases}$$

(b) Find $T(2)$, $T(3)$, and $T(5)$.

(c) What do your answers in part (b) represent?

79. Speeding Tickets In a certain state the maximum speed permitted on freeways is 65 mi/h, and the minimum is 40. The fine F for violating these limits is $15 for every mile above the maximum or below the minimum.

(a) Complete the expressions in the following piecewise defined function, where x is the speed at which you are driving.

$$F(x) = \begin{cases} \rule{1cm}{0.3mm} & \text{if } 0 < x < 40 \\ \rule{1cm}{0.3mm} & \text{if } 40 \le x \le 65 \\ \rule{1cm}{0.3mm} & \text{if } x > 65 \end{cases}$$

(b) Find $F(30)$, $F(50)$, and $F(75)$.

(c) What do your answers in part (b) represent?

80. Height of Grass A home owner mows the lawn every Wednesday afternoon. Sketch a rough graph of the height of the grass as a function of time over the course of a four-week period beginning on a Sunday.

81. Temperature Change You place a frozen pie in an oven and bake it for an hour. Then you take the pie out and let it cool before eating it. Sketch a rough graph of the temperature of the pie as a function of time.

82. Daily Temperature Change Temperature readings T (in °F) were recorded every 2 hours from midnight to noon in Atlanta, Georgia, on March 18, 1996. The time t was measured in hours from midnight. Sketch a rough graph of T as a function of t.

t	0	2	4	6	8	10	12
T	58	57	53	50	51	57	61

83. Population Growth The population P (in thousands) of San Jose, California, from 1988 to 2000 is shown in the table. (Midyear estimates are given.) Draw a rough graph of P as a function of time t.

t	1988	1990	1992	1994	1996	1998	2000
P	733	782	800	817	838	861	895

DISCOVERY ■ DISCUSSION ■ WRITING

84. Examples of Functions At the beginning of this section we discussed three examples of everyday, ordinary functions: Height is a function of age, temperature is a function of date, and postage cost is a function of weight. Give three other examples of functions from everyday life.

85. Four Ways to Represent a Function In the box on page 34 we represented four different functions verbally, algebraically, visually, and numerically. Think of a function that can be represented in all four ways, and write the four representations.

1.4 GRAPHS OF FUNCTIONS

Graphing Functions by Plotting Points ▶ Graphing Functions with a Graphing Calculator ▶ Graphing Piecewise Defined Functions ▶ The Vertical Line Test ▶ Equations That Define Functions

The most important way to visualize a function is through its graph. In this section we investigate in more detail the concept of graphing functions.

▼ Graphing Functions by Plotting Points

To graph a function f, we plot the points $(x, f(x))$ in a coordinate plane. In other words, we plot the points (x, y) whose x-coordinate is an input and whose y-coordinate is the corresponding output of the function.

> **THE GRAPH OF A FUNCTION**
>
> If f is a function with domain A, then the **graph** of f is the set of ordered pairs
>
> $$\{(x, f(x)) \mid x \in A\}$$
>
> plotted in a coordinate plane. In other words, the graph of f is the set of all points (x, y) such that $y = f(x)$; that is, the graph of f is the graph of the equation $y = f(x)$.

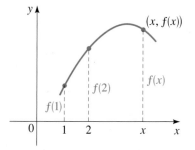

FIGURE 1 The height of the graph above the point x is the value of $f(x)$.

The graph of a function f gives a picture of the behavior or "life history" of the function. We can read the value of $f(x)$ from the graph as being the height of the graph above the point x (see Figure 1).

A function f of the form $f(x) = mx + b$ is called a **linear function** because its graph is the graph of the equation $y = mx + b$, which represents a line with slope m and y-intercept b. A special case of a linear function occurs when the slope is $m = 0$. The function $f(x) = b$, where b is a given number, is called a **constant function** because all its values are the same number, namely, b. Its graph is the horizontal line $y = b$. Figure 2 shows the graphs of the constant function $f(x) = 3$ and the linear function $f(x) = 2x + 1$.

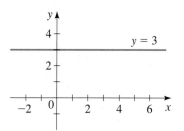

The constant function $f(x) = 3$

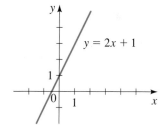

The linear function $f(x) = 2x + 1$

FIGURE 2

EXAMPLE 1 | Graphing Functions by Plotting Points

Sketch graphs of the following functions.

(a) $f(x) = x^2$ **(b)** $g(x) = x^3$ **(c)** $h(x) = \sqrt{x}$

SOLUTION We first make a table of values. Then we plot the points given by the table and join them by a smooth curve to obtain the graph. The graphs are sketched in Figure 3 on the next page.

x	$f(x) = x^2$
0	0
$\pm\frac{1}{2}$	$\frac{1}{4}$
± 1	1
± 2	4
± 3	9

x	$g(x) = x^3$
0	0
$\frac{1}{2}$	$\frac{1}{8}$
1	1
2	8
$-\frac{1}{2}$	$-\frac{1}{8}$
-1	-1
-2	-8

x	$h(x) = \sqrt{x}$
0	0
1	1
2	$\sqrt{2}$
3	$\sqrt{3}$
4	2
5	$\sqrt{5}$

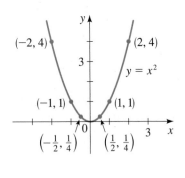

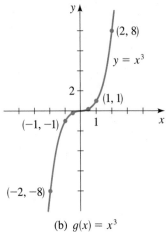

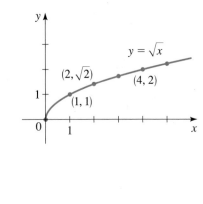

FIGURE 3   (a) $f(x) = x^2$ (b) $g(x) = x^3$ (c) $h(x) = \sqrt{x}$

✎ NOW TRY EXERCISES **11, 15**, AND **19**   ◼

▼ **Graphing Functions with a Graphing Calculator**

A convenient way to graph a function is to use a graphing calculator. Because the graph of a function f is the graph of the equation $y = f(x)$, we can use the methods of Appendix C to graph functions on a graphing calculator.

EXAMPLE 2 | Graphing a Function with a Graphing Calculator

Use a graphing calculator to graph the function $f(x) = x^3 - 8x^2$ in an appropriate viewing rectangle.

SOLUTION   To graph the function $f(x) = x^3 - 8x^2$, we must graph the equation $y = x^3 - 8x^2$. On the TI-83 graphing calculator the default viewing rectangle gives the graph in Figure 4(a). But this graph appears to spill over the top and bottom of the screen. We need to expand the vertical axis to get a better representation of the graph. The viewing rectangle $[-4, 10]$ by $[-100, 100]$ gives a more complete picture of the graph, as shown in Figure 4(b).

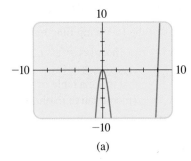

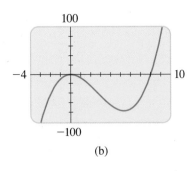

(a)   (b)

FIGURE 4 Graphing the function $f(x) = x^3 - 8x^2$

✎ NOW TRY EXERCISE **29**   ◼

EXAMPLE 3 | A Family of Power Functions

(a) Graph the functions $f(x) = x^n$ for $n = 2, 4,$ and 6 in the viewing rectangle $[-2, 2]$ by $[-1, 3]$.

(b) Graph the functions $f(x) = x^n$ for $n = 1, 3,$ and 5 in the viewing rectangle $[-2, 2]$ by $[-2, 2]$.

(c) What conclusions can you draw from these graphs?

SOLUTION To graph the function $f(x) = x^n$, we graph the equation $y = x^n$. The graphs for parts (a) and (b) are shown in Figure 5.

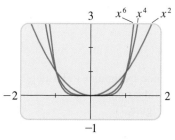

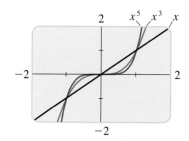

(a) Even powers of x (b) Odd powers of x

FIGURE 5 A family of power functions $f(x) = x^n$

(c) We see that the general shape of the graph of $f(x) = x^n$ depends on whether n is even or odd.

If n is even, the graph of $f(x) = x^n$ is similar to the parabola $y = x^2$.

If n is odd, the graph of $f(x) = x^n$ is similar to that of $y = x^3$.

✎ NOW TRY EXERCISE **69** ◼

Notice from Figure 5 that as n increases, the graph of $y = x^n$ becomes flatter near 0 and steeper when $x > 1$. When $0 < x < 1$, the lower powers of x are the "bigger" functions. But when $x > 1$, the higher powers of x are the dominant functions.

▼ Graphing Piecewise Defined Functions

A piecewise defined function is defined by different formulas on different parts of its domain. As you might expect, the graph of such a function consists of separate pieces.

EXAMPLE 4 | Graph of a Piecewise Defined Function

Sketch the graph of the function.

$$f(x) = \begin{cases} x^2 & \text{if } x \le 1 \\ 2x + 1 & \text{if } x > 1 \end{cases}$$

SOLUTION If $x \le 1$, then $f(x) = x^2$, so the part of the graph to the left of $x = 1$ coincides with the graph of $y = x^2$, which we sketched in Figure 3. If $x > 1$, then $f(x) = 2x + 1$, so the part of the graph to the right of $x = 1$ coincides with the line $y = 2x + 1$, which we graphed in Figure 2. This enables us to sketch the graph in Figure 6.

The solid dot at $(1, 1)$ indicates that this point is included in the graph; the open dot at $(1, 3)$ indicates that this point is excluded from the graph.

On many graphing calculators the graph in Figure 6 can be produced by using the logical functions in the calculator. For example, on the TI-83 the following equation gives the required graph:

$$Y_1 = (X \le 1)X^2 + (X > 1)(2X + 1)$$

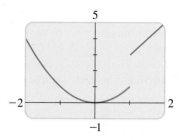

(To avoid the extraneous vertical line between the two parts of the graph, put the calculator in **D o t** mode.)

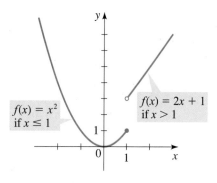

$f(x) = x^2$
if $x \le 1$

$f(x) = 2x + 1$
if $x > 1$

FIGURE 6

$$f(x) = \begin{cases} x^2 & \text{if } x \le 1 \\ 2x + 1 & \text{if } x > 1 \end{cases}$$

✎ NOW TRY EXERCISE **35** ◼

EXAMPLE 5 | Graph of the Absolute Value Function

Sketch a graph of the absolute value function $f(x) = |x|$.

SOLUTION Recall that

$$|x| = \begin{cases} x & \text{if } x \geq 0 \\ -x & \text{if } x < 0 \end{cases}$$

Using the same method as in Example 4, we note that the graph of f coincides with the line $y = x$ to the right of the y-axis and coincides with the line $y = -x$ to the left of the y-axis (see Figure 7).

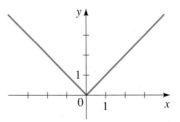

FIGURE 7 Graph of $f(x) = |x|$

✎ NOW TRY EXERCISE **23**

The **greatest integer function** is defined by

$$[\![x]\!] = \text{greatest integer less than or equal to } x$$

For example, $[\![2]\!] = 2$, $[\![2.3]\!] = 2$, $[\![1.999]\!] = 1$, $[\![0.002]\!] = 0$, $[\![-3.5]\!] = -4$, and $[\![-0.5]\!] = -1$.

EXAMPLE 6 | Graph of the Greatest Integer Function

Sketch a graph of $f(x) = [\![x]\!]$.

SOLUTION The table shows the values of f for some values of x. Note that $f(x)$ is constant between consecutive integers, so the graph between integers is a horizontal line segment, as shown in Figure 8.

x	$[\![x]\!]$
⋮	⋮
$-2 \leq x < -1$	-2
$-1 \leq x < 0$	-1
$0 \leq x < 1$	0
$1 \leq x < 2$	1
$2 \leq x < 3$	2
⋮	⋮

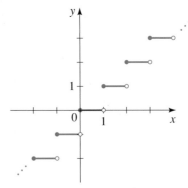

FIGURE 8 The greatest integer function, $y = [\![x]\!]$

The greatest integer function is an example of a **step function**. The next example gives a real-world example of a step function.

EXAMPLE 7 | The Cost Function for Long-Distance Phone Calls

The cost of a long-distance daytime phone call from Toronto, Canada, to Mumbai, India, is 69 cents for the first minute and 58 cents for each additional minute (or part of a minute). Draw the graph of the cost C (in dollars) of the phone call as a function of time t (in minutes).

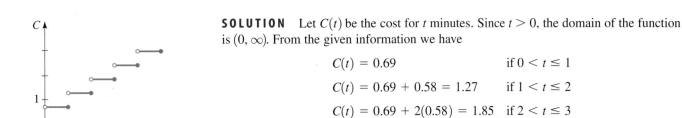

SOLUTION   Let $C(t)$ be the cost for t minutes. Since $t > 0$, the domain of the function is $(0, \infty)$. From the given information we have

$$C(t) = 0.69 \qquad\qquad\qquad \text{if } 0 < t \le 1$$
$$C(t) = 0.69 + 0.58 = 1.27 \qquad \text{if } 1 < t \le 2$$
$$C(t) = 0.69 + 2(0.58) = 1.85 \quad \text{if } 2 < t \le 3$$
$$C(t) = 0.69 + 3(0.58) = 2.43 \quad \text{if } 3 < t \le 4$$

FIGURE 9 Cost of a long-distance call

and so on. The graph is shown in Figure 9.

✎ NOW TRY EXERCISE **81**   ■

A function is called **continuous** if its graph has no "breaks" or "holes." The functions in Examples 1, 2, 3, and 5 are continuous; the functions in Examples 4, 6, and 7 are not continuous.

▼ The Vertical Line Test

The graph of a function is a curve in the xy-plane. But the question arises: Which curves in the xy-plane are graphs of functions? This is answered by the following test.

> ### THE VERTICAL LINE TEST
>
> A curve in the coordinate plane is the graph of a function if and only if no vertical line intersects the curve more than once.

We can see from Figure 10 why the Vertical Line Test is true. If each vertical line $x = a$ intersects a curve only once at (a, b), then exactly one functional value is defined by $f(a) = b$. But if a line $x = a$ intersects the curve twice, at (a, b) and at (a, c), then the curve cannot represent a function because a function cannot assign two different values to a.

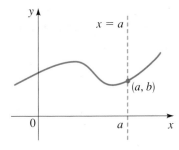
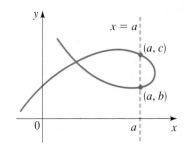

FIGURE 10 Vertical Line Test

Graph of a function   Not a graph of a function

EXAMPLE 8 | Using the Vertical Line Test

Using the Vertical Line Test, we see that the curves in parts (b) and (c) of Figure 11 represent functions, whereas those in parts (a) and (d) do not.

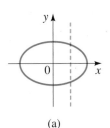
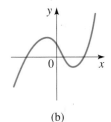
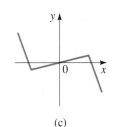
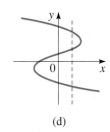

FIGURE 11   (a)   (b)   (c)   (d)

✎ NOW TRY EXERCISE **51**   ■

▼ Equations That Define Functions

Any equation in the variables x and y defines a relationship between these variables. For example, the equation

$$y - x^2 = 0$$

defines a relationship between y and x. Does this equation define y as a *function* of x? To find out, we solve for y and get

$$y = x^2$$

We see that the equation defines a rule, or function, that gives one value of y for each value of x. We can express this rule in function notation as

$$f(x) = x^2$$

But not every equation defines y as a function of x, as the following example shows.

EXAMPLE 9 | Equations That Define Functions

Does the equation define y as a function of x?

(a) $y - x^2 = 2$ **(b)** $x^2 + y^2 = 4$

SOLUTION

(a) Solving for y in terms of x gives

$$y - x^2 = 2$$
$$y = x^2 + 2 \qquad \text{Add } x^2$$

The last equation is a rule that gives one value of y for each value of x, so it defines y as a function of x. We can write the function as $f(x) = x^2 + 2$.

(b) We try to solve for y in terms of x:

$$x^2 + y^2 = 4$$
$$y^2 = 4 - x^2 \qquad \text{Subtract } x^2$$
$$y = \pm\sqrt{4 - x^2} \qquad \text{Take square roots}$$

The last equation gives two values of y for a given value of x. Thus, the equation does not define y as a function of x.

✎ NOW TRY EXERCISES **57** AND **61**

The graphs of the equations in Example 9 are shown in Figure 12. The Vertical Line Test shows graphically that the equation in Example 9(a) defines a function but the equation in Example 9(b) does not.

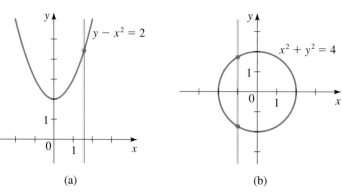

(a)   (b)

FIGURE 12

DONALD KNUTH was born in Milwaukee in 1938 and is Professor Emeritus of Computer Science at Stanford University. When Knuth was a high school student, he became fascinated with graphs of functions and laboriously drew many hundreds of them because he wanted to see the behavior of a great variety of functions. (Today, of course, it is far easier to use computers and graphing calculators to do this.) While still a graduate student at Caltech, he started writing a monumental series of books entitled *The Art of Computer Programming*.

Knuth is famous for his invention of TEX, a system of computer-assisted typesetting. This system was used in the preparation of the manuscript for this textbook.

Knuth has received numerous honors, among them election as an associate of the French Academy of Sciences, and as a Fellow of the Royal Society. President Carter awarded him the National Medal of Science in 1979.

The following table shows the graphs of some functions that you will see frequently in this book.

SOME FUNCTIONS AND THEIR GRAPHS

Linear functions
$f(x) = mx + b$

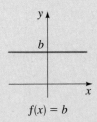

$f(x) = b$

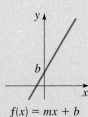

$f(x) = mx + b$

Power functions
$f(x) = x^n$

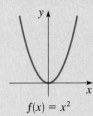

$f(x) = x^2$

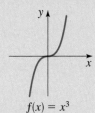

$f(x) = x^3$

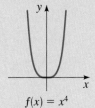

$f(x) = x^4$

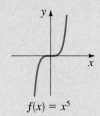

$f(x) = x^5$

Root functions
$f(x) = \sqrt[n]{x}$

$f(x) = \sqrt{x}$

$f(x) = \sqrt[3]{x}$

$f(x) = \sqrt[4]{x}$

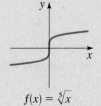

$f(x) = \sqrt[5]{x}$

Reciprocal functions
$f(x) = \dfrac{1}{x^n}$

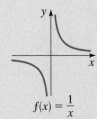

$f(x) = \dfrac{1}{x}$

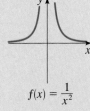

$f(x) = \dfrac{1}{x^2}$

Absolute value function
$f(x) = |x|$

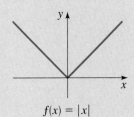

$f(x) = |x|$

Greatest integer function
$f(x) = [\![x]\!]$

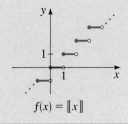

$f(x) = [\![x]\!]$

1.4 EXERCISES

CONCEPTS

1. To graph the function f, we plot the points $(x, \underline{\hspace{1cm}})$ in a coordinate plane. To graph $f(x) = x^3 + 2$, we plot the points $(x, \underline{\hspace{1cm}})$. So the point $(2, \underline{\hspace{1cm}})$ is on the graph of f.

The height of the graph of f above the x-axis when $x = 2$ is

$\underline{\hspace{1cm}}$.

2. If $f(2) = 3$, then the point $(2, \underline{\hspace{1cm}})$ is on the graph of f.

3. If the point $(2, 3)$ is on the graph of f, then $f(2) = $ _____.

4. Match the function with its graph.

(a) $f(x) = x^2$ (b) $f(x) = x^3$

(c) $f(x) = \sqrt{x}$ (d) $f(x) = |x|$

I
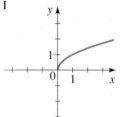

II
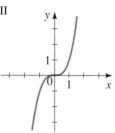

III
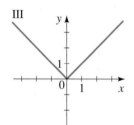

IV
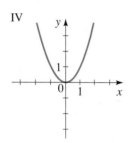

SKILLS

5–28 ■ Sketch the graph of the function by first making a table of values.

5. $f(x) = 2$ **6.** $f(x) = -3$

7. $f(x) = 2x - 4$ **8.** $f(x) = 6 - 3x$

9. $f(x) = -x + 3$, $-3 \le x \le 3$

10. $f(x) = \dfrac{x - 3}{2}$, $0 \le x \le 5$

 11. $f(x) = -x^2$ **12.** $f(x) = x^2 - 4$

13. $h(x) = 16 - x^2$ **14.** $g(x) = (x - 3)^2$

15. $g(x) = x^3 - 8$ **16.** $g(x) = (x + 2)^3$

17. $g(x) = x^2 - 2x$ **18.** $h(x) = 4x^2 - x^4$

19. $f(x) = 1 + \sqrt{x}$ **20.** $f(x) = \sqrt{x + 4}$

21. $g(x) = -\sqrt{x}$ **22.** $g(x) = \sqrt{-x}$

23. $H(x) = |2x|$ **24.** $H(x) = |x + 1|$

25. $G(x) = |x| + x$ **26.** $G(x) = |x| - x$

27. $f(x) = |2x - 2|$ **28.** $f(x) = \dfrac{x}{|x|}$

29–32 ■ Graph the function in each of the given viewing rectangles, and select the one that produces the most appropriate graph of the function.

29. $f(x) = 8x - x^2$

(a) $[-5, 5]$ by $[-5, 5]$

(b) $[-10, 10]$ by $[-10, 10]$

(c) $[-2, 10]$ by $[-5, 20]$

(d) $[-10, 10]$ by $[-100, 100]$

30. $g(x) = x^2 - x - 20$

(a) $[-2, 2]$ by $[-5, 5]$

(b) $[-10, 10]$ by $[-10, 10]$

(c) $[-7, 7]$ by $[-25, 20]$

(d) $[-10, 10]$ by $[-100, 100]$

31. $h(x) = x^3 - 5x - 4$

(a) $[-2, 2]$ by $[-2, 2]$

(b) $[-3, 3]$ by $[-10, 10]$

(c) $[-3, 3]$ by $[-10, 5]$

(d) $[-10, 10]$ by $[-10, 10]$

32. $k(x) = \frac{1}{32}x^4 - x^2 + 2$

(a) $[-1, 1]$ by $[-1, 1]$

(b) $[-2, 2]$ by $[-2, 2]$

(c) $[-5, 5]$ by $[-5, 5]$

(d) $[-10, 10]$ by $[-10, 10]$

33–46 ■ Sketch the graph of the piecewise defined function.

33. $f(x) = \begin{cases} 0 & \text{if } x < 2 \\ 1 & \text{if } x \ge 2 \end{cases}$

34. $f(x) = \begin{cases} 1 & \text{if } x \le 1 \\ x + 1 & \text{if } x > 1 \end{cases}$

35. $f(x) = \begin{cases} 3 & \text{if } x < 2 \\ x - 1 & \text{if } x \ge 2 \end{cases}$

36. $f(x) = \begin{cases} 1 - x & \text{if } x < -2 \\ 5 & \text{if } x \ge -2 \end{cases}$

37. $f(x) = \begin{cases} x & \text{if } x \le 0 \\ x + 1 & \text{if } x > 0 \end{cases}$

38. $f(x) = \begin{cases} 2x + 3 & \text{if } x < -1 \\ 3 - x & \text{if } x \ge -1 \end{cases}$

39. $f(x) = \begin{cases} -1 & \text{if } x < -1 \\ 1 & \text{if } -1 \le x \le 1 \\ -1 & \text{if } x > 1 \end{cases}$

40. $f(x) = \begin{cases} -1 & \text{if } x < -1 \\ x & \text{if } -1 \le x \le 1 \\ 1 & \text{if } x > 1 \end{cases}$

41. $f(x) = \begin{cases} 2 & \text{if } x \le -1 \\ x^2 & \text{if } x > -1 \end{cases}$

42. $f(x) = \begin{cases} 1 - x^2 & \text{if } x \le 2 \\ x & \text{if } x > 2 \end{cases}$

43. $f(x) = \begin{cases} 0 & \text{if } |x| \le 2 \\ 3 & \text{if } |x| > 2 \end{cases}$

44. $f(x) = \begin{cases} x^2 & \text{if } |x| \le 1 \\ 1 & \text{if } |x| > 1 \end{cases}$

45. $f(x) = \begin{cases} 4 & \text{if } x < -2 \\ x^2 & \text{if } -2 \le x \le 2 \\ -x + 6 & \text{if } x > 2 \end{cases}$

46. $f(x) = \begin{cases} -x & \text{if } x \le 0 \\ 9 - x^2 & \text{if } 0 < x \le 3 \\ x - 3 & \text{if } x > 3 \end{cases}$

47–48 ■ Use a graphing device to draw the graph of the piecewise defined function. (See the margin note on page 41.)

47. $f(x) = \begin{cases} x + 2 & \text{if } x \le -1 \\ x^2 & \text{if } x > -1 \end{cases}$

48. $f(x) = \begin{cases} 2x - x^2 & \text{if } x > 1 \\ (x - 1)^3 & \text{if } x \le 1 \end{cases}$

49–50 ■ The graph of a piecewise defined function is given. Find a formula for the function in the indicated form.

49.
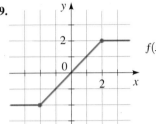

$f(x) = \begin{cases} \rule{2em}{0.6em} & \text{if } x < -2 \\ \rule{2em}{0.6em} & \text{if } -2 \le x \le 2 \\ \rule{2em}{0.6em} & \text{if } x > 2 \end{cases}$

50.
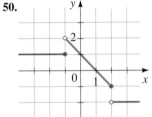

$f(x) = \begin{cases} \rule{2em}{0.6em} & \text{if } x \le -1 \\ \rule{2em}{0.6em} & \text{if } -1 < x \le 2 \\ \rule{2em}{0.6em} & \text{if } x > 2 \end{cases}$

51–52 ■ Use the Vertical Line Test to determine whether the curve is the graph of a function of x.

51. (a)

(b)
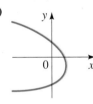

(c)

(d)

52. (a)

(b)

(c)

(d)

53–56 ■ Use the Vertical Line Test to determine whether the curve is the graph of a function of x. If it is, state the domain and range of the function.

53.
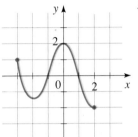

54.
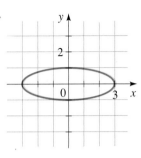

55.
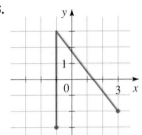

56.
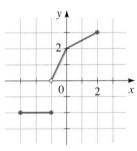

57–68 ■ Determine whether the equation defines y as a function of x. (See Example 9.)

57. $x^2 + 2y = 4$

58. $3x + 7y = 21$

59. $x = y^2$

60. $x^2 + (y - 1)^2 = 4$

61. $x + y^2 = 9$

62. $x^2 + y = 9$

63. $x^2 y + y = 1$

64. $\sqrt{x} + y = 12$

65. $2|x| + y = 0$

66. $2x + |y| = 0$

67. $x = y^3$

68. $x = y^4$

69–74 ■ A family of functions is given. In parts (a) and (b) graph all the given members of the family in the viewing rectangle indicated. In part (c) state the conclusions that you can make from your graphs.

69. $f(x) = x^2 + c$
(a) $c = 0, 2, 4, 6$; $[-5, 5]$ by $[-10, 10]$
(b) $c = 0, -2, -4, -6$; $[-5, 5]$ by $[-10, 10]$
(c) How does the value of c affect the graph?

70. $f(x) = (x - c)^2$
(a) $c = 0, 1, 2, 3$; $[-5, 5]$ by $[-10, 10]$
(b) $c = 0, -1, -2, -3$; $[-5, 5]$ by $[-10, 10]$
(c) How does the value of c affect the graph?

71. $f(x) = (x - c)^3$
(a) $c = 0, 2, 4, 6$; $[-10, 10]$ by $[-10, 10]$
(b) $c = 0, -2, -4, -6$; $[-10, 10]$ by $[-10, 10]$
(c) How does the value of c affect the graph?

72. $f(x) = cx^2$
(a) $c = 1, \frac{1}{2}, 2, 4$; $[-5, 5]$ by $[-10, 10]$
(b) $c = 1, -1, -\frac{1}{2}, -2$; $[-5, 5]$ by $[-10, 10]$
(c) How does the value of c affect the graph?

73. $f(x) = x^c$
(a) $c = \frac{1}{2}, \frac{1}{4}, \frac{1}{6}$; $[-1, 4]$ by $[-1, 3]$
(b) $c = 1, \frac{1}{3}, \frac{1}{5}$; $[-3, 3]$ by $[-2, 2]$
(c) How does the value of c affect the graph?

74. $f(x) = \dfrac{1}{x^n}$

   (a) $n = 1, 3$; $[-3, 3]$ by $[-3, 3]$
   (b) $n = 2, 4$; $[-3, 3]$ by $[-3, 3]$
   (c) How does the value of n affect the graph?

75–78 ■ Find a function whose graph is the given curve.

75. The line segment joining the points $(-2, 1)$ and $(4, -6)$

76. The line segment joining the points $(-3, -2)$ and $(6, 3)$

77. The top half of the circle $x^2 + y^2 = 9$

78. The bottom half of the circle $x^2 + y^2 = 9$

APPLICATIONS

 79. Weather Balloon As a weather balloon is inflated, the thickness T of its rubber skin is related to the radius of the balloon by

$$T(r) = \frac{0.5}{r^2}$$

where T and r are measured in centimeters. Graph the function T for values of r between 10 and 100.

 80. Power from a Wind Turbine The power produced by a wind turbine depends on the speed of the wind. If a windmill has blades 3 meters long, then the power P produced by the turbine is modeled by

$$P(v) = 14.1v^3$$

where P is measured in watts (W) and v is measured in meters per second (m/s). Graph the function P for wind speeds between 1 m/s and 10 m/s.

81. Utility Rates Westside Energy charges its electric customers a base rate of $6.00 per month, plus 10¢ per kilowatt-hour (kWh) for the first 300 kWh used and 6¢ per kWh for all usage over 300 kWh. Suppose a customer uses x kWh of electricity in one month.

   (a) Express the monthly cost E as a piecewise-defined function of x.
   (b) Graph the function E for $0 \le x \le 600$.

82. Taxicab Function A taxi company charges $2.00 for the first mile (or part of a mile) and 20 cents for each succeeding tenth of a mile (or part). Express the cost C (in dollars) of a ride as a piecewise-defined function of the distance x traveled (in miles) for $0 < x < 2$, and sketch the graph of this function.

83. Postage Rates The domestic postage rate for first-class letters weighing 3.5 oz or less is 44 cents for the first ounce (or less), plus 17 cents for each additional ounce (or part of an ounce). Express the postage P as a piecewise-defined function of the weight x of a letter, with $0 < x \le 3.5$, and sketch the graph of this function.

DISCOVERY ■ DISCUSSION ■ WRITING

84. When Does a Graph Represent a Function? For every integer n, the graph of the equation $y = x^n$ is the graph of a function, namely $f(x) = x^n$. Explain why the graph of $x = y^2$ is *not* the graph of a function of x. Is the graph of $x = y^3$ the graph of a function of x? If so, of what function of x is it the graph? Determine for what integers n the graph of $x = y^n$ is the graph of a function of x.

85. Step Functions In Example 7 and Exercises 82 and 83 we are given functions whose graphs consist of horizontal line segments. Such functions are often called *step functions*, because their graphs look like stairs. Give some other examples of step functions that arise in everyday life.

86. Stretched Step Functions Sketch graphs of the functions $f(x) = [\![x]\!]$, $g(x) = [\![2x]\!]$, and $h(x) = [\![3x]\!]$ on separate graphs. How are the graphs related? If n is a positive integer, what does the graph of $k(x) = [\![nx]\!]$ look like?

 87. Graph of the Absolute Value of a Function
   (a) Draw the graphs of the functions

$$f(x) = x^2 + x - 6$$

   and $\qquad g(x) = |x^2 + x - 6|$

   How are the graphs of f and g related?
   (b) Draw the graphs of the functions $f(x) = x^4 - 6x^2$ and $g(x) = |x^4 - 6x^2|$. How are the graphs of f and g related?
   (c) In general, if $g(x) = |f(x)|$, how are the graphs of f and g related? Draw graphs to illustrate your answer.

◯ DISCOVERY PROJECT **Relations and Functions**

In this project we explore the concept of function by comparing it with the concept of a *relation*. You can find the project at the book companion website:
www.stewartmath.com

1.5 GETTING INFORMATION FROM THE GRAPH OF A FUNCTION

Values of a Function; Domain and Range ▶ Increasing and Decreasing
Functions ▶ Local Maximum and Minimum Values of a Function

Many properties of a function are more easily obtained from a graph than from the rule
that describes the function. We will see in this section how a graph tells us whether the
values of a function are increasing or decreasing and also where the maximum and minimum values of a function are.

▼ Values of a Function; Domain and Range

A complete graph of a function contains all the information about a function, because the
graph tells us which input values correspond to which output values. To analyze the graph
of a function, we must keep in mind that *the height of the graph is the value of the function.* So we can read off the values of a function from its graph.

EXAMPLE 1 | Finding the Values of a Function from a Graph

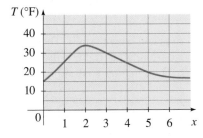

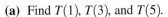

FIGURE 1   Temperature function

The function T graphed in Figure 1 gives the temperature between noon and 6:00 P.M. at
a certain weather station.

(a) Find $T(1)$, $T(3)$, and $T(5)$.

(b) Which is larger, $T(2)$ or $T(4)$?

(c) Find the value(s) of x for which $T(x) = 25$.

(d) Find the value(s) of x for which $T(x) \geq 25$.

SOLUTION

(a) $T(1)$ is the temperature at 1:00 P.M. It is represented by the height of the graph
above the x-axis at $x = 1$. Thus, $T(1) = 25$. Similarly, $T(3) = 30$ and $T(5) = 20$.

(b) Since the graph is higher at $x = 2$ than at $x = 4$, it follows that $T(2)$ is larger than $T(4)$.

(c) The height of the graph is 25 when x is 1 and when x is 4. In other words, the temperature is 25 at 1:00 P.M. and 4:00 P.M.

(d) The graph is higher than 25 for x between 1 and 4. In other words, the temperature
was 25 or greater between 1:00 P.M. and 4:00 P.M.

✎ NOW TRY EXERCISE **5**   ■

The graph of a function helps us to picture the domain and range of the function on the
x-axis and y-axis, as shown in Figure 2.

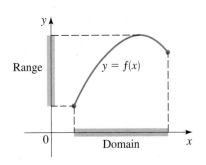

FIGURE 2   Domain and range of f

EXAMPLE 2 | Finding the Domain and Range from a Graph

(a) Use a graphing calculator to draw the graph of $f(x) = \sqrt{4 - x^2}$.

(b) Find the domain and range of f.

SOLUTION

(a) The graph is shown in Figure 3.

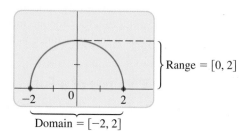

FIGURE 3 Graph of $f(x) = \sqrt{4 - x^2}$

(b) From the graph in Figure 3 we see that the domain is $[-2, 2]$ and the range is $[0, 2]$.

✎ NOW TRY EXERCISE **15**

▼ Increasing and Decreasing Functions

It is very useful to know where the graph of a function rises and where it falls. The graph shown in Figure 4 rises, falls, then rises again as we move from left to right: It rises from A to B, falls from B to C, and rises again from C to D. The function f is said to be *increasing* when its graph rises and *decreasing* when its graph falls.

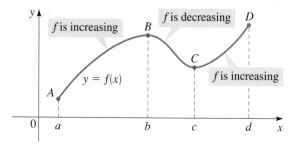

FIGURE 4 f is increasing on $[a, b]$ and $[c, d]$. f is decreasing on $[b, c]$.

We have the following definition.

DEFINITION OF INCREASING AND DECREASING FUNCTIONS

f is **increasing** on an interval I if $f(x_1) < f(x_2)$ whenever $x_1 < x_2$ in I.

f is **decreasing** on an interval I if $f(x_1) > f(x_2)$ whenever $x_1 < x_2$ in I.

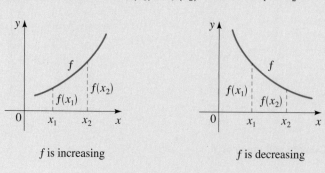

EXAMPLE 3 | Intervals on Which a Function Increases and Decreases

The graph in Figure 5 gives the weight W of a person at age x. Determine the intervals on which the function W is increasing and on which it is decreasing.

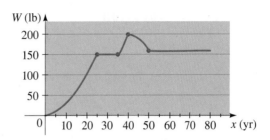

FIGURE 5 Weight as a function of age

SOLUTION   The function W is increasing on $[0, 25]$ and $[35, 40]$. It is decreasing on $[40, 50]$. The function W is constant (neither increasing nor decreasing) on $[25, 30]$ and $[50, 80]$. This means that the person gained weight until age 25, then gained weight again between ages 35 and 40. He lost weight between ages 40 and 50.

✎ NOW TRY EXERCISE **45** ∎

EXAMPLE 4 | Finding Intervals Where a Function Increases and Decreases

(a) Sketch a graph of the function $f(x) = 12x^2 + 4x^3 - 3x^4$.

(b) Find the domain and range of f.

(c) Find the intervals on which f increases and decreases.

SOLUTION

(a) We use a graphing calculator to sketch the graph in Figure 6.

(b) The domain of f is $\mathbb{R}$ because f is defined for all real numbers. Using the ⟨TRACE⟩ feature on the calculator, we find that the highest value is $f(2) = 32$. So the range of f is $(-\infty, 32]$.

(c) From the graph we see that f is increasing on the intervals $(-\infty, -1]$ and $[0, 2]$ and is decreasing on $[-1, 0]$ and $[2, \infty)$.

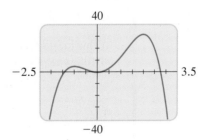

FIGURE 6 Graph of
$f(x) = 12x^2 + 4x^3 - 3x^4$

✎ NOW TRY EXERCISE **23** ∎

EXAMPLE 5 | Finding Intervals Where a Function Increases and Decreases

(a) Sketch the graph of the function $f(x) = x^{2/3}$.

(b) Find the domain and range of the function.

(c) Find the intervals on which f increases and decreases.

SOLUTION

(a) We use a graphing calculator to sketch the graph in Figure 7.

(b) From the graph we observe that the domain of f is $\mathbb{R}$ and the range is $[0, \infty)$.

(c) From the graph we see that f is decreasing on $(-\infty, 0]$ and increasing on $[0, \infty)$.

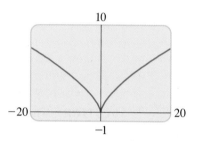

FIGURE 7 Graph of $f(x) = x^{2/3}$

✎ NOW TRY EXERCISE **29**

▼ Local Maximum and Minimum Values of a Function

Finding the largest or smallest values of a function is important in many applications. For example, if a function represents revenue or profit, then we are interested in its maximum value. For a function that represents cost, we would want to find its minimum value. We can easily find these values from the graph of a function. We first define what we mean by a local maximum or minimum.

LOCAL MAXIMA AND MINIMA OF A FUNCTION

1. The function value $f(a)$ is a **local maximum value** of f if

$$f(a) \geq f(x) \quad \text{when } x \text{ is near } a$$

(This means that $f(a) \geq f(x)$ for all x in some open interval containing a.)
In this case we say that f has a **local maximum** at $x = a$.

2. The function value $f(a)$ is a **local minimum** of f if

$$f(a) \leq f(x) \quad \text{when } x \text{ is near } a$$

(This means that $f(a) \leq f(x)$ for all x in some open interval containing a.)
In this case we say that f has a **local minimum** at $x = a$.

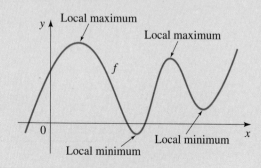

We can find the local maximum and minimum values of a function using a graphing calculator.

If there is a viewing rectangle such that the point $(a, f(a))$ is the highest point on the graph of f *within* the viewing rectangle (not on the edge), then the number $f(a)$ is a local maximum value of f (see Figure 8). Notice that $f(a) \geq f(x)$ for all numbers x that are close to a.

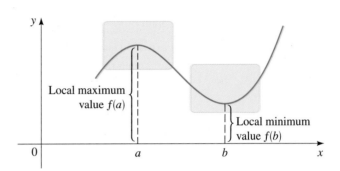

FIGURE 8

Similarly, if there is a viewing rectangle such that the point $(b, f(b))$ is the lowest point on the graph of f within the viewing rectangle, then the number $f(b)$ is a local minimum value of f. In this case, $f(b) \leq f(x)$ for all numbers x that are close to b.

EXAMPLE 6 | Finding Local Maxima and Minima from a Graph

Find the local maximum and minimum values of the function $f(x) = x^3 - 8x + 1$, correct to three decimal places.

SOLUTION   The graph of f is shown in Figure 9. There appears to be one local maximum between $x = -2$ and $x = -1$, and one local minimum between $x = 1$ and $x = 2$.

Let's find the coordinates of the local maximum point first. We zoom in to enlarge the area near this point, as shown in Figure 10. Using the $\boxed{\text{TRACE}}$ feature on the graphing device, we move the cursor along the curve and observe how the y-coordinates change. The local maximum value of y is 9.709, and this value occurs when x is -1.633, correct to three decimal places.

We locate the minimum value in a similar fashion. By zooming in to the viewing rectangle shown in Figure 11, we find that the local minimum value is about -7.709, and this value occurs when $x \approx 1.633$.

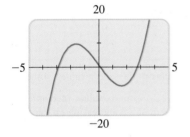

FIGURE 9 Graph of
$f(x) = x^3 - 8x + 1$

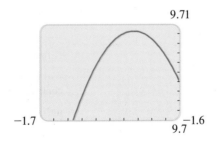

FIGURE 10

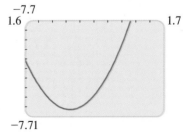

FIGURE 11

✎ NOW TRY EXERCISE **35**   ■

The `maximum` and `minimum` commands on a TI-83 or TI-84 calculator provide another method for finding extreme values of functions. We use this method in the next example.

EXAMPLE 7 | A Model for the Food Price Index

A model for the food price index (the price of a representative "basket" of foods) between 1990 and 2000 is given by the function

$$I(t) = -0.0113t^3 + 0.0681t^2 + 0.198t + 99.1$$

where t is measured in years since midyear 1990, so $0 \le t \le 10$, and $I(t)$ is scaled so that $I(3) = 100$. Estimate the time when food was most expensive during the period 1990–2000.

SOLUTION The graph of I as a function of t is shown in Figure 12(a). There appears to be a maximum between $t = 4$ and $t = 7$. Using the `maximum` command, as shown in Figure 12(b), we see that the maximum value of I is about 100.38, and it occurs when $t \approx 5.15$, which corresponds to August 1995.

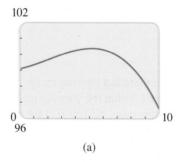

FIGURE 12   (a)   (b)

✎ NOW TRY EXERCISE **53**

1.5 EXERCISES

CONCEPTS

1–4 ■ These exercises refer to the graph of the function f shown below.

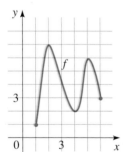

1. To find a function value $f(a)$ from the graph of f, we find the height of the graph above the x-axis at $x =$ _____. From the graph of f we see that $f(3) =$ _____.

2. The domain of the function f is all the _____-values of the points on the graph, and the range is all the corresponding _____-values. From the graph of f we see that the domain of f is the interval _____ and the range of f is the interval _____.

3. (a) If f is increasing on an interval, then the y-values of the points on the graph _____ as the x-values increase. From the graph of f we see that f is increasing on the intervals _____ and _____.

(b) If f is decreasing on an interval, then y-values of the points on the graph _____ as the x-values increase. From the graph of f we see that f is decreasing on the intervals _____ and _____.

4. (a) A function value $f(a)$ is a local maximum value of f if $f(a)$ is the _____ value of f on some interval containing a. From the graph of f we see that one local maximum value of f is _____ and that this value occurs when x is _____.

(b) The function value $f(a)$ is a local minimum value of f if $f(a)$ is the _____ value of f on some interval containing a. From the graph of f we see that one local minimum value of f is _____ and that this value occurs when x is _____.

SKILLS

5. The graph of a function h is given.
  (a) Find $h(-2)$, $h(0)$, $h(2)$, and $h(3)$.
  (b) Find the domain and range of h.
  (c) Find the values of x for which $h(x) = 3$.
  (d) Find the values of x for which $h(x) \le 3$.

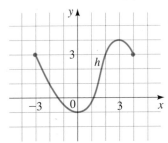

6. The graph of a function g is given.
  (a) Find $g(-2)$, $g(0)$, and $g(7)$.
  (b) Find the domain and range of g.
  (c) Find the values of x for which $g(x) = 4$.
  (d) Find the values of x for which $g(x) > 4$.

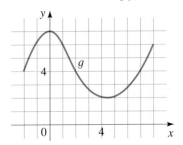

7. The graph of a function g is given.
  (a) Find $g(-4)$, $g(-2)$, $g(0)$, $g(2)$, and $g(4)$.
  (b) Find the domain and range of g.

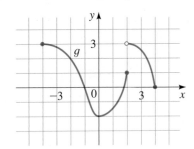

8. Graphs of the functions f and g are given.
  (a) Which is larger, $f(0)$ or $g(0)$?
  (b) Which is larger, $f(-3)$ or $g(-3)$?
  (c) For which values of x is $f(x) = g(x)$?

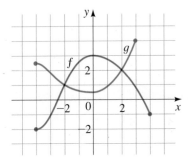

9–18 ■ A function f is given. **(a)** Use a graphing calculator to draw the graph of f. **(b)** Find the domain and range of f from the graph.

9. $f(x) = x - 1$ **10.** $f(x) = 2(x + 1)$

11. $f(x) = 4, \quad 1 \le x \le 3$ **12.** $f(x) = x^2, \quad -2 \le x \le 5$

13. $f(x) = 4 - x^2$ **14.** $f(x) = x^2 + 4$

15. $f(x) = \sqrt{16 - x^2}$ **16.** $f(x) = -\sqrt{25 - x^2}$

17. $f(x) = \sqrt{x - 1}$ **18.** $f(x) = \sqrt{x + 2}$

19–22 ■ The graph of a function is given. Determine the intervals on which the function is **(a)** increasing and **(b)** decreasing.

19.
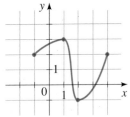

20.
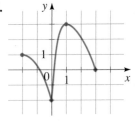

21.
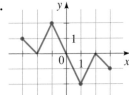

22.
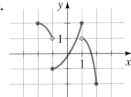

23–30 ■ A function f is given. **(a)** Use a graphing device to draw the graph of f. **(b)** State approximately the intervals on which f is increasing and on which f is decreasing.

23. $f(x) = x^2 - 5x$ **24.** $f(x) = x^3 - 4x$

25. $f(x) = 2x^3 - 3x^2 - 12x$ **26.** $f(x) = x^4 - 16x^2$

27. $f(x) = x^3 + 2x^2 - x - 2$

28. $f(x) = x^4 - 4x^3 + 2x^2 + 4x - 3$

29. $f(x) = x^{2/5}$ **30.** $f(x) = 4 - x^{2/3}$

31–34 ■ The graph of a function is given. **(a)** Find all the local maximum and minimum values of the function and the value of x at which each occurs. **(b)** Find the intervals on which the function is increasing and on which the function is decreasing.

31.
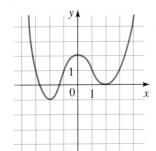

32.
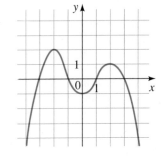

33.

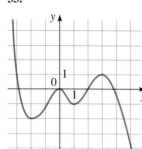

34.

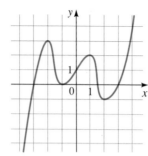

35–42 ■ A function is given. **(a)** Find all the local maximum and minimum values of the function and the value of x at which each occurs. State each answer correct to two decimal places. **(b)** Find the intervals on which the function is increasing and on which the function is decreasing. State each answer correct to two decimal places.

35. $f(x) = x^3 - x$

36. $f(x) = 3 + x + x^2 - x^3$

37. $g(x) = x^4 - 2x^3 - 11x^2$

38. $g(x) = x^5 - 8x^3 + 20x$

39. $U(x) = x\sqrt{6 - x}$

40. $U(x) = x\sqrt{x - x^2}$

41. $V(x) = \dfrac{1 - x^2}{x^3}$

42. $V(x) = \dfrac{1}{x^2 + x + 1}$

APPLICATIONS

43. Power Consumption The figure shows the power consumption in San Francisco for September 19, 1996 (P is measured in megawatts; t is measured in hours starting at midnight).
  (a) What was the power consumption at 6:00 A.M.? At 6:00 P.M.?
  (b) When was the power consumption the lowest?
  (c) When was the power consumption the highest?

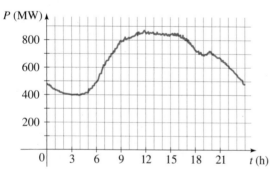

Source: Pacific Gas & Electric

44. Earthquake The graph shows the vertical acceleration of the ground from the 1994 Northridge earthquake in Los Angeles, as measured by a seismograph. (Here t represents the time in seconds.)
  (a) At what time t did the earthquake first make noticeable movements of the earth?
  (b) At what time t did the earthquake seem to end?
  (c) At what time t was the maximum intensity of the earthquake reached?

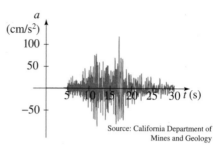

Source: California Department of
Mines and Geology

45. Weight Function The graph gives the weight W of a person at age x.
  (a) Determine the intervals on which the function W is increasing and those on which it is decreasing.
  (b) What do you think happened when this person was 30 years old?

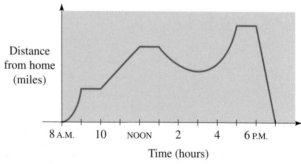

46. Distance Function The graph gives a sales representative's distance from his home as a function of time on a certain day.
  (a) Determine the time intervals on which his distance from home was increasing and those on which it was decreasing.
  (b) Describe in words what the graph indicates about his travels on this day.

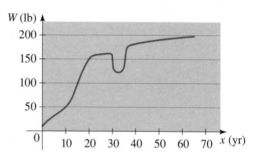

47. Changing Water Levels The graph shows the depth of water W in a reservoir over a one-year period as a function of the number of days x since the beginning of the year.
  (a) Determine the intervals on which the function W is increasing and on which it is decreasing.
  (b) At what value of x does W achieve a local maximum? A local minimum?

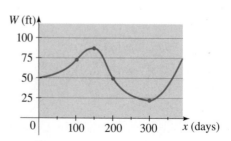

48. Population Growth and Decline The graph shows the population P in a small industrial city from 1950 to 2000. The variable x represents the number of years since 1950.

(a) Determine the intervals on which the function P is increasing and on which it is decreasing.

(b) What was the maximum population, and in what year was it attained?

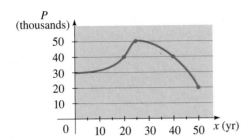

49. Hurdle Race Three runners compete in a 100-meter hurdle race. The graph depicts the distance run as a function of time for each runner. Describe in words what the graph tells you about this race. Who won the race? Did each runner finish the race? What do you think happened to runner B?

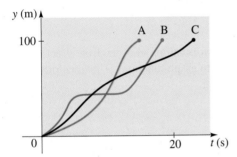

 50. Gravity Near the Moon We can use Newton's Law of Gravity to measure the gravitational attraction between the moon and an algebra student in a space ship located a distance x above the moon's surface:

$$F(x) = \frac{350}{x^2}$$

Here F is measured in newtons (N), and x is measured in millions of meters.

(a) Graph the function F for values of x between 0 and 10.

(b) Use the graph to describe the behavior of the gravitational attraction F as the distance x increases.

 51. Radii of Stars Astronomers infer the radii of stars using the Stefan Boltzmann Law:

$$E(T) = (5.67 \times 10^{-8})T^4$$

where E is the energy radiated per unit of surface area

measured in watts (W) and T is the absolute temperature measured in kelvins (K).

(a) Graph the function E for temperatures T between 100 K and 300 K.

(b) Use the graph to describe the change in energy E as the temperature T increases.

 52. Migrating Fish A fish swims at a speed v relative to the water, against a current of 5 mi/h. Using a mathematical model of energy expenditure, it can be shown that the total energy E required to swim a distance of 10 mi is given by

$$E(v) = 2.73v^3 \frac{10}{v - 5}$$

Biologists believe that migrating fish try to minimize the total energy required to swim a fixed distance. Find the value of v that minimizes energy required.

NOTE: This result has been verified; migrating fish swim against a current at a speed 50% greater than the speed of the current.

53. Highway Engineering A highway engineer wants to estimate the maximum number of cars that can safely travel a particular highway at a given speed. She assumes that each car is 17 ft long, travels at a speed s, and follows the car in front of it at the "safe following distance" for that speed. She finds that the number N of cars that can pass a given point per minute is modeled by the function

$$N(s) = \frac{88s}{17 + 17\left(\dfrac{s}{20}\right)^2}$$

At what speed can the greatest number of cars travel the highway safely?

 54. Volume of Water Between 0°C and 30°C, the volume V (in cubic centimeters) of 1 kg of water at a temperature T is given by the formula

$$V = 999.87 - 0.06426T + 0.0085043T^2 - 0.0000679T^3$$

Find the temperature at which the volume of 1 kg of water is a minimum.

55. Coughing When a foreign object that is lodged in the trachea (windpipe) forces a person to cough, the diaphragm thrusts upward, causing an increase in pressure in the lungs. At the same time, the trachea contracts, causing the expelled air to move faster and increasing the pressure on the foreign object. According to a mathematical model of coughing, the velocity v (in cm/s) of the airstream through an average-sized person's trachea is related to the radius r (in cm) of the trachea by the function

$$v(r) = 3.2(1 - r)r^2 \qquad \tfrac{1}{2} \le r \le 1$$

Determine the value of r for which v is a maximum.

56. Functions That Are Always Increasing or Decreasing Sketch rough graphs of functions that are defined for all real numbers and that exhibit the indicated behavior (or explain why the behavior is impossible).
  (a) f is always increasing, and $f(x) > 0$ for all x
  (b) f is always decreasing, and $f(x) > 0$ for all x
  (c) f is always increasing, and $f(x) < 0$ for all x
  (d) f is always decreasing, and $f(x) < 0$ for all x

57. Maxima and Minima In Example 7 we saw a real-world situation in which the maximum value of a function is important. Name several other everyday situations in which a maximum or minimum value is important.

58. Minimizing a Distance When we seek a minimum or maximum value of a function, it is sometimes easier to work with a simpler function instead.
  (a) Suppose

$$g(x) = \sqrt{f(x)}$$

  where $f(x) \geq 0$ for all x. Explain why the local minima and maxima of f and g occur at the same values of x.
  (b) Let $g(x)$ be the distance between the point $(3, 0)$ and the point (x, x^2) on the graph of the parabola $y = x^2$. Express g as a function of x.
  (c) Find the minimum value of the function g that you found in part (b). Use the principle described in part (a) to simplify your work.

1.6 TRANSFORMATIONS OF FUNCTIONS

Vertical Shifting ▶ Horizontal Shifting ▶ Reflecting Graphs ▶ Vertical Stretching and Shrinking ▶ Horizontal Stretching and Shrinking ▶ Even and Odd Functions

In this section we study how certain transformations of a function affect its graph. This will give us a better understanding of how to graph functions. The transformations that we study are shifting, reflecting, and stretching.

▼ Vertical Shifting

Adding a constant to a function shifts its graph vertically: upward if the constant is positive and downward if it is negative.

In general, suppose we know the graph of $y = f(x)$. How do we obtain from it the graphs of

$$y = f(x) + c \qquad \text{and} \qquad y = f(x) - c \qquad (c > 0)$$

The *y*-coordinate of each point on the graph of $y = f(x) + c$ is c units above the *y*-coordinate of the corresponding point on the graph of $y = f(x)$. So we obtain the graph of $y = f(x) + c$ simply by shifting the graph of $y = f(x)$ upward c units. Similarly, we obtain the graph of $y = f(x) - c$ by shifting the graph of $y = f(x)$ downward c units.

Recall that the graph of the function f is the same as the graph of the equation $y = f(x)$.

VERTICAL SHIFTS OF GRAPHS

Suppose $c > 0$.

To graph $y = f(x) + c$, shift the graph of $y = f(x)$ upward c units.
To graph $y = f(x) - c$, shift the graph of $y = f(x)$ downward c units.

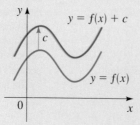

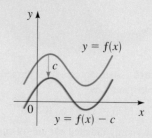

EXAMPLE 1 | Vertical Shifts of Graphs

Use the graph of $f(x) = x^2$ to sketch the graph of each function.

(a) $g(x) = x^2 + 3$ **(b)** $h(x) = x^2 - 2$

SOLUTION The function $f(x) = x^2$ was graphed in Example 1(a), Section 1.4. It is sketched again in Figure 1.

(a) Observe that

$$g(x) = x^2 + 3 = f(x) + 3$$

So the y-coordinate of each point on the graph of g is 3 units above the corresponding point on the graph of f. This means that to graph g, we shift the graph of f upward 3 units, as in Figure 1.

(b) Similarly, to graph h, we shift the graph of f downward 2 units, as shown in Figure 1.

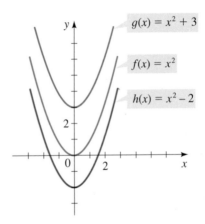

FIGURE 1

✎ NOW TRY EXERCISES **21** AND **23** ■

▼ Horizontal Shifting

Suppose that we know the graph of $y = f(x)$. How do we use it to obtain the graphs of

$$y = f(x + c) \quad \text{and} \quad y = f(x - c) \quad (c > 0)$$

The value of $f(x - c)$ at x is the same as the value of $f(x)$ at $x - c$. Since $x - c$ is c units to the left of x, it follows that the graph of $y = f(x - c)$ is just the graph of $y = f(x)$ shifted to the right c units. Similar reasoning shows that the graph of $y = f(x + c)$ is the graph of $y = f(x)$ shifted to the left c units. The following box summarizes these facts.

HORIZONTAL SHIFTS OF GRAPHS

Suppose $c > 0$.

To graph $y = f(x - c)$, shift the graph of $y = f(x)$ to the right c units.

To graph $y = f(x + c)$, shift the graph of $y = f(x)$ to the left c units.

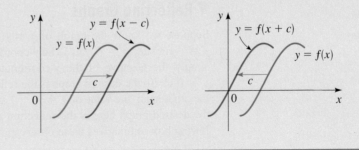

RENÉ DESCARTES (1596–1650) was born in the town of La Haye in southern France. From an early age Descartes liked mathematics because of "the certainty of its results and the clarity of its reasoning." He believed that to arrive at truth, one must begin by doubting everything, including one's own existence; this led him to formulate perhaps the best-known sentence in all of philosophy: "I think, therefore I am." In his book *Discourse on Method* he described what is now called the Cartesian plane. This idea of combining algebra and geometry enabled mathematicians for the first time to graph functions and thus "see" the equations they were studying. The philosopher John Stuart Mill called this invention "the greatest single step ever made in the progress of the exact sciences." Descartes liked to get up late and spend the morning in bed thinking and writing. He invented the coordinate plane while lying in bed watching a fly crawl on the ceiling, reasoning that he could describe the exact location of the fly by knowing its distance from two perpendicular walls. In 1649 Descartes became the tutor of Queen Christina of Sweden. She liked her lessons at 5 o'clock in the morning, when, she said, her mind was sharpest. However, the change from his usual habits and the ice-cold library where they studied proved too much for Descartes. In February 1650, after just two months of this, he caught pneumonia and died.

EXAMPLE 2 | Horizontal Shifts of Graphs

Use the graph of $f(x) = x^2$ to sketch the graph of each function.

(a) $g(x) = (x + 4)^2$ **(b)** $h(x) = (x - 2)^2$

SOLUTION

(a) To graph g, we shift the graph of f to the left 4 units.

(b) To graph h, we shift the graph of f to the right 2 units.

The graphs of g and h are sketched in Figure 2.

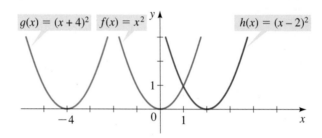

FIGURE 2

✎ NOW TRY EXERCISE **25** AND **27**

EXAMPLE 3 | Combining Horizontal and Vertical Shifts

Sketch the graph of $f(x) = \sqrt{x - 3} + 4$.

SOLUTION We start with the graph of $y = \sqrt{x}$ (Example 1(c), Section 1.4) and shift it to the right 3 units to obtain the graph of $y = \sqrt{x - 3}$. Then we shift the resulting graph upward 4 units to obtain the graph of $f(x) = \sqrt{x - 3} + 4$ shown in Figure 3.

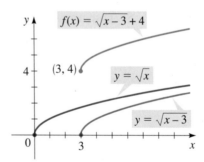

FIGURE 3

✎ NOW TRY EXERCISE **37**

▼ Reflecting Graphs

Suppose we know the graph of $y = f(x)$. How do we use it to obtain the graphs of $y = -f(x)$ and $y = f(-x)$? The y-coordinate of each point on the graph of $y = -f(x)$ is simply the negative of the y-coordinate of the corresponding point on the graph of $y = f(x)$. So the desired graph is the reflection of the graph of $y = f(x)$ in the x-axis. On the other hand, the value of $y = f(-x)$ at x is the same as the value of $y = f(x)$ at $-x$, so the desired graph here is the reflection of the graph of $y = f(x)$ in the y-axis. The following box summarizes these observations.

REFLECTING GRAPHS

To graph $y = -f(x)$, reflect the graph of $y = f(x)$ in the x-axis.

To graph $y = f(-x)$, reflect the graph of $y = f(x)$ in the y-axis.

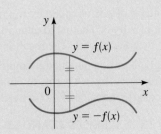

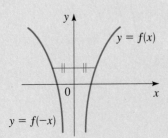

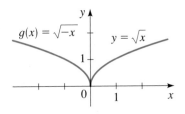

FIGURE 4

FIGURE 5

EXAMPLE 4 | Reflecting Graphs

Sketch the graph of each function.

(a) $f(x) = -x^2$ **(b)** $g(x) = \sqrt{-x}$

SOLUTION

(a) We start with the graph of $y = x^2$. The graph of $f(x) = -x^2$ is the graph of $y = x^2$ reflected in the x-axis (see Figure 4).

(b) We start with the graph of $y = \sqrt{x}$ (Example 1(c) in Section 1.4). The graph of $g(x) = \sqrt{-x}$ is the graph of $y = \sqrt{x}$ reflected in the y-axis (see Figure 5). Note that the domain of the function $g(x) = \sqrt{-x}$ is $\{x \mid x \leq 0\}$.

✎ NOW TRY EXERCISES **29** AND **31**

▼ Vertical Stretching and Shrinking

Suppose we know the graph of $y = f(x)$. How do we use it to obtain the graph of $y = cf(x)$? The y-coordinate of $y = cf(x)$ at x is the same as the corresponding y-coordinate of $y = f(x)$ multiplied by c. Multiplying the y-coordinates by c has the effect of vertically stretching or shrinking the graph by a factor of c.

VERTICAL STRETCHING AND SHRINKING OF GRAPHS

To graph $y = cf(x)$:

If $c > 1$, stretch the graph of $y = f(x)$ vertically by a factor of c.

If $0 < c < 1$, shrink the graph of $y = f(x)$ vertically by a factor of c.

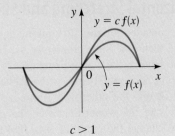

$c > 1$

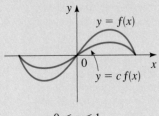

$0 < c < 1$

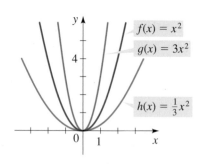

FIGURE 6

EXAMPLE 5 | Vertical Stretching and Shrinking of Graphs

Use the graph of $f(x) = x^2$ to sketch the graph of each function.

(a) $g(x) = 3x^2$ **(b)** $h(x) = \frac{1}{3}x^2$

SOLUTION

(a) The graph of g is obtained by multiplying the y-coordinate of each point on the graph of f by 3. That is, to obtain the graph of g, we stretch the graph of f vertically by a factor of 3. The result is the narrower parabola in Figure 6.

(b) The graph of h is obtained by multiplying the y-coordinate of each point on the graph of f by $\frac{1}{3}$. That is, to obtain the graph of h, we shrink the graph of f vertically by a factor of $\frac{1}{3}$. The result is the wider parabola in Figure 6.

✎. NOW TRY EXERCISES **33** AND **35**   ■

We illustrate the effect of combining shifts, reflections, and stretching in the following example.

EXAMPLE 6 | Combining Shifting, Stretching, and Reflecting

Sketch the graph of the function $f(x) = 1 - 2(x - 3)^2$.

SOLUTION   Starting with the graph of $y = x^2$, we first shift to the right 3 units to get the graph of $y = (x - 3)^2$. Then we reflect in the x-axis and stretch by a factor of 2 to get the graph of $y = -2(x - 3)^2$. Finally, we shift upward 1 unit to get the graph of $f(x) = 1 - 2(x - 3)^2$ shown in Figure 7.

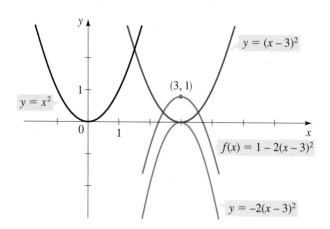

FIGURE 7

✎. NOW TRY EXERCISE **39**   ■

SONYA KOVALEVSKY (1850–1891) is considered the most important woman mathematician of the 19th century. She was born in Moscow to an aristocratic family. While a child, she was exposed to the principles of calculus in a very unusual fashion: Her bedroom was temporarily wallpapered with the pages of a calculus book. She later wrote that she "spent many hours in front of that wall, trying to understand it." Since Russian law forbade women from studying in universities, she entered a marriage of convenience, which allowed her to travel to Germany and obtain a doctorate in mathematics from the University of Göttingen. She eventually was awarded a full professorship at the University of Stockholm, where she taught for eight years before dying in an influenza epidemic at the age of 41. Her research was instrumental in helping to put the ideas and applications of functions and calculus on a sound and logical foundation. She received many accolades and prizes for her research work.

▼ Horizontal Stretching and Shrinking

Now we consider horizontal shrinking and stretching of graphs. If we know the graph of $y = f(x)$, then how is the graph of $y = f(cx)$ related to it? The y-coordinate of $y = f(cx)$ at x is the same as the y-coordinate of $y = f(x)$ at cx. Thus, the x-coordinates in the graph of $y = f(x)$ correspond to the x-coordinates in the graph of $y = f(cx)$ multiplied by c. Looking at this the other way around, we see that the x-coordinates in the graph of $y = f(cx)$ are the x-coordinates in the graph of $y = f(x)$ multiplied by $1/c$. In other words, to change the graph of $y = f(x)$ to the graph of $y = f(cx)$, we must shrink (or stretch) the graph horizontally by a factor of $1/c$, as summarized in the following box.

HORIZONTAL SHRINKING AND STRETCHING OF GRAPHS

To graph $y = f(cx)$:

If $c > 1$, shrink the graph of $y = f(x)$ horizontally by a factor of $1/c$.

If $0 < c < 1$, stretch the graph of $y = f(x)$ horizontally by a factor of $1/c$.

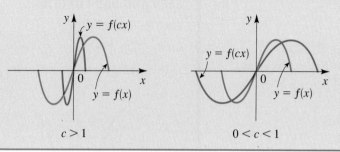

EXAMPLE 7 | Horizontal Stretching and Shrinking of Graphs

The graph of $y = f(x)$ is shown in Figure 8. Sketch the graph of each function.

(a) $y = f(2x)$ **(b)** $y = f\left(\tfrac{1}{2}x\right)$

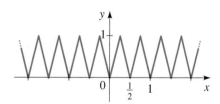

FIGURE 8 $y = f(x)$

SOLUTION Using the principles described in the preceding box, we obtain the graphs shown in Figures 9 and 10.

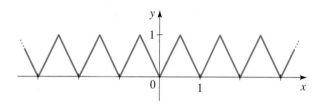

FIGURE 9 $y = f(2x)$ **FIGURE 10** $y = f\left(\tfrac{1}{2}x\right)$

✎ NOW TRY EXERCISE **63**

▼ **Even and Odd Functions**

If a function f satisfies $f(-x) = f(x)$ for every number x in its domain, then f is called an **even function**. For instance, the function $f(x) = x^2$ is even because

$$f(-x) = (-x)^2 = (-1)^2 x^2 = x^2 = f(x)$$

The graph of an even function is symmetric with respect to the y-axis (see Figure 11). This means that if we have plotted the graph of f for $x \geq 0$, then we can obtain the entire graph simply by reflecting this portion in the y-axis.

If f satisfies $f(-x) = -f(x)$ for every number x in its domain, then f is called an **odd function**. For example, the function $f(x) = x^3$ is odd because

$$f(-x) = (-x)^3 = (-1)^3 x^3 = -x^3 = -f(x)$$

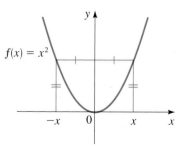

FIGURE 11 $f(x) = x^2$ is an even function.

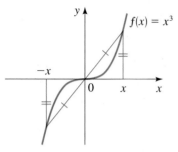

FIGURE 12 $f(x) = x^3$ is an odd function.

The graph of an odd function is symmetric about the origin (see Figure 12). If we have plotted the graph of f for $x \geq 0$, then we can obtain the entire graph by rotating this portion through $180°$ about the origin. (This is equivalent to reflecting first in the x-axis and then in the y-axis.)

EVEN AND ODD FUNCTIONS

Let f be a function.

f is **even** if $f(-x) = f(x)$ for all x in the domain of f.

f is **odd** if $f(-x) = -f(x)$ for all x in the domain of f.

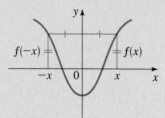

The graph of an even function is symmetric with respect to the y-axis.

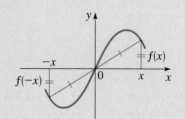

The graph of an odd function is symmetric with respect to the origin.

EXAMPLE 8 | Even and Odd Functions

Determine whether the functions are even, odd, or neither even nor odd.

(a) $f(x) = x^5 + x$ **(b)** $g(x) = 1 - x^4$ **(c)** $h(x) = 2x - x^2$

SOLUTION

(a) $\begin{aligned} f(-x) &= (-x)^5 + (-x) \\ &= -x^5 - x = -(x^5 + x) \\ &= -f(x) \end{aligned}$

Therefore, f is an odd function.

(b) $g(-x) = 1 - (-x)^4 = 1 - x^4 = g(x)$

So g is even.

(c) $h(-x) = 2(-x) - (-x)^2 = -2x - x^2$

Since $h(-x) \neq h(x)$ and $h(-x) \neq -h(x)$, we conclude that h is neither even nor odd.

✎ NOW TRY EXERCISES **75**, **77**, AND **79** ∎

The graphs of the functions in Example 8 are shown in Figure 13. The graph of f is symmetric about the origin, and the graph of g is symmetric about the y-axis. The graph of h is not symmetric either about the y-axis or the origin.

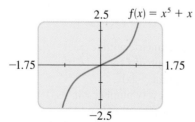

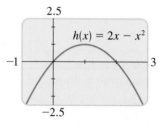

FIGURE 13   (a)   (b)   (c)

1.6 EXERCISES

CONCEPTS

1–2 ■ Fill in the blank with the appropriate direction (left, right, up, or down).

1. **(a)** The graph of $y = f(x) + 3$ is obtained from the graph of

$y = f(x)$ by shifting _____ 3 units.

(b) The graph of $y = f(x + 3)$ is obtained from the graph of

$y = f(x)$ by shifting _____ 3 units.

2. **(a)** The graph of $y = f(x) - 3$ is obtained from the graph of

$y = f(x)$ by shifting _____ 3 units.

(b) The graph of $y = f(x - 3)$ is obtained from the graph of

$y = f(x)$ by shifting _____ 3 units.

3. Fill in the blank with the appropriate axis (x-axis or y-axis).

(a) The graph of $y = -f(x)$ is obtained from the graph of

$y = f(x)$ by reflecting in the _____.

(b) The graph of $y = f(-x)$ is obtained from the graph of

$y = f(x)$ by reflecting in the _____.

4. Match the graph with the function.

(a) $y = |x + 1|$ **(b)** $y = |x - 1|$

(c) $y = |x| - 1$ **(d)** $y = -|x|$

I

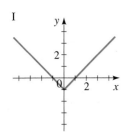

II

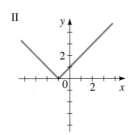

III

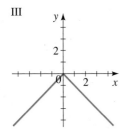

IV

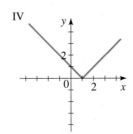

SKILLS

5–14 ■ Suppose the graph of f is given. Describe how the graph of each function can be obtained from the graph of f.

5. **(a)** $y = f(x) - 5$ **(b)** $y = f(x - 5)$

6. **(a)** $y = f(x + 7)$ **(b)** $y = f(x) + 7$

7. **(a)** $y = -f(x)$ **(b)** $y = f(-x)$

8. **(a)** $y = -2f(x)$ **(b)** $y = -\frac{1}{2}f(x)$

9. **(a)** $y = -f(x) + 5$ **(b)** $y = 3f(x) - 5$

10. **(a)** $y = f(x - 4) + \frac{3}{4}$ **(b)** $y = f(x + 4) - \frac{3}{4}$

11. **(a)** $y = 2f(x + 1) - 3$ **(b)** $y = 2f(x - 1) + 3$

12. **(a)** $y = 3 - 2f(x)$ **(b)** $y = 2 - f(-x)$

13. **(a)** $y = f(4x)$ **(b)** $y = f(\frac{1}{4}x)$

14. **(a)** $y = f(2x) - 1$ **(b)** $y = 2f(\frac{1}{2}x)$

15–18 ■ Explain how the graph of g is obtained from the graph of f.

15. **(a)** $f(x) = x^2$, $g(x) = (x + 2)^2$

    (b) $f(x) = x^2$, $g(x) = x^2 + 2$

16. **(a)** $f(x) = x^3$, $g(x) = (x - 4)^3$

    (b) $f(x) = x^3$, $g(x) = x^3 - 4$

17. **(a)** $f(x) = |x|$, $g(x) = |x + 2| - 2$

    (b) $f(x) = |x|$, $g(x) = |x - 2| + 2$

18. **(a)** $f(x) = \sqrt{x}$, $g(x) = -\sqrt{x} + 1$

    (b) $f(x) = \sqrt{x}$, $g(x) = \sqrt{-x} + 1$

19. Use the graph of $y = x^2$ in Figure 4 to graph the following.

    (a) $g(x) = x^2 + 1$

    (b) $g(x) = (x - 1)^2$

    (c) $g(x) = -x^2$

    (d) $g(x) = (x - 1)^2 + 3$

20. Use the graph of $y = \sqrt{x}$ in Figure 5 to graph the following.

    (a) $g(x) = \sqrt{x - 2}$

    (b) $g(x) = \sqrt{x} + 1$

    (c) $g(x) = \sqrt{x + 2} + 2$

    (d) $g(x) = -\sqrt{x} + 1$

21–44 ■ Sketch the graph of the function, not by plotting points, but by starting with the graph of a standard function and applying transformations.

21. $f(x) = x^2 - 1$

22. $f(x) = x^2 + 5$

23. $f(x) = \sqrt{x} + 1$

24. $f(x) = |x| - 1$

25. $f(x) = (x - 5)^2$

26. $f(x) = (x + 1)^2$

27. $f(x) = \sqrt{x + 4}$

28. $f(x) = |x - 3|$

29. $f(x) = -x^3$

30. $f(x) = -|x|$

31. $y = \sqrt[4]{-x}$

32. $y = \sqrt[3]{-x}$

33. $y = \frac{1}{4}x^2$

34. $y = -5\sqrt{x}$

35. $y = 3|x|$

36. $y = \frac{1}{2}|x|$

37. $y = (x - 3)^2 + 5$

38. $y = \sqrt{x + 4} - 3$

39. $y = 3 - \frac{1}{2}(x - 1)^2$

40. $y = 2 - \sqrt{x + 1}$

41. $y = |x + 2| + 2$

42. $y = 2 - |x|$

43. $y = \frac{1}{2}\sqrt{x + 4} - 3$

44. $y = 3 - 2(x - 1)^2$

45–54 ■ A function f is given, and the indicated transformations are applied to its graph (in the given order). Write the equation for the final transformed graph.

45. $f(x) = x^2$; shift upward 3 units

46. $f(x) = x^3$; shift downward 1 unit

47. $f(x) = \sqrt{x}$; shift 2 units to the left

48. $f(x) = \sqrt[3]{x}$; shift 1 unit to the right

49. $f(x) = |x|$; shift 3 units to the right and shift upward 1 unit

50. $f(x) = |x|$; shift 4 units to the left and shift downward 2 units

51. $f(x) = \sqrt[4]{x}$; reflect in the y-axis and shift upward 1 unit

52. $f(x) = x^2$; shift 2 units to the left and reflect in the x-axis

53. $f(x) = x^2$; stretch vertically by a factor of 2, shift downward 2 units, and shift 3 units to the right

54. $f(x) = |x|$; shrink vertically by a factor of $\frac{1}{2}$, shift to the left 1 unit, and shift upward 3 units

55–60 ■ The graphs of f and g are given. Find a formula for the function g.

55.

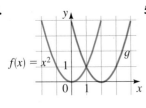

$f(x) = x^2$

56.

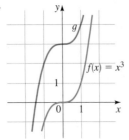

$f(x) = x^3$

57.

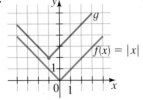

g

$f(x) = |x|$

58.

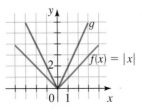

g

$f(x) = |x|$

59.

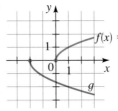

$f(x) = \sqrt{x}$

g

60.

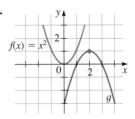

$f(x) = x^2$

g

61–62 ■ The graph of $y = f(x)$ is given. Match each equation with its graph.

61. (a) $y = f(x - 4)$ **(b)** $y = f(x) + 3$
    (c) $y = 2f(x + 6)$ **(d)** $y = -f(2x)$

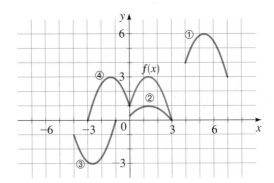

62. (a) $y = \frac{1}{3}f(x)$ **(b)** $y = -f(x + 4)$
    (c) $y = f(x - 4) + 3$ **(d)** $y = f(-x)$

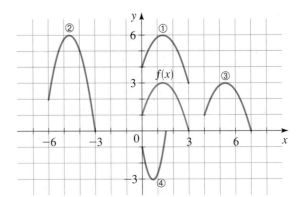

63. The graph of f is given. Sketch the graphs of the following functions.
    (a) $y = f(x - 2)$ **(b)** $y = f(x) - 2$
    (c) $y = 2f(x)$ **(d)** $y = -f(x) + 3$
    (e) $y = f(-x)$ **(f)** $y = \frac{1}{2}f(x - 1)$

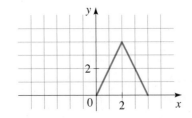

64. The graph of g is given. Sketch the graphs of the following functions.

(a) $y = g(x + 1)$ (b) $y = g(-x)$
(c) $y = g(x - 2)$ (d) $y = g(x) - 2$
(e) $y = -g(x)$ (f) $y = 2g(x)$

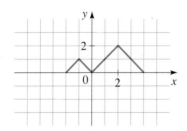

65. The graph of g is given. Use it to graph each of the following functions.

(a) $y = g(2x)$ (b) $y = g\left(\tfrac{1}{2}x\right)$

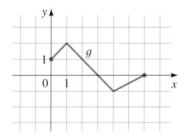

66. The graph of h is given. Use it to graph each of the following functions.

(a) $y = h(3x)$ (b) $y = h\left(\tfrac{1}{3}x\right)$

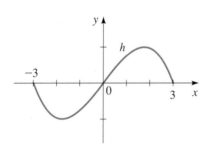

67–68 ■ Use the graph of $f(x) = [\![x]\!]$ described on page 42 to graph the indicated function.

67. $y = [\![2x]\!]$ **68.** $y = \left[\!\left[\tfrac{1}{4}x\right]\!\right]$

 69–72 ■ Graph the functions on the same screen using the given viewing rectangle. How is each graph related to the graph in part (a)?

69. Viewing rectangle $[-8, 8]$ by $[-2, 8]$
(a) $y = \sqrt[4]{x}$ (b) $y = \sqrt[4]{x + 5}$
(c) $y = 2\sqrt[4]{x + 5}$ (d) $y = 4 + 2\sqrt[4]{x + 5}$

70. Viewing rectangle $[-8, 8]$ by $[-6, 6]$
(a) $y = |x|$ (b) $y = -|x|$
(c) $y = -3|x|$ (d) $y = -3|x - 5|$

71. Viewing rectangle $[-4, 6]$ by $[-4, 4]$
(a) $y = x^6$ (b) $y = \tfrac{1}{3}x^6$
(c) $y = -\tfrac{1}{3}x^6$ (d) $y = -\tfrac{1}{3}(x - 4)^6$

72. Viewing rectangle $[-6, 6]$ by $[-4, 4]$
(a) $y = \dfrac{1}{\sqrt{x}}$ (b) $y = \dfrac{1}{\sqrt{x + 3}}$
(c) $y = \dfrac{1}{2\sqrt{x + 3}}$ (d) $y = \dfrac{1}{2\sqrt{x + 3}} - 3$

 73. If $f(x) = \sqrt{2x - x^2}$, graph the following functions in the viewing rectangle $[-5, 5]$ by $[-4, 4]$. How is each graph related to the graph in part (a)?
(a) $y = f(x)$ (b) $y = f(2x)$ (c) $y = f\left(\tfrac{1}{2}x\right)$

 74. If $f(x) = \sqrt{2x - x^2}$, graph the following functions in the viewing rectangle $[-5, 5]$ by $[-4, 4]$. How is each graph related to the graph in part (a)?
(a) $y = f(x)$ (b) $y = f(-x)$
(c) $y = -f(-x)$ (d) $y = f(-2x)$
(e) $y = f\left(-\tfrac{1}{2}x\right)$

75–82 ■ Determine whether the function f is even, odd, or neither. If f is even or odd, use symmetry to sketch its graph.

75. $f(x) = x^4$ **76.** $f(x) = x^3$

77. $f(x) = x^2 + x$ **78.** $f(x) = x^4 - 4x^2$

79. $f(x) = x^3 - x$ **80.** $f(x) = 3x^3 + 2x^2 + 1$

81. $f(x) = 1 - \sqrt[3]{x}$ **82.** $f(x) = x + \dfrac{1}{x}$

83–84 ■ The graph of a function defined for $x \geq 0$ is given. Complete the graph for $x < 0$ to make (a) an even function and (b) an odd function.

83.

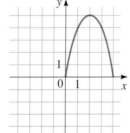

84.

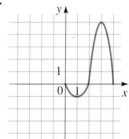

85–86 ■ These exercises show how the graph of $y = |f(x)|$ is obtained from the graph of $y = f(x)$.

85. The graphs of $f(x) = x^2 - 4$ and $g(x) = |x^2 - 4|$ are shown. Explain how the graph of g is obtained from the graph of f.

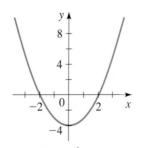

$f(x) = x^2 - 4$

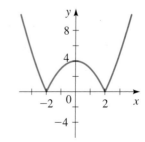

$g(x) = |x^2 - 4|$

86. The graph of $f(x) = x^4 - 4x^2$ is shown. Use this graph to sketch the graph of $g(x) = |x^4 - 4x^2|$.

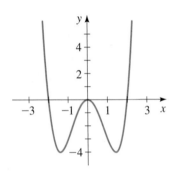

87–88 ■ Sketch the graph of each function.

87. (a) $f(x) = 4x - x^2$ **(b)** $g(x) = |4x - x^2|$

88. (a) $f(x) = x^3$ **(b)** $g(x) = |x^3|$

APPLICATIONS

89. Sales Growth The annual sales of a certain company can be modeled by the function $f(t) = 4 + 0.01t^2$, where t represents years since 1990 and $f(t)$ is measured in millions of dollars.
(a) What shifting and shrinking operations must be performed on the function $y = t^2$ to obtain the function $y = f(t)$?
(b) Suppose you want t to represent years since 2000 instead of 1990. What transformation would you have to apply to the function $y = f(t)$ to accomplish this? Write the new function $y = g(t)$ that results from this transformation.

90. Changing Temperature Scales The temperature on a certain afternoon is modeled by the function

$$C(t) = \tfrac{1}{2}t^2 + 2$$

where t represents hours after 12 noon ($0 \le t \le 6$) and C is measured in °C.
(a) What shifting and shrinking operations must be performed on the function $y = t^2$ to obtain the function $y = C(t)$?
(b) Suppose you want to measure the temperature in °F instead. What transformation would you have to apply to the function $y = C(t)$ to accomplish this? (Use the fact that the relationship between Celsius and Fahrenheit degrees is given by $F = \tfrac{9}{5}C + 32$.) Write the new function $y = F(t)$ that results from this transformation.

DISCOVERY ▪ DISCUSSION ▪ WRITING

91. Sums of Even and Odd Functions If f and g are both even functions, is $f + g$ necessarily even? If both are odd, is their sum necessarily odd? What can you say about the sum if one is odd and one is even? In each case, prove your answer.

92. Products of Even and Odd Functions Answer the same questions as in Exercise 91, except this time consider the product of f and g instead of the sum.

93. Even and Odd Power Functions What must be true about the integer n if the function

$$f(x) = x^n$$

is an even function? If it is an odd function? Why do you think the names "even" and "odd" were chosen for these function properties?

1.7 COMBINING FUNCTIONS

| Sums, Differences, Products, and Quotients ▶ Composition of Functions

In this section we study different ways to combine functions to make new functions.

▼ Sums, Differences, Products, and Quotients

The sum of f and g is defined by

$$(f + g)(x) = f(x) + g(x)$$

The name of the new function is "$f + g$." So this $+$ sign stands for the operation of addition of *functions*. The $+$ sign on the right side, however, stands for addition of the *numbers* $f(x)$ and $g(x)$.

Two functions f and g can be combined to form new functions $f + g$, $f - g$, fg, and f/g in a manner similar to the way we add, subtract, multiply, and divide real numbers. For example, we define the function $f + g$ by

$$(f + g)(x) = f(x) + g(x)$$

The new function $f + g$ is called the **sum** of the functions f and g; its value at x is $f(x) + g(x)$. Of course, the sum on the right-hand side makes sense only if both $f(x)$ and $g(x)$ are defined, that is, if x belongs to the domain of f and also to the domain of g. So if the domain of f is A and the domain of g is B, then the domain of $f + g$ is the intersection of these domains, that is, $A \cap B$. Similarly, we can define the **difference** $f - g$, the **product** fg, and the **quotient** f/g of the functions f and g. Their domains are $A \cap B$, but in the case of the quotient we must remember not to divide by 0.

ALGEBRA OF FUNCTIONS

Let f and g be functions with domains A and B. Then the functions $f + g$, $f - g$, fg, and f/g are defined as follows.

$$(f + g)(x) = f(x) + g(x) \qquad \text{Domain } A \cap B$$

$$(f - g)(x) = f(x) - g(x) \qquad \text{Domain } A \cap B$$

$$(fg)(x) = f(x)g(x) \qquad \text{Domain } A \cap B$$

$$\left(\frac{f}{g}\right)(x) = \frac{f(x)}{g(x)} \qquad \text{Domain } \{x \in A \cap B \mid g(x) \neq 0\}$$

EXAMPLE 1 | Combinations of Functions and Their Domains

Let $f(x) = \dfrac{1}{x - 2}$ and $g(x) = \sqrt{x}$.

(a) Find the functions $f + g$, $f - g$, fg, and f/g and their domains.

(b) Find $(f + g)(4)$, $(f - g)(4)$, $(fg)(4)$, and $(f/g)(4)$.

SOLUTION

(a) The domain of f is $\{x \mid x \neq 2\}$, and the domain of g is $\{x \mid x \geq 0\}$. The intersection of the domains of f and g is

$$\{x \mid x \geq 0 \text{ and } x \neq 2\} = [0, 2) \cup (2, \infty)$$

Thus, we have

$$(f + g)(x) = f(x) + g(x) = \frac{1}{x - 2} + \sqrt{x} \qquad \text{Domain } \{x \mid x \geq 0 \text{ and } x \neq 2\}$$

$$(f - g)(x) = f(x) - g(x) = \frac{1}{x - 2} - \sqrt{x} \qquad \text{Domain } \{x \mid x \geq 0 \text{ and } x \neq 2\}$$

To divide fractions, invert the denominator and multiply:

$$\frac{1/(x - 2)}{\sqrt{x}} = \frac{1/(x - 2)}{\sqrt{x}/1}$$

$$= \frac{1}{x - 2} \cdot \frac{1}{\sqrt{x}}$$

$$= \frac{1}{(x - 2)\sqrt{x}}$$

$$(fg)(x) = f(x)g(x) = \frac{\sqrt{x}}{x - 2} \qquad \text{Domain } \{x \mid x \geq 0 \text{ and } x \neq 2\}$$

$$\left(\frac{f}{g}\right)(x) = \frac{f(x)}{g(x)} = \frac{1}{(x - 2)\sqrt{x}} \qquad \text{Domain } \{x \mid x > 0 \text{ and } x \neq 2\}$$

Note that in the domain of f/g we exclude 0 because $g(0) = 0$.

(b) Each of these values exist because $x = 4$ is in the domain of each function.

$$(f + g)(4) = f(4) + g(4) = \frac{1}{4 - 2} + \sqrt{4} = \frac{5}{2}$$

$$(f - g)(4) = f(4) - g(4) = \frac{1}{4 - 2} - \sqrt{4} = -\frac{3}{2}$$

$$(fg)(4) = f(4)g(4) = \left(\frac{1}{4 - 2}\right)\sqrt{4} = 1$$

$$\left(\frac{f}{g}\right)(4) = \frac{f(4)}{g(4)} = \frac{1}{(4 - 2)\sqrt{4}} = \frac{1}{4}$$

✎ NOW TRY EXERCISE **5**

The graph of the function $f + g$ can be obtained from the graphs of f and g by **graphical addition**. This means that we add corresponding y-coordinates, as illustrated in the next example.

EXAMPLE 2 | Using Graphical Addition

The graphs of f and g are shown in Figure 1. Use graphical addition to graph the function $f + g$.

SOLUTION We obtain the graph of $f + g$ by "graphically adding" the value of $f(x)$ to $g(x)$ as shown in Figure 2. This is implemented by copying the line segment PQ on top of PR to obtain the point S on the graph of $f + g$.

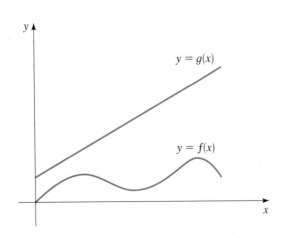

FIGURE 1

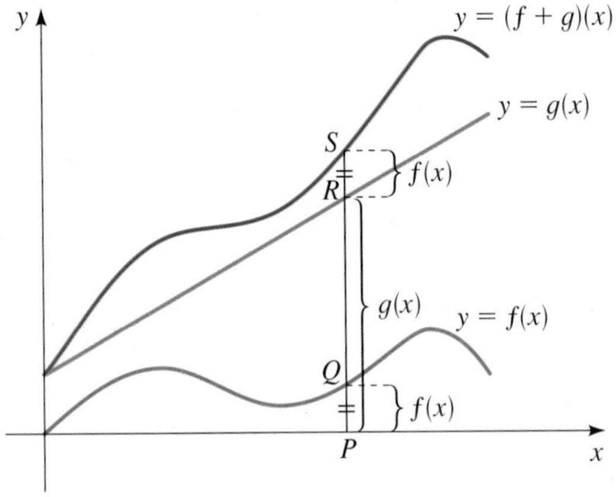

FIGURE 2 Graphical addition

✎ NOW TRY EXERCISE **15**

▼ Composition of Functions

Now let's consider a very important way of combining two functions to get a new function. Suppose $f(x) = \sqrt{x}$ and $g(x) = x^2 + 1$. We may define a new function h as

$$h(x) = f(g(x)) = f(x^2 + 1) = \sqrt{x^2 + 1}$$

The function h is made up of the functions f and g in an interesting way: Given a number x, we first apply the function g to it, then apply f to the result. In this case, f is the rule "take the square root," g is the rule "square, then add 1," and h is the rule "square, then add 1, then take the square root." In other words, we get the rule h by applying the rule g and then the rule f. Figure 3 shows a machine diagram for h.

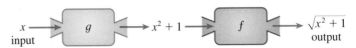

FIGURE 3 The h machine is composed of the g machine (first) and then the f machine.

In general, given any two functions f and g, we start with a number x in the domain of g and find its image $g(x)$. If this number $g(x)$ is in the domain of f, we can then calculate the value of $f(g(x))$. The result is a new function $h(x) = f(g(x))$ that is obtained by substituting g into f. It is called the *composition* (or *composite*) of f and g and is denoted by $f \circ g$ ("f composed with g").

COMPOSITION OF FUNCTIONS

Given two functions f and g, the **composite function** $f \circ g$ (also called the **composition** of f and g) is defined by

$$(f \circ g)(x) = f(g(x))$$

The domain of $f \circ g$ is the set of all x in the domain of g such that $g(x)$ is in the domain of f. In other words, $(f \circ g)(x)$ is defined whenever both $g(x)$ and $f(g(x))$ are defined. We can picture $f \circ g$ using an arrow diagram (Figure 4).

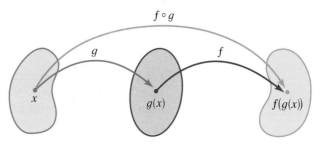

FIGURE 4 Arrow diagram for $f \circ g$

EXAMPLE 3 | Finding the Composition of Functions

Let $f(x) = x^2$ and $g(x) = x - 3$.

(a) Find the functions $f \circ g$ and $g \circ f$ and their domains.

(b) Find $(f \circ g)(5)$ and $(g \circ f)(7)$.

SOLUTION

In Example 3, f is the rule "square" and g is the rule "subtract 3." The function $f \circ g$ *first* subtracts 3 and *then* squares; the function $g \circ f$ *first* squares and *then* subtracts 3.

(a) We have

$$
\begin{aligned}
(f \circ g)(x) &= f(g(x)) && \text{Definition of } f \circ g \\
&= f(x - 3) && \text{Definition of } g \\
&= (x - 3)^2 && \text{Definition of } f
\end{aligned}
$$

and

$$
\begin{aligned}
(g \circ f)(x) &= g(f(x)) && \text{Definition of } g \circ f \\
&= g(x^2) && \text{Definition of } f \\
&= x^2 - 3 && \text{Definition of } g
\end{aligned}
$$

The domains of both $f \circ g$ and $g \circ f$ are $\mathbb{R}$.

(b) We have

$$(f \circ g)(5) = f(g(5)) = f(2) = 2^2 = 4$$

$$(g \circ f)(7) = g(f(7)) = g(49) = 49 - 3 = 46$$

✎ NOW TRY EXERCISES **21** AND **35**

You can see from Example 3 that, in general, $f \circ g \neq g \circ f$. Remember that the notation $f \circ g$ means that the function g is applied first and then f is applied second.

The graphs of f and g of Example 4, as well as those of $f \circ g$, $g \circ f$, $f \circ f$, and $g \circ g$, are shown below. These graphs indicate that the operation of composition can produce functions that are quite different from the original functions.

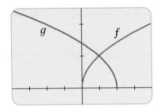

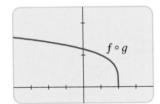

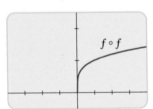

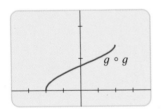

EXAMPLE 4 | Finding the Composition of Functions

If $f(x) = \sqrt{x}$ and $g(x) = \sqrt{2 - x}$, find the following functions and their domains.

(a) $f \circ g$ **(b)** $g \circ f$ **(c)** $f \circ f$ **(d)** $g \circ g$

SOLUTION

(a)
$$
\begin{aligned}
(f \circ g)(x) &= f(g(x)) && \text{Definition of } f \circ g \\
&= f(\sqrt{2 - x}) && \text{Definition of } g \\
&= \sqrt{\sqrt{2 - x}} && \text{Definition of } f \\
&= \sqrt[4]{2 - x}
\end{aligned}
$$

The domain of $f \circ g$ is $\{x \mid 2 - x \geq 0\} = \{x \mid x \leq 2\} = (-\infty, 2]$.

(b)
$$
\begin{aligned}
(g \circ f)(x) &= g(f(x)) && \text{Definition of } g \circ f \\
&= g(\sqrt{x}) && \text{Definition of } f \\
&= \sqrt{2 - \sqrt{x}} && \text{Definition of } g
\end{aligned}
$$

For $\sqrt{x}$ to be defined, we must have $x \geq 0$. For $\sqrt{2 - \sqrt{x}}$ to be defined, we must have $2 - \sqrt{x} \geq 0$, that is, $\sqrt{x} \leq 2$, or $x \leq 4$. Thus, we have $0 \leq x \leq 4$, so the domain of $g \circ f$ is the closed interval $[0, 4]$.

(c)
$$
\begin{aligned}
(f \circ f)(x) &= f(f(x)) && \text{Definition of } f \circ f \\
&= f(\sqrt{x}) && \text{Definition of } f \\
&= \sqrt{\sqrt{x}} && \text{Definition of } f \\
&= \sqrt[4]{x}
\end{aligned}
$$

The domain of $f \circ f$ is $[0, \infty)$.

(d)
$$
\begin{aligned}
(g \circ g)(x) &= g(g(x)) && \text{Definition of } g \circ g \\
&= g(\sqrt{2 - x}) && \text{Definition of } g \\
&= \sqrt{2 - \sqrt{2 - x}} && \text{Definition of } g
\end{aligned}
$$

This expression is defined when both $2 - x \geq 0$ and $2 - \sqrt{2 - x} \geq 0$. The first inequality means $x \leq 2$, and the second is equivalent to $\sqrt{2 - x} \leq 2$, or $2 - x \leq 4$, or $x \geq -2$. Thus, $-2 \leq x \leq 2$, so the domain of $g \circ g$ is $[-2, 2]$.

✎ NOW TRY EXERCISE **41** ■

It is possible to take the composition of three or more functions. For instance, the composite function $f \circ g \circ h$ is found by first applying h, then g, and then f as follows:

$$(f \circ g \circ h)(x) = f(g(h(x)))$$

EXAMPLE 5 | A Composition of Three Functions

Find $f \circ g \circ h$ if $f(x) = x/(x + 1)$, $g(x) = x^{10}$, and $h(x) = x + 3$.

SOLUTION

$$
\begin{aligned}
(f \circ g \circ h)(x) &= f(g(h(x))) && \text{Definition of } f \circ g \circ h \\
&= f(g(x + 3)) && \text{Definition of } h \\
&= f((x + 3)^{10}) && \text{Definition of } g \\
&= \frac{(x + 3)^{10}}{(x + 3)^{10} + 1} && \text{Definition of } f
\end{aligned}
$$

✎ NOW TRY EXERCISE **45** ■

So far, we have used composition to build complicated functions from simpler ones. But in calculus it is useful to be able to "decompose" a complicated function into simpler ones, as shown in the following example.

EXAMPLE 6 | Recognizing a Composition of Functions

Given $F(x) = \sqrt[4]{x + 9}$, find functions f and g such that $F = f \circ g$.

SOLUTION   Since the formula for F says to first add 9 and then take the fourth root, we let

$$g(x) = x + 9 \qquad \text{and} \qquad f(x) = \sqrt[4]{x}$$

Then

$$
\begin{aligned}
(f \circ g)(x) &= f(g(x)) && \text{Definition of } f \circ g \\
&= f(x + 9) && \text{Definition of } g \\
&= \sqrt[4]{x + 9} && \text{Definition of } f \\
&= F(x)
\end{aligned}
$$

✎ NOW TRY EXERCISE **49** ■

EXAMPLE 7 | An Application of Composition of Functions

A ship is traveling at 20 mi/h parallel to a straight shoreline. The ship is 5 mi from shore. It passes a lighthouse at noon.

(a) Express the distance s between the lighthouse and the ship as a function of d, the distance the ship has traveled since noon; that is, find f so that $s = f(d)$.

(b) Express d as a function of t, the time elapsed since noon; that is, find g so that $d = g(t)$.

(c) Find $f \circ g$. What does this function represent?

SOLUTION   We first draw a diagram as in Figure 5.

(a) We can relate the distances s and d by the Pythagorean Theorem. Thus, s can be expressed as a function of d by

$$s = f(d) = \sqrt{25 + d^2}$$

(b) Since the ship is traveling at 20 mi/h, the distance d it has traveled is a function of t as follows:

$$d = g(t) = 20t$$

(c) We have

$$
\begin{aligned}
(f \circ g)(t) &= f(g(t)) && \text{Definition of } f \circ g \\
&= f(20t) && \text{Definition of } g \\
&= \sqrt{25 + (20t)^2} && \text{Definition of } f
\end{aligned}
$$

The function $f \circ g$ gives the distance of the ship from the lighthouse as a function of time.

✎ NOW TRY EXERCISE **63** ■

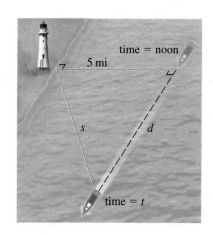

FIGURE 5

distance = rate × time

1.7 EXERCISES

CONCEPTS

1. From the graphs of f and g in the figure, we find

$(f + g)(2) = $ _____ $(f - g)(2) = $ _____

$(fg)(2) = $ _____ $\left(\dfrac{f}{g}\right)(2) = $ _____

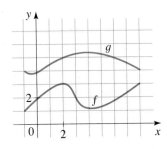

2. By definition, $f \circ g(x) = $ _____. So if $g(2) = 5$ and $f(5) = 12$, then $f \circ g(2) = $ _____.

3. If the rule of the function f is "add one" and the rule of the function g is "multiply by 2," then the rule of $f \circ g$ is

"_____,"

and the rule of $g \circ f$ is

"_____."

4. We can express the functions in Exercise 3 algebraically as

$f(x) = $ _____ $g(x) = $ _____

$f \circ g(x) = $ _____ $g \circ f(x) = $ _____

SKILLS

5–10 ■ Find $f + g$, $f - g$, fg, and f/g and their domains.

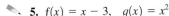

5. $f(x) = x - 3$, $g(x) = x^2$

6. $f(x) = x^2 + 2x$, $g(x) = 3x^2 - 1$

7. $f(x) = \sqrt{4 - x^2}$, $g(x) = \sqrt{1 + x}$

8. $f(x) = \sqrt{9 - x^2}$, $g(x) = \sqrt{x^2 - 4}$

9. $f(x) = \dfrac{2}{x}$, $g(x) = \dfrac{4}{x + 4}$

10. $f(x) = \dfrac{2}{x + 1}$, $g(x) = \dfrac{x}{x + 1}$

11–14 ■ Find the domain of the function.

11. $f(x) = \sqrt{x} + \sqrt{1 - x}$ **12.** $g(x) = \sqrt{x + 1} - \dfrac{1}{x}$

13. $h(x) = (x - 3)^{-1/4}$ **14.** $k(x) = \dfrac{\sqrt{x + 3}}{x - 1}$

15–16 ■ Use graphical addition to sketch the graph of $f + g$.

15. 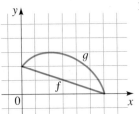   **16.**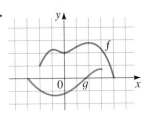

17–20 ■ Draw the graphs of f, g, and $f + g$ on a common screen to illustrate graphical addition.

17. $f(x) = \sqrt{1 + x}$, $g(x) = \sqrt{1 - x}$

18. $f(x) = x^2$, $g(x) = \sqrt{x}$

19. $f(x) = x^2$, $g(x) = \frac{1}{3}x^3$

20. $f(x) = \sqrt[4]{1 - x}$, $g(x) = \sqrt{1 - \dfrac{x^2}{9}}$

21–26 ■ Use $f(x) = 3x - 5$ and $g(x) = 2 - x^2$ to evaluate the expression.

21. (a) $f(g(0))$ (b) $g(f(0))$

22. (a) $f(f(4))$ (b) $g(g(3))$

23. (a) $(f \circ g)(-2)$ (b) $(g \circ f)(-2)$

24. (a) $(f \circ f)(-1)$ (b) $(g \circ g)(2)$

25. (a) $(f \circ g)(x)$ (b) $(g \circ f)(x)$

26. (a) $(f \circ f)(x)$ (b) $(g \circ g)(x)$

27–32 ■ Use the given graphs of f and g to evaluate the expression.

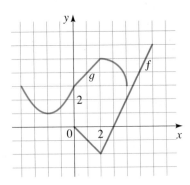

27. $f(g(2))$

28. $g(f(0))$

29. $(g \circ f)(4)$

30. $(f \circ g)(0)$

31. $(g \circ g)(-2)$

32. $(f \circ f)(4)$

33–44 ■ Find the functions $f \circ g$, $g \circ f$, $f \circ f$, and $g \circ g$ and their domains.

33. $f(x) = 2x + 3$, $g(x) = 4x - 1$

34. $f(x) = 6x - 5$, $g(x) = \dfrac{x}{2}$

35. $f(x) = x^2$, $g(x) = x + 1$

36. $f(x) = x^3 + 2$, $g(x) = \sqrt[3]{x}$

37. $f(x) = \dfrac{1}{x}$, $g(x) = 2x + 4$

38. $f(x) = x^2$, $g(x) = \sqrt{x - 3}$

39. $f(x) = |x|$, $g(x) = 2x + 3$

40. $f(x) = x - 4$, $g(x) = |x + 4|$

41. $f(x) = \dfrac{x}{x + 1}$, $g(x) = 2x - 1$

42. $f(x) = \dfrac{1}{\sqrt{x}}$, $g(x) = x^2 - 4x$

43. $f(x) = \dfrac{x}{x + 1}$, $g(x) = \dfrac{1}{x}$

44. $f(x) = \dfrac{2}{x}$, $g(x) = \dfrac{x}{x + 2}$

45–48 ■ Find $f \circ g \circ h$.

45. $f(x) = x - 1$, $g(x) = \sqrt{x}$, $h(x) = x - 1$

46. $f(x) = \dfrac{1}{x}$, $g(x) = x^3$, $h(x) = x^2 + 2$

47. $f(x) = x^4 + 1$, $g(x) = x - 5$, $h(x) = \sqrt{x}$

48. $f(x) = \sqrt{x}$, $g(x) = \dfrac{x}{x - 1}$, $h(x) = \sqrt[3]{x}$

49–54 ■ Express the function in the form $f \circ g$.

49. $F(x) = (x - 9)^5$

50. $F(x) = \sqrt{x} + 1$

51. $G(x) = \dfrac{x^2}{x^2 + 4}$

52. $G(x) = \dfrac{1}{x + 3}$

53. $H(x) = |1 - x^3|$

54. $H(x) = \sqrt{1 + \sqrt{x}}$

55–58 ■ Express the function in the form $f \circ g \circ h$.

55. $F(x) = \dfrac{1}{x^2 + 1}$

56. $F(x) = \sqrt[3]{\sqrt{x} - 1}$

57. $G(x) = (4 + \sqrt[3]{x})^9$

58. $G(x) = \dfrac{2}{(3 + \sqrt{x})^2}$

APPLICATIONS

59–60 ■ **Revenue, Cost, and Profit**   A print shop makes bumper stickers for election campaigns. If x stickers are ordered (where $x < 10{,}000$), then the price per bumper sticker is $0.15 - 0.000002x$ dollars, and the total cost of producing the order is $0.095x - 0.0000005x^2$ dollars.

59. Use the fact that

> revenue   =   price per item   ×   number of items sold

to express $R(x)$, the revenue from an order of x stickers, as a product of two functions of x.

60. Use the fact that

> profit   =   revenue   −   cost

to express $P(x)$, the profit on an order of x stickers, as a difference of two functions of x.

61. **Area of a Ripple**   A stone is dropped in a lake, creating a circular ripple that travels outward at a speed of 60 cm/s.
(a) Find a function g that models the radius as a function of time.
(b) Find a function f that models the area of the circle as a function of the radius.
(c) Find $f \circ g$. What does this function represent?

62. **Inflating a Balloon**   A spherical balloon is being inflated. The radius of the balloon is increasing at the rate of 1 cm/s.
(a) Find a function f that models the radius as a function of time.
(b) Find a function g that models the volume as a function of the radius.
(c) Find $g \circ f$. What does this function represent?

63. **Area of a Balloon**   A spherical weather balloon is being inflated. The radius of the balloon is increasing at the rate of 2 cm/s. Express the surface area of the balloon as a function of time t (in seconds).

64. **Multiple Discounts**   You have a $50 coupon from the manufacturer good for the purchase of a cell phone. The store where you are purchasing your cell phone is offering a 20% discount on all cell phones. Let x represent the regular price of the cell phone.
(a) Suppose only the 20% discount applies. Find a function f that models the purchase price of the cell phone as a function of the regular price x.

(b) Suppose only the $50 coupon applies. Find a function g that models the purchase price of the cell phone as a function of the sticker price x.

(c) If you can use the coupon and the discount, then the purchase price is either $f \circ g(x)$ or $g \circ f(x)$, depending on the order in which they are applied to the price. Find both $f \circ g(x)$ and $g \circ f(x)$. Which composition gives the lower price?

65. Multiple Discounts An appliance dealer advertises a 10% discount on all his washing machines. In addition, the manufacturer offers a $100 rebate on the purchase of a washing machine. Let x represent the sticker price of the washing machine.

(a) Suppose only the 10% discount applies. Find a function f that models the purchase price of the washer as a function of the sticker price x.

(b) Suppose only the $100 rebate applies. Find a function g that models the purchase price of the washer as a function of the sticker price x.

(c) Find $f \circ g$ and $g \circ f$. What do these functions represent? Which is the better deal?

66. Airplane Trajectory An airplane is flying at a speed of 350 mi/h at an altitude of one mile. The plane passes directly above a radar station at time $t = 0$.

(a) Express the distance s (in miles) between the plane and the radar station as a function of the horizontal distance d (in miles) that the plane has flown.

(b) Express d as a function of the time t (in hours) that the plane has flown.

(c) Use composition to express s as a function of t.

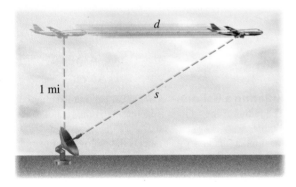

Find

$$A \circ A$$

$$A \circ A \circ A$$

$$A \circ A \circ A \circ A$$

What do these compositions represent? Find a formula for what you get when you compose n copies of A.

68. Composing Linear Functions The graphs of the functions

$$f(x) = m_1 x + b_1$$

$$g(x) = m_2 x + b_2$$

are lines with slopes m_1 and m_2, respectively. Is the graph of $f \circ g$ a line? If so, what is its slope?

69. Solving an Equation for an Unknown Function Suppose that

$$g(x) = 2x + 1$$

$$h(x) = 4x^2 + 4x + 7$$

Find a function f such that $f \circ g = h$. (Think about what operations you would have to perform on the formula for g to end up with the formula for h.) Now suppose that

$$f(x) = 3x + 5$$

$$h(x) = 3x^2 + 3x + 2$$

Use the same sort of reasoning to find a function g such that $f \circ g = h$.

70. Compositions of Odd and Even Functions Suppose that

$$h = f \circ g$$

If g is an even function, is h necessarily even? If g is odd, is h odd? What if g is odd and f is odd? What if g is odd and f is even?

DISCOVERY ▪ DISCUSSION ▪ WRITING

67. Compound Interest A savings account earns 5% interest compounded annually. If you invest x dollars in such an account, then the amount $A(x)$ of the investment after one year is the initial investment plus 5%; that is,

$$A(x) = x + 0.05x = 1.05x$$

 **DISCOVERY PROJECT** **Iteration and Chaos**

In this project we explore the process of repeatedly composing a function with itself; the result can be regular or chaotic. You can find the project at the book companion website: **www.stewartmath.com**

1.8 ONE-TO-ONE FUNCTIONS AND THEIR INVERSES

One-to-One Functions ▶ The Inverse of a Function ▶ Graphing the Inverse of a Function

The *inverse* of a function is a rule that acts on the output of the function and produces the corresponding input. So the inverse "undoes" or reverses what the function has done. Not all functions have inverses; those that do are called *one-to-one*.

▼ One-to-One Functions

Let's compare the functions f and g whose arrow diagrams are shown in Figure 1. Note that f never takes on the same value twice (any two numbers in A have different images), whereas g does take on the same value twice (both 2 and 3 have the same image, 4). In symbols, $g(2) = g(3)$ but $f(x_1) \neq f(x_2)$ whenever $x_1 \neq x_2$. Functions that have this latter property are called *one-to-one*.

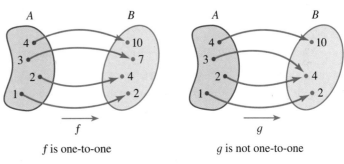

f is one-to-one g is not one-to-one

FIGURE 1

DEFINITION OF A ONE-TO-ONE FUNCTION

A function with domain A is called a **one-to-one function** if no two elements of A have the same image, that is,

$$f(x_1) \neq f(x_2) \quad \text{whenever } x_1 \neq x_2$$

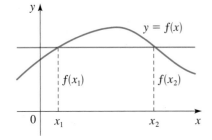

FIGURE 2 This function is not one-to-one because $f(x_1) = f(x_2)$.

An equivalent way of writing the condition for a one-to-one function is this:

$$\text{If } f(x_1) = f(x_2), \text{ then } x_1 = x_2.$$

If a horizontal line intersects the graph of f at more than one point, then we see from Figure 2 that there are numbers $x_1 \neq x_2$ such that $f(x_1) = f(x_2)$. This means that f is not one-to-one. Therefore, we have the following geometric method for determining whether a function is one-to-one.

HORIZONTAL LINE TEST

A function is one-to-one if and only if no horizontal line intersects its graph more than once.

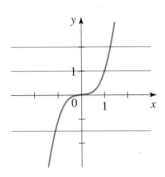

FIGURE 3 $f(x) = x^3$ is one-to-one.

EXAMPLE 1 | Deciding Whether a Function Is One-to-One

Is the function $f(x) = x^3$ one-to-one?

SOLUTION 1   If $x_1 \neq x_2$, then $x_1^3 \neq x_2^3$ (two different numbers cannot have the same cube). Therefore, $f(x) = x^3$ is one-to-one.

SOLUTION 2   From Figure 3 we see that no horizontal line intersects the graph of $f(x) = x^3$ more than once. Therefore, by the Horizontal Line Test, f is one-to-one.

✎ NOW TRY EXERCISE **13**   ◼

Notice that the function f of Example 1 is increasing and is also one-to-one. In fact, it can be proved that *every increasing function and every decreasing function is one-to-one.*

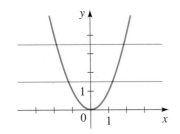

FIGURE 4 $f(x) = x^2$ is not one-to-one.

EXAMPLE 2 | Deciding Whether a Function Is One-to-One

Is the function $g(x) = x^2$ one-to-one?

SOLUTION 1   This function is not one-to-one because, for instance,

$$g(1) = 1 \qquad \text{and} \qquad g(-1) = 1$$

so 1 and -1 have the same image.

SOLUTION 2   From Figure 4 we see that there are horizontal lines that intersect the graph of g more than once. Therefore, by the Horizontal Line Test, g is not one-to-one.

✎ NOW TRY EXERCISE **15**   ◼

Although the function g in Example 2 is not one-to-one, it is possible to restrict its domain so that the resulting function is one-to-one. In fact, if we define

$$h(x) = x^2 \qquad x \geq 0$$

then h is one-to-one, as you can see from Figure 5 and the Horizontal Line Test.

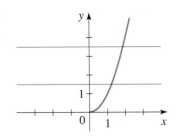

FIGURE 5 $f(x) = x^2$ $(x \geq 0)$ is one-to-one.

EXAMPLE 3 | Showing That a Function Is One-to-One

Show that the function $f(x) = 3x + 4$ is one-to-one.

SOLUTION   Suppose there are numbers x_1 and x_2 such that $f(x_1) = f(x_2)$. Then

$$3x_1 + 4 = 3x_2 + 4 \qquad \text{Suppose } f(x_1) = f(x_2)$$
$$3x_1 = 3x_2 \qquad \text{Subtract 4}$$
$$x_1 = x_2 \qquad \text{Divide by 3}$$

Therefore, f is one-to-one.

✎ NOW TRY EXERCISE **11**   ◼

▼ The Inverse of a Function

One-to-one functions are important because they are precisely the functions that possess inverse functions according to the following definition.

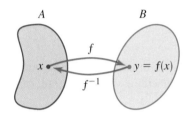

FIGURE 6

⊘ Don't mistake the -1 in f^{-1} for an exponent.

$$f^{-1}(x) \quad \textit{does not mean} \quad \frac{1}{f(x)}$$

The reciprocal $1/f(x)$ is written as $(f(x))^{-1}$.

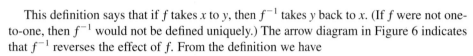

DEFINITION OF THE INVERSE OF A FUNCTION

Let f be a one-to-one function with domain A and range B. Then its **inverse function** f^{-1} has domain B and range A and is defined by

$$f^{-1}(y) = x \quad \Leftrightarrow \quad f(x) = y$$

for any y in B.

This definition says that if f takes x to y, then f^{-1} takes y back to x. (If f were not one-to-one, then f^{-1} would not be defined uniquely.) The arrow diagram in Figure 6 indicates that f^{-1} reverses the effect of f. From the definition we have

$$\text{domain of } f^{-1} = \text{range of } f$$
$$\text{range of } f^{-1} = \text{domain of } f$$

EXAMPLE 4 | Finding f^{-1} for Specific Values

If $f(1) = 5$, $f(3) = 7$, and $f(8) = -10$, find $f^{-1}(5)$, $f^{-1}(7)$, and $f^{-1}(-10)$.

SOLUTION   From the definition of f^{-1} we have

$$f^{-1}(5) = 1 \quad \text{because} \quad f(1) = 5$$
$$f^{-1}(7) = 3 \quad \text{because} \quad f(3) = 7$$
$$f^{-1}(-10) = 8 \quad \text{because} \quad f(8) = -10$$

Figure 7 shows how f^{-1} reverses the effect of f in this case.

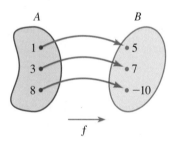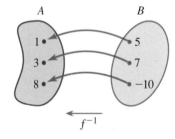

FIGURE 7

✎ NOW TRY EXERCISE **21**

By definition the inverse function f^{-1} undoes what f does: If we start with x, apply f, and then apply f^{-1}, we arrive back at x, where we started. Similarly, f undoes what f^{-1} does. In general, any function that reverses the effect of f in this way must be the inverse of f. These observations are expressed precisely as follows.

INVERSE FUNCTION PROPERTY

Let f be a one-to-one function with domain A and range B. The inverse function f^{-1} satisfies the following cancellation properties:

$$f^{-1}(f(x)) = x \quad \text{for every } x \text{ in } A$$
$$f(f^{-1}(x)) = x \quad \text{for every } x \text{ in } B$$

Conversely, any function f^{-1} satisfying these equations is the inverse of f.

These properties indicate that f is the inverse function of f^{-1}, so we say that f and f^{-1} are *inverses of each other*.

EXAMPLE 5 | Verifying That Two Functions Are Inverses

Show that $f(x) = x^3$ and $g(x) = x^{1/3}$ are inverses of each other.

SOLUTION Note that the domain and range of both f and g is $\mathbb{R}$. We have

$$g(f(x)) = g(x^3) = (x^3)^{1/3} = x$$
$$f(g(x)) = f(x^{1/3}) = (x^{1/3})^3 = x$$

So by the Property of Inverse Functions, f and g are inverses of each other. These equations simply say that the cube function and the cube root function, when composed, cancel each other.

✎ **NOW TRY EXERCISE 27**

Now let's examine how we compute inverse functions. We first observe from the definition of f^{-1} that

$$y = f(x) \quad \Leftrightarrow \quad f^{-1}(y) = x$$

So if $y = f(x)$ and if we are able to solve this equation for x in terms of y, then we must have $x = f^{-1}(y)$. If we then interchange x and y, we have $y = f^{-1}(x)$, which is the desired equation.

HOW TO FIND THE INVERSE OF A ONE-TO-ONE FUNCTION

1. Write $y = f(x)$.

2. Solve this equation for x in terms of y (if possible).

3. Interchange x and y. The resulting equation is $y = f^{-1}(x)$.

Note that Steps 2 and 3 can be reversed. In other words, we can interchange x and y first and then solve for y in terms of x.

In Example 6 note how f^{-1} reverses the effect of f. The function f is the rule "Multiply by 3, then subtract 2," whereas f^{-1} is the rule "Add 2, then divide by 3."

EXAMPLE 6 | Finding the Inverse of a Function

Find the inverse of the function $f(x) = 3x - 2$.

SOLUTION First we write $y = f(x)$.

$$y = 3x - 2$$

Then we solve this equation for x.

$$3x = y + 2 \qquad \text{Add 2}$$
$$x = \frac{y + 2}{3} \qquad \text{Divide by 3}$$

Finally, we interchange x and y.

$$y = \frac{x + 2}{3}$$

Therefore, the inverse function is $f^{-1}(x) = \dfrac{x + 2}{3}$.

✎ **NOW TRY EXERCISE 37**

CHECK YOUR ANSWER

We use the Inverse Function Property.

$$f^{-1}(f(x)) = f^{-1}(3x - 2)$$
$$= \frac{(3x - 2) + 2}{3}$$
$$= \frac{3x}{3} = x$$

$$f(f^{-1}(x)) = f\left(\frac{x + 2}{3}\right)$$
$$= 3\left(\frac{x + 2}{3}\right) - 2$$
$$= x + 2 - 2 = x \quad ✔$$

EXAMPLE 7 | Finding the Inverse of a Function

In Example 7 note how f^{-1} reverses the effect of f. The function f is the rule "Take the fifth power, subtract 3, then divide by 2," whereas f^{-1} is the rule "Multiply by 2, add 3, then take the fifth root."

Find the inverse of the function $f(x) = \dfrac{x^5 - 3}{2}$.

SOLUTION We first write $y = (x^5 - 3)/2$ and solve for x.

$$y = \frac{x^5 - 3}{2} \qquad \text{Equation defining function}$$

$$2y = x^5 - 3 \qquad \text{Multiply by 2}$$

$$x^5 = 2y + 3 \qquad \text{Add 3 (and switch sides)}$$

$$x = (2y + 3)^{1/5} \qquad \text{Take fifth root of each side}$$

Then we interchange x and y to get $y = (2x + 3)^{1/5}$. Therefore, the inverse function is $f^{-1}(x) = (2x + 3)^{1/5}$.

➤ **NOW TRY EXERCISE 53**

CHECK YOUR ANSWER

We use the Inverse Function Property.

$$f^{-1}(f(x)) = f^{-1}\left(\frac{x^5 - 3}{2}\right)$$

$$= \left[2\left(\frac{x^5 - 3}{2}\right) + 3\right]^{1/5}$$

$$= (x^5 - 3 + 3)^{1/5}$$

$$= (x^5)^{1/5} = x$$

$$f(f^{-1}(x)) = f((2x + 3)^{1/5})$$

$$= \frac{[(2x + 3)^{1/5}]^5 - 3}{2}$$

$$= \frac{2x + 3 - 3}{2}$$

$$= \frac{2x}{2} = x \quad ✔$$

A **rational function** is a function defined by a rational expression. In the next example we find the inverse of a rational function.

EXAMPLE 8 | Finding the Inverse of a Rational Function

Find the inverse of the function $f(x) = \dfrac{2x + 3}{x - 1}$.

SOLUTION We first write $y = (2x + 3)/(x - 1)$ and solve for x.

$$y = \frac{2x + 3}{x - 1} \qquad \text{Equation defining function}$$

$$y(x - 1) = 2x + 3 \qquad \text{Multiply by } x - 1$$

$$yx - y = 2x + 3 \qquad \text{Expand}$$

$$yx - 2x = y + 3 \qquad \text{Bring } x\text{-terms to LHS}$$

$$x(y - 2) = y + 3 \qquad \text{Factor } x$$

$$x = \frac{y + 3}{y - 2} \qquad \text{Divide by } y - 2$$

Therefore the inverse function is $f^{-1}(x) = \dfrac{x + 3}{x - 2}$.

➤ **NOW TRY EXERCISE 45**

▼ Graphing the Inverse of a Function

The principle of interchanging x and y to find the inverse function also gives us a method for obtaining the graph of f^{-1} from the graph of f. If $f(a) = b$, then $f^{-1}(b) = a$. Thus, the point (a, b) is on the graph of f if and only if the point (b, a) is on the graph of f^{-1}. But we get the point (b, a) from the point (a, b) by reflecting in the line $y = x$ (see Figure 8 on the next page). Therefore, as Figure 9 on the next page illustrates, the following is true.

> The graph of f^{-1} is obtained by reflecting the graph of f in the line $y = x$.

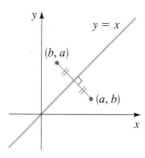

FIGURE 8

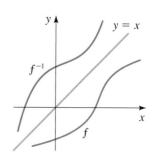

FIGURE 9

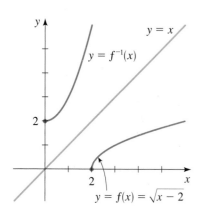

FIGURE 10

In Example 9 note how f^{-1} reverses the effect of f. The function f is the rule "Subtract 2, then take the square root," whereas f^{-1} is the rule "Square, then add 2."

EXAMPLE 9 | Graphing the Inverse of a Function

(a) Sketch the graph of $f(x) = \sqrt{x - 2}$.

(b) Use the graph of f to sketch the graph of f^{-1}.

(c) Find an equation for f^{-1}.

SOLUTION

(a) Using the transformations from Section 1.6, we sketch the graph of $y = \sqrt{x - 2}$ by plotting the graph of the function $y = \sqrt{x}$ (Example 1(c) in Section 1.4) and moving it to the right 2 units.

(b) The graph of f^{-1} is obtained from the graph of f in part (a) by reflecting it in the line $y = x$, as shown in Figure 10.

(c) Solve $y = \sqrt{x - 2}$ for x, noting that $y \geq 0$.

$$\sqrt{x - 2} = y$$

$$x - 2 = y^2 \qquad \text{Square each side}$$

$$x = y^2 + 2 \qquad y \geq 0 \qquad \text{Add 2}$$

Interchange x and y:

$$y = x^2 + 2 \qquad x \geq 0$$

Thus $\qquad\qquad f^{-1}(x) = x^2 + 2 \qquad x \geq 0$

This expression shows that the graph of f^{-1} is the right half of the parabola $y = x^2 + 2$, and from the graph shown in Figure 10, this seems reasonable.

✎ NOW TRY EXERCISE **63**

1.8 EXERCISES

CONCEPTS

1. A function f is one-to-one if different inputs produce

   _____ outputs. You can tell from the graph that a function

   is one-to-one by using the _____ Test.

2. (a) For a function to have an inverse, it must be _____.

   So which one of the following functions has an inverse?

   $$f(x) = x^2 \qquad g(x) = x^3$$

   (b) What is the inverse of the function that you chose in part (a)?

3. A function f has the following verbal description: "Multiply by 3, add 5, and then take the third power of the result."

   (a) Write a verbal description for f^{-1}.

   (b) Find algebraic formulas that express f and f^{-1} in terms of the input x.

4. *True or false?*

   (a) If f has an inverse, then $f^{-1}(x)$ is the same as $\dfrac{1}{f(x)}$.

   (b) If f has an inverse, then $f^{-1}(f(x)) = x$.

SKILLS

5–10 ■ The graph of a function f is given. Determine whether f is one-to-one.

5.
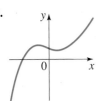

6.

7.

8.

9.

10.

11–20 ■ Determine whether the function is one-to-one.

11. $f(x) = -2x + 4$

12. $f(x) = 3x - 2$

13. $g(x) = \sqrt{x}$

14. $g(x) = |x|$

15. $h(x) = x^2 - 2x$

16. $h(x) = x^3 + 8$

17. $f(x) = x^4 + 5$

18. $f(x) = x^4 + 5, \quad 0 \le x \le 2$

19. $f(x) = \dfrac{1}{x^2}$

20. $f(x) = \dfrac{1}{x}$

21–22 ■ Assume that f is a one-to-one function.

21. (a) If $f(2) = 7$, find $f^{-1}(7)$.
 (b) If $f^{-1}(3) = -1$, find $f(-1)$.

22. (a) If $f(5) = 18$, find $f^{-1}(18)$.
 (b) If $f^{-1}(4) = 2$, find $f(2)$.

23. If $f(x) = 5 - 2x$, find $f^{-1}(3)$.

24. If $g(x) = x^2 + 4x$ with $x \ge -2$, find $g^{-1}(5)$.

25–36 ■ Use the Inverse Function Property to show that f and g are inverses of each other.

25. $f(x) = x - 6; \quad g(x) = x + 6$

26. $f(x) = 3x; \quad g(x) = \dfrac{x}{3}$

27. $f(x) = 2x - 5; \quad g(x) = \dfrac{x + 5}{2}$

28. $f(x) = \dfrac{3 - x}{4}; \quad g(x) = 3 - 4x$

29. $f(x) = \dfrac{1}{x}; \quad g(x) = \dfrac{1}{x}$

30. $f(x) = x^5; \quad g(x) = \sqrt[5]{x}$

31. $f(x) = x^2 - 4, \quad x \ge 0;$
 $g(x) = \sqrt{x + 4}, \quad x \ge -4$

32. $f(x) = x^3 + 1; \quad g(x) = (x - 1)^{1/3}$

33. $f(x) = \dfrac{1}{x - 1}, \quad x \ne 1; \quad g(x) = \dfrac{1}{x} + 1, \quad x \ne 0$

34. $f(x) = \sqrt{4 - x^2}, \quad 0 \le x \le 2;$
 $g(x) = \sqrt{4 - x^2}, \quad 0 \le x \le 2$

35. $f(x) = \dfrac{x + 2}{x - 2}; \quad g(x) = \dfrac{2x + 2}{x - 1}$

36. $f(x) = \dfrac{x - 5}{3x + 4}; \quad g(x) = \dfrac{5 + 4x}{1 - 3x}$

37–60 ■ Find the inverse function of f.

37. $f(x) = 2x + 1$

38. $f(x) = 6 - x$

39. $f(x) = 4x + 7$

40. $f(x) = 3 - 5x$

41. $f(x) = 5 - 4x^3$

42. $f(x) = \dfrac{1}{x^2}, \quad x > 0$

43. $f(x) = \dfrac{1}{x + 2}$

44. $f(x) = \dfrac{x - 2}{x + 2}$

45. $f(x) = \dfrac{x}{x + 4}$

46. $f(x) = \dfrac{3x}{x - 2}$

47. $f(x) = \dfrac{2x + 5}{x - 7}$

48. $f(x) = \dfrac{4x - 2}{3x + 1}$

49. $f(x) = \dfrac{1 + 3x}{5 - 2x}$

50. $f(x) = \dfrac{2x - 1}{x - 3}$

51. $f(x) = \sqrt{2 + 5x}$

52. $f(x) = x^2 + x, \quad x \ge -\tfrac{1}{2}$

53. $f(x) = 4 - x^2, \quad x \ge 0$

54. $f(x) = \sqrt{2x - 1}$

55. $f(x) = 4 + \sqrt[3]{x}$

56. $f(x) = (2 - x^3)^5$

57. $f(x) = 1 + \sqrt{1 + x}$

58. $f(x) = \sqrt{9 - x^2}, \quad 0 \le x \le 3$

59. $f(x) = x^4, \quad x \ge 0$

60. $f(x) = 1 - x^3$

61–64 ■ A function f is given. (a) Sketch the graph of f. (b) Use the graph of f to sketch the graph of f^{-1}. (c) Find f^{-1}.

61. $f(x) = 3x - 6$

62. $f(x) = 16 - x^2, \quad x \ge 0$

63. $f(x) = \sqrt{x + 1}$

64. $f(x) = x^3 - 1$

65–70 ■ Draw the graph of f and use it to determine whether the function is one-to-one.

65. $f(x) = x^3 - x$

66. $f(x) = x^3 + x$

67. $f(x) = \dfrac{x + 12}{x - 6}$

68. $f(x) = \sqrt{x^3 - 4x + 1}$

69. $f(x) = |x| - |x - 6|$

70. $f(x) = x \cdot |x|$

71–74 ■ A one-to-one function is given. **(a)** Find the inverse of the function. **(b)** Graph both the function and its inverse on the same screen to verify that the graphs are reflections of each other in the line $y = x$.

71. $f(x) = 2 + x$　　　　**72.** $f(x) = 2 - \frac{1}{2}x$

73. $g(x) = \sqrt{x + 3}$　　　**74.** $g(x) = x^2 + 1, \quad x \geq 0$

75–78 ■ The given function is not one-to-one. Restrict its domain so that the resulting function *is* one-to-one. Find the inverse of the function with the restricted domain. (There is more than one correct answer.)

75. $f(x) = 4 - x^2$　　　**76.** $g(x) = (x - 1)^2$

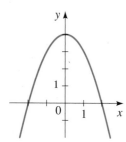　　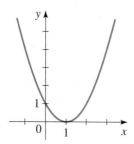

77. $h(x) = (x + 2)^2$　　　**78.** $k(x) = |x - 3|$

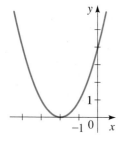　　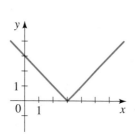

79–80 ■ Use the graph of f to sketch the graph of f^{-1}.

79.　　　　　　　　　　**80.**

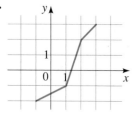　　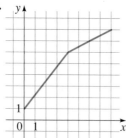

APPLICATIONS

81. Fee for Service For his services, a private investigator requires a $500 retention fee plus $80 per hour. Let x represent the number of hours the investigator spends working on a case.
  (a) Find a function f that models the investigator's fee as a function of x.
  (b) Find f^{-1}. What does f^{-1} represent?
  (c) Find $f^{-1}(1220)$. What does your answer represent?

82. Toricelli's Law A tank holds 100 gallons of water, which drains from a leak at the bottom, causing the tank to empty in 40 minutes. Toricelli's Law gives the volume of water remaining in the tank after t minutes as

$$V(t) = 100\left(1 - \frac{t}{40}\right)^2$$

  (a) Find V^{-1}. What does V^{-1} represent?
  (b) Find $V^{-1}(15)$. What does your answer represent?

83. Blood Flow As blood moves through a vein or artery, its velocity v is greatest along the central axis and decreases as the distance r from the central axis increases (see the figure below). For an artery with radius 0.5 cm, v (in cm/s) is given as a function of r (in cm) by

$$v(r) = 18,500(0.25 - r^2)$$

  (a) Find v^{-1}. What does v^{-1} represent?
  (b) Find $v^{-1}(30)$. What does your answer represent?

84. Demand Function The amount of a commodity that is sold is called the *demand* for the commodity. The demand D for a certain commodity is a function of the price given by

$$D(p) = -3p + 150$$

  (a) Find D^{-1}. What does D^{-1} represent?
  (b) Find $D^{-1}(30)$. What does your answer represent?

85. Temperature Scales The relationship between the Fahrenheit (F) and Celsius (C) scales is given by

$$F(C) = \tfrac{9}{5}C + 32$$

  (a) Find F^{-1}. What does F^{-1} represent?
  (b) Find $F^{-1}(86)$. What does your answer represent?

86. Exchange Rates The relative value of currencies fluctuates every day. When this problem was written, one Canadian dollar was worth 1.0573 U.S. dollar.
  (a) Find a function f that gives the U.S. dollar value $f(x)$ of x Canadian dollars.
  (b) Find f^{-1}. What does f^{-1} represent?
  (c) How much Canadian money would $12,250 in U.S. currency be worth?

87. Income Tax In a certain country, the tax on incomes less than or equal to €20,000 is 10%. For incomes that are more than €20,000, the tax is €2000 plus 20% of the amount over €20,000.
  (a) Find a function f that gives the income tax on an income x. Express f as a piecewise defined function.
  (b) Find f^{-1}. What does f^{-1} represent?
  (c) How much income would require paying a tax of €10,000?

88. Multiple Discounts A car dealership advertises a 15% discount on all its new cars. In addition, the manufacturer offers a $1000 rebate on the purchase of a new car. Let x represent the sticker price of the car.
  (a) Suppose only the 15% discount applies. Find a function f that models the purchase price of the car as a function of the sticker price x.

(b) Suppose only the $1000 rebate applies. Find a function g that models the purchase price of the car as a function of the sticker price x.

(c) Find a formula for $H = f \circ g$.

(d) Find H^{-1}. What does H^{-1} represent?

(e) Find $H^{-1}(13,000)$. What does your answer represent?

89. Pizza Cost Marcello's Pizza charges a base price of $7 for a large pizza plus $2 for each topping. Thus, if you order a large pizza with x toppings, the price of your pizza is given by the function $f(x) = 7 + 2x$. Find f^{-1}. What does the function f^{-1} represent?

DISCOVERY ■ DISCUSSION ■ WRITING

90. Determining When a Linear Function Has an Inverse For the linear function $f(x) = mx + b$ to be one-to-one, what must be true about its slope? If it is one-to-one, find its inverse. Is the inverse linear? If so, what is its slope?

91. Finding an Inverse "in Your Head" In the margin notes in this section we pointed out that the inverse of a function can be found by simply reversing the operations that make up the function. For instance, in Example 6 we saw that the inverse of

$$f(x) = 3x - 2 \quad \text{is} \quad f^{-1}(x) = \frac{x + 2}{3}$$

because the "reverse" of "Multiply by 3 and subtract 2" is "Add 2 and divide by 3." Use the same procedure to find the inverse of the following functions.

(a) $f(x) = \dfrac{2x + 1}{5}$ **(b)** $f(x) = 3 - \dfrac{1}{x}$

(c) $f(x) = \sqrt{x^3 + 2}$ **(d)** $f(x) = (2x - 5)^3$

Now consider another function:

$$f(x) = x^3 + 2x + 6$$

Is it possible to use the same sort of simple reversal of operations to find the inverse of this function? If so, do it. If not, explain what is different about this function that makes this task difficult.

92. The Identity Function The function $I(x) = x$ is called the **identity function**. Show that for any function f we have $f \circ I = f$, $I \circ f = f$, and $f \circ f^{-1} = f^{-1} \circ f = I$. (This means that the identity function I behaves for functions and composition just the way the number 1 behaves for real numbers and multiplication.)

93. Solving an Equation for an Unknown Function In Exercise 69 of Section 1.7 you were asked to solve equations in which the unknowns were functions. Now that we know about inverses and the identity function (see Exercise 92), we can use algebra to solve such equations. For instance, to solve $f \circ g = h$ for the unknown function f, we perform the following steps:

$f \circ g = h$	Problem: Solve for f
$f \circ g \circ g^{-1} = h \circ g^{-1}$	Compose with g^{-1} on the right
$f \circ I = h \circ g^{-1}$	Because $g \circ g^{-1} = I$
$f = h \circ g^{-1}$	Because $f \circ I = f$

So the solution is $f = h \circ g^{-1}$. Use this technique to solve the equation $f \circ g = h$ for the indicated unknown function.

(a) Solve for f, where $g(x) = 2x + 1$ and $h(x) = 4x^2 + 4x + 7$.

(b) Solve for g, where $f(x) = 3x + 5$ and $h(x) = 3x^2 + 3x + 2$.

CHAPTER 1 | REVIEW

■ CONCEPT CHECK

1. (a) Describe the coordinate plane.

   (b) How do you locate points in the coordinate plane?

2. State each formula.

   (a) The Distance Formula

   (b) The Midpoint Formula

3. Given an equation, what is its graph?

4. How do you find the x-intercepts and y-intercepts of a graph?

5. Write an equation of the circle with center (h, k) and radius r.

6. Explain the meaning of each type of symmetry. How do you test for it?

   (a) Symmetry with respect to the x-axis

   (b) Symmetry with respect to the y-axis

   (c) Symmetry with respect to the origin

7. Define the slope of a line.

8. Write each form of the equation of a line.

   (a) The point-slope form

   (b) The slope-intercept form

9. (a) What is the equation of a vertical line?

   (b) What is the equation of a horizontal line?

10. What is the general equation of a line?

11. Given lines with slopes m_1 and m_2, explain how you can tell if the lines are

   (a) parallel   **(b)** perpendicular

12. Define each concept in your own words. (Check by referring to the definition in the text.)

   (a) Function

   (b) Domain and range of a function

   (c) Graph of a function

   (d) Independent and dependent variables

13. Sketch by hand, on the same axes, the graphs of the following functions.

   (a) $f(x) = x$ **(b)** $g(x) = x^2$

   (c) $h(x) = x^3$ **(d)** $j(x) = x^4$

14. **(a)** State the Vertical Line Test.

   (b) State the Horizontal Line Test.

15. Define each concept in your own words.

   (a) Increasing function

   (b) Decreasing function

   (c) Constant function

16. Suppose the graph of f is given. Write an equation for each graph that is obtained from the graph of f as follows.

   (a) Shift 3 units upward.

   (b) Shift 3 units downward.

   (c) Shift 3 units to the right.

   (d) Shift 3 units to the left.

   (e) Reflect in the x-axis.

   (f) Reflect in the y-axis.

   (g) Stretch vertically by a factor of 3.

   (h) Shrink vertically by a factor of $\frac{1}{3}$.

   (i) Stretch horizontally by a factor of 2.

   (j) Shrink horizontally by a factor of $\frac{1}{2}$.

17. **(a)** What is an even function? What symmetry does its graph possess? Give an example of an even function.

   (b) What is an odd function? What symmetry does its graph possess? Give an example of an odd function.

18. What does it mean to say that $f(3)$ is a local maximum value of f?

19. Suppose that f has domain A and g has domain B.

   (a) What is the domain of $f + g$?

   (b) What is the domain of fg?

   (c) What is the domain of f/g?

20. How is the composite function $f \circ g$ defined?

21. **(a)** What is a one-to-one function?

   (b) How can you tell from the graph of a function whether it is one-to-one?

   (c) Suppose f is a one-to-one function with domain A and range B. How is the inverse function f^{-1} defined? What is the domain of f^{-1}? What is the range of f^{-1}?

   (d) If you are given a formula for f, how do you find a formula for f^{-1}?

   (e) If you are given the graph of f, how do you find the graph of f^{-1}?

■ EXERCISES

1–2 ■ Two points P and Q are given.

   (a) Plot P and Q on a coordinate plane.

   (b) Find the distance from P to Q.

   (c) Find the midpoint of the segment PQ.

   (d) Sketch the line determined by P and Q, and find its equation in slope-intercept form.

   (e) Sketch the circle that passes through Q and has center P, and find the equation of this circle.

1. $P(2, 0)$, $Q(-5, 12)$ **2.** $P(7, -1)$, $Q(2, -11)$

3–4 ■ Sketch the region given by the set.

3. $\{(x, y) \mid -4 < x < 4 \quad \text{and} \quad -2 < y < 2\}$

4. $\{(x, y) \mid x \geq 4 \quad \text{or} \quad y \geq 2\}$

5. Which of the points $A(4, 4)$ or $B(5, 3)$ is closer to the point $C(-1, -3)$?

6. Find an equation of the circle that has center $(2, -5)$ and radius $\sqrt{2}$.

7. Find an equation of the circle that has center $(-5, -1)$ and passes through the origin.

8. Find an equation of the circle that contains the points $P(2, 3)$ and $Q(-1, 8)$ and has the midpoint of the segment PQ as its center.

9–12 ■ Determine whether the equation represents a circle, represents a point, or has no graph. If the equation is that of a circle, find its center and radius.

9. $x^2 + y^2 + 2x - 6y + 9 = 0$

10. $2x^2 + 2y^2 - 2x + 8y = \frac{1}{2}$

11. $x^2 + y^2 + 72 = 12x$

12. $x^2 + y^2 - 6x - 10y + 34 = 0$

13–20 ■ Test the equation for symmetry, and sketch its graph.

13. $y = 2 - 3x$ **14.** $2x - y + 1 = 0$

15. $x + 3y = 21$ **16.** $x = 2y + 12$

17. $y = 16 - x^2$ **18.** $8x + y^2 = 0$

19. $x = \sqrt{y}$ **20.** $y = -\sqrt{1 - x^2}$

21. Find an equation for the line that passes through the points $(-1, -6)$ and $(2, -4)$.

22. Find an equation for the line that passes through the point $(6, -3)$ and has slope $-\frac{1}{2}$.

23. Find an equation for the line that has x-intercept 4 and y-intercept 12.

24. Find an equation for the line that passes through the point $(1, 7)$ and is perpendicular to the line $x - 3y + 16 = 0$.

25. Find an equation for the line that passes through the origin and is parallel to the line $3x + 15y = 22$.

26. Find an equation for the line that passes through the point $(5, 2)$ and is parallel to the line passing through $(-1, -3)$ and $(3, 2)$.

27–28 ■ Find equations for the circle and the line in the figure.

27.

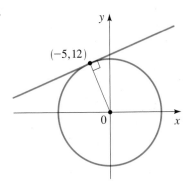

28.

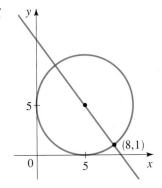

29. Hooke's Law states that if a weight w is attached to a hanging spring, then the stretched length s of the spring is linearly related to w. For a particular spring we have

$$s = 0.3w + 2.5$$

where s is measured in inches and w in pounds.
   (a) What do the slope and s-intercept in this equation represent?
   (b) How long is the spring when a 5-lb weight is attached?

30. Margarita is hired by an accounting firm at a salary of $60,000 per year. Three years later her annual salary has increased to $70,500. Assume that her salary increases linearly.
   (a) Find an equation that relates her annual salary S and the number of years t that she has worked for the firm.
   (b) What do the slope and S-intercept of her salary equation represent?
   (c) What will her salary be after 12 years with the firm?

31–32 ■ A verbal description of a function f is given. Find a formula that expresses f in function notation.

31. "Square, then subtract 5."

32. "Divide by 2, then add 9."

33–34 ■ A formula for a function f is given. Give a verbal description of the function.

33. $f(x) = 3(x + 10)$

34. $f(x) = \sqrt{6x - 10}$

35–36 ■ Complete the table of values for the given function.

35. $g(x) = x^2 - 4x$

x	$g(x)$
-1	
0	
1	
2	
3	

36. $h(x) = 3x^2 + 2x - 5$

x	$h(x)$
-2	
-1	
0	
1	
2	

37. A publisher estimates that the cost $C(x)$ of printing a run of x copies of a certain mathematics textbook is given by the function $C(x) = 5000 + 30x - 0.001x^2$.
   (a) Find $C(1000)$ and $C(10,000)$.
   (b) What do your answers in part (a) represent?
   (c) Find $C(0)$. What does this number represent?

38. Reynalda works as a salesperson in the electronics division of a department store. She earns a base weekly salary plus a commission based on the retail price of the goods she has sold. If she sells x dollars worth of goods in a week, her earnings for that week are given by the function $E(x) = 400 + 0.03x$.
   (a) Find $E(2000)$ and $E(15,000)$.
   (b) What do your answers in part (a) represent?
   (c) Find $E(0)$. What does this number represent?
   (d) From the formula for E, determine what percentage Reynalda earns on the goods that she sells.

39. If $f(x) = x^2 - 4x + 6$, find $f(0)$, $f(2)$, $f(-2)$, $f(a)$, $f(-a)$, $f(x + 1)$, $f(2x)$, and $2f(x) - 2$.

40. If $f(x) = 4 - \sqrt{3x - 6}$, find $f(5)$, $f(9)$, $f(a + 2)$, $f(-x)$, $f(x^2)$, and $[f(x)]^2$.

41. Which of the following figures are graphs of functions? Which of the functions are one-to-one?

(a) 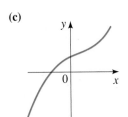 **(b)**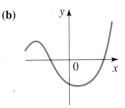

(c) 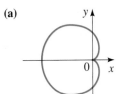 **(d)**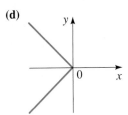

42. The graph of a function f is given.
  (a) Find $f(-2)$ and $f(2)$.
  (b) Find the domain of f.
  (c) Find the range of f.
  (d) On what intervals is f increasing? On what intervals is f decreasing?
  (e) What are the local maximum values of f?
  (f) Is f one-to-one?

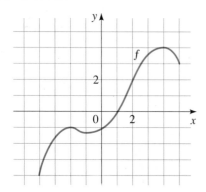

43–44 ■ Find the domain and range of the function.

43. $f(x) = \sqrt{x + 3}$ **44.** $F(t) = t^2 + 2t + 5$

45–52 ■ Find the domain of the function.

45. $f(x) = 7x + 15$ **46.** $f(x) = \dfrac{2x + 1}{2x - 1}$

47. $f(x) = \sqrt{x + 4}$ **48.** $f(x) = 3x - \dfrac{2}{\sqrt{x + 1}}$

49. $f(x) = \dfrac{1}{x} + \dfrac{1}{x + 1} + \dfrac{1}{x + 2}$ **50.** $g(x) = \dfrac{2x^2 + 5x + 3}{2x^2 - 5x - 3}$

51. $h(x) = \sqrt{4 - x} + \sqrt{x^2 - 1}$ **52.** $f(x) = \dfrac{\sqrt[3]{2x + 1}}{\sqrt[3]{2x + 2}}$

53–70 ■ Sketch the graph of the function.

53. $f(x) = 1 - 2x$

54. $f(x) = \frac{1}{3}(x - 5),\ 2 \le x \le 8$

55. $f(t) = 1 - \frac{1}{2}t^2$ **56.** $g(t) = t^2 - 2t$

57. $f(x) = x^2 - 6x + 6$ **58.** $f(x) = 3 - 8x - 2x^2$

59. $g(x) = 1 - \sqrt{x}$ **60.** $g(x) = -|x|$

61. $h(x) = \frac{1}{2}x^3$ **62.** $h(x) = \sqrt{x + 3}$

63. $h(x) = \sqrt[3]{x}$ **64.** $H(x) = x^3 - 3x^2$

65. $g(x) = \dfrac{1}{x^2}$ **66.** $G(x) = \dfrac{1}{(x - 3)^2}$

67. $f(x) = \begin{cases} 1 - x & \text{if } x < 0 \\ 1 & \text{if } x \ge 0 \end{cases}$

68. $f(x) = \begin{cases} 1 - 2x & \text{if } x \le 0 \\ 2x - 1 & \text{if } x > 0 \end{cases}$

69. $f(x) = \begin{cases} x + 6 & \text{if } x < -2 \\ x^2 & \text{if } x \ge -2 \end{cases}$

70. $f(x) = \begin{cases} -x & \text{if } x < 0 \\ x^2 & \text{if } 0 \le x < 2 \\ 1 & \text{if } x \ge 2 \end{cases}$

71–74 ■ Determine whether the equation defines y as a function of x.

71. $x + y^2 = 14$ **72.** $3x - \sqrt{y} = 8$

73. $x^3 - y^3 = 27$ **74.** $2x = y^4 - 16$

 75. Find, approximately, the domain of the function
$$f(x) = \sqrt{x^3 - 4x + 1}$$

76. Find, approximately, the range of the function
$$f(x) = x^4 - x^3 + x^2 + 3x - 6$$

77–78 ■ Draw a graph of the function f, and determine the intervals on which f is increasing and on which f is decreasing.

77. $f(x) = x^3 - 4x^2$ **78.** $f(x) = |x^4 - 16|$

79. Suppose the graph of f is given. Describe how the graphs of the following functions can be obtained from the graph of f.
  (a) $y = f(x) + 8$ **(b)** $y = f(x + 8)$
  (c) $y = 1 + 2f(x)$ **(d)** $y = f(x - 2) - 2$
  (e) $y = f(-x)$ **(f)** $y = -f(-x)$
  (g) $y = -f(x)$ **(h)** $y = f^{-1}(x)$

80. The graph of f is given. Draw the graphs of the following functions.
  (a) $y = f(x - 2)$ **(b)** $y = -f(x)$
  (c) $y = 3 - f(x)$ **(d)** $y = \frac{1}{2}f(x) - 1$
  (e) $y = f^{-1}(x)$ **(f)** $y = f(-x)$

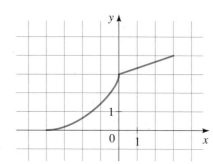

81. Determine whether f is even, odd, or neither.
  (a) $f(x) = 2x^5 - 3x^2 + 2$ **(b)** $f(x) = x^3 - x^7$
  (c) $f(x) = \dfrac{1 - x^2}{1 + x^2}$ **(d)** $f(x) = \dfrac{1}{x + 2}$

82. Determine whether the function in the figure is even, odd, or neither.

(a) 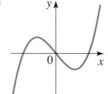   **(b)**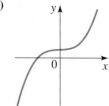

(c)   **(d)**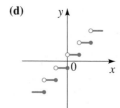

83. Find the minimum value of the function $g(x) = 2x^2 + 4x - 5$.

84. Find the maximum value of the function $f(x) = 1 - x - x^2$.

85. A stone is thrown upward from the top of a building. Its height (in feet) above the ground after t seconds is given by

$$h(t) = -16t^2 + 48t + 32$$

What maximum height does it reach?

86. The profit P (in dollars) generated by selling x units of a certain commodity is given by

$$P(x) = -1500 + 12x - 0.0004x^2$$

What is the maximum profit, and how many units must be sold to generate it?

87–88 ■ Find the local maximum and minimum values of the function and the values of x at which they occur. State each answer correct to two decimal places.

87. $f(x) = 3.3 + 1.6x - 2.5x^3$ **88.** $f(x) = x^{2/3}(6 - x)^{1/3}$

89–90 ■ Two functions, f and g, are given. Draw graphs of f, g, and $f + g$ on the same graphing calculator screen to illustrate the concept of graphical addition.

89. $f(x) = x + 2$, $g(x) = x^2$

90. $f(x) = x^2 + 1$, $g(x) = 3 - x^2$

91. If $f(x) = x^2 - 3x + 2$ and $g(x) = 4 - 3x$, find the following functions.
(a) $f + g$ **(b)** $f - g$ **(c)** fg
(d) f/g **(e)** $f \circ g$ **(f)** $g \circ f$

92. If $f(x) = 1 + x^2$ and $g(x) = \sqrt{x - 1}$, find the following.
(a) $f \circ g$ **(b)** $g \circ f$ **(c)** $(f \circ g)(2)$
(d) $(f \circ f)(2)$ **(e)** $f \circ g \circ f$ **(f)** $g \circ f \circ g$

93–94 ■ Find the functions $f \circ g, g \circ f, f \circ f$, and $g \circ g$ and their domains.

93. $f(x) = 3x - 1$, $g(x) = 2x - x^2$

94. $f(x) = \sqrt{x}$, $g(x) = \dfrac{2}{x - 4}$

95. Find $f \circ g \circ h$, where $f(x) = \sqrt{1 - x}, g(x) = 1 - x^2$, and $h(x) = 1 + \sqrt{x}$.

96. If $T(x) = \dfrac{1}{\sqrt{1 + \sqrt{x}}}$, find functions f, g, and h such that $f \circ g \circ h = T$.

97–102 ■ Determine whether the function is one-to-one.

97. $f(x) = 3 + x^3$ **98.** $g(x) = 2 - 2x + x^2$

99. $h(x) = \dfrac{1}{x^4}$ **100.** $r(x) = 2 + \sqrt{x + 3}$

101. $p(x) = 3.3 + 1.6x - 2.5x^3$

102. 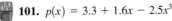 $q(x) = 3.3 + 1.6x + 2.5x^3$

103–106 ■ Find the inverse of the function.

103. $f(x) = 3x - 2$ **104.** $f(x) = \dfrac{2x + 1}{3}$

105. $f(x) = (x + 1)^3$ **106.** $f(x) = 1 + \sqrt[5]{x - 2}$

107. **(a)** Sketch the graph of the function

$$f(x) = x^2 - 4 \qquad x \geq 0$$

(b) Use part (a) to sketch the graph of f^{-1}.
(c) Find an equation for f^{-1}.

108. **(a)** Show that the function $f(x) = 1 + \sqrt[4]{x}$ is one-to-one.
(b) Sketch the graph of f.
(c) Use part (b) to sketch the graph of f^{-1}.
(d) Find an equation for f^{-1}.

1. **(a)** Plot the points $P(0, 3)$, $Q(3, 0)$, and $R(6, 3)$ in the coordinate plane. Where must the point S be located so that $PQRS$ is a square?

   (b) Find the area of $PQRS$.

2. **(a)** Sketch the graph of $y = x^2 - 4$.

   (b) Find the x- and y-intercepts of the graph.

   (c) Is the graph symmetric about the x-axis, the y-axis, or the origin?

3. Let $P(-3, 1)$ and $Q(5, 6)$ be two points in the coordinate plane.

   (a) Plot P and Q in the coordinate plane.

   (b) Find the distance between P and Q.

   (c) Find the midpoint of the segment PQ.

   (d) Find the slope of the line that contains P and Q.

   (e) Find the perpendicular bisector of the line that contains P and Q.

   (f) Find an equation for the circle for which the segment PQ is a diameter.

4. Find the center and radius of each circle and sketch its graph.

   (a) $x^2 + y^2 = 25$ **(b)** $(x - 2)^2 + (y + 1)^2 = 9$ **(c)** $x^2 + 6x + y^2 - 2y + 6 = 0$

5. Write the linear equation $2x - 3y = 15$ in slope-intercept form, and sketch its graph. What are the slope and y-intercept?

6. Find an equation for the line with the given property.

   (a) It passes through the point $(3, -6)$ and is parallel to the line $3x + y - 10 = 0$.

   (b) It has x-intercept 6 and y-intercept 4.

7. Which of the following are graphs of functions? If the graph is that of a function, is it one-to-one?

(a)
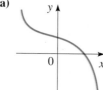

(b)
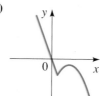

(c)

(d)
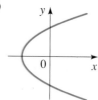

8. Let $f(x) = \dfrac{\sqrt{x + 1}}{x}$.

   (a) Evaluate $f(3)$, $f(5)$, and $f(a - 1)$.

   (b) Find the domain of f.

9. A function f has the following verbal description: "Subtract 2, then cube the result."

   (a) Find a formula that expresses f algebraically.

   (b) Make a table of values of f, for the inputs -1, 0, 1, 2, 3, and 4.

   (c) Sketch a graph of f, using the table of values from part (b) to help you.

   (d) How do we know that f has an inverse? Give a verbal description for f^{-1}.

   (e) Find a formula that expresses f^{-1} algebraically.

10. A school fund-raising group sells chocolate bars to help finance a swimming pool for their physical education program. The group finds that when they set their price at x dollars per bar (where $0 < x \le 5$), their total sales revenue (in dollars) is given by the function $R(x) = -500x^2 + 3000x$.

(a) Evaluate $R(2)$ and $R(4)$. What do these values represent?

(b) Use a graphing calculator to draw a graph of R. What does the graph tell us about what happens to revenue as the price increases from 0 to 5 dollars?

(c) What is the maximum revenue, and at what price is it achieved?

11. (a) Sketch the graph of the function $f(x) = x^3$.

(b) Use part (a) to graph the function $g(x) = (x - 1)^3 - 2$.

12. (a) How is the graph of $y = f(x - 3) + 2$ obtained from the graph of f?

(b) How is the graph of $y = f(-x)$ obtained from the graph of f?

13. Let $f(x) = \begin{cases} 1 - x & \text{if } x \le 1 \\ 2x + 1 & \text{if } x > 1 \end{cases}$

(a) Evaluate $f(-2)$ and $f(1)$.

(b) Sketch the graph of f.

14. If $f(x) = x^2 + 1$ and $g(x) = x - 3$, find the following.

(a) $f \circ g$ (b) $g \circ f$

(c) $f(g(2))$ (d) $g(f(2))$

(e) $g \circ g \circ g$

15. (a) If $f(x) = \sqrt{3 - x}$, find the inverse function f^{-1}.

(b) Sketch the graphs of f and f^{-1} on the same coordinate axes.

16. The graph of a function f is given.

(a) Find the domain and range of f.

(b) Sketch the graph of f^{-1}.

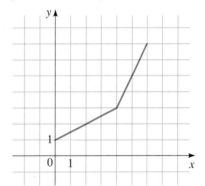

17. Let $f(x) = 3x^4 - 14x^2 + 5x - 3$.

(a) Draw the graph of f in an appropriate viewing rectangle.

(b) Is f one-to-one?

(c) Find the local maximum and minimum values of f and the values of x at which they occur. State each answer correct to two decimal places.

(d) Use the graph to determine the range of f.

(e) Find the intervals on which f is increasing and on which f is decreasing.

A model is a representation of an object or process. For example, a toy Ferrari is a model of the actual car; a road map is a model of the streets in a city. A **mathematical model** is a mathematical representation (usually an equation) of an object or process. Once a mathematical model is made it can be used to obtain useful information or make predictions about the thing being modeled. In these *Focus on Modeling* sections we explore different ways in which mathematics is used to model real-world phenomena.

▼ The Line That Best Fits the Data

In Section 1.2 we used linear equations to model relationships between varying quantities. In practice, such relationships are discovered by collecting data. But real-world data seldom fall into a precise line. The **scatter plot** in Figure 1(a) shows the result of a study on childhood obesity. The graph plots the body mass index (BMI) versus the number of hours of television watched per day for 25 adolescent subjects. Of course, we would not expect the data to be exactly linear as in Figure 1(b). But there is a linear *trend* indicated by the blue line in Figure 1(a): The more hours a subject watches TV the higher the BMI. In this section we learn how to find the line that best fits the data.

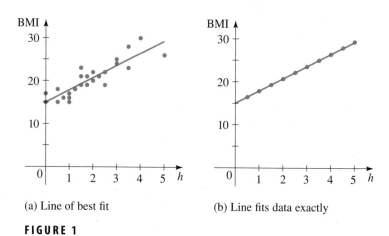

(a) Line of best fit

(b) Line fits data exactly

FIGURE 1

Table 1 gives the nationwide infant mortality rate for the period from 1950 to 2000. The *rate* is the number of infants who die before reaching their first birthday, out of every 1000 live births.

TABLE 1

U.S. Infant Mortality

Year	Rate
1950	29.2
1960	26.0
1970	20.0
1980	12.6
1990	9.2
2000	6.9

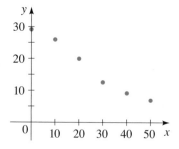

FIGURE 2 U.S. infant mortality rate

The scatter plot in Figure 2 shows that the data lie roughly on a straight line. We can try to fit a line visually to approximate the data points, but since the data aren't *exactly* linear, there are many lines that might seem to work. Figure 3 shows two attempts at "eyeballing" a line to fit the data.

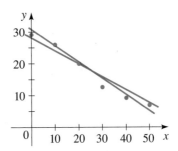

FIGURE 3 Visual attempts to fit line to data

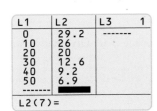

FIGURE 4 Distance from the data points to the line

Of all the lines that run through these data points, there is one that "best" fits the data, in the sense that it provides the most accurate linear model for the data. We now describe how to find this line.

It seems reasonable that the line of best fit is the line that is as close as possible to all the data points. This is the line for which the sum of the vertical distances from the data points to the line is as small as possible (see Figure 4). For technical reasons it is better to use the line where the sum of the squares of these distances is smallest. This is called the **regression line**. The formula for the regression line is found by using calculus, but fortunately, the formula is programmed into most graphing calculators. In Example 1 we see how to use a TI-83 calculator to find the regression line for the infant mortality data described above. (The process for other calculators is similar.)

EXAMPLE 1 | Regression Line for U.S. Infant Mortality Rates

(a) Find the regression line for the infant mortality data in Table 1.

(b) Graph the regression line on a scatter plot of the data.

(c) Use the regression line to estimate the infant mortality rates in 1995 and 2006.

SOLUTION

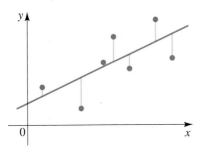

FIGURE 5 Entering the data

(a) To find the regression line using a TI-83 calculator, we must first enter the data into the lists L_1 and L_2, which are accessed by pressing the $\boxed{\text{STAT}}$ key and selecting Edit. Figure 5 shows the calculator screen after the data have been entered. (Note that we are letting $x = 0$ correspond to the year 1950, so that $x = 50$ corresponds to 2000. This makes the equations easier to work with.) We then press the $\boxed{\text{STAT}}$ key again and select Calc, then 4:LinReg(ax+b), which provides the output shown in Figure 6(a). This tells us that the regression line is

$$y = -0.48x + 29.4$$

Here x represents the number of years since 1950, and y represents the corresponding infant mortality rate.

(b) The scatter plot and the regression line have been plotted on a graphing calculator screen in Figure 6(b).

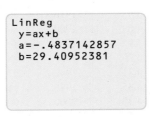

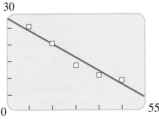

(a) Output of the LinReg command

(b) Scatter plot and regression line

FIGURE 6

(c) The year 1995 is 45 years after 1950, so substituting 45 for x, we find that $y = -0.48(45) + 29.4 = 7.8$. So the infant mortality rate in 1995 was about 7.8. Similarly, substituting 56 for x, we find that the infant mortality rate predicted for 2006 was about $-0.48(56) + 29.4 \approx 2.5$.   ∎

An Internet search shows that the actual infant mortality rate was 7.6 in 1995 and 6.4 in 2006. So the regression line is fairly accurate for 1995 (the actual rate was slightly lower than the predicted rate), but it is considerably off for 2006 (the actual rate was more than twice the predicted rate). The reason is that infant mortality in the United States stopped declining and actually started rising in 2002, for the first time in more than a century. This shows that we have to be very careful about extrapolating linear models outside the domain over which the data are spread.

▼ Examples of Regression Analysis

Since the modern Olympic Games began in 1896, achievements in track and field events have been improving steadily. One example in which the winning records have shown an upward linear trend is the pole vault. Pole vaulting began in the northern Netherlands as a practical activity: When traveling from village to village, people would vault across the many canals that crisscrossed the area to avoid having to go out of their way to find a bridge. Households maintained a supply of wooden poles of lengths appropriate for each member of the family. Pole vaulting for height rather than distance became a collegiate track and field event in the mid-1800s and was one of the events in the first modern Olympics. In the next example we find a linear model for the gold-medal-winning records in the men's Olympic pole vault.

Steven Hooker, 2008 Olympic gold medal winner, men's pole vault

EXAMPLE 2 | Regression Line for Olympic Pole Vault Records

Table 2 gives the men's Olympic pole vault records up to 2004.

(a) Find the regression line for the data.

(b) Make a scatter plot of the data, and graph the regression line. Does the regression line appear to be a suitable model for the data?

(c) What does the slope of the regression line represent?

(d) Use the model to predict the winning pole vault height for the 2008 Olympics.

TABLE 2
Men's Olympic Pole Vault Records

Year	x	Gold medalist	Height (m)	Year	x	Gold medalist	Height (m)
1896	−4	William Hoyt, USA	3.30	1956	56	Robert Richards, USA	4.56
1900	0	Irving Baxter, USA	3.30	1960	60	Don Bragg, USA	4.70
1904	4	Charles Dvorak, USA	3.50	1964	64	Fred Hansen, USA	5.10
1906	6	Fernand Gonder, France	3.50	1968	68	Bob Seagren, USA	5.40
1908	8	A. Gilbert, E. Cook, USA	3.71	1972	72	W. Nordwig, E. Germany	5.64
1912	12	Harry Babcock, USA	3.95	1976	76	Tadeusz Slusarski, Poland	5.64
1920	20	Frank Foss, USA	4.09	1980	80	W. Kozakiewicz, Poland	5.78
1924	24	Lee Barnes, USA	3.95	1984	84	Pierre Quinon, France	5.75
1928	28	Sabin Can, USA	4.20	1988	88	Sergei Bubka, USSR	5.90
1932	32	William Miller, USA	4.31	1992	92	M. Tarassob, Unified Team	5.87
1936	36	Earle Meadows, USA	4.35	1996	96	Jean Jaffione, France	5.92
1948	48	Guinn Smith, USA	4.30	2000	100	Nick Hysong, USA	5.90
1952	52	Robert Richards, USA	4.55	2004	104	Timothy Mack, USA	5.95

```
LinReg
  y=ax+b
  a=.0265652857
  b=3.400989881
```

Output of the `LinReg`
function on the TI-83

FIGURE 7 Scatter plot and regression line for pole vault data

SOLUTION

(a) Let x = year $-$ 1900, so 1896 corresponds to $x = -4$, 1900 to $x = 0$, and so on. Using a calculator, we find the following regression line:

$$y = 0.0266x + 3.40$$

(b) The scatter plot and the regression line are shown in Figure 7. The regression line appears to be a good model for the data.

(c) The slope is the average rate of increase in the pole vault record per year. So on average, the pole vault record increased by 0.0266 m/yr.

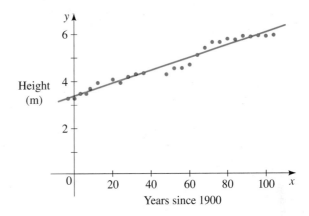

(d) The year 2008 corresponds to $x = 108$ in our model. The model gives

$$y = 0.0266(108) + 3.40$$

$$\approx 6.27$$

So the model predicts that in 2008 the winning pole vault will be 6.27 m. ■

At the 2008 Olympics in Beijing, China, the men's Olympic gold medal in the pole vault was won by Steven Hooker of Australia, with a vault of 5.96 m. Although this height set an Olympic record, it was considerably lower than the 6.27 m predicted by the model of Example 2. In Problem 10 we find a regression line for the pole vault data from 1972 to 2004. Do the problem to see whether this restricted set of more recent data provides a better predictor for the 2008 record.

Is a linear model really appropriate for the data of Example 2? In subsequent *Focus on Modeling* sections, we study regression models that use other types of functions, and we learn how to choose the best model for a given set of data.

In the next example we see how linear regression is used in medical research to investigate potential causes of diseases such as cancer.

TABLE 3
Asbestos–Tumor Data

Asbestos exposure (fibers/mL)	Percent that develop lung tumors
50	2
400	6
500	5
900	10
1100	26
1600	42
1800	37
2000	28
3000	50

EXAMPLE 3 | Regression Line for Links Between Asbestos and Cancer

When laboratory rats are exposed to asbestos fibers, some of the rats develop lung tumors. Table 3 lists the results of several experiments by different scientists.

(a) Find the regression line for the data.

(b) Make a scatter plot and graph the regression line. Does the regression line appear to be a suitable model for the data?

(c) What does the y-intercept of the regression line represent?

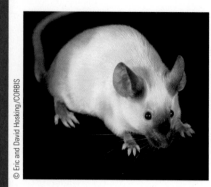

SOLUTION

(a) Using a calculator, we find the following regression line (see Figure 8(a)):

$$y = 0.0177x + 0.5405$$

(b) The scatter plot and regression line are graphed in Figure 8(b). The regression line appears to be a reasonable model for the data.

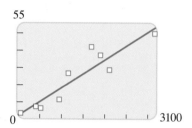

(a) Output of the LinReg command

(b) Scatter plot and regression line

FIGURE 8 Linear regression for the asbestos–tumor data

(c) The y-intercept is the percentage of rats that develop tumors when no asbestos fibers are present. In other words, this is the percentage that normally develop lung tumors (for reasons other than asbestos). ■

▼ How Good Is the Fit? The Correlation Coefficient

For any given set of two-variable data it is always possible to find a regression line, even if the data points do not tend to lie on a line and even if the variables don't seem to be related at all. Look at the three scatter plots in Figure 9. In the first scatter plot, the data points lie close to a line. In the second plot, there is still a linear trend but the points are more scattered. In the third plot there doesn't seem to be any trend at all, linear or otherwise.

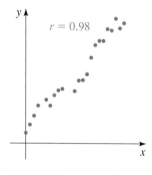

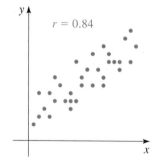

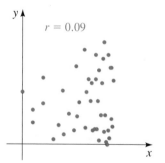

FIGURE 9

A graphing calculator can give us a regression line for each of these scatter plots. But how well do these lines represent or "fit" the data? To answer this question, statisticians have invented the **correlation coefficient**, usually denoted r. The correlation coefficient is a number between -1 and 1 that measures how closely the data follow the regression line—or, in other words, how strongly the variables are **correlated**. Many graphing calculators give the value of r when they compute a regression line. If r is close to -1

or 1, then the variables are strongly correlated—that is, the scatter plot follows the regression line closely. If r is close to 0, then the variables are weakly correlated or not correlated at all. (The sign of r depends on the slope of the regression line.) The correlation coefficients of the scatter plots in Figure 9 are indicated on the graphs. For the first plot, r is close to 1 because the data are very close to linear. The second plot also has a relatively large r, but it is not as large as the first, because the data, while fairly linear, are more diffuse. The third plot has an r close to 0, since there is virtually no linear trend in the data.

There are no hard and fast rules for deciding what values of r are sufficient for deciding that a linear correlation is "significant." The correlation coefficient is only a rough guide in helping us decide how much faith to put into a given regression line. In Example 1 the correlation coefficient is -0.99, indicating a very high level of correlation, so we can safely say that the drop in infant mortality rates from 1950 to 2000 was strongly linear. (The value of r is negative, since infant mortality trended *down* over this period.) In Example 3 the correlation coefficient is 0.92, which also indicates a strong correlation between the variables. So exposure to asbestos is clearly associated with the growth of lung tumors in rats. Does this mean that asbestos *causes* lung cancer?

If two variables are correlated, it does not necessarily mean that a change in one variable *causes* a change in the other. For example, the mathematician John Allen Paulos points out that shoe size is strongly correlated to mathematics scores among schoolchildren. Does this mean that big feet cause high math scores? Certainly not—both shoe size and math skills increase independently as children get older. So it is important not to jump to conclusions: Correlation and causation are not the same thing. Correlation is a useful tool in bringing important cause-and-effect relationships to light; but to prove causation, we must explain the mechanism by which one variable affects the other. For example, the link between smoking and lung cancer was observed as a correlation long before science found the mechanism through which smoking causes lung cancer.

PROBLEMS

1. Femur Length and Height Anthropologists use a linear model that relates femur length to height. The model allows an anthropologist to determine the height of an individual when only a partial skeleton (including the femur) is found. In this problem we find the model by analyzing the data on femur length and height for the eight males given in the table.

(a) Make a scatter plot of the data.

(b) Find and graph a linear function that models the data.

(c) An anthropologist finds a femur of length 58 cm. How tall was the person?

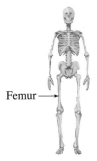

Femur →

Femur length (cm)	Height (cm)
50.1	178.5
48.3	173.6
45.2	164.8
44.7	163.7
44.5	168.3
42.7	165.0
39.5	155.4
38.0	155.8

2. Demand for Soft Drinks A convenience store manager notices that sales of soft drinks are higher on hotter days, so he assembles the data in the table.

(a) Make a scatter plot of the data.

(b) Find and graph a linear function that models the data.

(c) Use the model to predict soft drink sales if the temperature is 95°F.

High temperature (°F)	Number of cans sold
55	340
58	335
64	410
68	460
70	450
75	610
80	735
84	780

3. Tree Diameter and Age   To estimate ages of trees, forest rangers use a linear model that relates tree diameter to age. The model is useful because tree diameter is much easier to measure than tree age (which requires special tools for extracting a representative cross section of the tree and counting the rings). To find the model, use the data in the table, which were collected for a certain variety of oaks.

(a) Make a scatter plot of the data.

(b) Find and graph a linear function that models the data.

(c) Use the model to estimate the age of an oak whose diameter is 18 in.

Diameter (in.)	Age (years)
2.5	15
4.0	24
6.0	32
8.0	56
9.0	49
9.5	76
12.5	90
15.5	89

4. Carbon Dioxide Levels   The Mauna Loa Observatory, located on the island of Hawaii, has been monitoring carbon dioxide (CO_2) levels in the atmosphere since 1958. The table lists the average annual CO_2 levels measured in parts per million (ppm) from 1984 to 2006.

(a) Make a scatter plot of the data.

(b) Find and graph the regression line.

(c) Use the linear model in part (b) to estimate the CO_2 level in the atmosphere in 2005. Compare your answer with the actual CO_2 level of 379.7 that was measured in 2005.

Year	CO_2 level (ppm)
1984	344.3
1986	347.0
1988	351.3
1990	354.0
1992	356.3
1994	358.9
1996	362.7
1998	366.5
2000	369.4
2002	372.0
2004	377.5
2006	380.9

Temperature (°F)	Chirping rate (chirps/min)
50	20
55	46
60	79
65	91
70	113
75	140
80	173
85	198
90	211

5. Temperature and Chirping Crickets   Biologists have observed that the chirping rate of crickets of a certain species appears to be related to temperature. The table shows the chirping rates for various temperatures.

(a) Make a scatter plot of the data.

(b) Find and graph the regression line.

(c) Use the linear model in part (b) to estimate the chirping rate at 100°F.

6. Extent of Arctic Sea Ice   The National Snow and Ice Data Center monitors the amount of ice in the Arctic year round. The table gives approximate values for the sea ice extent in millions of square kilometers from 1980 to 2006, in two-year intervals.

(a) Make a scatter plot of the data.

(b) Find and graph the regression line.

(c) Use the linear model in part (b) to estimate the ice extent in the year 2010.

Year	Ice extent (million km^2)	Year	Ice extent (million km^2)
1980	7.9	1994	7.1
1982	7.4	1996	7.9
1984	7.2	1998	6.6
1986	7.6	2000	6.3
1988	7.5	2002	6.0
1990	6.2	2004	6.1
1992	7.6	2006	5.7

Flow rate (%)	Mosquito positive rate (%)
0	22
10	16
40	12
60	11
90	6
100	2

7. Mosquito Prevalence   The table lists the relative abundance of mosquitoes (as measured by the mosquito positive rate) versus the flow rate (measured as a percentage of maximum flow) of canal networks in Saga City, Japan.

(a) Make a scatter plot of the data.

(b) Find and graph the regression line.

(c) Use the linear model in part (b) to estimate the mosquito positive rate if the canal flow is 70% of maximum.

8. Noise and Intelligibility   Audiologists study the intelligibility of spoken sentences under different noise levels. Intelligibility, the MRT score, is measured as the percent of a spoken sentence that the listener can decipher at a certain noise level in decibels (dB). The table shows the results of one such test.

(a) Make a scatter plot of the data.

(b) Find and graph the regression line.

(c) Find the correlation coefficient. Is a linear model appropriate?

(d) Use the linear model in part (b) to estimate the intelligibility of a sentence at a 94-dB noise level.

Noise level (dB)	MRT score (%)
80	99
84	91
88	84
92	70
96	47
100	23
104	11

9. **Life Expectancy** The average life expectancy in the United States has been rising steadily over the past few decades, as shown in the table.

   (a) Make a scatter plot of the data.

   (b) Find and graph the regression line.

   (c) Use the linear model you found in part (b) to predict the life expectancy in the year 2006.

   (d) Search the Internet or your campus library to find the actual 2006 average life expectancy. Compare to your answer in part (c).

Year	Life expectancy
1920	54.1
1930	59.7
1940	62.9
1950	68.2
1960	69.7
1970	70.8
1980	73.7
1990	75.4
2000	76.9

10. **Olympic Pole Vault** The graph in Figure 7 indicates that in recent years the winning Olympic men's pole vault height has fallen below the value predicted by the regression line in Example 2. This might have occurred because when the pole vault was a new event, there was much room for improvement in vaulters' performances, whereas now even the best training can produce only incremental advances. Let's see whether concentrating on more recent results gives a better predictor of future records.

   (a) Use the data in Table 2 to complete the table of winning pole vault heights. (Note that we are using $x = 0$ to correspond to the year 1972, where this restricted data set begins.)

   (b) Find the regression line for the data in part (a).

   (c) Plot the data and the regression line on the same axes. Does the regression line seem to provide a good model for the data?

   (d) What does the regression line predict as the winning pole vault height for the 2008 Olympics? Compare this predicted value to the actual 2008 winning height of 5.96 m, as described on page 95. Has this new regression line provided a better prediction than the line in Example 2?

Year	x	Height (m)
1972	0	5.64
1976	4	
1980	8	
1984		
1988		
1992		
1996		
2000		
2004		

11. **Olympic Swimming Records**   The tables give the gold medal times in the men's and women's 100-m freestyle Olympic swimming event.

(a) Find the regression lines for the men's data and the women's data.

(b) Sketch both regression lines on the same graph. When do these lines predict that the women will overtake the men in the event? Does this conclusion seem reasonable?

MEN

Year	Gold medalist	Time (s)
1908	C. Daniels, USA	65.6
1912	D. Kahanamoku, USA	63.4
1920	D. Kahanamoku, USA	61.4
1924	J. Weissmuller, USA	59.0
1928	J. Weissmuller, USA	58.6
1932	Y. Miyazaki, Japan	58.2
1936	F. Csik, Hungary	57.6
1948	W. Ris, USA	57.3
1952	C. Scholes, USA	57.4
1956	J. Henricks, Australia	55.4
1960	J. Devitt, Australia	55.2
1964	D. Schollander, USA	53.4
1968	M. Wenden, Australia	52.2
1972	M. Spitz, USA	51.22
1976	J. Montgomery, USA	49.99
1980	J. Woithe, E. Germany	50.40
1984	R. Gaines, USA	49.80
1988	M. Biondi, USA	48.63
1992	A. Popov, Russia	49.02
1996	A. Popov, Russia	48.74
2000	P. van den Hoogenband, Netherlands	48.30
2004	P. van den Hoogenband, Netherlands	48.17
2008	A. Bernard, France	47.21

WOMEN

Year	Gold medalist	Time (s)
1912	F. Durack, Australia	82.2
1920	E. Bleibtrey, USA	73.6
1924	E. Lackie, USA	72.4
1928	A. Osipowich, USA	71.0
1932	H. Madison, USA	66.8
1936	H. Mastenbroek, Holland	65.9
1948	G. Andersen, Denmark	66.3
1952	K. Szoke, Hungary	66.8
1956	D. Fraser, Australia	62.0
1960	D. Fraser, Australia	61.2
1964	D. Fraser, Australia	59.5
1968	J. Henne, USA	60.0
1972	S. Nielson, USA	58.59
1976	K. Ender, E. Germany	55.65
1980	B. Krause, E. Germany	54.79
1984	(Tie) C. Steinseifer, USA	55.92
	N. Hogshead, USA	55.92
1988	K. Otto, E. Germany	54.93
1992	Z. Yong, China	54.64
1996	L. Jingyi, China	54.50
2000	I. DeBruijn, Netherlands	53.83
2004	J. Henry, Australia	53.84
2008	B. Steffen, Germany	53.12

12. **Shoe Size and Height**   Do you think that shoe size and height are correlated? Find out by surveying the shoe sizes and heights of people in your class. (Of course, the data for men and women should be separate.) Find the correlation coefficient.

13. **Demand for Candy Bars**   In this problem you will determine a linear demand equation that describes the demand for candy bars in your class. Survey your classmates to determine what price they would be willing to pay for a candy bar. Your survey form might look like the sample to the left.

(a) Make a table of the number of respondents who answered "yes" at each price level.

(b) Make a scatter plot of your data.

(c) Find and graph the regression line $y = mp + b$, which gives the number of responents y who would buy a candy bar if the price were p cents. This is the *demand equation*. Why is the slope m negative?

(d) What is the p-intercept of the demand equation? What does this intercept tell you about pricing candy bars?

Would you buy a candy bar from the vending machine in the hallway if the price is as indicated?

Price	Yes or No
30¢	
40¢	
50¢	
60¢	
70¢	
80¢	
90¢	
$1.00	
$1.10	
$1.20	

Image copyright © Kristian Peetz.
Used under license from Shutterstock.com

TRIGONOMETRIC FUNCTIONS: UNIT CIRCLE APPROACH

If you've ever taken a ferris wheel ride, then you know about periodic motion—that is, motion that repeats over and over. Periodic motion is common in nature. Think about the daily rising and setting of the sun (day, night, day, night, . . .), the daily variation in tide levels (high, low, high, low, . . .), or the vibrations of a leaf in the wind (left, right, left, right, . . .). To model such motion, we need a function whose values increase, then decrease, then increase, and so on. To understand how to define such a function, let's look at a person riding on a ferris wheel. The graph shows the height of the person above the center of the wheel at time *t*. Notice that the graph goes up and down repeatedly.

The trigonometric function *sine* is defined in a similar way, using the unit circle (in place of the ferris wheel). The trigonometric functions can be defined in two different but equivalent ways: as functions of real numbers (Chapter 2) or as functions of angles (Chapter 3). The two approaches are independent of each other, so either Chapter 2 or Chapter 3 may be studied first. We study both approaches because the different approaches are required for different applications.

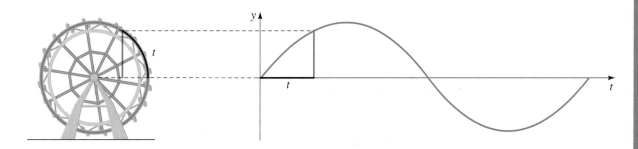

2.1 THE UNIT CIRCLE

| The Unit Circle ▶ Terminal Points on the Unit Curcle ▶ The Reference Number

In this section we explore some properties of the circle of radius 1 centered at the origin. These properties are used in the next section to define the trigonometric functions.

▼ The Unit Circle

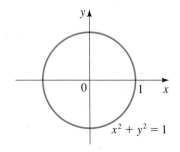

FIGURE 1   The unit circle

The set of points at a distance 1 from the origin is a circle of radius 1 (see Figure 1). In Section 1.1 we learned that the equation of this circle is $x^2 + y^2 = 1$.

> **THE UNIT CIRCLE**
>
> The **unit circle** is the circle of radius 1 centered at the origin in the xy-plane. Its equation is
>
> $$x^2 + y^2 = 1$$

EXAMPLE 1 | A Point on the Unit Circle

Show that the point $P\left(\dfrac{\sqrt{3}}{3}, \dfrac{\sqrt{6}}{3}\right)$ is on the unit circle.

SOLUTION   We need to show that this point satisfies the equation of the unit circle, that is, $x^2 + y^2 = 1$. Since

$$\left(\frac{\sqrt{3}}{3}\right)^2 + \left(\frac{\sqrt{6}}{3}\right)^2 = \frac{3}{9} + \frac{6}{9} = 1$$

P is on the unit circle.

✎ NOW TRY EXERCISE **3**   ■

EXAMPLE 2 | Locating a Point on the Unit Circle

The point $P(\sqrt{3}/2, y)$ is on the unit circle in Quadrant IV. Find its y-coordinate.

SOLUTION   Since the point is on the unit circle, we have

$$\left(\frac{\sqrt{3}}{2}\right)^2 + y^2 = 1$$

$$y^2 = 1 - \frac{3}{4} = \frac{1}{4}$$

$$y = \pm\frac{1}{2}$$

Since the point is in Quadrant IV, its y-coordinate must be negative, so $y = -\frac{1}{2}$.

✎ NOW TRY EXERCISE **9**   ■

▼ Terminal Points on the Unit Circle

Suppose t is a real number. Let's mark off a distance t along the unit circle, starting at the point $(1, 0)$ and moving in a counterclockwise direction if t is positive or in a clockwise direction if t is negative (Figure 2). In this way we arrive at a point $P(x, y)$ on the unit circle. The point $P(x, y)$ obtained in this way is called the **terminal point** determined by the real number t.

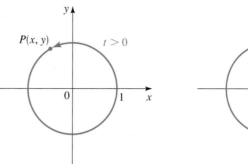

FIGURE 2

(a) Terminal point $P(x, y)$ determined by $t > 0$

(b) Terminal point $P(x, y)$ determined by $t < 0$

The circumference of the unit circle is $C = 2\pi(1) = 2\pi$. So if a point starts at $(1, 0)$ and moves counterclockwise all the way around the unit circle and returns to $(1, 0)$, it travels a distance of 2π. To move halfway around the circle, it travels a distance of $\frac{1}{2}(2\pi) = \pi$. To move a quarter of the distance around the circle, it travels a distance of $\frac{1}{4}(2\pi) = \pi/2$. Where does the point end up when it travels these distances along the circle? From Figure 3 we see, for example, that when it travels a distance of π starting at $(1, 0)$, its terminal point is $(-1, 0)$.

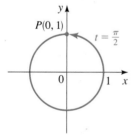

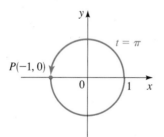

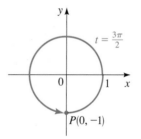

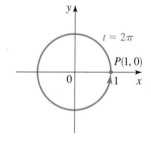

FIGURE 3 Terminal points determined by $t = \frac{\pi}{2}, \pi, \frac{3\pi}{2}$, and 2π

EXAMPLE 3 | Finding Terminal Points

Find the terminal point on the unit circle determined by each real number t.

(a) $t = 3\pi$ **(b)** $t = -\pi$ **(c)** $t = -\dfrac{\pi}{2}$

SOLUTION From Figure 4 we get the following:

(a) The terminal point determined by 3π is $(-1, 0)$.

(b) The terminal point determined by $-\pi$ is $(-1, 0)$.

(c) The terminal point determined by $-\pi/2$ is $(0, -1)$.

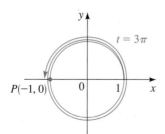

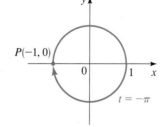

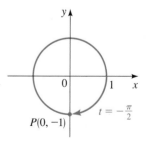

FIGURE 4

Notice that different values of t can determine the same terminal point.

✎ NOW TRY EXERCISE **23**

The terminal point $P(x, y)$ determined by $t = \pi/4$ is the same distance from $(1, 0)$ as from $(0, 1)$ along the unit circle (see Figure 5).

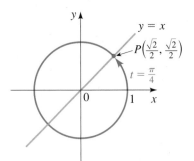

FIGURE 5

Since the unit circle is symmetric with respect to the line $y = x$, it follows that P lies on the line $y = x$. So P is the point of intersection (in the first quadrant) of the circle $x^2 + y^2 = 1$ and the line $y = x$. Substituting x for y in the equation of the circle, we get

$$x^2 + x^2 = 1$$

$$2x^2 = 1 \qquad \text{Combine like terms}$$

$$x^2 = \frac{1}{2} \qquad \text{Divide by 2}$$

$$x = \pm\frac{1}{\sqrt{2}} \qquad \text{Take square roots}$$

Since P is in the first quadrant, $x = 1/\sqrt{2}$ and since $y = x$, we have $y = 1/\sqrt{2}$ also. Thus, the terminal point determined by $\pi/4$ is

$$P\left(\frac{1}{\sqrt{2}}, \frac{1}{\sqrt{2}}\right) = P\left(\frac{\sqrt{2}}{2}, \frac{\sqrt{2}}{2}\right)$$

Similar methods can be used to find the terminal points determined by $t = \pi/6$ and $t = \pi/3$ (see Exercises 57 and 58). Table 1 and Figure 6 give the terminal points for some special values of t.

TABLE 1

t	Terminal point determined by t
0	$(1, 0)$
$\frac{\pi}{6}$	$\left(\frac{\sqrt{3}}{2}, \frac{1}{2}\right)$
$\frac{\pi}{4}$	$\left(\frac{\sqrt{2}}{2}, \frac{\sqrt{2}}{2}\right)$
$\frac{\pi}{3}$	$\left(\frac{1}{2}, \frac{\sqrt{3}}{2}\right)$
$\frac{\pi}{2}$	$(0, 1)$

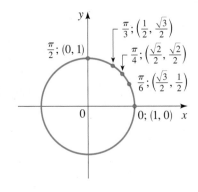

FIGURE 6

EXAMPLE 4 | Finding Terminal Points

Find the terminal point determined by each given real number t.

(a) $t = -\dfrac{\pi}{4}$ **(b)** $t = \dfrac{3\pi}{4}$ **(c)** $t = -\dfrac{5\pi}{6}$

SOLUTION

(a) Let P be the terminal point determined by $-\pi/4$, and let Q be the terminal point determined by $\pi/4$. From Figure 7(a) we see that the point P has the same coordinates as Q except for sign. Since P is in quadrant IV, its x-coordinate is positive and its y-coordinate is negative. Thus, the terminal point is $P\left(\sqrt{2}/2, -\sqrt{2}/2\right)$.

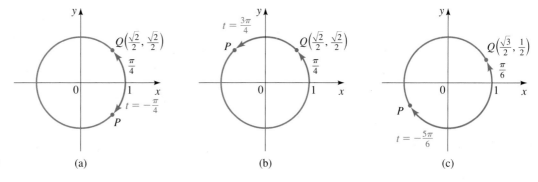

FIGURE 7   (a)   (b)   (c)

(b) Let P be the terminal point determined by $3\pi/4$, and let Q be the terminal point determined by $\pi/4$. From Figure 7(b) we see that the point P has the same coordinates as Q except for sign. Since P is in quadrant II, its x-coordinate is negative and its y-coordinate is positive. Thus, the terminal point is $P\left(-\sqrt{2}/2, \sqrt{2}/2\right)$.

(c) Let P be the terminal point determined by $-5\pi/6$, and let Q be the terminal point determined by $\pi/6$. From Figure 7(c) we see that the point P has the same coordinates as Q except for sign. Since P is in quadrant III, its coordinates are both negative. Thus, the terminal point is $P\left(-\sqrt{3}/2, -\frac{1}{2}\right)$.

✎ NOW TRY EXERCISE **25**   ■

▼ The Reference Number

From Examples 3 and 4 we see that to find a terminal point in any quadrant we need only know the "corresponding" terminal point in the first quadrant. We use the idea of the *reference number* to help us find terminal points.

> ### REFERENCE NUMBER
>
> Let t be a real number. The **reference number $\bar{t}$** associated with t is the shortest distance along the unit circle between the terminal point determined by t and the x-axis.

Figure 8 shows that to find the reference number $\bar{t}$, it's helpful to know the quadrant in which the terminal point determined by t lies. If the terminal point lies in quadrants I or IV, where x is positive, we find $\bar{t}$ by moving along the circle to the *positive* x-axis. If it lies in quadrants II or III, where x is negative, we find $\bar{t}$ by moving along the circle to the *negative* x-axis.

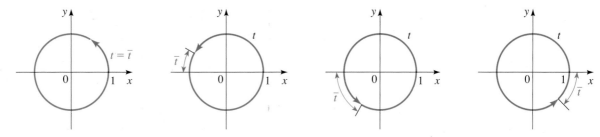

FIGURE 8   The reference number $\bar{t}$ for t

EXAMPLE 5 | Finding Reference Numbers

Find the reference number for each value of t.

(a) $t = \dfrac{5\pi}{6}$ **(b)** $t = \dfrac{7\pi}{4}$ **(c)** $t = -\dfrac{2\pi}{3}$ **(d)** $t = 5.80$

SOLUTION From Figure 9 we find the reference numbers as follows:

(a) $\bar{t} = \pi - \dfrac{5\pi}{6} = \dfrac{\pi}{6}$

(b) $\bar{t} = 2\pi - \dfrac{7\pi}{4} = \dfrac{\pi}{4}$

(c) $\bar{t} = \pi - \dfrac{2\pi}{3} = \dfrac{\pi}{3}$

(d) $\bar{t} = 2\pi - 5.80 \approx 0.48$

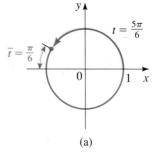

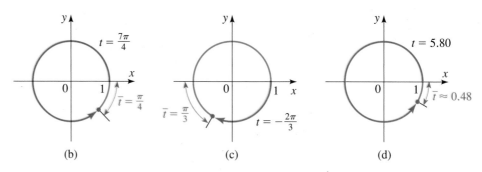

(a)   (b)   (c)   (d)

FIGURE 9

✎ NOW TRY EXERCISE **35**

USING REFERENCE NUMBERS TO FIND TERMINAL POINTS

To find the terminal point P determined by any value of t, we use the following steps:

1. Find the reference number $\bar{t}$.

2. Find the terminal point $Q(a, b)$ determined by $\bar{t}$.

3. The terminal point determined by t is $P(\pm a, \pm b)$, where the signs are chosen according to the quadrant in which this terminal point lies.

EXAMPLE 6 | Using Reference Numbers to Find Terminal Points

Find the terminal point determined by each given real number t.

(a) $t = \dfrac{5\pi}{6}$ **(b)** $t = \dfrac{7\pi}{4}$ **(c)** $t = -\dfrac{2\pi}{3}$

SOLUTION The reference numbers associated with these values of t were found in Example 5.

(a) The reference number is $\bar{t} = \pi/6$, which determines the terminal point $\left(\sqrt{3}/2, \tfrac{1}{2}\right)$ from Table 1. Since the terminal point determined by t is in Quadrant II, its x-coordinate is negative and its y-coordinate is positive. Thus, the desired terminal point is

$$\left(-\dfrac{\sqrt{3}}{2}, \dfrac{1}{2}\right)$$

(b) The reference number is $\bar{t} = \pi/4$, which determines the terminal point $(\sqrt{2}/2, \sqrt{2}/2)$ from Table 1. Since the terminal point is in Quadrant IV, its x-coordinate is positive and its y-coordinate is negative. Thus, the desired terminal point is

$$\left(\frac{\sqrt{2}}{2}, -\frac{\sqrt{2}}{2} \right)$$

(c) The reference number is $\bar{t} = \pi/3$, which determines the terminal point $(\frac{1}{2}, \sqrt{3}/2)$ from Table 1. Since the terminal point determined by t is in Quadrant III, its coordinates are both negative. Thus, the desired terminal point is

$$\left(-\frac{1}{2}, -\frac{\sqrt{3}}{2} \right)$$

✎ NOW TRY EXERCISE **39** ■

Since the circumference of the unit circle is 2π, the terminal point determined by t is the same as that determined by $t + 2\pi$ or $t - 2\pi$. In general, we can add or subtract 2π any number of times without changing the terminal point determined by t. We use this observation in the next example to find terminal points for large t.

EXAMPLE 7 | Finding the Terminal Point for Large t

Find the terminal point determined by $t = \dfrac{29\pi}{6}$.

SOLUTION Since

$$t = \frac{29\pi}{6} = 4\pi + \frac{5\pi}{6}$$

we see that the terminal point of t is the same as that of $5\pi/6$ (that is, we subtract 4π). So by Example 6(a) the terminal point is $\left(-\sqrt{3}/2, \frac{1}{2} \right)$. (See Figure 10.)

✎ NOW TRY EXERCISE **45** ■

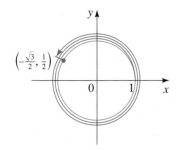

FIGURE 10

2.1 EXERCISES

CONCEPTS

1. (a) The unit circle is the circle centered at _____ with radius _____.

(b) The equation of the unit circle is _____.

(c) Suppose the point $P(x, y)$ is on the unit circle. Find the missing coordinate:

   (i) $P(1, \rule{1cm}{0.4pt})$ (ii) $P(\rule{1cm}{0.4pt}, 1)$

   (iii) $P(-1, \rule{1cm}{0.4pt})$ (iv) $P(\rule{1cm}{0.4pt}, -1)$

2. (a) If we mark off a distance t along the unit circle, starting at $(1, 0)$ and moving in a counterclockwise direction, we arrive at the _____ point determined by t.

(b) The terminal points determined by $\pi/2, \pi, -\pi/2, 2\pi$ are _____, _____, _____, and _____, respectively.

SKILLS

3–8 ■ Show that the point is on the unit circle.

3. $\left(\dfrac{4}{5}, -\dfrac{3}{5} \right)$ **4.** $\left(-\dfrac{5}{13}, \dfrac{12}{13} \right)$ **5.** $\left(\dfrac{7}{25}, \dfrac{24}{25} \right)$

6. $\left(-\dfrac{5}{7}, -\dfrac{2\sqrt{6}}{7} \right)$ **7.** $\left(-\dfrac{\sqrt{5}}{3}, \dfrac{2}{3} \right)$ **8.** $\left(\dfrac{\sqrt{11}}{6}, \dfrac{5}{6} \right)$

9–14 ■ Find the missing coordinate of P, using the fact that P lies on the unit circle in the given quadrant.

Coordinates	Quadrant
9. $P\left(-\dfrac{3}{5}, \rule{1cm}{0.4pt}\right)$	III
10. $P\left(\rule{1cm}{0.4pt}, -\dfrac{7}{25}\right)$	IV
11. $P\left(\rule{1cm}{0.4pt}, \dfrac{1}{3}\right)$	II

Coordinates	Quadrant
12. $P\left(\frac{2}{5},\ \_\_\_\right)$	I
13. $P\left(\_\_\_,\ -\frac{2}{7}\right)$	IV
14. $P\left(-\frac{2}{3},\ \_\_\_\right)$	II

15–20 ■ The point P is on the unit circle. Find $P(x, y)$ from the given information.

15. The x-coordinate of P is $\frac{4}{5}$, and the y-coordinate is positive.

16. The y-coordinate of P is $-\frac{1}{3}$, and the x-coordinate is positive.

17. The y-coordinate of P is $\frac{2}{3}$, and the x-coordinate is negative.

18. The x-coordinate of P is positive, and the y-coordinate of P is $-\sqrt{5}/5$.

19. The x-coordinate of P is $-\sqrt{2}/3$, and P lies below the x-axis.

20. The x-coordinate of P is $-\frac{2}{5}$, and P lies above the x-axis.

21–22 ■ Find t and the terminal point determined by t for each point in the figure. In Exercise 21, t increases in increments of $\pi/4$; in Exercise 22, t increases in increments of $\pi/6$.

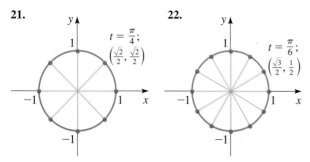

21. $t = \frac{\pi}{4}$; $\left(\frac{\sqrt{2}}{2}, \frac{\sqrt{2}}{2}\right)$

22. $t = \frac{\pi}{6}$; $\left(\frac{\sqrt{3}}{2}, \frac{1}{2}\right)$

23–32 ■ Find the terminal point $P(x, y)$ on the unit circle determined by the given value of t.

23. $t = \dfrac{\pi}{2}$

24. $t = \dfrac{3\pi}{2}$

25. $t = \dfrac{5\pi}{6}$

26. $t = \dfrac{7\pi}{6}$

27. $t = -\dfrac{\pi}{3}$

28. $t = \dfrac{5\pi}{3}$

29. $t = \dfrac{2\pi}{3}$

30. $t = -\dfrac{\pi}{2}$

31. $t = -\dfrac{3\pi}{4}$

32. $t = \dfrac{11\pi}{6}$

★33. Suppose that the terminal point determined by t is the point $\left(\frac{3}{5}, \frac{4}{5}\right)$ on the unit circle. Find the terminal point determined by each of the following.
  (a) $\pi - t$ **(b)** $-t$
  (c) $\pi + t$ **(d)** $2\pi + t$

34. Suppose that the terminal point determined by t is the point $\left(\frac{3}{4}, \sqrt{7}/4\right)$ on the unit circle. Find the terminal point determined by each of the following.
  (a) $-t$ **(b)** $4\pi + t$
  (c) $\pi - t$ **(d)** $t - \pi$

35–38 ■ Find the reference number for each value of t.

35. (a) $t = \dfrac{5\pi}{4}$ **(b)** $t = \dfrac{7\pi}{3}$

   (c) $t = -\dfrac{4\pi}{3}$ **(d)** $t = \dfrac{\pi}{6}$

36. (a) $t = \dfrac{5\pi}{6}$ **(b)** $t = \dfrac{7\pi}{6}$

   (c) $t = \dfrac{11\pi}{3}$ **(d)** $t = -\dfrac{7\pi}{4}$

37. (a) $t = \dfrac{5\pi}{7}$ **(b)** $t = -\dfrac{7\pi}{9}$

   (c) $t = -3$ **(d)** $t = 5$

38. (a) $t = \dfrac{11\pi}{5}$ **(b)** $t = -\dfrac{9\pi}{7}$

   (c) $t = 6$ **(d)** $t = -7$

39–52 ■ Find **(a)** the reference number for each value of t and **(b)** the terminal point determined by t.

39. $t = \dfrac{2\pi}{3}$ **40.** $t = \dfrac{4\pi}{3}$

41. $t = \dfrac{3\pi}{4}$ **42.** $t = \dfrac{7\pi}{3}$

43. $t = -\dfrac{2\pi}{3}$ **44.** $t = -\dfrac{7\pi}{6}$

45. $t = \dfrac{13\pi}{4}$ **46.** $t = \dfrac{13\pi}{6}$

47. $t = \dfrac{7\pi}{6}$ **48.** $t = \dfrac{17\pi}{4}$

49. $t = -\dfrac{11\pi}{3}$ **50.** $t = \dfrac{31\pi}{6}$

51. $t = \dfrac{16\pi}{3}$ **52.** $t = -\dfrac{41\pi}{4}$

53–56 ■ Use the figure to find the terminal point determined by the real number t, with coordinates rounded to one decimal place.

★53. $t \doteq 1$

54. $t = 2.5$

55. $t = -1.1$

56. $t = 4.2$

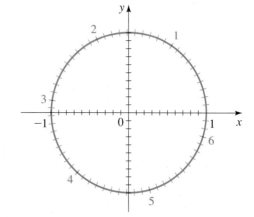

57. Finding the Terminal Point for $\pi/6$ Suppose the terminal point determined by $t = \pi/6$ is $P(x, y)$ and the points Q and R are as shown in the figure. Why are the distances PQ and PR the same? Use this fact, together with the Distance Formula, to show that the coordinates of P satisfy the equation $2y = \sqrt{x^2 + (y - 1)^2}$. Simplify this equation using the fact that $x^2 + y^2 = 1$. Solve the simplified equation to find $P(x, y)$.

58. Finding the Terminal Point for $\pi/3$ Now that you know the terminal point determined by $t = \pi/6$, use symmetry to find the terminal point determined by $t = \pi/3$ (see the figure). Explain your reasoning.

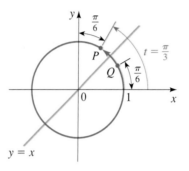

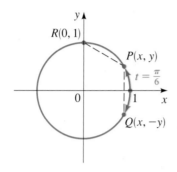

2.2 TRIGONOMETRIC FUNCTIONS OF REAL NUMBERS

The Trigonometric Functions ▶ Values of the Trigonometric Functions ▶ Fundamental Identities

A function is a rule that assigns to each real number another real number. In this section we use properties of the unit circle from the preceding section to define the trigonometric functions.

▼ The Trigonometric Functions

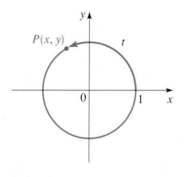

FIGURE 1

Recall that to find the terminal point $P(x, y)$ for a given real number t, we move a distance t along the unit circle, starting at the point $(1, 0)$. We move in a counterclockwise direction if t is positive and in a clockwise direction if t is negative (see Figure 1). We now use the x- and y-coordinates of the point $P(x, y)$ to define several functions. For instance, we define the function called *sine* by assigning to each real number t the y-coordinate of the terminal point $P(x, y)$ determined by t. The functions *cosine*, *tangent*, *cosecant*, *secant*, and *cotangent* are also defined by using the coordinates of $P(x, y)$.

DEFINITION OF THE TRIGONOMETRIC FUNCTIONS

Let t be any real number and let $P(x, y)$ be the terminal point on the unit circle determined by t. We define

$$\sin t = y \qquad \cos t = x \qquad \tan t = \frac{y}{x} \;\; (x \neq 0)$$

$$\csc t = \frac{1}{y} \;\; (y \neq 0) \qquad \sec t = \frac{1}{x} \;\; (x \neq 0) \qquad \cot t = \frac{x}{y} \;\; (y \neq 0)$$

Because the trigonometric functions can be defined in terms of the unit circle, they are sometimes called the **circular functions**.

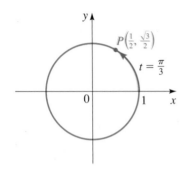

FIGURE 2

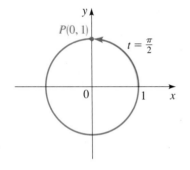

FIGURE 3

We can easily remember the sines and cosines of the basic angles by writing them in the form $\sqrt{\blacksquare}/2$:

t	$\sin t$	$\cos t$
0	$\sqrt{0}/2$	$\sqrt{4}/2$
$\pi/6$	$\sqrt{1}/2$	$\sqrt{3}/2$
$\pi/4$	$\sqrt{2}/2$	$\sqrt{2}/2$
$\pi/3$	$\sqrt{3}/2$	$\sqrt{1}/2$
$\pi/2$	$\sqrt{4}/2$	$\sqrt{0}/2$

EXAMPLE 1 | Evaluating Trigonometric Functions

Find the six trigonometric functions of each given real number t.

(a) $t = \dfrac{\pi}{3}$ **(b)** $t = \dfrac{\pi}{2}$

SOLUTION

(a) From Table 1 on page 106, we see that the terminal point determined by $t = \pi/3$ is $P(\frac{1}{2}, \sqrt{3}/2)$. (See Figure 2.) Since the coordinates are $x = \frac{1}{2}$ and $y = \sqrt{3}/2$, we have

$$\sin \frac{\pi}{3} = \frac{\sqrt{3}}{2} \qquad \cos \frac{\pi}{3} = \frac{1}{2} \qquad \tan \frac{\pi}{3} = \frac{\sqrt{3}/2}{1/2} = \sqrt{3}$$

$$\csc \frac{\pi}{3} = \frac{2\sqrt{3}}{3} \qquad \sec \frac{\pi}{3} = 2 \qquad \cot \frac{\pi}{3} = \frac{1/2}{\sqrt{3}/2} = \frac{\sqrt{3}}{3}$$

(b) The terminal point determined by $\pi/2$ is $P(0, 1)$. (See Figure 3.) So

$$\sin \frac{\pi}{2} = 1 \qquad \cos \frac{\pi}{2} = 0 \qquad \csc \frac{\pi}{2} = \frac{1}{1} = 1 \qquad \cot \frac{\pi}{2} = \frac{0}{1} = 0$$

But $\tan \pi/2$ and $\sec \pi/2$ are undefined because $x = 0$ appears in the denominator in each of their definitions.

✎. NOW TRY EXERCISE **3**   ■

Some special values of the trigonometric functions are listed in Table 1. This table is easily obtained from Table 1 of Section 2.1, together with the definitions of the trigonometric functions.

TABLE 1
Special values of the trigonometric functions

t	$\sin t$	$\cos t$	$\tan t$	$\csc t$	$\sec t$	$\cot t$
0	0	1	0	—	1	—
$\dfrac{\pi}{6}$	$\dfrac{1}{2}$	$\dfrac{\sqrt{3}}{2}$	$\dfrac{\sqrt{3}}{3}$	2	$\dfrac{2\sqrt{3}}{3}$	$\sqrt{3}$
$\dfrac{\pi}{4}$	$\dfrac{\sqrt{2}}{2}$	$\dfrac{\sqrt{2}}{2}$	1	$\sqrt{2}$	$\sqrt{2}$	1
$\dfrac{\pi}{3}$	$\dfrac{\sqrt{3}}{2}$	$\dfrac{1}{2}$	$\sqrt{3}$	$\dfrac{2\sqrt{3}}{3}$	2	$\dfrac{\sqrt{3}}{3}$
$\dfrac{\pi}{2}$	1	0	—	1	—	0

Example 1 shows that some of the trigonometric functions fail to be defined for certain real numbers. So we need to determine their domains. The functions sine and cosine are defined for all values of t. Since the functions cotangent and cosecant have y in the denominator of their definitions, they are not defined whenever the y-coordinate of the terminal point $P(x, y)$ determined by t is 0. This happens when $t = n\pi$ for any integer n, so their domains do not include these points. The functions tangent and secant have x in the denominator in their definitions, so they are not defined whenever $x = 0$. This happens when $t = (\pi/2) + n\pi$ for any integer n.

Relationship to the Trigonometric Functions of Angles

If you have previously studied trigonometry of right triangles (Chapter 3), you are probably wondering how the sine and cosine of an *angle* relate to those of this section. To see how, let's start with a right triangle, ΔOPQ.

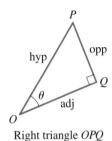

Right triangle OPQ

Place the triangle in the coordinate plane as shown, with angle θ in standard position.

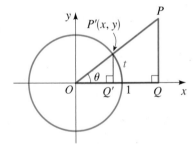

$P'(x, y)$ is the terminal point determined by t.

The point $P'(x, y)$ in the figure is the terminal point determined by the arc t. Note that triangle OPQ is similar to the small triangle $OP'Q'$ whose legs have lengths x and y.

Now, by the definition of the trigonometric functions of the *angle* θ we have

$$\sin \theta = \frac{\text{opp}}{\text{hyp}} = \frac{PQ}{OP} = \frac{P'Q'}{OP'}$$

$$= \frac{y}{1} = y$$

$$\cos \theta = \frac{\text{adj}}{\text{hyp}} = \frac{OQ}{OP} = \frac{OQ'}{OP'}$$

$$= \frac{x}{1} = x$$

By the definition of the trigonometric functions of the *real number t*, we have

$$\sin t = y \qquad \cos t = x$$

Now, if θ is measured in radians, then $\theta = t$ (see the figure). So the trigonometric functions of the angle with radian measure θ are exactly the same as the trigonometric functions defined in terms of the terminal point determined by the real number t.

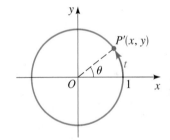

The radian measure of angle θ is t.

Why then study trigonometry in two different ways? Because different applications require that we view the trigonometric functions differently. (Compare Section 2.6 with Sections 3.2, 3.5, and 3.6.)

DOMAINS OF THE TRIGONOMETRIC FUNCTIONS

Function	Domain
sin, cos	All real numbers
tan, sec	All real numbers other than $\dfrac{\pi}{2} + n\pi$ for any integer n
cot, csc	All real numbers other than $n\pi$ for any integer, n

---

▼ Values of the Trigonometric Functions

To compute other values of the trigonometric functions, we first determine their signs. The signs of the trigonometric functions depend on the quadrant in which the terminal point of t lies. For example, if the terminal point $P(x, y)$ determined by t lies in Quadrant III, then its coordinates are both negative. So sin t, cos t, csc t, and sec t are all negative, whereas tan t and cot t are positive. You can check the other entries in the following box.

The following mnemonic device will help you remember which trigonometric functions are positive in each quadrant: **A**ll of them, **S**ine, **T**angent, or **C**osine.

You can remember this as "All Students Take Calculus."

SIGNS OF THE TRIGONOMETRIC FUNCTIONS

Quadrant	Positive Functions	Negative Functions
I	all	none
II	sin, csc	cos, sec, tan, cot
III	tan, cot	sin, csc, cos, sec
IV	cos, sec	sin, csc, tan, cot

For example $\cos(2\pi/3) < 0$ because the terminal point of $t = 2\pi/3$ is in Quadrant II, whereas $\tan 4 > 0$ because the terminal point of $t = 4$ is in Quadrant III.

In Section 2.1 we used the reference number to find the terminal point determined by a real number t. Since the trigonometric functions are defined in terms of the coordinates of terminal points, we can use the reference number to find values of the trigonometric functions. Suppose that $\bar{t}$ is the reference number for t. Then the terminal point of $\bar{t}$ has the same coordinates, except possibly for sign, as the terminal point of t. So the values of the trigonometric functions at t are the same, except possibly for sign, as their values at $\bar{t}$. We illustrate this procedure in the next example.

EXAMPLE 2 | Evaluating Trigonometric Functions

Find each value.

(a) $\cos \dfrac{2\pi}{3}$ **(b)** $\tan\left(-\dfrac{\pi}{3}\right)$ **(c)** $\sin \dfrac{19\pi}{4}$

SOLUTION

(a) The reference number for $2\pi/3$ is $\pi/3$ (see Figure 4(a)). Since the terminal point of $2\pi/3$ is in Quadrant II, $\cos(2\pi/3)$ is negative. Thus,

$$\cos \frac{2\pi}{3} = -\cos \frac{\pi}{3} = -\frac{1}{2}$$

Sign   Reference number   From Table 1

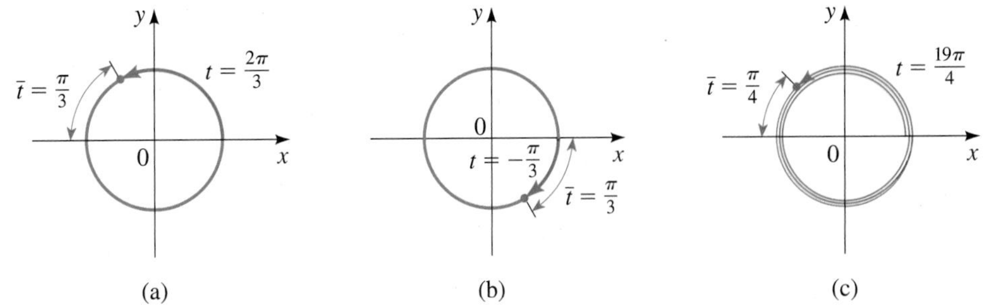

FIGURE 4

(a)   (b)   (c)

(b) The reference number for $-\pi/3$ is $\pi/3$ (see Figure 4(b)). Since the terminal point of $-\pi/3$ is in Quadrant IV, $\tan(-\pi/3)$ is negative. Thus,

$$\tan\left(-\frac{\pi}{3}\right) = -\tan\frac{\pi}{3} = -\sqrt{3}$$

Sign   Reference number   From Table 1

(c) Since $(19\pi/4) - 4\pi = 3\pi/4$, the terminal points determined by $19\pi/4$ and $3\pi/4$ are the same. The reference number for $3\pi/4$ is $\pi/4$ (see Figure 4(c)). Since the terminal point of $3\pi/4$ is in Quadrant II, $\sin(3\pi/4)$ is positive. Thus,

$$\sin\frac{19\pi}{4} = \sin\frac{3\pi}{4} = +\sin\frac{\pi}{4} = \frac{\sqrt{2}}{2}$$

Subtract 4π Sign   Reference number   From Table 1

✎ NOW TRY EXERCISE **7**

So far, we have been able to compute the values of the trigonometric functions only for certain values of t. In fact, we can compute the values of the trigonometric functions whenever t is a multiple of $\pi/6$, $\pi/4$, $\pi/3$, and $\pi/2$. How can we compute the trigonometric functions for other values of t? For example, how can we find $\sin 1.5$? One way is to carefully sketch a diagram and read the value (see Exercises 39–46); however, this method is not very accurate. Fortunately, programmed directly into scientific calculators are mathematical procedures (see the margin note on page 134) that find the values of *sine*, *cosine*, and *tangent* correct to the number of digits in the display. The calculator must be put in *radian mode* to evaluate these functions. To find values of cosecant, secant, and cotangent using a calculator, we need to use the following *reciprocal relations*:

$$\csc t = \frac{1}{\sin t} \qquad \sec t = \frac{1}{\cos t} \qquad \cot t = \frac{1}{\tan t}$$

These identities follow from the definitions of the trigonometric functions. For instance, since $\sin t = y$ and $\csc t = 1/y$, we have $\csc t = 1/y = 1/(\sin t)$. The others follow similarly.

EXAMPLE 3 | Using a Calculator to Evaluate Trigonometric Functions

Making sure our calculator is set to radian mode and rounding the results to six decimal places, we get

(a) $\sin 2.2 \approx 0.808496$

(b) $\cos 1.1 \approx 0.453596$

(c) $\cot 28 = \dfrac{1}{\tan 28} \approx -3.553286$

(d) $\csc 0.98 = \dfrac{1}{\sin 0.98} \approx 1.204098$

✎ NOW TRY EXERCISES **41** AND **43**

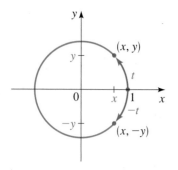

FIGURE 5

Even and odd functions are defined in Section 1.6.

Let's consider the relationship between the trigonometric functions of t and those of $-t$. From Figure 5 we see that

$$\sin(-t) = -y = -\sin t$$

$$\cos(-t) = x = \cos t$$

$$\tan(-t) = \frac{-y}{x} = -\frac{y}{x} = -\tan t$$

These equations show that sine and tangent are odd functions, whereas cosine is an even function. It's easy to see that the reciprocal of an even function is even and the reciprocal of an odd function is odd. This fact, together with the reciprocal relations, completes our knowledge of the even-odd properties for all the trigonometric functions.

EVEN-ODD PROPERTIES

Sine, cosecant, tangent, and cotangent are odd functions; cosine and secant are even functions.

$$\sin(-t) = -\sin t \qquad \cos(-t) = \cos t \qquad \tan(-t) = -\tan t$$

$$\csc(-t) = -\csc t \qquad \sec(-t) = \sec t \qquad \cot(-t) = -\cot t$$

EXAMPLE 4 | Even and Odd Trigonometric Functions

Use the even-odd properties of the trigonometric functions to determine each value.

(a) $\sin\left(-\dfrac{\pi}{6}\right)$ **(b)** $\cos\left(-\dfrac{\pi}{4}\right)$

SOLUTION   By the even-odd properties and Table 1 we have

(a) $\sin\left(-\dfrac{\pi}{6}\right) = -\sin\dfrac{\pi}{6} = -\dfrac{1}{2}$ Sine is odd

(b) $\cos\left(-\dfrac{\pi}{4}\right) = \cos\dfrac{\pi}{4} = \dfrac{\sqrt{2}}{2}$ Cosine is even

✎ NOW TRY EXERCISE **13**   ■

▼ Fundamental Identities

The trigonometric functions are related to each other through equations called **trigonometric identities**. We give the most important ones in the following box.*

FUNDAMENTAL IDENTITIES

Reciprocal Identities

$$\csc t = \frac{1}{\sin t} \qquad \sec t = \frac{1}{\cos t} \qquad \cot t = \frac{1}{\tan t} \qquad \tan t = \frac{\sin t}{\cos t} \qquad \cot t = \frac{\cos t}{\sin t}$$

Pythagorean Identities

$$\sin^2 t + \cos^2 t = 1 \qquad \tan^2 t + 1 = \sec^2 t \qquad 1 + \cot^2 t = \csc^2 t$$

---

*We follow the usual convention of writing $\sin^2 t$ for $(\sin t)^2$. In general, we write $\sin^n t$ for $(\sin t)^n$ for all integers n except $n = -1$. The exponent $n = -1$ will be assigned another meaning in Section 2.5. Of course, the same convention applies to the other five trigonometric functions.

PROOF The reciprocal identities follow immediately from the definitions on page 111. We now prove the Pythagorean identities. By definition, $\cos t = x$ and $\sin t = y$, where x and y are the coordinates of a point $P(x, y)$ on the unit circle. Since $P(x, y)$ is on the unit circle, we have $x^2 + y^2 = 1$. Thus

$$\sin^2 t + \cos^2 t = 1$$

Dividing both sides by $\cos^2 t$ (provided that $\cos t \neq 0$), we get

$$\frac{\sin^2 t}{\cos^2 t} + \frac{\cos^2 t}{\cos^2 t} = \frac{1}{\cos^2 t}$$

$$\left(\frac{\sin t}{\cos t}\right)^2 + 1 = \left(\frac{1}{\cos t}\right)^2$$

$$\tan^2 t + 1 = \sec^2 t$$

We have used the reciprocal identities $\sin t/\cos t = \tan t$ and $1/\cos t = \sec t$. Similarly, dividing both sides of the first Pythagorean identity by $\sin^2 t$ (provided that $\sin t \neq 0$) gives us $1 + \cot^2 t = \csc^2 t$. ∎

As their name indicates, the fundamental identities play a central role in trigonometry because we can use them to relate any trigonometric function to any other. So, if we know the value of any one of the trigonometric functions at t, then we can find the values of all the others at t.

EXAMPLE 5 | Finding All Trigonometric Functions from the Value of One

If $\cos t = \frac{3}{5}$ and t is in Quadrant IV, find the values of all the trigonometric functions at t.

SOLUTION From the Pythagorean identities we have

$$\sin^2 t + \cos^2 t = 1$$

$$\sin^2 t + \left(\tfrac{3}{5}\right)^2 = 1 \qquad \text{Substitute } \cos t = \tfrac{3}{5}$$

$$\sin^2 t = 1 - \tfrac{9}{25} = \tfrac{16}{25} \qquad \text{Solve for } \sin^2 t$$

$$\sin t = \pm\tfrac{4}{5} \qquad \text{Take square roots}$$

Since this point is in Quadrant IV, $\sin t$ is negative, so $\sin t = -\frac{4}{5}$. Now that we know both $\sin t$ and $\cos t$, we can find the values of the other trigonometric functions using the reciprocal identities:

$$\sin t = -\frac{4}{5} \qquad \cos t = \frac{3}{5} \qquad \tan t = \frac{\sin t}{\cos t} = \frac{-\frac{4}{5}}{\frac{3}{5}} = -\frac{4}{3}$$

$$\csc t = \frac{1}{\sin t} = -\frac{5}{4} \qquad \sec t = \frac{1}{\cos t} = \frac{5}{3} \qquad \cot t = \frac{1}{\tan t} = -\frac{3}{4}$$

✎ NOW TRY EXERCISE **65**

EXAMPLE 6 | Writing One Trigonometric Function in Terms of Another

Write $\tan t$ in terms of $\cos t$, where t is in Quadrant III.

SOLUTION   Since $\tan t = \sin t/\cos t$, we need to write $\sin t$ in terms of $\cos t$. By the Pythagorean identities we have

$$\sin^2 t + \cos^2 t = 1$$

$$\sin^2 t = 1 - \cos^2 t \qquad \text{Solve for } \sin^2 t$$

$$\sin t = \pm\sqrt{1 - \cos^2 t} \qquad \text{Take square roots}$$

Since $\sin t$ is negative in Quadrant III, the negative sign applies here. Thus,

$$\tan t = \frac{\sin t}{\cos t} = \frac{-\sqrt{1 - \cos^2 t}}{\cos t}$$

✎. **NOW TRY EXERCISE 55**

2.2 EXERCISES

CONCEPTS

1. Let $P(x, y)$ be the terminal point on the unit circle determined by t. Then $\sin t =$ _____, $\cos t =$ _____, and $\tan t =$ _____.

2. If $P(x, y)$ is on the unit circle, then $x^2 + y^2 =$ _____. So for all t we have $\sin^2 t + \cos^2 t =$ _____.

SKILLS

3–4 ■ Find $\sin t$ and $\cos t$ for the values of t whose terminal points are shown on the unit circle in the figure. In Exercise 3, t increases in increments of $\pi/4$; in Exercise 4, t increases in increments of $\pi/6$. (See Exercises 21 and 22 in Section 2.1.)

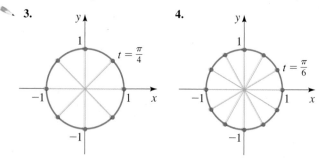

5–24 ■ Find the exact value of the trigonometric function at the given real number.

5. (a) $\sin\dfrac{2\pi}{3}$ (b) $\cos\dfrac{2\pi}{3}$ (c) $\tan\dfrac{2\pi}{3}$

6. (a) $\sin\dfrac{5\pi}{6}$ (b) $\cos\dfrac{5\pi}{6}$ (c) $\tan\dfrac{5\pi}{6}$

7. (a) $\sin\dfrac{7\pi}{6}$ (b) $\sin\left(-\dfrac{\pi}{6}\right)$ (c) $\sin\dfrac{11\pi}{6}$

8. (a) $\cos\dfrac{5\pi}{3}$ (b) $\cos\left(-\dfrac{5\pi}{3}\right)$ (c) $\cos\dfrac{7\pi}{3}$

9. (a) $\cos\dfrac{3\pi}{4}$ (b) $\cos\dfrac{5\pi}{4}$ (c) $\cos\dfrac{7\pi}{4}$

10. (a) $\sin\dfrac{3\pi}{4}$ (b) $\sin\dfrac{5\pi}{4}$ (c) $\sin\dfrac{7\pi}{4}$

11. (a) $\sin\dfrac{7\pi}{3}$ (b) $\csc\dfrac{7\pi}{3}$ (c) $\cot\dfrac{7\pi}{3}$

12. (a) $\cos\left(-\dfrac{\pi}{3}\right)$ (b) $\sec\left(-\dfrac{\pi}{3}\right)$ (c) $\tan\left(-\dfrac{\pi}{3}\right)$

13. (a) $\sin\left(-\dfrac{\pi}{2}\right)$ (b) $\cos\left(-\dfrac{\pi}{2}\right)$ (c) $\cot\left(-\dfrac{\pi}{2}\right)$

14. (a) $\sin\left(-\dfrac{3\pi}{2}\right)$ (b) $\cos\left(-\dfrac{3\pi}{2}\right)$ (c) $\cot\left(-\dfrac{3\pi}{2}\right)$

15. (a) $\sec\dfrac{11\pi}{3}$ (b) $\csc\dfrac{11\pi}{3}$ (c) $\sec\left(-\dfrac{\pi}{3}\right)$

16. (a) $\cos\dfrac{7\pi}{6}$ (b) $\sec\dfrac{7\pi}{6}$ (c) $\csc\dfrac{7\pi}{6}$

17. (a) $\tan\dfrac{5\pi}{6}$ (b) $\tan\dfrac{7\pi}{6}$ (c) $\tan\dfrac{11\pi}{6}$

18. (a) $\cot\left(-\dfrac{\pi}{3}\right)$ (b) $\cot\dfrac{2\pi}{3}$ (c) $\cot\dfrac{5\pi}{3}$

19. (a) $\cos\left(-\dfrac{\pi}{4}\right)$ (b) $\csc\left(-\dfrac{\pi}{4}\right)$ (c) $\cot\left(-\dfrac{\pi}{4}\right)$

20. (a) $\sin\dfrac{5\pi}{4}$ (b) $\sec\dfrac{5\pi}{4}$ (c) $\tan\dfrac{5\pi}{4}$

21. (a) $\csc\left(-\dfrac{\pi}{2}\right)$ (b) $\csc\dfrac{\pi}{2}$ (c) $\csc\dfrac{3\pi}{2}$

22. (a) $\sec(-\pi)$ (b) $\sec\pi$ (c) $\sec 4\pi$

23. (a) $\sin 13\pi$ (b) $\cos 14\pi$ (c) $\tan 15\pi$

24. (a) $\sin\dfrac{25\pi}{2}$ (b) $\cos\dfrac{25\pi}{2}$ (c) $\cot\dfrac{25\pi}{2}$

25–28 ■ Find the value of each of the six trigonometric functions (if it is defined) at the given real number t. Use your answers to complete the table.

25. $t = 0$ **26.** $t = \dfrac{\pi}{2}$ **27.** $t = \pi$ **28.** $t = \dfrac{3\pi}{2}$

t	$\sin t$	$\cos t$	$\tan t$	$\csc t$	$\sec t$	$\cot t$
0	0	1		undefined		
$\dfrac{\pi}{2}$						
π		0			undefined	
$\dfrac{3\pi}{2}$						

29–38 ■ The terminal point $P(x, y)$ determined by a real number t is given. Find $\sin t$, $\cos t$, and $\tan t$.

29. $\left(\dfrac{3}{5}, \dfrac{4}{5}\right)$ **30.** $\left(-\dfrac{3}{5}, \dfrac{4}{5}\right)$

31. $\left(\dfrac{\sqrt{5}}{4}, -\dfrac{\sqrt{11}}{4}\right)$ **32.** $\left(-\dfrac{1}{3}, -\dfrac{2\sqrt{2}}{3}\right)$

33. $\left(-\dfrac{6}{7}, \dfrac{\sqrt{13}}{7}\right)$ **34.** $\left(\dfrac{40}{41}, \dfrac{9}{41}\right)$

35. $\left(-\dfrac{5}{13}, -\dfrac{12}{13}\right)$ **36.** $\left(\dfrac{\sqrt{5}}{5}, \dfrac{2\sqrt{5}}{5}\right)$

37. $\left(-\dfrac{20}{29}, \dfrac{21}{29}\right)$ **38.** $\left(\dfrac{24}{25}, -\dfrac{7}{25}\right)$

39–46 ■ Find an approximate value of the given trigonometric function by using **(a)** the figure and **(b)** a calculator. Compare the two values.

39. $\sin 1$

40. $\cos 0.8$

41. $\sin 1.2$

42. $\cos 5$

43. $\tan 0.8$

44. $\tan(-1.3)$

45. $\cos 4.1$

46. $\sin(-5.2)$

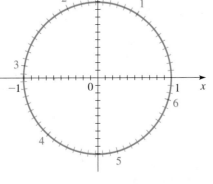

47–50 ■ Find the sign of the expression if the terminal point determined by t is in the given quadrant.

47. $\sin t \cos t$,   Quadrant II   **48.** $\tan t \sec t$,   Quadrant IV

49. $\dfrac{\tan t \sin t}{\cot t}$,   Quadrant III   **50.** $\cos t \sec t$,   any quadrant

51–54 ■ From the information given, find the quadrant in which the terminal point determined by t lies.

51. $\sin t > 0$ and $\cos t < 0$ **52.** $\tan t > 0$ and $\sin t < 0$

53. $\csc t > 0$ and $\sec t < 0$ **54.** $\cos t < 0$ and $\cot t < 0$

55–64 ■ Write the first expression in terms of the second if the terminal point determined by t is in the given quadrant.

55. $\sin t$, $\cos t$;   Quadrant II   **56.** $\cos t$, $\sin t$;   Quadrant IV

57. $\tan t$, $\sin t$;   Quadrant IV   **58.** $\tan t$, $\cos t$;   Quadrant III

59. $\sec t$, $\tan t$;   Quadrant II   **60.** $\csc t$, $\cot t$;   Quadrant III

61. $\tan t$, $\sec t$;   Quadrant III   **62.** $\sin t$, $\sec t$;   Quadrant IV

63. $\tan^2 t$, $\sin t$;   any quadrant

64. $\sec^2 t \sin^2 t$, $\cos t$;   any quadrant

65–72 ■ Find the values of the trigonometric functions of t from the given information.

65. $\sin t = \tfrac{3}{5}$,   terminal point of t is in Quadrant II

66. $\cos t = -\tfrac{4}{5}$,   terminal point of t is in Quadrant III

67. $\sec t = 3$,   terminal point of t is in Quadrant IV

68. $\tan t = \tfrac{1}{4}$,   terminal point of t is in Quadrant III

69. $\tan t = -\tfrac{3}{4}$, $\cos t > 0$

70. $\sec t = 2$, $\sin t < 0$

71. $\sin t = -\tfrac{1}{4}$, $\sec t < 0$

72. $\tan t = -4$, $\csc t > 0$

73–80 ■ Determine whether the function is even, odd, or neither.

73. $f(x) = x^2 \sin x$ **74.** $f(x) = x^2 \cos 2x$

75. $f(x) = \sin x \cos x$ **76.** $f(x) = \sin x + \cos x$

77. $f(x) = |x| \cos x$ **78.** $f(x) = x \sin^3 x$

79. $f(x) = x^3 + \cos x$ **80.** $f(x) = \cos(\sin x)$

APPLICATIONS

81. Harmonic Motion   The displacement from equilibrium of an oscillating mass attached to a spring is given by $y(t) = 4 \cos 3\pi t$ where y is measured in inches and t in seconds. Find the displacement at the times indicated in the table.

t	$y(t)$
0	
0.25	
0.50	
0.75	
1.00	
1.25	

$y > 0$

Equilibrium, $y = 0$

$y < 0$

82. Circadian Rhythms   Everybody's blood pressure varies over the course of the day. In a certain individual the resting diastolic blood pressure at time t is given by $B(t) = 80 + 7 \sin(\pi t/12)$, where t is measured in hours since midnight and $B(t)$ in mmHg (millimeters of mercury). Find this person's diastolic blood pressure at
(a) 6:00 A.M.   **(b)** 10:30 A.M.   **(c)** Noon   **(d)** 8:00 P.M.

83. Electric Circuit   After the switch is closed in the circuit shown, the current t seconds later is $I(t) = 0.8e^{-3t}\sin 10t$. Find the current at the times
(a) $t = 0.1$ s and **(b)** $t = 0.5$ s.

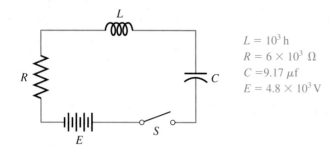

$$L = 10^3 \, h$$
$$R = 6 \times 10^3 \, \Omega$$
$$C = 9.17 \, \mu f$$
$$E = 4.8 \times 10^3 \, V$$

84. Bungee Jumping   A bungee jumper plummets from a high bridge to the river below and then bounces back over and over again. At time t seconds after her jump, her height H (in meters) above the river is given by $H(t) = 100 + 75 \cdot 2^{-t/20}\cos\left(\frac{\pi}{4}t\right)$. Find her height at the times indicated in the table.

t	$H(t)$
0	
1	
2	
4	
6	
8	
12	

DISCOVERY ▪ DISCUSSION ▪ WRITING

85. Reduction Formulas   A *reduction formula* is one that can be used to "reduce" the number of terms in the input for a

trigonometric function. Explain how the figure shows that the following reduction formulas are valid:

$$\sin(t + \pi) = -\sin t \qquad \cos(t + \pi) = -\cos t$$

$$\tan(t + \pi) = \tan t$$

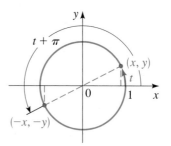

86. More Reduction Formulas   By the "Angle-Side-Angle" theorem from elementary geometry, triangles CDO and AOB in the figure are congruent. Explain how this proves that if B has coordinates (x, y), then D has coordinates $(-y, x)$. Then explain how the figure shows that the following reduction formulas are valid:

$$\sin\left(t + \frac{\pi}{2}\right) = \cos t$$

$$\cos\left(t + \frac{\pi}{2}\right) = -\sin t$$

$$\tan\left(t + \frac{\pi}{2}\right) = -\cot t$$

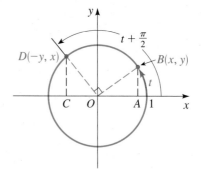

2.3 TRIGONOMETRIC GRAPHS

| Graphs of Sine and Cosine ▶ Graphs of Transformations of Sine and Cosine ▶ Using Graphing Devices to Graph Trigonometric Functions

The graph of a function gives us a better idea of its behavior. So, in this section we graph the sine and cosine functions and certain transformations of these functions. The other trigonometric functions are graphed in the next section.

▼ Graphs of Sine and Cosine

To help us graph the sine and cosine functions, we first observe that these functions repeat their values in a regular fashion. To see exactly how this happens, recall that the circumference of the unit circle is 2π. It follows that the terminal point $P(x, y)$ determined by the real number t is the same as that determined by $t + 2\pi$. Since the sine and cosine functions are defined in terms of the coordinates of $P(x, y)$, it follows that their values are unchanged by the addition of any integer multiple of 2π. In other words,

$$\sin(t + 2n\pi) = \sin t \qquad \text{for any integer } n$$

$$\cos(t + 2n\pi) = \cos t \qquad \text{for any integer } n$$

Thus, the sine and cosine functions are *periodic* according to the following definition: A function f is **periodic** if there is a positive number p such that $f(t + p) = f(t)$ for every t. The least such positive number (if it exists) is the **period** of f. If f has period p, then the graph of f on any interval of length p is called **one complete period** of f.

PERIODIC PROPERTIES OF SINE AND COSINE

The functions sine and cosine have period 2π:

$$\sin(t + 2\pi) = \sin t \qquad\qquad \cos(t + 2\pi) = \cos t$$

TABLE 1

t	$\sin t$	$\cos t$
$0 \to \dfrac{\pi}{2}$	$0 \to 1$	$1 \to 0$
$\dfrac{\pi}{2} \to \pi$	$1 \to 0$	$0 \to -1$
$\pi \to \dfrac{3\pi}{2}$	$0 \to -1$	$-1 \to 0$
$\dfrac{3\pi}{2} \to 2\pi$	$-1 \to 0$	$0 \to 1$

So the sine and cosine functions repeat their values in any interval of length 2π. To sketch their graphs, we first graph one period. To sketch the graphs on the interval $0 \le t \le 2\pi$, we could try to make a table of values and use those points to draw the graph. Since no such table can be complete, let's look more closely at the definitions of these functions.

Recall that $\sin t$ is the y-coordinate of the terminal point $P(x, y)$ on the unit circle determined by the real number t. How does the y-coordinate of this point vary as t increases? It's easy to see that the y-coordinate of $P(x, y)$ increases to 1, then decreases to -1 repeatedly as the point $P(x, y)$ travels around the unit circle. (See Figure 1.) In fact, as t increases from 0 to $\pi/2$, $y = \sin t$ increases from 0 to 1. As t increases from $\pi/2$ to π, the value of $y = \sin t$ decreases from 1 to 0. Table 1 shows the variation of the sine and cosine functions for t between 0 and 2π.

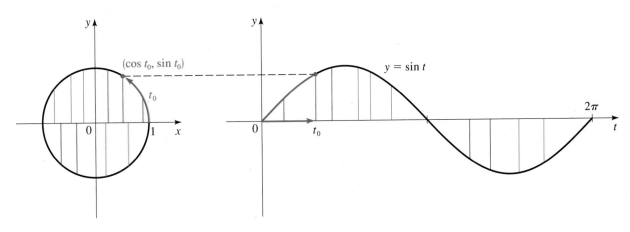

FIGURE 1

To draw the graphs more accurately, we find a few other values of $\sin t$ and $\cos t$ in Table 2. We could find still other values with the aid of a calculator.

TABLE 2

t	0	$\dfrac{\pi}{6}$	$\dfrac{\pi}{3}$	$\dfrac{\pi}{2}$	$\dfrac{2\pi}{3}$	$\dfrac{5\pi}{6}$	π	$\dfrac{7\pi}{6}$	$\dfrac{4\pi}{3}$	$\dfrac{3\pi}{2}$	$\dfrac{5\pi}{3}$	$\dfrac{11\pi}{6}$	2π
$\sin t$	0	$\dfrac{1}{2}$	$\dfrac{\sqrt{3}}{2}$	1	$\dfrac{\sqrt{3}}{2}$	$\dfrac{1}{2}$	0	$-\dfrac{1}{2}$	$-\dfrac{\sqrt{3}}{2}$	-1	$-\dfrac{\sqrt{3}}{2}$	$-\dfrac{1}{2}$	0
$\cos t$	1	$\dfrac{\sqrt{3}}{2}$	$\dfrac{1}{2}$	0	$-\dfrac{1}{2}$	$-\dfrac{\sqrt{3}}{2}$	-1	$-\dfrac{\sqrt{3}}{2}$	$-\dfrac{1}{2}$	0	$\dfrac{1}{2}$	$\dfrac{\sqrt{3}}{2}$	1

Now we use this information to graph the functions sin t and cos t for t between 0 and 2π in Figures 2 and 3. These are the graphs of one period. Using the fact that these functions are periodic with period 2π, we get their complete graphs by continuing the same pattern to the left and to the right in every successive interval of length 2π.

The graph of the sine function is symmetric with respect to the origin. This is as expected, since sine is an odd function. Since the cosine function is an even function, its graph is symmetric with respect to the y-axis.

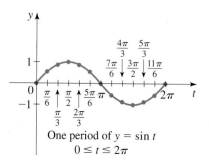

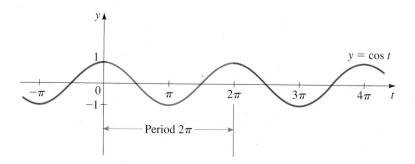

FIGURE 2 Graph of sin t

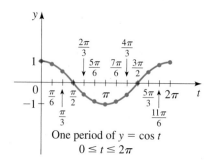

FIGURE 3 Graph of cos t

▼ Graphs of Transformations of Sine and Cosine

We now consider graphs of functions that are transformations of the sine and cosine functions. Thus, the graphing techniques of Section 1.6 are very useful here. The graphs we obtain are important for understanding applications to physical situations such as harmonic motion (see Section 2.6), but some of them are beautiful graphs that are interesting in their own right.

It's traditional to use the letter x to denote the variable in the domain of a function. So from here on we use the letter x and write $y = \sin x$, $y = \cos x$, $y = \tan x$, and so on to denote these functions.

EXAMPLE 1 | Cosine Curves

Sketch the graph of each function.

(a) $f(x) = 2 + \cos x$ **(b)** $g(x) = -\cos x$

SOLUTION

(a) The graph of $y = 2 + \cos x$ is the same as the graph of $y = \cos x$, but shifted up 2 units (see Figure 4(a)).

(b) The graph of $y = -\cos x$ in Figure 4(b) is the reflection of the graph of $y = \cos x$ in the x-axis.

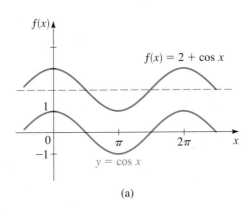

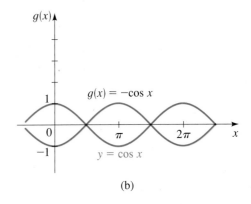

FIGURE 4   (a)   (b)

✎. NOW TRY EXERCISES **3** AND **5**

Let's graph $y = 2 \sin x$. We start with the graph of $y = \sin x$ and multiply the y-coordinate of each point by 2. This has the effect of stretching the graph vertically by a factor of 2. To graph $y = \frac{1}{2} \sin x$, we start with the graph of $y = \sin x$ and multiply the y-coordinate of each point by $\frac{1}{2}$. This has the effect of shrinking the graph vertically by a factor of $\frac{1}{2}$ (see Figure 5).

Vertical stretching and shrinking of graphs is discussed in Section 1.6.

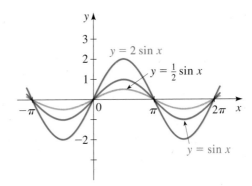

FIGURE 5

In general, for the functions

$$y = a \sin x \qquad \text{and} \qquad y = a \cos x$$

the number $|a|$ is called the **amplitude** and is the largest value these functions attain. Graphs of $y = a \sin x$ for several values of a are shown in Figure 6.

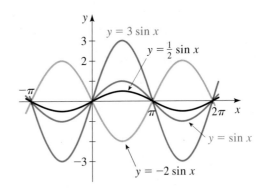

FIGURE 6

EXAMPLE 2 | Stretching a Cosine Curve

Find the amplitude of $y = -3 \cos x$, and sketch its graph.

SOLUTION The amplitude is $|-3| = 3$, so the largest value the graph attains is 3 and the smallest value is -3. To sketch the graph, we begin with the graph of $y = \cos x$, stretch the graph vertically by a factor of 3, and reflect in the x-axis, arriving at the graph in Figure 7.

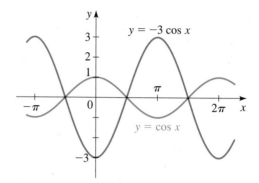

FIGURE 7

✎. NOW TRY EXERCISE **9**

Since the sine and cosine functions have period 2π, the functions

$$y = a \sin kx \qquad \text{and} \qquad y = a \cos kx \qquad (k > 0)$$

complete one period as kx varies from 0 to 2π, that is, for $0 \le kx \le 2\pi$ or for $0 \le x \le 2\pi/k$. So these functions complete one period as x varies between 0 and $2\pi/k$ and thus have period $2\pi/k$. The graphs of these functions are called **sine curves** and **cosine curves**, respectively. (Collectively, sine and cosine curves are often referred to as **sinusoidal** curves.)

SINE AND COSINE CURVES

The sine and cosine curves

$$y = a \sin kx \qquad \text{and} \qquad y = a \cos kx \qquad (k > 0)$$

have **amplitude** $|a|$ and **period** $2\pi/k$.

An appropriate interval on which to graph one complete period is $[0, 2\pi/k]$.

To see how the value of k affects the graph of $y = \sin kx$, let's graph the sine curve $y = \sin 2x$. Since the period is $2\pi/2 = \pi$, the graph completes one period in the interval $0 \le x \le \pi$ (see Figure 8(a)). For the sine curve $y = \sin \frac{1}{2}x$, the period is $2\pi \div \frac{1}{2} = 4\pi$, so the graph completes one period in the interval $0 \le x \le 4\pi$ (see Figure 8(b)). We see that the effect is to *shrink* the graph horizontally if $k > 1$ or to *stretch* the graph horizontally if $k < 1$.

Horizontal stretching and shrinking of graphs is discussed in Section 1.6.

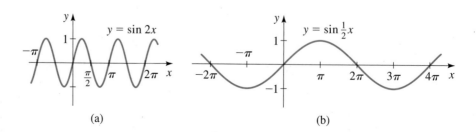

FIGURE 8 (a) (b)

For comparison, in Figure 9 we show the graphs of one period of the sine curve $y = a \sin kx$ for several values of k.

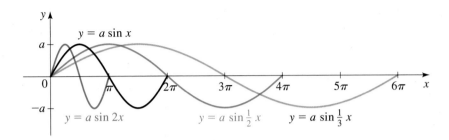

FIGURE 9

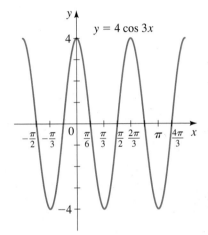

FIGURE 10

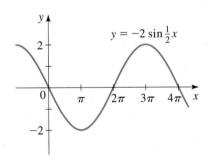

FIGURE 11

EXAMPLE 3 | Amplitude and Period

Find the amplitude and period of each function, and sketch its graph.

(a) $y = 4 \cos 3x$ **(b)** $y = -2 \sin \frac{1}{2}x$

SOLUTION

(a) We get the amplitude and period from the form of the function as follows:

$$\text{amplitude} = |a| = 4$$

$$y = 4 \cos 3x$$

$$\text{period} = \frac{2\pi}{k} = \frac{2\pi}{3}$$

The amplitude is 4 and the period is $2\pi/3$. The graph is shown in Figure 10.

(b) For $y = -2 \sin \frac{1}{2}x$,

$$\text{amplitude} = |a| = |-2| = 2$$

$$\text{period} = \frac{2\pi}{\frac{1}{2}} = 4\pi$$

The graph is shown in Figure 11.

◆ NOW TRY EXERCISES **19** AND **21**

The graphs of functions of the form $y = a \sin k(x - b)$ and $y = a \cos k(x - b)$ are simply sine and cosine curves shifted horizontally by an amount $|b|$. They are shifted to the right if $b > 0$ or to the left if $b < 0$. The number b is the *phase shift*. We summarize the properties of these functions in the following box.

SHIFTED SINE AND COSINE CURVES

The sine and cosine curves

$$y = a \sin k(x - b) \quad \text{and} \quad y = a \cos k(x - b) \quad (k > 0)$$

have **amplitude** $|a|$, **period** $2\pi/k$, and **phase shift** b.

An appropriate interval on which to graph one complete period is $[b, b + (2\pi/k)]$.

The graphs of $y = \sin\left(x - \dfrac{\pi}{3}\right)$ and $y = \sin\left(x + \dfrac{\pi}{6}\right)$ are shown in Figure 12.

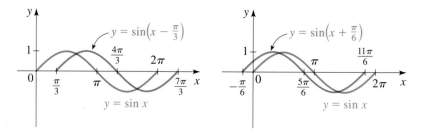

FIGURE 12

EXAMPLE 4 | A Shifted Sine Curve

Find the amplitude, period, and phase shift of $y = 3\sin 2\left(x - \dfrac{\pi}{4}\right)$, and graph one complete period.

SOLUTION We get the amplitude, period, and phase shift from the form of the function as follows:

$$\text{amplitude} = |a| = 3 \qquad \text{period} = \frac{2\pi}{k} = \frac{2\pi}{2} = \pi$$

$$y = 3\sin 2\left(x - \frac{\pi}{4}\right)$$

$$\text{phase shift} = \frac{\pi}{4} \text{ (to the right)}$$

Since the phase shift is $\pi/4$ and the period is π, one complete period occurs on the interval

$$\left[\frac{\pi}{4}, \frac{\pi}{4} + \pi\right] = \left[\frac{\pi}{4}, \frac{5\pi}{4}\right]$$

As an aid in sketching the graph, we divide this interval into four equal parts, then graph a sine curve with amplitude 3 as in Figure 13.

Here is another way to find an appropriate interval on which to graph one complete period. Since the period of $y = \sin x$ is 2π, the function $y = 3\sin 2(x - \frac{\pi}{4})$ will go through one complete period as $2(x - \frac{\pi}{4})$ varies from 0 to 2π.

Start of period: End of period:

$2(x - \frac{\pi}{4}) = 0$ $2(x - \frac{\pi}{4}) = 2\pi$

$x - \frac{\pi}{4} = 0$ $x - \frac{\pi}{4} = \pi$

$x = \frac{\pi}{4}$ $x = \frac{5\pi}{4}$

So we graph one period on the interval $\left[\frac{\pi}{4}, \frac{5\pi}{4}\right]$.

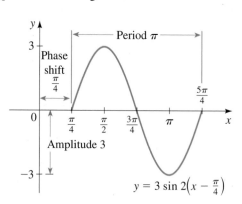

FIGURE 13 $y = 3\sin 2\left(x - \frac{\pi}{4}\right)$

✎ NOW TRY EXERCISE **33**

EXAMPLE 5 | A Shifted Cosine Curve

Find the amplitude, period, and phase shift of

$$y = \frac{3}{4}\cos\left(2x + \frac{2\pi}{3}\right)$$

and graph one complete period.

SOLUTION   We first write this function in the form $y = a \cos k(x - b)$. To do this, we factor 2 from the expression $2x + \dfrac{2\pi}{3}$ to get

$$y = \frac{3}{4} \cos 2\left[x - \left(-\frac{\pi}{3}\right)\right]$$

Thus we have

$$\text{amplitude} = |a| = \frac{3}{4}$$

$$\text{period} = \frac{2\pi}{k} = \frac{2\pi}{2} = \pi$$

$$\text{phase shift} = b = -\frac{\pi}{3} \qquad \text{Shift } \frac{\pi}{3} \text{ to the } left$$

We can also find one complete period as follows:

Start of period:

$2x + \frac{2\pi}{3} = 0$

$2x = -\frac{2\pi}{3}$

$x = -\frac{\pi}{3}$

End of period:

$2x + \frac{2\pi}{3} = 2\pi$

$2x = \frac{4\pi}{3}$

$x = \frac{2\pi}{3}$

So we graph one period on the interval $\left[-\frac{\pi}{3}, \frac{2\pi}{3}\right]$.

From this information it follows that one period of this cosine curve begins at $-\pi/3$ and ends at $(-\pi/3) + \pi = 2\pi/3$. To sketch the graph over the interval $\left[-\pi/3, 2\pi/3\right]$, we divide this interval into four equal parts and graph a cosine curve with amplitude $\frac{3}{4}$ as shown in Figure 14.

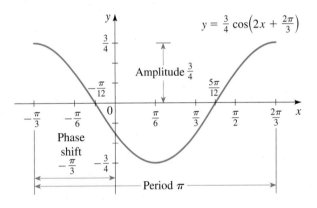

FIGURE 14

◥ NOW TRY EXERCISE **35**

▼ Using Graphing Devices to Graph Trigonometric Functions

When using a graphing calculator or a computer to graph a function, it is important to choose the viewing rectangle carefully in order to produce a reasonable graph of the function. This is especially true for trigonometric functions; the next example shows that, if care is not taken, it's easy to produce a very misleading graph of a trigonometric function.

See Appendix C.1 for guidelines on choosing an appropriate viewing rectangle.

EXAMPLE 6 | Choosing the Viewing Rectangle

Graph the function $f(x) = \sin 50x$ in an appropriate viewing rectangle.

SOLUTION   Figure 15(a) shows the graph of f produced by a graphing calculator using the viewing rectangle $[-12, 12]$ by $[-1.5, 1.5]$. At first glance the graph appears to be reasonable. But if we change the viewing rectangle to the ones shown in Figure 15, the graphs look very different. Something strange is happening.

The appearance of the graphs in Figure 15 depends on the machine used. The graphs you get with your own graphing device might not look like these figures, but they will also be quite inaccurate.

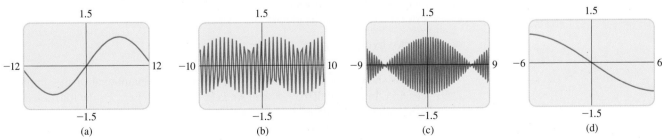

(a)   (b)   (c)   (d)

FIGURE 15   Graphs of $f(x) = \sin 50x$ in different viewing rectangles

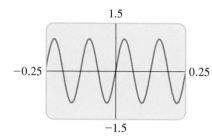

FIGURE 16 $f(x) = \sin 50x$

To explain the big differences in appearance of these graphs and to find an appropriate viewing rectangle, we need to find the period of the function $y = \sin 50x$:

$$\text{period} = \frac{2\pi}{50} = \frac{\pi}{25} \approx 0.126$$

This suggests that we should deal only with small values of x in order to show just a few oscillations of the graph. If we choose the viewing rectangle $[-0.25, 0.25]$ by $[-1.5, 1.5]$, we get the graph shown in Figure 16.

Now we see what went wrong in Figure 15. The oscillations of $y = \sin 50x$ are so rapid that when the calculator plots points and joins them, it misses most of the maximum and minimum points and therefore gives a very misleading impression of the graph.

✎ NOW TRY EXERCISE **51**

The function h in Example 7 is **periodic** with period 2π. In general, functions that are sums of functions from the following list are periodic:

$$1, \cos kx, \cos 2kx, \cos 3kx, \dots$$

$$\sin kx, \sin 2kx, \sin 3kx, \dots$$

Although these functions appear to be special, they are actually fundamental to describing all periodic functions that arise in practice. The French mathematician J. B. J. Fourier (see page 233) discovered that nearly every periodic function can be written as a sum (usually an infinite sum) of these functions. This is remarkable because it means that any situation in which periodic variation occurs can be described mathematically using the functions sine and cosine. A modern application of Fourier's discovery is the digital encoding of sound on compact discs.

EXAMPLE 7 | A Sum of Sine and Cosine Curves

Graph $f(x) = 2 \cos x$, $g(x) = \sin 2x$, and $h(x) = 2 \cos x + \sin 2x$ on a common screen to illustrate the method of graphical addition.

SOLUTION Notice that $h = f + g$, so its graph is obtained by adding the corresponding y-coordinates of the graphs of f and g. The graphs of f, g, and h are shown in Figure 17.

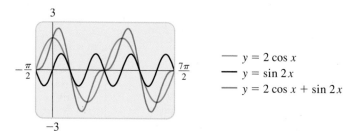

FIGURE 17

✎ NOW TRY EXERCISE **59**

EXAMPLE 8 | A Cosine Curve with Variable Amplitude

Graph the functions $y = x^2$, $y = -x^2$, and $y = x^2 \cos 6\pi x$ on a common screen. Comment on and explain the relationship among the graphs.

SOLUTION Figure 18 shows all three graphs in the viewing rectangle $[-1.5, 1.5]$ by $[-2, 2]$. It appears that the graph of $y = x^2 \cos 6\pi x$ lies between the graphs of the functions $y = x^2$ and $y = -x^2$.

To understand this, recall that the values of $\cos 6\pi x$ lie between -1 and 1, that is,

$$-1 \le \cos 6\pi x \le 1$$

for all values of x. Multiplying the inequalities by x^2 and noting that $x^2 \ge 0$, we get

$$-x^2 \le x^2 \cos 6\pi x \le x^2$$

This explains why the functions $y = x^2$ and $y = -x^2$ form a boundary for the graph of $y = x^2 \cos 6\pi x$. (Note that the graphs touch when $\cos 6\pi x = \pm 1$.)

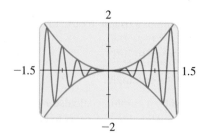

FIGURE 18 $y = x^2 \cos 6\pi x$

✎ NOW TRY EXERCISE **63**

Example 8 shows that the function $y = x^2$ controls the amplitude of the graph of $y = x^2 \cos 6\pi x$. In general, if $f(x) = a(x) \sin kx$ or $f(x) = a(x) \cos kx$, the function a determines how the amplitude of f varies, and the graph of f lies between the graphs of $y = -a(x)$ and $y = a(x)$. Here is another example.

AM and FM Radio

Radio transmissions consist of sound waves superimposed on a harmonic electromagnetic wave form called the **carrier signal.**

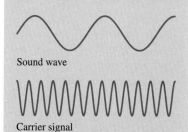

Sound wave

Carrier signal

There are two types of radio transmission, called **amplitude modulation (AM)** and **frequency modulation (FM).** In AM broadcasting, the sound wave changes, or **modulates,** the amplitude of the carrier, but the frequency remains unchanged.

AM signal

In FM broadcasting, the sound wave modulates the frequency, but the amplitude remains the same.

FM signal

EXAMPLE 9 | A Cosine Curve with Variable Amplitude

Graph the function $f(x) = \cos 2\pi x \cos 16\pi x$.

SOLUTION The graph is shown in Figure 19. Although it was drawn by a computer, we could have drawn it by hand, by first sketching the boundary curves $y = \cos 2\pi x$ and $y = -\cos 2\pi x$. The graph of f is a cosine curve that lies between the graphs of these two functions.

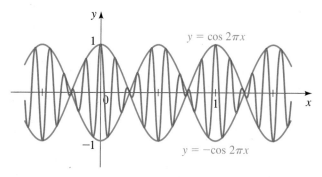

FIGURE 19 $f(x) = \cos 2\pi x \cos 16\pi x$

✎ NOW TRY EXERCISE **65**

EXAMPLE 10 | A Sine Curve with Decaying Amplitude

The function $f(x) = \dfrac{\sin x}{x}$ is important in calculus. Graph this function and comment on its behavior when x is close to 0.

SOLUTION The viewing rectangle $[-15, 15]$ by $[-0.5, 1.5]$ shown in Figure 20(a) gives a good global view of the graph of f. The viewing rectangle $[-1, 1]$ by $[-0.5, 1.5]$ in Figure 20(b) focuses on the behavior of f when $x \approx 0$. Notice that although $f(x)$ is not defined when $x = 0$ (in other words, 0 is not in the domain of f), the values of f seem to approach 1 when x gets close to 0. This fact is crucial in calculus.

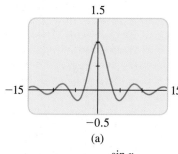
(a)

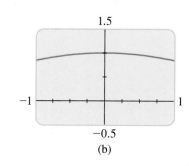
(b)

FIGURE 20 $f(x) = \dfrac{\sin x}{x}$

✎ NOW TRY EXERCISE **75**

The function in Example 10 can be written as

$$f(x) = \frac{1}{x} \sin x$$

and may thus be viewed as a sine function whose amplitude is controlled by the function $a(x) = 1/x$.

2.3 EXERCISES

CONCEPTS

1. The trigonometric functions $y = \sin x$ and $y = \cos x$ have

amplitude _____ and period _____. Sketch a graph of
each function on the interval $[0, 2\pi]$.

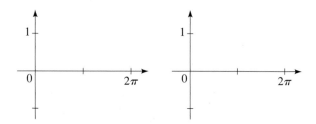

2. The trigonometric function $y = 3 \sin 2x$ has amplitude

_____ and period _____.

SKILLS

3–16 ■ Graph the function.

   **3.** $f(x) = 1 + \cos x$ **4.** $f(x) = 3 + \sin x$

5. $f(x) = -\sin x$ **6.** $f(x) = 2 - \cos x$

7. $f(x) = -2 + \sin x$ **8.** $f(x) = -1 + \cos x$

9. $g(x) = 3 \cos x$ **10.** $g(x) = 2 \sin x$

11. $g(x) = -\frac{1}{2} \sin x$ **12.** $g(x) = -\frac{2}{3} \cos x$

13. $g(x) = 3 + 3 \cos x$ **14.** $g(x) = 4 - 2 \sin x$

15. $h(x) = |\cos x|$ **16.** $h(x) = |\sin x|$

17–28 ■ Find the amplitude and period of the function, and
sketch its graph.

17. $y = \cos 2x$ **18.** $y = -\sin 2x$

19. $y = -3 \sin 3x$ **20.** $y = \frac{1}{2} \cos 4x$

21. $y = 10 \sin \frac{1}{2}x$ **22.** $y = 5 \cos \frac{1}{4}x$

23. $y = -\frac{1}{3} \cos \frac{1}{3}x$ **24.** $y = 4 \sin(-2x)$

25. $y = -2 \sin 2\pi x$ **26.** $y = -3 \sin \pi x$

27. $y = 1 + \frac{1}{2} \cos \pi x$ **28.** $y = -2 + \cos 4\pi x$

29–42 ■ Find the amplitude, period, and phase shift of the function, and graph one complete period.

29. $y = \cos\left(x - \dfrac{\pi}{2}\right)$ **30.** $y = 2 \sin\left(x - \dfrac{\pi}{3}\right)$

31. $y = -2 \sin\left(x - \dfrac{\pi}{6}\right)$ **32.** $y = 3 \cos\left(x + \dfrac{\pi}{4}\right)$

33. $y = -4 \sin 2\left(x + \dfrac{\pi}{2}\right)$ **34.** $y = \sin \dfrac{1}{2}\left(x + \dfrac{\pi}{4}\right)$

35. $y = 5 \cos\left(3x - \dfrac{\pi}{4}\right)$ **36.** $y = 2 \sin\left(\dfrac{2}{3}x - \dfrac{\pi}{6}\right)$

37. $y = \dfrac{1}{2} - \dfrac{1}{2} \cos\left(2x - \dfrac{\pi}{3}\right)$ **38.** $y = 1 + \cos\left(3x + \dfrac{\pi}{2}\right)$

39. $y = 3 \cos \pi\left(x + \dfrac{1}{2}\right)$ **40.** $y = 3 + 2 \sin 3(x + 1)$

41. $y = \sin(\pi + 3x)$ **42.** $y = \cos\left(\dfrac{\pi}{2} - x\right)$

43–50 ■ The graph of one complete period of a sine or cosine
curve is given.

(a) Find the amplitude, period, and phase shift.

(b) Write an equation that represents the curve in the form

$$y = a \sin k(x - b) \qquad \text{or} \qquad y = a \cos k(x - b)$$

43.

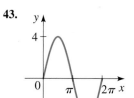

44.

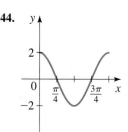

45.

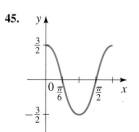

46.

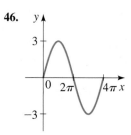

47.

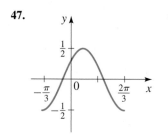

48.

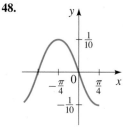

49.

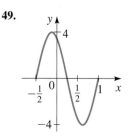

50.

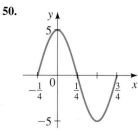

51–58 ■ Determine an appropriate viewing rectangle for each
function, and use it to draw the graph.

51. $f(x) = \cos 100x$ **52.** $f(x) = 3 \sin 120x$

53. $f(x) = \sin(x/40)$ **54.** $f(x) = \cos(x/80)$

55. $y = \tan 25x$ **56.** $y = \csc 40x$

57. $y = \sin^2 20x$ **58.** $y = \sqrt{\tan 10\pi x}$

59–60 ■ Graph f, g, and $f + g$ on a common screen to illustrate graphical addition.

 59. $f(x) = x$, $g(x) = \sin x$

60. $f(x) = \sin x$, $g(x) = \sin 2x$

 61–66 ■ Graph the three functions on a common screen. How are the graphs related?

61. $y = x^2$, $y = -x^2$, $y = x^2 \sin x$

62. $y = x$, $y = -x$, $y = x \cos x$

63. $y = \sqrt{x}$, $y = -\sqrt{x}$, $y = \sqrt{x} \sin 5\pi x$

64. $y = \dfrac{1}{1 + x^2}$, $y = -\dfrac{1}{1 + x^2}$, $y = \dfrac{\cos 2\pi x}{1 + x^2}$

65. $y = \cos 3\pi x$, $y = -\cos 3\pi x$, $y = \cos 3\pi x \cos 21\pi x$

66. $y = \sin 2\pi x$, $y = -\sin 2\pi x$, $y = \sin 2\pi x \sin 10\pi x$

 67–70 ■ Find the maximum and minimum values of the function.

67. $y = \sin x + \sin 2x$

68. $y = x - 2 \sin x, 0 \le x \le 2\pi$

69. $y = 2 \sin x + \sin^2 x$

70. $y = \dfrac{\cos x}{2 + \sin x}$

 71–74 ■ Find all solutions of the equation that lie in the interval $[0, \pi]$. State each answer correct to two decimal places.

71. $\cos x = 0.4$

72. $\tan x = 2$

73. $\csc x = 3$

74. $\cos x = x$

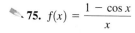 **75–76** ■ A function f is given.

(a) Is f even, odd, or neither?

(b) Find the x-intercepts of the graph of f.

(c) Graph f in an appropriate viewing rectangle.

(d) Describe the behavior of the function as $x \to \pm\infty$.

(e) Notice that $f(x)$ is not defined when $x = 0$. What happens as x approaches 0?

75. $f(x) = \dfrac{1 - \cos x}{x}$

76. $f(x) = \dfrac{\sin 4x}{2x}$

APPLICATIONS

77. Height of a Wave   As a wave passes by an offshore piling, the height of the water is modeled by the function

$$h(t) = 3 \cos\left(\frac{\pi}{10} t\right)$$

where $h(t)$ is the height in feet above mean sea level at time t seconds.

(a) Find the period of the wave.

(b) Find the wave height, that is, the vertical distance between the trough and the crest of the wave.

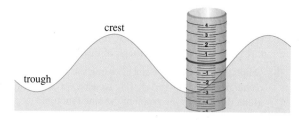

78. Sound Vibrations   A tuning fork is struck, producing a pure tone as its tines vibrate. The vibrations are modeled by the function

$$v(t) = 0.7 \sin(880\pi t)$$

where $v(t)$ is the displacement of the tines in millimeters at time t seconds.

(a) Find the period of the vibration.

(b) Find the frequency of the vibration, that is, the number of times the fork vibrates per second.

(c) Graph the function v.

79. Blood Pressure   Each time your heart beats, your blood pressure first increases and then decreases as the heart rests between beats. The maximum and minimum blood pressures are called the *systolic* and *diastolic* pressures, respectively. Your *blood pressure reading* is written as systolic/diastolic. A reading of 120/80 is considered normal.

A certain person's blood pressure is modeled by the function

$$p(t) = 115 + 25 \sin(160\pi t)$$

where $p(t)$ is the pressure in mmHg (millimeters of mercury), at time t measured in minutes.

(a) Find the period of p.

(b) Find the number of heartbeats per minute.

(c) Graph the function p.

(d) Find the blood pressure reading. How does this compare to normal blood pressure?

80. Variable Stars   Variable stars are ones whose brightness varies periodically. One of the most visible is R Leonis; its brightness is modeled by the function

$$b(t) = 7.9 - 2.1 \cos\left(\frac{\pi}{156} t\right)$$

where t is measured in days.

(a) Find the period of R Leonis.

(b) Find the maximum and minimum brightness.

(c) Graph the function b.

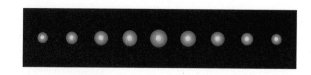

DISCOVERY ■ DISCUSSION ■ WRITING

81. Compositions Involving Trigonometric Functions
This exercise explores the effect of the inner function g on a composite function $y = f(g(x))$.

(a) Graph the function $y = \sin \sqrt{x}$ using the viewing rectangle $[0, 400]$ by $[-1.5, 1.5]$. In what ways does this graph differ from the graph of the sine function?

(b) Graph the function $y = \sin(x^2)$ using the viewing rectangle $[-5, 5]$ by $[-1.5, 1.5]$. In what ways does this graph differ from the graph of the sine function?

82. Periodic Functions I Recall that a function f is *periodic* if there is a positive number p such that $f(t + p) = f(t)$ for every t, and the least such p (if it exists) is the *period* of f. The graph of a function of period p looks the same on each interval of length p, so we can easily determine the period from the graph. Determine whether the function whose graph is shown is periodic; if it is periodic, find the period.

(a)

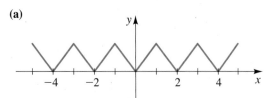

(b)

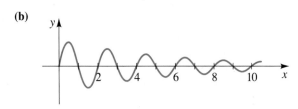

(c)

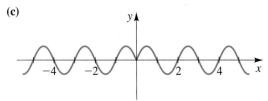

(d)

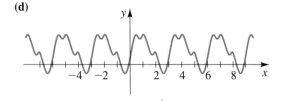

 83. Periodic Functions II Use a graphing device to graph the following functions. From the graph, determine whether the function is periodic; if it is periodic, find the period. (See page 42 for the definition of $[\![x]\!]$.)

(a) $y = |\sin x|$

(b) $y = \sin |x|$

(c) $y = 2^{\cos x}$

(d) $y = x - [\![x]\!]$

(e) $y = \cos(\sin x)$

(f) $y = \cos(x^2)$

84. Sinusoidal Curves The graph of $y = \sin x$ is the same as the graph of $y = \cos x$ shifted to the right $\pi/2$ units. So the sine curve $y = \sin x$ is also at the same time a cosine curve: $y = \cos(x - \pi/2)$. In fact, any sine curve is also a cosine curve with a different phase shift, and any cosine curve is also a sine curve. Sine and cosine curves are collectively referred to as *sinusoidal*. For the curve whose graph is shown, find all possible ways of expressing it as a sine curve $y = a \sin(x - b)$ or as a cosine curve $y = a \cos(x - b)$. Explain why you think you have found all possible choices for a and b in each case.

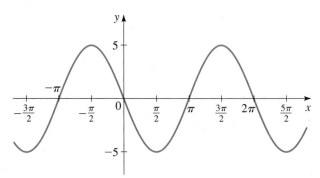

DISCOVERY PROJECT | **Predator/Prey Models**

In this project we explore the use of sine functions in modeling the population of a predator and its prey. You can find the project at the book companion website: **www.stewartmath.com**

2.4 MORE TRIGONOMETRIC GRAPHS

Graphs of Tangent, Cotangent, Secant, and Cosecant ▶ Graphs of Transformations of Tangent and Cotangent ▶ Graphs of Transformations of Cosecant and Secant

In this section we graph the tangent, cotangent, secant, and cosecant functions and transformations of these functions.

▼ Graphs of Tangent, Cotangent, Secant, and Cosecant

We begin by stating the periodic properties of these functions. Recall that sine and cosine have period 2π. Since cosecant and secant are the reciprocals of sine and cosine, respectively, they also have period 2π (see Exercise 55). Tangent and cotangent, however, have period π (see Exercise 85 of Section 2.2).

PERIODIC PROPERTIES

The functions tangent and cotangent have period π:

$$\tan(x + \pi) = \tan x \qquad \cot(x + \pi) = \cot x$$

The functions cosecant and secant have period 2π:

$$\csc(x + 2\pi) = \csc x \qquad \sec(x + 2\pi) = \sec x$$

---

x	$\tan x$
0	0
$\pi/6$	0.58
$\pi/4$	1.00
$\pi/3$	1.73
1.4	5.80
1.5	14.10
1.55	48.08
1.57	1,255.77
1.5707	10,381.33

We first sketch the graph of tangent. Since it has period π, we need only sketch the graph on any interval of length π and then repeat the pattern to the left and to the right. We sketch the graph on the interval $(-\pi/2, \pi/2)$. Since $\tan(\pi/2)$ and $\tan(-\pi/2)$ aren't defined, we need to be careful in sketching the graph at points near $\pi/2$ and $-\pi/2$. As x gets near $\pi/2$ through values less than $\pi/2$, the value of $\tan x$ becomes large. To see this, notice that as x gets close to $\pi/2$, $\cos x$ approaches 0 and $\sin x$ approaches 1 and so $\tan x = \sin x/\cos x$ is large. A table of values of $\tan x$ for x close to $\pi/2$ (≈ 1.570796) is shown in the margin.

Thus, by choosing x close enough to $\pi/2$ through values less than $\pi/2$, we can make the value of $\tan x$ larger than any given positive number. We express this by writing

$$\tan x \to \infty \qquad \text{as} \qquad x \to \frac{\pi^-}{2}$$

This is read "$\tan x$ approaches infinity as x approaches $\pi/2$ from the left."

In a similar way, by choosing x close to $-\pi/2$ through values greater than $-\pi/2$, we can make $\tan x$ smaller than any given negative number. We write this as

$$\tan x \to -\infty \qquad \text{as} \qquad x \to -\frac{\pi^+}{2}$$

This is read "$\tan x$ approaches negative infinity as x approaches $-\pi/2$ from the right."

Thus, the graph of $y = \tan x$ approaches the vertical lines $x = \pi/2$ and $x = -\pi/2$. So these lines are **vertical asymptotes**. With the information we have so far, we sketch the graph of

$y = \tan x$ for $-\pi/2 < x < \pi/2$ in Figure 1. The complete graph of tangent (see Figure 5(a) on the next page) is now obtained using the fact that tangent is periodic with period π.

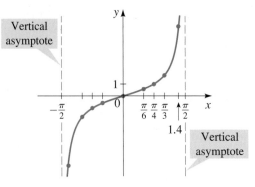

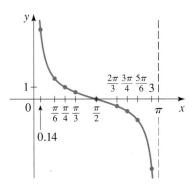

FIGURE 1
One period of $y = \tan x$

FIGURE 2
One period of $y = \cot x$

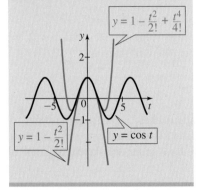
The function $y = \cot x$ is graphed on the interval $(0, \pi)$ by a similar analysis (see Figure 2). Since $\cot x$ is undefined for $x = n\pi$ with n an integer, its complete graph (in Figure 5(b) on the next page) has vertical asymptotes at these values.

To graph the cosecant and secant functions, we use the reciprocal identities

$$\csc x = \frac{1}{\sin x} \qquad \text{and} \qquad \sec x = \frac{1}{\cos x}$$

So to graph $y = \csc x$, we take the reciprocals of the y-coordinates of the points of the graph of $y = \sin x$. (See Figure 3.) Similarly, to graph $y = \sec x$, we take the reciprocals of the y-coordinates of the points of the graph of $y = \cos x$. (See Figure 4.)

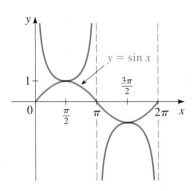

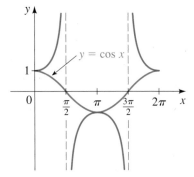

FIGURE 3
One period of $y = \csc x$

FIGURE 4
One period of $y = \sec x$

Let's consider more closely the graph of the function $y = \csc x$ on the interval $0 < x < \pi$. We need to examine the values of the function near 0 and π, since at these values $\sin x = 0$, and $\csc x$ is thus undefined. We see that

$$\csc x \to \infty \qquad \text{as} \qquad x \to 0^+$$

$$\csc x \to \infty \qquad \text{as} \qquad x \to \pi^-$$

Thus, the lines $x = 0$ and $x = \pi$ are vertical asymptotes. In the interval $\pi < x < 2\pi$ the graph is sketched in the same way. The values of $\csc x$ in that interval are the same as those in the interval $0 < x < \pi$ except for sign (see Figure 3). The complete graph in Figure 5(c) is now obtained from the fact that the function cosecant is periodic with period

2π. Note that the graph has vertical asymptotes at the points where $\sin x = 0$, that is, at $x = n\pi$, for n an integer.

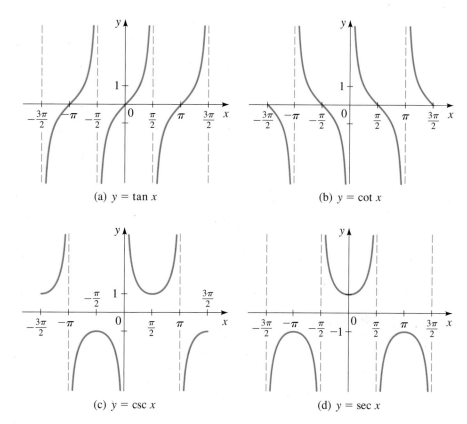

(a) $y = \tan x$

(b) $y = \cot x$

(c) $y = \csc x$

(d) $y = \sec x$

FIGURE 5

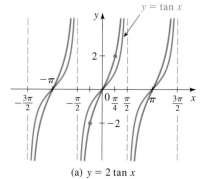

(a) $y = 2 \tan x$

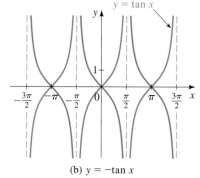

(b) $y = -\tan x$

FIGURE 6

The graph of $y = \sec x$ is sketched in a similar manner. Observe that the domain of $\sec x$ is the set of all real numbers other than $x = (\pi/2) + n\pi$, for n an integer, so the graph has vertical asymptotes at those points. The complete graph is shown in Figure 5(d).

It is apparent that the graphs of $y = \tan x$, $y = \cot x$, and $y = \csc x$ are symmetric about the origin, whereas that of $y = \sec x$ is symmetric about the y-axis. This is because tangent, cotangent, and cosecant are odd functions, whereas secant is an even function.

▼ Graphs of Transformations of Tangent and Cotangent

We now consider graphs of transformations of the tangent and cotangent functions.

EXAMPLE 1 | Graphing Tangent Curves

Graph each function.

(a) $y = 2 \tan x$ (b) $y = -\tan x$

SOLUTION We first graph $y = \tan x$ and then transform it as required.

(a) To graph $y = 2 \tan x$, we multiply the y-coordinate of each point on the graph of $y = \tan x$ by 2. The resulting graph is shown in Figure 6(a).

(b) The graph of $y = -\tan x$ in Figure 6(b) is obtained from that of $y = \tan x$ by reflecting in the x-axis.

✎ NOW TRY EXERCISES **9** AND **11**

Since the tangent and cotangent functions have period π, the functions

$$y = a \tan kx \quad \text{and} \quad y = a \cot kx \quad (k > 0)$$

complete one period as kx varies from 0 to π, that is, for $0 \le kx \le \pi$. Solving this inequality, we get $0 \le x \le \pi/k$. So they each have period π/k.

TANGENT AND COTANGENT CURVES

The functions

$$y = a \tan kx \quad \text{and} \quad y = a \cot kx \quad (k > 0)$$

have period π/k.

Thus, one complete period of the graphs of these functions occurs on any interval of length π/k. To sketch a complete period of these graphs, it's convenient to select an interval between vertical asymptotes:

To graph one period of $y = a \tan kx$, an appropriate interval is $\left(-\dfrac{\pi}{2k}, \dfrac{\pi}{2k} \right)$.

To graph one period of $y = a \cot kx$, an appropriate interval is $\left(0, \dfrac{\pi}{k} \right)$.

EXAMPLE 2 | Graphing Tangent Curves

Graph each function.

(a) $y = \tan 2x$ **(b)** $y = \tan 2\left(x - \dfrac{\pi}{4} \right)$

SOLUTION

(a) The period is $\pi/2$ and an appropriate interval is $(-\pi/4, \pi/4)$. The endpoints $x = -\pi/4$ and $x = \pi/4$ are vertical asymptotes. Thus, we graph one complete period of the function on $(-\pi/4, \pi/4)$. The graph has the same shape as that of the tangent function, but is shrunk horizontally by a factor of $\frac{1}{2}$. We then repeat that portion of the graph to the left and to the right. See Figure 7(a).

(b) The graph is the same as that in part (a), but it is shifted to the right $\pi/4$, as shown in Figure 7(b).

Since $y = \tan x$ completes one period between $x = -\frac{\pi}{2}$ and $x = \frac{\pi}{2}$, the function $y = \tan 2(x - \frac{\pi}{4})$ completes one period as $2(x - \frac{\pi}{4})$ varies from $-\frac{\pi}{2}$ to $\frac{\pi}{2}$.

Start of period: End of period:

$2\left(x - \frac{\pi}{4}\right) = -\frac{\pi}{2}$ $2\left(x - \frac{\pi}{4}\right) = \frac{\pi}{2}$

$x - \frac{\pi}{4} = -\frac{\pi}{4}$ $x - \frac{\pi}{4} = \frac{\pi}{4}$

$x = 0$ $x = \frac{\pi}{2}$

So we graph one period on the interval $(0, \frac{\pi}{2})$.

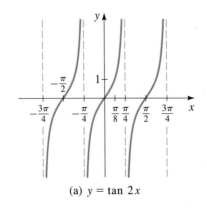

(a) $y = \tan 2x$

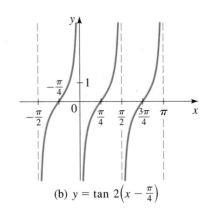

(b) $y = \tan 2\left(x - \frac{\pi}{4}\right)$

FIGURE 7

NOW TRY EXERCISES **27** AND **39**

EXAMPLE 3 | A Shifted Cotangent Curve

Graph $y = 2 \cot\left(3x - \dfrac{\pi}{2}\right)$.

SOLUTION We first put this in the form $y = a \cot k(x - b)$ by factoring 3 from the expression $3x - \dfrac{\pi}{2}$:

$$y = 2 \cot\left(3x - \frac{\pi}{2}\right) = 2 \cot 3\left(x - \frac{\pi}{6}\right)$$

Thus the graph is the same as that of $y = 2 \cot 3x$ but is shifted to the right $\pi/6$. The period of $y = 2 \cot 3x$ is $\pi/3$, and an appropriate interval is $(0, \pi/3)$. To get the corresponding interval for the desired graph, we shift this interval to the right $\pi/6$. This gives

$$\left(0 + \frac{\pi}{6}, \frac{\pi}{3} + \frac{\pi}{6}\right) = \left(\frac{\pi}{6}, \frac{\pi}{2}\right)$$

Finally, we graph one period in the shape of cotangent on the interval $(\pi/6, \pi/2)$ and repeat that portion of the graph to the left and to the right. (See Figure 8.)

Since $y = \cot x$ completes one period between $x = 0$ and $x = \pi$, the function $y = 2 \cot(3x - \frac{\pi}{2})$ completes one period as $3x - \frac{\pi}{2}$ varies from 0 to π.

Start of period: End of period:

$3x - \frac{\pi}{2} = 0$ $3x - \frac{\pi}{2} = \pi$

$\qquad 3x = \frac{\pi}{2}$ $\qquad 3x = \frac{3\pi}{2}$

$\qquad\quad x = \frac{\pi}{6}$ $\qquad\quad x = \frac{\pi}{2}$

So we graph one period on the interval $(\frac{\pi}{6}, \frac{\pi}{2})$.

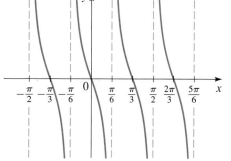

FIGURE 8

$y = 2 \cot\left(3x - \dfrac{\pi}{2}\right)$

✎ NOW TRY EXERCISE **43**

▼ Graphs of Transformations of Cosecant and Secant

We have already observed that the cosecant and secant functions are the reciprocals of the sine and cosine functions. Thus, the following result is the counterpart of the result for sine and cosine curves in Section 2.3.

COSECANT AND SECANT CURVES

The functions

$$y = a \csc kx \qquad \text{and} \qquad y = a \sec kx \qquad (k > 0)$$

have period $2\pi/k$.

An appropriate interval on which to graph one complete period is $[0, 2\pi/k]$.

EXAMPLE 4 | Graphing Cosecant Curves

Graph each function.

(a) $y = \dfrac{1}{2} \csc 2x$ **(b)** $y = \dfrac{1}{2} \csc\left(2x + \dfrac{\pi}{2}\right)$

SOLUTION

(a) The period is $2\pi/2 = \pi$. An appropriate interval is $[0, \pi]$, and the asymptotes occur in this interval whenever $\sin 2x = 0$. So the asymptotes in this interval are $x = 0$, $x = \pi/2$, and $x = \pi$. With this information we sketch on the interval $[0, \pi]$ a graph with the same general shape as that of one period of the cosecant function. The complete graph in Figure 9(a) is obtained by repeating this portion of the graph to the left and to the right.

Since $y = \csc x$ completes one period between $x = 0$ and $x = 2\pi$, the function $y = \frac{1}{2}\csc(2x + \frac{\pi}{2})$ completes one period as $2x + \frac{\pi}{2}$ varies from 0 to 2π.

Start of period: End of period:

$2x + \frac{\pi}{2} = 0$ $2x + \frac{\pi}{2} = 2\pi$

$2x = -\frac{\pi}{2}$ $2x = \frac{3\pi}{2}$

$x = -\frac{\pi}{4}$ $x = \frac{3\pi}{4}$

So we graph one period on the interval $\left(-\frac{\pi}{4}, \frac{3\pi}{4}\right)$.

(b) We first write

$$y = \frac{1}{2}\csc\left(2x + \frac{\pi}{2}\right) = \frac{1}{2}\csc 2\left(x + \frac{\pi}{4}\right)$$

From this we see that the graph is the same as that in part (a) but shifted to the left $\pi/4$. The graph is shown in Figure 9(b).

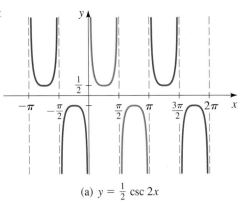

(a) $y = \frac{1}{2}\csc 2x$

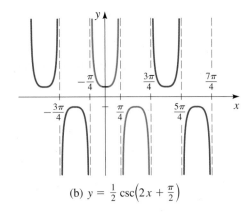

(b) $y = \frac{1}{2}\csc\left(2x + \frac{\pi}{2}\right)$

FIGURE 9

✎. NOW TRY EXERCISES **33** AND **45**

EXAMPLE 5 | Graphing a Secant Curve

Graph $y = 3 \sec \frac{1}{2}x$.

SOLUTION   The period is $2\pi \div \frac{1}{2} = 4\pi$. An appropriate interval is $[0, 4\pi]$, and the asymptotes occur in this interval wherever $\cos \frac{1}{2}x = 0$. Thus, the asymptotes in this interval are $x = \pi$, $x = 3\pi$. With this information we sketch on the interval $[0, 4\pi]$ a graph with the same general shape as that of one period of the secant function. The complete graph in Figure 10 is obtained by repeating this portion of the graph to the left and to the right.

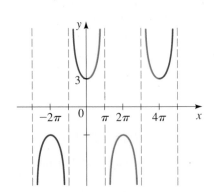

FIGURE 10
$y = 3 \sec \frac{1}{2}x$

✎. NOW TRY EXERCISE **31**

2.4 EXERCISES

CONCEPTS

1. The trigonometric function $y = \tan x$ has period _____
and asymptotes $x =$ _____ . Sketch a graph of this function on the interval $(-\pi/2, \pi/2)$.

2. The trigonometric function $y = \csc x$ has period _____
and asymptotes $x =$ _____ . Sketch a graph of this function on the interval $(-\pi, \pi)$.

SKILLS

3–8 ■ Match the trigonometric function with one of the graphs I–VI.

3. $f(x) = \tan\left(x + \dfrac{\pi}{4}\right)$

4. $f(x) = \sec 2x$

5. $f(x) = \cot 2x$

6. $f(x) = -\tan x$

7. $f(x) = 2 \sec x$

8. $f(x) = 1 + \csc x$

I
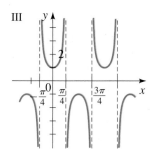

II
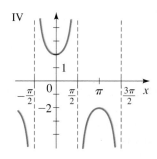

III
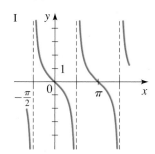

IV
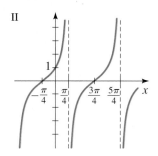

V
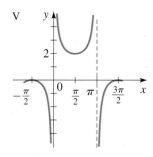

VI
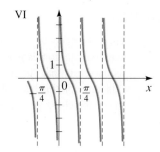

9–54 ■ Find the period and graph the function.

9. $y = 4 \tan x$

10. $y = -4 \tan x$

11. $y = -\frac{1}{2} \tan x$

12. $y = \frac{1}{2} \tan x$

13. $y = -\cot x$

14. $y = 2 \cot x$

15. $y = 2 \csc x$

16. $y = \frac{1}{2} \csc x$

17. $y = 3 \sec x$

18. $y = -3 \sec x$

19. $y = \tan\left(x + \dfrac{\pi}{2}\right)$

20. $y = \tan\left(x - \dfrac{\pi}{4}\right)$

21. $y = \csc\left(x - \dfrac{\pi}{2}\right)$

22. $y = \sec\left(x + \dfrac{\pi}{4}\right)$

23. $y = \cot\left(x + \dfrac{\pi}{4}\right)$

24. $y = 2 \csc\left(x - \dfrac{\pi}{3}\right)$

25. $y = \dfrac{1}{2} \sec\left(x - \dfrac{\pi}{6}\right)$

26. $y = 3 \csc\left(x + \dfrac{\pi}{2}\right)$

27. $y = \tan 4x$

28. $y = \tan \frac{1}{2}x$

29. $y = \tan \dfrac{\pi}{4}x$

30. $y = \cot \dfrac{\pi}{2}x$

31. $y = \sec 2x$

32. $y = 5 \csc 3x$

33. $y = \csc 4x$

34. $y = \csc \frac{1}{2}x$

35. $y = 2 \tan 3\pi x$

36. $y = 2 \tan \dfrac{\pi}{2}x$

37. $y = 5 \csc \dfrac{3\pi}{2}x$

38. $y = 5 \sec 2\pi x$

39. $y = \tan 2\left(x + \dfrac{\pi}{2}\right)$

40. $y = \csc 2\left(x + \dfrac{\pi}{2}\right)$

41. $y = \tan 2(x - \pi)$

42. $y = \sec 2\left(x - \dfrac{\pi}{2}\right)$

43. $y = \cot\left(2x - \dfrac{\pi}{2}\right)$

44. $y = \frac{1}{2} \tan(\pi x - \pi)$

45. $y = 2 \csc\left(\pi x - \dfrac{\pi}{3}\right)$

46. $y = 2 \sec\left(\dfrac{1}{2}x - \dfrac{\pi}{3}\right)$

47. $y = 5 \sec\left(3x - \dfrac{\pi}{2}\right)$

48. $y = \frac{1}{2} \sec(2\pi x - \pi)$

49. $y = \tan\left(\dfrac{2}{3}x - \dfrac{\pi}{6}\right)$

50. $y = \tan \dfrac{1}{2}\left(x + \dfrac{\pi}{4}\right)$

51. $y = 3 \sec \pi\left(x + \dfrac{1}{2}\right)$

52. $y = \sec\left(3x + \dfrac{\pi}{2}\right)$

53. $y = -2 \tan\left(2x - \dfrac{\pi}{3}\right)$

54. $y = 2 \csc(3x + 3)$

55. (a) Prove that if f is periodic with period p, then $1/f$ is also periodic with period p.
(b) Prove that cosecant and secant each have period 2π.

56. Prove that if f and g are periodic with period p, then f/g is also periodic, but its period could be smaller than p.

APPLICATIONS

57. Lighthouse The beam from a lighthouse completes one rotation every two minutes. At time t, the distance d shown in the figure on the next page is

$$d(t) = 3 \tan \pi t$$

where t is measured in minutes and d in miles.
(a) Find $d(0.15)$, $d(0.25)$, and $d(0.45)$.

(b) Sketch a graph of the function d for $0 \leq t < \frac{1}{2}$.

(c) What happens to the distance d as t approaches $\frac{1}{2}$?

(d) Explain what happens to the shadow as the time approaches 6 P.M. (that is, as $t \to 12^-$).

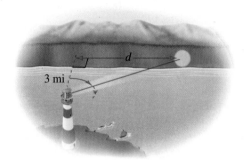

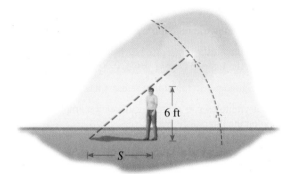

58. Length of a Shadow On a day when the sun passes directly overhead at noon, a six-foot-tall man casts a shadow of length

$$S(t) = 6 \left| \cot \frac{\pi}{12} t \right|$$

where S is measured in feet and t is the number of hours since 6 A.M.

(a) Find the length of the shadow at 8:00 A.M., noon, 2:00 P.M., and 5:45 P.M.

(b) Sketch a graph of the function S for $0 < t < 12$.

(c) From the graph determine the values of t at which the length of the shadow equals the man's height. To what time of day does each of these values correspond?

DISCOVERY ▪ DISCUSSION ▪ WRITING

59. Reduction Formulas Use the graphs in Figure 5 to explain why the following formulas are true.

$$\tan\left(x - \frac{\pi}{2}\right) = -\cot x$$

$$\sec\left(x - \frac{\pi}{2}\right) = \csc x$$

2.5 INVERSE TRIGONOMETRIC FUNCTIONS AND THEIR GRAPHS

▶ The Inverse Sine Function ▶ The Inverse Cosine Function ▶ The Inverse Tangent Function ▶ The Inverse Secant, Cosecant, and Cotangent Functions

We study applications of inverse trigonometric functions to triangles in Sections 3.4–3.6.

Recall from Section 1.8 that the inverse of a function f is a function f^{-1} that reverses the rule of f. For a function to have an inverse, it must be one-to-one. Since the trigonometric functions are not one-to-one, they do not have inverses. It is possible, however, to restrict the domains of the trigonometric functions in such a way that the resulting functions are one-to-one.

▼ The Inverse Sine Function

Let's first consider the sine function. There are many ways to restrict the domain of sine so that the new function is one-to-one. A natural way to do this is to restrict the domain to the interval $[-\pi/2, \pi/2]$. The reason for this choice is that sine is one-to-one on this interval and moreover attains each of the values in its range on this interval. From Figure 1 we see that sine is one-to-one on this restricted domain (by the Horizontal Line Test) and so has an inverse.

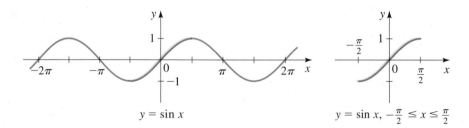

FIGURE 1 Graphs of the sine function and the restricted sine function

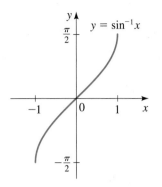

FIGURE 2   Graph of $y = \sin^{-1} x$

We can now define an inverse sine function on this restricted domain. The graph of $y = \sin^{-1} x$ is shown in Figure 2; it is obtained by reflecting the graph of $y = \sin x$, $-\pi/2 \le x \le \pi/2$, in the line $y = x$.

DEFINITION OF THE INVERSE SINE FUNCTION

The **inverse sine function** is the function $\sin^{-1}$ with domain $[-1, 1]$ and range $[-\pi/2, \pi/2]$ defined by

$$\sin^{-1} x = y \quad \Leftrightarrow \quad \sin y = x$$

The inverse sine function is also called **arcsine**, denoted by **arcsin**.

Thus, $y = \sin^{-1} x$ *is the number in the interval* $[-\pi/2, \pi/2]$ *whose sine is x.* In other words, $\sin(\sin^{-1} x) = x$. In fact, from the general properties of inverse functions studied in Section 1.8, we have the following **cancellation properties**.

$$\sin(\sin^{-1} x) = x \quad \text{for} \quad -1 \le x \le 1$$
$$\sin^{-1}(\sin x) = x \quad \text{for} \quad -\frac{\pi}{2} \le x \le \frac{\pi}{2}$$

EXAMPLE 1 │ Evaluating the Inverse Sine Function

Find each value.

(a) $\sin^{-1} \dfrac{1}{2}$ 　　　 **(b)** $\sin^{-1}\left(-\dfrac{1}{2}\right)$ 　　　 **(c)** $\sin^{-1} \dfrac{3}{2}$

SOLUTION

(a) The number in the interval $[-\pi/2, \pi/2]$ whose sine is $\frac{1}{2}$ is $\pi/6$. Thus, $\sin^{-1}\frac{1}{2} = \pi/6$.

(b) The number in the interval $[-\pi/2, \pi/2]$ whose sine is $-\frac{1}{2}$ is $-\pi/6$. Thus, $\sin^{-1}(-\frac{1}{2}) = -\pi/6$.

(c) Since $\frac{3}{2} > 1$, it is not in the domain of $\sin^{-1} x$, so $\sin^{-1} \frac{3}{2}$ is not defined.

✎ NOW TRY EXERCISE **3**　　　　　　　　　　　　　　　　　　　　　■

EXAMPLE 2 │ Using a Calculator to Evaluate Inverse Sine

Find approximate values for **(a)** $\sin^{-1}(0.82)$ and **(b)** $\sin^{-1}\frac{1}{3}$.

SOLUTION

We use a calculator to approximate these values. Using the $\boxed{\text{SIN}^{-1}}$, or $\boxed{\text{INV}}\,\boxed{\text{SIN}}$, or $\boxed{\text{ARC}}\,\boxed{\text{SIN}}$ key(s) on the calculator (with the calculator in radian mode), we get

(a) $\sin^{-1}(0.82) \approx 0.96141$ 　　　　 **(b)** $\sin^{-1}\frac{1}{3} \approx 0.33984$

✎ NOW TRY EXERCISES **11** AND **21**　　　　　　　　　　　　　　　　■

When evaluating expressions involving $\sin^{-1}$, we need to remember that the range of $\sin^{-1}$ is the interval $[-\pi/2, \pi/2]$.

EXAMPLE 3 │ Evaluating Expressions with Inverse Sine

Find each value.

(a) $\sin^{-1}\left(\sin \dfrac{\pi}{3}\right)$ 　　　　 **(b)** $\sin^{-1}\left(\sin \dfrac{2\pi}{3}\right)$

SOLUTION

(a) Since $\pi/3$ is in the interval $[-\pi/2, \pi/2]$, we can use the above cancellation properties of inverse functions:

$$\sin^{-1}\left(\sin\frac{\pi}{3}\right) = \frac{\pi}{3} \qquad \text{Cancellation property: } -\frac{\pi}{2} \le \frac{\pi}{3} \le \frac{\pi}{2}$$

(b) We first evaluate the expression in the parentheses:

$$\sin^{-1}\left(\sin\frac{2\pi}{3}\right) = \sin^{-1}\left(\frac{\sqrt{3}}{2}\right) \qquad \text{Evaluate}$$

$$= \frac{\pi}{3} \qquad \text{Because } \sin\frac{\pi}{3} = \frac{\sqrt{3}}{2}$$

⊘ Note: $\sin^{-1}(\sin x) = x$ only if $-\frac{\pi}{2} \le x \le \frac{\pi}{2}$.

✎ NOW TRY EXERCISES **31** AND **33** ■

▼ The Inverse Cosine Function

If the domain of the cosine function is restricted to the interval $[0, \pi]$, the resulting function is one-to-one and so has an inverse. We choose this interval because on it, cosine attains each of its values exactly once (see Figure 3).

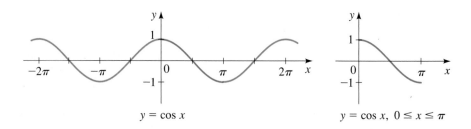

FIGURE 3 Graphs of the cosine function and the restricted cosine function

$y = \cos x$

$y = \cos x, \ 0 \le x \le \pi$

DEFINITION OF THE INVERSE COSINE FUNCTION

The **inverse cosine function** is the function $\cos^{-1}$ with domain $[-1, 1]$ and range $[0, \pi]$ defined by

$$\cos^{-1}x = y \quad \Leftrightarrow \quad \cos y = x$$

The inverse cosine function is also called **arccosine**, denoted by **arccos**.

Thus, $y = \cos^{-1}x$ *is the number in the interval* $[0, \pi]$ *whose cosine is x.* The following **cancellation properties** follow from the inverse function properties.

$$\cos(\cos^{-1}x) = x \quad \text{for} \quad -1 \le x \le 1$$
$$\cos^{-1}(\cos x) = x \quad \text{for} \quad 0 \le x \le \pi$$

The graph of $y = \cos^{-1}x$ is shown in Figure 4; it is obtained by reflecting the graph of $y = \cos x, 0 \le x \le \pi$, in the line $y = x$.

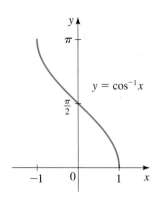

$y = \cos^{-1}x$

FIGURE 4 Graph of $y = \cos^{-1}x$

EXAMPLE 4 | Evaluating the Inverse Cosine Function

Find each value.

(a) $\cos^{-1}\dfrac{\sqrt{3}}{2}$ **(b)** $\cos^{-1}0$ **(c)** $\cos^{-1}\dfrac{5}{7}$

SOLUTION

(a) The number in the interval $[0, \pi]$ whose cosine is $\sqrt{3}/2$ is $\pi/6$. Thus, $\cos^{-1}(\sqrt{3}/2) = \pi/6$.

(b) The number in the interval $[0, \pi]$ whose cosine is 0 is $\pi/2$. Thus, $\cos^{-1} 0 = \pi/2$.

(c) Since no rational multiple of π has cosine $\frac{5}{7}$, we use a calculator (in radian mode) to find this value approximately:

$$\cos^{-1} \frac{5}{7} \approx 0.77519$$

✎ NOW TRY EXERCISES **5** AND **13** ▪

EXAMPLE 5 | Evaluating Expressions with Inverse Cosine

Find each value.

(a) $\cos^{-1}\left(\cos \dfrac{2\pi}{3}\right)$ (b) $\cos^{-1}\left(\cos \dfrac{5\pi}{3}\right)$

SOLUTION

(a) Since $2\pi/3$ is in the interval $[0, \pi]$ we can use the above cancellation properties:

$$\cos^{-1}\left(\cos \frac{2\pi}{3}\right) = \frac{2\pi}{3} \qquad \text{Cancellation property: } 0 \le \frac{2\pi}{3} \le \pi$$

(b) We first evaluate the expression in the parentheses:

$$\cos^{-1}\left(\cos \frac{5\pi}{3}\right) = \cos^{-1}\left(\tfrac{1}{2}\right) \qquad \text{Evaluate}$$

$$= \frac{\pi}{3} \qquad \text{Because } \cos \frac{\pi}{3} = \frac{1}{2}$$

⊘ Note: $\cos^{-1}(\cos x) = x$ only if $0 \le x \le \pi$.

✎ NOW TRY EXERCISES **29** AND **35** ▪

▼ The Inverse Tangent Function

We restrict the domain of the tangent function to the interval $(-\pi/2, \pi/2)$ in order to obtain a one-to-one function.

DEFINITION OF THE INVERSE TANGENT FUNCTION

The **inverse tangent function** is the function $\tan^{-1}$ with domain $\mathbb{R}$ and range $(-\pi/2, \pi/2)$ defined by

$$\tan^{-1} x = y \quad \Leftrightarrow \quad \tan y = x$$

The inverse tangent function is also called **arctangent**, denoted by **arctan**.

Thus, $y = \tan^{-1} x$ *is the number in the interval* $(-\pi/2, \pi/2)$ *whose tangent is* x. The following **cancellation properties** follow from the inverse function properties.

$$\tan(\tan^{-1} x) = x \quad \text{for} \quad x \in \mathbb{R}$$

$$\tan^{-1}(\tan x) = x \quad \text{for} \quad -\frac{\pi}{2} < x < \frac{\pi}{2}$$

Figure 5 shows the graph of $y = \tan x$ on the interval $(-\pi/2, \pi/2)$ and the graph of its inverse function, $y = \tan^{-1} x$.

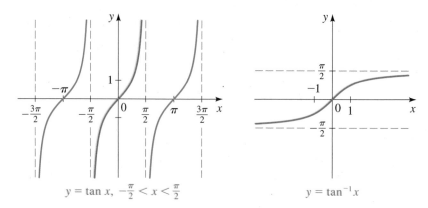

FIGURE 5 Graphs of the restricted tangent function and the inverse tangent function

$y = \tan x,\ -\frac{\pi}{2} < x < \frac{\pi}{2}$
$y = \tan^{-1} x$

EXAMPLE 6 | Evaluating the Inverse Tangent Function

Find each value.

(a) $\tan^{-1} 1$ **(b)** $\tan^{-1} \sqrt{3}$ **(c)** $\tan^{-1}(20)$

SOLUTION

(a) The number in the interval $(-\pi/2, \pi/2)$ with tangent 1 is $\pi/4$. Thus, $\tan^{-1} 1 = \pi/4$.

(b) The number in the interval $(-\pi/2, \pi/2)$ with tangent $\sqrt{3}$ is $\pi/3$. Thus, $\tan^{-1}\sqrt{3} = \pi/3$.

(c) We use a calculator (in radian mode) to find that $\tan^{-1}(20) \approx -1.52084$.

✎ NOW TRY EXERCISES **7** AND **17** ∎

▼ The Inverse Secant, Cosecant, and Cotangent Functions

See Exercise 44 in Section 3.4 (page 201) for a way of finding the values of these inverse trigonometric functions on a calculator.

To define the inverse functions of the secant, cosecant, and cotangent functions, we restrict the domain of each function to a set on which it is one-to-one and on which it attains all its values. Although any interval satisfying these criteria is appropriate, we choose to restrict the domains in a way that simplifies the choice of sign in computations involving inverse trigonometric functions. The choices we make are also appropriate for calculus. This explains the seemingly strange restriction for the domains of the secant and cosecant functions. We end this section by displaying the graphs of the secant, cosecant, and cotangent functions with their restricted domains and the graphs of their inverse functions (Figures 6–8).

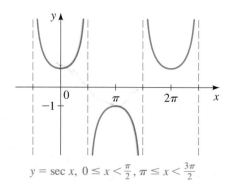
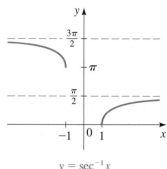

FIGURE 6 The inverse secant function

$y = \sec x,\ 0 \le x < \frac{\pi}{2},\ \pi \le x < \frac{3\pi}{2}$
$y = \sec^{-1} x$

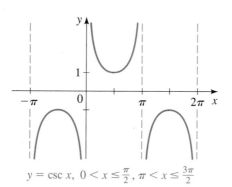

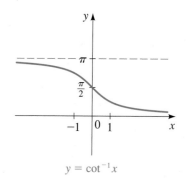

FIGURE 7 The inverse cosecant function

$y = \csc x, \ 0 < x \le \frac{\pi}{2}, \ \pi < x \le \frac{3\pi}{2}$

$y = \csc^{-1}x$

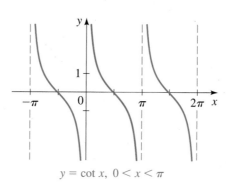

FIGURE 8 The inverse cotangent function

$y = \cot x, \ 0 < x < \pi$

$y = \cot^{-1}x$

2.5 EXERCISES

CONCEPTS

1. **(a)** To define the inverse sine function, we restrict the domain of sine to the interval _____. On this interval the sine function is one-to-one, and its inverse function $\sin^{-1}$ is defined by $\sin^{-1}x = y \Leftrightarrow \sin$ _____ $=$ _____. For example, $\sin^{-1}\frac{1}{2} =$ _____ because $\sin$ _____ $=$ _____.

(b) To define the inverse cosine function we restrict the domain of cosine to the interval _____. On this interval the cosine function is one-to-one and its inverse function $\cos^{-1}$ is defined by $\cos^{-1}x = y \Leftrightarrow$ $\cos$ _____ $=$ _____. For example, $\cos^{-1}\frac{1}{2} =$ _____ because $\cos$ _____ $=$ _____.

2. The cancellation property $\sin^{-1}(\sin x) = x$ is valid for x in the interval _____. Which of the following is not true?

(a) $\sin^{-1}\left(\sin\frac{\pi}{3}\right) = \frac{\pi}{3}$

(b) $\sin^{-1}\left(\sin\frac{10\pi}{3}\right) = \frac{10\pi}{3}$

SKILLS

3–10 ■ Find the exact value of each expression, if it is defined.

3. **(a)** $\sin^{-1} 1$ **(b)** $\sin^{-1}\dfrac{\sqrt{3}}{2}$ **(c)** $\sin^{-1} 2$

4. **(a)** $\sin^{-1}(-1)$ **(b)** $\sin^{-1}\dfrac{\sqrt{2}}{2}$ **(c)** $\sin^{-1}(-2)$

5. **(a)** $\cos^{-1}(-1)$ **(b)** $\cos^{-1}\frac{1}{2}$ **(c)** $\cos^{-1}\left(-\dfrac{\sqrt{3}}{2}\right)$

6. **(a)** $\cos^{-1}\left(\dfrac{\sqrt{2}}{2}\right)$ **(b)** $\cos^{-1} 1$ **(c)** $\cos^{-1}\left(-\dfrac{\sqrt{2}}{2}\right)$

7. **(a)** $\tan^{-1}(-1)$ **(b)** $\tan^{-1}\sqrt{3}$ **(c)** $\tan^{-1}\dfrac{\sqrt{3}}{3}$

8. **(a)** $\tan^{-1} 0$ **(b)** $\tan^{-1}(-\sqrt{3})$ **(c)** $\tan^{-1}\left(-\dfrac{\sqrt{3}}{3}\right)$

9. **(a)** $\cos^{-1}(-\frac{1}{2})$ **(b)** $\sin^{-1}\left(-\dfrac{\sqrt{2}}{2}\right)$ **(c)** $\tan^{-1} 1$

10. **(a)** $\cos^{-1} 0$ **(b)** $\sin^{-1} 0$ **(c)** $\sin^{-1}(-\frac{1}{2})$

11–22 ■ Use a calculator to find an approximate value of each expression correct to five decimal places, if it is defined.

11. $\sin^{-1}\frac{2}{3}$ **12.** $\sin^{-1}(-\frac{8}{9})$

13. $\cos^{-1}(-\frac{3}{7})$ **14.** $\cos^{-1}(\frac{4}{9})$

15. $\cos^{-1}(-0.92761)$ **16.** $\sin^{-1}(0.13844)$

17. $\tan^{-1} 10$ **18.** $\tan^{-1}(-26)$

19. $\tan^{-1}(1.23456)$ **20.** $\cos^{-1}(1.23456)$

21. $\sin^{-1}(-0.25713)$ **22.** $\tan^{-1}(-0.25713)$

23–44 ■ Find the exact value of the expression, if it is defined.

23. $\sin\left(\sin^{-1}\frac{1}{4}\right)$

24. $\cos\left(\cos^{-1}\frac{2}{3}\right)$

25. $\tan(\tan^{-1}5)$

26. $\sin(\sin^{-1}5)$

27. $\sin\left(\sin^{-1}\left(\frac{3}{2}\right)\right)$

28. $\tan\left(\tan^{-1}\left(\frac{3}{2}\right)\right)$

29. $\cos^{-1}\left(\cos\frac{5\pi}{6}\right)$

30. $\tan^{-1}\left(\tan\left(\frac{\pi}{4}\right)\right)$

31. $\sin^{-1}\left(\sin\left(-\frac{\pi}{6}\right)\right)$

32. $\tan^{-1}\left(\tan\left(-\frac{\pi}{4}\right)\right)$

33. $\sin^{-1}\left(\sin\left(\frac{5\pi}{6}\right)\right)$

34. $\cos^{-1}\left(\cos\left(-\frac{\pi}{6}\right)\right)$

35. $\cos^{-1}\left(\cos\left(\frac{17\pi}{6}\right)\right)$

36. $\tan^{-1}\left(\tan\left(\frac{4\pi}{3}\right)\right)$

37. $\tan^{-1}\left(\tan\left(\frac{2\pi}{3}\right)\right)$

38. $\sin^{-1}\left(\sin\left(\frac{11\pi}{4}\right)\right)$

39. $\tan\left(\sin^{-1}\frac{1}{2}\right)$

40. $\cos(\sin^{-1}0)$

41. $\cos\left(\sin^{-1}\frac{\sqrt{3}}{2}\right)$

42. $\tan\left(\sin^{-1}\frac{\sqrt{2}}{2}\right)$

43. $\sin(\tan^{-1}(-1))$

44. $\sin(\tan^{-1}(-\sqrt{3}))$

DISCOVERY ■ DISCUSSION ■ WRITING

45. Two Different Compositions Let f and g be the functions

$$f(x) = \sin(\sin^{-1}x)$$

and

$$g(x) = \sin^{-1}(\sin x)$$

By the cancellation properties, $f(x) = x$ and $g(x) = x$ for suitable values of x. But these functions are not the same for all x. Graph both f and g to show how the functions differ. (Think carefully about the domain and range of $\sin^{-1}$.)

46–47 ■ **Graphing Inverse Trigonometric Functions**
(a) Graph the function and make a conjecture, and **(b)** prove that your conjecture is true.

46. $y = \sin^{-1}x + \cos^{-1}x$

47. $y = \tan^{-1}x + \tan^{-1}\frac{1}{x}$

2.6 MODELING HARMONIC MOTION

| Simple Harmonic Motion

Periodic behavior—behavior that repeats over and over again—is common in nature. Perhaps the most familiar example is the daily rising and setting of the sun, which results in the repetitive pattern of day, night, day, night, ... Another example is the daily variation of tide levels at the beach, which results in the repetitive pattern of high tide, low tide, high tide, low tide, ... Certain animal populations increase and decrease in a predictable periodic pattern: A large population exhausts the food supply, which causes the population to dwindle; this in turn results in a more plentiful food supply, which makes it possible for the population to increase; and the pattern then repeats over and over (see the Discovery Project *Predator/Prey Models* referenced on page 132).

Other common examples of periodic behavior involve motion that is caused by vibration or oscillation. A mass suspended from a spring that has been compressed and then allowed to vibrate vertically is a simple example. This "back and forth" motion also occurs in such diverse phenomena as sound waves, light waves, alternating electrical current, and pulsating stars, to name a few. In this section we consider the problem of modeling periodic behavior.

▼ Simple Harmonic Motion

The trigonometric functions are ideally suited for modeling periodic behavior. A glance at the graphs of the sine and cosine functions, for instance, tells us that these functions themselves exhibit periodic behavior. Figure 1 shows the graph of $y = \sin t$. If we think of *t* as time, we see that as time goes on, $y = \sin t$ increases and decreases over and over again. Figure 2 shows that the motion of a vibrating mass on a spring is modeled very accurately by $y = \sin t$.

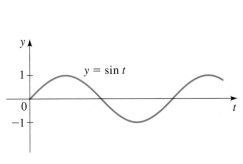

FIGURE 1 $y = \sin t$

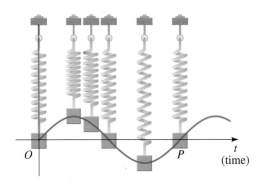

FIGURE 2 Motion of a vibrating spring is modeled by $y = \sin t$.

Notice that the mass returns to its original position over and over again. A **cycle** is one complete vibration of an object, so the mass in Figure 2 completes one cycle of its motion between *O* and *P*. Our observations about how the sine and cosine functions model periodic behavior are summarized in the following box.

The main difference between the two equations describing simple harmonic motion is the starting point. At $t = 0$ we get

$$y = a \sin \omega \cdot 0 = 0$$
$$y = a \cos \omega \cdot 0 = a$$

In the first case the motion "starts" with zero displacement, whereas in the second case the motion "starts" with the displacement at maximum (at the amplitude *a*).

SIMPLE HARMONIC MOTION

If the equation describing the displacement *y* of an object at time *t* is

$$y = a \sin \omega t \qquad \text{or} \qquad y = a \cos \omega t$$

then the object is in **simple harmonic motion**. In this case,

amplitude $= |a|$ Maximum displacement of the object

period $= \dfrac{2\pi}{\omega}$ Time required to complete one cycle

frequency $= \dfrac{\omega}{2\pi}$ Number of cycles per unit of time

---

The symbol ω is the lowercase Greek letter "omega," and ν is the letter "nu."

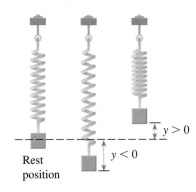

Rest position

$y > 0$

$y < 0$

FIGURE 3

Notice that the functions

$$y = a \sin 2\pi\nu t \qquad \text{and} \qquad y = a \cos 2\pi\nu t$$

have frequency ν, because $2\pi\nu/(2\pi) = \nu$. Since we can immediately read the frequency from these equations, we often write equations of simple harmonic motion in this form.

EXAMPLE 1 | A Vibrating Spring

The displacement of a mass suspended by a spring is modeled by the function

$$y = 10 \sin 4\pi t$$

where *y* is measured in inches and *t* in seconds (see Figure 3).

(a) Find the amplitude, period, and frequency of the motion of the mass.

(b) Sketch a graph of the displacement of the mass.

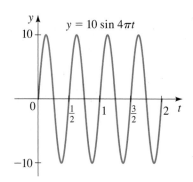

FIGURE 4

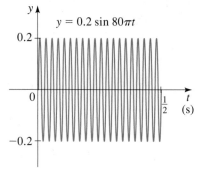

FIGURE 5

SOLUTION

(a) From the formulas for amplitude, period, and frequency we get

$$\text{amplitude} = |a| = 10 \text{ in.}$$

$$\text{period} = \frac{2\pi}{\omega} = \frac{2\pi}{4\pi} = \frac{1}{2} \text{ s}$$

$$\text{frequency} = \frac{\omega}{2\pi} = \frac{4\pi}{2\pi} = 2 \text{ cycles per second (Hz)}$$

(b) The graph of the displacement of the mass at time t is shown in Figure 4.

✎ NOW TRY EXERCISE **3** ■

An important situation in which simple harmonic motion occurs is in the production of sound. Sound is produced by a regular variation in air pressure from the normal pressure. If the pressure varies in simple harmonic motion, then a pure sound is produced. The tone of the sound depends on the frequency, and the loudness depends on the amplitude.

EXAMPLE 2 | Vibrations of a Musical Note

A tuba player plays the note E and sustains the sound for some time. For a pure E the variation in pressure from normal air pressure is given by

$$V(t) = 0.2 \sin 80\pi t$$

where V is measured in pounds per square inch and t is measured in seconds.

(a) Find the amplitude, period, and frequency of V.

(b) Sketch a graph of V.

(c) If the tuba player increases the loudness of the note, how does the equation for V change?

(d) If the player is playing the note incorrectly and it is a little flat, how does the equation for V change?

SOLUTION

(a) From the formulas for amplitude, period, and frequency we get

$$\text{amplitude} = |0.2| = 0.2$$

$$\text{period} = \frac{2\pi}{80\pi} = \frac{1}{40}$$

$$\text{frequency} = \frac{80\pi}{2\pi} = 40$$

(b) The graph of V is shown in Figure 5.

(c) If the player increases the loudness the amplitude increases. So the number 0.2 is replaced by a larger number.

(d) If the note is flat, then the frequency is decreased. Thus, the coefficient of t is less than 80π.

✎ NOW TRY EXERCISE **21** ■

EXAMPLE 3 | Modeling a Vibrating Spring

A mass is suspended from a spring. The spring is compressed a distance of 4 cm and then released. It is observed that the mass returns to the compressed position after $\frac{1}{3}$ s.

(a) Find a function that models the displacement of the mass.

(b) Sketch the graph of the displacement of the mass.

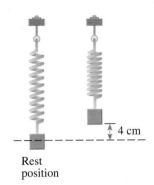

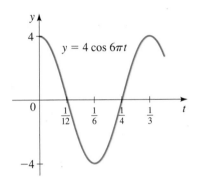

FIGURE 6

SOLUTION

(a) The motion of the mass is given by one of the equations for simple harmonic motion. The amplitude of the motion is 4 cm. Since this amplitude is reached at time $t = 0$, an appropriate function that models the displacement is of the form

$$y = a \cos \omega t$$

Since the period is $p = \frac{1}{3}$, we can find ω from the following equation:

$$\text{period} = \frac{2\pi}{\omega}$$

$$\frac{1}{3} = \frac{2\pi}{\omega} \qquad \text{Period} = \frac{1}{3}$$

$$\omega = 6\pi \qquad \text{Solve for } \omega$$

So the motion of the mass is modeled by the function

$$y = 4 \cos 6\pi t$$

where y is the displacement from the rest position at time t. Notice that when $t = 0$, the displacement is $y = 4$, as we expect.

(b) The graph of the displacement of the mass at time t is shown in Figure 6.

✎ NOW TRY EXERCISES **15** AND **27**

In general, the sine or cosine functions representing harmonic motion may be shifted horizontally or vertically. In this case, the equations take the form

$$y = a \sin(\omega(t - c)) + b \qquad \text{or} \qquad y = a \cos(\omega(t - c)) + b$$

The vertical shift b indicates that the variation occurs around an average value b. The horizontal shift c indicates the position of the object at $t = 0$. (See Figure 7.)

FIGURE 7

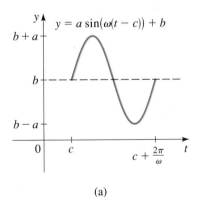

(a)

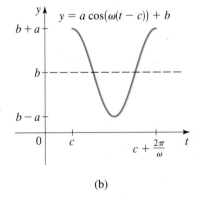

(b)

EXAMPLE 4 | Modeling the Brightness of a Variable Star

A variable star is one whose brightness alternately increases and decreases. For the variable star Delta Cephei, the time between periods of maximum brightness is 5.4 days. The average brightness (or magnitude) of the star is 4.0, and its brightness varies by ±0.35 magnitude.

(a) Find a function that models the brightness of Delta Cephei as a function of time.

(b) Sketch a graph of the brightness of Delta Cephei as a function of time.

SOLUTION

(a) Let's find a function in the form

$$y = a \cos(\omega(t - c)) + b$$

The amplitude is the maximum variation from average brightness, so the amplitude is $a = 0.35$ magnitude. We are given that the period is 5.4 days, so

$$\omega = \frac{2\pi}{5.4} \approx 1.164$$

Since the brightness varies from an average value of 4.0 magnitudes, the graph is shifted upward by $b = 4.0$. If we take $t = 0$ to be a time when the star is at maximum brightness, there is no horizontal shift, so $c = 0$ (because a cosine curve achieves its maximum at $t = 0$). Thus, the function we want is

$$y = 0.35 \cos(1.16t) + 4.0$$

where t is the number of days from a time when the star is at maximum brightness.

(b) The graph is sketched in Figure 8.

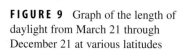

FIGURE 8

🔸 NOW TRY EXERCISE **31** ▪

The number of hours of daylight varies throughout the course of a year. In the Northern Hemisphere, the longest day is June 21, and the shortest is December 21. The average length of daylight is 12 h, and the variation from this average depends on the latitude. (For example, Fairbanks, Alaska, experiences more than 20 h of daylight on the longest day and less than 4 h on the shortest day!) The graph in Figure 9 shows the number of hours of daylight at different times of the year for various latitudes. It's apparent from the graph that the variation in hours of daylight is simple harmonic.

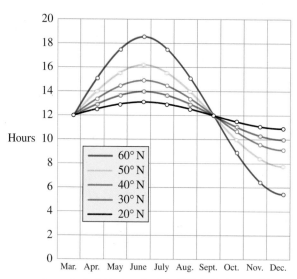

Hours

FIGURE 9 Graph of the length of daylight from March 21 through December 21 at various latitudes

Source: Lucia C. Harrison, Daylight, Twilight, Darkness and Time (New York: Silver, Burdett, 1935), page 40.

EXAMPLE 5 | Modeling the Number of Hours of Daylight

In Philadelphia ($40°$ N latitude) the longest day of the year has 14 h 50 min of daylight, and the shortest day has 9 h 10 min of daylight.

(a) Find a function L that models the length of daylight as a function of t, the number of days from January 1.

(b) An astronomer needs at least 11 hours of darkness for a long exposure astronomical photograph. On what days of the year are such long exposures possible?

SOLUTION

(a) We need to find a function in the form

$$y = a \sin(\omega(t - c)) + b$$

whose graph is the 40° N latitude curve in Figure 9. From the information given, we see that the amplitude is

$$a = \tfrac{1}{2}\left(14\tfrac{5}{6} - 9\tfrac{1}{6}\right) \approx 2.83 \text{ h}$$

Since there are 365 days in a year, the period is 365, so

$$\omega = \frac{2\pi}{365} \approx 0.0172$$

Since the average length of daylight is 12 h, the graph is shifted upward by 12, so $b = 12$. Since the curve attains the average value (12) on March 21, the 80th day of the year, the curve is shifted 80 units to the right. Thus, $c = 80$. So a function that models the number of hours of daylight is

$$y = 2.83 \sin(0.0172(t - 80)) + 12$$

where t is the number of days from January 1.

(b) A day has 24 h, so 11 h of night correspond to 13 h of daylight. So we need to solve the inequality $y \leq 13$. To solve this inequality graphically, we graph $y = 2.83 \sin 0.0172(t - 80) + 12$ and $y = 13$ on the same graph. From the graph in Figure 10 we see that there are fewer than 13 h of daylight between day 1 (January 1) and day 101 (April 11) and from day 241 (August 29) to day 365 (December 31).

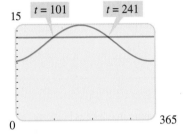

FIGURE 10

✎ NOW TRY EXERCISE **33**   ■

Another situation in which simple harmonic motion occurs is in alternating current (AC) generators. Alternating current is produced when an armature rotates about its axis in a magnetic field.

Figure 11 represents a simple version of such a generator. As the wire passes through the magnetic field, a voltage E is generated in the wire. It can be shown that the voltage generated is given by

$$E(t) = E_0 \cos \omega t$$

where E_0 is the maximum voltage produced (which depends on the strength of the magnetic field) and $\omega/(2\pi)$ is the number of revolutions per second of the armature (the frequency).

Why do we say that household current is 110 V when the maximum voltage produced is 155 V? From the symmetry of the cosine function we see that the average voltage produced is zero. This average value would be the same for all AC generators and so gives no information about the voltage generated. To obtain a more informative measure of voltage, engineers use the **root-mean-square** (rms) method. It can be shown that the rms voltage is $1/\sqrt{2}$ times the maximum voltage. So for household current the rms voltage is

$$155 \times \frac{1}{\sqrt{2}} \approx 110 \text{ V}$$

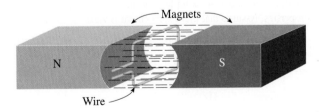

FIGURE 11

EXAMPLE 6 | Modeling Alternating Current

Ordinary 110-V household alternating current varies from $+155$ V to -155 V with a frequency of 60 Hz (cycles per second). Find an equation that describes this variation in voltage.

SOLUTION The variation in voltage is simple harmonic. Since the frequency is 60 cycles per second, we have

$$\frac{\omega}{2\pi} = 60 \qquad \text{or} \qquad \omega = 120\pi$$

Let's take $t = 0$ to be a time when the voltage is $+155$ V. Then

$$E(t) = a \cos \omega t = 155 \cos 120\pi t$$

✎ NOW TRY EXERCISE **35**   ■

2.6 EXERCISES

CONCEPTS

1–2 ■ An object is in simple harmonic motion with amplitude a and period $2\pi/\omega$. Find an equation that models the displacement y at time t under the given condition.

1. $y = 0$ at time $t = 0$: $y = $ _____.

2. $y = a$ at time $t = 0$: $y = $ _____.

SKILLS

3–10 ■ The given function models the displacement of an object moving in simple harmonic motion.
(a) Find the amplitude, period, and frequency of the motion.
(b) Sketch a graph of the displacement of the object over one complete period.

3. $y = 2 \sin 3t$ **4.** $y = 3 \cos \frac{1}{2}t$

5. $y = -\cos 0.3t$ **6.** $y = 2.4 \sin 3.6t$

7. $y = -0.25 \cos\left(1.5t - \dfrac{\pi}{3}\right)$ **8.** $y = -\frac{3}{2} \sin(0.2t + 1.4)$

9. $y = 5 \cos\left(\frac{2}{3}t + \frac{3}{4}\right)$ **10.** $y = 1.6 \sin(t - 1.8)$

11–14 ■ Find a function that models the simple harmonic motion having the given properties. Assume that the displacement is zero at time $t = 0$.

11. amplitude 10 cm,   period 3 s

12. amplitude 24 ft,   period 2 min

13. amplitude 6 in.,   frequency $5/\pi$ Hz

14. amplitude 1.2 m,   frequency 0.5 Hz

15–18 ■ Find a function that models the simple harmonic motion having the given properties. Assume that the displacement is at its maximum at time $t = 0$.

15. amplitude 60 ft,   period 0.5 min

16. amplitude 35 cm,   period 8 s

17. amplitude 2.4 m,   frequency 750 Hz

18. amplitude 6.25 in.,   frequency 60 Hz

APPLICATIONS

19. A Bobbing Cork   A cork floating in a lake is bobbing in simple harmonic motion. Its displacement above the bottom of the lake is modeled by

$$y = 0.2 \cos 20\pi t + 8$$

where y is measured in meters and t is measured in minutes.
(a) Find the frequency of the motion of the cork.
(b) Sketch a graph of y.
(c) Find the maximum displacement of the cork above the lake bottom.

20. FM Radio Signals   The carrier wave for an FM radio signal is modeled by the function

$$y = a \sin(2\pi(9.15 \times 10^7)t)$$

where t is measured in seconds. Find the period and frequency of the carrier wave.

21. Blood Pressure   Each time your heart beats, your blood pressure increases, then decreases as the heart rests between beats. A certain person's blood pressure is modeled by the function

$$p(t) = 115 + 25 \sin(160\pi t)$$

where $p(t)$ is the pressure in mmHg at time t, measured in minutes.
(a) Find the amplitude, period, and frequency of p.
(b) Sketch a graph of p.
(c) If a person is exercising, his or her heart beats faster. How does this affect the period and frequency of p?

22. Predator Population Model   In a predator/prey model the predator population is modeled by the function

$$y = 900 \cos 2t + 8000$$

where t is measured in years.
(a) What is the maximum population?
(b) Find the length of time between successive periods of maximum population.

23. Spring–Mass System A mass attached to a spring is moving up and down in simple harmonic motion. The graph gives its displacement $d(t)$ from equilibrium at time t. Express the function d in the form $d(t) = a \sin \omega t$.

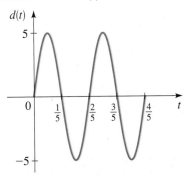

24. Tides The graph shows the variation of the water level relative to mean sea level in Commencement Bay at Tacoma, Washington, for a particular 24-hour period. Assuming that this variation is modeled by simple harmonic motion, find an equation of the form $y = a \sin \omega t$ that describes the variation in water level as a function of the number of hours after midnight.

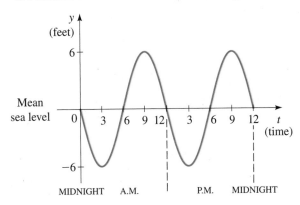

25. Tides The Bay of Fundy in Nova Scotia has the highest tides in the world. In one 12-hour period the water starts at mean sea level, rises to 21 ft above, drops to 21 ft below, then returns to mean sea level. Assuming that the motion of the tides is simple harmonic, find an equation that describes the height of the tide in the Bay of Fundy above mean sea level. Sketch a graph that shows the level of the tides over a 12-hour period.

26. Spring–Mass System A mass suspended from a spring is pulled down a distance of 2 ft from its rest position, as shown in the figure. The mass is released at time $t = 0$ and allowed to oscillate. If the mass returns to this position after 1 s, find an equation that describes its motion.

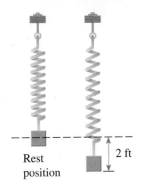

27. Spring–Mass System A mass is suspended on a spring. The spring is compressed so that the mass is located 5 cm above its rest position. The mass is released at time $t = 0$ and allowed to oscillate. It is observed that the mass reaches its lowest point $\frac{1}{2}$ s after it is released. Find an equation that describes the motion of the mass.

28. Spring–Mass System The frequency of oscillation of an object suspended on a spring depends on the stiffness k of the spring (called the *spring constant*) and the mass m of the object. If the spring is compressed a distance a and then allowed to oscillate, its displacement is given by

$$f(t) = a \cos \sqrt{k/m}\; t$$

(a) A 10-g mass is suspended from a spring with stiffness $k = 3$. If the spring is compressed a distance 5 cm and then released, find the equation that describes the oscillation of the spring.

(b) Find a general formula for the frequency (in terms of k and m).

(c) How is the frequency affected if the mass is increased? Is the oscillation faster or slower?

(d) How is the frequency affected if a stiffer spring is used (larger k)? Is the oscillation faster or slower?

29. Ferris Wheel A ferris wheel has a radius of 10 m, and the bottom of the wheel passes 1 m above the ground. If the ferris wheel makes one complete revolution every 20 s, find an equation that gives the height above the ground of a person on the ferris wheel as a function of time.

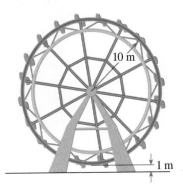

30. Clock Pendulum The pendulum in a grandfather clock makes one complete swing every 2 s. The maximum angle that the pendulum makes with respect to its rest position is 10°. We know from physical principles that the angle θ between the pendulum and its rest position changes in simple harmonic fashion. Find an equation that describes the size of the angle θ as a function of time. (Take $t = 0$ to be a time when the pendulum is vertical.)

31. Variable Stars The variable star Zeta Gemini has a period of 10 days. The average brightness of the star is 3.8 magnitudes, and the maximum variation from the average is 0.2 magnitude. Assuming that the variation in brightness is simple harmonic, find an equation that gives the brightness of the star as a function of time.

32. Variable Stars Astronomers believe that the radius of a variable star increases and decreases with the brightness of the star. The variable star Delta Cephei (Example 4) has an average radius of 20 million miles and changes by a maximum of 1.5 million miles from this average during a single pulsation. Find an equation that describes the radius of this star as a function of time.

33. Biological Clocks *Circadian rhythms* are biological processes that oscillate with a period of approximately 24 hours. That is, a circadian rhythm is an internal daily biological clock. Blood pressure appears to follow such a rhythm. For a certain individual the average resting blood pressure varies from a maximum of 100 mmHg at 2:00 P.M. to a minimum of 80 mmHg at 2:00 A.M. Find a sine function of the form

$$f(t) = a \sin(\omega(t - c)) + b$$

that models the blood pressure at time t, measured in hours from midnight.

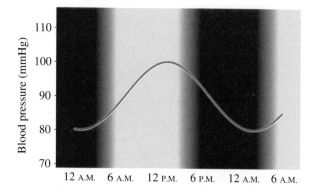

34. Electric Generator The armature in an electric generator is rotating at the rate of 100 revolutions per second (rps). If the maximum voltage produced is 310 V, find an equation that describes this variation in voltage? What is the rms voltage? (See Example 6 and the margin note adjacent to it.)

35. Electric Generator The graph shows an oscilloscope reading of the variation in voltage of an AC current produced by a simple generator.
  (a) Find the maximum voltage produced.
  (b) Find the frequency (cycles per second) of the generator.

(c) How many revolutions per second does the armature in the generator make?
(d) Find a formula that describes the variation in voltage as a function of time.

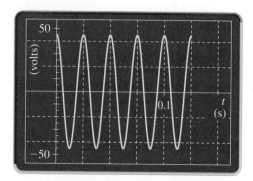

36. Doppler Effect When a car with its horn blowing drives by an observer, the pitch of the horn seems higher as it approaches and lower as it recedes (see the figure below). This phenomenon is called the **Doppler effect**. If the sound source is moving at speed v relative to the observer and if the speed of sound is v_0, then the perceived frequency f is related to the actual frequency f_0 as follows:

$$f = f_0 \left(\frac{v_0}{v_0 \pm v} \right)$$

We choose the minus sign if the source is moving toward the observer and the plus sign if it is moving away.

Suppose that a car drives at 110 ft/s past a woman standing on the shoulder of a highway, blowing its horn, which has a frequency of 500 Hz. Assume that the speed of sound is 1130 ft/s. (This is the speed in dry air at 70°F.)
  (a) What are the frequencies of the sounds that the woman hears as the car approaches her and as it moves away from her?
  (b) Let A be the amplitude of the sound. Find functions of the form

$$y = A \sin \omega t$$

that model the perceived sound as the car approaches the woman and as it recedes.

CHAPTER 2 | REVIEW

■ CONCEPT CHECK

1. (a) What is the unit circle?
  (b) Use a diagram to explain what is meant by the terminal point determined by a real number t.
  (c) What is the reference number $\bar{t}$ associated with t?
  (d) If t is a real number and $P(x, y)$ is the terminal point determined by t, write equations that define $\sin t$, $\cos t$, $\tan t$, $\cot t$, $\sec t$, and $\csc t$.
  (e) What are the domains of the six functions that you defined in part (d)?
  (f) Which trigonometric functions are positive in Quadrants I, II, III, and IV?

2. (a) What is an even function?
  (b) Which trigonometric functions are even?
  (c) What is an odd function?
  (d) Which trigonometric functions are odd?

3. (a) State the reciprocal identities.
  (b) State the Pythagorean identities.

4. (a) What is a periodic function?
  (b) What are the periods of the six trigonometric functions?

5. Graph the sine and cosine functions. How is the graph of cosine related to the graph of sine?

6. Write expressions for the amplitude, period, and phase shift of the sine curve $y = a \sin k(x - b)$ and the cosine curve $y = a \cos k(x - b)$.

7. (a) Graph the tangent and cotangent functions.
  (b) State the periods of the tangent curve $y = a \tan kx$ and the cotangent curve $y = a \cot kx$.

8. (a) Graph the secant and cosecant functions.
  (b) State the periods of the secant curve $y = a \sec kx$ and the cosecant curve $y = a \csc kx$.

9. (a) Define the inverse sine function $\sin^{-1}$. What are its domain and range?
  (b) For what values of x is the equation $\sin(\sin^{-1} x) = x$ true?
  (c) For what values of x is the equation $\sin^{-1}(\sin x) = x$ true?

10. (a) Define the inverse cosine function $\cos^{-1}$. What are its domain and range?
  (b) For what values of x is the equation $\cos(\cos^{-1} x) = x$ true?
  (c) For what values of x is the equation $\cos^{-1}(\cos x) = x$ true?

11. (a) Define the inverse tangent function $\tan^{-1}$. What are its domain and range?
  (b) For what values of x is the equation $\tan(\tan^{-1} x) = x$ true?
  (c) For what values of x is the equation $\tan^{-1}(\tan x) = x$ true?

12. (a) What is simple harmonic motion?
  (b) Give three real-life examples of simple harmonic motion.

■ EXERCISES

1–2 ■ A point $P(x, y)$ is given.
(a) Show that P is on the unit circle.
(b) Suppose that P is the terminal point determined by t. Find $\sin t$, $\cos t$, and $\tan t$.

1. $P\left(-\dfrac{\sqrt{3}}{2}, \dfrac{1}{2}\right)$ **2.** $P\left(\dfrac{3}{5}, -\dfrac{4}{5}\right)$

3–6 ■ A real number t is given.
(a) Find the reference number for t.
(b) Find the terminal point $P(x, y)$ on the unit circle determined by t.
(c) Find the six trigonometric functions of t.

3. $t = \dfrac{2\pi}{3}$ **4.** $t = \dfrac{5\pi}{3}$

5. $t = -\dfrac{11\pi}{4}$ **6.** $t = -\dfrac{7\pi}{6}$

7–16 ■ Find the value of the trigonometric function. If possible, give the exact value; otherwise, use a calculator to find an approximate value correct to five decimal places.

7. (a) $\sin \dfrac{3\pi}{4}$ **(b)** $\cos \dfrac{3\pi}{4}$

8. (a) $\tan \dfrac{\pi}{3}$ **(b)** $\tan\left(-\dfrac{\pi}{3}\right)$

9. (a) $\sin 1.1$ **(b)** $\cos 1.1$

10. (a) $\cos \dfrac{\pi}{5}$ **(b)** $\cos\left(-\dfrac{\pi}{5}\right)$

11. (a) $\cos \dfrac{9\pi}{2}$ **(b)** $\sec \dfrac{9\pi}{2}$

12. (a) $\sin \dfrac{\pi}{7}$ **(b)** $\csc \dfrac{\pi}{7}$

13. (a) $\tan \dfrac{5\pi}{2}$ **(b)** $\cot \dfrac{5\pi}{2}$

14. (a) $\sin 2\pi$ **(b)** $\csc 2\pi$

15. (a) $\tan \dfrac{5\pi}{6}$ **(b)** $\cot \dfrac{5\pi}{6}$

16. (a) $\cos \dfrac{\pi}{3}$ **(b)** $\sin \dfrac{\pi}{6}$

17–20 ■ Use the fundamental identities to write the first expression in terms of the second.

17. $\dfrac{\tan t}{\cos t}$, $\sin t$ **18.** $\tan^2 t \sec t$, $\cos t$

19. $\tan t$, $\sin t$; t in Quadrant IV

20. $\sec t$, $\sin t$; t in Quadrant II

21–24 ■ Find the values of the remaining trigonometric functions at t from the given information.

21. $\sin t = \frac{5}{13}$, $\cos t = -\frac{12}{13}$

22. $\sin t = -\frac{1}{2}$, $\cos t > 0$

23. $\cot t = -\frac{1}{2}$, $\csc t = \sqrt{5}/2$

24. $\cos t = -\frac{3}{5}$, $\tan t < 0$

25. If $\tan t = \frac{1}{4}$ and the terminal point for t is in Quadrant III, find $\sec t + \cot t$.

26. If $\sin t = -\frac{8}{17}$ and the terminal point for t is in Quadrant IV, find $\csc t + \sec t$.

27. If $\cos t = \frac{3}{5}$ and the terminal point for t is in Quadrant I, find $\tan t + \sec t$.

28. If $\sec t = -5$ and the terminal point for t is in Quadrant II, find $\sin^2 t + \cos^2 t$.

29–36 ■ A trigonometric function is given.
(a) Find the amplitude, period, and phase shift of the function.
(b) Sketch the graph.

29. $y = 10 \cos \frac{1}{2}x$

30. $y = 4 \sin 2\pi x$

31. $y = -\sin \frac{1}{2}x$

32. $y = 2 \sin\left(x - \frac{\pi}{4}\right)$

33. $y = 3 \sin(2x - 2)$

34. $y = \cos 2\left(x - \frac{\pi}{2}\right)$

35. $y = -\cos\left(\frac{\pi}{2}x + \frac{\pi}{6}\right)$

36. $y = 10 \sin\left(2x - \frac{\pi}{2}\right)$

37–40 ■ The graph of one period of a function of the form $y = a \sin k(x - b)$ or $y = a \cos k(x - b)$ is shown. Determine the function.

37.

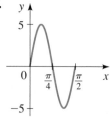

38.

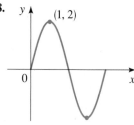

(1, 2)

39.

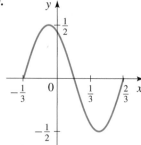

40.

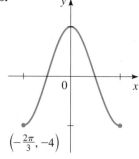

$\left(-\frac{2\pi}{3}, -4\right)$

41–48 ■ Find the period, and sketch the graph.

41. $y = 3 \tan x$

42. $y = \tan \pi x$

43. $y = 2 \cot\left(x - \frac{\pi}{2}\right)$

44. $y = \sec\left(\frac{1}{2}x - \frac{\pi}{2}\right)$

45. $y = 4 \csc(2x + \pi)$

46. $y = \tan\left(x + \frac{\pi}{6}\right)$

47. $y = \tan\left(\frac{1}{2}x - \frac{\pi}{8}\right)$

48. $y = -4 \sec 4\pi x$

49–52 ■ Find the exact value of each expression, if it is defined.

49. $\sin^{-1} 1$

50. $\cos^{-1}\left(-\frac{1}{2}\right)$

51. $\sin^{-1}\left(\sin \frac{13\pi}{6}\right)$

52. $\tan\left(\cos^{-1}\left(\frac{1}{2}\right)\right)$

 53–58 ■ A function is given.
(a) Use a graphing device to graph the function.
(b) Determine from the graph whether the function is periodic and, if so, determine the period.
(c) Determine from the graph whether the function is odd, even, or neither.

53. $y = |\cos x|$

54. $y = \sin(\cos x)$

55. $y = \cos(2^{0.1x})$

56. $y = 1 + 2^{\cos x}$

57. $y = |x| \cos 3x$

58. $y = \sqrt{x} \sin 3x$ $(x > 0)$

 59–62 ■ Graph the three functions on a common screen. How are the graphs related?

59. $y = x$, $y = -x$, $y = x \sin x$

60. $y = 2^{-x}$, $y = -2^{-x}$, $y = 2^{-x} \cos 4\pi x$

61. $y = x$, $y = \sin 4x$, $y = x + \sin 4x$

62. $y = \sin^2 x$, $y = \cos^2 x$, $y = \sin^2 x + \cos^2 x$

 63–64 ■ Find the maximum and minimum values of the function.

63. $y = \cos x + \sin 2x$

64. $y = \cos x + \sin^2 x$

 65. Find the solutions of $\sin x = 0.3$ in the interval $[0, 2\pi]$.

66. Find the solutions of $\cos 3x = x$ in the interval $[0, \pi]$.

67. Let $f(x) = \dfrac{\sin^2 x}{x}$.
(a) Is the function f even, odd, or neither?
(b) Find the x-intercepts of the graph of f.
(c) Graph f in an appropriate viewing rectangle.
(d) Describe the behavior of the function as x becomes large.
(e) Notice that $f(x)$ is not defined when $x = 0$. What happens as x approaches 0?

68. Let $y_1 = \cos(\sin x)$ and $y_2 = \sin(\cos x)$.
(a) Graph y_1 and y_2 in the same viewing rectangle.
(b) Determine the period of each of these functions from its graph.
(c) Find an inequality between $\sin(\cos x)$ and $\cos(\sin x)$ that is valid for all x.

69. A point P moving in simple harmonic motion completes 8 cycles every second. If the amplitude of the motion is 50 cm, find an equation that describes the motion of P as a function of time. Assume that the point P is at its maximum displacement when $t = 0$.

70. A mass suspended from a spring oscillates in simple harmonic motion at a frequency of 4 cycles per second. The distance from the highest to the lowest point of the oscillation is 100 cm. Find an equation that describes the distance of the mass from its rest position as a function of time. Assume that the mass is at its lowest point when $t = 0$.

71. The graph shows the variation of the water level relative to mean sea level in the Long Beach harbor for a particular 24-hour period. Assuming that this variation is simple harmonic, find an equation of the form $y = a \cos \omega t$ that describes the variation in water level as a function of the number of hours after midnight.

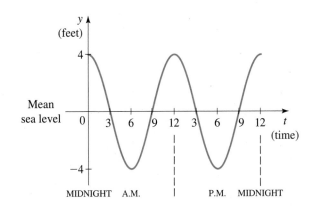

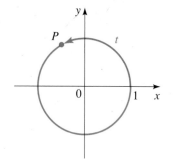

1. The point $P(x, y)$ is on the unit circle in Quadrant IV. If $x = \sqrt{11}/6$, find y.

2. The point P in the figure at the left has y-coordinate $\frac{4}{5}$. Find:

   (a) $\sin t$ (b) $\cos t$

   (c) $\tan t$ (d) $\sec t$

3. Find the exact value.

   (a) $\sin \dfrac{7\pi}{6}$ (b) $\cos \dfrac{13\pi}{4}$

   (c) $\tan\left(-\dfrac{5\pi}{3}\right)$ (d) $\csc \dfrac{3\pi}{2}$

4. Express $\tan t$ in terms of $\sin t$, if the terminal point determined by t is in Quadrant II.

5. If $\cos t = -\frac{8}{17}$ and if the terminal point determined by t is in Quadrant III, find $\tan t \cot t + \csc t$.

6–7 ■ A trigonometric function is given.

    (a) Find the amplitude, period, and phase shift of the function.

    (b) Sketch the graph.

6. $y = -5 \cos 4x$ 7. $y = 2 \sin\left(\dfrac{1}{2}x - \dfrac{\pi}{6}\right)$

8–9 ■ Find the period, and graph the function.

8. $y = -\csc 2x$ 9. $y = \tan\left(2x - \dfrac{\pi}{2}\right)$

10. Find the exact value of each expression, if it is defined.

    (a) $\tan^{-1} 1$ (b) $\cos^{-1}\left(-\dfrac{\sqrt{3}}{2}\right)$

    (c) $\tan^{-1}(\tan 3\pi)$ (d) $\cos(\tan^{-1}(-\sqrt{3}))$

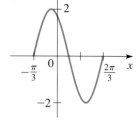

11. The graph shown at left is one period of a function of the form $y = a \sin k(x - b)$. Determine the function.

12. Let $f(x) = \dfrac{\cos x}{1 + x^2}$.

    (a) Use a graphing device to graph f in an appropriate viewing rectangle.

    (b) Determine from the graph if f is even, odd, or neither.

    (c) Find the minimum and maximum values of f.

13. A mass suspended from a spring oscillates in simple harmonic motion. The mass completes 2 cycles every second, and the distance between the highest point and the lowest point of the oscillation is 10 cm. Find an equation of the form $y = a \sin \omega t$ that gives the distance of the mass from its rest position as a function of time.

In the *Focus on Modeling* section at the end of Chapter 1, we learned how to fit linear models to data. Figure 1 shows some scatter plots of data. The scatter plots can help guide us in choosing an appropriate model. (Try to determine what type of function would best model the data in each graph.) If the scatter plot indicates simple harmonic motion, then we might try to model the data with a sine or cosine function. The next example illustrates this process.

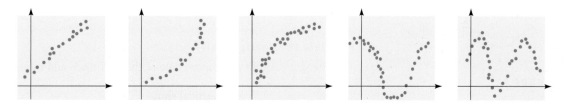

FIGURE 1

EXAMPLE 1 │ Modeling the Height of a Tide

The water depth in a narrow channel varies with the tides. Table 1 shows the water depth over a 12-hour period.

(a) Make a scatter plot of the water depth data.

(b) Find a function that models the water depth with respect to time.

(c) If a boat needs at least 11 ft of water to cross the channel, during which times can it safely do so?

TABLE 1

Time	Depth (ft)
12:00 A.M.	9.8
1:00 A.M.	11.4
2:00 A.M.	11.6
3:00 A.M.	11.2
4:00 A.M.	9.6
5:00 A.M.	8.5
6:00 A.M.	6.5
7:00 A.M.	5.7
8:00 A.M.	5.4
9:00 A.M.	6.0
10:00 A.M.	7.0
11:00 A.M.	8.6
12:00 P.M.	10.0

SOLUTION

(a) A scatter plot of the data is shown in Figure 2.

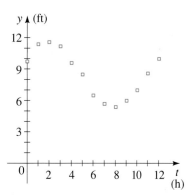

FIGURE 2

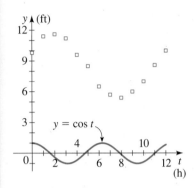

FIGURE 3

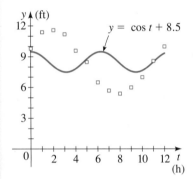

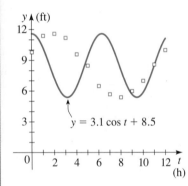

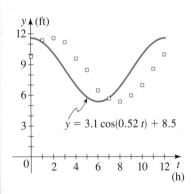

(b) The data appear to lie on a cosine (or sine) curve. But if we graph $y = \cos t$ on the same graph as the scatter plot, the result in Figure 3 is not even close to the data. To fit the data, we need to adjust the vertical shift, amplitude, period, and phase shift of the cosine curve. In other words, we need to find a function of the form

$$y = a\cos(\omega(t - c)) + b$$

We use the following steps, which are illustrated by the graphs in the margin.

▶ **Adjust the Vertical Shift**

The vertical shift b is the average of the maximum and minimum values:

$$b = \text{vertical shift}$$

$$= \frac{1}{2} \cdot (\text{maximum value} + \text{minimum value})$$

$$= \frac{1}{2}(11.6 + 5.4) = 8.5$$

▶ **Adjust the Amplitude**

The amplitude a is half of the difference between the maximum and minimum values:

$$a = \text{amplitude}$$

$$= \frac{1}{2} \cdot (\text{maximum value} - \text{minimum value})$$

$$= \frac{1}{2}(11.6 - 5.4) = 3.1$$

▶ **Adjust the Period**

The time between consecutive maximum and minimum values is half of one period. Thus

$$\frac{2\pi}{\omega} = \text{period}$$

$$= 2 \cdot (\text{time of maximum value} - \text{time of minimum value})$$

$$= 2(8 - 2) = 12$$

Thus, $\omega = 2\pi/12 = 0.52$.

▶ **Adjust the Horizontal Shift**

Since the maximum value of the data occurs at approximately $t = 2.0$, it represents a cosine curve shifted 2 h to the right. So

$$c = \text{phase shift}$$

$$= \text{time of maximum value}$$

$$= 2.0$$

▶ **The Model**

We have shown that a function that models the tides over the given time period is given by

$$y = 3.1\cos(0.52(t - 2.0)) + 8.5$$

A graph of the function and the scatter plot are shown in Figure 4. It appears that the model we found is a good approximation to the data.

(c) We need to solve the inequality $y \geq 11$. We solve this inequality graphically by graphing $y = 3.1 \cos 0.52(t - 2.0) + 8.5$ and $y = 11$ on the same graph. From the graph in Figure 5 we see the water depth is higher than 11 ft between $t \approx 0.8$ and $t \approx 3.2$. This corresponds to the times 12:48 A.M. to 3:12 A.M.

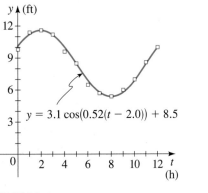

FIGURE 4

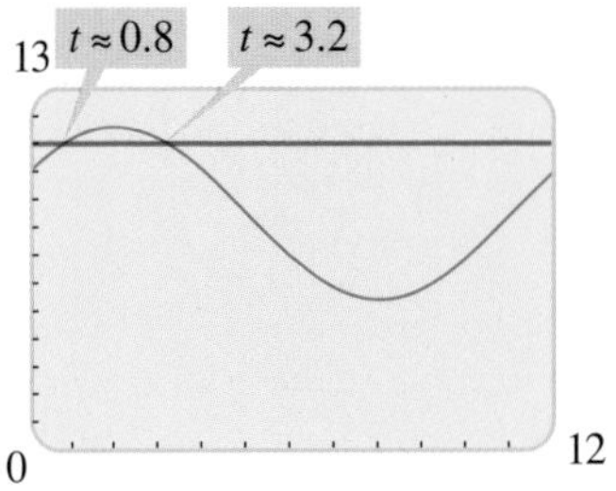

FIGURE 5

For the TI-83 and TI-84 the command SinReg (for sine regression) finds the sine curve that best fits the given data.

In Example 1 we used the scatter plot to guide us in finding a cosine curve that gives an approximate model of the data. Some graphing calculators are capable of finding a sine or cosine curve that best fits a given set of data points. The method these calculators use is similar to the method of finding a line of best fit, as explained on page 93.

EXAMPLE 2 | Fitting a Sine Curve to Data

(a) Use a graphing device to find the sine curve that best fits the depth of water data in Table 1 on page 159.

(b) Compare your result to the model found in Example 1.

SOLUTION

(a) Using the data in Table 1 and the SinReg command on the TI-83 calculator, we get a function of the form

$$y = a \sin(bt + c) + d$$

where

$$a = 3.1 \qquad b = 0.53$$
$$c = 0.55 \qquad d = 8.42$$

So the sine function that best fits the data is

$$y = 3.1 \sin(0.53t + 0.55) + 8.42$$

(b) To compare this with the function in Example 1, we change the sine function to a cosine function by using the reduction formula $\sin u = \cos(u - \pi/2)$.

$$y = 3.1 \sin(0.53t + 0.55) + 8.42$$

$$= 3.1 \cos\left(0.53t + 0.55 - \frac{\pi}{2}\right) + 8.42 \qquad \text{Reduction formula}$$

$$= 3.1 \cos(0.53t - 1.02) + 8.42$$

$$= 3.1 \cos(0.53(t - 1.92)) + 8.42 \qquad \text{Factor 0.53}$$

```
SinReg
y=a*sin(bx+c)+d
a=3.097877596
b=.5268322697
c=.5493035195
d=8.424021899
```

Output of the SinReg function on the TI-83.

Comparing this with the function we obtained in Example 1, we see that there are small differences in the coefficients. In Figure 6 we graph a scatter plot of the data together with the sine function of best fit.

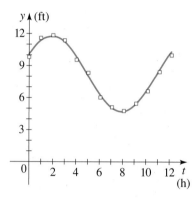

FIGURE 6

In Example 1 we estimated the values of the amplitude, period, and shifts from the data. In Example 2 the calculator computed the sine curve that best fits the data (that is, the curve that deviates least from the data as explained on page 93). The different ways of obtaining the model account for the differences in the functions.

PROBLEMS

1–4 ■ Modeling Periodic Data   A set of data is given.

(a) Make a scatter plot of the data.

(b) Find a cosine function of the form $y = a \cos(\omega(t - c)) + b$ that models the data, as in Example 1.

(c) Graph the function you found in part (b) together with the scatter plot. How well does the curve fit the data?

(d) Use a graphing calculator to find the sine function that best fits the data, as in Example 2.

(e) Compare the functions you found in parts (b) and (d). [Use the reduction formula $\sin u = \cos(u - \pi/2)$.]

1.

t	y
0	2.1
2	1.1
4	−0.8
6	−2.1
8	−1.3
10	0.6
12	1.9
14	1.5

2.

t	y
0	190
25	175
50	155
75	125
100	110
125	95
150	105
175	120
200	140
225	165
250	185
275	200
300	195
325	185
350	165

3.

t	y
0.1	21.1
0.2	23.6
0.3	24.5
0.4	21.7
0.5	17.5
0.6	12.0
0.7	5.6
0.8	2.2
0.9	1.0
1.0	3.5
1.1	7.6
1.2	13.2
1.3	18.4
1.4	23.0
1.5	25.1

4.

t	y
0.0	0.56
0.5	0.45
1.0	0.29
1.5	0.13
2.0	0.05
2.5	−0.10
3.0	0.02
3.5	0.12
4.0	0.26
4.5	0.43
5.0	0.54
5.5	0.63
6.0	0.59

5. Annual Temperature Change   The table gives the average monthly temperature in Montgomery County, Maryland.

(a) Make a scatter plot of the data.

(b) Find a cosine curve that models the data (as in Example 1).

(c) Graph the function you found in part (b) together with the scatter plot.

(d) Use a graphing calculator to find the sine curve that best fits the data (as in Example 2).

Month	Average temperature (°F)	Month	Average temperature (°F)
January	40.0	July	85.8
February	43.1	August	83.9
March	54.6	September	76.9
April	64.2	October	66.8
May	73.8	November	55.5
June	81.8	December	44.5

6. Circadian Rhythms   Circadian rhythm (from the Latin *circa*—about, and *diem*—day) is the daily biological pattern by which body temperature, blood pressure, and other physiological variables change. The data in the table below show typical changes in human body temperature over a 24-hour period ($t = 0$ corresponds to midnight).

(a) Make a scatter plot of the data.

(b) Find a cosine curve that models the data (as in Example 1).

(c) Graph the function you found in part (b) together with the scatter plot.

(d) Use a graphing calculator to find the sine curve that best fits the data (as in Example 2).

Time	Body temperature (°C)	Time	Body temperature (°C)
0	36.8	14	37.3
2	36.7	16	37.4
4	36.6	18	37.3
6	36.7	20	37.2
8	36.8	22	37.0
10	37.0	24	36.8
12	37.2		

7. Predator Population   When two species interact in a predator/prey relationship, the populations of both species tend to vary in a sinusoidal fashion. (See the Discovery Project *Predator/Prey Models* referenced on page 132). In a certain midwestern county, the main food source for barn owls consists of field mice and other small mammals. The table gives the population of barn owls in this county every July 1 over a 12-year period.

(a) Make a scatter plot of the data.

(b) Find a sine curve that models the data (as in Example 1).

(c) Graph the function you found in part (b) together with the scatter plot.

(d) Use a graphing calculator to find the sine curve that best fits the data (as in Example 2). Compare to your answer from part (b).

Year	Owl population
0	50
1	62
2	73
3	80
4	71
5	60
6	51
7	43
8	29
9	20
10	28
11	41
12	49

8. Salmon Survival For reasons not yet fully understood, the number of fingerling salmon that survive the trip from their riverbed spawning grounds to the open ocean varies approximately sinusoidally from year to year. The table shows the number of salmon that hatch in a certain British Columbia creek and then make their way to the Strait of Georgia. The data is given in thousands of fingerlings, over a period of 16 years.

(a) Make a scatter plot of the data.

(b) Find a sine curve that models the data (as in Example 1).

(c) Graph the function you found in part (b) together with the scatter plot.

(d) Use a graphing calculator to find the sine curve that best fits the data (as in Example 2). Compare to your answer from part (b).

Year	Salmon ($\times$ 1000)	Year	Salmon ($\times$ 1000)
1985	43	1993	56
1986	36	1994	63
1987	27	1995	57
1988	23	1996	50
1989	26	1997	44
1990	33	1998	38
1991	43	1999	30
1992	50	2000	22

9. Sunspot Activity Sunspots are relatively "cool" regions on the sun that appear as dark spots when observed through special solar filters. The number of sunspots varies in an 11-year cycle. The table gives the average daily sunspot count for the years 1975–2004.

(a) Make a scatter plot of the data.

(b) Find a cosine curve that models the data (as in Example 1).

(c) Graph the function you found in part (b) together with the scatter plot.

(d) Use a graphing calculator to find the sine curve that best fits the data (as in Example 2). Compare to your answer in part (b).

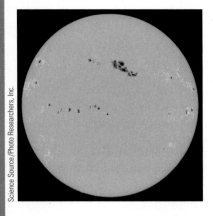

Year	Sunspots	Year	Sunspots
1975	16	1990	143
1976	13	1991	146
1977	28	1992	94
1978	93	1993	55
1979	155	1994	30
1980	155	1995	18
1981	140	1996	9
1982	116	1997	21
1983	67	1998	64
1984	46	1999	93
1985	18	2000	119
1986	13	2001	111
1987	29	2002	104
1988	100	2003	64
1989	158	2004	40

TRIGONOMETRIC FUNCTIONS: RIGHT TRIANGLE APPROACH

© age fotostock /SuperStock

Suppose we want to find the distance from the earth to the sun. Using a tape measure is obviously impractical, so we need something other than simple measurements to tackle this problem. Angles are easier to measure than distances. For example, we can find the angle formed by the sun, earth, and moon by simply pointing to the sun with one arm and to the moon with the other and estimating the angle between them. The key idea is to find relationships between angles and distances. So if we had a way of determining distances from angles, we would be able to find the distance to the sun without having to go there. The trigonometric functions provide us with just the tools we need.

If θ is an angle in a right triangle, then the trigonometric ratio $\sin \theta$ is defined as the length of the side opposite θ divided by the length of the hypotenuse. This ratio is the same in *any* similar right triangle, including the huge triangle formed by the sun, earth, and moon! (See Section 3.2, Exercise 61.)

The trigonometric functions can be defined in two different but equivalent ways: as functions of real numbers (Chapter 2) or as functions of angles (Chapter 3). The two approaches are independent of each other, so either Chapter 2 or Chapter 3 may be studied first. We study both approaches because the different approaches are required for different applications.

3.1 ANGLE MEASURE

> Angle Measure ▶ Angles in Standard Position ▶ Length of a Circular Arc
> ▶ Area of a Circular Sector ▶ Circular Motion

An **angle** AOB consists of two rays R_1 and R_2 with a common vertex O (see Figure 1). We often interpret an angle as a rotation of the ray R_1 onto R_2. In this case, R_1 is called the **initial side**, and R_2 is called the **terminal side** of the angle. If the rotation is counterclockwise, the angle is considered **positive**, and if the rotation is clockwise, the angle is considered **negative**.

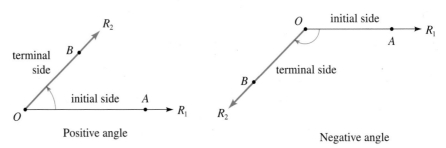

FIGURE 1

▼ Angle Measure

The **measure** of an angle is the amount of rotation about the vertex required to move R_1 onto R_2. Intuitively, this is how much the angle "opens." One unit of measurement for angles is the **degree**. An angle of measure 1 degree is formed by rotating the initial side $\frac{1}{360}$ of a complete revolution. In calculus and other branches of mathematics, a more natural method of measuring angles is used—*radian measure*. The amount an angle opens is measured along the arc of a circle of radius 1 with its center at the vertex of the angle.

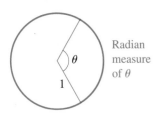

Radian measure of θ

FIGURE 2

DEFINITION OF RADIAN MEASURE

If a circle of radius 1 is drawn with the vertex of an angle at its center, then the measure of this angle in **radians** (abbreviated **rad**) is the length of the arc that subtends the angle (see Figure 2).

The circumference of the circle of radius 1 is 2π and so a complete revolution has measure 2π rad, a straight angle has measure π rad, and a right angle has measure $\pi/2$ rad. An angle that is subtended by an arc of length 2 along the unit circle has radian measure 2 (see Figure 3).

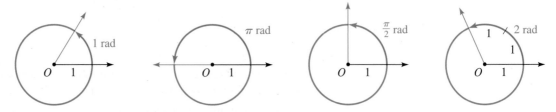

FIGURE 3 Radian measure

Since a complete revolution measured in degrees is $360°$ and measured in radians is 2π rad, we get the following simple relationship between these two methods of angle measurement.

RELATIONSHIP BETWEEN DEGREES AND RADIANS

$$180° = \pi \text{ rad} \qquad 1 \text{ rad} = \left(\frac{180}{\pi}\right)° \qquad 1° = \frac{\pi}{180} \text{ rad}$$

1. To convert degrees to radians, multiply by $\dfrac{\pi}{180}$.

2. To convert radians to degrees, multiply by $\dfrac{180}{\pi}$.

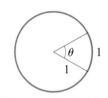

Measure of $\theta = 1$ rad
Measure of $\theta \approx 57.296°$

FIGURE 4

To get some idea of the size of a radian, notice that

$$1 \text{ rad} \approx 57.296° \qquad \text{and} \qquad 1° \approx 0.01745 \text{ rad}$$

An angle θ of measure 1 rad is shown in Figure 4.

EXAMPLE 1 | Converting Between Radians and Degrees

(a) Express 60° in radians. **(b)** Express $\dfrac{\pi}{6}$ rad in degrees.

SOLUTION The relationship between degrees and radians gives

(a) $60° = 60\left(\dfrac{\pi}{180}\right) \text{ rad} = \dfrac{\pi}{3} \text{ rad}$ **(b)** $\dfrac{\pi}{6} \text{ rad} = \left(\dfrac{\pi}{6}\right)\left(\dfrac{180}{\pi}\right) = 30°$

✎ NOW TRY EXERCISES **3** AND **15** ∎

A note on terminology: We often use a phrase such as "a 30° angle" to mean *an angle whose measure is 30°*. Also, for an angle θ, we write $\theta = 30°$ or $\theta = \pi/6$ to mean *the measure of θ is 30° or $\pi/6$ rad*. When no unit is given, the angle is assumed to be measured in radians.

▼ Angles in Standard Position

An angle is in **standard position** if it is drawn in the *xy*-plane with its vertex at the origin and its initial side on the positive *x*-axis. Figure 5 gives examples of angles in standard position.

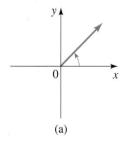 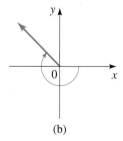 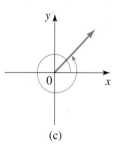 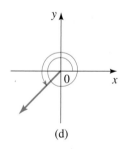

(a) (b) (c) (d)

FIGURE 5 Angles in standard position

Two angles in standard position are **coterminal** if their sides coincide. In Figure 5 the angles in (a) and (c) are coterminal.

EXAMPLE 2 | Coterminal Angles

(a) Find angles that are coterminal with the angle $\theta = 30°$ in standard position.

(b) Find angles that are coterminal with the angle $\theta = \dfrac{\pi}{3}$ in standard position.

SOLUTION

(a) To find positive angles that are coterminal with θ, we add any multiple of 360°. Thus

$$30° + 360° = 390° \qquad \text{and} \qquad 30° + 720° = 750°$$

are coterminal with $\theta = 30°$. To find negative angles that are coterminal with θ, we subtract any multiple of 360°. Thus

$$30° - 360° = -330° \qquad \text{and} \qquad 30° - 720° = -690°$$

are coterminal with θ. (See Figure 6.)

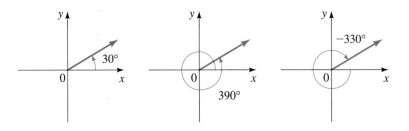

FIGURE 6

(b) To find positive angles that are coterminal with θ, we add any multiple of 2π. Thus

$$\frac{\pi}{3} + 2\pi = \frac{7\pi}{3} \qquad \text{and} \qquad \frac{\pi}{3} + 4\pi = \frac{13\pi}{3}$$

are coterminal with $\theta = \pi/3$. To find negative angles that are coterminal with θ, we subtract any multiple of 2π. Thus

$$\frac{\pi}{3} - 2\pi = -\frac{5\pi}{3} \qquad \text{and} \qquad \frac{\pi}{3} - 4\pi = -\frac{11\pi}{3}$$

are coterminal with θ. (See Figure 7.)

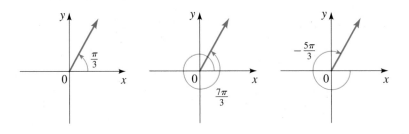

FIGURE 7

✎ NOW TRY EXERCISES **27** AND **29**   ■

EXAMPLE 3 | Coterminal Angles

Find an angle with measure between 0° and 360° that is coterminal with the angle of measure 1290° in standard position.

SOLUTION   We can subtract 360° as many times as we wish from 1290°, and the resulting angle will be coterminal with 1290°. Thus, $1290° - 360° = 930°$ is coterminal with 1290°, and so is the angle $1290° - 2(360)° = 570°$.

To find the angle we want between 0° and 360°, we subtract 360° from 1290° as many times as necessary. An efficient way to do this is to determine how many times 360° goes into 1290°, that is, divide 1290 by 360, and the remainder will be the angle we are looking for. We see that 360 goes into 1290 three times with a remainder of 210. Thus, 210° is the desired angle (see Figure 8).

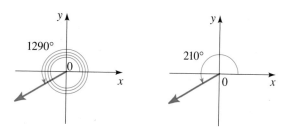

FIGURE 8

✎. NOW TRY EXERCISE **39**   ■

▼ Length of a Circular Arc

An angle whose radian measure is θ is subtended by an arc that is the fraction $\theta/(2\pi)$ of the circumference of a circle. Thus, in a circle of radius r, the length s of an arc that subtends the angle θ (see Figure 9) is

$$s = \frac{\theta}{2\pi} \times \text{circumference of circle}$$

$$= \frac{\theta}{2\pi}(2\pi r) = \theta r$$

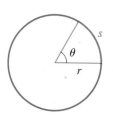

FIGURE 9 $s = \theta r$

LENGTH OF A CIRCULAR ARC

In a circle of radius r, the length s of an arc that subtends a central angle of θ radians is

$$s = r\theta$$

Solving for θ, we get the important formula

$$\theta = \frac{s}{r}$$

This formula allows us to define radian measure using a circle of any radius r: The radian measure of an angle θ is s/r, where s is the length of the circular arc that subtends θ in a circle of radius r (see Figure 10).

FIGURE 10 The radian measure of θ is the number of "radiuses" that can fit in the arc that subtends θ; hence the term *radian*.

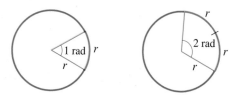

EXAMPLE 4 | Arc Length and Angle Measure

(a) Find the length of an arc of a circle with radius 10 m that subtends a central angle of 30°.

(b) A central angle θ in a circle of radius 4 m is subtended by an arc of length 6 m. Find the measure of θ in radians.

SOLUTION

(a) From Example 1(b) we see that $30° = \pi/6$ rad. So the length of the arc is

$$s = r\theta = (10)\frac{\pi}{6} = \frac{5\pi}{3}\,\text{m}$$

> The formula $s = r\theta$ is true only when θ is measured in radians.

(b) By the formula $\theta = s/r$, we have

$$\theta = \frac{s}{r} = \frac{6}{4} = \frac{3}{2}\,\text{rad}$$

✎. NOW TRY EXERCISES **55** AND **57**

▼ Area of a Circular Sector

The area of a circle of radius r is $A = \pi r^2$. A sector of this circle with central angle θ has an area that is the fraction $\theta/(2\pi)$ of the area of the entire circle (see Figure 11). So the area of this sector is

$$A = \frac{\theta}{2\pi} \times \text{area of circle}$$

$$= \frac{\theta}{2\pi}(\pi r^2) = \frac{1}{2}r^2\theta$$

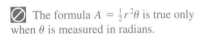

FIGURE 11
$A = \frac{1}{2}r^2\theta$

AREA OF A CIRCULAR SECTOR

In a circle of radius r, the area A of a sector with a central angle of θ radians is

$$A = \frac{1}{2}r^2\theta$$

EXAMPLE 5 │ Area of a Sector

Find the area of a sector of a circle with central angle 60° if the radius of the circle is 3 m.

SOLUTION   To use the formula for the area of a circular sector, we must find the central angle of the sector in radians: $60° = 60(\pi/180)$ rad $= \pi/3$ rad. Thus, the area of the sector is

> The formula $A = \frac{1}{2}r^2\theta$ is true only when θ is measured in radians.

$$A = \frac{1}{2}r^2\theta = \frac{1}{2}(3)^2\left(\frac{\pi}{3}\right) = \frac{3\pi}{2}\,\text{m}^2$$

✎. NOW TRY EXERCISE **61**

▼ Circular Motion

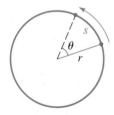

FIGURE 12

Suppose a point moves along a circle as shown in Figure 12. There are two ways to describe the motion of the point: linear speed and angular speed. **Linear speed** is the rate at which the distance traveled is changing, so linear speed is the distance traveled divided by the time elapsed. **Angular speed** is the rate at which the central angle θ is changing, so angular speed is the number of radians this angle changes divided by the time elapsed.

LINEAR SPEED AND ANGULAR SPEED

Suppose a point moves along a circle of radius r and the ray from the center of the circle to the point traverses θ radians in time t. Let $s = r\theta$ be the distance the point travels in time t. Then the speed of the object is given by

$$\text{Angular speed} \qquad \omega = \frac{\theta}{t}$$

$$\text{Linear speed} \qquad v = \frac{s}{t}$$

The symbol ω is the Greek letter "omega."

EXAMPLE 6 | Finding Linear and Angular Speed

A boy rotates a stone in a 3-ft-long sling at the rate of 15 revolutions every 10 seconds. Find the angular and linear velocities of the stone.

SOLUTION In 10 s, the angle θ changes by $15 \cdot 2\pi = 30\pi$ radians. So the *angular speed* of the stone is

$$\omega = \frac{\theta}{t} = \frac{30\pi \text{ rad}}{10 \text{ s}} = 3\pi \text{ rad/s}$$

The distance traveled by the stone in 10 s is $s = 15 \cdot 2\pi r = 15 \cdot 2\pi \cdot 3 = 90\pi$ ft. So the *linear speed* of the stone is

$$v = \frac{s}{t} = \frac{90\pi \text{ ft}}{10 \text{ s}} = 9\pi \text{ ft/s}$$

✎. NOW TRY EXERCISE **79**   ■

Notice that angular speed does *not* depend on the radius of the circle, but only on the angle θ. However, if we know the angular speed ω and the radius r, we can find linear speed as follows: $v = s/t = r\theta/t = r(\theta/t) = r\omega$.

RELATIONSHIP BETWEEN LINEAR AND ANGULAR SPEED

If a point moves along a circle of radius r with angular speed ω, then its linear speed v is given by

$$v = r\omega$$

EXAMPLE 7 | Finding Linear Speed from Angular Speed

A woman is riding a bicycle whose wheels are 26 inches in diameter. If the wheels rotate at 125 revolutions per minute (rpm), find the speed at which she is traveling, in mi/h.

SOLUTION The angular speed of the wheels is $2\pi \cdot 125 = 250\pi$ rad/min. Since the wheels have radius 13 in. (half the diameter), the linear speed is

$$v = r\omega = 13 \cdot 250\pi \approx 10{,}210.2 \text{ in./min}$$

Since there are 12 inches per foot, 5280 feet per mile, and 60 minutes per hour, her speed in miles per hour is

$$\frac{10{,}210.2 \text{ in./min} \times 60 \text{ min/h}}{12 \text{ in./ft} \times 5280 \text{ ft/mi}} = \frac{612{,}612 \text{ in./h}}{63{,}360 \text{ in./mi}}$$

$$\approx 9.7 \text{ mi/h}$$

✎. NOW TRY EXERCISE **81**   ■

3.1 EXERCISES

CONCEPTS

1. (a) The radian measure of an angle θ is the length of the _____ that subtends the angle in a circle of radius _____.

   (b) To convert degrees to radians, we multiply by _____.

   (c) To convert radians to degrees, we multiply by _____.

2. A central angle θ is drawn in a circle of radius r.

   (a) The length of the arc subtended by θ is $s =$ _____.

   (b) The area of the sector with central angle θ is

   $A =$ _____.

SKILLS

3–14 ■ Find the radian measure of the angle with the given degree measure.

3. $72°$ 4. $54°$ 5. $-45°$

6. $-60°$ 7. $-75°$ 8. $-300°$

9. $1080°$ 10. $3960°$ 11. $96°$

12. $15°$ 13. $7.5°$ 14. $202.5°$

15–26 ■ Find the degree measure of the angle with the given radian measure.

15. $\dfrac{7\pi}{6}$ 16. $\dfrac{11\pi}{3}$ 17. $-\dfrac{5\pi}{4}$

18. $-\dfrac{3\pi}{2}$ 19. 3 20. -2

21. -1.2 22. 3.4 23. $\dfrac{\pi}{10}$

24. $\dfrac{5\pi}{18}$ 25. $-\dfrac{2\pi}{15}$ 26. $-\dfrac{13\pi}{12}$

27–32 ■ The measure of an angle in standard position is given. Find two positive angles and two negative angles that are coterminal with the given angle.

27. $50°$ 28. $135°$ 29. $\dfrac{3\pi}{4}$

30. $\dfrac{11\pi}{6}$ 31. $-\dfrac{\pi}{4}$ 32. $-45°$

33–38 ■ The measures of two angles in standard position are given. Determine whether the angles are coterminal.

33. $70°, \quad 430°$ 34. $-30°, \quad 330°$

35. $\dfrac{5\pi}{6}, \dfrac{17\pi}{6}$ 36. $\dfrac{32\pi}{3}, \dfrac{11\pi}{3}$

37. $155°, \quad 875°$ 38. $50°, \quad 340°$

39–44 ■ Find an angle between $0°$ and $360°$ that is coterminal with the given angle.

39. $733°$ 40. $361°$ 41. $1110°$

42. $-100°$ 43. $-800°$ 44. $1270°$

45–50 ■ Find an angle between 0 and 2π that is coterminal with the given angle.

45. $\dfrac{17\pi}{6}$ 46. $-\dfrac{7\pi}{3}$ 47. 87π

48. 10 49. $\dfrac{17\pi}{4}$ 50. $\dfrac{51\pi}{2}$

51. Find the length of the arc s in the figure.

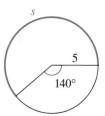

52. Find the angle θ in the figure.

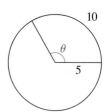

53. Find the radius r of the circle in the figure.

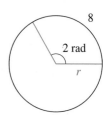

54. Find the length of an arc that subtends a central angle of $45°$ in a circle of radius 10 m.

55. Find the length of an arc that subtends a central angle of 2 rad in a circle of radius 2 mi.

56. A central angle θ in a circle of radius 5 m is subtended by an arc of length 6 m. Find the measure of θ in degrees and in radians.

57. An arc of length 100 m subtends a central angle θ in a circle of radius 50 m. Find the measure of θ in degrees and in radians.

58. A circular arc of length 3 ft subtends a central angle of $25°$. Find the radius of the circle.

59. Find the radius of the circle if an arc of length 6 m on the circle subtends a central angle of $\pi/6$ rad.

60. Find the radius of the circle if an arc of length 4 ft on the circle subtends a central angle of $135°$.

61. Find the area of the sector shown in each figure.

(a)

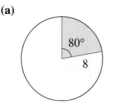

(b)

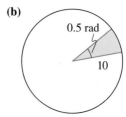

62. Find the radius of each circle if the area of the sector is 12.

(a) **(b)**

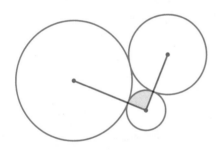

0.7 rad   150°

63. Find the area of a sector with central angle 1 rad in a circle of radius 10 m.

64. A sector of a circle has a central angle of 60°. Find the area of the sector if the radius of the circle is 3 mi.

65. The area of a sector of a circle with a central angle of 2 rad is 16 m^2. Find the radius of the circle.

66. A sector of a circle of radius 24 mi has an area of 288 mi^2. Find the central angle of the sector.

67. The area of a circle is 72 cm^2. Find the area of a sector of this circle that subtends a central angle of $\pi/6$ rad.

68. Three circles with radii 1, 2, and 3 ft are externally tangent to one another, as shown in the figure. Find the area of the sector of the circle of radius 1 that is cut off by the line segments joining the center of that circle to the centers of the other two circles.

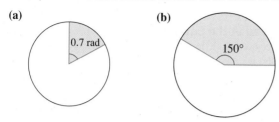

APPLICATIONS

69. Travel Distance A car's wheels are 28 in. in diameter. How far (in miles) will the car travel if its wheels revolve 10,000 times without slipping?

70. Wheel Revolutions How many revolutions will a car wheel of diameter 30 in. make as the car travels a distance of one mile?

71. Latitudes Pittsburgh, Pennsylvania, and Miami, Florida, lie approximately on the same meridian. Pittsburgh has a latitude of 40.5° N, and Miami has a latitude of 25.5° N. Find the distance between these two cities. (The radius of the earth is 3960 mi.)

Pittsburgh
Miami

72. Latitudes Memphis, Tennessee, and New Orleans, Louisiana, lie approximately on the same meridian. Memphis has a latitude of 35° N, and New Orleans has a latitude of 30° N. Find the distance between these two cities. (The radius of the earth is 3960 mi.)

73. Orbit of the Earth Find the distance that the earth travels in one day in its path around the sun. Assume that a year has 365 days and that the path of the earth around the sun is a circle of radius 93 million miles. [The path of the earth around the sun is actually an *ellipse* with the sun at one focus (see Section 7.2). This ellipse, however, has very small eccentricity, so it is nearly circular.]

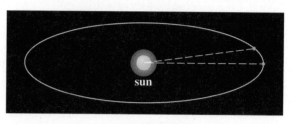

sun

74. Circumference of the Earth The Greek mathematician Eratosthenes (ca. 276–195 B.C.) measured the circumference of the earth from the following observations. He noticed that on a certain day the sun shone directly down a deep well in Syene (modern Aswan). At the same time in Alexandria, 500 miles north (on the same meridian), the rays of the sun shone at an angle of 7.2° to the zenith. Use this information and the figure to find the radius and circumference of the earth.

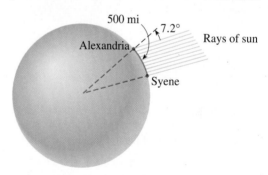

500 mi   7.2°   Rays of sun
Alexandria
Syene

75. Nautical Miles Find the distance along an arc on the surface of the earth that subtends a central angle of 1 minute (1 minute = $\frac{1}{60}$ degree). This distance is called a *nautical mile*. (The radius of the earth is 3960 mi.)

76. Irrigation An irrigation system uses a straight sprinkler pipe 300 ft long that pivots around a central point as shown. Due to an obstacle the pipe is allowed to pivot through 280° only. Find the area irrigated by this system.

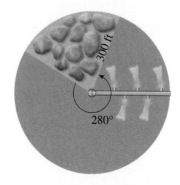

300 ft

280°

77. Windshield Wipers The top and bottom ends of a windshield wiper blade are 34 in. and 14 in., respectively, from the pivot point. While in operation, the wiper sweeps through 135°. Find the area swept by the blade.

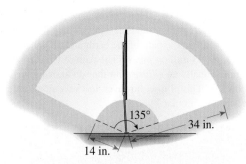

78. The Tethered Cow A cow is tethered by a 100-ft rope to the inside corner of an L-shaped building, as shown in the figure. Find the area that the cow can graze.

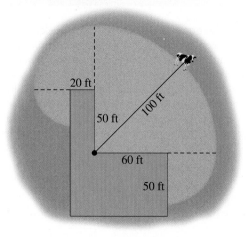

79. Fan A ceiling fan with 16-in. blades rotates at 45 rpm.
(a) Find the angular speed of the fan in rad/min.
(b) Find the linear speed of the tips of the blades in in./min.

80. Radial Saw A radial saw has a blade with a 6-in. radius. Suppose that the blade spins at 1000 rpm.
(a) Find the angular speed of the blade in rad/min.
(b) Find the linear speed of the sawteeth in ft/s.

81. Winch A winch of radius 2 ft is used to lift heavy loads. If the winch makes 8 revolutions every 15 s, find the speed at which the load is rising.

82. Speed of a Car The wheels of a car have radius 11 in. and are rotating at 600 rpm. Find the speed of the car in mi/h.

83. Speed at the Equator The earth rotates about its axis once every 23 h 56 min 4 s, and the radius of the earth is 3960 mi. Find the linear speed of a point on the equator in mi/h.

84. Truck Wheels A truck with 48-in.-diameter wheels is traveling at 50 mi/h.
(a) Find the angular speed of the wheels in rad/min.
(b) How many revolutions per minute do the wheels make?

85. Speed of a Current To measure the speed of a current, scientists place a paddle wheel in the stream and observe the rate at which it rotates. If the paddle wheel has radius 0.20 m and rotates at 100 rpm, find the speed of the current in m/s.

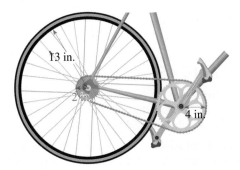

86. Bicycle Wheel The sprockets and chain of a bicycle are shown in the figure. The pedal sprocket has a radius of 4 in., the wheel sprocket a radius of 2 in., and the wheel a radius of 13 in. The cyclist pedals at 40 rpm.
(a) Find the angular speed of the wheel sprocket.
(b) Find the speed of the bicycle. (Assume that the wheel turns at the same rate as the wheel sprocket.)

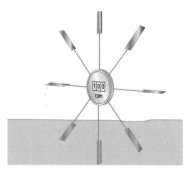

87. Conical Cup A conical cup is made from a circular piece of paper with radius 6 cm by cutting out a sector and joining the edges as shown on the next page. Suppose $\theta = 5\pi/3$.
(a) Find the circumference C of the opening of the cup.
(b) Find the radius r of the opening of the cup. [*Hint:* Use $C = 2\pi r$.]
(c) Find the height h of the cup. [*Hint:* Use the Pythagorean Theorem.]

(d) Find the volume of the cup.

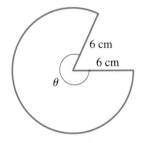

88. Conical Cup   In this exercise we find the volume of the conical cup in Exercise 87 for any angle θ.

(a) Follow the steps in Exercise 87 to show that the volume of the cup as a function of θ is

$$V(\theta) = \frac{9}{\pi^2}\theta^2\sqrt{4\pi^2 - \theta^2}, \qquad 0 < \theta < 2\pi$$

 (b) Graph the function V.

 (c) For what angle θ is the volume of the cup a maximum?

89. Different Ways of Measuring Angles   The custom of measuring angles using degrees, with 360° in a circle, dates back to the ancient Babylonians, who used a number system based on groups of 60. Another system of measuring angles divides the circle into 400 units, called *grads*. In this system a right angle is 100 grad, so this fits in with our base 10 number system.

Write a short essay comparing the advantages and disadvantages of these two systems and the radian system of measuring angles. Which system do you prefer? Why?

90. Clocks and Angles   In one hour, the minute hand on a clock moves through a complete circle, and the hour hand moves through $\frac{1}{12}$ of a circle. Through how many radians do the minute and the hour hand move between 1:00 P.M. and 6:45 P.M. (on the same day)?

3.2 TRIGONOMETRY OF RIGHT TRIANGLES

Trigonometric Ratios ▶ Special Triangles ▶ Applications of Trigonometry of Right Triangles

In this section we study certain ratios of the sides of right triangles, called trigonometric ratios, and give several applications.

▼ Trigonometric Ratios

Consider a right triangle with θ as one of its acute angles. The trigonometric ratios are defined as follows (see Figure 1).

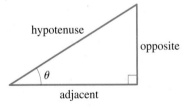

FIGURE 1

THE TRIGONOMETRIC RATIOS

$$\sin\theta = \frac{\text{opposite}}{\text{hypotenuse}} \qquad \cos\theta = \frac{\text{adjacent}}{\text{hypotenuse}} \qquad \tan\theta = \frac{\text{opposite}}{\text{adjacent}}$$

$$\csc\theta = \frac{\text{hypotenuse}}{\text{opposite}} \qquad \sec\theta = \frac{\text{hypotenuse}}{\text{adjacent}} \qquad \cot\theta = \frac{\text{adjacent}}{\text{opposite}}$$

The symbols we use for these ratios are abbreviations for their full names: **sine**, **cosine**, **tangent**, **cosecant**, **secant**, **cotangent**. Since any two right triangles with angle θ are similar, these ratios are the same, regardless of the size of the triangle; the trigonometric ratios depend only on the angle θ (see Figure 2).

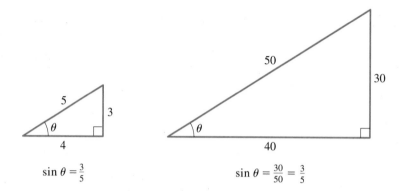

FIGURE 2

$$\sin \theta = \frac{3}{5} \qquad\qquad \sin \theta = \frac{30}{50} = \frac{3}{5}$$

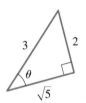

FIGURE 3

EXAMPLE 1 | Finding Trigonometric Ratios

Find the six trigonometric ratios of the angle θ in Figure 3.

SOLUTION

$$\sin \theta = \frac{2}{3} \qquad \cos \theta = \frac{\sqrt{5}}{3} \qquad \tan \theta = \frac{2}{\sqrt{5}}$$

$$\csc \theta = \frac{3}{2} \qquad \sec \theta = \frac{3}{\sqrt{5}} \qquad \cot \theta = \frac{\sqrt{5}}{2}$$

✎ NOW TRY EXERCISE **3**

EXAMPLE 2 | Finding Trigonometric Ratios

If $\cos \alpha = \frac{3}{4}$, sketch a right triangle with acute angle α, and find the other five trigonometric ratios of α.

SOLUTION Since $\cos \alpha$ is defined as the ratio of the adjacent side to the hypotenuse, we sketch a triangle with hypotenuse of length 4 and a side of length 3 adjacent to α. If the opposite side is x, then by the Pythagorean Theorem, $3^2 + x^2 = 4^2$ or $x^2 = 7$, so $x = \sqrt{7}$. We then use the triangle in Figure 4 to find the ratios.

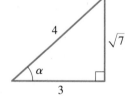

FIGURE 4

$$\sin \alpha = \frac{\sqrt{7}}{4} \qquad \cos \alpha = \frac{3}{4} \qquad \tan \alpha = \frac{\sqrt{7}}{3}$$

$$\csc \alpha = \frac{4}{\sqrt{7}} \qquad \sec \alpha = \frac{4}{3} \qquad \cot \alpha = \frac{3}{\sqrt{7}}$$

✎ NOW TRY EXERCISE **19**

▼ Special Triangles

Certain right triangles have ratios that can be calculated easily from the Pythagorean Theorem. Since they are used frequently, we mention them here.

The first triangle is obtained by drawing a diagonal in a square of side 1 (see Figure 5). By the Pythagorean Theorem this diagonal has length $\sqrt{2}$. The resulting triangle has angles 45°, 45°, and 90° (or $\pi/4$, $\pi/4$, and $\pi/2$). To get the second triangle, we start with an equilateral triangle ABC of side 2 and draw the perpendicular bisector DB of the base, as in Figure 6. By the Pythagorean Theorem the length of DB is $\sqrt{3}$. Since DB bisects angle ABC, we obtain a triangle with angles 30°, 60°, and 90° (or $\pi/6$, $\pi/3$, and $\pi/2$).

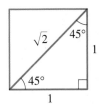

FIGURE 5

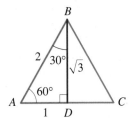

FIGURE 6

For an explanation of numerical methods, see the margin note on page 134.

We can now use the special triangles in Figures 5 and 6 to calculate the trigonometric ratios for angles with measures 30°, 45°, and 60° (or $\pi/6$, $\pi/4$, and $\pi/3$). These are listed in Table 1.

TABLE 1
Values of the trigonometric ratios for special angles

θ in degrees	θ in radians	$\sin \theta$	$\cos \theta$	$\tan \theta$	$\csc \theta$	$\sec \theta$	$\cot \theta$
30°	$\dfrac{\pi}{6}$	$\dfrac{1}{2}$	$\dfrac{\sqrt{3}}{2}$	$\dfrac{\sqrt{3}}{3}$	2	$\dfrac{2\sqrt{3}}{3}$	$\sqrt{3}$
45°	$\dfrac{\pi}{4}$	$\dfrac{\sqrt{2}}{2}$	$\dfrac{\sqrt{2}}{2}$	1	$\sqrt{2}$	$\sqrt{2}$	1
60°	$\dfrac{\pi}{3}$	$\dfrac{\sqrt{3}}{2}$	$\dfrac{1}{2}$	$\sqrt{3}$	$\dfrac{2\sqrt{3}}{3}$	2	$\dfrac{\sqrt{3}}{3}$

It's useful to remember these special trigonometric ratios because they occur often. Of course, they can be recalled easily if we remember the triangles from which they are obtained.

To find the values of the trigonometric ratios for other angles, we use a calculator. Mathematical methods (called *numerical methods*) used in finding the trigonometric ratios are programmed directly into scientific calculators. For instance, when the $\boxed{\text{SIN}}$ key is pressed, the calculator computes an approximation to the value of the sine of the given angle. Calculators give the values of sine, cosine, and tangent; the other ratios can be easily calculated from these by using the following *reciprocal relations*:

$$\csc t = \frac{1}{\sin t} \qquad \sec t = \frac{1}{\cos t} \qquad \cot t = \frac{1}{\tan t}$$

You should check that these relations follow immediately from the definitions of the trigonometric ratios.

We follow the convention that when we write $\sin t$, *we mean the sine of the angle whose radian measure is t.* For instance, $\sin 1$ means the sine of the angle whose radian measure is 1. When using a calculator to find an approximate value for this number, set your calculator to radian mode; you will find that

$$\sin 1 \approx 0.841471$$

If you want to find the sine of the angle whose measure is 1°, set your calculator to degree mode; you will find that

$$\sin 1° \approx 0.0174524$$

▼ Applications of Trigonometry of Right Triangles

A triangle has six parts: three angles and three sides. To **solve a triangle** means to determine all of its parts from the information known about the triangle, that is, to determine the lengths of the three sides and the measures of the three angles.

EXAMPLE 3 | Solving a Right Triangle

Solve triangle *ABC*, shown in Figure 7 on the next page.

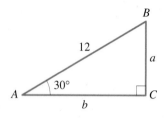

FIGURE 7

SOLUTION It's clear that $\angle B = 60°$. To find a, we look for an equation that relates a to the lengths and angles we already know. In this case, we have $\sin 30° = a/12$, so

$$a = 12 \sin 30° = 12\left(\tfrac{1}{2}\right) = 6$$

Similarly, $\cos 30° = b/12$, so

$$b = 12 \cos 30° = 12\left(\frac{\sqrt{3}}{2}\right) = 6\sqrt{3}$$

✎. NOW TRY EXERCISE **31**

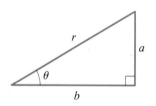

FIGURE 8
$a = r \sin \theta$
$b = r \cos \theta$

Figure 8 shows that if we know the hypotenuse r and an acute angle θ in a right triangle, then the legs a and b are given by

$$a = r \sin \theta \qquad \text{and} \qquad b = r \cos \theta$$

The ability to solve right triangles by using the trigonometric ratios is fundamental to many problems in navigation, surveying, astronomy, and the measurement of distances. The applications we consider in this section always involve right triangles, but as we will see in the next three sections, trigonometry is also useful in solving triangles that are not right triangles.

To discuss the next examples, we need some terminology. If an observer is looking at an object, then the line from the eye of the observer to the object is called the **line of sight** (Figure 9). If the object being observed is above the horizontal, then the angle between the line of sight and the horizontal is called the **angle of elevation**. If the object is below the horizontal, then the angle between the line of sight and the horizontal is called the **angle of depression**. In many of the examples and exercises in this chapter, angles of elevation and depression will be given for a hypothetical observer at ground level. If the line of sight follows a physical object, such as an inclined plane or a hillside, we use the term **angle of inclination**.

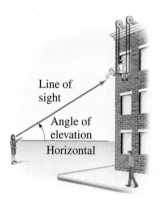 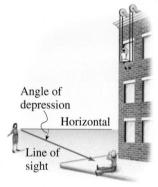

FIGURE 9

The next example gives an important application of trigonometry to the problem of measurement: We measure the height of a tall tree without having to climb it! Although the example is simple, the result is fundamental to understanding how the trigonometric ratios are applied to such problems.

EXAMPLE 4 | Finding the Height of a Tree

A giant redwood tree casts a shadow 532 ft long. Find the height of the tree if the angle of elevation of the sun is 25.7°.

SOLUTION   Let the height of the tree be h. From Figure 10 we see that

$$\frac{h}{532} = \tan 25.7° \qquad \text{Definition of tangent}$$

$$h = 532 \tan 25.7° \qquad \text{Multiply by 532}$$

$$\approx 532(0.48127) \approx 256 \qquad \text{Use a calculator}$$

Therefore, the height of the tree is about 256 ft.

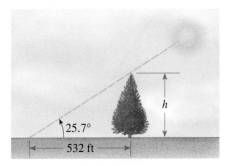

FIGURE 10

✎. NOW TRY EXERCISE **47**

EXAMPLE 5 | A Problem Involving Right Triangles

From a point on the ground 500 ft from the base of a building, an observer finds that the angle of elevation to the top of the building is 24° and that the angle of elevation to the top of a flagpole atop the building is 27°. Find the height of the building and the length of the flagpole.

SOLUTION   Figure 11 illustrates the situation. The height of the building is found in the same way that we found the height of the tree in Example 4.

$$\frac{h}{500} = \tan 24° \qquad \text{Definition of tangent}$$

$$h = 500 \tan 24° \qquad \text{Multiply by 500}$$

$$\approx 500(0.4452) \approx 223 \qquad \text{Use a calculator}$$

The height of the building is approximately 223 ft.

To find the length of the flagpole, let's first find the height from the ground to the top of the pole:

$$\frac{k}{500} = \tan 27°$$

$$k = 500 \tan 27°$$

$$\approx 500(0.5095)$$

$$\approx 255$$

To find the length of the flagpole, we subtract h from k. So the length of the pole is approximately $255 - 223 = 32$ ft.

✎ NOW TRY EXERCISE **55**

FIGURE 11

3.2 EXERCISES

CONCEPTS

1. A right triangle with an angle θ is shown in the figure.

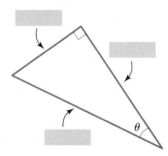

(a) Label the "opposite" and "adjacent" sides of θ and the hypotenuse of the triangle.

(b) The trigonometric functions of the angle θ are defined as follows:

$$\sin \theta = \frac{\rule{1cm}{0.4pt}}{\rule{1cm}{0.4pt}} \qquad \cos \theta = \frac{\rule{1cm}{0.4pt}}{\rule{1cm}{0.4pt}} \qquad \tan \theta = \frac{\rule{1cm}{0.4pt}}{\rule{1cm}{0.4pt}}$$

(c) The trigonometric ratios do not depend on the size of the triangle. This is because all right triangles with an acute angle θ are _____.

2. The reciprocal identities state that

$$\csc \theta = \frac{1}{\rule{1cm}{0.4pt}} \qquad \sec \theta = \frac{1}{\rule{1cm}{0.4pt}} \qquad \cot \theta = \frac{1}{\rule{1cm}{0.4pt}}$$

SKILLS

3–8 ■ Find the exact values of the six trigonometric ratios of the angle θ in the triangle.

3.

4.

5.

6.

7.

8.

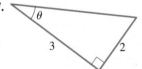
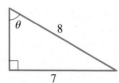

9–10 ■ Find (a) $\sin \alpha$ and $\cos \beta$, (b) $\tan \alpha$ and $\cot \beta$, and (c) $\sec \alpha$ and $\csc \beta$.

9.
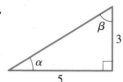

10.
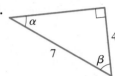

11–16 ■ Find the side labeled x. In Exercises 13 and 14 state your answer rounded to five decimal places.

11.
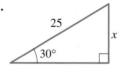

12.

13.
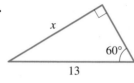

14.
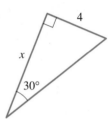

15.
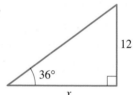

16.
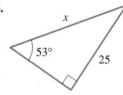

17–18 ■ Express x and y in terms of trigonometric ratios of θ.

17.
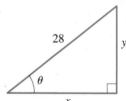

18.
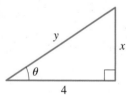

19–24 ■ Sketch a triangle that has acute angle θ, and find the other five trigonometric ratios of θ.

19. $\sin \theta = \frac{3}{5}$ **20.** $\cos \theta = \frac{9}{40}$

21. $\cot \theta = 1$ **22.** $\tan \theta = \sqrt{3}$

23. $\sec \theta = \frac{7}{2}$ **24.** $\csc \theta = \frac{13}{12}$

25–30 ■ Evaluate the expression without using a calculator.

25. $\sin \dfrac{\pi}{6} + \cos \dfrac{\pi}{6}$

26. $\sin 30° \csc 30°$

27. $\sin 30° \cos 60° + \sin 60° \cos 30°$

28. $(\sin 60°)^2 + (\cos 60°)^2$

29. $(\cos 30°)^2 - (\sin 30°)^2$

30. $\left(\sin \dfrac{\pi}{3} \cos \dfrac{\pi}{4} - \sin \dfrac{\pi}{4} \cos \dfrac{\pi}{3} \right)^2$

31–38 ■ Solve the right triangle.

31.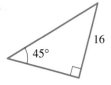

32.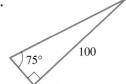

33.

34.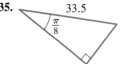

35.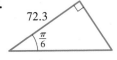

36.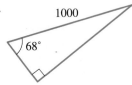

37.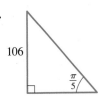

38.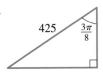

39. Use a ruler to carefully measure the sides of the triangle, and then use your measurements to estimate the six trigonometric ratios of θ.

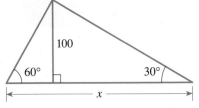

40. Using a protractor, sketch a right triangle that has the acute angle 40°. Measure the sides carefully, and use your results to estimate the six trigonometric ratios of 40°.

41–44 ■ Find x rounded to one decimal place.

41.

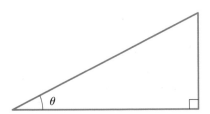

42.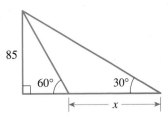

43.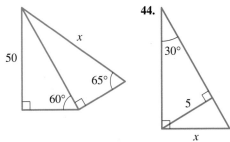

44.

45. Express the length x in terms of the trigonometric ratios of θ.

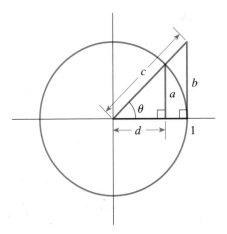

46. Express the length a, b, c, and d in the figure in terms of the trigonometric ratios of θ.

APPLICATIONS

47. Height of a Building The angle of elevation to the top of the Empire State Building in New York is found to be 11° from the ground at a distance of 1 mi from the base of the building. Using this information, find the height of the Empire State Building.

48. Gateway Arch A plane is flying within sight of the Gateway Arch in St. Louis, Missouri, at an elevation of 35,000 ft. The pilot would like to estimate her distance from the Gateway Arch. She finds that the angle of depression to a point on the ground below the arch is 22°.
(a) What is the distance between the plane and the arch?
(b) What is the distance between a point on the ground directly below the plane and the arch?

49. Deviation of a Laser Beam A laser beam is to be directed toward the center of the moon, but the beam strays 0.5° from its intended path.
(a) How far has the beam diverged from its assigned target when it reaches the moon? (The distance from the earth to the moon is 240,000 mi.)
(b) The radius of the moon is about 1000 mi. Will the beam strike the moon?

50. Distance at Sea From the top of a 200-ft lighthouse, the angle of depression to a ship in the ocean is 23°. How far is the ship from the base of the lighthouse?

51. Leaning Ladder A 20-ft ladder leans against a building so that the angle between the ground and the ladder is 72°. How high does the ladder reach on the building?

52. Height of a Tower A 600-ft guy wire is attached to the top of a communications tower. If the wire makes an angle of 65° with the ground, how tall is the communications tower?

53. Elevation of a Kite A man is lying on the beach, flying a kite. He holds the end of the kite string at ground level, and estimates the angle of elevation of the kite to be 50°. If the string is 450 ft long, how high is the kite above the ground?

54. Determining a Distance A woman standing on a hill sees a flagpole that she knows is 60 ft tall. The angle of depression to the bottom of the pole is 14°, and the angle of elevation to the top of the pole is 18°. Find her distance x from the pole.

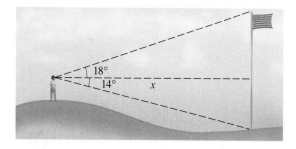

55. Height of a Tower A water tower is located 325 ft from a building (see the figure). From a window in the building, an observer notes that the angle of elevation to the top of the tower is 39° and that the angle of depression to the bottom of the tower is 25°. How tall is the tower? How high is the window?

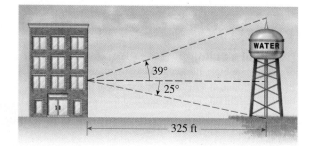

56. Determining a Distance An airplane is flying at an elevation of 5150 ft, directly above a straight highway. Two motorists are driving cars on the highway on opposite sides of the plane, and the angle of depression to one car is 35° and to the other is 52°. How far apart are the cars?

57. Determining a Distance If both cars in Exercise 56 are on one side of the plane and if the angle of depression to one car is 38° and to the other car is 52°, how far apart are the cars?

58. Height of a Balloon A hot-air balloon is floating above a straight road. To estimate their height above the ground, the balloonists simultaneously measure the angle of depression to two consecutive mileposts on the road on the same side of the balloon. The angles of depression are found to be 20° and 22°. How high is the balloon?

59. Height of a Mountain To estimate the height of a mountain above a level plain, the angle of elevation to the top of the mountain is measured to be 32°. One thousand feet closer to the mountain along the plain, it is found that the angle of elevation is 35°. Estimate the height of the mountain.

60. Height of Cloud Cover To measure the height of the cloud cover at an airport, a worker shines a spotlight upward at an angle 75° from the horizontal. An observer 600 m away measures the angle of elevation to the spot of light to be 45°. Find the height h of the cloud cover.

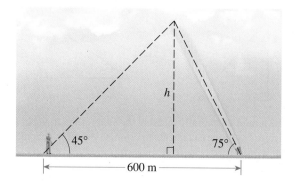

61. Distance to the Sun When the moon is exactly half full, the earth, moon, and sun form a right angle (see the figure). At that time the angle formed by the sun, earth, and moon is measured to be 89.85°. If the distance from the earth to the moon is 240,000 mi, estimate the distance from the earth to the sun.

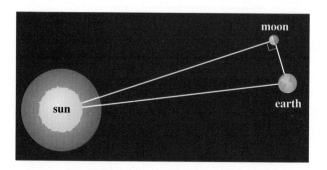

62. Distance to the Moon To find the distance to the sun as in Exercise 61, we needed to know the distance to the moon. Here is a way to estimate that distance: When the moon is seen at its zenith at a point A on the earth, it is observed to be at the horizon from point B (see the following figure). Points A and B are 6155 mi apart, and the radius of the earth is 3960 mi.

(a) Find the angle θ in degrees.

(b) Estimate the distance from point A to the moon.

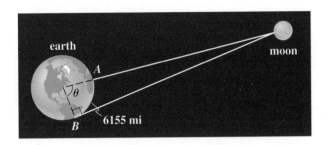

63. Radius of the Earth In Exercise 74 of Section 3.1 a method was given for finding the radius of the earth. Here is a more modern method: From a satellite 600 mi above the earth, it is observed that the angle formed by the vertical and the line of sight to the horizon is 60.276°. Use this information to find the radius of the earth.

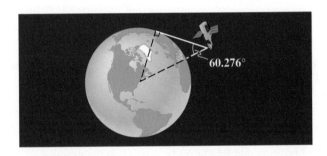

64. Parallax To find the distance to nearby stars, the method of parallax is used. The idea is to find a triangle with the star at one vertex and with a base as large as possible. To do this, the star is observed at two different times exactly 6 months apart, and its apparent change in position is recorded. From these two observations, $\angle E_1SE_2$ can be calculated. (The times are chosen so that $\angle E_1SE_2$ is as large as possible, which guarantees that $\angle E_1OS$ is 90°.) The angle E_1SO is called the *parallax* of the star. Alpha Centauri, the star nearest the earth, has a

parallax of 0.000211°. Estimate the distance to this star. (Take the distance from the earth to the sun to be 9.3×10^7 mi.)

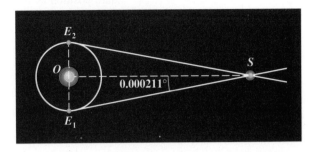

65. Distance from Venus to the Sun The **elongation** α of a planet is the angle formed by the planet, earth, and sun (see the figure). When Venus achieves its maximum elongation of 46.3°, the earth, Venus, and the sun form a triangle with a right angle at Venus. Find the distance between Venus and the sun in astronomical units (AU). (By definition the distance between the earth and the sun is 1 AU.)

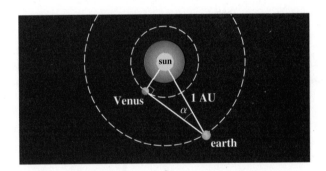

DISCOVERY ▪ DISCUSSION ▪ WRITING

66. Similar Triangles If two triangles are similar, what properties do they share? Explain how these properties make it possible to define the trigonometric ratios without regard to the size of the triangle.

3.3 TRIGONOMETRIC FUNCTIONS OF ANGLES

| Trigonometric Functions of Angles ▶ Evaluating Trigonometric Functions at Any Angle ▶ Trigonometric Identities ▶ Areas of Triangles

In the preceding section we defined the trigonometric ratios for acute angles. Here we extend the trigonometric ratios to all angles by defining the trigonometric functions of angles. With these functions we can solve practical problems that involve angles that are not necessarily acute.

▼ Trigonometric Functions of Angles

Let *POQ* be a right triangle with acute angle θ as shown in Figure 1(a). Place θ in standard position as shown in Figure 1(b).

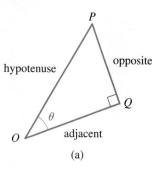

FIGURE 1

Then $P = P(x, y)$ is a point on the terminal side of θ. In triangle *POQ*, the opposite side has length y and the adjacent side has length x. Using the Pythagorean Theorem, we see that the hypotenuse has length $r = \sqrt{x^2 + y^2}$. So

$$\sin \theta = \frac{y}{r} \qquad \cos \theta = \frac{x}{r} \qquad \tan \theta = \frac{y}{x}$$

The other trigonometric ratios can be found in the same way.

These observations allow us to extend the trigonometric ratios to any angle. We define the trigonometric functions of angles as follows (see Figure 2).

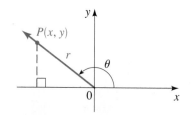

FIGURE 2

DEFINITION OF THE TRIGONOMETRIC FUNCTIONS

Let θ be an angle in standard position and let $P(x, y)$ be a point on the terminal side. If $r = \sqrt{x^2 + y^2}$ is the distance from the origin to the point $P(x, y)$, then

$$\sin \theta = \frac{y}{r} \qquad \cos \theta = \frac{x}{r} \qquad \tan \theta = \frac{y}{x} \quad (x \neq 0)$$

$$\csc \theta = \frac{r}{y} \quad (y \neq 0) \qquad \sec \theta = \frac{r}{x} \quad (x \neq 0) \qquad \cot \theta = \frac{x}{y} \quad (y \neq 0)$$

Since division by 0 is an undefined operation, certain trigonometric functions are not defined for certain angles. For example, $\tan 90° = y/x$ is undefined because $x = 0$. The angles for which the trigonometric functions may be undefined are the angles for which either the *x*- or *y*-coordinate of a point on the terminal side of the angle is 0. These are **quadrantal angles**—angles that are coterminal with the coordinate axes.

It is a crucial fact that the values of the trigonometric functions do *not* depend on the choice of the point $P(x, y)$. This is because if $P'(x', y')$ is any other point on the terminal side, as in Figure 3, then triangles *POQ* and $P'OQ'$ are similar.

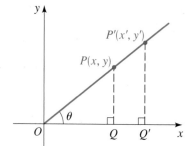

FIGURE 3

▼ Evaluating Trigonometric Functions at Any Angle

From the definition we see that the values of the trigonometric functions are all positive if the angle θ has its terminal side in Quadrant I. This is because x and y are positive in this quadrant. [Of course, r is always positive, since it is simply the distance from the origin to the point $P(x, y)$.] If the terminal side of θ is in Quadrant II, however, then x is

Relationship to the Trigonometric Functions of Real Numbers

You may have already studied the trigonometric functions defined using the unit circle (Chapter 2). To see how they relate to the trigonometric functions of an *angle*, let's start with the unit circle in the coordinate plane.

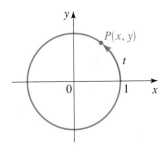

$P(x, y)$ is the terminal
point determined by t.

Let $P(x, y)$ be the terminal point determined by an arc of length t on the unit circle. Then t subtends an angle θ at the center of the circle. If we drop a perpendicular from P onto the point Q on the x-axis, then triangle $\triangle OPQ$ is a right triangle with legs of length x and y, as shown in the figure.

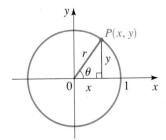

Triangle OPQ is
a right triangle.

Now, by the definition of the trigonometric functions of the *real number t* we have

$$\sin t = y$$

$$\cos t = x$$

By the definition of the trigonometric functions of the *angle θ* we have

$$\sin \theta = \frac{\text{opp}}{\text{hyp}} = \frac{y}{1} = y$$

$$\cos \theta = \frac{\text{adj}}{\text{hyp}} = \frac{x}{1} = x$$

If θ is measured in radians, then $\theta = t$. (See the figure below.) Comparing the two ways of defining the trigonometric functions, we see that they are identical. In other words, as functions, they assign identical values to a given real number. (The real number is the radian measure of θ in one case or the length t of an arc in the other.)

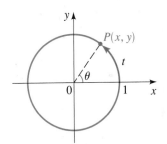

The radian measure
of angle θ is t.

Why then do we study trigonometry in two different ways? Because different applications require that we view the trigonometric functions differently. (See *Focus on Modeling*, pages 159, 221, and 265, and Sections 3.2, 3.5, and 3.6.)

The following mnemonic device can be used to remember which trigonometric functions are positive in each quadrant: **A**ll of them, **S**ine, **T**angent, or **C**osine.

Sine	**A**ll
Tangent	**C**osine

You can remember this as "All Students Take Calculus."

negative and y is positive. Thus, in Quadrant II the functions $\sin \theta$ and $\csc \theta$ are positive, and all the other trigonometric functions have negative values. You can check the other entries in the following table.

SIGNS OF THE TRIGONOMETRIC FUNCTIONS

Quadrant	Positive Functions	Negative Functions
I	all	none
II	sin, csc	cos, sec, tan, cot
III	tan, cot	sin, csc, cos, sec
IV	cos, sec	sin, csc, tan, cot

We now turn our attention to finding the values of the trigonometric functions for angles that are not acute.

EXAMPLE 1 | Finding Trigonometric Functions of Angles

Find **(a)** $\cos 135°$ and **(b)** $\tan 390°$.

SOLUTION

(a) From Figure 4 we see that $\cos 135° = -x/r$. But $\cos 45° = x/r$, and since $\cos 45° = \sqrt{2}/2$, we have

$$\cos 135° = -\frac{\sqrt{2}}{2}$$

(b) The angles $390°$ and $30°$ are coterminal. From Figure 5 it's clear that $\tan 390° = \tan 30°$ and, since $\tan 30° = \sqrt{3}/3$, we have

$$\tan 390° = \frac{\sqrt{3}}{3}$$

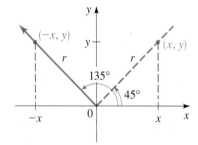

FIGURE 4

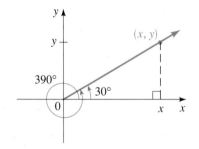

FIGURE 5

✎. NOW TRY EXERCISES **11** AND **13**

From Example 1 we see that the trigonometric functions for angles that aren't acute have the same value, except possibly for sign, as the corresponding trigonometric functions of an acute angle. That acute angle will be called the *reference angle*.

REFERENCE ANGLE

Let θ be an angle in standard position. The **reference angle** $\bar{\theta}$ associated with θ is the acute angle formed by the terminal side of θ and the x-axis.

Figure 6 shows that to find a reference angle $\bar{\theta}$, it's useful to know the quadrant in which the terminal side of the angle θ lies.

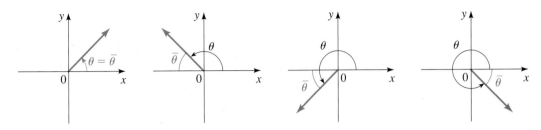

FIGURE 6 The reference angle $\bar{\theta}$ for an angle θ

EXAMPLE 2 | Finding Reference Angles

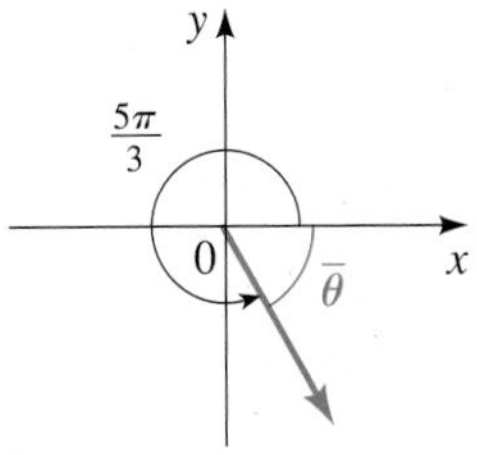

Find the reference angle for **(a)** $\theta = \dfrac{5\pi}{3}$ and **(b)** $\theta = 870°$.

SOLUTION

(a) The reference angle is the acute angle formed by the terminal side of the angle $5\pi/3$ and the x-axis (see Figure 7). Since the terminal side of this angle is in Quadrant IV, the reference angle is

$$\bar{\theta} = 2\pi - \frac{5\pi}{3} = \frac{\pi}{3}$$

(b) The angles $870°$ and $150°$ are coterminal [because $870 - 2(360) = 150$]. Thus, the terminal side of this angle is in Quadrant II (see Figure 8). So the reference angle is

$$\bar{\theta} = 180° - 150° = 30°$$

🖎 NOW TRY EXERCISES **3** AND **7**   ■

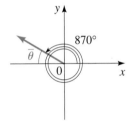

FIGURE 7

FIGURE 8

EVALUATING TRIGONOMETRIC FUNCTIONS FOR ANY ANGLE

To find the values of the trigonometric functions for any angle θ, we carry out the following steps.

1. Find the reference angle $\bar{\theta}$ associated with the angle θ.

2. Determine the sign of the trigonometric function of θ by noting the quadrant in which θ lies.

3. The value of the trigonometric function of θ is the same, except possibly for sign, as the value of the trigonometric function of $\bar{\theta}$.

EXAMPLE 3 | Using the Reference Angle to Evaluate Trigonometric Functions

Find **(a)** $\sin 240°$ and **(b)** $\cot 495°$.

SOLUTION

(a) This angle has its terminal side in Quadrant III, as shown in Figure 9. The reference angle is therefore $240° - 180° = 60°$, and the value of $\sin 240°$ is negative. Thus

$$\sin 240° = -\sin 60° = -\frac{\sqrt{3}}{2}$$

Sign   Reference angle

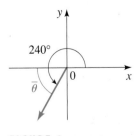

FIGURE 9

$$\begin{array}{c|c} S & A \\ \hline T & C \end{array}$$ $\sin 240°$ is *negative*.

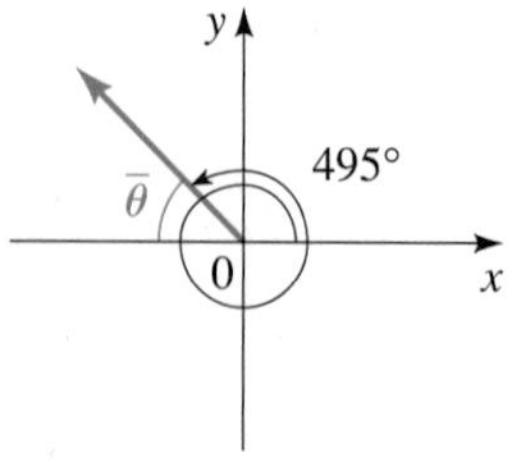

FIGURE 10

$\frac{S\ |\ A}{T\ |\ C}$ *tan 495° is negative,*
so cot 495° *is negative.*

(b) The angle 495° is coterminal with the angle 135°, and the terminal side of this angle is in Quadrant II, as shown in Figure 10. So the reference angle is $180° - 135° = 45°$, and the value of cot 495° is negative. We have

$$\cot 495° = \cot 135° = -\cot 45° = -1$$

Coterminal angles   Sign   Reference angle

✎. NOW TRY EXERCISES **17** AND **19**

EXAMPLE 4 | Using the Reference Angle to Evaluate Trigonometric Functions

Find **(a)** $\sin \dfrac{16\pi}{3}$ and **(b)** $\sec\left(-\dfrac{\pi}{4}\right)$.

SOLUTION

(a) The angle $16\pi/3$ is coterminal with $4\pi/3$, and these angles are in Quadrant III (see Figure 11). Thus, the reference angle is $(4\pi/3) - \pi = \pi/3$. Since the value of sine is negative in Quadrant III, we have

$$\sin \frac{16\pi}{3} = \sin \frac{4\pi}{3} = -\sin \frac{\pi}{3} = -\frac{\sqrt{3}}{2}$$

Coterminal angles   Sign   Reference angle

(b) The angle $-\pi/4$ is in Quadrant IV, and its reference angle is $\pi/4$ (see Figure 12). Since secant is positive in this quadrant, we get

$$\sec\left(-\frac{\pi}{4}\right) = +\sec \frac{\pi}{4} = \sqrt{2}$$

Sign   Reference angle

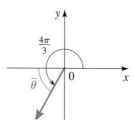

FIGURE 11

$\frac{S\ |\ A}{T\ |\ C}$ $\sin \frac{16\pi}{3}$ *is negative.*

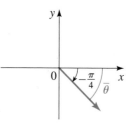

FIGURE 12

$\frac{S\ |\ A}{T\ |\ C}$ $\cos(-\frac{\pi}{4})$ *is positive,*
so $\sec(-\frac{\pi}{4})$ *is positive.*

✎. NOW TRY EXERCISES **23** AND **25**

▼ Trigonometric Identities

The trigonometric functions of angles are related to each other through several important equations called **trigonometric identities**. We've already encountered the reciprocal identities. These identities continue to hold for any angle θ, provided that both sides of the equation are defined. The Pythagorean identities are a consequence of the Pythagorean Theorem.*

FUNDAMENTAL IDENTITIES

Reciprocal Identities

$$\csc \theta = \frac{1}{\sin \theta} \qquad \sec \theta = \frac{1}{\cos \theta} \qquad \cot \theta = \frac{1}{\tan \theta}$$

$$\tan \theta = \frac{\sin \theta}{\cos \theta} \qquad \cot \theta = \frac{\cos \theta}{\sin \theta}$$

Pythagorean Identities

$$\sin^2\theta + \cos^2\theta = 1 \qquad \tan^2\theta + 1 = \sec^2\theta \qquad 1 + \cot^2\theta = \csc^2\theta$$

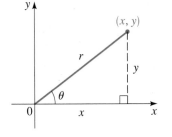

FIGURE 13

PROOF Let's prove the first Pythagorean identity. Using $x^2 + y^2 = r^2$ (the Pythagorean Theorem) in Figure 13, we have

$$\sin^2\theta + \cos^2\theta = \left(\frac{y}{r}\right)^2 + \left(\frac{x}{r}\right)^2 = \frac{x^2 + y^2}{r^2} = \frac{r^2}{r^2} = 1$$

Thus, $\sin^2\theta + \cos^2\theta = 1$. (Although the figure indicates an acute angle, you should check that the proof holds for all angles θ.) ∎

See Exercises 61 and 62 for the proofs of the other two Pythagorean identities.

EXAMPLE 5 | Expressing One Trigonometric Function in Terms of Another

(a) Express $\sin \theta$ in terms of $\cos \theta$.

(b) Express $\tan \theta$ in terms of $\sin \theta$, where θ is in Quadrant II.

SOLUTION

(a) From the first Pythagorean identity we get

$$\sin \theta = \pm\sqrt{1 - \cos^2\theta}$$

where the sign depends on the quadrant. If θ is in Quadrant I or II, then $\sin \theta$ is positive, and so

$$\sin \theta = \sqrt{1 - \cos^2\theta}$$

whereas if θ is in Quadrant III or IV, $\sin \theta$ is negative, and so

$$\sin \theta = -\sqrt{1 - \cos^2\theta}$$

(b) Since $\tan \theta = \sin \theta/\cos \theta$, we need to write $\cos \theta$ in terms of $\sin \theta$. By part (a)

$$\cos \theta = \pm\sqrt{1 - \sin^2\theta}$$

and since $\cos \theta$ is negative in Quadrant II, the negative sign applies here. Thus

$$\tan \theta = \frac{\sin \theta}{\cos \theta} = \frac{\sin \theta}{-\sqrt{1 - \sin^2\theta}}$$

✎. NOW TRY EXERCISE 39   ∎

---

* We follow the usual convention of writing $\sin^2\theta$ for $(\sin \theta)^2$. In general, we write $\sin^n\theta$ for $(\sin \theta)^n$ for all integers n except $n = -1$. The exponent $n = -1$ will be assigned another meaning in Section 3.4. Of course, the same convention applies to the other five trigonometric functions.

EXAMPLE 6 | Evaluating a Trigonometric Function

If $\tan \theta = \frac{2}{3}$ and θ is in Quadrant III, find $\cos \theta$.

SOLUTION 1 We need to write $\cos \theta$ in terms of $\tan \theta$. From the identity $\tan^2\theta + 1 = \sec^2\theta$ we get $\sec \theta = \pm\sqrt{\tan^2\theta + 1}$. In Quadrant III, $\sec \theta$ is negative, so

$$\sec \theta = -\sqrt{\tan^2\theta + 1}$$

If you wish to rationalize the denominator, you can express $\cos \theta$ as

$$-\frac{3}{\sqrt{13}} \cdot \frac{\sqrt{13}}{\sqrt{13}} = -\frac{3\sqrt{3}}{13}$$

Thus,
$$\cos \theta = \frac{1}{\sec \theta} = \frac{1}{-\sqrt{\tan^2\theta + 1}}$$

$$= \frac{1}{-\sqrt{\left(\frac{2}{3}\right)^2 + 1}} = \frac{1}{-\sqrt{\frac{13}{9}}} = -\frac{3}{\sqrt{13}}$$

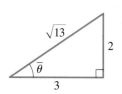

FIGURE 14

SOLUTION 2 This problem can be solved more easily by using the method of Example 2 of Section 3.2. Recall that, except for sign, the values of the trigonometric functions of any angle are the same as those of an acute angle (the reference angle). So, ignoring the sign for the moment, let's sketch a right triangle with an acute angle $\bar{\theta}$ satisfying $\tan \bar{\theta} = \frac{2}{3}$ (see Figure 14). By the Pythagorean Theorem the hypotenuse of this triangle has length $\sqrt{13}$. From the triangle in Figure 14 we immediately see that $\cos \bar{\theta} = 3/\sqrt{13}$. Since θ is in Quadrant III, $\cos \theta$ is negative, so

$$\cos \theta = -\frac{3}{\sqrt{13}}$$

✎ NOW TRY EXERCISE **45**   ■

EXAMPLE 7 | Evaluating Trigonometric Functions

If $\sec \theta = 2$ and θ is in Quadrant IV, find the other five trigonometric functions of θ.

SOLUTION We sketch a triangle as in Figure 15 so that $\sec \bar{\theta} = 2$. Taking into account the fact that θ is in Quadrant IV, we get

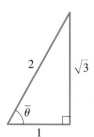

FIGURE 15

$$\sin \theta = -\frac{\sqrt{3}}{2} \qquad \cos \theta = \frac{1}{2} \qquad \tan \theta = -\sqrt{3}$$

$$\csc \theta = -\frac{2}{\sqrt{3}} \qquad \sec \theta = 2 \qquad \cot \theta = -\frac{1}{\sqrt{3}}$$

✎ NOW TRY EXERCISE **47**   ■

▼ Areas of Triangles

We conclude this section with an application of the trigonometric functions that involves angles that are not necessarily acute. More extensive applications appear in the next two sections.

The area of a triangle is $\mathcal{A} = \frac{1}{2} \times$ base $\times$ height. If we know two sides and the included angle of a triangle, then we can find the height using the trigonometric functions, and from this we can find the area.

If θ is an acute angle, then the height of the triangle in Figure 16(a) is given by $h = b \sin \theta$. Thus the area is

$$\mathcal{A} = \frac{1}{2} \times \text{base} \times \text{height} = \frac{1}{2}ab \sin \theta$$

If the angle θ is not acute, then from Figure 16(b) we see that the height of the triangle is

$$h = b \sin(180° - \theta) = b \sin \theta$$

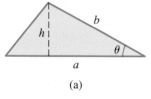

(a)

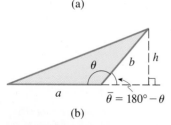

(b)

FIGURE 16

This is so because the reference angle of θ is the angle $180° - \theta$. Thus, in this case also, the area of the triangle is

$$\mathcal{A} = \tfrac{1}{2} \times \text{base} \times \text{height} = \tfrac{1}{2}ab \sin \theta$$

> ### AREA OF A TRIANGLE
>
> The area $\mathcal{A}$ of a triangle with sides of lengths a and b and with included angle θ is
>
> $$\mathcal{A} = \tfrac{1}{2}ab \sin \theta$$

EXAMPLE 8 | Finding the Area of a Triangle

Find the area of triangle ABC shown in Figure 17.

SOLUTION The triangle has sides of length 10 cm and 3 cm, with included angle 120°. Therefore

$$\mathcal{A} = \tfrac{1}{2}ab \sin \theta$$
$$= \tfrac{1}{2}(10)(3) \sin 120°$$
$$= 15 \sin 60° \qquad \text{Reference angle}$$
$$= 15\frac{\sqrt{3}}{2} \approx 13 \text{ cm}^2$$

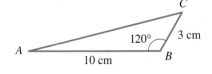

FIGURE 17

✎ NOW TRY EXERCISE **55**

3.3 EXERCISES

CONCEPTS

1. If the angle θ is in standard position and $P(x, y)$ is a point on the terminal side of θ, and r is the distance from the origin to P, then

$$\sin \theta = \underline{} \qquad \cos \theta = \underline{} \qquad \tan \theta = \underline{}$$

2. The sign of a trigonometric function of θ depends on the

_____ in which the terminal side of the angle θ lies.

In Quadrant II, $\sin \theta$ is _____ (positive / negative).

In Quadrant III, $\cos \theta$ is _____ (positive / negative).

In Quadrant IV, $\sin \theta$ is _____ (positive / negative).

SKILLS

3–10 ■ Find the reference angle for the given angle.

3. (a) 150° (b) 330° (c) −30°

4. (a) 120° (b) −210° (c) 780°

5. (a) 225° (b) 810° (c) −105°

6. (a) 99° (b) −199° (c) 359°

✎ **7.** (a) $\dfrac{11\pi}{4}$ (b) $-\dfrac{11\pi}{6}$ (c) $\dfrac{11\pi}{3}$

8. (a) $\dfrac{4\pi}{3}$ (b) $\dfrac{33\pi}{4}$ (c) $-\dfrac{23\pi}{6}$

9. (a) $\dfrac{5\pi}{7}$ (b) -1.4π (c) 1.4

10. (a) 2.3π (b) 2.3 (c) -10π

11–34 ■ Find the exact value of the trigonometric function.

✎ **11.** $\sin 150°$ **12.** $\sin 225°$ ✎ **13.** $\cos 210°$

14. $\cos(-60°)$ **15.** $\tan(-60°)$ **16.** $\sec 300°$

✎ **17.** $\csc(-630°)$ **18.** $\cot 210°$ ✎ **19.** $\cos 570°$

20. $\sec 120°$ **21.** $\tan 750°$ **22.** $\cos 660°$

✎ **23.** $\sin \dfrac{2\pi}{3}$ **24.** $\sin \dfrac{5\pi}{3}$ ✎ **25.** $\sin \dfrac{3\pi}{2}$

26. $\cos \dfrac{7\pi}{3}$ **27.** $\cos\left(-\dfrac{7\pi}{3}\right)$ **28.** $\tan \dfrac{5\pi}{6}$

29. $\sec \dfrac{17\pi}{3}$ **30.** $\csc \dfrac{5\pi}{4}$ **31.** $\cot\left(-\dfrac{\pi}{4}\right)$

32. $\cos \dfrac{7\pi}{4}$ **33.** $\tan \dfrac{5\pi}{2}$ **34.** $\sin \dfrac{11\pi}{6}$

35–38 ■ Find the quadrant in which θ lies from the information given.

35. $\sin \theta < 0$ and $\cos \theta < 0$

36. $\tan \theta < 0$ and $\sin \theta < 0$

37. $\sec \theta > 0$ and $\tan \theta < 0$

38. $\csc \theta > 0$ and $\cos \theta < 0$

39–44 ■ Write the first trigonometric function in terms of the second for θ in the given quadrant.

39. $\tan \theta$, $\cos \theta$; θ in Quadrant III

40. $\cot \theta$, $\sin \theta$; θ in Quadrant II

41. $\cos \theta$, $\sin \theta$; θ in Quadrant IV

42. $\sec \theta$, $\sin \theta$; θ in Quadrant I

43. $\sec \theta$, $\tan \theta$; θ in Quadrant II

44. $\csc \theta$, $\cot \theta$; θ in Quadrant III

45–52 ■ Find the values of the trigonometric functions of θ from the information given.

45. $\sin \theta = \frac{3}{5}$, θ in Quadrant II

46. $\cos \theta = -\frac{7}{12}$, θ in Quadrant III

47. $\tan \theta = -\frac{3}{4}$, $\cos \theta > 0$

48. $\sec \theta = 5$, $\sin \theta < 0$

49. $\csc \theta = 2$, θ in Quadrant I

50. $\cot \theta = \frac{1}{4}$, $\sin \theta < 0$

51. $\cos \theta = -\frac{2}{7}$, $\tan \theta < 0$

52. $\tan \theta = -4$, $\sin \theta > 0$

53. If $\theta = \pi/3$, find the value of each expression.
 (a) $\sin 2\theta$, $2 \sin \theta$ **(b)** $\sin \frac{1}{2}\theta$, $\frac{1}{2} \sin \theta$
 (c) $\sin^2 \theta$, $\sin(\theta^2)$

54. Find the area of a triangle with sides of length 7 and 9 and included angle 72°.

55. Find the area of a triangle with sides of length 10 and 22 and included angle 10°.

56. Find the area of an equilateral triangle with side of length 10.

57. A triangle has an area of 16 in², and two of the sides of the triangle have lengths 5 in. and 7 in. Find the angle included by these two sides.

58. An isosceles triangle has an area of 24 cm², and the angle between the two equal sides is $5\pi/6$. What is the length of the two equal sides?

59–60 ■ Find the area of the shaded region in the figure.

59. 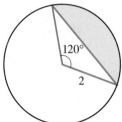 **60.**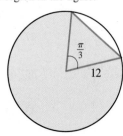

61. Use the first Pythagorean identity to prove the second. [*Hint:* Divide by $\cos^2 \theta$.]

62. Use the first Pythagorean identity to prove the third.

APPLICATIONS

63. Height of a Rocket   A rocket fired straight up is tracked by an observer on the ground a mile away.
 (a) Show that when the angle of elevation is θ, the height of the rocket in feet is $h = 5280 \tan \theta$.
 (b) Complete the table to find the height of the rocket at the given angles of elevation.

θ	20°	60°	80°	85°
h				

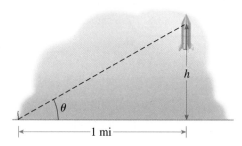

64. Rain Gutter   A rain gutter is to be constructed from a metal sheet of width 30 cm by bending up one-third of the sheet on each side through an angle θ.
 (a) Show that the cross-sectional area of the gutter is modeled by the function

$$A(\theta) = 100 \sin \theta + 100 \sin \theta \cos \theta$$

 (b) Graph the function A for $0 \le \theta \le \pi/2$.
 (c) For what angle θ is the largest cross-sectional area achieved?

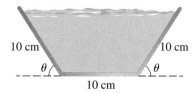

65. Wooden Beam   A rectangular beam is to be cut from a cylindrical log of diameter 20 cm. The figures show different ways this can be done.
 (a) Express the cross-sectional area of the beam as a function of the angle θ in the figures.
 (b) Graph the function you found in part (a).
 (c) Find the dimensions of the beam with largest cross-sectional area.

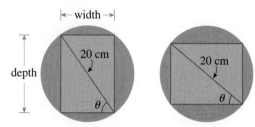

66. Strength of a Beam The strength of a beam is proportional to the width and the square of the depth. A beam is cut from a log as in Exercise 65. Express the strength of the beam as a function of the angle θ in the figures.

67. Throwing a Shot Put The range R and height H of a shot put thrown with an initial velocity of v_0 ft/s at an angle θ are given by

$$R = \frac{v_0^2 \sin(2\theta)}{g}$$

$$H = \frac{v_0^2 \sin^2\theta}{2g}$$

On the earth $g = 32$ ft/s^2 and on the moon $g = 5.2$ ft/s^2. Find the range and height of a shot put thrown under the given conditions.
(a) On the earth with $v_0 = 12$ ft/s and $\theta = \pi/6$
(b) On the moon with $v_0 = 12$ ft/s and $\theta = \pi/6$

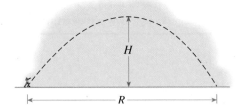

68. Sledding The time in seconds that it takes for a sled to slide down a hillside inclined at an angle θ is

$$t = \sqrt{\frac{d}{16 \sin \theta}}$$

where d is the length of the slope in feet. Find the time it takes to slide down a 2000-ft slope inclined at 30°.

69. Beehives In a beehive each cell is a regular hexagonal prism, as shown in the figure. The amount of wax W in the cell depends on the apex angle θ and is given by

$$W = 3.02 - 0.38 \cot \theta + 0.65 \csc \theta$$

Bees instinctively choose θ so as to use the least amount of wax possible.
(a) Use a graphing device to graph W as a function of θ for $0 < \theta < \pi$.

(b) For what value of θ does W have its minimum value? [*Note:* Biologists have discovered that bees rarely deviate from this value by more than a degree or two.]

70. Turning a Corner A steel pipe is being carried down a hallway that is 9 ft wide. At the end of the hall there is a right-angled turn into a narrower hallway 6 ft wide.
(a) Show that the length of the pipe in the figure is modeled by the function

$$L(\theta) = 9 \csc \theta + 6 \sec \theta$$

(b) Graph the function L for $0 < \theta < \pi/2$.
(c) Find the minimum value of the function L.
(d) Explain why the value of L you found in part (c) is the length of the longest pipe that can be carried around the corner.

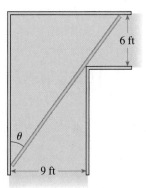

71. Rainbows Rainbows are created when sunlight of different wavelengths (colors) is refracted and reflected in raindrops. The angle of elevation θ of a rainbow is always the same. It can be shown that $\theta = 4\beta - 2\alpha$, where

$$\sin \alpha = k \sin \beta$$

and $\alpha = 59.4°$ and $k = 1.33$ is the index of refraction of water. Use the given information to find the angle of elevation θ of a rainbow. (For a mathematical explanation of rainbows see *Calculus Early Transcendentals,* 7th Edition, by James Stewart, page 282.)

DISCOVERY • DISCUSSION • WRITING

72. Using a Calculator To solve a certain problem, you need to find the sine of 4 rad. Your study partner uses his calculator and tells you that

$$\sin 4 = 0.0697564737$$

On your calculator you get

$$\sin 4 = -0.7568024953$$

What is wrong? What mistake did your partner make?

73. Viète's Trigonometric Diagram In the 16th century the French mathematician François Viète (see page 515) published the following remarkable diagram. Each of the six trigonometric functions of θ is equal to the length of a line segment in the figure. For instance, $\sin \theta = |PR|$, since from $\triangle OPR$ we see that

$$\sin \theta = \frac{\text{opp}}{\text{hyp}} = \frac{|PR|}{|OR|}$$

$$= \frac{|PR|}{1} = |PR|$$

For each of the five other trigonometric functions, find a line segment in the figure whose length equals the value of the function at θ. (*Note:* The radius of the circle is 1, the center is O, segment QS is tangent to the circle at R, and $\angle SOQ$ is a right angle.)

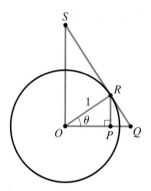

DISCOVERY PROJECT | **Similarity**

In this project we explore the idea of similarity and some of its consequences for any type of figure. You can find the project at the book companion website: **www.stewartmath.com**

3.4 INVERSE TRIGONOMETRIC FUNCTIONS AND RIGHT TRIANGLES

The Inverse Sine, Inverse Cosine, and Inverse Tangent Functions ▶ Solving for Angles in Right Triangles ▶ Evaluating Expressions Involving Inverse Trigonometric Functions

The graphs of the inverse trigonometric functions are studied in Section 2.5.

Recall that for a function to have an inverse, it must be one-to-one. Since the trigonometric functions are not one-to-one, they do not have inverses. So we restrict the domain of each of the trigonometric functions to intervals on which they attain all their values and on which they are one-to-one. The resulting functions have the same range as the original functions but are one-to-one.

▼ The Inverse Sine, Inverse Cosine, and Inverse Tangent Functions

Let's first consider the sine function. We restrict the domain of the sine function to angles θ with $-\pi/2 \le \theta \le \pi/2$. From Figure 1 we see that on this domain the sine function attains each of the values in the interval $[-1, 1]$ exactly once and so is one-to-one. Similarly, we restrict the domains of cosine and tangent as shown in Figure 1.

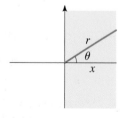
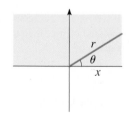
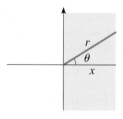

FIGURE 1 Restricted domains of the sine, cosine, and tangent functions

$$\sin \theta = \frac{y}{r}$$
$$-\frac{\pi}{2} \le \theta \le \frac{\pi}{2}$$

$$\cos \theta = \frac{x}{r}$$
$$0 \le \theta \le \pi$$

$$\tan \theta = \frac{x}{y}$$
$$-\frac{\pi}{2} < \theta < \frac{\pi}{2}$$

On these restricted domains we can define an inverse for each of these functions. By the definition of inverse function we have

$$
\begin{aligned}
\sin^{-1} x = y &\iff \sin y = x \\
\cos^{-1} x = y &\iff \cos y = x \\
\tan^{-1} x = y &\iff \tan y = x
\end{aligned}
$$

We summarize the domains and ranges of the inverse trigonometric functions in the following box.

THE INVERSE SINE, INVERSE COSINE, AND INVERSE TANGENT FUNCTIONS

The sine, cosine, and tangent functions on the restricted domains $[-\pi/2, \pi/2]$, $[0, \pi]$, and $(-\pi/2, \pi/2)$, respectively, are one-to one and so have inverses. The inverse functions have domain and range as follows.

Function	Domain	Range
$\sin^{-1}$	$[-1, 1]$	$[-\pi/2, \pi/2]$
$\cos^{-1}$	$[-1, 1]$	$[0, \pi]$
$\tan^{-1}$	$\mathbb{R}$	$(-\pi/2, \pi/2)$

The functions $\sin^{-1}$, $\cos^{-1}$, and $\tan^{-1}$ are sometimes called **arcsine**, **arccosine**, and **arctangent**, respectively.

Since these are inverse functions, they reverse the rule of the original function. For example, since $\sin \pi/6 = \frac{1}{2}$, it follows that $\sin^{-1} \frac{1}{2} = \pi/6$. The following example gives further illustrations.

EXAMPLE 1 | Evaluating Inverse Trigonometric Functions

Find the exact value.

(a) $\sin^{-1} \dfrac{\sqrt{3}}{2}$ **(b)** $\cos^{-1}\left(-\frac{1}{2}\right)$ **(c)** $\tan^{-1} 1$

SOLUTION

(a) The angle in the interval $[-\pi/2, \pi/2]$ whose sine is $\sqrt{3}/2$ is $\pi/3$. Thus $\sin^{-1}(\sqrt{3}/2) = \pi/3$.

(b) The angle in the interval $[0, \pi]$ whose cosine is $-\frac{1}{2}$ is $2\pi/3$. Thus $\cos^{-1}(-\frac{1}{2}) = 2\pi/3$.

(c) The angle in the interval $[-\pi/2, \pi/2]$ whose tangent is 1 is $\pi/4$. Thus $\tan^{-1} 1 = \pi/4$.

✎ NOW TRY EXERCISE **3**

EXAMPLE 2 | Evaluating Inverse Trigonometric Functions

Find approximate values for the given expression.

(a) $\sin^{-1}(0.71)$ **(b)** $\tan^{-1}(2)$ **(c)** $\cos^{-1} 2$

SOLUTION We use a calculator to approximate these values.

(a) Using the $\boxed{\text{INV}}\,\boxed{\text{SIN}}$, or $\boxed{\text{SIN}^{-1}}$, or $\boxed{\text{ARC}}\,\boxed{\text{SIN}}$ key(s) on the calculator (with the calculator in radian mode), we get

$$\sin^{-1}(0.71) \approx 0.78950$$

(b) Using the $\boxed{\text{INV}}$ $\boxed{\text{TAN}}$, or $\boxed{\text{TAN}^{-1}}$, or $\boxed{\text{ARC}}$ $\boxed{\text{TAN}}$ key(s) on the calculator (with the calculator in radian mode), we get

$$\tan^{-1} 2 \approx 1.10715$$

(c) Since $2 > 1$, it is not in the domain of $\cos^{-1}$, so $\cos^{-1} 2$ is not defined.

✎ NOW TRY EXERCISES **7, 11,** AND **13** ■

▼ Solving for Angles in Right Triangles

In Section 3.2 we solved triangles by using the trigonometric functions to find the unknown sides. We now use the inverse trigonometric functions to solve for *angles* in a right triangle.

EXAMPLE 3 | Finding an Angle in a Right Triangle

Find the angle θ in the triangle shown in Figure 2.

SOLUTION Since θ is the angle opposite the side of length 10 and the hypotenuse has length 50, we have

$$\sin \theta = \frac{10}{50} = \frac{1}{5} \qquad \sin \theta = \frac{\text{opp}}{\text{hyp}}$$

Now we can use $\sin^{-1}$ to find θ:

$$\theta = \sin^{-1} \tfrac{1}{5} \qquad \text{Definition of } \sin^{-1}$$

$$\theta \approx 11.5° \qquad \text{Calculator (in degree mode)}$$

✎ NOW TRY EXERCISE **15** ■

FIGURE 2

EXAMPLE 4 | Solving for an Angle in a Right Triangle

A 40-ft ladder leans against a building. If the base of the ladder is 6 ft from the base of the building, what is the angle formed by the ladder and the building?

SOLUTION First we sketch a diagram as in Figure 3. If θ is the angle between the ladder and the building, then

$$\sin \theta = \frac{6}{40} = 0.15 \qquad \sin \theta = \frac{\text{opp}}{\text{hyp}}$$

Now we use $\sin^{-1}$ to find θ:

$$\theta = \sin^{-1}(0.15) \qquad \text{Definition of } \sin^{-1}$$

$$\theta \approx 8.6° \qquad \text{Calculator (in degree mode)}$$

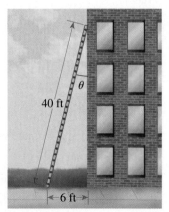

FIGURE 3

✎ NOW TRY EXERCISE **37** ■

EXAMPLE 5 │ The Angle of a Beam of Light

A lighthouse is located on an island that is 2 mi off a straight shoreline (see Figure 4). Express the angle formed by the beam of light and the shoreline in terms of the distance d in the figure.

SOLUTION   From the figure we see that

$$\tan \theta = \frac{2}{d} \qquad\qquad \tan \theta = \frac{\text{opp}}{\text{adj}}$$

Taking the inverse tangent of both sides, we get

$$\tan^{-1}(\tan \theta) = \tan^{-1}\left(\frac{2}{d}\right) \qquad \text{Take } \tan^{-1} \text{ of both sides}$$

$$\theta = \tan^{-1}\left(\frac{2}{d}\right) \qquad \text{Property of inverse functions: } \tan^{-1}(\tan \theta) = \theta$$

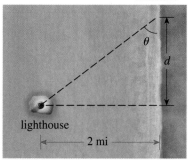

FIGURE 4

shoreline

lighthouse

2 mi

d

θ

✎ NOW TRY EXERCISE **39**   ■

In Section 3.5 we learn how to solve any triangle (not necessarily a right triangle). The angles in a triangle are always in the interval $[0, \pi]$ (or between 0° and 180°). We'll see that to solve such triangles we need to find all angles in the interval $[0, \pi]$ that have a specified sine or cosine. We do this in the next example.

EXAMPLE 6 │ Solving a Basic Trigonometric Equation on an Interval

Find all angles θ between 0° and 180° satisfying the given equation.

(a) $\sin \theta = 0.4$ **(b)** $\cos \theta = 0.4$

SOLUTION

(a) We use $\sin^{-1}$ to find one solution in the interval $[-\pi/2, \pi/2]$.

$$\sin \theta = 0.4 \qquad \text{Equation}$$

$$\theta = \sin^{-1}(0.4) \qquad \text{Take } \sin^{-1} \text{ of each side}$$

$$\theta \approx 23.6° \qquad \text{Calculator (in degree mode)}$$

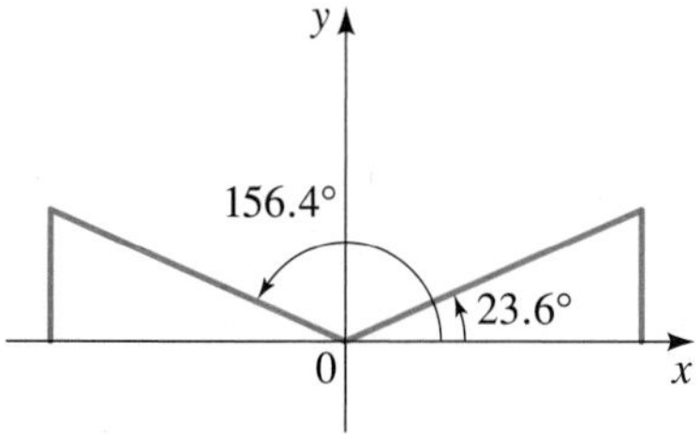

Another solution with θ between 0° and 180° is obtained by taking the supplement of the angle: $180° - 23.6° = 156.4°$ (see Figure 5). So the solutions of the equation with θ between 0° and 180° are

$$\theta \approx 23.6° \qquad \text{and} \qquad \theta \approx 156.4°$$

(b) The cosine function is one-to-one on the interval $[0, \pi]$, so there is only one solution of the equation with θ between 0° and 180°. We find that solution by taking $\cos^{-1}$ of each side.

$$\cos \theta = 0.4$$

$$\theta = \cos^{-1}(0.4) \qquad \text{Take } \cos^{-1} \text{ of each side}$$

$$\theta \approx 66.4° \qquad \text{Calculator (in degree mode)}$$

The solution is $\theta \approx 66.4°$

FIGURE 5

156.4°   23.6°

✎ NOW TRY EXERCISES **23** AND **25**   ■

▼ Evaluating Expressions Involving Inverse Trigonometric Functions

Expressions like $\cos(\sin^{-1} x)$ arise in calculus. We find exact values of such expressions using trigonometric identities or right triangles.

EXAMPLE 7 | Composing Trigonometric Functions and Their Inverses

Find $\cos(\sin^{-1}\frac{3}{5})$.

SOLUTION 1

Let $\theta = \sin^{-1}\frac{3}{5}$. Then θ is the number in the interval $[-\pi/2, \pi/2]$ whose sine is $\frac{3}{5}$. Let's interpret θ as an angle and draw a right triangle with θ as one of its acute angles, with opposite side 3 and hypotenuse 5 (see Figure 6). The remaining leg of the triangle is found by the Pythagorean Theorem to be 4. From the figure we get

$$\cos(\sin^{-1}\tfrac{3}{5}) = \cos\theta \qquad \theta = \sin^{-1}\tfrac{3}{5}$$

$$= \frac{4}{5} \qquad \cos\theta = \frac{\text{adj}}{\text{hyp}}$$

So $\cos(\sin^{-1}\frac{3}{5}) = \frac{4}{5}$.

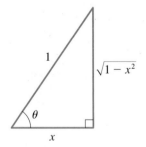

FIGURE 6 $\cos\theta = \dfrac{4}{5}$

SOLUTION 2

It's easy to find $\sin(\sin^{-1}\frac{3}{5})$. In fact, by the cancellation properties of inverse functions, this value is exactly $\frac{3}{5}$. To find $\cos(\sin^{-1}\frac{3}{5})$, we first write the cosine function in terms of the sine function. Let $u = \sin^{-1}\frac{3}{5}$. Since $-\pi/2 \le u \le \pi/2$, $\cos u$ is positive, and we can write the following:

$$\cos u = +\sqrt{1 - \sin^2 u} \qquad \cos^2 u + \sin^2 u = 1$$

$$= \sqrt{1 - \sin^2(\sin^{-1}\tfrac{3}{5})} \qquad u = \sin^{-1}\tfrac{3}{5}$$

$$= \sqrt{1 - \left(\tfrac{3}{5}\right)^2} \qquad \text{Property of inverse functions: } \sin\!\left(\sin^{-1}\tfrac{3}{5}\right) = \tfrac{3}{5}$$

$$= \sqrt{1 - \tfrac{9}{25}} = \sqrt{\tfrac{16}{25}} = \tfrac{4}{5} \qquad \text{Calculate}$$

So $\cos\!\left(\sin^{-1}\frac{3}{5}\right) = \frac{4}{5}$.

✎ NOW TRY EXERCISE **27**

EXAMPLE 8 | Composing Trigonometric Functions and Their Inverses

Write $\sin(\cos^{-1}x)$ and $\tan(\cos^{-1}x)$ as algebraic expressions in x for $-1 \le x \le 1$.

SOLUTION 1

Let $\theta = \cos^{-1}x$; then $\cos\theta = x$. In Figure 7 we sketch a right triangle with an acute angle θ, adjacent side x, and hypotenuse 1. By the Pythagorean Theorem the remaining leg is $\sqrt{1 - x^2}$. From the figure we have

$$\sin(\cos^{-1}x) = \sin\theta = \sqrt{1 - x^2} \qquad \text{and} \qquad \tan(\cos^{-1}x) = \tan\theta = \frac{\sqrt{1 - x^2}}{x}$$

FIGURE 7 $\cos\theta = \dfrac{x}{1} = x$

SOLUTION 2

Let $u = \cos^{-1}x$. We need to find $\sin u$ and $\tan u$ in terms of x. As in Example 5 the idea here is to write sine and tangent in terms of cosine. Note that $0 \le u \le \pi$ because $u = \cos^{-1}x$. We have

$$\sin u = \pm\sqrt{1 - \cos^2 u} \qquad \text{and} \qquad \tan u = \frac{\sin u}{\cos u} = \frac{\pm\sqrt{1 - \cos^2 u}}{\cos u}$$

To choose the proper signs, note that u lies in the interval $[0, \pi]$ because $u = \cos^{-1} x$. Since $\sin u$ is positive on this interval, the $+$ sign is the correct choice. Substituting $u = \cos^{-1} x$ in the displayed equations and using the cancellation property $\cos(\cos^{-1} x) = x$, we get

$$\sin(\cos^{-1} x) = \sqrt{1 - x^2} \quad \text{and} \quad \tan(\cos^{-1} x) = \frac{\sqrt{1 - x^2}}{x}$$

✎ NOW TRY EXERCISES **33** AND **35** ∎

Note: In Solution 1 of Example 8 it might seem that because we are sketching a triangle, the angle $\theta = \cos^{-1} x$ must be acute. But it turns out that the triangle method works for any x. The domains and ranges of all six inverse trigonometric functions have been chosen in such a way that we can always use a triangle to find $S(T^{-1}(x))$, where S and T are any trigonometric functions.

3.4 EXERCISES

CONCEPTS

1. The inverse sine, inverse cosine, and inverse tangent functions have the followings domains and ranges.

(a) The function $\sin^{-1}$ has domain _____ and range

_____.

(b) The function $\cos^{-1}$ has domain _____ and range

_____.

(c) The function $\tan^{-1}$ has domain _____ and range

_____.

2. In the triangle shown, we can find the angle θ as follows:

(a) $\theta = \sin^{-1} \dfrac{\blacksquare}{\blacksquare}$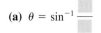

(b) $\theta = \cos^{-1} \dfrac{\blacksquare}{\blacksquare}$

(c) $\theta = \tan^{-1} \dfrac{\blacksquare}{\blacksquare}$

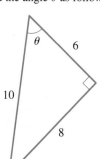

SKILLS

3–6 ■ Find the exact value of each expression, if it is defined.

3. **(a)** $\sin^{-1} \dfrac{1}{2}$ **(b)** $\cos^{-1}\left(-\dfrac{\sqrt{3}}{2}\right)$ **(c)** $\tan^{-1}(-1)$

4. **(a)** $\sin^{-1}\left(-\dfrac{\sqrt{3}}{2}\right)$ **(b)** $\cos^{-1}\left(-\dfrac{\sqrt{2}}{2}\right)$ **(c)** $\tan^{-1}(-\sqrt{3})$

5. **(a)** $\sin^{-1}\left(-\dfrac{1}{2}\right)$ **(b)** $\cos^{-1}\dfrac{1}{2}$ **(c)** $\tan^{-1}\left(\dfrac{\sqrt{3}}{3}\right)$

6. **(a)** $\sin^{-1}(-1)$ **(b)** $\cos^{-1} 1$ **(c)** $\tan^{-1} 0$

7–14 ■ Use a calculator to find an approximate value of each expression rounded to five decimal places, if it is defined.

✎ **7.** $\sin^{-1}(0.45)$ **8.** $\cos^{-1}(-0.75)$

9. $\cos^{-1}\left(-\dfrac{1}{4}\right)$ **10.** $\sin^{-1} \dfrac{1}{3}$

✎ **11.** $\tan^{-1} 3$ **12.** $\tan^{-1}(-4)$

✎ **13.** $\cos^{-1} 3$ **14.** $\sin^{-1}(-2)$

15–20 ■ Find the angle θ in degrees, rounded to one decimal.

✎ **15.**   **16.**

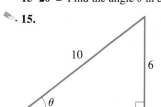   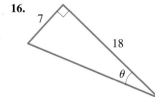

17.   **18.**

   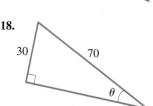

19.   **20.**

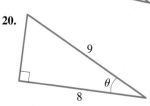

21–26 ■ Find all angles θ between $0°$ and $180°$ satisfying the given equation.

21. $\sin \theta = \dfrac{1}{2}$ **22.** $\sin \theta = \dfrac{\sqrt{3}}{2}$

23. $\sin \theta = 0.7$

24. $\sin \theta = \dfrac{1}{4}$

25. $\cos \theta = 0.7$

26. $\cos \theta = \dfrac{1}{9}$

27–32 ■ Find the exact value of the expression.

27. $\sin\left(\cos^{-1}\frac{3}{5}\right)$

28. $\tan\left(\sin^{-1}\frac{4}{5}\right)$

29. $\sec\left(\sin^{-1}\frac{12}{13}\right)$

30. $\csc\left(\cos^{-1}\frac{7}{25}\right)$

31. $\tan\left(\sin^{-1}\frac{12}{13}\right)$

32. $\cot\left(\sin^{-1}\frac{2}{3}\right)$

33–36 ■ Rewrite the expression as an algebraic expression in x.

33. $\cos(\sin^{-1} x)$

34. $\sin(\tan^{-1} x)$

35. $\tan(\sin^{-1} x)$

36. $\cos(\tan^{-1} x)$

APPLICATIONS

37. Leaning Ladder A 20-ft ladder is leaning against a building. If the base of the ladder is 6 ft from the base of the building, what is the angle of elevation of the ladder? How high does the ladder reach on the building?

38. Angle of the Sun A 96-ft tree casts a shadow that is 120 ft long. What is the angle of elevation of the sun?

39. Height of the Space Shuttle An observer views the space shuttle from a distance of 2 mi from the launch pad.
(a) Express the height of the space shuttle as a function of the angle of elevation θ.
(b) Express the angle of elevation θ as a function of the height h of the space shuttle.

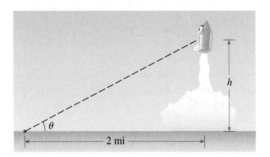

40. Height of a Pole A 50-ft pole casts a shadow as shown in the figure.
(a) Express the angle of elevation θ of the sun as a function of the length s of the shadow.
(b) Find the angle θ of elevation of the sun when the shadow is 20 ft long.

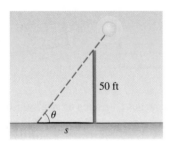

41. Height of a Balloon A 680-ft rope anchors a hot-air balloon as shown in the figure.
(a) Express the angle θ as a function of the height h of the balloon.

(b) Find the angle θ if the balloon is 500 ft high.

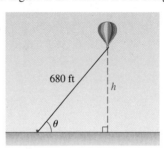

42. View from a Satellite The figures indicate that the higher the orbit of a satellite, the more of the earth the satellite can "see." Let θ, s, and h be as in the figure, and assume the earth is a sphere of radius 3960 mi.
(a) Express the angle θ as a function of h.
(b) Express the distance s as a function of θ.
(c) Express the distance s as a function of h. [*Hint:* Find the composition of the functions in parts (a) and (b).]
(d) If the satellite is 100 mi above the earth, what is the distance s that it can see?
(e) How high does the satellite have to be to see both Los Angeles and New York, 2450 mi apart?

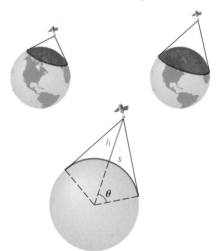

43. Surfing the Perfect Wave For a wave to be surfable, it can't break all at once. Robert Guza and Tony Bowen have shown that a wave has a surfable shoulder if it hits the shoreline at an angle θ given by

$$\theta = \sin^{-1}\left(\frac{1}{(2n+1)\tan \beta}\right)$$

where β is the angle at which the beach slopes down and where $n = 0, 1, 2, \ldots$.
(a) For $\beta = 10°$, find θ when $n = 3$.
(b) For $\beta = 15°$, find θ when $n = 2, 3$, and 4. Explain why the formula does not give a value for θ when $n = 0$ or 1.

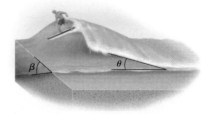

DISCOVERY ▪ DISCUSSION ▪ WRITING

44. Inverse Trigonometric Functions on a Calculator
Most calculators do not have keys for $\sec^{-1}$, $\csc^{-1}$, or $\cot^{-1}$.
Prove the following identities, and then use these identities
and a calculator to find $\sec^{-1} 2$, $\csc^{-1} 3$, and $\cot^{-1} 4$.

$$\sec^{-1} x = \cos^{-1}\left(\frac{1}{x}\right), \qquad x \geq 1$$

$$\csc^{-1} x = \sin^{-1}\left(\frac{1}{x}\right), \qquad x \geq 1$$

$$\cot^{-1} x = \tan^{-1}\left(\frac{1}{x}\right), \qquad x > 0$$

3.5 THE LAW OF SINES

| The Law of Sines ▶ The Ambiguous Case

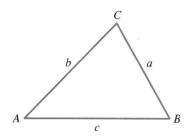

FIGURE 1

In Section 3.2 we used the trigonometric ratios to solve right triangles. The trigonometric functions can also be used to solve *oblique triangles*, that is, triangles with no right angles. To do this, we first study the Law of Sines here and then the Law of Cosines in the next section. To state these laws (or formulas) more easily, we follow the convention of labeling the angles of a triangle as A, B, C and the lengths of the corresponding opposite sides as a, b, c, as in Figure 1.

To solve a triangle, we need to know certain information about its sides and angles. To decide whether we have enough information, it's often helpful to make a sketch. For instance, if we are given two angles and the included side, then it's clear that one and only one triangle can be formed (see Figure 2(a)). Similarly, if two sides and the included angle are known, then a unique triangle is determined (Figure 2(c)). But if we know all three angles and no sides, we cannot uniquely determine the triangle because many triangles can have the same three angles. (All these triangles would be similar, of course.) So we won't consider this last case.

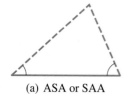   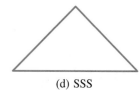

(a) ASA or SAA   (b) SSA   (c) SAS   (d) SSS

FIGURE 2

In general, a triangle is determined by three of its six parts (angles and sides) as long as at least one of these three parts is a side. So the possibilities, illustrated in Figure 2, are as follows.

Case 1 One side and two angles (ASA or SAA)

Case 2 Two sides and the angle opposite one of those sides (SSA)

Case 3 Two sides and the included angle (SAS)

Case 4 Three sides (SSS)

Cases 1 and 2 are solved by using the Law of Sines; Cases 3 and 4 require the Law of Cosines.

▼ The Law of Sines

The **Law of Sines** says that in any triangle the lengths of the sides are proportional to the sines of the corresponding opposite angles.

THE LAW OF SINES

In triangle ABC we have

$$\frac{\sin A}{a} = \frac{\sin B}{b} = \frac{\sin C}{c}$$

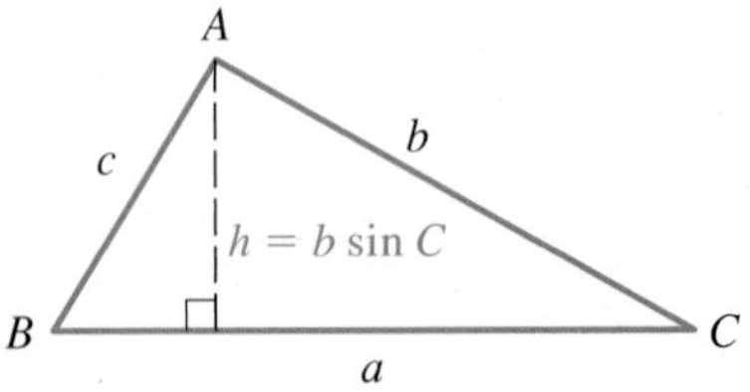

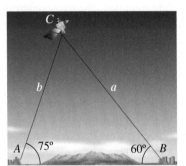

FIGURE 3

PROOF To see why the Law of Sines is true, refer to Figure 3. By the formula in Section 3.3 the area of triangle ABC is $\frac{1}{2}ab \sin C$. By the same formula the area of this triangle is also $\frac{1}{2}ac \sin B$ and $\frac{1}{2}bc \sin A$. Thus,

$$\tfrac{1}{2}bc \sin A = \tfrac{1}{2}ac \sin B = \tfrac{1}{2}ab \sin C$$

Multiplying by $2/(abc)$ gives the Law of Sines. ■

EXAMPLE 1 | Tracking a Satellite (ASA)

A satellite orbiting the earth passes directly overhead at observation stations in Phoenix and Los Angeles, 340 mi apart. At an instant when the satellite is between these two stations, its angle of elevation is simultaneously observed to be 60° at Phoenix and 75° at Los Angeles. How far is the satellite from Los Angeles?

SOLUTION We need to find the distance b in Figure 4. Since the sum of the angles in any triangle is 180°, we see that $\angle C = 180° - (75° + 60°) = 45°$ (see Figure 4), so we have

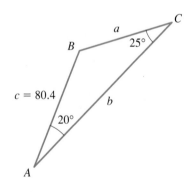

Los Angeles $c = 340$ mi   Phoenix

FIGURE 4

$$\frac{\sin B}{b} = \frac{\sin C}{c} \qquad \text{Law of Sines}$$

$$\frac{\sin 60°}{b} = \frac{\sin 45°}{340} \qquad \text{Substitute}$$

$$b = \frac{340 \sin 60°}{\sin 45°} \approx 416 \qquad \text{Solve for } b$$

The distance of the satellite from Los Angeles is approximately 416 mi.

✎ NOW TRY EXERCISES **5** AND **33** ■

EXAMPLE 2 | Solving a Triangle (SAA)

Solve the triangle in Figure 5.

SOLUTION First, $\angle B = 180° - (20° + 25°) = 135°$. Since side c is known, to find side a we use the relation

$$\frac{\sin A}{a} = \frac{\sin C}{c} \qquad \text{Law of Sines}$$

$$a = \frac{c \sin A}{\sin C} = \frac{80.4 \sin 20°}{\sin 25°} \approx 65.1 \qquad \text{Solve for } a$$

Similarly, to find b, we use

$$\frac{\sin B}{b} = \frac{\sin C}{c} \qquad \text{Law of Sines}$$

$$b = \frac{c \sin B}{\sin C} = \frac{80.4 \sin 135°}{\sin 25°} \approx 134.5 \qquad \text{Solve for } b$$

✎ NOW TRY EXERCISE **13** ■

FIGURE 5

▼ The Ambiguous Case

In Examples 1 and 2 a unique triangle was determined by the information given. This is always true of Case 1 (ASA or SAA). But in Case 2 (SSA) there may be two triangles, one triangle, or no triangle with the given properties. For this reason, Case 2 is sometimes called the **ambiguous case**. To see why this is so, we show in Figure 6 the possibilities when angle A and sides a and b are given. In part (a) no solution is possible, since side a is too short to complete the triangle. In part (b) the solution is a right triangle. In part (c) two solutions are possible, and in part (d) there is a unique triangle with the given properties. We illustrate the possibilities of Case 2 in the following examples.

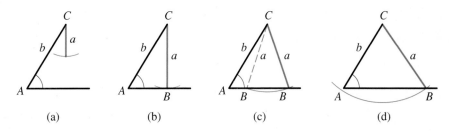

(a)   (b)   (c)   (d)

FIGURE 6 The ambiguous case

EXAMPLE 3 | SSA, the One-Solution Case

Solve triangle ABC, where $\angle A = 45°$, $a = 7\sqrt{2}$, and $b = 7$.

SOLUTION We first sketch the triangle with the information we have (see Figure 7). Our sketch is necessarily tentative, since we don't yet know the other angles. Nevertheless, we can now see the possibilities.

We first find $\angle B$.

$$\frac{\sin A}{a} = \frac{\sin B}{b} \qquad \text{Law of Sines}$$

$$\sin B = \frac{b \sin A}{a} = \frac{7}{7\sqrt{2}} \sin 45° = \left(\frac{1}{\sqrt{2}}\right)\left(\frac{\sqrt{2}}{2}\right) = \frac{1}{2} \qquad \text{Solve for } \sin B$$

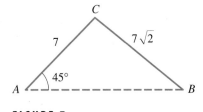

FIGURE 7

We consider only angles smaller than 180°, since no triangle can contain an angle of 180° or larger.

Which angles B have $\sin B = \frac{1}{2}$? From the preceding section we know that there are two such angles smaller than 180° (they are 30° and 150°). Which of these angles is compatible with what we know about triangle ABC? Since $\angle A = 45°$, we cannot have $\angle B = 150°$, because $45° + 150° > 180°$. So $\angle B = 30°$, and the remaining angle is $\angle C = 180° - (30° + 45°) = 105°$.

Now we can find side c.

$$\frac{\sin B}{b} = \frac{\sin C}{c} \qquad \text{Law of Sines}$$

$$c = \frac{b \sin C}{\sin B} = \frac{7 \sin 105°}{\sin 30°} = \frac{7 \sin 105°}{\frac{1}{2}} \approx 13.5 \qquad \text{Solve for } c$$

 NOW TRY EXERCISE 19

In Example 3 there were two possibilities for angle B, and one of these was not compatible with the rest of the information. In general, if $\sin A < 1$, we must check the angle and its supplement as possibilities, because any angle smaller than 180° can be in the triangle. To decide whether either possibility works, we check to see whether the resulting sum of the angles exceeds 180°. It can happen, as in Figure 6(c), that both possibilities are compatible with the given information. In that case, two different triangles are solutions to the problem.

The *supplement* of an angle θ (where $0 \le \theta \le 180°$) is the angle $180° - \theta$.

EXAMPLE 4 | SSA, the Two-Solution Case

Solve triangle ABC if $\angle A = 43.1°$, $a = 186.2$, and $b = 248.6$.

Surveying is a method of land measurement used for mapmaking. Surveyors use a process called *triangulation* in which a network of thousands of interlocking triangles is created on the area to be mapped. The process is started by measuring the length of a *baseline* between two surveying stations. Then, with the use of an instrument called a *theodolite*, the angles between these two stations and a third station are measured. The Law of Sines is then used to calculate the two other sides of the triangle formed by the three stations. The calculated sides are used as baselines, and the process is repeated over and over to create a network of triangles. In this method the only distance measured is the initial baseline; all other distances are calculated from the Law of Sines. This method is practical because it is much easier to measure angles than distances.

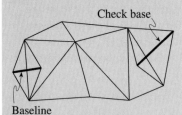

One of the most ambitious mapmaking efforts of all time was the Great Trigonometric Survey of India (see Problem 8, page 224) which required several expeditions and took over a century to complete. The famous expedition of 1823, led by **Sir George Everest**, lasted 20 years. Ranging over treacherous terrain and encountering the dreaded malaria-carrying mosquitoes, this expedition reached the foothills of the Himalayas. A later expedition, using triangulation, calculated the height of the highest peak of the Himalayas to be 29,002 ft. The peak was named in honor of Sir George Everest.

Today, with the use of satellites, the height of Mt. Everest is estimated to be 29,028 ft. The very close agreement of these two estimates shows the great accuracy of the trigonometric method.

SOLUTION From the given information we sketch the triangle shown in Figure 8. Note that side a may be drawn in two possible positions to complete the triangle. From the Law of Sines

$$\sin B = \frac{b \sin A}{a} = \frac{248.6 \sin 43.1°}{186.2} \approx 0.91225$$

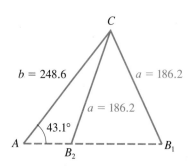

FIGURE 8

There are two possible angles B between $0°$ and $180°$ such that $\sin B = 0.91225$. Using a calculator, we find that one of the angles is $\sin^{-1}(0.91225) \approx 65.8°$. The other angle is approximately $180° - 65.8° = 114.2°$. We denote these two angles by B_1 and B_2 so that

$$\angle B_1 \approx 65.8° \qquad \text{and} \qquad \angle B_2 \approx 114.2°$$

Thus two triangles satisfy the given conditions: triangle $A_1B_1C_1$ and triangle $A_2B_2C_2$.

Solve triangle $A_1B_1C_1$:

$$\angle C_1 \approx 180° - (43.1° + 65.8°) = 71.1° \qquad \text{Find } \angle C_1$$

Thus $\qquad c_1 = \dfrac{a_1 \sin C_1}{\sin A_1} \approx \dfrac{186.2 \sin 71.1°}{\sin 43.1°} \approx 257.8 \qquad$ Law of Sines

Solve triangle $A_2B_2C_2$:

$$\angle C_2 \approx 180° - (43.1° + 114.2°) = 22.7° \qquad \text{Find } \angle C_2$$

Thus $\qquad c_2 = \dfrac{a_2 \sin C_2}{\sin A_2} \approx \dfrac{186.2 \sin 22.7°}{\sin 43.1°} \approx 105.2 \qquad$ Law of Sines

Triangles $A_1B_1C_1$ and $A_2B_2C_2$ are shown in Figure 9.

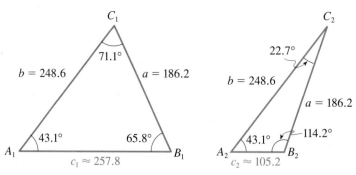

FIGURE 9

◥ **NOW TRY EXERCISE 23**

The next example presents a situation for which no triangle is compatible with the given data.

EXAMPLE 5 | SSA, the No-Solution Case

Solve triangle ABC, where $\angle A = 42°$, $a = 70$, and $b = 122$.

SOLUTION   To organize the given information, we sketch the diagram in Figure 10. Let's try to find $\angle B$. We have

$$\frac{\sin A}{a} = \frac{\sin B}{b} \qquad \text{Law of Sines}$$

$$\sin B = \frac{b \sin A}{a} = \frac{122 \sin 42°}{70} \approx 1.17 \qquad \text{Solve for } \sin B$$

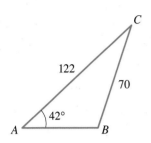

FIGURE 10

Since the sine of an angle is never greater than 1, we conclude that no triangle satisfies the conditions given in this problem.

✎ NOW TRY EXERCISE **21**

3.5 EXERCISES

CONCEPTS

1. In triangle ABC with sides a, b, and c the Law of Sines states that

$$\frac{\blacksquare}{\blacksquare} = \frac{\blacksquare}{\blacksquare} = \frac{\blacksquare}{\blacksquare}$$

2. In which of the following cases can we use the Law of Sines to solve a triangle?

> ASA   SSS   SAS   SSA

SKILLS

3–8 ■ Use the Law of Sines to find the indicated side x or angle θ.

3.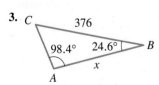

4.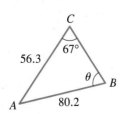

5.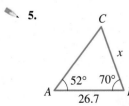

6.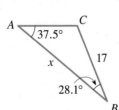

7.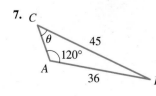

8.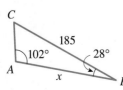

9–12 ■ Solve the triangle using the Law of Sines.

9.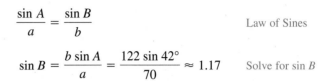

10.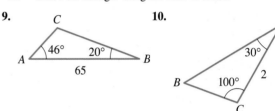

11.

12.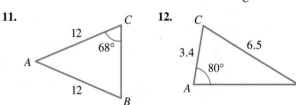

13–18 ■ Sketch each triangle, and then solve the triangle using the Law of Sines.

✎ **13.** $\angle A = 50°$, $\angle B = 68°$, $c = 230$

14. $\angle A = 23°$, $\angle B = 110°$, $c = 50$

15. $\angle A = 30°$, $\angle C = 65°$, $b = 10$

16. $\angle A = 22°$, $\angle B = 95°$, $a = 420$

17. $\angle B = 29°$, $\angle C = 51°$, $b = 44$

18. $\angle B = 10°$, $\angle C = 100°$, $c = 115$

19–28 ■ Use the Law of Sines to solve for all possible triangles that satisfy the given conditions.

✎ **19.** $a = 28$, $b = 15$, $\angle A = 110°$

20. $a = 30$, $c = 40$, $\angle A = 37°$

✎ **21.** $a = 20$, $c = 45$, $\angle A = 125°$

22. $b = 45$, $c = 42$, $\angle C = 38°$

23. $b = 25$, $c = 30$, $\angle B = 25°$

24. $a = 75$, $b = 100$, $\angle A = 30°$

25. $a = 50$, $b = 100$, $\angle A = 50°$

26. $a = 100$, $b = 80$, $\angle A = 135°$

27. $a = 26$, $c = 15$, $\angle C = 29°$

28. $b = 73$, $c = 82$, $\angle B = 58°$

29. For the triangle shown, find
   (a) $\angle BCD$ and
   (b) $\angle DCA$.

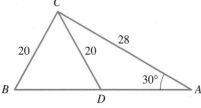

30. For the triangle shown, find the length AD.

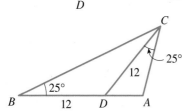

31. In triangle ABC, $\angle A = 40°$, $a = 15$, and $b = 20$.
   (a) Show that there are two triangles, ABC and $A'B'C'$, that satisfy these conditions.
   (b) Show that the areas of the triangles in part (a) are proportional to the sines of the angles C and C', that is,

$$\frac{\text{area of } \triangle ABC}{\text{area of } \triangle A'B'C'} = \frac{\sin C}{\sin C'}$$

32. Show that, given the three angles A, B, C of a triangle and one side, say a, the area of the triangle is

$$\text{area} = \frac{a^2 \sin B \sin C}{2 \sin A}$$

APPLICATIONS

33. Tracking a Satellite   The path of a satellite orbiting the earth causes it to pass directly over two tracking stations A and B, which are 50 mi apart. When the satellite is on one side of the two stations, the angles of elevation at A and B are measured to be $87.0°$ and $84.2°$, respectively.
   (a) How far is the satellite from station A?
   (b) How high is the satellite above the ground?

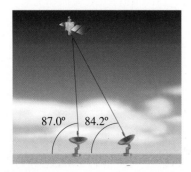

34. Flight of a Plane   A pilot is flying over a straight highway. He determines the angles of depression to two mileposts, 5 mi apart, to be $32°$ and $48°$, as shown in the figure.
   (a) Find the distance of the plane from point A.
   (b) Find the elevation of the plane.

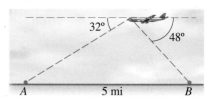

35. Distance Across a River   To find the distance across a river, a surveyor chooses points A and B, which are 200 ft apart on one side of the river (see the figure). She then chooses a reference point C on the opposite side of the river and finds that $\angle BAC \approx 82°$ and $\angle ABC \approx 52°$. Approximate the distance from A to C.

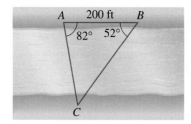

36. Distance Across a Lake   Points A and B are separated by a lake. To find the distance between them, a surveyor locates a point C on land such that $\angle CAB = 48.6°$. He also measures CA as 312 ft and CB as 527 ft. Find the distance between A and B.

37. The Leaning Tower of Pisa   The bell tower of the cathedral in Pisa, Italy, leans $5.6°$ from the vertical. A tourist stands 105 m from its base, with the tower leaning directly toward her. She measures the angle of elevation to the top of the tower to be $29.2°$. Find the length of the tower to the nearest meter.

38. Radio Antenna   A short-wave radio antenna is supported by two guy wires, 165 ft and 180 ft long. Each wire is attached to the top of the antenna and anchored to the ground, at two anchor points on opposite sides of the antenna. The shorter wire makes an angle of $67°$ with the ground. How far apart are the anchor points?

39. Height of a Tree   A tree on a hillside casts a shadow 215 ft down the hill. If the angle of inclination of the hillside is $22°$ to the horizontal and the angle of elevation of the sun is $52°$, find the height of the tree.

40. Length of a Guy Wire A communications tower is located at the top of a steep hill, as shown. The angle of inclination of the hill is 58°. A guy wire is to be attached to the top of the tower and to the ground, 100 m downhill from the base of the tower. The angle α in the figure is determined to be 12°. Find the length of cable required for the guy wire.

41. Calculating a Distance Observers at P and Q are located on the side of a hill that is inclined 32° to the horizontal, as shown. The observer at P determines the angle of elevation to a hot-air balloon to be 62°. At the same instant the observer at Q measures the angle of elevation to the balloon to be 71°. If P is 60 m down the hill from Q, find the distance from Q to the balloon.

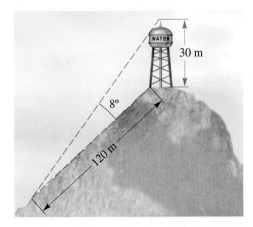

42. Calculating an Angle A water tower 30 m tall is located at the top of a hill. From a distance of 120 m down the hill, it is observed that the angle formed between the top and base of the tower is 8°. Find the angle of inclination of the hill.

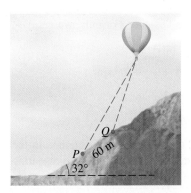

43. Distances to Venus The *elongation* α of a planet is the angle formed by the planet, earth, and sun (see the figure). It is known that the distance from the sun to Venus is 0.723 AU (see Exercise 65 in Section 3.2). At a certain time the elongation of

Venus is found to be 39.4°. Find the possible distances from the earth to Venus at that time in astronomical units (AU).

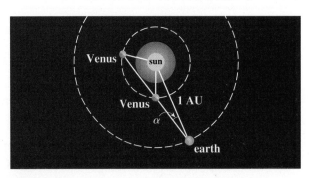

44. Soap Bubbles When two bubbles cling together in midair, their common surface is part of a sphere whose center D lies on the line passing throught the centers of the bubbles (see the figure). Also, angles ACB and ACD each have measure 60°.

(a) Show that the radius r of the common face is given by

$$r = \frac{ab}{a - b}$$

[*Hint:* Use the Law of Sines together with the fact that an angle θ and its supplement $180° - \theta$ have the same sine.]

(b) Find the radius of the common face if the radii of the bubbles are 4 cm and 3 cm.

(c) What shape does the common face take if the two bubbles have equal radii?

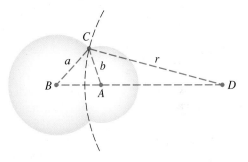

DISCOVERY ▪ DISCUSSION ▪ WRITING

45. Number of Solutions in the Ambiguous Case We have seen that when the Law of Sines is used to solve a triangle in the SSA case, there may be two, one, or no solution(s). Sketch triangles like those in Figure 6 to verify the criteria in the table for the number of solutions if you are given $\angle A$ and sides a and b.

Criterion	Number of solutions
$a \geq b$	1
$b > a > b \sin A$	2
$a = b \sin A$	1
$a < b \sin A$	0

If $\angle A = 30°$ and $b = 100$, use these criteria to find the range of values of a for which the triangle ABC has two solutions, one solution, or no solution.

3.6 THE LAW OF COSINES

> | The Law of Cosines ▶ Navigation: Heading and Bearing ▶ The Area of a Triangle

▼ The Law of Cosines

The Law of Sines cannot be used directly to solve triangles if we know two sides and the angle between them or if we know all three sides (these are Cases 3 and 4 of the preceding section). In these two cases the **Law of Cosines** applies.

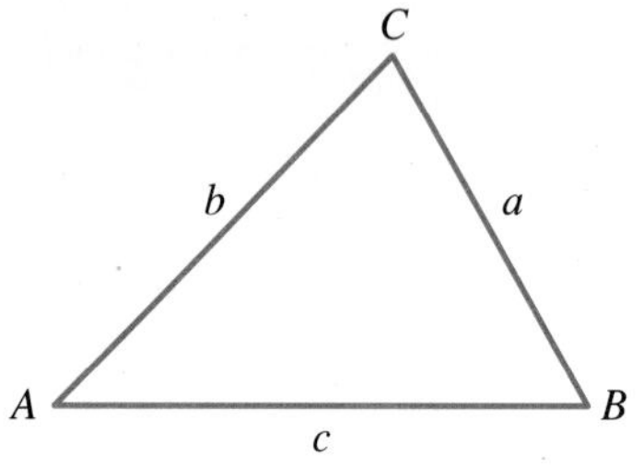

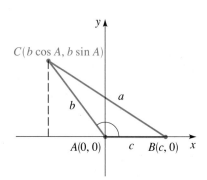

FIGURE 1

THE LAW OF COSINES

In any triangle ABC (see Figure 1), we have

$$a^2 = b^2 + c^2 - 2bc \cos A$$

$$b^2 = a^2 + c^2 - 2ac \cos B$$

$$c^2 = a^2 + b^2 - 2ab \cos C$$

PROOF   To prove the Law of Cosines, place triangle ABC so that $\angle A$ is at the origin, as shown in Figure 2. The coordinates of vertices B and C are $(c, 0)$ and $(b \cos A, b \sin A)$, respectively. (You should check that the coordinates of these points will be the same if we draw angle A as an acute angle.) Using the Distance Formula, we get

$$a^2 = (b \cos A - c)^2 + (b \sin A - 0)^2$$

$$= b^2 \cos^2 A - 2bc \cos A + c^2 + b^2 \sin^2 A$$

$$= b^2(\cos^2 A + \sin^2 A) - 2bc \cos A + c^2$$

$$= b^2 + c^2 - 2bc \cos A \qquad \text{Because } \sin^2 A + \cos^2 A = 1$$

FIGURE 2

This proves the first formula. The other two formulas are obtained in the same way by placing each of the other vertices of the triangle at the origin and repeating the preceding argument. ∎

In words, the Law of Cosines says that the square of any side of a triangle is equal to the sum of the squares of the other two sides, minus twice the product of those two sides times the cosine of the included angle.

If one of the angles of a triangle, say $\angle C$, is a right angle, then $\cos C = 0$, and the Law of Cosines reduces to the Pythagorean Theorem, $c^2 = a^2 + b^2$. Thus the Pythagorean Theorem is a special case of the Law of Cosines.

EXAMPLE 1 | Length of a Tunnel

A tunnel is to be built through a mountain. To estimate the length of the tunnel, a surveyor makes the measurements shown in Figure 3. Use the surveyor's data to approximate the length of the tunnel.

SOLUTION   To approximate the length c of the tunnel, we use the Law of Cosines:

$$c^2 = a^2 + b^2 - 2ab \cos C \qquad \text{Law of Cosines}$$

$$= 388^2 + 212^2 - 2(388)(212) \cos 82.4° \qquad \text{Substitute}$$

$$\approx 173730.2367 \qquad \text{Use a calculator}$$

$$c \approx \sqrt{173730.2367} \approx 416.8 \qquad \text{Take square roots}$$

FIGURE 3

Thus the tunnel will be approximately 417 ft long.

✎ NOW TRY EXERCISES **3** AND **39**

EXAMPLE 2 | SSS, the Law of Cosines

The sides of a triangle are $a = 5$, $b = 8$, and $c = 12$ (see Figure 4). Find the angles of the triangle.

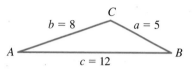

FIGURE 4

SOLUTION   We first find $\angle A$. From the Law of Cosines, $a^2 = b^2 + c^2 - 2bc \cos A$. Solving for $\cos A$, we get

$$\cos A = \frac{b^2 + c^2 - a^2}{2bc} = \frac{8^2 + 12^2 - 5^2}{2(8)(12)} = \frac{183}{192} = 0.953125$$

Using a calculator, we find that $\angle A = \cos^{-1}(0.953125) \approx 18°$. In the same way we get

$$\cos B = \frac{a^2 + c^2 - b^2}{2ac} = \frac{5^2 + 12^2 - 8^2}{2(5)(12)} = 0.875$$

$$\cos C = \frac{a^2 + b^2 - c^2}{2ab} = \frac{5^2 + 8^2 - 12^2}{2(5)(8)} = -0.6875$$

Using a calculator, we find that

$$\angle B = \cos^{-1}(0.875) \approx 29° \quad \text{and} \quad \angle C = \cos^{-1}(-0.6875) \approx 133°$$

Of course, once two angles have been calculated, the third can more easily be found from the fact that the sum of the angles of a triangle is 180°. However, it's a good idea to calculate all three angles using the Law of Cosines and add the three angles as a check on your computations.

✎ NOW TRY EXERCISE **7**

EXAMPLE 3 | SAS, the Law of Cosines

Solve triangle ABC, where $\angle A = 46.5°$, $b = 10.5$, and $c = 18.0$.

SOLUTION   We can find a using the Law of Cosines.

$$a^2 = b^2 + c^2 - 2bc \cos A$$

$$= (10.5)^2 + (18.0)^2 - 2(10.5)(18.0)(\cos 46.5°) \approx 174.05$$

Thus, $a \approx \sqrt{174.05} \approx 13.2$. We also use the Law of Cosines to find $\angle B$ and $\angle C$, as in Example 2.

$$\cos B = \frac{a^2 + c^2 - b^2}{2ac} = \frac{13.2^2 + 18.0^2 - 10.5^2}{2(13.2)(18.0)} \approx 0.816477$$

$$\cos C = \frac{a^2 + b^2 - c^2}{2ab} = \frac{13.2^2 + 10.5^2 - 18.0^2}{2(13.2)(10.5)} \approx -0.142532$$

Using a calculator, we find that

$$\angle B = \cos^{-1}(0.816477) \approx 35.3° \quad \text{and} \quad \angle C = \cos^{-1}(-0.142532) \approx 98.2°$$

To summarize: $\angle B \approx 35.3°$, $\angle C \approx 98.2°$, and $a \approx 13.2$. (See Figure 5.)

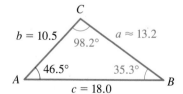

FIGURE 5

✎ NOW TRY EXERCISE **13**

We could have used the Law of Sines to find $\angle B$ and $\angle C$ in Example 3, since we knew all three sides and an angle in the triangle. But knowing the sine of an angle does not uniquely specify the angle, since an angle θ and its supplement $180° - \theta$ both have the same sine. Thus, we would need to decide which of the two angles is the correct choice. This ambiguity does not arise when we use the Law of Cosines, because every angle between 0° and 180° has a unique cosine. So using only the Law of Cosines is preferable in problems like Example 3.

▼ Navigation: Heading and Bearing

In navigation a direction is often given as a **bearing**, that is, as an acute angle measured from due north or due south. The bearing N 30° E, for example, indicates a direction that points 30° to the east of due north (see Figure 6).

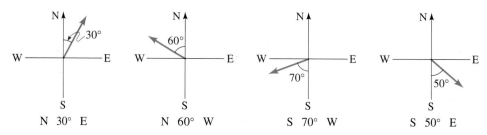

FIGURE 6

EXAMPLE 4 | Navigation

A pilot sets out from an airport and heads in the direction N 20° E, flying at 200 mi/h. After one hour, he makes a course correction and heads in the direction N 40° E. Half an hour after that, engine trouble forces him to make an emergency landing.

(a) Find the distance between the airport and his final landing point.

(b) Find the bearing from the airport to his final landing point.

SOLUTION

(a) In one hour the plane travels 200 mi, and in half an hour it travels 100 mi, so we can plot the pilot's course as in Figure 7. When he makes his course correction, he turns 20° to the right, so the angle between the two legs of his trip is $180° - 20° = 160°$. So by the Law of Cosines we have

$$b^2 = 200^2 + 100^2 - 2 \cdot 200 \cdot 100 \cos 160°$$

$$\approx 87{,}587.70$$

Thus, $b \approx 295.95$. The pilot lands about 296 mi from his starting point.

(b) We first use the Law of Sines to find $\angle A$.

$$\frac{\sin A}{100} = \frac{\sin 160°}{295.95}$$

$$\sin A = 100 \cdot \frac{\sin 160°}{295.95}$$

$$\approx 0.11557$$

Another angle with sine 0.11557 is $180° - 6.636° = 173.364°$. But this is clearly too large to be $\angle A$ in $\angle ABC$.

Using the $\boxed{\text{SIN}^{-1}}$ key on a calculator, we find that $\angle A \approx 6.636°$. From Figure 7 we see that the line from the airport to the final landing site points in the direction $20° + 6.636° = 26.636°$ east of due north. Thus, the bearing is about N 26.6° E.

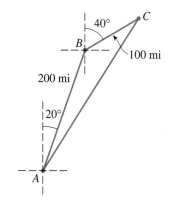

FIGURE 7

✎ NOW TRY EXERCISE **45**

▼ The Area of a Triangle

An interesting application of the Law of Cosines involves a formula for finding the area of a triangle from the lengths of its three sides (see Figure 8).

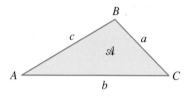

FIGURE 8

> **HERON'S FORMULA**
>
> The area $\mathcal{A}$ of triangle ABC is given by
> $$\mathcal{A} = \sqrt{s(s-a)(s-b)(s-c)}$$
> where $s = \frac{1}{2}(a+b+c)$ is the **semiperimeter** of the triangle; that is, s is half the perimeter.

PROOF We start with the formula $\mathcal{A} = \frac{1}{2}ab \sin C$ from Section 3.3. Thus

$$\mathcal{A}^2 = \tfrac{1}{4}a^2b^2 \sin^2 C$$

$$= \tfrac{1}{4}a^2b^2(1 - \cos^2 C) \qquad \text{Pythagorean identity}$$

$$= \tfrac{1}{4}a^2b^2(1 - \cos C)(1 + \cos C) \qquad \text{Factor}$$

Next, we write the expressions $1 - \cos C$ and $1 + \cos C$ in terms of a, b, and c. By the Law of Cosines we have

$$\cos C = \frac{a^2 + b^2 - c^2}{2ab} \qquad \text{Law of Cosines}$$

$$1 + \cos C = 1 + \frac{a^2 + b^2 - c^2}{2ab} \qquad \text{Add 1}$$

$$= \frac{2ab + a^2 + b^2 - c^2}{2ab} \qquad \text{Common denominator}$$

$$= \frac{(a + b)^2 - c^2}{2ab} \qquad \text{Factor}$$

$$= \frac{(a + b + c)(a + b - c)}{2ab} \qquad \text{Difference of squares}$$

Similarly

$$1 - \cos C = \frac{(c + a - b)(c - a + b)}{2ab}$$

Substituting these expressions in the formula we obtained for $\mathcal{A}^2$ gives

$$\mathcal{A}^2 = \tfrac{1}{4}a^2b^2 \frac{(a + b + c)(a + b - c)}{2ab} \cdot \frac{(c + a - b)(c - a + b)}{2ab}$$

$$= \frac{(a + b + c)}{2} \cdot \frac{(a + b - c)}{2} \cdot \frac{(c + a - b)}{2} \cdot \frac{(c - a + b)}{2}$$

$$= s(s - c)(s - b)(s - a)$$

To see that the factors in the last two products are equal, note for example that

$$\frac{a + b - c}{2} = \frac{a + b + c}{2} - c$$

$$= s - c$$

Heron's Formula now follows by taking the square root of each side. ■

FIGURE 9

EXAMPLE 5 | Area of a Lot

A businessman wishes to buy a triangular lot in a busy downtown location (see Figure 9). The lot frontages on the three adjacent streets are 125, 280, and 315 ft. Find the area of the lot.

SOLUTION　The semiperimeter of the lot is

$$s = \frac{125 + 280 + 315}{2} = 360$$

By Heron's Formula the area is

$$\mathcal{A} = \sqrt{360(360 - 125)(360 - 280)(360 - 315)} \approx 17{,}451.6$$

Thus, the area is approximately 17,452 ft^2.

✎ NOW TRY EXERCISES **29** AND **53** ■

3.6 EXERCISES

CONCEPTS

1. For triangle ABC with sides a, b, and c the Law of Cosines states

$c^2 = $ _____

2. In which of the following cases must the Law of Cosines be used to solve a triangle?

ASA　SSS　SAS　SSA

SKILLS

3–10 ■ Use the Law of Cosines to determine the indicated side x or angle θ.

3.

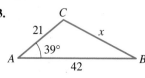

4.

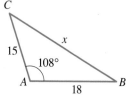

5.

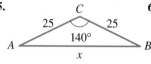

6.

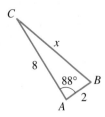

7.

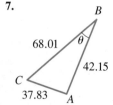

8.

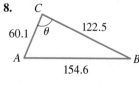

9.

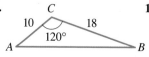

10.

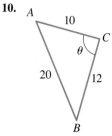

11–20 ■ Solve triangle ABC.

11.

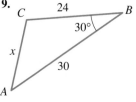

12.

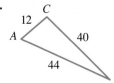

13. $a = 3.0$,　$b = 4.0$,　$\angle C = 53°$

14. $b = 60$,　$c = 30$,　$\angle A = 70°$

15. $a = 20$,　$b = 25$,　$c = 22$

16. $a = 10$,　$b = 12$,　$c = 16$

17. $b = 125$,　$c = 162$,　$\angle B = 40°$

18. $a = 65$,　$c = 50$,　$\angle C = 52°$

19. $a = 50$,　$b = 65$,　$\angle A = 55°$

20. $a = 73.5$,　$\angle B = 61°$,　$\angle C = 83°$

21–28 ■ Find the indicated side x or angle θ. (Use either the Law of Sines or the Law of Cosines, as appropriate.)

21.

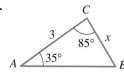

22.

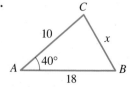

23.

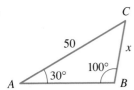

24.

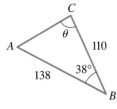

25.

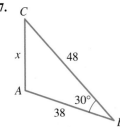

26.

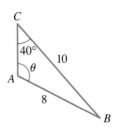

27.

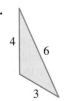

28.

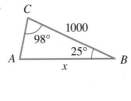

29–32 ■ Find the area of the triangle whose sides have the given lengths.

29. $a = 9$, $b = 12$, $c = 15$ **30.** $a = 1$, $b = 2$, $c = 2$

31. $a = 7$, $b = 8$, $c = 9$

32. $a = 11$, $b = 100$, $c = 101$

33–36 ■ Find the area of the shaded figure, rounded to two decimals.

33.

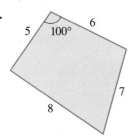

34.

35.

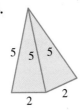

36.

37. Three circles of radii 4, 5, and 6 cm are mutually tangent. Find the shaded area enclosed between the circles.

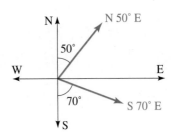

38. Prove that in triangle ABC

$$a = b \cos C + c \cos B$$
$$b = c \cos A + a \cos C$$
$$c = a \cos B + b \cos A$$

These are called the *Projection Laws*.   [*Hint:* To get the first equation, add the second and third equations in the Law of Cosines and solve for a.]

APPLICATIONS

39. Surveying   To find the distance across a small lake, a surveyor has taken the measurements shown. Find the distance across the lake using this information.

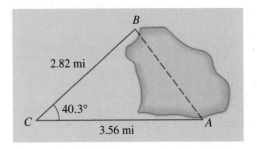

40. Geometry   A parallelogram has sides of lengths 3 and 5, and one angle is 50°. Find the lengths of the diagonals.

41. Calculating Distance   Two straight roads diverge at an angle of 65°. Two cars leave the intersection at 2:00 P.M., one traveling at 50 mi/h and the other at 30 mi/h. How far apart are the cars at 2:30 P.M.?

42. Calculating Distance   A car travels along a straight road, heading east for 1 h, then traveling for 30 min on another road that leads northeast. If the car has maintained a constant speed of 40 mi/h, how far is it from its starting position?

43. Dead Reckoning   A pilot flies in a straight path for 1 h 30 min. She then makes a course correction, heading 10° to the right of her original course, and flies 2 h in the new direction. If she maintains a constant speed of 625 mi/h, how far is she from her starting position?

44. Navigation   Two boats leave the same port at the same time. One travels at a speed of 30 mi/h in the direction N 50° E and the other travels at a speed of 26 mi/h in a direction S 70° E (see the figure). How far apart are the two boats after one hour?

45. Navigation A fisherman leaves his home port and heads in the direction N 70° W. He travels 30 mi and reaches Egg Island. The next day he sails N 10° E for 50 mi, reaching Forrest Island.

(a) Find the distance between the fisherman's home port and Forrest Island.

(b) Find the bearing from Forrest Island back to his home port.

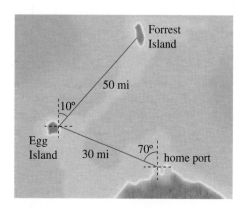

46. Navigation Airport B is 300 mi from airport A at a bearing N 50° E (see the figure). A pilot wishing to fly from A to B mistakenly flies due east at 200 mi/h for 30 minutes, when he notices his error.

(a) How far is the pilot from his destination at the time he notices the error?

(b) What bearing should he head his plane in order to arrive at airport B?

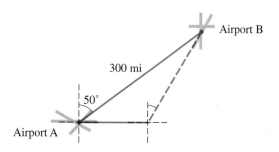

47. Triangular Field A triangular field has sides of lengths 22, 36, and 44 yd. Find the largest angle.

48. Towing a Barge Two tugboats that are 120 ft apart pull a barge, as shown. If the length of one cable is 212 ft and the length of the other is 230 ft, find the angle formed by the two cables.

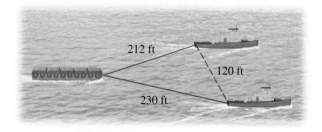

49. Flying Kites A boy is flying two kites at the same time. He has 380 ft of line out to one kite and 420 ft to the other. He estimates the angle between the two lines to be 30°. Approximate the distance between the kites.

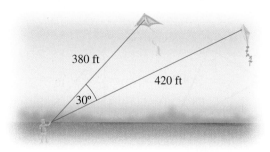

50. Securing a Tower A 125-ft tower is located on the side of a mountain that is inclined 32° to the horizontal. A guy wire is to be attached to the top of the tower and anchored at a point 55 ft downhill from the base of the tower. Find the shortest length of wire needed.

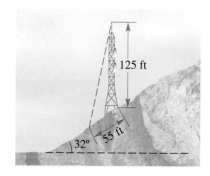

51. Cable Car A steep mountain is inclined 74° to the horizontal and rises 3400 ft above the surrounding plain. A cable car is to be installed from a point 800 ft from the base to the top of the mountain, as shown. Find the shortest length of cable needed.

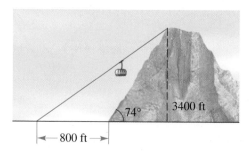

52. CN Tower The CN Tower in Toronto, Canada, is the tallest free-standing structure in North America. A woman on the observation deck, 1150 ft above the ground, wants to determine the distance between two landmarks on the ground below. She observes that the angle formed by the lines of sight to these two landmarks is 43°. She also observes that the angle between the vertical and the line of sight to one of the landmarks is 62° and to the other landmark is 54°. Find the distance between the two landmarks.

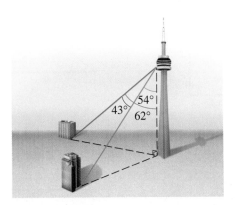

53. Land Value Land in downtown Columbia is valued at $20 a square foot. What is the value of a triangular lot with sides of lengths 112, 148, and 190 ft?

DISCOVERY ▪ DISCUSSION ▪ WRITING

54. Solving for the Angles in a Triangle The paragraph that follows the solution of Example 3 on page 209 explains an alternative method for finding $\angle B$ and $\angle C$, using the Law of Sines. Use this method to solve the triangle in the example, finding $\angle B$ first and then $\angle C$. Explain how you chose the appropriate value for the measure of $\angle B$. Which method do you prefer for solving an SAS triangle problem, the one explained in Example 3 or the one you used in this exercise?

CHAPTER 3 | REVIEW

■ CONCEPT CHECK

1. (a) Explain the difference between a positive angle and a negative angle.
  (b) How is an angle of measure 1 degree formed?
  (c) How is an angle of measure 1 radian formed?
  (d) How is the radian measure of an angle θ defined?
  (e) How do you convert from degrees to radians?
  (f) How do you convert from radians to degrees?

2. (a) When is an angle in standard position?
  (b) When are two angles coterminal?

3. (a) What is the length s of an arc of a circle with radius r that subtends a central angle of θ radians?
  (b) What is the area A of a sector of a circle with radius r and central angle θ radians?

4. If θ is an acute angle in a right triangle, define the six trigonometric ratios in terms of the adjacent and opposite sides and the hypotenuse.

5. What does it mean to solve a triangle?

6. If θ is an angle in standard position, $P(x, y)$ is a point on the terminal side, and r is the distance from the origin to P, write expressions for the six trigonometric functions of θ.

7. Which trigonometric functions are positive in Quadrants I, II, III, and IV?

8. If θ is an angle in standard position, what is its reference angle $\bar{\theta}$?

9. (a) State the reciprocal identities.
  (b) State the Pythagorean identities.

10. (a) What is the area of a triangle with sides of length a and b and with included angle θ?
  (b) What is the area of a triangle with sides of length a, b, and c?

11. Define the inverse sine function $\sin^{-1}$. What are its domain and range?

12. Define the inverse cosine function $\cos^{-1}$. What are its domain and range?

13. Define the inverse tangent function $\tan^{-1}$. What are its domain and range?

14. (a) State the Law of Sines.
  (b) State the Law of Cosines.

15. Explain the ambiguous case in the Law of Sines.

■ EXERCISES

1–2 ■ Find the radian measure that corresponds to the given degree measure.

1. (a) 60°   **(b)** 330°   **(c)** −135°   **(d)** −90°

2. (a) 24°   **(b)** −330°   **(c)** 750°   **(d)** 5°

3–4 ■ Find the degree measure that corresponds to the given radian measure.

3. (a) $\dfrac{5\pi}{2}$ **(b)** $-\dfrac{\pi}{6}$ **(c)** $\dfrac{9\pi}{4}$ **(d)** 3.1

4. (a) 8   **(b)** $-\dfrac{5}{2}$ **(c)** $\dfrac{11\pi}{6}$ **(d)** $\dfrac{3\pi}{5}$

5. Find the length of an arc of a circle of radius 8 m if the arc subtends a central angle of 1 rad.

6. Find the measure of a central angle θ in a circle of radius 5 ft if the angle is subtended by an arc of length 7 ft.

7. A circular arc of length 100 ft subtends a central angle of 70°. Find the radius of the circle.

8. How many revolutions will a car wheel of diameter 28 in. make over a period of half an hour if the car is traveling at 60 mi/h?

9. New York and Los Angeles are 2450 mi apart. Find the angle that the arc between these two cities subtends at the center of the earth. (The radius of the earth is 3960 mi.)

10. Find the area of a sector with central angle 2 rad in a circle of radius 5 m.

11. Find the area of a sector with central angle 52° in a circle of radius 200 ft.

12. A sector in a circle of radius 25 ft has an area of 125 ft². Find the central angle of the sector.

13. A potter's wheel with radius 8 in. spins at 150 rpm. Find the angular and linear speeds of a point on the rim of the wheel.

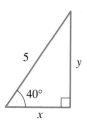

14. In an automobile transmission a *gear ratio g* is the ratio

$$g = \frac{\text{angular speed of engine}}{\text{angular speed of wheels}}$$

The angular speed of the engine is shown on the tachometer (in rpm).

A certain sports car has wheels with radius 11 in. Its gear ratios are shown in the following table. Suppose the car is in fourth gear and the tachometer reads 3500 rpm.
(a) Find the angular speed of the engine.
(b) Find the angular speed of the wheels.
(c) How fast (in mi/h) is the car traveling?

Gear	Ratio
1st	4.1
2nd	3.0
3rd	1.6
4th	0.9
5th	0.7

15–16 ■ Find the values of the six trigonometric ratios of θ.

15.

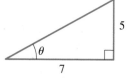

16.

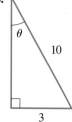

17–20 ■ Find the sides labeled x and y, rounded to two decimal places.

17.

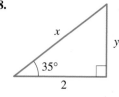

18.

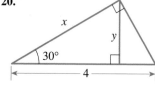

19.

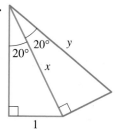

20.

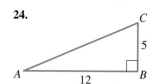

21–24 ■ Solve the triangle.

21.

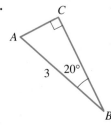

22.

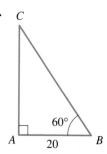

23.

24.

25. Express the lengths a and b in the figure in terms of the trigonometric ratios of θ.

26. The highest free-standing tower in North America is the CN Tower in Toronto, Canada. From a distance of 1 km from its base, the angle of elevation to the top of the tower is 28.81°. Find the height of the tower.

27. Find the perimeter of a regular hexagon that is inscribed in a circle of radius 8 m.

28. The pistons in a car engine move up and down repeatedly to turn the crankshaft, as shown. Find the height of the point P above the center O of the crankshaft in terms of the angle θ.

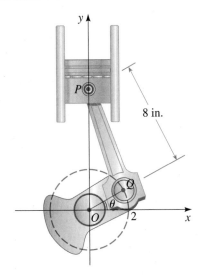

29. As viewed from the earth, the angle subtended by the full moon is $0.518°$. Use this information and the fact that the distance AB from the earth to the moon is 236,900 mi to find the radius of the moon.

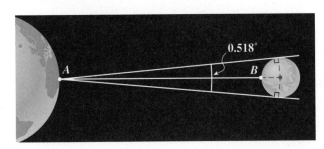

30. A pilot measures the angles of depression to two ships to be $40°$ and $52°$ (see the figure). If the pilot is flying at an elevation of 35,000 ft, find the distance between the two ships.

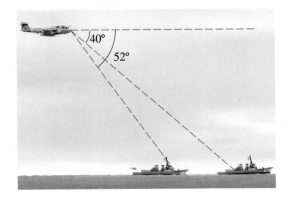

31–42 ■ Find the exact value.

31. $\sin 315°$

32. $\csc \dfrac{9\pi}{4}$

33. $\tan(-135°)$

34. $\cos \dfrac{5\pi}{6}$

35. $\cot\left(-\dfrac{22\pi}{3}\right)$

36. $\sin 405°$

37. $\cos 585°$

38. $\sec \dfrac{22\pi}{3}$

39. $\csc \dfrac{8\pi}{3}$

40. $\sec \dfrac{13\pi}{6}$

41. $\cot(-390°)$

42. $\tan \dfrac{23\pi}{4}$

43. Find the values of the six trigonometric ratios of the angle θ in standard position if the point $(-5, 12)$ is on the terminal side of θ.

44. Find $\sin \theta$ if θ is in standard position and its terminal side intersects the circle of radius 1 centered at the origin at the point $(-\sqrt{3}/2, \tfrac{1}{2})$.

45. Find the acute angle that is formed by the line $y - \sqrt{3}x + 1 = 0$ and the x-axis.

46. Find the six trigonometric ratios of the angle θ in standard position if its terminal side is in Quadrant III and is parallel to the line $4y - 2x - 1 = 0$.

47–50 ■ Write the first expression in terms of the second, for θ in the given quadrant.

47. $\tan \theta$, $\cos \theta$; θ in Quadrant II

48. $\sec \theta$, $\sin \theta$; θ in Quadrant III

49. $\tan^2\theta$, $\sin \theta$; θ in any quadrant

50. $\csc^2\theta \cos^2\theta$, $\sin \theta$; θ in any quadrant

51–54 ■ Find the values of the six trigonometric functions of θ from the information given.

51. $\tan \theta = \sqrt{7}/3$, $\sec \theta = \tfrac{4}{3}$

52. $\sec \theta = \tfrac{41}{40}$, $\csc \theta = -\tfrac{41}{9}$

53. $\sin \theta = \tfrac{3}{5}$, $\cos \theta < 0$

54. $\sec \theta = -\tfrac{13}{5}$, $\tan \theta > 0$

55. If $\tan \theta = -\tfrac{1}{2}$ for θ in Quadrant II, find $\sin \theta + \cos \theta$.

56. If $\sin \theta = \tfrac{1}{2}$ for θ in Quadrant I, find $\tan \theta + \sec \theta$.

57. If $\tan \theta = -1$, find $\sin^2\theta + \cos^2\theta$.

58. If $\cos \theta = -\sqrt{3}/2$ and $\pi/2 < \theta < \pi$, find $\sin 2\theta$.

59–62 ■ Find the exact value of the expression.

59. $\sin^{-1}(\sqrt{3}/2)$

60. $\tan^{-1}(\sqrt{3}/3)$

61. $\tan(\sin^{-1}\tfrac{2}{5})$

62. $\sin(\cos^{-1}\tfrac{3}{8})$

63–64 ■ Rewrite the expression as an algebraic expression in x.

63. $\sin(\tan^{-1}x)$

64. $\sec(\sin^{-1}x)$

65–66 ■ Express θ in terms of x.

65.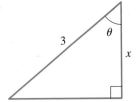

66.

67–76 ■ Find the side labeled *x* or the angle labeled *θ*.

67.

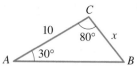

68.

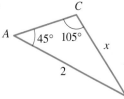

69.

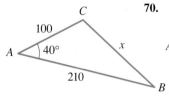

70.

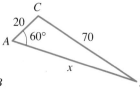

71.

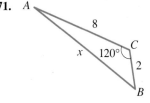

72.

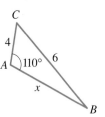

73.

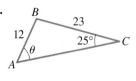

74.

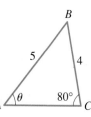

75.

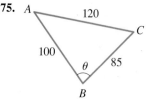

76.

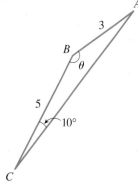

77. Two ships leave a port at the same time. One travels at 20 mi/h in a direction N 32° E, and the other travels at 28 mi/h in a direction S 42° E (see the figure). How far apart are the two ships after 2 h?

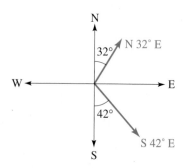

78. From a point *A* on the ground, the angle of elevation to the top of a tall building is 24.1°. From a point *B*, which is 600 ft closer to the building, the angle of elevation is measured to be 30.2°. Find the height of the building.

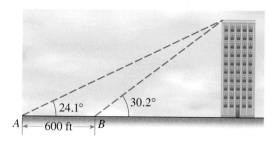

79. Find the distance between points *A* and *B* on opposite sides of a lake from the information shown.

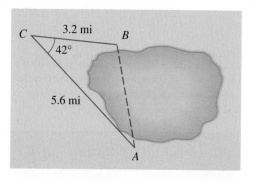

80. A boat is cruising the ocean off a straight shoreline. Points *A* and *B* are 120 mi apart on the shore, as shown. It is found that ∠*A* = 42.3° and ∠*B* = 68.9°. Find the shortest distance from the boat to the shore.

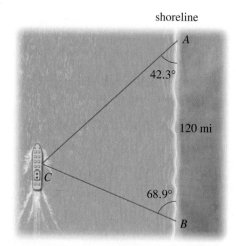

81. Find the area of a triangle with sides of length 8 and 14 and included angle 35°.

82. Find the area of a triangle with sides of length 5, 6, and 8.

1. Find the radian measures that correspond to the degree measures $330°$ and $-135°$.

2. Find the degree measures that correspond to the radian measures $\dfrac{4\pi}{3}$ and -1.3.

3. The rotor blades of a helicopter are 16 ft long and are rotating at 120 rpm.
    (a) Find the angular speed of the rotor.
    (b) Find the linear speed of a point on the tip of a blade.

4. Find the exact value of each of the following.
    (a) $\sin 405°$ (b) $\tan(-150°)$ (c) $\sec \dfrac{5\pi}{3}$ (d) $\csc \dfrac{5\pi}{2}$

5. Find $\tan \theta + \sin \theta$ for the angle θ shown.

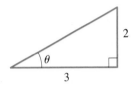

6. Express the lengths a and b shown in the figure in terms of θ.

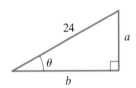

7. If $\cos \theta = -\frac{1}{3}$ and θ is in Quadrant III, find $\tan \theta \cot \theta + \csc \theta$.

8. If $\sin \theta = \frac{5}{13}$ and $\tan \theta = -\frac{5}{12}$, find $\sec \theta$.

9. Express $\tan \theta$ in terms of $\sec \theta$ for θ in Quadrant II.

10. The base of the ladder in the figure is 6 ft from the building, and the angle formed by the ladder and the ground is $73°$. How high up the building does the ladder touch?

11. Express θ in each figure in terms of x.

    (a)

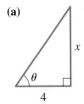

    (b)

12. Find the exact value of $\cos\left(\tan^{-1} \frac{9}{40}\right)$.

13–18 ■ Find the side labeled x or the angle labeled θ.

13.

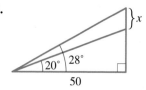

14.

15.

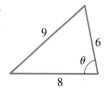

16.

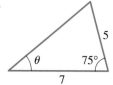

17.

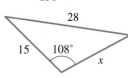

18.

19. Refer to the figure below.

   (a) Find the area of the shaded region.

   (b) Find the perimeter of the shaded region.

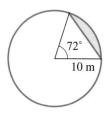

20. Refer to the figure below.

   (a) Find the angle opposite the longest side.

   (b) Find the area of the triangle.

21. Two wires tether a balloon to the ground, as shown. How high is the balloon above the ground?

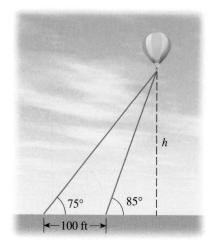

How can we measure the height of a mountain or the distance across a lake? Obviously, it may be difficult, inconvenient, or impossible to measure these distances directly (that is, by using a tape measure or a yardstick). On the other hand, it is easy to measure *angles* involving distant objects. That's where trigonometry comes in: the trigonometric ratios relate angles to distances, so they can be used to *calculate* distances from the *measured* angles. In this *Focus* we examine how trigonometry is used to map a town. Modern mapmaking methods use satellites and the Global Positioning System, but mathematics remains at the core of the process (see page 331).

▼ Mapping a Town

A student wants to draw a map of his hometown. To construct an accurate map (or scale model), he needs to find distances between various landmarks in the town. The student makes the measurements shown in Figure 1. Note that only one distance is measured, between City Hall and the first bridge. All other measurements are angles.

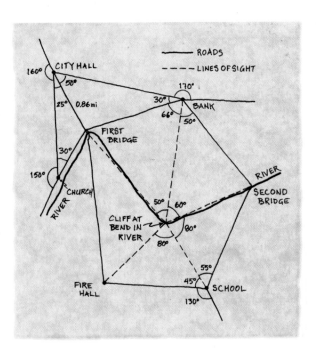

FIGURE 1

The distances between other landmarks can now be found by using the Law of Sines. For example, the distance x from the bank to the first bridge is calculated by applying the Law of Sines to the triangle with vertices at City Hall, the bank, and the first bridge:

$$\frac{x}{\sin 50°} = \frac{0.86}{\sin 30°} \qquad \text{Law of Sines}$$

$$x = \frac{0.86 \sin 50°}{\sin 30°} \qquad \text{Solve for } x$$

$$\approx 1.32 \text{ mi} \qquad \text{Calculator}$$

So the distance between the bank and the first bridge is 1.32 mi.

The distance we just found can now be used to find other distances. For instance, we find the distance y between the bank and the cliff as follows:

$$\frac{y}{\sin 64°} = \frac{1.32}{\sin 50°} \qquad \text{Law of Sines}$$

$$y = \frac{1.32 \sin 64°}{\sin 50°} \qquad \text{Solve for } y$$

$$\approx 1.55 \text{ mi} \qquad \text{Calculator}$$

Continuing in this fashion, we can calculate all the distances between the landmarks shown in the rough sketch in Figure 1. We can use this information to draw the map shown in Figure 2.

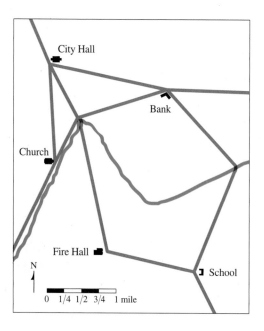

FIGURE 2

To make a topographic map, we need to measure elevation. This concept is explored in Problems 4–6.

PROBLEMS

1. **Completing the Map**   Find the distance between the church and City Hall.

2. **Completing the Map**   Find the distance between the fire hall and the school. (You will need to find other distances first.)

3. **Determining a Distance**   A surveyor on one side of a river wishes to find the distance between points A and B on the opposite side of the river. On her side, she chooses points C and D, which are 20 m apart, and measures the angles shown in the figure below. Find the distance between A and B.

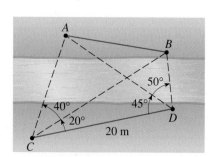

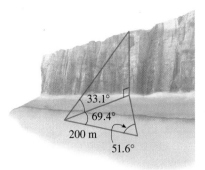

4. **Height of a Cliff**   To measure the height of an inaccessible cliff on the opposite side of a river, a surveyor makes the measurements shown in the figure at the left. Find the height of the cliff.

5. **Height of a Mountain**   To calculate the height h of a mountain, angle α, β, and distance d are measured, as shown in the figure below.

   (a) Show that

   $$h = \frac{d}{\cot \alpha - \cot \beta}$$

   (b) Show that

   $$h = d \frac{\sin \alpha \sin \beta}{\sin(\beta - \alpha)}$$

   (c) Use the formulas from parts (a) and (b) to find the height of a mountain if $\alpha = 25°$, $\beta = 29°$, and $d = 800$ ft. Do you get the same answer from each formula?

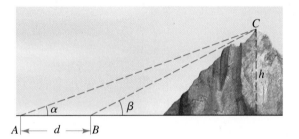

6. **Determining a Distance**   A surveyor has determined that a mountain is 2430 ft high. From the top of the mountain he measures the angles of depression to two landmarks at the base of the mountain and finds them to be 42° and 39°. (Observe that these are the same as the angles of elevation from the landmarks as shown in the figure at the left.) The angle between the lines of sight to the landmarks is 68°. Calculate the distance between the two landmarks.

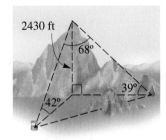

7. **Surveying Building Lots**   A surveyor surveys two adjacent lots and makes the following rough sketch showing his measurements. Calculate all the distances shown in the figure, and use your result to draw an accurate map of the two lots.

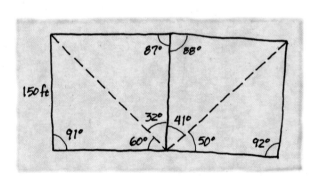

8. Great Survey of India The Great Trigonometric Survey of India was one of the most massive mapping projects ever undertaken (see the margin note on page 204). Do some research at your library or on the Internet to learn more about the Survey, and write a report on your findings.

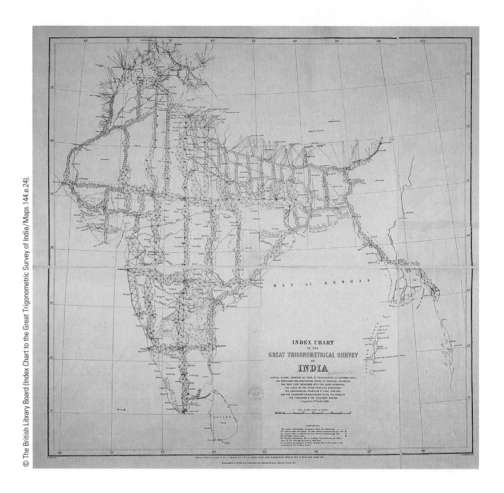

ANALYTIC TRIGONOMETRY

In Chapters 2 and 3 we studied graphical and geometric properties of the trigonometric functions. In this chapter we study algebraic properties of these functions, that is, simplifying and factoring expressions and solving equations that involve trigonometric functions.

We have used the trigonometric functions to model different real-world phenomena, including periodic motion (such as the motion of an ocean wave). To obtain information from a model, we often need to solve equations. If the model involves trigonometric functions, we need to solve trigonometric equations. Solving trigonometric equations often involves using trigonometric identities. We've already encountered some basic trigonometric identities in the preceding chapters. We begin this chapter by finding many new identities.

4.1 TRIGONOMETRIC IDENTITIES

│ Simplifying Trigonometric Expressions ▶ Proving Trigonometric Identities

We begin by listing some of the basic trigonometric identities. We studied most of these in Chapters 2 and 3; you are asked to prove the cofunction identities in Exercise 102.

FUNDAMENTAL TRIGONOMETRIC IDENTITIES

Reciprocal Identities

$$\csc x = \frac{1}{\sin x} \qquad \sec x = \frac{1}{\cos x} \qquad \cot x = \frac{1}{\tan x}$$

$$\tan x = \frac{\sin x}{\cos x} \qquad \cot x = \frac{\cos x}{\sin x}$$

Pythagorean Identities

$$\sin^2 x + \cos^2 x = 1 \qquad \tan^2 x + 1 = \sec^2 x \qquad 1 + \cot^2 x = \csc^2 x$$

Even-Odd Identities

$$\sin(-x) = -\sin x \qquad \cos(-x) = \cos x \qquad \tan(-x) = -\tan x$$

Cofunction Identities

$$\sin\left(\frac{\pi}{2} - u\right) = \cos u \qquad \tan\left(\frac{\pi}{2} - u\right) = \cot u \qquad \sec\left(\frac{\pi}{2} - u\right) = \csc u$$

$$\cos\left(\frac{\pi}{2} - u\right) = \sin u \qquad \cot\left(\frac{\pi}{2} - u\right) = \tan u \qquad \csc\left(\frac{\pi}{2} - u\right) = \sec u$$

▼ Simplifying Trigonometric Expressions

Identities enable us to write the same expression in different ways. It is often possible to rewrite a complicated-looking expression as a much simpler one. To simplify algebraic expressions, we used factoring, common denominators, and the Special Product Formulas. To simplify trigonometric expressions, we use these same techniques together with the fundamental trigonometric identities.

EXAMPLE 1 │ Simplifying a Trigonometric Expression

Simplify the expression $\cos t + \tan t \sin t$.

SOLUTION We start by rewriting the expression in terms of sine and cosine:

$$\cos t + \tan t \sin t = \cos t + \left(\frac{\sin t}{\cos t}\right) \sin t \qquad \text{Reciprocal identity}$$

$$= \frac{\cos^2 t + \sin^2 t}{\cos t} \qquad \text{Common denominator}$$

$$= \frac{1}{\cos t} \qquad \text{Pythagorean identity}$$

$$= \sec t \qquad \text{Reciprocal identity}$$

✎ NOW TRY EXERCISE **3**

EXAMPLE 2 | Simplifying by Combining Fractions

Simplify the expression $\dfrac{\sin\theta}{\cos\theta} + \dfrac{\cos\theta}{1+\sin\theta}$.

SOLUTION   We combine the fractions by using a common denominator:

$$\frac{\sin\theta}{\cos\theta} + \frac{\cos\theta}{1+\sin\theta} = \frac{\sin\theta\,(1+\sin\theta) + \cos^2\theta}{\cos\theta\,(1+\sin\theta)} \qquad \text{Common denominator}$$

$$= \frac{\sin\theta + \sin^2\theta + \cos^2\theta}{\cos\theta\,(1+\sin\theta)} \qquad \text{Distribute } \sin\theta$$

$$= \frac{\sin\theta + 1}{\cos\theta\,(1+\sin\theta)} \qquad \text{Pythagorean identity}$$

$$= \frac{1}{\cos\theta} = \sec\theta \qquad \text{Cancel and use reciprocal identity}$$

✎ NOW TRY EXERCISE **21**   ■

▼ Proving Trigonometric Identities

Many identities follow from the fundamental identities. In the examples that follow, we learn how to prove that a given trigonometric equation is an identity, and in the process we will see how to discover new identities.

First, it's easy to decide when a given equation is *not* an identity. All we need to do is show that the equation does not hold for some value of the variable (or variables). Thus the equation

$$\sin x + \cos x = 1$$

is not an identity, because when $x = \pi/4$, we have

$$\sin\frac{\pi}{4} + \cos\frac{\pi}{4} = \frac{\sqrt{2}}{2} + \frac{\sqrt{2}}{2} = \sqrt{2} \neq 1$$

To verify that a trigonometric equation is an identity, we transform one side of the equation into the other side by a series of steps, each of which is itself an identity.

GUIDELINES FOR PROVING TRIGONOMETRIC IDENTITIES

1. **Start with one side.**   Pick one side of the equation and write it down. Your goal is to transform it into the other side. It's usually easier to start with the more complicated side.

2. **Use known identities.**   Use algebra and the identities you know to change the side you started with. Bring fractional expressions to a common denominator, factor, and use the fundamental identities to simplify expressions.

3. **Convert to sines and cosines.**   If you are stuck, you may find it helpful to rewrite all functions in terms of sines and cosines.

 Warning: To prove an identity, we do *not* just perform the same operations on both sides of the equation. For example, if we start with an equation that is not an identity, such as

(1) $\sin x = -\sin x$

and square both sides, we get the equation

(2) $\sin^2 x = \sin^2 x$

which is clearly an identity. Does this mean that the original equation is an identity? Of course not. The problem here is that the operation of squaring is not **reversible** in the sense that we cannot arrive back at (1) from (2) by taking square roots (reversing the procedure). Only operations that are reversible will necessarily transform an identity into an identity.

EXAMPLE 3 | Proving an Identity by Rewriting in Terms of Sine and Cosine

Consider the equation $\cos \theta (\sec \theta - \cos \theta) = \sin^2\theta$.

(a) Verify algebraically that the equation is an identity.

(b) Confirm graphically that the equation is an identity.

SOLUTION

(a) The left-hand side looks more complicated, so we start with it and try to transform it into the right-hand side:

$$\text{LHS} = \cos \theta (\sec \theta - \cos \theta)$$

$$= \cos \theta \left(\frac{1}{\cos \theta} - \cos \theta \right) \qquad \text{Reciprocal identity}$$

$$= 1 - \cos^2\theta \qquad\qquad \text{Expand}$$

$$= \sin^2\theta = \text{RHS} \qquad \text{Pythagorean identity}$$

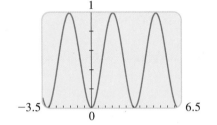

FIGURE 1

(b) We graph each side of the equation to see whether the graphs coincide. From Figure 1 we see that the graphs of $y = \cos \theta (\sec \theta - \cos \theta)$ and $y = \sin^2\theta$ are identical. This confirms that the equation is an identity.

✎ NOW TRY EXERCISE **27**   ■

In Example 3 it isn't easy to see how to change the right-hand side into the left-hand side, but it's definitely possible. Simply notice that each step is reversible. In other words, if we start with the last expression in the proof and work backward through the steps, the right-hand side is transformed into the left-hand side. You will probably agree, however, that it's more difficult to prove the identity this way. That's why it's often better to change the more complicated side of the identity into the simpler side.

EXAMPLE 4 | Proving an Identity by Combining Fractions

Verify the identity

$$2 \tan x \sec x = \frac{1}{1 - \sin x} - \frac{1}{1 + \sin x}$$

SOLUTION   Finding a common denominator and combining the fractions on the right-hand side of this equation, we get

$$\text{RHS} = \frac{1}{1 - \sin x} - \frac{1}{1 + \sin x}$$

$$= \frac{(1 + \sin x) - (1 - \sin x)}{(1 - \sin x)(1 + \sin x)} \qquad \text{Common denominator}$$

$$= \frac{2 \sin x}{1 - \sin^2 x} \qquad\qquad \text{Simplify}$$

$$= \frac{2 \sin x}{\cos^2 x} \qquad\qquad \text{Pythagorean identity}$$

$$= 2 \frac{\sin x}{\cos x} \left(\frac{1}{\cos x} \right) \qquad \text{Factor}$$

$$= 2 \tan x \sec x = \text{LHS} \qquad \text{Reciprocal identities}$$

✎ NOW TRY EXERCISE **79**   ■

See the Prologue: *Principles of Problem Solving,* pages P1–P4.

In Example 5 we introduce "something extra" to the problem by multiplying the numerator and the denominator by a trigonometric expression, chosen so that we can simplify the result.

EXAMPLE 5 | Proving an Identity by Introducing Something Extra

Verify the identity $\dfrac{\cos u}{1 - \sin u} = \sec u + \tan u$.

SOLUTION   We start with the left-hand side and multiply the numerator and denominator by $1 + \sin u$:

We multiply by $1 + \sin u$ because we know by the difference of squares formula that $(1 - \sin u)(1 + \sin u) = 1 - \sin^2 u$, and this is just $\cos^2 u$, a simpler expression.

$$\begin{aligned}
\text{LHS} &= \frac{\cos u}{1 - \sin u} \\[2mm]
&= \frac{\cos u}{1 - \sin u} \cdot \frac{1 + \sin u}{1 + \sin u} && \text{Multiply numerator and denominator by } 1 + \sin u \\[2mm]
&= \frac{\cos u\,(1 + \sin u)}{1 - \sin^2 u} && \text{Expand denominator} \\[2mm]
&= \frac{\cos u\,(1 + \sin u)}{\cos^2 u} && \text{Pythagorean identity} \\[2mm]
&= \frac{1 + \sin u}{\cos u} && \text{Cancel common factor} \\[2mm]
&= \frac{1}{\cos u} + \frac{\sin u}{\cos u} && \text{Separate into two fractions} \\[2mm]
&= \sec u + \tan u && \text{Reciprocal identities}
\end{aligned}$$

✎ **NOW TRY EXERCISE 53** ∎

Here is another method for proving that an equation is an identity. If we can transform each side of the equation *separately*, by way of identities, to arrive at the same result, then the equation is an identity. Example 6 illustrates this procedure.

EXAMPLE 6 | Proving an Identity by Working with Both Sides Separately

Verify the identity $\dfrac{1 + \cos\theta}{\cos\theta} = \dfrac{\tan^2\theta}{\sec\theta - 1}$.

SOLUTION   We prove the identity by changing each side separately into the same expression. Supply the reasons for each step:

$$\text{LHS} = \frac{1 + \cos\theta}{\cos\theta} = \frac{1}{\cos\theta} + \frac{\cos\theta}{\cos\theta} = \sec\theta + 1$$

$$\text{RHS} = \frac{\tan^2\theta}{\sec\theta - 1} = \frac{\sec^2\theta - 1}{\sec\theta - 1} = \frac{(\sec\theta - 1)(\sec\theta + 1)}{\sec\theta - 1} = \sec\theta + 1$$

It follows that $\text{LHS} = \text{RHS}$, so the equation is an identity.

✎ **NOW TRY EXERCISE 81** ∎

We conclude this section by describing the technique of *trigonometric substitution,* which we use to convert algebraic expressions to trigonometric ones. This is often useful in calculus, for instance, in finding the area of a circle or an ellipse.

EUCLID (circa 300 B.C.) taught in Alexandria. His *Elements* is the most widely influential scientific book in history. For 2000 years it was the standard introduction to geometry in the schools, and for many generations it was considered the best way to develop logical reasoning. Abraham Lincoln, for instance, studied the *Elements* as a way to sharpen his mind. The story is told that King Ptolemy once asked Euclid if there was a faster way to learn geometry than through the *Elements.* Euclid replied that there is "no royal road to geometry"—meaning by this that mathematics does not respect wealth or social status. Euclid was revered in his own time and was referred to by the title "The Geometer" or "The Writer of the *Elements.*" The greatness of the *Elements* stems from its precise, logical, and systematic treatment of geometry. For dealing with equality, Euclid lists the following rules, which he calls "common notions."

1. Things that are equal to the same thing are equal to each other.

2. If equals are added to equals, the sums are equal.

3. If equals are subtracted from equals, the remainders are equal.

4. Things that coincide with one another are equal.

5. The whole is greater than the part.

EXAMPLE 7 | Trigonometric Substitution

Substitute $\sin\theta$ for x in the expression $\sqrt{1-x^2}$ and simplify. Assume that $0 \le \theta \le \pi/2$.

SOLUTION Setting $x = \sin\theta$, we have

$$\sqrt{1-x^2} = \sqrt{1-\sin^2\theta} \qquad \text{Substitute } x = \sin\theta$$

$$= \sqrt{\cos^2\theta} \qquad \text{Pythagorean identity}$$

$$= \cos\theta \qquad \text{Take square root}$$

The last equality is true because $\cos\theta \ge 0$ for the values of θ in question.

✎ **NOW TRY EXERCISE 91** ∎

4.1 EXERCISES

CONCEPTS

1. An equation is called an identity if it is valid for _____ values of the variable. The equation $2x = x + x$ is an algebraic identity, and the equation $\sin^2 x + \cos^2 x =$ _____ is a trigonometric identity.

2. For any x it is true that $\cos(-x)$ has the same value as $\cos x$.
   We express this fact as the identity _____.

SKILLS

3–12 ■ Write the trigonometric expression in terms of sine and cosine, and then simplify.

✎ **3.** $\cos t \tan t$

4. $\cos t \csc t$

5. $\sin\theta \sec\theta$

6. $\tan\theta \csc\theta$

7. $\tan^2 x - \sec^2 x$

8. $\dfrac{\sec x}{\csc x}$

9. $\sin u + \cot u \cos u$

10. $\cos^2\theta \,(1 + \tan^2\theta)$

11. $\dfrac{\sec\theta - \cos\theta}{\sin\theta}$

12. $\dfrac{\cot\theta}{\csc\theta - \sin\theta}$

13–26 ■ Simplify the trigonometric expression.

13. $\dfrac{\sin x \sec x}{\tan x}$

14. $\cos^3 x + \sin^2 x \cos x$

15. $\dfrac{1 + \cos y}{1 + \sec y}$

16. $\dfrac{\tan x}{\sec(-x)}$

17. $\dfrac{\sec^2 x - 1}{\sec^2 x}$

18. $\dfrac{\sec x - \cos x}{\tan x}$

19. $\dfrac{1 + \csc x}{\cos x + \cot x}$

20. $\dfrac{\sin x}{\csc x} + \dfrac{\cos x}{\sec x}$

✎ **21.** $\dfrac{1 + \sin u}{\cos u} + \dfrac{\cos u}{1 + \sin u}$

22. $\tan x \cos x \csc x$

23. $\dfrac{2 + \tan^2 x}{\sec^2 x} - 1$

24. $\dfrac{1 + \cot A}{\csc A}$

25. $\tan\theta + \cos(-\theta) + \tan(-\theta)$

26. $\dfrac{\cos x}{\sec x + \tan x}$

27–28 ■ Consider the given equation. **(a)** Verify algebraically that the equation is an identity. **(b)** Confirm graphically that the equation is an identity.

✎ **27.** $\dfrac{\cos x}{\sec x \sin x} = \csc x - \sin x$ **28.** $\dfrac{\tan y}{\csc y} = \sec y - \cos y$

29–90 ■ Verify the identity.

29. $\dfrac{\sin\theta}{\tan\theta} = \cos\theta$

30. $\dfrac{\tan x}{\sec x} = \sin x$

31. $\dfrac{\cos u \sec u}{\tan u} = \cot u$

32. $\dfrac{\cot x \sec x}{\csc x} = 1$

33. $\sin B + \cos B \cot B = \csc B$

34. $\cos(-x) - \sin(-x) = \cos x + \sin x$

35. $\cot(-\alpha)\cos(-\alpha) + \sin(-\alpha) = -\csc\alpha$

36. $\csc x\,[\csc x + \sin(-x)] = \cot^2 x$

37. $\tan\theta + \cot\theta = \sec\theta \csc\theta$

38. $(\sin x + \cos x)^2 = 1 + 2\sin x \cos x$

39. $(1 - \cos\beta)(1 + \cos\beta) = \dfrac{1}{\csc^2\beta}$

40. $\dfrac{\cos x}{\sec x} + \dfrac{\sin x}{\csc x} = 1$

41. $\dfrac{(\sin x + \cos x)^2}{\sin^2 x - \cos^2 x} = \dfrac{\sin^2 x - \cos^2 x}{(\sin x - \cos x)^2}$

42. $(\sin x + \cos x)^4 = (1 + 2\sin x \cos x)^2$

43. $\dfrac{\sec t - \cos t}{\sec t} = \sin^2 t$

44. $\dfrac{1 - \sin x}{1 + \sin x} = (\sec x - \tan x)^2$

45. $\dfrac{1}{1 - \sin^2 y} = 1 + \tan^2 y$ **46.** $\csc x - \sin x = \cos x \cot x$

47. $(\cot x - \csc x)(\cos x + 1) = -\sin x$

48. $\sin^4\theta - \cos^4\theta = \sin^2\theta - \cos^2\theta$

49. $(1 - \cos^2 x)(1 + \cot^2 x) = 1$

50. $\cos^2 x - \sin^2 x = 2\cos^2 x - 1$

51. $2\cos^2 x - 1 = 1 - 2\sin^2 x$

52. $(\tan y + \cot y)\sin y \cos y = 1$

53. $\dfrac{1 - \cos \alpha}{\sin \alpha} = \dfrac{\sin \alpha}{1 + \cos \alpha}$

54. $\sin^2\alpha + \cos^2\alpha + \tan^2\alpha = \sec^2\alpha$

55. $\tan^2\theta - \sin^2\theta = \tan^2\theta \sin^2\theta$

56. $\cot^2\theta \cos^2\theta = \cot^2\theta - \cos^2\theta$

57. $\dfrac{\sin x - 1}{\sin x + 1} = \dfrac{-\cos^2 x}{(\sin x + 1)^2}$

58. $\dfrac{\sin w}{\sin w + \cos w} = \dfrac{\tan w}{1 + \tan w}$

59. $\dfrac{(\sin t + \cos t)^2}{\sin t \cos t} = 2 + \sec t \csc t$

60. $\sec t \csc t\,(\tan t + \cot t) = \sec^2 t + \csc^2 t$

61. $\dfrac{1 + \tan^2 u}{1 - \tan^2 u} = \dfrac{1}{\cos^2 u - \sin^2 u}$

62. $\dfrac{1 + \sec^2 x}{1 + \tan^2 x} = 1 + \cos^2 x$

63. $\dfrac{\sec x}{\sec x - \tan x} = \sec x\,(\sec x + \tan x)$

64. $\dfrac{\sec x + \csc x}{\tan x + \cot x} = \sin x + \cos x$

65. $\sec v - \tan v = \dfrac{1}{\sec v + \tan v}$

66. $\dfrac{\sin A}{1 - \cos A} - \cot A = \csc A$

67. $\dfrac{\sin x + \cos x}{\sec x + \csc x} = \sin x \cos x$

68. $\dfrac{1 - \cos x}{\sin x} + \dfrac{\sin x}{1 - \cos x} = 2 \csc x$

69. $\dfrac{\csc x - \cot x}{\sec x - 1} = \cot x$ **70.** $\dfrac{\csc^2 x - \cot^2 x}{\sec^2 x} = \cos^2 x$

71. $\tan^2 u - \sin^2 u = \tan^2 u \sin^2 u$

72. $\dfrac{\tan v \sin v}{\tan v + \sin v} = \dfrac{\tan v - \sin v}{\tan v \sin v}$

73. $\sec^4 x - \tan^4 x = \sec^2 x + \tan^2 x$

74. $\dfrac{\cos \theta}{1 - \sin \theta} = \sec \theta + \tan \theta$

75. $\dfrac{\cos \theta}{1 - \sin \theta} = \dfrac{\sin \theta - \csc \theta}{\cos \theta - \cot \theta}$

76. $\dfrac{1 + \tan x}{1 - \tan x} = \dfrac{\cos x + \sin x}{\cos x - \sin x}$

77. $\dfrac{\cos^2 t + \tan^2 t - 1}{\sin^2 t} = \tan^2 t$

78. $\dfrac{1}{1 - \sin x} - \dfrac{1}{1 + \sin x} = 2 \sec x \tan x$

79. $\dfrac{1}{\sec x + \tan x} + \dfrac{1}{\sec x - \tan x} = 2 \sec x$

80. $\dfrac{1 + \sin x}{1 - \sin x} - \dfrac{1 - \sin x}{1 + \sin x} = 4 \tan x \sec x$

81. $(\tan x + \cot x)^2 = \sec^2 x + \csc^2 x$

82. $\tan^2 x - \cot^2 x = \sec^2 x - \csc^2 x$

83. $\dfrac{\sec u - 1}{\sec u + 1} = \dfrac{1 - \cos u}{1 + \cos u}$ **84.** $\dfrac{\cot x + 1}{\cot x - 1} = \dfrac{1 + \tan x}{1 - \tan x}$

85. $\dfrac{\sin^3 x + \cos^3 x}{\sin x + \cos x} = 1 - \sin x \cos x$

86. $\dfrac{\tan v - \cot v}{\tan^2 v - \cot^2 v} = \sin v \cos v$

87. $\dfrac{1 + \sin x}{1 - \sin x} = (\tan x + \sec x)^2$

88. $\dfrac{\tan x + \tan y}{\cot x + \cot y} = \tan x \tan y$

89. $(\tan x + \cot x)^4 = \csc^4 x \sec^4 x$

90. $(\sin \alpha - \tan \alpha)(\cos \alpha - \cot \alpha) = (\cos \alpha - 1)(\sin \alpha - 1)$

91–96 ■ Make the indicated trigonometric substitution in the given algebraic expression and simplify (see Example 7). Assume that $0 \le \theta < \pi/2$.

91. $\dfrac{x}{\sqrt{1 - x^2}}$, $\quad x = \sin \theta$ **92.** $\sqrt{1 + x^2}$, $\quad x = \tan \theta$

93. $\sqrt{x^2 - 1}$, $\quad x = \sec \theta$ **94.** $\dfrac{1}{x^2\sqrt{4 + x^2}}$, $\quad x = 2 \tan \theta$

95. $\sqrt{9 - x^2}$, $\quad x = 3 \sin \theta$ **96.** $\dfrac{\sqrt{x^2 - 25}}{x}$, $\quad x = 5 \sec \theta$

97–100 ■ Graph f and g in the same viewing rectangle. Do the graphs suggest that the equation $f(x) = g(x)$ is an identity? Prove your answer.

97. $f(x) = \cos^2 x - \sin^2 x$, $\quad g(x) = 1 - 2\sin^2 x$

98. $f(x) = \tan x\,(1 + \sin x)$, $\quad g(x) = \dfrac{\sin x \cos x}{1 + \sin x}$

99. $f(x) = (\sin x + \cos x)^2$, $\quad g(x) = 1$

100. $f(x) = \cos^4 x - \sin^4 x$, $\quad g(x) = 2\cos^2 x - 1$

101. Show that the equation is not an identity.

(a) $\sin 2x = 2 \sin x$ (b) $\sin(x + y) = \sin x + \sin y$

(c) $\sec^2 x + \csc^2 x = 1$

(d) $\dfrac{1}{\sin x + \cos x} = \csc x + \sec x$

DISCOVERY ■ DISCUSSION ■ WRITING

102. Cofunction Identities In the right triangle shown, explain why $v = (\pi/2) - u$. Explain how you can obtain all six cofunction identities from this triangle for $0 < u < \pi/2$.

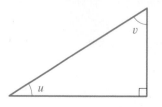

103. Graphs and Identities Suppose you graph two functions, f and g, on a graphing device and their graphs appear identical in the viewing rectangle. Does this prove that the equation $f(x) = g(x)$ is an identity? Explain.

104. Making Up Your Own Identity If you start with a trigonometric expression and rewrite it or simplify it, then setting the original expression equal to the rewritten expression yields a trigonometric identity. For instance, from Example 1 we get the identity

$$\cos t + \tan t \sin t = \sec t$$

Use this technique to make up your own identity, then give it to a classmate to verify.

4.2 ADDITION AND SUBTRACTION FORMULAS

Addition and Subtraction Formulas ▶ Evaluating Expressions Involving Inverse Trigonometric Functions ▶ Expressions of the form $A \sin x + B \cos x$

▼ Addition and Subtraction Formulas

We now derive identities for trigonometric functions of sums and differences.

ADDITION AND SUBTRACTION FORMULAS

Formulas for sine:
$$\sin(s + t) = \sin s \cos t + \cos s \sin t$$
$$\sin(s - t) = \sin s \cos t - \cos s \sin t$$

Formulas for cosine:
$$\cos(s + t) = \cos s \cos t - \sin s \sin t$$
$$\cos(s - t) = \cos s \cos t + \sin s \sin t$$

Formulas for tangent:
$$\tan(s + t) = \frac{\tan s + \tan t}{1 - \tan s \tan t}$$

$$\tan(s - t) = \frac{\tan s - \tan t}{1 + \tan s \tan t}$$

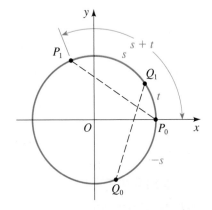

FIGURE 1

PROOF OF ADDITION FORMULA FOR COSINE To prove the formula $\cos(s + t) = \cos s \cos t - \sin s \sin t$, we use Figure 1. In the figure, the distances t, $s + t$, and $-s$ have been marked on the unit circle, starting at $P_0(1, 0)$ and terminating at Q_1, P_1, and Q_0, respectively. The coordinates of these points are

$$P_0(1, 0) \qquad\qquad Q_0(\cos(-s), \sin(-s))$$
$$P_1(\cos(s + t), \sin(s + t)) \qquad Q_1(\cos t, \sin t)$$

Since $\cos(-s) = \cos s$ and $\sin(-s) = -\sin s$, it follows that the point Q_0 has the coordinates $Q_0(\cos s, -\sin s)$. Notice that the distances between P_0 and P_1 and between Q_0 and Q_1 measured along the arc of the circle are equal. Since equal arcs are subtended by equal chords, it follows that $d(P_0, P_1) = d(Q_0, Q_1)$. Using the Distance Formula, we get

$$\sqrt{[\cos(s + t) - 1]^2 + [\sin(s + t) - 0]^2} = \sqrt{(\cos t - \cos s)^2 + (\sin t + \sin s)^2}$$

Squaring both sides and expanding, we have

$$\overbrace{\cos^2(s + t) - 2\cos(s + t) + 1 + \sin^2(s + t)}^{\text{These add to 1}}$$

$$= \underbrace{\cos^2 t - 2\cos s \cos t + \cos^2 s + \sin^2 t + 2\sin s \sin t + \sin^2 s}_{\text{These add to 1}}$$

Using the Pythagorean identity $\sin^2\theta + \cos^2\theta = 1$ three times gives

$$2 - 2\cos(s + t) = 2 - 2\cos s \cos t + 2\sin s \sin t$$

Finally, subtracting 2 from each side and dividing both sides by -2, we get

$$\cos(s + t) = \cos s \cos t - \sin s \sin t$$

which proves the Addition Formula for Cosine.   ■

PROOF OF SUBTRACTION FORMULA FOR COSINE   Replacing t with $-t$ in the Addition Formula for Cosine, we get

$$\cos(s - t) = \cos(s + (-t))$$

$$= \cos s \cos(-t) - \sin s \sin(-t) \qquad \text{Addition Formula for Cosine}$$

$$= \cos s \cos t + \sin s \sin t \qquad \text{Even-odd identities}$$

This proves the Subtraction Formula for Cosine.   ■

See Exercises 70 and 71 for proofs of the other Addition Formulas.

EXAMPLE 1 | Using the Addition and Subtraction Formulas

Find the exact value of each expression.

(a) $\cos 75°$ **(b)** $\cos \dfrac{\pi}{12}$

SOLUTION

(a) Notice that $75° = 45° + 30°$. Since we know the exact values of sine and cosine at $45°$ and $30°$, we use the Addition Formula for Cosine to get

$$\cos 75° = \cos(45° + 30°)$$

$$= \cos 45° \cos 30° - \sin 45° \sin 30°$$

$$= \frac{\sqrt{2}}{2}\frac{\sqrt{3}}{2} - \frac{\sqrt{2}}{2}\frac{1}{2} = \frac{\sqrt{2}\sqrt{3} - \sqrt{2}}{4} = \frac{\sqrt{6} - \sqrt{2}}{4}$$

(b) Since $\dfrac{\pi}{12} = \dfrac{\pi}{4} - \dfrac{\pi}{6}$, the Subtraction Formula for Cosine gives

$$\cos \frac{\pi}{12} = \cos\left(\frac{\pi}{4} - \frac{\pi}{6}\right)$$

$$= \cos \frac{\pi}{4} \cos \frac{\pi}{6} + \sin \frac{\pi}{4} \sin \frac{\pi}{6}$$

$$= \frac{\sqrt{2}}{2}\frac{\sqrt{3}}{2} + \frac{\sqrt{2}}{2}\frac{1}{2} = \frac{\sqrt{6} + \sqrt{2}}{4}$$

✎ NOW TRY EXERCISES **3** AND **9**   ■

EXAMPLE 2 | Using the Addition Formula for Sine

Find the exact value of the expression $\sin 20° \cos 40° + \cos 20° \sin 40°$.

SOLUTION   We recognize the expression as the right-hand side of the Addition Formula for Sine with $s = 20°$ and $t = 40°$. So we have

$$\sin 20° \cos 40° + \cos 20° \sin 40° = \sin(20° + 40°) = \sin 60° = \frac{\sqrt{3}}{2}$$

✎ NOW TRY EXERCISE **15**   ■

EXAMPLE 3 | Proving a Cofunction Identity

Prove the cofunction identity $\cos\left(\dfrac{\pi}{2} - u\right) = \sin u$.

SOLUTION   By the Subtraction Formula for Cosine, we have

$$\cos\left(\frac{\pi}{2} - u\right) = \cos\frac{\pi}{2}\cos u + \sin\frac{\pi}{2}\sin u$$

$$= 0 \cdot \cos u + 1 \cdot \sin u = \sin u$$

✎ NOW TRY EXERCISE **21**

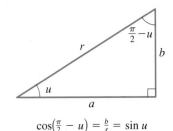

$$\cos\left(\tfrac{\pi}{2} - u\right) = \tfrac{b}{r} = \sin u$$

The cofunction identity in Example 3, as well as the other cofunction identities, can also be derived from the figure in the margin.

EXAMPLE 4 | Proving an Identity

Verify the identity $\dfrac{1 + \tan x}{1 - \tan x} = \tan\left(\dfrac{\pi}{4} + x\right)$.

SOLUTION   Starting with the right-hand side and using the Addition Formula for Tangent, we get

$$\text{RHS} = \tan\left(\frac{\pi}{4} + x\right) = \frac{\tan\dfrac{\pi}{4} + \tan x}{1 - \tan\dfrac{\pi}{4}\tan x}$$

$$= \frac{1 + \tan x}{1 - \tan x} = \text{LHS}$$

✎ NOW TRY EXERCISE **25**

The next example is a typical use of the Addition and Subtraction Formulas in calculus.

EXAMPLE 5 | An Identity from Calculus

If $f(x) = \sin x$, show that

$$\frac{f(x + h) - f(x)}{h} = \sin x\left(\frac{\cos h - 1}{h}\right) + \cos x\left(\frac{\sin h}{h}\right)$$

SOLUTION

$$\frac{f(x + h) - f(x)}{h} = \frac{\sin(x + h) - \sin x}{h} \qquad \text{Definition of } f$$

$$= \frac{\sin x\cos h + \cos x\sin h - \sin x}{h} \qquad \text{Addition Formula for Sine}$$

$$= \frac{\sin x\,(\cos h - 1) + \cos x\sin h}{h} \qquad \text{Factor}$$

$$= \sin x\left(\frac{\cos h - 1}{h}\right) + \cos x\left(\frac{\sin h}{h}\right) \qquad \text{Separate the fraction}$$

✎ NOW TRY EXERCISE **61**

▼ Evaluating Expressions Involving Inverse Trigonometric Functions

Expressions involving trigonometric functions and their inverses arise in calculus. In the next examples we illustrate how to evaluate such expressions.

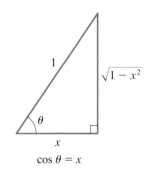

$$\cos \theta = x$$

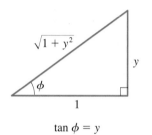

$$\tan \phi = y$$

FIGURE 2

EXAMPLE 6 | Simplifying an Expression Involving Inverse Trigonometric Functions

Write $\sin(\cos^{-1}x + \tan^{-1}y)$ as an algebraic expression in x and y, where $-1 \le x \le 1$ and y is any real number.

SOLUTION Let $\theta = \cos^{-1}x$ and $\phi = \tan^{-1}y$. Using the methods of Section 3.4, we sketch triangles with angles θ and ϕ such that $\cos \theta = x$ and $\tan \phi = y$ (see Figure 2). From the triangles we have

$$\sin \theta = \sqrt{1 - x^2} \qquad \cos \phi = \frac{1}{\sqrt{1 + y^2}} \qquad \sin \phi = \frac{y}{\sqrt{1 + y^2}}$$

From the Addition Formula for Sine we have

$$\sin(\cos^{-1}x + \tan^{-1}y) = \sin(\theta + \phi)$$

$$= \sin \theta \cos \phi + \cos \theta \sin \phi \qquad \text{Addition Formula for Sine}$$

$$= \sqrt{1 - x^2}\,\frac{1}{\sqrt{1 + y^2}} + x\,\frac{y}{\sqrt{1 + y^2}} \qquad \text{From triangles}$$

$$= \frac{1}{\sqrt{1 + y^2}}(\sqrt{1 - x^2} + xy) \qquad \text{Factor } \frac{1}{\sqrt{1 + y^2}}$$

NOW TRY EXERCISES 43 AND 47 ■

EXAMPLE 7 | Evaluating an Expression Involving Trigonometric Functions

Evaluate $\sin(\theta + \phi)$, where $\sin \theta = \frac{12}{13}$ with θ in Quadrant II and $\tan \phi = \frac{3}{4}$ with ϕ in Quadrant III.

SOLUTION We first sketch the angles θ and ϕ in standard position with terminal sides in the appropriate quadrants as in Figure 3. Since $\sin \theta = y/r = \frac{12}{13}$ we can label a side and the hypotenuse in the triangle in Figure 3(a). To find the remaining side, we use the Pythagorean Theorem:

$$x^2 + y^2 = r^2 \qquad \text{Pythagorean Theorem}$$

$$x^2 + 12^2 = 13^2 \qquad y = 12, \quad r = 13$$

$$x^2 = 25 \qquad \text{Solve for } x^2$$

$$x = -5 \qquad \text{Because } x < 0$$

Similarly, since $\tan \phi = y/x = \frac{3}{4}$, we can label two sides of the triangle in Figure 3(b) and then use the Pythagorean Theorem to find the hypotenuse.

FIGURE 3

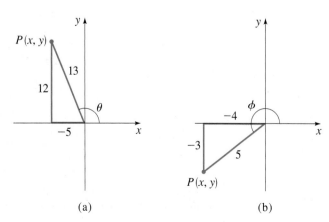

(a)   (b)

Now, to find $\sin(\theta + \phi)$, we use the Addition Formula for Sine and the triangles in Figure 3:

$$\sin(\theta + \phi) = \sin\theta\cos\phi + \cos\theta\sin\phi \qquad \text{Addition Formula}$$

$$= \left(\tfrac{12}{13}\right)\left(-\tfrac{4}{5}\right) + \left(-\tfrac{5}{13}\right)\left(-\tfrac{3}{5}\right) \qquad \text{From triangles}$$

$$= -\tfrac{33}{65} \qquad \text{Calculate}$$

✎ NOW TRY EXERCISE **51**   ■

▼ Expressions of the Form $A\sin x + B\cos x$

We can write expressions of the form $A\sin x + B\cos x$ in terms of a single trigonometric function using the Addition Formula for Sine. For example, consider the expression

$$\frac{1}{2}\sin x + \frac{\sqrt{3}}{2}\cos x$$

If we set $\phi = \pi/3$, then $\cos\phi = \frac{1}{2}$ and $\sin\phi = \sqrt{3}/2$, and we can write

$$\frac{1}{2}\sin x + \frac{\sqrt{3}}{2}\cos x = \cos\phi\sin x + \sin\phi\cos x$$

$$= \sin(x + \phi) = \sin\left(x + \frac{\pi}{3}\right)$$

We are able to do this because the coefficients $\frac{1}{2}$ and $\sqrt{3}/2$ are precisely the cosine and sine of a particular number, in this case, $\pi/3$. We can use this same idea in general to write $A\sin x + B\cos x$ in the form $k\sin(x + \phi)$. We start by multiplying the numerator and denominator by $\sqrt{A^2 + B^2}$ to get

$$A\sin x + B\cos x = \sqrt{A^2 + B^2}\left(\frac{A}{\sqrt{A^2 + B^2}}\sin x + \frac{B}{\sqrt{A^2 + B^2}}\cos x\right)$$

We need a number ϕ with the property that

$$\cos\phi = \frac{A}{\sqrt{A^2 + B^2}} \qquad \text{and} \qquad \sin\phi = \frac{B}{\sqrt{A^2 + B^2}}$$

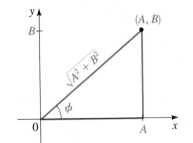

FIGURE 4

Figure 4 shows that the point (A, B) in the plane determines a number ϕ with precisely this property. With this ϕ we have

$$A\sin x + B\cos x = \sqrt{A^2 + B^2}(\cos\phi\sin x + \sin\phi\cos x)$$

$$= \sqrt{A^2 + B^2}\sin(x + \phi)$$

We have proved the following theorem.

SUMS OF SINES AND COSINES

If A and B are real numbers, then

$$A\sin x + B\cos x = k\sin(x + \phi)$$

where $k = \sqrt{A^2 + B^2}$ and ϕ satisfies

$$\cos\phi = \frac{A}{\sqrt{A^2 + B^2}} \qquad \text{and} \qquad \sin\phi = \frac{B}{\sqrt{A^2 + B^2}}$$

EXAMPLE 8 | A Sum of Sine and Cosine Terms

Express $3 \sin x + 4 \cos x$ in the form $k \sin(x + \phi)$.

SOLUTION By the preceding theorem, $k = \sqrt{A^2 + B^2} = \sqrt{3^2 + 4^2} = 5$. The angle ϕ has the property that $\sin \phi = \frac{4}{5}$ and $\cos \phi = \frac{3}{5}$. Using a calculator, we find $\phi \approx 53.1°$. Thus

$$3 \sin x + 4 \cos x \approx 5 \sin(x + 53.1°)$$

✎ NOW TRY EXERCISE **55**   ■

EXAMPLE 9 | Graphing a Trigonometric Function

Write the function $f(x) = -\sin 2x + \sqrt{3} \cos 2x$ in the form $k \sin(2x + \phi)$, and use the new form to graph the function.

SOLUTION Since $A = -1$ and $B = \sqrt{3}$, we have $k = \sqrt{A^2 + B^2} = \sqrt{1 + 3} = 2$. The angle ϕ satisfies $\cos \phi = -\frac{1}{2}$ and $\sin \phi = \sqrt{3}/2$. From the signs of these quantities we conclude that ϕ is in Quadrant II. Thus $\phi = 2\pi/3$. By the preceding theorem we can write

$$f(x) = -\sin 2x + \sqrt{3} \cos 2x = 2 \sin\left(2x + \frac{2\pi}{3}\right)$$

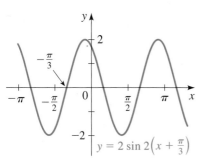

FIGURE 5

Using the form

$$f(x) = 2 \sin 2\left(x + \frac{\pi}{3}\right)$$

we see that the graph is a sine curve with amplitude 2, period $2\pi/2 = \pi$, and phase shift $-\pi/3$. The graph is shown in Figure 5.

✎ NOW TRY EXERCISE **59**   ■

4.2 EXERCISES

CONCEPTS

1. If we know the values of the sine and cosine of x and y, we can find the value of $\sin(x + y)$ by using the _____ Formula for Sine. State the formula: $\sin(x + y) = $ _____.

2. If we know the values of the sine and cosine of x and y, we can find the value of $\cos(x - y)$ by using the _____ Formula for Cosine. State the formula: $\cos(x - y) = $ _____.

SKILLS

3–14 ■ Use an Addition or Subtraction Formula to find the exact value of the expression, as demonstrated in Example 1.

✎ **3.** $\sin 75°$ **4.** $\sin 15°$

5. $\cos 105°$ **6.** $\cos 195°$

7. $\tan 15°$ **8.** $\tan 165°$

✎ **9.** $\sin \dfrac{19\pi}{12}$ **10.** $\cos \dfrac{17\pi}{12}$

11. $\tan\left(-\dfrac{\pi}{12}\right)$ **12.** $\sin\left(-\dfrac{5\pi}{12}\right)$

13. $\cos \dfrac{11\pi}{12}$ **14.** $\tan \dfrac{7\pi}{12}$

15–20 ■ Use an Addition or Subtraction Formula to write the expression as a trigonometric function of one number, and then find its exact value.

✎ **15.** $\sin 18° \cos 27° + \cos 18° \sin 27°$

16. $\cos 10° \cos 80° - \sin 10° \sin 80°$

17. $\cos \dfrac{3\pi}{7} \cos \dfrac{2\pi}{21} + \sin \dfrac{3\pi}{7} \sin \dfrac{2\pi}{21}$

18. $\dfrac{\tan \dfrac{\pi}{18} + \tan \dfrac{\pi}{9}}{1 - \tan \dfrac{\pi}{18} \tan \dfrac{\pi}{9}}$

19. $\dfrac{\tan 73° - \tan 13°}{1 + \tan 73° \tan 13°}$

20. $\cos \dfrac{13\pi}{15} \cos\left(-\dfrac{\pi}{5}\right) - \sin \dfrac{13\pi}{15} \sin\left(-\dfrac{\pi}{5}\right)$

21–24 ■ Prove the cofunction identity using the Addition and Subtraction Formulas.

21. $\tan\left(\dfrac{\pi}{2} - u\right) = \cot u$ **22.** $\cot\left(\dfrac{\pi}{2} - u\right) = \tan u$

23. $\sec\left(\dfrac{\pi}{2} - u\right) = \csc u$ **24.** $\csc\left(\dfrac{\pi}{2} - u\right) = \sec u$

25–42 ■ Prove the identity.

25. $\sin\left(x - \dfrac{\pi}{2}\right) = -\cos x$

26. $\cos\left(x - \dfrac{\pi}{2}\right) = \sin x$

27. $\sin(x - \pi) = -\sin x$ **28.** $\cos(x - \pi) = -\cos x$

29. $\tan(x - \pi) = \tan x$

30. $\sin\left(\dfrac{\pi}{2} - x\right) = \sin\left(\dfrac{\pi}{2} + x\right)$

31. $\cos\left(x + \dfrac{\pi}{6}\right) + \sin\left(x - \dfrac{\pi}{3}\right) = 0$

32. $\tan\left(x - \dfrac{\pi}{4}\right) = \dfrac{\tan x - 1}{\tan x + 1}$

33. $\sin(x + y) - \sin(x - y) = 2 \cos x \sin y$

34. $\cos(x + y) + \cos(x - y) = 2 \cos x \cos y$

35. $\cot(x - y) = \dfrac{\cot x \cot y + 1}{\cot y - \cot x}$

36. $\cot(x + y) = \dfrac{\cot x \cot y - 1}{\cot x + \cot y}$

37. $\tan x - \tan y = \dfrac{\sin(x - y)}{\cos x \cos y}$

38. $1 - \tan x \tan y = \dfrac{\cos(x + y)}{\cos x \cos y}$

39. $\dfrac{\sin(x + y) - \sin(x - y)}{\cos(x + y) + \cos(x - y)} = \tan y$

40. $\cos(x + y) \cos(x - y) = \cos^2 x - \sin^2 y$

41. $\sin(x + y + z) = \sin x \cos y \cos z + \cos x \sin y \cos z$
$\qquad\qquad + \cos x \cos y \sin z - \sin x \sin y \sin z$

42. $\tan(x - y) + \tan(y - z) + \tan(z - x)$
$\qquad\qquad = \tan(x - y) \tan(y - z) \tan(z - x)$

43–46 ■ Write the given expression in terms of x and y only.

43. $\cos(\sin^{-1}x - \tan^{-1}y)$ **44.** $\tan(\sin^{-1}x + \cos^{-1}y)$

45. $\sin(\tan^{-1}x - \tan^{-1}y)$ **46.** $\sin(\sin^{-1}x + \cos^{-1}y)$

47–50 ■ Find the exact value of the expression.

47. $\sin\left(\cos^{-1}\tfrac{1}{2} + \tan^{-1} 1\right)$ **48.** $\cos\left(\sin^{-1}\tfrac{\sqrt{3}}{2} + \cot^{-1}\sqrt{3}\right)$

49. $\tan\left(\sin^{-1}\tfrac{3}{4} - \cos^{-1}\tfrac{1}{3}\right)$ **50.** $\sin\left(\cos^{-1}\tfrac{2}{3} - \tan^{-1}\tfrac{1}{2}\right)$

51–54 ■ Evaluate each expression under the given conditions.

51. $\cos(\theta - \phi)$; $\cos \theta = \tfrac{3}{5}$, θ in Quadrant IV, $\tan \phi = -\sqrt{3}$, ϕ in Quadrant II.

52. $\sin(\theta - \phi)$; $\tan \theta = \tfrac{4}{3}$, θ in Quadrant III, $\sin \phi = -\sqrt{10}/10$, ϕ in Quadrant IV

53. $\sin(\theta + \phi)$; $\sin \theta = \tfrac{5}{13}$, θ in Quadrant I, $\cos \phi = -2\sqrt{5}/5$, ϕ in Quadrant II

54. $\tan(\theta + \phi)$; $\cos \theta = -\tfrac{1}{3}$, θ in Quadrant III, $\sin \phi = \tfrac{1}{4}$, ϕ in Quadrant II

55–58 ■ Write the expression in terms of sine only.

55. $-\sqrt{3} \sin x + \cos x$ **56.** $\sin x + \cos x$

57. $5(\sin 2x - \cos 2x)$ **58.** $3 \sin \pi x + 3\sqrt{3} \cos \pi x$

59–60 ■ **(a)** Express the function in terms of sine only. **(b)** Graph the function.

59. $g(x) = \cos 2x + \sqrt{3} \sin 2x$ **60.** $f(x) = \sin x + \cos x$

61. Let $g(x) = \cos x$. Show that

$$\dfrac{g(x + h) - g(x)}{h} = -\cos x \left(\dfrac{1 - \cos h}{h}\right) - \sin x \left(\dfrac{\sin h}{h}\right)$$

62. Show that if $\beta - \alpha = \pi/2$, then

$$\sin(x + \alpha) + \cos(x + \beta) = 0$$

63. Refer to the figure. Show that $\alpha + \beta = \gamma$, and find $\tan \gamma$.

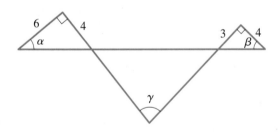

64. **(a)** If L is a line in the plane and θ is the angle formed by the line and the x-axis as shown in the figure, show that the slope m of the line is given by

$$m = \tan \theta$$

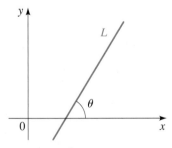

(b) Let L_1 and L_2 be two nonparallel lines in the plane with slopes m_1 and m_2, respectively. Let ψ be the acute angle formed by the two lines (see the following figure). Show that

$$\tan \psi = \dfrac{m_2 - m_1}{1 + m_1 m_2}$$

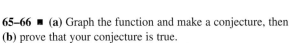

(c) Find the acute angle formed by the two lines

$$y = \tfrac{1}{3}x + 1 \qquad \text{and} \qquad y = -\tfrac{1}{2}x - 3$$

(d) Show that if two lines are perpendicular, then the slope of one is the negative reciprocal of the slope of the other. [*Hint:* First find an expression for cot ψ.]

 65–66 ■ **(a)** Graph the function and make a conjecture, then **(b)** prove that your conjecture is true.

65. $y = \sin^2\left(x + \dfrac{\pi}{4}\right) + \sin^2\left(x - \dfrac{\pi}{4}\right)$

66. $y = -\tfrac{1}{2}[\cos(x + \pi) + \cos(x - \pi)]$

67. Find $\angle A + \angle B + \angle C$ in the figure. [*Hint:* First use an addition formula to find tan$(A + B)$.]

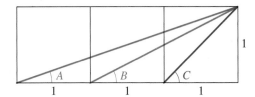

APPLICATIONS

 68. Adding an Echo A digital delay device echoes an input signal by repeating it a fixed length of time after it is received. If such a device receives the pure note $f_1(t) = 5 \sin t$ and echoes the pure note $f_2(t) = 5 \cos t$, then the combined sound is $f(t) = f_1(t) + f_2(t)$.
(a) Graph $y = f(t)$ and observe that the graph has the form of a sine curve $y = k \sin(t + \phi)$.
(b) Find k and ϕ.

69. Interference Two identical tuning forks are struck, one a fraction of a second after the other. The sounds produced are modeled by $f_1(t) = C \sin \omega t$ and $f_2(t) = C \sin(\omega t + \alpha)$. The two sound waves interfere to produce a single sound modeled by the sum of these functions

$$f(t) = C \sin \omega t + C \sin(\omega t + \alpha)$$

(a) Use the Addition Formula for Sine to show that f can be written in the form $f(t) = A \sin \omega t + B \cos \omega t$, where A and B are constants that depend on α.

(b) Suppose that $C = 10$ and $\alpha = \pi/3$. Find constants k and ϕ so that $f(t) = k \sin(\omega t + \phi)$.

DISCOVERY ■ DISCUSSION ■ WRITING

70. Addition Formula for Sine In the text we proved only the Addition and Subtraction Formulas for Cosine. Use these formulas and the cofunction identities

$$\sin x = \cos\left(\dfrac{\pi}{2} - x\right)$$

$$\cos x = \sin\left(\dfrac{\pi}{2} - x\right)$$

to prove the Addition Formula for Sine. [*Hint:* To get started, use the first cofunction identity to write

$$\sin(s + t) = \cos\left(\dfrac{\pi}{2} - (s + t)\right)$$

$$= \cos\left(\left(\dfrac{\pi}{2} - s\right) - t\right)$$

and use the Subtraction Formula for Cosine.]

71. Addition Formula for Tangent Use the Addition Formulas for Cosine and Sine to prove the Addition Formula for Tangent. [*Hint:* Use

$$\tan(s + t) = \dfrac{\sin(s + t)}{\cos(s + t)}$$

and divide the numerator and denominator by cos s cos t.]

4.3 DOUBLE-ANGLE, HALF-ANGLE, AND PRODUCT-SUM FORMULAS

Double-Angle Formulas ▶ Half-Angle Formulas ▶ Simplifying Expressions Involving Inverse Trigonometric Functions ▶ Product-Sum Formulas

The identities we consider in this section are consequences of the addition formulas. The **Double-Angle Formulas** allow us to find the values of the trigonometric functions at $2x$ from their values at x. The **Half-Angle Formulas** relate the values of the trigonometric functions at $\tfrac{1}{2}x$ to their values at x. The **Product-Sum Formulas** relate products of sines and cosines to sums of sines and cosines.

▼ Double-Angle Formulas

The formulas in the following box are immediate consequences of the addition formulas, which we proved in the preceding section.

DOUBLE-ANGLE FORMULAS

Formula for sine:	$\sin 2x = 2 \sin x \cos x$
Formulas for cosine:	$\cos 2x = \cos^2 x - \sin^2 x$
	$= 1 - 2 \sin^2 x$
	$= 2 \cos^2 x - 1$
Formula for tangent:	$\tan 2x = \dfrac{2 \tan x}{1 - \tan^2 x}$

The proofs for the formulas for cosine are given here. You are asked to prove the remaining formulas in Exercises 35 and 36.

PROOF OF DOUBLE-ANGLE FORMULAS FOR COSINE

$$\cos 2x = \cos(x + x)$$
$$= \cos x \cos x - \sin x \sin x$$
$$= \cos^2 x - \sin^2 x$$

The second and third formulas for $\cos 2x$ are obtained from the formula we just proved and the Pythagorean identity. Substituting $\cos^2 x = 1 - \sin^2 x$ gives

$$\cos 2x = \cos^2 x - \sin^2 x$$
$$= (1 - \sin^2 x) - \sin^2 x$$
$$= 1 - 2 \sin^2 x$$

The third formula is obtained in the same way, by substituting $\sin^2 x = 1 - \cos^2 x$. ■

EXAMPLE 1 | Using the Double-Angle Formulas

If $\cos x = -\frac{2}{3}$ and x is in Quadrant II, find $\cos 2x$ and $\sin 2x$.

SOLUTION Using one of the Double-Angle Formulas for Cosine, we get

$$\cos 2x = 2 \cos^2 x - 1$$
$$= 2\left(-\frac{2}{3}\right)^2 - 1 = \frac{8}{9} - 1 = -\frac{1}{9}$$

To use the formula $\sin 2x = 2 \sin x \cos x$, we need to find $\sin x$ first. We have

$$\sin x = \sqrt{1 - \cos^2 x} = \sqrt{1 - \left(-\frac{2}{3}\right)^2} = \frac{\sqrt{5}}{3}$$

where we have used the positive square root because $\sin x$ is positive in Quadrant II. Thus

$$\sin 2x = 2 \sin x \cos x$$
$$= 2\left(\frac{\sqrt{5}}{3}\right)\left(-\frac{2}{3}\right) = -\frac{4\sqrt{5}}{9}$$

✎ NOW TRY EXERCISE **3** ■

EXAMPLE 2 | A Triple-Angle Formula

Write $\cos 3x$ in terms of $\cos x$.

SOLUTION

$$
\begin{aligned}
\cos 3x &= \cos(2x + x) \\
&= \cos 2x \cos x - \sin 2x \sin x && \text{Addition Formula} \\
&= (2\cos^2 x - 1)\cos x - (2\sin x \cos x)\sin x && \text{Double-Angle Formulas} \\
&= 2\cos^3 x - \cos x - 2\sin^2 x \cos x && \text{Expand} \\
&= 2\cos^3 x - \cos x - 2\cos x\,(1 - \cos^2 x) && \text{Pythagorean identity} \\
&= 2\cos^3 x - \cos x - 2\cos x + 2\cos^3 x && \text{Expand} \\
&= 4\cos^3 x - 3\cos x && \text{Simplify}
\end{aligned}
$$

✎ NOW TRY EXERCISE **101**

Example 2 shows that $\cos 3x$ can be written as a polynomial of degree 3 in $\cos x$. The identity $\cos 2x = 2\cos^2 x - 1$ shows that $\cos 2x$ is a polynomial of degree 2 in $\cos x$. In fact, for any natural number n, we can write $\cos nx$ as a polynomial in $\cos x$ of degree n (see the note following Exercise 101). The analogous result for $\sin nx$ is not true in general.

EXAMPLE 3 | Proving an Identity

Prove the identity $\dfrac{\sin 3x}{\sin x \cos x} = 4\cos x - \sec x$.

SOLUTION   We start with the left-hand side:

$$
\begin{aligned}
\frac{\sin 3x}{\sin x \cos x} &= \frac{\sin(x + 2x)}{\sin x \cos x} \\[2mm]
&= \frac{\sin x \cos 2x + \cos x \sin 2x}{\sin x \cos x} && \text{Addition Formula} \\[2mm]
&= \frac{\sin x\,(2\cos^2 x - 1) + \cos x\,(2\sin x \cos x)}{\sin x \cos x} && \text{Double-Angle Formulas} \\[2mm]
&= \frac{\sin x\,(2\cos^2 x - 1)}{\sin x \cos x} + \frac{\cos x\,(2\sin x \cos x)}{\sin x \cos x} && \text{Separate fraction} \\[2mm]
&= \frac{2\cos^2 x - 1}{\cos x} + 2\cos x && \text{Cancel} \\[2mm]
&= 2\cos x - \frac{1}{\cos x} + 2\cos x && \text{Separate fraction} \\[2mm]
&= 4\cos x - \sec x && \text{Reciprocal identity}
\end{aligned}
$$

✎ NOW TRY EXERCISE **81**

▼ Half-Angle Formulas

The following formulas allow us to write any trigonometric expression involving even powers of sine and cosine in terms of the first power of cosine only. This technique is important in calculus. The Half-Angle Formulas are immediate consequences of these formulas.

PIERRE DE FERMAT (1601–1665) was a French lawyer who became interested in mathematics at the age of 30. Because of his job as a magistrate, Fermat had little time to write complete proofs of his discoveries and often wrote them in the margin of whatever book he was reading at the time. After his death, his copy of Diophantus' *Arithmetica* (see page 506) was found to contain a particularly tantalizing comment. Where Diophantus discusses the solutions of $x^2 + y^2 = z^2$ (for example, $x = 3, y = 4,$ and $z = 5$), Fermat states in the margin that for $n \geq 3$ there are no natural number solutions to the equation $x^n + y^n = z^n$. In other words, it's impossible for a cube to equal the sum of two cubes, a fourth power to equal the sum of two fourth powers, and so on. Fermat writes, "I have discovered a truly wonderful proof for this but the margin is too small to contain it." All the other margin comments in Fermat's copy of *Arithmetica* have been proved. This one, however, remained unproved, and it came to be known as "Fermat's Last Theorem."

In 1994, Andrew Wiles of Princeton University announced a proof of Fermat's Last Theorem, an astounding 350 years after it was conjectured. His proof is one of the most widely reported mathematical results in the popular press.

FORMULAS FOR LOWERING POWERS

$$\sin^2 x = \frac{1 - \cos 2x}{2} \qquad \cos^2 x = \frac{1 + \cos 2x}{2}$$

$$\tan^2 x = \frac{1 - \cos 2x}{1 + \cos 2x}$$

PROOF   The first formula is obtained by solving for $\sin^2 x$ in the double-angle formula $\cos 2x = 1 - 2\sin^2 x$. Similarly, the second formula is obtained by solving for $\cos^2 x$ in the Double-Angle Formula $\cos 2x = 2\cos^2 x - 1$.

The last formula follows from the first two and the reciprocal identities:

$$\tan^2 x = \frac{\sin^2 x}{\cos^2 x} = \frac{\dfrac{1 - \cos 2x}{2}}{\dfrac{1 + \cos 2x}{2}} = \frac{1 - \cos 2x}{1 + \cos 2x}$$

EXAMPLE 4 | Lowering Powers in a Trigonometric Expression

Express $\sin^2 x \cos^2 x$ in terms of the first power of cosine.

SOLUTION   We use the formulas for lowering powers repeatedly:

$$\sin^2 x \cos^2 x = \left(\frac{1 - \cos 2x}{2} \right)\left(\frac{1 + \cos 2x}{2} \right)$$

$$= \frac{1 - \cos^2 2x}{4} = \frac{1}{4} - \frac{1}{4}\cos^2 2x$$

$$= \frac{1}{4} - \frac{1}{4}\left(\frac{1 + \cos 4x}{2} \right) = \frac{1}{4} - \frac{1}{8} - \frac{\cos 4x}{8}$$

$$= \frac{1}{8} - \frac{1}{8}\cos 4x = \frac{1}{8}(1 - \cos 4x)$$

Another way to obtain this identity is to use the Double-Angle Formula for Sine in the form $\sin x \cos x = \frac{1}{2}\sin 2x$. Thus

$$\sin^2 x \cos^2 x = \frac{1}{4}\sin^2 2x = \frac{1}{4}\left(\frac{1 - \cos 4x}{2} \right)$$

$$= \frac{1}{8}(1 - \cos 4x)$$

✎ NOW TRY EXERCISE **11**

HALF-ANGLE FORMULAS

$$\sin \frac{u}{2} = \pm\sqrt{\frac{1 - \cos u}{2}} \qquad \cos \frac{u}{2} = \pm\sqrt{\frac{1 + \cos u}{2}}$$

$$\tan \frac{u}{2} = \frac{1 - \cos u}{\sin u} = \frac{\sin u}{1 + \cos u}$$

The choice of the $+$ or $-$ sign depends on the quadrant in which $u/2$ lies.

PROOF   We substitute $x = u/2$ in the formulas for lowering powers and take the square root of each side. This gives the first two Half-Angle Formulas. In the case of the Half-Angle Formula for Tangent we get

$$\tan \frac{u}{2} = \pm\sqrt{\frac{1 - \cos u}{1 + \cos u}}$$

$$= \pm\sqrt{\left(\frac{1 - \cos u}{1 + \cos u}\right)\left(\frac{1 - \cos u}{1 - \cos u}\right)} \qquad \text{Multiply numerator and denominator by } 1 - \cos u$$

$$= \pm\sqrt{\frac{(1 - \cos u)^2}{1 - \cos^2 u}} \qquad \text{Simplify}$$

$$= \pm\frac{|1 - \cos u|}{|\sin u|} \qquad \begin{array}{l}\sqrt{A^2} = |A| \\ \text{and} \quad 1 - \cos^2 u = \sin^2 u\end{array}$$

Now, $1 - \cos u$ is nonnegative for all values of u. It is also true that $\sin u$ and $\tan(u/2)$ always have the same sign. (Verify this.) It follows that

$$\tan \frac{u}{2} = \frac{1 - \cos u}{\sin u}$$

The other Half-Angle Formula for Tangent is derived from this by multiplying the numerator and denominator by $1 + \cos u$. ∎

EXAMPLE 5 | Using a Half-Angle Formula

Find the exact value of $\sin 22.5°$.

SOLUTION   Since $22.5°$ is half of $45°$, we use the Half-Angle Formula for Sine with $u = 45°$. We choose the $+$ sign because $22.5°$ is in the first quadrant:

$$\sin \frac{45°}{2} = \sqrt{\frac{1 - \cos 45°}{2}} \qquad \text{Half-Angle Formula}$$

$$= \sqrt{\frac{1 - \sqrt{2}/2}{2}} \qquad \cos 45° = \sqrt{2}/2$$

$$= \sqrt{\frac{2 - \sqrt{2}}{4}} \qquad \text{Common denominator}$$

$$= \tfrac{1}{2}\sqrt{2 - \sqrt{2}} \qquad \text{Simplify}$$

🔖 NOW TRY EXERCISE **17**   ∎

EXAMPLE 6 | Using a Half-Angle Formula

Find $\tan(u/2)$ if $\sin u = \frac{2}{5}$ and u is in Quadrant II.

SOLUTION   To use the Half-Angle Formula for Tangent, we first need to find $\cos u$. Since cosine is negative in Quadrant II, we have

$$\cos u = -\sqrt{1 - \sin^2 u}$$

$$= -\sqrt{1 - \left(\tfrac{2}{5}\right)^2} = -\frac{\sqrt{21}}{5}$$

Thus

$$\tan \frac{u}{2} = \frac{1 - \cos u}{\sin u}$$

$$= \frac{1 + \sqrt{21}/5}{\frac{2}{5}} = \frac{5 + \sqrt{21}}{2}$$

🔖 NOW TRY EXERCISE **37**

▼ Evaluating Expressions Involving Inverse Trigonometric Functions

Expressions involving trigonometric functions and their inverses arise in calculus. In the next examples we illustrate how to evaluate such expressions.

EXAMPLE 7 | Simplifying an Expression Involving an Inverse Trigonometric Function

Write $\sin(2\cos^{-1}x)$ as an algebraic expression in x only, where $-1 \le x \le 1$.

SOLUTION   Let $\theta = \cos^{-1}x$, and sketch a triangle as in Figure 1. We need to find $\sin 2\theta$, but from the triangle we can find trigonometric functions of θ only, not 2θ. So we use the Double-Angle Formula for Sine.

$$\sin(2\cos^{-1}x) = \sin 2\theta \qquad \cos^{-1}x = \theta$$
$$= 2\sin\theta\cos\theta \qquad \text{Double-Angle Formula}$$
$$= 2x\sqrt{1-x^2} \qquad \text{From the triangle}$$

✎ NOW TRY EXERCISES **43** AND **47**

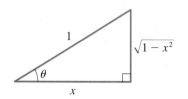

FIGURE 1

EXAMPLE 8 | Evaluating an Expression Involving Inverse Trigonometric Functions

Evaluate $\sin 2\theta$, where $\cos\theta = -\frac{2}{5}$ with θ in Quadrant II.

SOLUTION   We first sketch the angle θ in standard position with terminal side in Quadrant II as in Figure 2. Since $\cos\theta = x/r = -\frac{2}{5}$, we can label a side and the hypotenuse of the triangle in Figure 2. To find the remaining side, we use the Pythagorean Theorem:

$$x^2 + y^2 = r^2 \qquad \text{Pythagorean Theorem}$$
$$(-2)^2 + y^2 = 5^2 \qquad x = -2, \quad r = 5$$
$$y = \pm\sqrt{21} \qquad \text{Solve for } y^2$$
$$y = +\sqrt{21} \qquad \text{Because } y > 0$$

We can now use the Double-Angle Formula for Sine:

$$\sin 2\theta = 2\sin\theta\cos\theta \qquad \text{Double-Angle Formula}$$
$$= 2\left(\frac{\sqrt{21}}{5}\right)\left(-\frac{2}{5}\right) \qquad \text{From the triangle}$$
$$= -\frac{4\sqrt{21}}{25} \qquad \text{Simplify}$$

✎ NOW TRY EXERCISE **51**

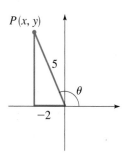

FIGURE 2

▼ Product-Sum Formulas

It is possible to write the product $\sin u \cos v$ as a sum of trigonometric functions. To see this, consider the addition and subtraction formulas for the sine function:

$$\sin(u + v) = \sin u \cos v + \cos u \sin v$$
$$\sin(u - v) = \sin u \cos v - \cos u \sin v$$

Adding the left- and right-hand sides of these formulas gives

$$\sin(u + v) + \sin(u - v) = 2\sin u \cos v$$

Dividing by 2 gives the formula

$$\sin u \cos v = \tfrac{1}{2}[\sin(u + v) + \sin(u - v)]$$

The other three **Product-to-Sum Formulas** follow from the addition formulas in a similar way.

PRODUCT-TO-SUM FORMULAS

$$\sin u \cos v = \tfrac{1}{2}[\sin(u + v) + \sin(u - v)]$$

$$\cos u \sin v = \tfrac{1}{2}[\sin(u + v) - \sin(u - v)]$$

$$\cos u \cos v = \tfrac{1}{2}[\cos(u + v) + \cos(u - v)]$$

$$\sin u \sin v = \tfrac{1}{2}[\cos(u - v) - \cos(u + v)]$$

EXAMPLE 9 | Expressing a Trigonometric Product as a Sum

Express $\sin 3x \sin 5x$ as a sum of trigonometric functions.

SOLUTION   Using the fourth Product-to-Sum Formula with $u = 3x$ and $v = 5x$ and the fact that cosine is an even function, we get

$$\sin 3x \sin 5x = \tfrac{1}{2}[\cos(3x - 5x) - \cos(3x + 5x)]$$

$$= \tfrac{1}{2}\cos(-2x) - \tfrac{1}{2}\cos 8x$$

$$= \tfrac{1}{2}\cos 2x - \tfrac{1}{2}\cos 8x$$

NOW TRY EXERCISE 55

The Product-to-Sum Formulas can also be used as Sum-to-Product Formulas. This is possible because the right-hand side of each Product-to-Sum Formula is a sum and the left side is a product. For example, if we let

$$u = \frac{x + y}{2} \qquad \text{and} \qquad v = \frac{x - y}{2}$$

in the first Product-to-Sum Formula, we get

$$\sin \frac{x + y}{2} \cos \frac{x - y}{2} = \tfrac{1}{2}(\sin x + \sin y)$$

so

$$\sin x + \sin y = 2 \sin \frac{x + y}{2} \cos \frac{x - y}{2}$$

The remaining three of the following **Sum-to-Product Formulas** are obtained in a similar manner.

SUM-TO-PRODUCT FORMULAS

$$\sin x + \sin y = 2 \sin \frac{x + y}{2} \cos \frac{x - y}{2}$$

$$\sin x - \sin y = 2 \cos \frac{x + y}{2} \sin \frac{x - y}{2}$$

$$\cos x + \cos y = 2 \cos \frac{x + y}{2} \cos \frac{x - y}{2}$$

$$\cos x - \cos y = -2 \sin \frac{x + y}{2} \sin \frac{x - y}{2}$$

EXAMPLE 10 | Expressing a Trigonometric Sum as a Product

Write $\sin 7x + \sin 3x$ as a product.

SOLUTION The first Sum-to-Product Formula gives

$$\sin 7x + \sin 3x = 2 \sin \frac{7x + 3x}{2} \cos \frac{7x - 3x}{2}$$

$$= 2 \sin 5x \cos 2x$$

✎ NOW TRY EXERCISE **61**

EXAMPLE 11 | Proving an Identity

Verify the identity $\dfrac{\sin 3x - \sin x}{\cos 3x + \cos x} = \tan x$.

SOLUTION We apply the second Sum-to-Product Formula to the numerator and the third formula to the denominator:

$$\text{LHS} = \frac{\sin 3x - \sin x}{\cos 3x + \cos x} = \frac{2 \cos \dfrac{3x + x}{2} \sin \dfrac{3x - x}{2}}{2 \cos \dfrac{3x + x}{2} \cos \dfrac{3x - x}{2}} \qquad \text{Sum-to-Product Formulas}$$

$$= \frac{2 \cos 2x \sin x}{2 \cos 2x \cos x} \qquad \text{Simplify}$$

$$= \frac{\sin x}{\cos x} = \tan x = \text{RHS} \qquad \text{Cancel}$$

✎ NOW TRY EXERCISE **89**

4.3 EXERCISES

CONCEPTS

1. If we know the values of $\sin x$ and $\cos x$, we can find the value of $\sin 2x$ by using the _____ Formula for Sine. State the formula: $\sin 2x =$ _____.

2. If we know the value of $\cos x$ and the quadrant in which $x/2$ lies, we can find the value of $\sin(x/2)$ by using the _____ Formula for Sine. State the formula:

$\sin(x/2) =$ _____.

SKILLS

3–10 ■ Find $\sin 2x$, $\cos 2x$, and $\tan 2x$ from the given information.

3. $\sin x = \frac{5}{13}$, x in Quadrant I

4. $\tan x = -\frac{4}{3}$, x in Quadrant II

5. $\cos x = \frac{4}{5}$, $\csc x < 0$ **6.** $\csc x = 4$, $\tan x < 0$

7. $\sin x = -\frac{3}{5}$, x in Quadrant III

8. $\sec x = 2$, x in Quadrant IV

9. $\tan x = -\frac{1}{3}$, $\cos x > 0$

10. $\cot x = \frac{2}{3}$, $\sin x > 0$

11–16 ■ Use the formulas for lowering powers to rewrite the expression in terms of the first power of cosine, as in Example 4.

11. $\sin^4 x$ **12.** $\cos^4 x$

13. $\cos^2 x \sin^4 x$ **14.** $\cos^4 x \sin^2 x$

15. $\cos^4 x \sin^4 x$ **16.** $\cos^6 x$

17–28 ■ Use an appropriate Half-Angle Formula to find the exact value of the expression.

17. $\sin 15°$ **18.** $\tan 15°$

19. $\tan 22.5°$ **20.** $\sin 75°$

21. $\cos 165°$ **22.** $\cos 112.5°$

23. $\tan \dfrac{\pi}{8}$

24. $\cos \dfrac{3\pi}{8}$

25. $\cos \dfrac{\pi}{12}$

26. $\tan \dfrac{5\pi}{12}$

27. $\sin \dfrac{9\pi}{8}$

28. $\sin \dfrac{11\pi}{12}$

29–34 ■ Simplify the expression by using a Double-Angle Formula or a Half-Angle Formula.

29. (a) $2 \sin 18° \cos 18°$ (b) $2 \sin 3\theta \cos 3\theta$

30. (a) $\dfrac{2 \tan 7°}{1 - \tan^2 7°}$ (b) $\dfrac{2 \tan 7\theta}{1 - \tan^2 7\theta}$

31. (a) $\cos^2 34° - \sin^2 34°$ (b) $\cos^2 5\theta - \sin^2 5\theta$

32. (a) $\cos^2 \dfrac{\theta}{2} - \sin^2 \dfrac{\theta}{2}$ (b) $2 \sin \dfrac{\theta}{2} \cos \dfrac{\theta}{2}$

33. (a) $\dfrac{\sin 8°}{1 + \cos 8°}$ (b) $\dfrac{1 - \cos 4\theta}{\sin 4\theta}$

34. (a) $\sqrt{\dfrac{1 - \cos 30°}{2}}$ (b) $\sqrt{\dfrac{1 - \cos 8\theta}{2}}$

35. Use the Addition Formula for Sine to prove the Double-Angle Formula for Sine.

36. Use the Addition Formula for Tangent to prove the Double-Angle Formula for Tangent.

37–42 ■ Find $\sin \dfrac{x}{2}$, $\cos \dfrac{x}{2}$, and $\tan \dfrac{x}{2}$ from the given information.

37. $\sin x = \frac{3}{5}$, $0° < x < 90°$

38. $\cos x = -\frac{4}{5}$, $180° < x < 270°$

39. $\csc x = 3$, $90° < x < 180°$

40. $\tan x = 1$, $0° < x < 90°$

41. $\sec x = \frac{3}{2}$, $270° < x < 360°$

42. $\cot x = 5$, $180° < x < 270°$

43–46 ■ Write the given expression as an algebraic expression in x.

43. $\sin(2 \tan^{-1} x)$

44. $\tan(2 \cos^{-1} x)$

45. $\sin(\frac{1}{2} \cos^{-1} x)$

46. $\cos(2 \sin^{-1} x)$

47–50 ■ Find the exact value of the given expression.

47. $\sin\left(2 \cos^{-1} \frac{7}{25}\right)$

48. $\cos\left(2 \tan^{-1} \frac{12}{5}\right)$

49. $\sec\left(2 \sin^{-1} \frac{1}{4}\right)$

50. $\tan\left(\frac{1}{2} \cos^{-1} \frac{2}{3}\right)$

51–54 ■ Evaluate each expression under the given conditions.

51. $\cos 2\theta$; $\sin \theta = -\frac{3}{5}$, θ in Quadrant III

52. $\sin(\theta/2)$; $\tan \theta = -\frac{5}{12}$, θ in Quadrant IV

53. $\sin 2\theta$; $\sin \theta = \frac{1}{7}$, θ in Quadrant II

54. $\tan 2\theta$; $\cos \theta = \frac{3}{5}$, θ in Quadrant I

55–60 ■ Write the product as a sum.

55. $\sin 2x \cos 3x$

56. $\sin x \sin 5x$

57. $\cos x \sin 4x$

58. $\cos 5x \cos 3x$

59. $3 \cos 4x \cos 7x$

60. $11 \sin \dfrac{x}{2} \cos \dfrac{x}{4}$

61–66 ■ Write the sum as a product.

61. $\sin 5x + \sin 3x$

62. $\sin x - \sin 4x$

63. $\cos 4x - \cos 6x$

64. $\cos 9x + \cos 2x$

65. $\sin 2x - \sin 7x$

66. $\sin 3x + \sin 4x$

67–72 ■ Find the value of the product or sum.

67. $2 \sin 52.5° \sin 97.5°$

68. $3 \cos 37.5° \cos 7.5°$

69. $\cos 37.5° \sin 7.5°$

70. $\sin 75° + \sin 15°$

71. $\cos 255° - \cos 195°$

72. $\cos \dfrac{\pi}{12} + \cos \dfrac{5\pi}{12}$

73–90 ■ Prove the identity.

73. $\cos^2 5x - \sin^2 5x = \cos 10x$

74. $\sin 8x = 2 \sin 4x \cos 4x$

75. $(\sin x + \cos x)^2 = 1 + \sin 2x$

76. $\dfrac{2 \tan x}{1 + \tan^2 x} = \sin 2x$

77. $\dfrac{\sin 4x}{\sin x} = 4 \cos x \cos 2x$

78. $\dfrac{1 + \sin 2x}{\sin 2x} = 1 + \frac{1}{2} \sec x \csc x$

79. $\dfrac{2(\tan x - \cot x)}{\tan^2 x - \cot^2 x} = \sin 2x$

80. $\cot 2x = \dfrac{1 - \tan^2 x}{2 \tan x}$

81. $\tan 3x = \dfrac{3 \tan x - \tan^3 x}{1 - 3 \tan^2 x}$

82. $4(\sin^6 x + \cos^6 x) = 4 - 3 \sin^2 2x$

83. $\cos^4 x - \sin^4 x = \cos 2x$

84. $\tan^2\left(\dfrac{x}{2} + \dfrac{\pi}{4}\right) = \dfrac{1 + \sin x}{1 - \sin x}$

85. $\dfrac{\sin x + \sin 5x}{\cos x + \cos 5x} = \tan 3x$

86. $\dfrac{\sin 3x + \sin 7x}{\cos 3x - \cos 7x} = \cot 2x$

87. $\dfrac{\sin 10x}{\sin 9x + \sin x} = \dfrac{\cos 5x}{\cos 4x}$

88. $\dfrac{\sin x + \sin 3x + \sin 5x}{\cos x + \cos 3x + \cos 5x} = \tan 3x$

89. $\dfrac{\sin x + \sin y}{\cos x + \cos y} = \tan\left(\dfrac{x + y}{2}\right)$

90. $\tan y = \dfrac{\sin(x + y) - \sin(x - y)}{\cos(x + y) + \cos(x - y)}$

91. Show that $\sin 130° - \sin 110° = -\sin 10°$.

92. Show that $\cos 100° - \cos 200° = \sin 50°$.

93. Show that $\sin 45° + \sin 15° = \sin 75°$.

94. Show that $\cos 87° + \cos 33° = \sin 63°$.

95. Prove the identity

$$\frac{\sin x + \sin 2x + \sin 3x + \sin 4x + \sin 5x}{\cos x + \cos 2x + \cos 3x + \cos 4x + \cos 5x} = \tan 3x$$

96. Use the identity

$$\sin 2x = 2 \sin x \cos x$$

n times to show that

$$\sin(2^n x) = 2^n \sin x \cos x \cos 2x \cos 4x \cdots \cos 2^{n-1}x$$

 97. **(a)** Graph $f(x) = \dfrac{\sin 3x}{\sin x} - \dfrac{\cos 3x}{\cos x}$ and make a conjecture.

    (b) Prove the conjecture you made in part (a).

 98. **(a)** Graph $f(x) = \cos 2x + 2 \sin^2 x$ and make a conjecture.

    (b) Prove the conjecture you made in part (a).

 99. Let $f(x) = \sin 6x + \sin 7x$.

    (a) Graph $y = f(x)$.

    (b) Verify that $f(x) = 2 \cos \frac{1}{2}x \sin \frac{13}{2}x$.

    (c) Graph $y = 2 \cos \frac{1}{2}x$ and $y = -2 \cos \frac{1}{2}x$, together with the graph in part (a), in the same viewing rectangle. How are these graphs related to the graph of f?

100. Let $3x = \pi/3$ and let $y = \cos x$. Use the result of Example 2 to show that y satisfies the equation

$$8y^3 - 6y - 1 = 0$$

NOTE This equation has roots of a certain kind that are used to show that the angle $\pi/3$ cannot be trisected by using a ruler and compass only.

101. **(a)** Show that there is a polynomial $P(t)$ of degree 4 such that $\cos 4x = P(\cos x)$ (see Example 2).

    (b) Show that there is a polynomial $Q(t)$ of degree 5 such that $\cos 5x = Q(\cos x)$.

NOTE In general, there is a polynomial $P_n(t)$ of degree n such that $\cos nx = P_n(\cos x)$. These polynomials are called *Tchebycheff polynomials*, after the Russian mathematician P. L. Tchebycheff (1821–1894).

102. In triangle ABC (see the figure) the line segment s bisects angle C. Show that the length of s is given by

$$s = \frac{2ab \cos x}{a + b}$$

[*Hint:* Use the Law of Sines.]

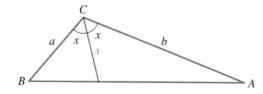

103. If A, B, and C are the angles in a triangle, show that

$$\sin 2A + \sin 2B + \sin 2C = 4 \sin A \sin B \sin C$$

104. A rectangle is to be inscribed in a semicircle of radius 5 cm as shown in the following figure.

    (a) Show that the area of the rectangle is modeled by the function

$$A(\theta) = 25 \sin 2\theta$$

    (b) Find the largest possible area for such an inscribed rectangle.

    (c) Find the dimensions of the inscribed rectangle with the largest possible area.

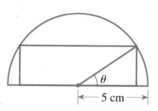

APPLICATIONS

105. **Sawing a Wooden Beam** A rectangular beam is to be cut from a cylindrical log of diameter 20 in.

    (a) Show that the cross-sectional area of the beam is modeled by the function

$$A(\theta) = 200 \sin 2\theta$$

    where θ is as shown in the figure.

    (b) Show that the maximum cross-sectional area of such a beam is 200 in². [*Hint:* Use the fact that $\sin u$ achieves its maximum value at $u = \pi/2$.]

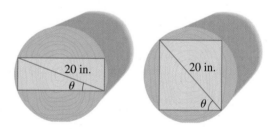

106. **Length of a Fold** The lower right-hand corner of a long piece of paper 6 in. wide is folded over to the left-hand edge as shown. The length L of the fold depends on the angle θ. Show that

$$L = \frac{3}{\sin \theta \cos^2 \theta}$$

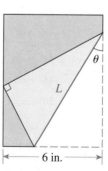

 107. **Sound Beats** When two pure notes that are close in frequency are played together, their sounds interfere to produce *beats*; that is, the loudness (or amplitude) of the sound alternately increases and decreases. If the two notes are given by

$$f_1(t) = \cos 11t \qquad \text{and} \qquad f_2(t) = \cos 13t$$

the resulting sound is $f(t) = f_1(t) + f_2(t)$.

    (a) Graph the function $y = f(t)$.

    (b) Verify that $f(t) = 2 \cos t \cos 12t$.

(c) Graph $y = 2 \cos t$ and $y = -2 \cos t$, together with the graph in part (a), in the same viewing rectangle. How do these graphs describe the variation in the loudness of the sound?

108. Touch-Tone Telephones When a key is pressed on a touch-tone telephone, the keypad generates two pure tones, which combine to produce a sound that uniquely identifies the key. The figure shows the low frequency f_1 and the high frequency f_2 associated with each key. Pressing a key produces the sound wave $y = \sin(2\pi f_1 t) + \sin(2\pi f_2 t)$.

(a) Find the function that models the sound produced when the 4 key is pressed.

(b) Use a Sum-to-Product Formula to express the sound generated by the 4 key as a product of a sine and a cosine function.

(c) Graph the sound wave generated by the 4 key, from $t = 0$ to $t = 0.006$ s.

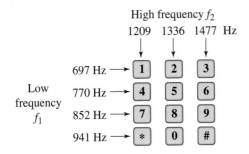

DISCOVERY ■ DISCUSSION ■ WRITING

109. Geometric Proof of a Double-Angle Formula
Use the figure to prove that $\sin 2\theta = 2 \sin \theta \cos \theta$.

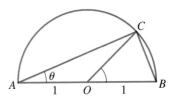

[*Hint:* Find the area of triangle ABC in two different ways. You will need the following facts from geometry:

An angle inscribed in a semicircle is a right angle, so $\angle ACB$ is a right angle.

The central angle subtended by the chord of a circle is twice the angle subtended by the chord on the circle, so $\angle BOC$ is 2θ.]

DISCOVERY PROJECT Where to Sit at the Movies

In this project we use trigonometry to find the best location to observe such things as a painting or a movie. You can find the project at the book companion website:
www.stewartmath.com

4.4 BASIC TRIGONOMETRIC EQUATIONS

| Basic Trigonometric Equations ▶ Solving Trigonometric Equations by Factoring

An equation that contains trigonometric functions is called a **trigonometric equation**. For example, the following are trigonometric equations:

$$\sin^2\theta + \cos^2\theta = 1 \qquad 2\sin\theta - 1 = 0 \qquad \tan 2\theta - 1 = 0$$

The first equation is an *identity*—that is, it is true for every value of the variable θ. The other two equations are true only for certain values of θ. To solve a trigonometric equation, we find all the values of the variable that make the equation true.

▼ Basic Trigonometric Equations

Solving any trigonometric equation always reduces to solving a **basic trigonometric equation**—an equation of the form $T(\theta) = c$, where T is a trigonometric function and c is a constant. In the next three examples we solve such basic equations.

EXAMPLE 1 | Solving a Basic Trigonometric Equation

Solve the equation $\sin\theta = \dfrac{1}{2}$.

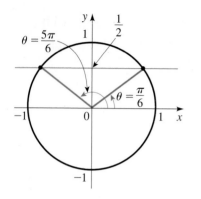

FIGURE 1

SOLUTION

Find the solutions in one period. Because sine has period 2π, we first find the solutions in any interval of length 2π. To find these solutions, we look at the unit circle in Figure 1. We see that $\sin \theta = \frac{1}{2}$ in Quadrants I and II, so the solutions in the interval $[0, 2\pi)$ are

$$\theta = \frac{\pi}{6} \qquad \theta = \frac{5\pi}{6}$$

Find all solutions. Because the sine function repeats its values every 2π units, we get all solutions of the equation by adding integer multiples of 2π to these solutions:

$$\theta = \frac{\pi}{6} + 2k\pi \qquad \theta = \frac{5\pi}{6} + 2k\pi$$

where k is any integer. Figure 2 gives a graphical representation of the solutions.

FIGURE 2

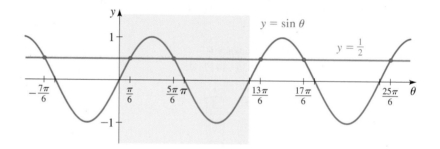

✏ NOW TRY EXERCISE **5**

EXAMPLE 2 | Solving a Basic Trigonometric Equation

Solve the equation $\cos \theta = -\dfrac{\sqrt{2}}{2}$, and list eight specific solutions.

SOLUTION

Find the solutions in one period. Because cosine has period 2π, we first find the solutions in any interval of length 2π. From the unit circle in Figure 3 we see that $\cos \theta = -\sqrt{2}/2$ in Quadrants II and III, so the solutions in the interval $[0, 2\pi)$ are

$$\theta = \frac{3\pi}{4} \qquad \theta = \frac{5\pi}{4}$$

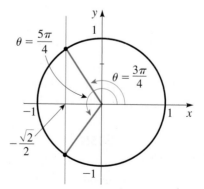

FIGURE 3

Find all solutions. Because the cosine function repeats its values every 2π units, we get all solutions of the equation by adding integer multiples of 2π to these solutions:

$$\theta = \frac{3\pi}{4} + 2k\pi \qquad \theta = \frac{5\pi}{4} + 2k\pi$$

where k is any integer. You can check that for $k = -1, 0, 1, 2$ we get the following specific solutions:

$$\theta = \underbrace{-\frac{5\pi}{4},\, -\frac{3\pi}{4}}_{k=-1},\, \underbrace{\frac{3\pi}{4},\, \frac{5\pi}{4}}_{k=0},\, \underbrace{\frac{11\pi}{4},\, \frac{13\pi}{4}}_{k=1},\, \underbrace{\frac{19\pi}{4},\, \frac{21\pi}{4}}_{k=2}$$

Figure 4 gives a graphical representation of the solutions.

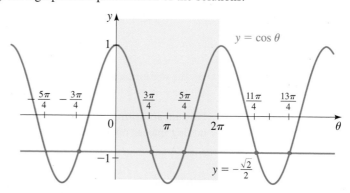

FIGURE 4

✎ NOW TRY EXERCISE **17**

EXAMPLE 3 | Solving a Basic Trigonometric Equation

Solve the equation $\cos \theta = 0.65$.

SOLUTION

Find the solutions in one period.   We first find one solution by taking $\cos^{-1}$ of each side of the equation.

$\cos \theta = 0.65$	Given equation
$\theta = \cos^{-1}(0.65)$	Take $\cos^{-1}$ of each side
$\theta \approx 0.86$	Calculator (in radian mode)

Because cosine has period 2π, we next find the solutions in any interval of length 2π. To find these solutions, we look at the unit circle in Figure 5. We see that $\cos \theta = 0.85$ in Quadrants I and IV, so the solutions are

$$\theta \approx 0.86 \qquad \theta \approx 2\pi - 0.86 \approx 5.42$$

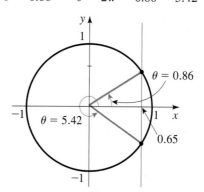

FIGURE 5

Find all solutions.   To get all solutions of the equation, we add integer multiples of 2π to these solutions:

$$\theta \approx 0.86 + 2k\pi \qquad \theta \approx 5.42 + 2k\pi$$

where k is any integer.

✎ NOW TRY EXERCISE **21**

EXAMPLE 4 | Solving a Basic Trigonometric Equation

Solve the equation $\tan \theta = 2$.

SOLUTION

Find the solutions in one period. We first find one solution by taking $\tan^{-1}$ of each side of the equation:

$$\tan \theta = 2 \qquad \text{Given equation}$$

$$\theta = \tan^{-1}(2) \qquad \text{Take } \tan^{-1} \text{ of each side}$$

$$\theta \approx 1.12 \qquad \text{Calculator (in radian mode)}$$

By the definition of $\tan^{-1}$, the solution that we obtained is the only solution in the interval $(-\pi/2, \pi/2)$ (which is an interval of length π).

Find all solutions. Since tangent has period π, we get all solutions of the equation by adding integer multiples of π:

$$\theta \approx 1.12 + k\pi$$

where k is any integer. A graphical representation of the solutions is shown in Figure 6. You can check that the solutions shown in the graph correspond to $k = -1, 0, 1, 2, 3$.

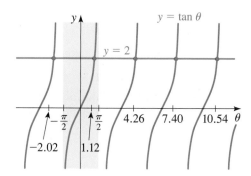

FIGURE 6

✎ NOW TRY EXERCISE **23**

In the next example we solve trigonometric equations that are algebraically equivalent to basic trigonometric equations.

EXAMPLE 5 | Solving Trigonometric Equations

Find all solutions of the equation.
(a) $2 \sin \theta - 1 = 0$ **(b)** $\tan^2 \theta - 3 = 0$

SOLUTION

(a) We start by isolating $\sin \theta$:

$$2 \sin \theta - 1 = 0 \qquad \text{Given equation}$$

$$2 \sin \theta = 1 \qquad \text{Add 1}$$

$$\sin \theta = \frac{1}{2} \qquad \text{Divide by 2}$$

This last equation is the same as that in Example 1. The solutions are

$$\theta = \frac{\pi}{6} + 2k\pi \qquad \theta = \frac{5\pi}{6} + 2k\pi$$

where k is any integer.

(b) We start by isolating $\tan \theta$:

$$\tan^2\theta - 3 = 0 \qquad \text{Given equation}$$

$$\tan^2\theta = 3 \qquad \text{Add 3}$$

$$\tan\theta = \pm\sqrt{3} \qquad \text{Take the square root}$$

Because tangent has period π, we first find the solutions in any interval of length π. In the interval $(-\pi/2, \pi/2)$ the solutions are $\theta = \pi/3$ and $\theta = -\pi/3$. To get all solutions, we add integer multiples of π to these solutions:

$$\theta = \frac{\pi}{3} + k\pi \qquad \theta = -\frac{\pi}{3} + k\pi$$

where k is any integer.

✎. NOW TRY EXERCISES **27** AND **33**   ■

▼ Solving Trigonometric Equations by Factoring

Zero-Product Property
If $AB = 0$, then $A = 0$ or $B = 0$.

Factoring is one of the most useful techniques for solving equations, including trigonometric equations. The idea is to move all terms to one side of the equation, factor, and then use the Zero-Product Property (see Appendix A.4).

EXAMPLE 6 | A Trigonometric Equation of Quadratic Type

Solve the equation $2\cos^2\theta - 7\cos\theta + 3 = 0$.

SOLUTION We factor the left-hand side of the equation.

Equation of Quadratic Type

$$2C^2 - 7C + 3 = 0$$

$$(2C - 1)(C - 3) = 0$$

$$2\cos^2\theta - 7\cos\theta + 3 = 0 \qquad \text{Given equation}$$

$$(2\cos\theta - 1)(\cos\theta - 3) = 0 \qquad \text{Factor}$$

$$2\cos\theta - 1 = 0 \quad \text{or} \quad \cos\theta - 3 = 0 \qquad \text{Set each factor equal to 0}$$

$$\cos\theta = \frac{1}{2} \quad \text{or} \quad \cos\theta = 3 \qquad \text{Solve for } \cos\theta$$

Because cosine has period 2π, we first find the solutions in the interval $[0, 2\pi)$. For the first equation the solutions are $\theta = \pi/3$ and $\theta = 5\pi/3$ (see Figure 7). The second equation has no solution because $\cos\theta$ is never greater than 1. Thus the solutions are

$$\theta = \frac{\pi}{3} + 2k\pi \qquad \theta = \frac{5\pi}{3} + 2k\pi$$

where k is any integer.

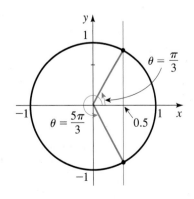

FIGURE 7

✎. NOW TRY EXERCISE **41**   ■

EXAMPLE 7 | Solving a Trigonometric Equation by Factoring

Solve the equation $5 \sin \theta \cos \theta + 4 \cos \theta = 0$.

SOLUTION We factor the left-hand side of the equation:

$$5 \sin \theta \cos \theta + 2 \cos \theta = 0 \qquad \text{Given equation}$$

$$\cos \theta (5 \sin \theta + 2) = 0 \qquad \text{Factor}$$

$$\cos \theta = 0 \qquad \text{or} \qquad 5 \sin \theta + 4 = 0 \qquad \text{Set each factor equal to 0}$$

$$\sin \theta = -0.8 \qquad \text{Solve for } \sin \theta$$

Because sine and cosine have period 2π, we first find the solutions of these equations in an interval of length 2π. For the first equation the solutions in the interval $[0, 2\pi)$ are $\theta = \pi/2$ and $\theta = 3\pi/2$. To solve the second equation, we take $\sin^{-1}$ of each side:

$$\sin \theta = -0.80 \qquad \text{Second equation}$$

$$\theta = \sin^{-1}(-0.80) \qquad \text{Take } \sin^{-1} \text{ of each side}$$

$$\theta \approx -0.93 \qquad \text{Calculator (in radian mode)}$$

So the solutions in an interval of length 2π are $\theta = -0.93$ and $\theta = \pi + 0.93 \approx 4.07$ (see Figure 8). We get all the solutions of the equation by adding integer multiples of 2π to these solutions.

$$\theta = \frac{\pi}{2} + 2k\pi, \qquad \theta = \frac{3\pi}{2} + 2k\pi, \qquad \theta \approx -0.93 + 2k\pi, \qquad \theta \approx 4.07 + 2k\pi$$

where k is any integer.

✎ NOW TRY EXERCISE **53**

FIGURE 8

$\theta = \pi + 0.93$

$\theta = -0.93$

4.4 EXERCISES

CONCEPTS

1. Because the trigonometric functions are periodic, if a basic trigonometric equation has one solution, it has _____ (several/infinitely many) solutions.

2. The basic equation $\sin x = 2$ has _____ (no/one/infinitely many) solutions, whereas the basic equation $\sin x = 0.3$ has

_____ (no/one/infinitely many) solutions.

3. We can find some of the solutions of $\sin x = 0.3$ graphically by graphing $y = \sin x$ and $y = $ _____. Use the graph below to estimate some of the solutions.

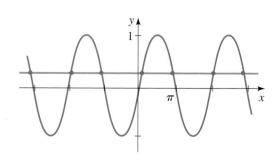

4. We can find the solutions of $\sin x = 0.3$ algebraically.

  (a) First we find the solutions in the interval $[-\pi, \pi]$. We get one such solution by taking $\sin^{-1}$ to get $x = $ _____.

  The other solution in this interval is $x = $ _____.

  (b) We find all solutions by adding multiples of _____ to the solutions in $[-\pi, \pi]$. The solutions are

  $x = $ _____ and $x = $ _____.

SKILLS

5–16 ■ Solve the given equation.

✎ **5.** $\sin \theta = \dfrac{\sqrt{3}}{2}$ **6.** $\sin \theta = -\dfrac{\sqrt{2}}{2}$

7. $\cos \theta = -1$ **8.** $\cos \theta = \dfrac{\sqrt{3}}{2}$

9. $\cos \theta = \frac{1}{4}$ **10.** $\sin \theta = -0.3$

11. $\sin \theta = -0.45$ **12.** $\cos \theta = 0.32$

13. $\tan \theta = -\sqrt{3}$ **14.** $\tan \theta = 1$

15. $\tan \theta = 5$ **16.** $\tan \theta = -\frac{1}{3}$

17–24 ■ Solve the given equation, and list six specific solutions.

17. $\cos\theta = -\dfrac{\sqrt{3}}{2}$

18. $\cos\theta = \dfrac{1}{2}$

19. $\sin\theta = \dfrac{\sqrt{2}}{2}$

20. $\sin\theta = -\dfrac{\sqrt{3}}{2}$

21. $\cos\theta = 0.28$

22. $\tan\theta = 2.5$

23. $\tan\theta = -10$

24. $\sin\theta = -0.9$

25–38 ■ Find all solutions of the given equation.

25. $\cos\theta + 1 = 0$

26. $\sin\theta + 1 = 0$

27. $\sqrt{2}\sin\theta + 1 = 0$

28. $\sqrt{2}\cos\theta - 1 = 0$

29. $5\sin\theta - 1 = 0$

30. $4\cos\theta + 1 = 0$

31. $3\tan^2\theta - 1 = 0$

32. $\cot\theta + 1 = 0$

33. $2\cos^2\theta - 1 = 0$

34. $4\sin^2\theta - 3 = 0$

35. $\tan^2\theta - 4 = 0$

36. $9\sin^2\theta - 1 = 0$

37. $\sec^2\theta - 2 = 0$

38. $\csc^2\theta - 4 = 0$

39–56 ■ Solve the given equation.

39. $(\tan^2\theta - 4)(2\cos\theta + 1) = 0$

40. $(\tan\theta - 2)(16\sin^2\theta - 1) = 0$

41. $4\cos^2\theta - 4\cos\theta + 1 = 0$

42. $2\sin^2\theta - \sin\theta - 1 = 0$

43. $3\sin^2\theta - 7\sin\theta + 2 = 0$

44. $\tan^4\theta - 13\tan^2\theta + 36 = 0$

45. $2\cos^2\theta - 7\cos\theta + 3 = 0$

46. $\sin^2\theta - \sin\theta - 2 = 0$

47. $\cos^2\theta - \cos\theta - 6 = 0$

48. $2\sin^2\theta + 5\sin\theta - 12 = 0$

49. $\sin^2\theta = 2\sin\theta + 3$

50. $3\tan^3\theta = \tan\theta$

51. $\cos\theta(2\sin\theta + 1) = 0$

52. $\sec\theta(2\cos\theta - \sqrt{2}) = 0$

53. $\cos\theta\sin\theta - 2\cos\theta = 0$

54. $\tan\theta\sin\theta + \sin\theta = 0$

55. $3\tan\theta\sin\theta - 2\tan\theta = 0$

56. $4\cos\theta\sin\theta + 3\cos\theta = 0$

APPLICATIONS

57. Refraction of Light It has been observed since ancient times that light refracts or "bends" as it travels from one medium to another (from air to water, for example). If v_1 is the speed of light in one medium and v_2 its speed in another medium, then according to **Snell's Law**,

$$\frac{\sin\theta_1}{\sin\theta_2} = \frac{v_1}{v_2}$$

where θ_1 is the *angle of incidence* and θ_2 is the *angle of refraction* (see the figure). The number v_1/v_2 is called the *index of refraction*. The index of refraction for several substances is given in the table.

If a ray of light passes through the surface of a lake at an angle of incidence of 70°, what is the angle of refraction?

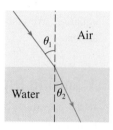

Substance	Refraction from air to substance
Water	1.33
Alcohol	1.36
Glass	1.52
Diamond	2.41

58. Total Internal Reflection When light passes from a more-dense to a less-dense medium—from glass to air, for example—the angle of refraction predicted by Snell's Law (see Exercise 57) can be 90° or larger. In this case the light beam is actually reflected back into the denser medium. This phenomenon, called *total internal reflection*, is the principle behind fiber optics. Set $\theta_2 = 90°$ in Snell's Law, and solve for θ_1 to determine the critical angle of incidence at which total internal reflection begins to occur when light passes from glass to air. (Note that the index of refraction from glass to air is the reciprocal of the index from air to glass.)

59. Phases of the Moon As the moon revolves around the earth, the side that faces the earth is usually just partially illuminated by the sun. The phases of the moon describe how much of the surface appears to be in sunlight. An astronomical measure of phase is given by the fraction F of the lunar disc that is lit. When the angle between the sun, earth, and moon is θ ($0 \le \theta \le 360°$), then

$$F = \frac{1}{2}(1 - \cos\theta)$$

Determine the angles θ that correspond to the following phases:
(a) $F = 0$ (new moon)
(b) $F = 0.25$ (a crescent moon)
(c) $F = 0.5$ (first or last quarter)
(d) $F = 1$ (full moon)

DISCOVERY ■ DISCUSSION ■ WRITING

60. Equations and Identities Which of the following statements is true?

  A. Every identity is an equation.

  B. Every equation is an identity.

Give examples to illustrate your answer. Write a short paragraph to explain the difference between an equation and an identity.

4.5 MORE TRIGONOMETRIC EQUATIONS

Solving Trigonometric Equations by Using Identities ▶ Equations with Trigonometric Functions of Multiples of Angles

In this section we solve trigonometric equations by first using identities to simplify the equation. We also solve trigonometric equations in which the terms contain multiples of angles.

▼ Solving Trigonometric Equations by Using Identities

In the next two examples we use trigonometric identities to express a trigonometric equation in a form in which it can be factored.

EXAMPLE 1 | Using a Trigonometric Identity

Solve the equation $1 + \sin \theta = 2 \cos^2 \theta$.

SOLUTION We first need to rewrite this equation so that it contains only one trigonometric function. To do this, we use a trigonometric identity:

$$1 + \sin \theta = 2 \cos^2 \theta \qquad \text{Given equation}$$

$$1 + \sin \theta = 2(1 - \sin^2 \theta) \qquad \text{Pythagorean identity}$$

$$2 \sin^2 \theta + \sin \theta - 1 = 0 \qquad \text{Put all terms on one side}$$

$$(2 \sin \theta - 1)(\sin \theta + 1) = 0 \qquad \text{Factor}$$

$$2 \sin \theta - 1 = 0 \quad \text{or} \quad \sin \theta + 1 = 0 \qquad \text{Set each factor equal to 0}$$

$$\sin \theta = \frac{1}{2} \quad \text{or} \quad \sin \theta = -1 \qquad \text{Solve for } \sin \theta$$

$$\theta = \frac{\pi}{6}, \frac{5\pi}{6} \quad \text{or} \quad \theta = \frac{3\pi}{2} \qquad \begin{array}{l}\text{Solve for } \theta \text{ in the} \\ \text{interval } [0, 2\pi)\end{array}$$

Because sine has period 2π, we get all the solutions of the equation by adding integer multiples of 2π to these solutions. Thus the solutions are

$$\theta = \frac{\pi}{6} + 2k\pi \qquad \theta = \frac{5\pi}{6} + 2k\pi \qquad \theta = \frac{3\pi}{2} + 2k\pi$$

where k is any integer.

✎ NOW TRY EXERCISE **3** ∎

EXAMPLE 2 | Using a Trigonometric Identity

Solve the equation $\sin 2\theta - \cos \theta = 0$.

SOLUTION The first term is a function of 2θ, and the second is a function of θ, so we begin by using a trigonometric identity to rewrite the first term as a function of θ only:

$$\sin 2\theta - \cos \theta = 0 \qquad \text{Given equation}$$

$$2 \sin \theta \cos \theta - \cos \theta = 0 \qquad \text{Double-Angle Formula}$$

$$\cos \theta \, (2 \sin \theta - 1) = 0 \qquad \text{Factor}$$

$$\cos\theta = 0 \quad \text{or} \quad 2\sin\theta - 1 = 0 \qquad \text{Set each factor equal to 0}$$

$$\sin\theta = \frac{1}{2} \qquad \text{Solve for } \sin\theta$$

$$\theta = \frac{\pi}{2}, \frac{3\pi}{2} \quad \text{or} \quad \theta = \frac{\pi}{6}, \frac{5\pi}{6} \qquad \text{Solve for } \theta \text{ in } [0, 2\pi)$$

Both sine and cosine have period 2π, so we get all the solutions of the equation by adding integer multiples of 2π to these solutions. Thus the solutions are

$$\theta = \frac{\pi}{2} + 2k\pi \qquad \theta = \frac{3\pi}{2} + 2k\pi \qquad \theta = \frac{\pi}{6} + 2k\pi \qquad \theta = \frac{5\pi}{6} + 2k\pi$$

where k is any integer.

✎ **NOW TRY EXERCISES 7 AND 11** ∎

EXAMPLE 3 | Squaring and Using an Identity

Solve the equation $\cos\theta + 1 = \sin\theta$ in the interval $[0, 2\pi)$.

SOLUTION To get an equation that involves either sine only or cosine only, we square both sides and use a Pythagorean identity:

$$\cos\theta + 1 = \sin\theta \qquad \text{Given equation}$$

$$\cos^2\theta + 2\cos\theta + 1 = \sin^2\theta \qquad \text{Square both sides}$$

$$\cos^2\theta + 2\cos\theta + 1 = 1 - \cos^2\theta \qquad \text{Pythagorean identity}$$

$$2\cos^2\theta + 2\cos\theta = 0 \qquad \text{Simplify}$$

$$2\cos\theta(\cos\theta + 1) = 0 \qquad \text{Factor}$$

$$2\cos\theta = 0 \quad \text{or} \quad \cos\theta + 1 = 0 \qquad \text{Set each factor equal to 0}$$

$$\cos\theta = 0 \quad \text{or} \quad \cos\theta = -1 \qquad \text{Solve for } \cos\theta$$

$$\theta = \frac{\pi}{2}, \frac{3\pi}{2} \quad \text{or} \quad \theta = \pi \qquad \text{Solve for } \theta \text{ in } [0, 2\pi)$$

Because we squared both sides, we need to check for extraneous solutions. From *Check Your Answers* we see that the solutions of the given equation are $\pi/2$ and π.

CHECK YOUR ANSWER

$$\theta = \frac{\pi}{2} \qquad\qquad \theta = \frac{3\pi}{2} \qquad\qquad \theta = \pi$$

$$\cos\frac{\pi}{2} + 1 = \sin\frac{\pi}{2} \qquad \cos\frac{3\pi}{2} + 1 = \sin\frac{3\pi}{2} \qquad \cos\pi + 1 = \sin\pi$$

$$0 + 1 = 1 \quad ✔ \qquad\qquad 0 + 1 \overset{?}{=} -1 \quad ✗ \qquad\qquad -1 + 1 = 0 \quad ✔$$

✎ **NOW TRY EXERCISE 13** ∎

EXAMPLE 4 | Finding Intersection Points

Find the values of x for which the graphs of $f(x) = \sin x$ and $g(x) = \cos x$ intersect.

MATHEMATICS IN THE MODERN WORLD

Weather Prediction

Modern meteorologists do much more than predict tomorrow's weather. They research long-term weather patterns, depletion of the ozone layer, global warming, and other effects of human activity on the weather. But daily weather prediction is still a major part of meteorology; its value is measured by the innumerable human lives that are saved each year through accurate prediction of hurricanes, blizzards, and other catastrophic weather phenomena. Early in the 20th century mathematicians proposed to model weather with equations that used the current values of hundreds of atmospheric variables. Although this model worked in principle, it was impossible to predict future weather patterns with it because of the difficulty of measuring all the variables accurately and solving all the equations. Today, new mathematical models combined with high-speed computer simulations and better data have vastly improved weather prediction. As a result, many human as well as economic disasters have been averted. Mathematicians at the National Oceanographic and Atmospheric Administration (NOAA) are continually researching better methods of weather prediction.

SOLUTION 1: Graphical

The graphs intersect where $f(x) = g(x)$. In Figure 1 we graph $y_1 = \sin x$ and $y_2 = \cos x$ on the same screen, for x between 0 and 2π. Using TRACE or the intersect command on the graphing calculator, we see that the two points of intersection in this interval occur where $x \approx 0.785$ and $x \approx 3.927$. Since sine and cosine are periodic with period 2π, the intersection points occur where

$$x \approx 0.785 + 2k\pi \qquad \text{and} \qquad x \approx 3.927 + 2k\pi$$

where k is any integer.

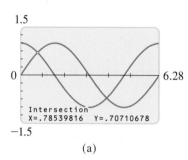
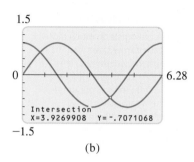

FIGURE 1   (a)   (b)

SOLUTION 2: Algebraic

To find the exact solution, we set $f(x) = g(x)$ and solve the resulting equation algebraically:

$$\sin x = \cos x \qquad \text{Equate functions}$$

Since the numbers x for which $\cos x = 0$ are not solutions of the equation, we can divide both sides by $\cos x$:

$$\frac{\sin x}{\cos x} = 1 \qquad \text{Divide by } \cos x$$

$$\tan x = 1 \qquad \text{Reciprocal identity}$$

The only solution of this equation in the interval $(-\pi/2, \pi/2)$ is $x = \pi/4$. Since tangent has period π, we get all solutions of the equation by adding integer multiples of π:

$$x = \frac{\pi}{4} + k\pi$$

where k is any integer. The graphs intersect for these values of x. You should use your calculator to check that, rounded to three decimals, these are the same values that we obtained in Solution 1.

✎. NOW TRY EXERCISE **35**   ■

▼ Equations with Trigonometric Functions of Multiples of Angles

When solving trigonometric equations that involve functions of multiples of angles, we first solve for the multiple of the angle, then divide to solve for the angle.

EXAMPLE 5 | A Trigonometric Equation Involving a Multiple of an Angle

Consider the equation $2 \sin 3\theta - 1 = 0$.

(a) Find all solutions of the equation.

(b) Find the solutions in the interval $[0, 2\pi)$.

SOLUTION

(a) We first isolate $\sin 3\theta$ and then solve for the angle 3θ.

$$2 \sin 3\theta - 1 = 0 \qquad \text{Given equation}$$

$$2 \sin 3\theta = 1 \qquad \text{Add 1}$$

$$\sin 3\theta = \frac{1}{2} \qquad \text{Divide by 2}$$

$$3\theta = \frac{\pi}{6}, \frac{5\pi}{6} \qquad \text{Solve for } 3\theta \text{ in the interval } [0, 2\pi) \text{ (see Figure 2)}$$

To get all solutions, we add integer multiples of 2π to these solutions. So the solutions are of the form

$$3\theta = \frac{\pi}{6} + 2k\pi \qquad 3\theta = \frac{5\pi}{6} + 2k\pi$$

To solve for θ, we divide by 3 to get the solutions

$$\theta = \frac{\pi}{18} + \frac{2k\pi}{3} \qquad \theta = \frac{5\pi}{18} + \frac{2k\pi}{3}$$

where k is any integer.

(b) The solutions from part (a) that are in the interval $[0, 2\pi)$ correspond to $k = 0, 1,$ and 2. For all other values of k the corresponding values of θ lie outside this interval. So the solutions in the interval $[0, 2\pi)$ are

$$\theta = \underbrace{\frac{\pi}{18}, \frac{5\pi}{18}}_{k = 0}, \underbrace{\frac{13\pi}{18}, \frac{17\pi}{18}}_{k = 1}, \underbrace{\frac{25\pi}{18}, \frac{29\pi}{18}}_{k = 2}$$

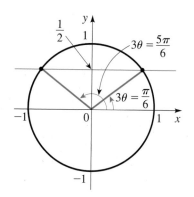

FIGURE 2

✎. NOW TRY EXERCISE **17**

EXAMPLE 6 | A Trigonometric Equation Involving a Half Angle

Consider the equation $\sqrt{3} \tan \dfrac{\theta}{2} - 1 = 0$.

(a) Find all solutions of the equation.

(b) Find the solutions in the interval $[0, 4\pi)$.

SOLUTION

(a) We start by isolating $\tan\dfrac{\theta}{2}$:

$$\sqrt{3}\,\tan\frac{\theta}{2} - 1 = 0 \qquad \text{Given equation}$$

$$\sqrt{3}\,\tan\frac{\theta}{2} = 1 \qquad \text{Add 1}$$

$$\tan\frac{\theta}{2} = \frac{1}{\sqrt{3}} \qquad \text{Divide by } \sqrt{3}$$

$$\frac{\theta}{2} = \frac{\pi}{6} \qquad \text{Solve for } \frac{\theta}{2} \text{ in the interval } \left(-\frac{\pi}{2}, \frac{\pi}{2}\right)$$

Since tangent has period π, to get all solutions we add integer multiples of π to this solution. So the solutions are of the form

$$\frac{\theta}{2} = \frac{\pi}{6} + k\pi$$

Multiplying by 2, we get the solutions

$$\theta = \frac{\pi}{3} + 2k\pi$$

where k is any integer.

(b) The solutions from part (a) that are in the interval $[0, 4\pi)$ correspond to $k = 0$ and $k = 1$. For all other values of k the corresponding values of x lie outside this interval. Thus the solutions in the interval $[0, 4\pi)$ are

$$x = \frac{\pi}{3}, \frac{7\pi}{3}$$

✎ NOW TRY EXERCISE **23** ∎

4.5 EXERCISES

CONCEPTS

1–2 ■ We can use identities to help us solve trigonometric equations.

1. Using a Pythagorean identity we see that the equation $\sin x + \sin^2 x + \cos^2 x = 1$ is equivalent to the basic equation

_____ whose solutions are $x = $ _____.

2. Using a Double-Angle Formula we see that the equation

$\sin x + \sin 2x = 0$ is equivalent to the equation _____.
Factoring we see that solving this equation is equivalent to

solving the two basic equations _____ and _____.

SKILLS

3–16 ■ Solve the given equation.

3. $2\cos^2\theta + \sin\theta = 1$

4. $\sin^2\theta = 4 - 2\cos^2\theta$

5. $\tan^2\theta - 2\sec\theta = 2$

6. $\csc^2\theta = \cot\theta + 3$

7. $2\sin 2\theta - 3\sin\theta = 0$

8. $3\sin 2\theta - 2\sin\theta = 0$

9. $\cos 2\theta = 3\sin\theta - 1$

10. $\cos 2\theta = \cos^2\theta - \frac{1}{2}$

11. $2\sin^2\theta - \cos\theta = 1$

12. $\tan\theta - 3\cot\theta = 0$

13. $\sin\theta - 1 = \cos\theta$

14. $\cos\theta - \sin\theta = 1$

15. $\tan\theta + 1 = \sec\theta$

16. $2\tan\theta + \sec^2\theta = 4$

17–34 ■ An equation is given. **(a)** Find all solutions of the equation. **(b)** Find the solutions in the interval $[0, 2\pi)$.

17. $2\cos 3\theta = 1$

18. $3\csc^2\theta = 4$

19. $2\cos 2\theta + 1 = 0$

20. $2\sin 3\theta + 1 = 0$

21. $\sqrt{3}\tan 3\theta + 1 = 0$

22. $\sec 4\theta - 2 = 0$

23. $\cos\dfrac{\theta}{2} - 1 = 0$

24. $\tan\dfrac{\theta}{4} + \sqrt{3} = 0$

25. $2\sin\dfrac{\theta}{3} + \sqrt{3} = 0$

26. $\sec\dfrac{\theta}{2} = \cos\dfrac{\theta}{2}$

27. $\sin 2\theta = 3 \cos 2\theta$

28. $\csc 3\theta = 5 \sin 3\theta$

29. $\sec \theta - \tan \theta = \cos \theta$

30. $\tan 3\theta + 1 = \sec 3\theta$

31. $3 \tan^3\theta - 3 \tan^2\theta - \tan \theta + 1 = 0$

32. $4 \sin \theta \cos \theta + 2 \sin \theta - 2 \cos \theta - 1 = 0$

33. $2 \sin \theta \tan \theta - \tan \theta = 1 - 2 \sin \theta$

34. $\sec \theta \tan \theta - \cos \theta \cot \theta = \sin \theta$

35–38 ■ **(a)** Graph f and g in the given viewing rectangle and find the intersection points graphically, rounded to two decimal places. **(b)** Find the intersection points of f and g algebraically. Give exact answers.

35. $f(x) = 3 \cos x + 1$, $g(x) = \cos x - 1$; $[-2\pi, 2\pi]$ by $[-2.5, 4.5]$

36. $f(x) = \sin 2x + 1$, $g(x) = 2 \sin 2x + 1$; $[-2\pi, 2\pi]$ by $[-1.5, 3.5]$

37. $f(x) = \tan x$, $g(x) = \sqrt{3}$; $\left[-\dfrac{\pi}{2}, \dfrac{\pi}{2}\right]$ by $[-10, 10]$

38. $f(x) = \sin x - 1$, $g(x) = \cos x$; $[-2\pi, 2\pi]$ by $[-2.5, 1.5]$

39–42 ■ Use an Addition or Subtraction Formula to simplify the equation. Then find all solutions in the interval $[0, 2\pi)$.

39. $\cos \theta \cos 3\theta - \sin \theta \sin 3\theta = 0$

40. $\cos \theta \cos 2\theta + \sin \theta \sin 2\theta = \frac{1}{2}$

41. $\sin 2\theta \cos \theta - \cos 2\theta \sin \theta = \sqrt{3}/2$

42. $\sin 3\theta \cos \theta - \cos 3\theta \sin \theta = 0$

43–52 ■ Use a Double- or Half-Angle Formula to solve the equation in the interval $[0, 2\pi)$.

43. $\sin 2\theta + \cos \theta = 0$

44. $\tan \dfrac{\theta}{2} - \sin \theta = 0$

45. $\cos 2\theta + \cos \theta = 2$

46. $\tan \theta + \cot \theta = 4 \sin 2\theta$

47. $\cos 2\theta - \cos^2\theta = 0$

48. $2 \sin^2\theta = 2 + \cos 2\theta$

49. $\cos 2\theta - \cos 4\theta = 0$

50. $\sin 3\theta - \sin 6\theta = 0$

51. $\cos \theta - \sin \theta = \sqrt{2} \sin \dfrac{\theta}{2}$

52. $\sin \theta - \cos \theta = \frac{1}{2}$

53–56 ■ Solve the equation by first using a Sum-to-Product Formula.

53. $\sin \theta + \sin 3\theta = 0$

54. $\cos 5\theta - \cos 7\theta = 0$

55. $\cos 4\theta + \cos 2\theta = \cos \theta$

56. $\sin 5\theta - \sin 3\theta = \cos 4\theta$

57–62 ■ Use a graphing device to find the solutions of the equation, correct to two decimal places.

57. $\sin 2x = x$

58. $\cos x = \dfrac{x}{3}$

59. $2^{\sin x} = x$

60. $\sin x = x^3$

61. $\dfrac{\cos x}{1 + x^2} = x^2$

62. $\cos x = \frac{1}{2}(e^x + e^{-x})$

APPLICATIONS

63. Range of a Projectile If a projectile is fired with velocity v_0 at an angle θ, then its *range*, the horizontal distance it travels (in feet), is modeled by the function

$$R(\theta) = \dfrac{v_0^2 \sin 2\theta}{32}$$

(See page 306.) If $v_0 = 2200$ ft/s, what angle (in degrees) should be chosen for the projectile to hit a target on the ground 5000 ft away?

64. Damped Vibrations The displacement of a spring vibrating in damped harmonic motion is given by

$$y = 4e^{-3t} \sin 2\pi t$$

Find the times when the spring is at its equilibrium position $(y = 0)$.

65. Hours of Daylight In Philadelphia the number of hours of daylight on day t (where t is the number of days after January 1) is modeled by the function

$$L(t) = 12 + 2.83 \sin\left(\dfrac{2\pi}{365}(t - 80)\right)$$

(a) Which days of the year have about 10 hours of daylight?

(b) How many days of the year have more than 10 hours of daylight?

66. Belts and Pulleys A thin belt of length L surrounds two pulleys of radii R and r, as shown in the figure.

(a) Show that the angle θ (in radians) where the belt crosses itself satisfies the equation

$$\theta + 2 \cot \dfrac{\theta}{2} = \dfrac{L}{R + r} - \pi$$

[*Hint:* Express L in terms of R, r, and θ by adding up the lengths of the curved and straight parts of the belt.]

(b) Suppose that $R = 2.42$ ft, $r = 1.21$ ft, and $L = 27.78$ ft. Find θ by solving the equation in part (a) graphically. Express your answer both in radians and in degrees.

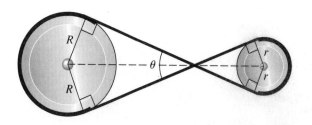

DISCOVERY ■ DISCUSSION ■ WRITING

67. A Special Trigonometric Equation What makes the equation $\sin(\cos x) = 0$ different from all the other equations we've looked at in this section? Find all solutions of this equation.

CHAPTER 4 | REVIEW

■ CONCEPT CHECK

1. (a) State the reciprocal identities.
   (b) State the Pythagorean identities.
   (c) State the even-odd identities.
   (d) State the cofunction identities.

2. Explain the difference between an equation and an identity.

3. How do you prove a trigonometric identity?

4. (a) State the Addition Formulas for Sine, Cosine, and Tangent.
   (b) State the Subtraction Formulas for Sine, Cosine, and Tangent.

5. (a) State the Double-Angle Formulas for Sine, Cosine, and Tangent.
   (b) State the formulas for lowering powers.
   (c) State the Half-Angle Formulas.

6. (a) State the Product-to-Sum Formulas.
   (b) State the Sum-to-Product Formulas.

7. Explain how you solve a trigonometric equation by factoring.

8. What identity would you use to solve the equation $\cos x - \sin 2x = 0$?

■ EXERCISES

1–24 ■ Verify the identity.

1. $\sin \theta \, (\cot \theta + \tan \theta) = \sec \theta$

2. $(\sec \theta - 1)(\sec \theta + 1) = \tan^2 \theta$

3. $\cos^2 x \csc x - \csc x = -\sin x$

4. $\dfrac{1}{1 - \sin^2 x} = 1 + \tan^2 x$

5. $\dfrac{\cos^2 x - \tan^2 x}{\sin^2 x} = \cot^2 x - \sec^2 x$

6. $\dfrac{1 + \sec x}{\sec x} = \dfrac{\sin^2 x}{1 - \cos x}$

7. $\dfrac{\cos^2 x}{1 - \sin x} = \dfrac{\cos x}{\sec x - \tan x}$

8. $(1 - \tan x)(1 - \cot x) = 2 - \sec x \csc x$

9. $\sin^2 x \cot^2 x + \cos^2 x \tan^2 x = 1$

10. $(\tan x + \cot x)^2 = \csc^2 x \sec^2 x$

11. $\dfrac{\sin 2x}{1 + \cos 2x} = \tan x$

12. $\dfrac{\cos(x + y)}{\cos x \sin y} = \cot y - \tan x$

13. $\tan \dfrac{x}{2} = \csc x - \cot x$

14. $\dfrac{\sin(x + y) + \sin(x - y)}{\cos(x + y) + \cos(x - y)} = \tan x$

15. $\sin(x + y) \sin(x - y) = \sin^2 x - \sin^2 y$

16. $\csc x - \tan \dfrac{x}{2} = \cot x$

17. $1 + \tan x \tan \dfrac{x}{2} = \sec x$

18. $\dfrac{\sin 3x + \cos 3x}{\cos x - \sin x} = 1 + 2 \sin 2x$

19. $\left(\cos \dfrac{x}{2} - \sin \dfrac{x}{2}\right)^2 = 1 - \sin x$

20. $\dfrac{\cos 3x - \cos 7x}{\sin 3x + \sin 7x} = \tan 2x$

21. $\dfrac{\sin 2x}{\sin x} - \dfrac{\cos 2x}{\cos x} = \sec x$

22. $(\cos x + \cos y)^2 + (\sin x - \sin y)^2 = 2 + 2 \cos(x + y)$

23. $\tan\left(x + \dfrac{\pi}{4}\right) = \dfrac{1 + \tan x}{1 - \tan x}$

24. $\dfrac{\sec x - 1}{\sin x \sec x} = \tan \dfrac{x}{2}$

25–28 ■ (a) Graph f and g. (b) Do the graphs suggest that the equation $f(x) = g(x)$ is an identity? Prove your answer.

25. $f(x) = 1 - \left(\cos \dfrac{x}{2} - \sin \dfrac{x}{2}\right)^2, \quad g(x) = \sin x$

26. $f(x) = \sin x + \cos x, \quad g(x) = \sqrt{\sin^2 x + \cos^2 x}$

27. $f(x) = \tan x \tan \dfrac{x}{2}, \quad g(x) = \dfrac{1}{\cos x}$

28. $f(x) = 1 - 8 \sin^2 x + 8 \sin^4 x, \quad g(x) = \cos 4x$

29–30 ■ (a) Graph the function(s) and make a conjecture, and (b) prove your conjecture.

29. $f(x) = 2 \sin^2 3x + \cos 6x$

30. $f(x) = \sin x \cot \dfrac{x}{2}, \quad g(x) = \cos x$

31–48 ■ Solve the equation in the interval $[0, 2\pi)$.

31. $4 \sin \theta - 3 = 0$

32. $5 \cos \theta + 3 = 0$

33. $\cos x \sin x - \sin x = 0$

34. $\sin x - 2 \sin^2 x = 0$

35. $2 \sin^2 x - 5 \sin x + 2 = 0$

36. $\sin x - \cos x - \tan x = -1$

37. $2 \cos^2 x - 7 \cos x + 3 = 0$

38. $4 \sin^2 x + 2 \cos^2 x = 3$

39. $\dfrac{1 - \cos x}{1 + \cos x} = 3$

40. $\sin x = \cos 2x$

41. $\tan^3 x + \tan^2 x - 3 \tan x - 3 = 0$

42. $\cos 2x \csc^2 x = 2 \cos 2x$

43. $\tan \frac{1}{2} x + 2 \sin 2x = \csc x$

44. $\cos 3x + \cos 2x + \cos x = 0$

45. $\tan x + \sec x = \sqrt{3}$

46. $2 \cos x - 3 \tan x = 0$

 47. $\cos x = x^2 - 1$

 48. $e^{\sin x} = x$

49. If a projectile is fired with velocity v_0 at an angle θ, then the maximum height it reaches (in feet) is modeled by the function

$$M(\theta) = \dfrac{v_0^2 \sin^2 \theta}{64}$$

Suppose $v_0 = 400$ ft/s.

(a) At what angle θ should the projectile be fired so that the maximum height it reaches is 2000 ft?

(b) Is it possible for the projectile to reach a height of 3000 ft?

(c) Find the angle θ for which the projectile will travel highest.

50. The displacement of an automobile shock absorber is modeled by the function

$$f(t) = 2^{-0.2t} \sin 4\pi t$$

Find the times when the shock absorber is at its equilibrium position (that is, when $f(t) = 0$). [*Hint:* $2^x > 0$ for all real x.]

51–60 ■ Find the exact value of the expression.

51. $\cos 15°$

52. $\sin \dfrac{5\pi}{12}$

53. $\tan \dfrac{\pi}{8}$

54. $2 \sin \dfrac{\pi}{12} \cos \dfrac{\pi}{12}$

55. $\sin 5° \cos 40° + \cos 5° \sin 40°$

56. $\dfrac{\tan 66° - \tan 6°}{1 + \tan 66° \tan 6°}$

57. $\cos^2 \dfrac{\pi}{8} - \sin^2 \dfrac{\pi}{8}$

58. $\dfrac{1}{2} \cos \dfrac{\pi}{12} + \dfrac{\sqrt{3}}{2} \sin \dfrac{\pi}{12}$

59. $\cos 37.5° \cos 7.5°$

60. $\cos 67.5° + \cos 22.5°$

61–66 ■ Find the exact value of the expression given that $\sec x = \frac{3}{2}$, $\csc y = 3$, and x and y are in Quadrant I.

61. $\sin(x + y)$

62. $\cos(x - y)$

63. $\tan(x + y)$

64. $\sin 2x$

65. $\cos \dfrac{y}{2}$

66. $\tan \dfrac{y}{2}$

67–68 ■ Find the exact value of the expression.

67. $\tan\left(2 \cos^{-1} \frac{3}{7}\right)$

68. $\sin\left(\tan^{-1} \frac{3}{4} + \cos^{-1} \frac{5}{13}\right)$

69–70 ■ Write the expression as an algebraic expression in the variable(s).

69. $\tan(2 \tan^{-1} x)$

70. $\cos(\sin^{-1} x + \cos^{-1} y)$

71. A 10-ft-wide highway sign is adjacent to a roadway, as shown in the figure. As a driver approaches the sign, the viewing angle θ changes.

(a) Express viewing angle θ as a function of the distance x between the driver and the sign.

(b) The sign is legible when the viewing angle is 2° or greater. At what distance x does the sign first become legible?

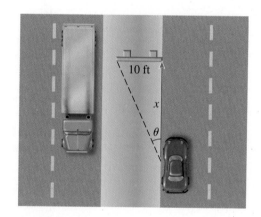

72. A 380-ft-tall building supports a 40-ft communications tower (see the figure). As a driver approaches the building, the viewing angle θ of the tower changes.

(a) Express the viewing angle θ as a function of the distance x between the driver and the building.

 (b) At what distance from the building is the viewing angle θ as large as possible?

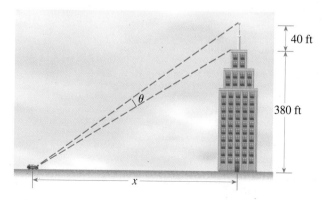

1. Verify each identity.

   (a) $\tan \theta \sin \theta + \cos \theta = \sec \theta$

   (b) $\dfrac{\tan x}{1 - \cos x} = \csc x\,(1 + \sec x)$

   (c) $\dfrac{2 \tan x}{1 + \tan^2 x} = \sin 2x$

2. Let $x = 2 \sin \theta$, $-\pi/2 < \theta < \pi/2$. Simplify the expression

$$\frac{x}{\sqrt{4 - x^2}}$$

3. Find the exact value of each expression.

   (a) $\sin 8° \cos 22° + \cos 8° \sin 22°$ (b) $\sin 75°$ (c) $\sin \dfrac{\pi}{12}$

4. For the angles α and β in the figures, find $\cos(\alpha + \beta)$.

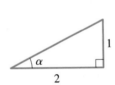 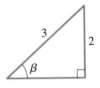

5. (a) Write $\sin 3x \cos 5x$ as a sum of trigonometric functions.

   (b) Write $\sin 2x - \sin 5x$ as a product of trigonometric functions.

6. If $\sin \theta = -\frac{4}{5}$ and θ is in Quadrant III, find $\tan(\theta/2)$.

7. Solve each trigonometric equation in the interval $[0, 2\pi)$, rounded to two decimal places.

   (a) $3 \sin \theta - 1 = 0$

   (b) $(2 \cos \theta - 1)(\sin \theta - 1) = 0$

   (c) $2 \cos^2\theta + 5 \cos \theta + 2 = 0$

   (d) $\sin 2\theta - \cos \theta = 0$

8. Find all solutions in the interval $[0, 2\pi)$, rounded to five decimal places:

$$5 \cos 2\theta = 2$$

9. Find the exact value of $\cos\left(2 \tan^{-1} \frac{9}{40}\right)$.

10. Rewrite the expression as an algebraic function of x and y:

$$\sin(\cos^{-1}x - \tan^{-1}y)$$

We've learned that the position of a particle in simple harmonic motion is described by a function of the form $y = A \sin \omega t$ (see Section 2.6). For example, if a string is moved up and down as in Figure 1, then the red dot on the string moves up and down in simple harmonic motion. Of course, the same holds true for each point on the string.

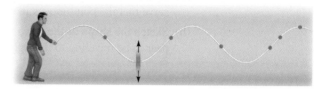

FIGURE 1

What function describes the shape of the whole string? If we fix an instant in time ($t = 0$) and snap a photograph of the string, we get the shape in Figure 2, which is modeled by

$$y = A \sin kx$$

where y is the height of the string above the x-axis at the point x.

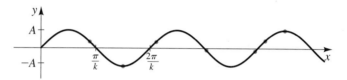

FIGURE 2 $y = A \sin kx$

▼ Traveling Waves

If we snap photographs of the string at other instants, as in Figure 3, it appears that the waves in the string "travel" or shift to the right.

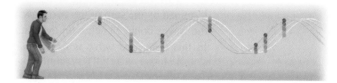

FIGURE 3

The **velocity** of the wave is the rate at which it moves to the right. If the wave has velocity v, then it moves to the right a distance vt in time t. So the graph of the shifted wave at time t is

$$y(x, t) = A \sin k(x - vt)$$

This function models the position of any point x on the string at any time t. We use the notation $y(x, t)$ to indicate that the function depends on the *two* variables x and t. Here is how this function models the motion of the string.

- **If we fix x**, then $y(x, t)$ is a function of t only, which gives the position of the fixed point x at time t.

- **If we fix t**, then $y(x, t)$ is a function of x only, whose graph is the shape of the string at the fixed time t.

EXAMPLE 1 | A Traveling Wave

A traveling wave is described by the function

$$y(x, t) = 3 \sin\left(2x - \frac{\pi}{2}t\right), \qquad x \geq 0$$

(a) Find the function that models the position of the point $x = \pi/6$ at any time t. Observe that the point moves in simple harmonic motion.

(b) Sketch the shape of the wave when $t = 0, 0.5, 1.0, 1.5,$ and 2.0. Does the wave appear to be traveling to the right?

(c) Find the velocity of the wave.

SOLUTION

(a) Substituting $x = \pi/6$ we get

$$y\left(\frac{\pi}{6}, t\right) = 3 \sin\left(2 \cdot \frac{\pi}{6} - \frac{\pi}{2}t\right) = 3 \sin\left(\frac{\pi}{3} - \frac{\pi}{2}t\right)$$

The function $y = 3 \sin\left(\frac{\pi}{3} - \frac{\pi}{2}t\right)$ describes simple harmonic motion with amplitude 3 and period $2\pi/(\pi/2) = 4$.

(b) The graphs are shown in Figure 4. As t increases, the wave moves to the right.

(c) We express the given function in the standard form $y(x, t) = A \sin k(x - vt)$:

$$y(x, t) = 3 \sin\left(2x - \frac{\pi}{2}t\right) \qquad \text{Given}$$

$$= 3 \sin 2\left(x - \frac{\pi}{4}t\right) \qquad \text{Factor 2}$$

Comparing this to the standard form, we see that the wave is moving with velocity $v = \pi/4$. ∎

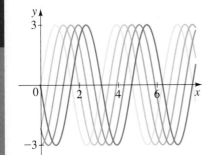

FIGURE 4 Traveling wave

▼ Standing Waves

If two waves are traveling along the same string, then the movement of the string is determined by the sum of the two waves. For example, if the string is attached to a wall, then the waves bounce back with the same amplitude and speed but in the opposite direction. In this case, one wave is described by $y = A \sin k(x - vt)$ and the reflected wave by $y = A \sin k(x + vt)$. The resulting wave is

$$y(x, t) = A \sin k(x - vt) + A \sin k(x + vt) \qquad \text{Add the two waves}$$

$$= 2A \sin kx \cos kvt \qquad \text{Sum-to-Product Formula}$$

The points where kx is a multiple of 2π are special, because at these points $y = 0$ for any time t. In other words, these points never move. Such points are called **nodes**. Figure 5 shows the graph of the wave for several values of t. We see that the wave does not travel, but simply vibrates up and down. Such a wave is called a **standing wave**.

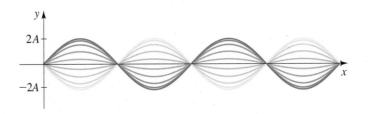

FIGURE 5 A standing wave

EXAMPLE 2 | A Standing Wave

Traveling waves are generated at each end of a wave tank 30 ft long, with equations

$$y = 1.5 \sin\left(\frac{\pi}{5}x - 3t\right)$$

and

$$y = 1.5 \sin\left(\frac{\pi}{5}x + 3t\right)$$

(a) Find the equation of the combined wave, and find the nodes.

(b) Sketch the graph for $t = 0, 0.17, 0.34, 0.51, 0.68, 0.85,$ and 1.02. Is this a standing wave?

SOLUTION

(a) The combined wave is obtained by adding the two equations:

$$y = 1.5 \sin\left(\frac{\pi}{5}x - 3t\right) + 1.5 \sin\left(\frac{\pi}{5}x + 3t\right) \qquad \text{Add the two waves}$$

$$= 3 \sin\frac{\pi}{5}x \cos 3t \qquad\qquad\qquad \text{Sum-to-Product Formula}$$

The nodes occur at the values of x for which $\sin\frac{\pi}{5}x = 0$, that is, where $\frac{\pi}{5}x = k\pi$ (k an integer). Solving for x, we get $x = 5k$. So the nodes occur at

$$x = 0, 5, 10, 15, 20, 25, 30$$

(b) The graphs are shown in Figure 6. From the graphs we see that this is a standing wave.

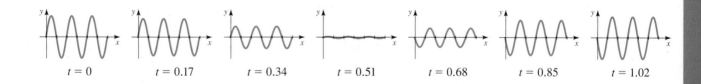

| $t = 0$ | $t = 0.17$ | $t = 0.34$ | $t = 0.51$ | $t = 0.68$ | $t = 0.85$ | $t = 1.02$ |

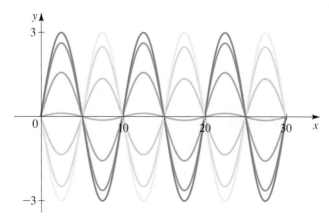

FIGURE 6 $y(x, t) = 3 \sin\dfrac{\pi}{5}x \cos 3t$

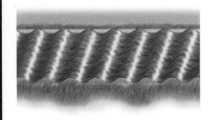

PROBLEMS

1. **Wave on a Canal** A wave on the surface of a long canal is described by the function

$$y(x, t) = 5 \sin\left(2x - \frac{\pi}{2}t \right), \quad x \geq 0$$

   (a) Find the function that models the position of the point $x = 0$ at any time t.
   (b) Sketch the shape of the wave when $t = 0, 0.4, 0.8, 1.2,$ and 1.6. Is this a traveling wave?
   (c) Find the velocity of the wave.

2. **Wave in a Rope** Traveling waves are generated at each end of a tightly stretched rope 24 ft long, with equations

$$y = 0.2 \sin(1.047x - 0.524t) \quad \text{and} \quad y = 0.2 \sin(1.047x + 0.524t)$$

   (a) Find the equation of the combined wave, and find the nodes.
   (b) Sketch the graph for $t = 0, 1, 2, 3, 4, 5,$ and 6. Is this a standing wave?

3. **Traveling Wave** A traveling wave is graphed at the instant $t = 0$. If it is moving to the right with velocity 6, find an equation of the form $y(x, t) = A \sin(kx - kvt)$ for this wave.

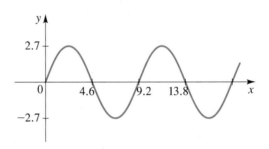

4. **Traveling Wave** A traveling wave has period $2\pi/3$, amplitude 5, and velocity 0.5.
   (a) Find the equation of the wave.
   (b) Sketch the graph for $t = 0, 0.5, 1, 1.5,$ and 2.

5. **Standing Wave** A standing wave with amplitude 0.6 is graphed at several times t as shown in the figure. If the vibration has a frequency of 20 Hz, find an equation of the form $y(x, t) = A \sin \alpha x \cos \beta t$ that models this wave.

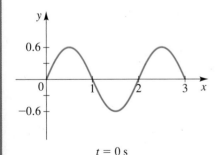

$t = 0$ s

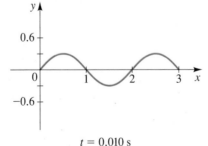

$t = 0.010$ s

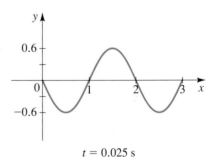

$t = 0.025$ s

6. **Standing Wave**   A standing wave has maximum amplitude 7 and nodes at 0, $\pi/2$, π, $3\pi/2$, 2π, as shown in the figure. Each point that is not a node moves up and down with period 4π. Find a function of the form $y(x, t) = A \sin \alpha x \cos \beta t$ that models this wave.

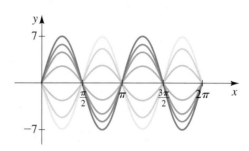

7. **Vibrating String**   When a violin string vibrates, the sound produced results from a combination of standing waves that have evenly placed nodes. The figure illustrates some of the possible standing waves. Let's assume that the string has length π.

   (a) For fixed t, the string has the shape of a sine curve $y = A \sin \alpha x$. Find the appropriate value of α for each of the illustrated standing waves.

   (b) Do you notice a pattern in the values of α that you found in part (a)? What would the next two values of α be? Sketch rough graphs of the standing waves associated with these new values of α.

   (c) Suppose that for fixed t, each point on the string that is not a node vibrates with frequency 440 Hz. Find the value of β for which an equation of the form $y = A \cos \beta t$ would model this motion.

   (d) Combine your answers for parts (a) and (c) to find functions of the form $y(x, t) = A \sin \alpha x \cos \beta t$ that model each of the standing waves in the figure. (Assume that $A = 1$.)

8. **Waves in a Tube**   Standing waves in a violin string must have nodes at the ends of the string because the string is fixed at its endpoints. But this need not be the case with sound waves in a tube (such as a flute or an organ pipe). The figure shows some possible standing waves in a tube.

   Suppose that a standing wave in a tube 37.7 ft long is modeled by the function

   $$y(x, t) = 0.3 \cos \tfrac{1}{2}x \cos 50\pi t$$

   Here $y(x, t)$ represents the variation from normal air pressure at the point x feet from the end of the tube, at time t seconds.

   (a) At what points x are the nodes located? Are the endpoints of the tube nodes?

   (b) At what frequency does the air vibrate at points that are not nodes?

POLAR COORDINATES AND PARAMETRIC EQUATIONS

In Section 1.1 we learned how to graph points in rectangular coordinates. In this chapter we study a different way of locating points in the plane, called *polar coordinates.* Using rectangular coordinates is like describing a location in a city by saying that it's at the corner of 2nd Street and 4th Avenue; these directions would help a taxi driver find the location. But we may also describe this same location "as the crow flies"; we can say that it's 1.5 miles northeast of City Hall. So instead of specifying the location with respect to a grid of streets and avenues, we specify it by giving its distance and direction from a fixed reference point. That's what we do in the polar coordinate system. In polar coordinates the location of a point is given by an ordered pair of numbers: the distance of the point from the origin (or pole) and the angle from the positive *x*-axis.

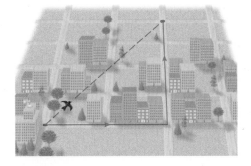

Why do we study different coordinate systems? It's because certain curves are more naturally described in one coordinate system rather than another. For example, in rectangular coordinates lines and parabolas have simple equations, but equations of circles are rather complicated. We'll see that in polar coordinates circles have very simple equations.

5.1 POLAR COORDINATES

Definition of Polar Coordinates ▶ Relationship Between Polar
and Rectangular Coordinates ▶ Polar Equations

In this section we define polar coordinates, and we learn how polar coordinates are related to rectangular coordinates.

▼ Definition of Polar Coordinates

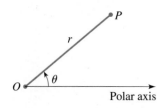

FIGURE 1

The **polar coordinate system** uses distances and directions to specify the location of a point in the plane. To set up this system, we choose a fixed point O in the plane called the **pole** (or **origin**) and draw from O a ray (half-line) called the **polar axis** as in Figure 1. Then each point P can be assigned polar coordinates $P(r, \theta)$ where

$$r \text{ is the } distance \text{ from } O \text{ to } P$$

$$\theta \text{ is the } angle \text{ between the polar axis and the segment } \overline{OP}$$

We use the convention that θ is positive if measured in a counterclockwise direction from the polar axis or negative if measured in a clockwise direction. If r is negative, then $P(r, \theta)$ is defined to be the point that lies $|r|$ units from the pole in the direction opposite to that given by θ (see Figure 2).

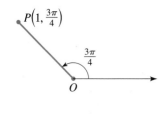

FIGURE 2

EXAMPLE 1 | Plotting Points in Polar Coordinates

Plot the points whose polar coordinates are given.

(a) $(1, 3\pi/4)$ **(b)** $(3, -\pi/6)$ **(c)** $(3, 3\pi)$ **(d)** $(-4, \pi/4)$

SOLUTION The points are plotted in Figure 3. Note that the point in part (d) lies 4 units from the origin along the angle $5\pi/4$, because the given value of r is negative.

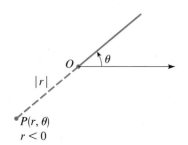

(a) (b) (c) (d)

FIGURE 3

✎ NOW TRY EXERCISES **3** AND **5**

Note that the coordinates (r, θ) and $(-r, \theta + \pi)$ represent the same point, as shown in Figure 4. Moreover, because the angles $\theta + 2n\pi$ (where n is any integer) all have the same terminal side as the angle θ, each point in the plane has infinitely many representations in polar coordinates. In fact, any point $P(r, \theta)$ can also be represented by

$$P(r, \theta + 2n\pi) \qquad \text{and} \qquad P(-r, \theta + (2n + 1)\pi)$$

for any integer n.

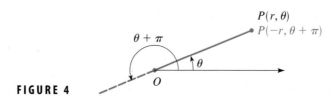

FIGURE 4

EXAMPLE 2 | Different Polar Coordinates for the Same Point

(a) Graph the point with polar coordinates $P(2, \pi/3)$.

(b) Find two other polar coordinate representations of P with $r > 0$ and two with $r < 0$.

SOLUTION

(a) The graph is shown in Figure 5(a).

(b) Other representations with $r > 0$ are

$$\left(2, \frac{\pi}{3} + 2\pi\right) = \left(2, \frac{7\pi}{3}\right) \qquad \text{Add } 2\pi \text{ to } \theta$$

$$\left(2, \frac{\pi}{3} - 2\pi\right) = \left(2, -\frac{5\pi}{3}\right) \qquad \text{Add } -2\pi \text{ to } \theta$$

Other representations with $r < 0$ are

$$\left(-2, \frac{\pi}{3} + \pi\right) = \left(-2, \frac{4\pi}{3}\right) \qquad \text{Replace } r \text{ by } -r \text{ and add } \pi \text{ to } \theta$$

$$\left(-2, \frac{\pi}{3} - \pi\right) = \left(-2, -\frac{2\pi}{3}\right) \qquad \text{Replace } r \text{ by } -r \text{ and add } -\pi \text{ to } \theta$$

The graphs in Figure 5 explain why these coordinates represent the same point.

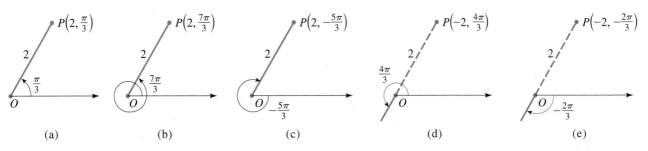

(a)   (b)   (c)   (d)   (e)

FIGURE 5

✎ NOW TRY EXERCISE **9**   ∎

▼ Relationship Between Polar and Rectangular Coordinates

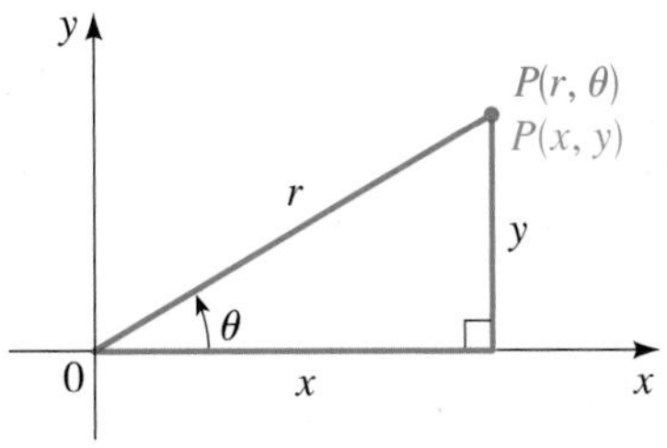

Situations often arise in which we need to consider polar and rectangular coordinates simultaneously. The connection between the two systems is illustrated in Figure 6, where the polar axis coincides with the positive x-axis. The formulas in the following box are obtained from the figure using the definitions of the trigonometric functions and the Pythagorean Theorem. (Although we have pictured the case where $r > 0$ and θ is acute, the formulas hold for any angle θ and for any value of r.)

FIGURE 6

RELATIONSHIP BETWEEN POLAR AND RECTANGULAR COORDINATES

1. To change from polar to rectangular coordinates, use the formulas

$$x = r \cos \theta \qquad \text{and} \qquad y = r \sin \theta$$

2. To change from rectangular to polar coordinates, use the formulas

$$r^2 = x^2 + y^2 \qquad \text{and} \qquad \tan \theta = \frac{y}{x} \quad (x \neq 0)$$

EXAMPLE 3 | Converting Polar Coordinates to Rectangular Coordinates

Find rectangular coordinates for the point that has polar coordinates $(4, 2\pi/3)$.

SOLUTION Since $r = 4$ and $\theta = 2\pi/3$, we have

$$x = r \cos \theta = 4 \cos \frac{2\pi}{3} = 4 \cdot \left(-\frac{1}{2}\right) = -2$$

$$y = r \sin \theta = 4 \sin \frac{2\pi}{3} = 4 \cdot \frac{\sqrt{3}}{2} = 2\sqrt{3}$$

Thus the point has rectangular coordinates $(-2, 2\sqrt{3})$.

✎ NOW TRY EXERCISE **27** ■

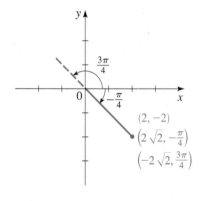

FIGURE 7

EXAMPLE 4 | Converting Rectangular Coordinates to Polar Coordinates

Find polar coordinates for the point that has rectangular coordinates $(2, -2)$.

SOLUTION Using $x = 2$, $y = -2$, we get

$$r^2 = x^2 + y^2 = 2^2 + (-2)^2 = 8$$

so $r = 2\sqrt{2}$ or $-2\sqrt{2}$. Also

$$\tan \theta = \frac{y}{x} = \frac{-2}{2} = -1$$

so $\theta = 3\pi/4$ or $-\pi/4$. Since the point $(2, -2)$ lies in Quadrant IV (see Figure 7), we can represent it in polar coordinates as $(2\sqrt{2}, -\pi/4)$ or $(-2\sqrt{2}, 3\pi/4)$.

✎ NOW TRY EXERCISE **35** ■

  Note that the equations relating polar and rectangular coordinates do not uniquely determine r or θ. When we use these equations to find the polar coordinates of a point, we must be careful that the values we choose for r and θ give us a point in the correct quadrant, as we did in Example 4.

▼ Polar Equations

In Examples 3 and 4 we converted points from one coordinate system to the other. Now we consider the same problem for equations.

EXAMPLE 5 | Converting an Equation from Rectangular to Polar Coordinates

Express the equation $x^2 = 4y$ in polar coordinates.

SOLUTION We use the formulas $x = r \cos \theta$ and $y = r \sin \theta$:

$$x^2 = 4y \qquad \text{Rectangular equation}$$

$$(r \cos \theta)^2 = 4(r \sin \theta) \qquad \text{Substitute } x = r \cos \theta, \ y = r \sin \theta$$

$$r^2 \cos^2 \theta = 4r \sin \theta \qquad \text{Expand}$$

$$r = 4 \frac{\sin \theta}{\cos^2 \theta} \qquad \text{Divide by } r \cos^2 \theta$$

$$r = 4 \sec \theta \tan \theta \qquad \text{Simplify}$$

✎ NOW TRY EXERCISE **45** ■

Art Wolfe/Stone/Getty Images

Mathematical Ecology
In the 1970s humpback whales became a
center of controversy. Environmentalists
believed that whaling threatened the
whales with imminent extinction; whalers
saw their livelihood threatened by any attempt to stop whaling. Are whales really
threatened to extinction by whaling?
What level of whaling is safe to guarantee
survival of the whales? These questions
motivated mathematicians to study population patterns of whales and other
species more closely.

As early as the 1920s Lotka and
Volterra had founded the field of
mathematical biology by creating
predator-prey models. Their models,
which draw on a branch of mathematics
called differential equations, take into
account the rates at which predator eats
prey and the rates of growth of each
population. Note that as predator eats
prey, the prey population decreases; this
means less food supply for the predators, so their population begins to
decrease; with fewer predators the prey
population begins to increase, and so
on. Normally, a state of equilibrium
develops, and the two populations
alternate between a minimum and a
maximum. Notice that if the predators
eat the prey too fast, they will be left
without food and will thus ensure their
own extinction.

Since Lotka and Volterra's time,
more detailed mathematical models of
animal populations have been developed. For many species the population is divided into several stages:
immature, juvenile, adult, and so on.
The proportion of each stage that
survives or reproduces in a given time
period is entered into a matrix (called a
transition matrix); matrix multiplication
is then used to predict the population
in succeeding time periods. (See the
Discovery Project *Will the Species
Survive?* at the book companion
website: www.stewartmath.com.)

As you can see, the power of mathematics to model and predict is an invaluable tool in the ongoing debate
over the environment.

As Example 5 shows, converting from rectangular to polar coordinates is straightfor-
ward: Just replace x by $r \cos \theta$ and y by $r \sin \theta$, and then simplify. But converting polar
equations to rectangular form often requires more thought.

EXAMPLE 6 | Converting Equations from Polar
to Rectangular Coordinates

Express the polar equation in rectangular coordinates. If possible, determine the graph of
the equation from its rectangular form.

(a) $r = 5 \sec \theta$ **(b)** $r = 2 \sin \theta$ **(c)** $r = 2 + 2 \cos \theta$

SOLUTION

(a) Since $\sec \theta = 1/\cos \theta$, we multiply both sides by $\cos \theta$:

$$r = 5 \sec \theta \qquad \text{Polar equation}$$
$$r \cos \theta = 5 \qquad \text{Multiply by } \cos \theta$$
$$x = 5 \qquad \text{Substitute } x = r \cos \theta$$

The graph of $x = 5$ is the vertical line in Figure 8.

(b) We multiply both sides of the equation by r, because then we can use the
formulas $r^2 = x^2 + y^2$ and $r \sin \theta = y$:

$$r = 2 \sin \theta \qquad \text{Polar equation}$$
$$r^2 = 2r \sin \theta \qquad \text{Multiply by } r$$
$$x^2 + y^2 = 2y \qquad r^2 = x^2 + y^2 \text{ and } r \sin \theta = y$$
$$x^2 + y^2 - 2y = 0 \qquad \text{Subtract } 2y$$
$$x^2 + (y - 1)^2 = 1 \qquad \text{Complete the square in } y$$

This is the equation of a circle of radius 1 centered at the point $(0, 1)$. It is graphed
in Figure 9.

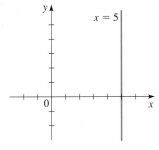

FIGURE 8

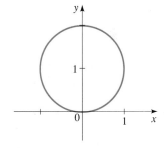

FIGURE 9

(c) We first multiply both sides of the equation by r:

$$r^2 = 2r + 2r \cos \theta$$

Using $r^2 = x^2 + y^2$ and $x = r \cos \theta$, we can convert two terms in the equation into
rectangular coordinates, but eliminating the remaining r requires more work:

$$x^2 + y^2 = 2r + 2x \qquad r^2 = x^2 + y^2 \text{ and } r \cos \theta = x$$
$$x^2 + y^2 - 2x = 2r \qquad \text{Subtract } 2x$$
$$(x^2 + y^2 - 2x)^2 = 4r^2 \qquad \text{Square both sides}$$
$$(x^2 + y^2 - 2x)^2 = 4(x^2 + y^2) \qquad r^2 = x^2 + y^2$$

In this case the rectangular equation looks more complicated than the polar equation.
Although we cannot easily determine the graph of the equation from its rectangular
form, we will see in the next section how to graph it using the polar equation.

➤ NOW TRY EXERCISES **53, 55,** AND **57**

5.1 EXERCISES

CONCEPTS

1. We can describe the location of a point in the plane using different _____ systems. The point P shown in the figure has rectangular coordinates (▢ , ▢) and polar coordinates (▢ , ▢).

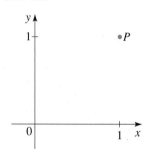

2. Let P be a point in the plane.
   (a) If P has polar coordinates (r, θ) then it has rectangular coordinates (x, y) where $x =$ _____ and $y =$ _____.

   (b) If P has rectangular coordinates (x, y) then it has polar coordinates (r, θ) where $r^2 =$ _____ and $\tan \theta =$ _____.

SKILLS

3–8 ■ Plot the point that has the given polar coordinates.

3. $(4, \pi/4)$ 4. $(1, 0)$ 5. $(6, -7\pi/6)$
6. $(3, -2\pi/3)$ 7. $(-2, 4\pi/3)$ 8. $(-5, -17\pi/6)$

9–14 ■ Plot the point that has the given polar coordinates. Then give two other polar coordinate representations of the point, one with $r < 0$ and the other with $r > 0$.

9. $(3, \pi/2)$ 10. $(2, 3\pi/4)$ 11. $(-1, 7\pi/6)$
12. $(-2, -\pi/3)$ 13. $(-5, 0)$ 14. $(3, 1)$

15–22 ■ Determine which point in the figure, P, Q, R, or S, has the given polar coordinates.

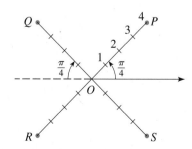

15. $(4, 3\pi/4)$ 16. $(4, -3\pi/4)$
17. $(-4, -\pi/4)$ 18. $(-4, 13\pi/4)$

19. $(4, -23\pi/4)$ 20. $(-4, 23\pi/4)$
21. $(-4, 101\pi/4)$ 22. $(4, 103\pi/4)$

23–24 ■ A point is graphed in rectangular form. Find polar coordinates for the point, with $r > 0$ and $0 < \theta < 2\pi$.

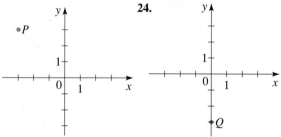

25–26 ■ A point is graphed in polar form. Find its rectangular coordinates.

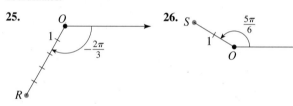

27–34 ■ Find the rectangular coordinates for the point whose polar coordinates are given.

27. $(4, \pi/6)$ 28. $(6, 2\pi/3)$
29. $(\sqrt{2}, -\pi/4)$ 30. $(-1, 5\pi/2)$
31. $(5, 5\pi)$ 32. $(0, 13\pi)$
33. $(6\sqrt{2}, 11\pi/6)$ 34. $(\sqrt{3}, -5\pi/3)$

35–42 ■ Convert the rectangular coordinates to polar coordinates with $r > 0$ and $0 \leq \theta < 2\pi$.

35. $(-1, 1)$ 36. $(3\sqrt{3}, -3)$
37. $(\sqrt{8}, \sqrt{8})$ 38. $(-\sqrt{6}, -\sqrt{2})$
39. $(3, 4)$ 40. $(1, -2)$
41. $(-6, 0)$ 42. $(0, -\sqrt{3})$

43–48 ■ Convert the equation to polar form.

43. $x = y$ 44. $x^2 + y^2 = 9$
45. $y = x^2$ 46. $y = 5$
47. $x = 4$ 48. $x^2 - y^2 = 1$

49–68 ■ Convert the polar equation to rectangular coordinates.

49. $r = 7$ 50. $r = -3$
51. $\theta = -\dfrac{\pi}{2}$ 52. $\theta = \pi$
53. $r \cos \theta = 6$ 54. $r = 2 \csc \theta$

55. $r = 4 \sin \theta$

56. $r = 6 \cos \theta$

57. $r = 1 + \cos \theta$

58. $r = 3(1 - \sin \theta)$

59. $r = 1 + 2 \sin \theta$

60. $r = 2 - \cos \theta$

61. $r = \dfrac{1}{\sin \theta - \cos \theta}$

62. $r = \dfrac{1}{1 + \sin \theta}$

63. $r = \dfrac{4}{1 + 2 \sin \theta}$

64. $r = \dfrac{2}{1 - \cos \theta}$

65. $r^2 = \tan \theta$

66. $r^2 = \sin 2\theta$

67. $\sec \theta = 2$

68. $\cos 2\theta = 1$

DISCOVERY ▪ DISCUSSION ▪ WRITING

69. The Distance Formula in Polar Coordinates

(a) Use the Law of Cosines to prove that the distance between the polar points (r_1, θ_1) and (r_2, θ_2) is

$$d = \sqrt{r_1^2 + r_2^2 - 2r_1 r_2 \cos(\theta_2 - \theta_1)}$$

(b) Find the distance between the points whose polar coordinates are $(3, 3\pi/4)$ and $(1, 7\pi/6)$, using the formula from part (a).

(c) Now convert the points in part (b) to rectangular coordinates. Find the distance between them using the usual Distance Formula. Do you get the same answer?

5.2 GRAPHS OF POLAR EQUATIONS

| Graphing Polar Equations ▶ Symmetry ▶ Graphing Polar Equations with Graphing Devices

The **graph of a polar equation** $r = f(\theta)$ consists of all points P that have at least one polar representation (r, θ) whose coordinates satisfy the equation. Many curves that arise in mathematics and its applications are more easily and naturally represented by polar equations than by rectangular equations.

▼ Graphing Polar Equations

A rectangular grid is helpful for plotting points in rectangular coordinates (see Figure 1(a)). To plot points in polar coordinates, it is convenient to use a grid consisting of circles centered at the pole and rays emanating from the pole, as in Figure 1(b). We will use such grids to help us sketch polar graphs.

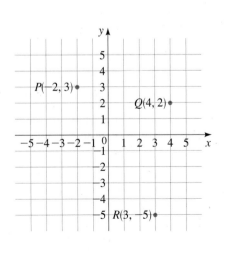

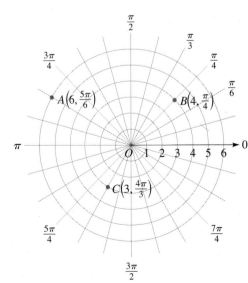

FIGURE 1   (a) Grid for rectangular coordinates   (b) Grid for polar coordinates

In Examples 1 and 2 we see that circles centered at the origin and lines that pass through the origin have particularly simple equations in polar coordinates.

EXAMPLE 1 | Sketching the Graph of a Polar Equation

Sketch a graph of the equation $r = 3$, and express the equation in rectangular coordinates.

SOLUTION The graph consists of all points whose r-coordinate is 3, that is, all points that are 3 units away from the origin. So the graph is a circle of radius 3 centered at the origin, as shown in Figure 2.

Squaring both sides of the equation, we get

$$r^2 = 3^2 \qquad \text{Square both sides}$$

$$x^2 + y^2 = 9 \qquad \text{Substitute } r^2 = x^2 + y^2$$

So the equivalent equation in rectangular coordinates is $x^2 + y^2 = 9$.

✎. NOW TRY EXERCISE **17**

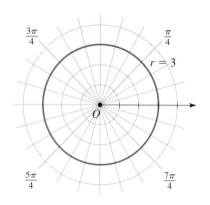

FIGURE 2

In general, the graph of the equation $r = a$ is a circle of radius $|a|$ centered at the origin. Squaring both sides of this equation, we see that the equivalent equation in rectangular coordinates is $x^2 + y^2 = a^2$.

EXAMPLE 2 | Sketching the Graph of a Polar Equation

Sketch a graph of the equation $\theta = \pi/3$, and express the equation in rectangular coordinates.

SOLUTION The graph consists of all points whose θ-coordinate is $\pi/3$. This is the straight line that passes through the origin and makes an angle of $\pi/3$ with the polar axis (see Figure 3). Note that the points $(r, \pi/3)$ on the line with $r > 0$ lie in Quadrant I, whereas those with $r < 0$ lie in Quadrant III. If the point (x, y) lies on this line, then

$$\frac{y}{x} = \tan \theta = \tan \frac{\pi}{3} = \sqrt{3}$$

Thus, the rectangular equation of this line is $y = \sqrt{3}x$.

✎. NOW TRY EXERCISE **19**

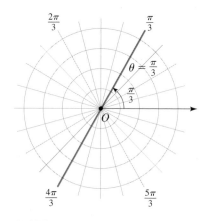

FIGURE 3

To sketch a polar curve whose graph isn't as obvious as the ones in the preceding examples, we plot points calculated for sufficiently many values of θ and then join them in a continuous curve. (This is what we did when we first learned to graph functions in rectangular coordinates.)

EXAMPLE 3 | Sketching the Graph of a Polar Equation

Sketch a graph of the polar equation $r = 2 \sin \theta$.

SOLUTION We first use the equation to determine the polar coordinates of several points on the curve. The results are shown in the following table.

θ	0	$\pi/6$	$\pi/4$	$\pi/3$	$\pi/2$	$2\pi/3$	$3\pi/4$	$5\pi/6$	π
$r = 2 \sin \theta$	0	1	$\sqrt{2}$	$\sqrt{3}$	2	$\sqrt{3}$	$\sqrt{2}$	1	0

We plot these points in Figure 4 and then join them to sketch the curve. The graph appears to be a circle. We have used values of θ only between 0 and π, since the same points (this time expressed with negative r-coordinates) would be obtained if we allowed θ to range from π to 2π.

The polar equation $r = 2 \sin \theta$ in rectangular coordinates is

$$x^2 + (y - 1)^2 = 1$$

(see Section 5.1, Example 6(b)). From the rectangular form of the equation we see that the graph is a circle of radius 1 centered at $(0, 1)$.

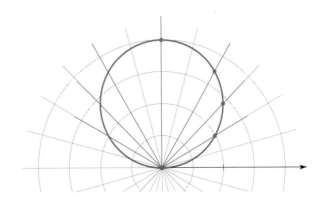

FIGURE 4 $r = 2 \sin \theta$

✎. NOW TRY EXERCISE **21**

In general, the graphs of equations of the form

$$r = 2a \sin \theta \qquad \text{and} \qquad r = 2a \cos \theta$$

are **circles** with radius $|a|$ centered at the points with polar coordinates $(a, \pi/2)$ and $(a, 0)$, respectively.

EXAMPLE 4 | Sketching the Graph of a Cardioid

Sketch a graph of $r = 2 + 2 \cos \theta$.

SOLUTION Instead of plotting points as in Example 3, we first sketch the graph of $r = 2 + 2 \cos \theta$ in *rectangular* coordinates in Figure 5. We can think of this graph as a table of values that enables us to read at a glance the values of r that correspond to increasing values of θ. For instance, we see that as θ increases from 0 to $\pi/2$, r (the distance from O) decreases from 4 to 2, so we sketch the corresponding part of the polar graph in Figure 6(a). As θ increases from $\pi/2$ to π, Figure 5 shows that r decreases from 2 to 0, so we sketch the next part of the graph as in Figure 6(b). As θ increases from π to $3\pi/2$, r increases from 0 to 2, as shown in part (c). Finally, as θ increases from $3\pi/2$ to 2π, r increases from 2 to 4, as shown in part (d). If we let θ increase beyond 2π or decrease beyond 0, we would simply retrace our path. Combining the portions of the graph from parts (a) through (d) of Figure 6, we sketch the complete graph in part (e).

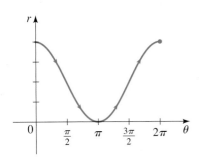

FIGURE 5 $r = 2 + 2 \cos \theta$

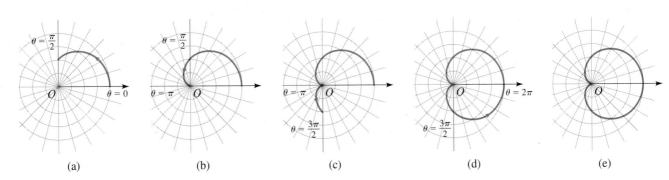

(a)   (b)   (c)   (d)   (e)

FIGURE 6 Steps in sketching $r = 2 + 2 \cos \theta$

The polar equation $r = 2 + 2 \cos \theta$ in rectangular coordinates is

$$(x^2 + y^2 - 2x)^2 = 4(x^2 + y^2)$$

(see Section 5.1, Example 6(c)). The simpler form of the polar equation shows that it is more natural to describe cardioids using polar coordinates.

✎. NOW TRY EXERCISE **25**

The curve in Figure 6 is called a **cardioid** because it is heart-shaped. In general, the graph of any equation of the form

$$r = a(1 \pm \cos \theta) \qquad \text{or} \qquad r = a(1 \pm \sin \theta)$$

is a cardioid.

EXAMPLE 5 | Sketching the Graph of a Four-Leaved Rose

Sketch the curve $r = \cos 2\theta$.

SOLUTION As in Example 4, we first sketch the graph of $r = \cos 2\theta$ in *rectangular* coordinates, as shown in Figure 7. As θ increases from 0 to $\pi/4$, Figure 7 shows that r decreases from 1 to 0, so we draw the corresponding portion of the polar curve in Figure 8 (indicated by ①). As θ increases from $\pi/4$ to $\pi/2$, the value of r goes from 0 to -1. This means that the distance from the origin increases from 0 to 1, but instead of being in Quadrant I, this portion of the polar curve (indicated by ②) lies on the opposite side of the origin in Quadrant III. The remainder of the curve is drawn in a similar fashion, with the arrows and numbers indicating the order in which the portions are traced out. The resulting curve has four petals and is called a **four-leaved rose**.

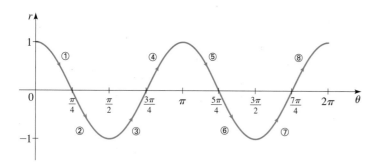

FIGURE 7 Graph of $r = \cos 2\theta$ sketched in rectangular coordinates

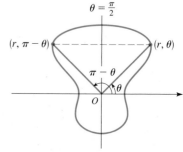

FIGURE 8 Four-leaved rose $r = \cos 2\theta$ sketched in polar coordinates

✎. NOW TRY EXERCISE **29**

In general, the graph of an equation of the form

$$r = a \cos n\theta \qquad \text{or} \qquad r = a \sin n\theta$$

is an **n-leaved rose** if n is odd or a $2n$-leaved rose if n is even (as in Example 5).

▼ Symmetry

In graphing a polar equation, it's often helpful to take advantage of symmetry. We list three tests for symmetry; Figure 9 shows why these tests work.

TESTS FOR SYMMETRY

1. If a polar equation is unchanged when we replace θ by $-\theta$, then the graph is symmetric about the polar axis (Figure 9(a)).

2. If the equation is unchanged when we replace r by $-r$, then the graph is symmetric about the pole (Figure 9(b)).

3. If the equation is unchanged when we replace θ by $\pi - \theta$, the graph is symmetric about the vertical line $\theta = \pi/2$ (the y-axis) (Figure 9(c)).

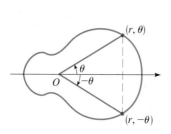

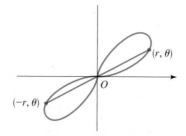

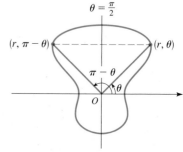

FIGURE 9   (a) Symmetry about the polar axis   (b) Symmetry about the pole   (c) Symmetry about the line $\theta = \frac{\pi}{2}$

The graphs in Figures 2, 6(e), and 8 are symmetric about the polar axis. The graph in Figure 8 is also symmetric about the pole. Figures 4 and 8 show graphs that are symmetric about $\theta = \pi/2$. Note that the four-leaved rose in Figure 8 meets all three tests for symmetry.

In rectangular coordinates, the zeros of the function $y = f(x)$ correspond to the x-intercepts of the graph. In polar coordinates, the zeros of the function $r = f(\theta)$ are the angles θ at which the curve crosses the pole. The zeros help us sketch the graph, as is illustrated in the next example.

EXAMPLE 6 | Using Symmetry to Sketch a Limaçon

Sketch a graph of the equation $r = 1 + 2 \cos \theta$.

SOLUTION   We use the following as aids in sketching the graph:

Symmetry: Since the equation is unchanged when θ is replaced by $-\theta$, the graph is symmetric about the polar axis.

Zeros: To find the zeros, we solve

$$0 = 1 + 2 \cos \theta$$
$$\cos \theta = -\frac{1}{2}$$
$$\theta = \frac{2\pi}{3}, \frac{4\pi}{3}$$

Table of values: As in Example 4, we sketch the graph of $r = 1 + 2 \cos \theta$ in *rectangular* coordinates to serve as a table of values (Figure 10).

Now we sketch the polar graph of $r = 1 + 2 \cos \theta$ from $\theta = 0$ to $\theta = \pi$, and then use symmetry to complete the graph in Figure 11.

✎ NOW TRY EXERCISE **35**

The curve in Figure 11 is called a **limaçon**, after the Middle French word for snail. In general, the graph of an equation of the form

$$r = a \pm b \cos \theta \qquad \text{or} \qquad r = a \pm b \sin \theta$$

is a limaçon. The shape of the limaçon depends on the relative size of a and b (see the table on the next page).

▼ Graphing Polar Equations with Graphing Devices

Although it's useful to be able to sketch simple polar graphs by hand, we need a graphing calculator or computer when the graph is as complicated as the one in Figure 12. Fortunately, most graphing calculators are capable of graphing polar equations directly.

EXAMPLE 7 | Drawing the Graph of a Polar Equation

Graph the equation $r = \cos(2\theta/3)$.

SOLUTION   We need to determine the domain for θ. So we ask ourselves: How many times must θ go through a complete rotation (2π radians) before the graph starts to repeat itself? The graph repeats itself when the same value of r is obtained at θ and $\theta + 2n\pi$. Thus we need to find an integer n, so that

$$\cos \frac{2(\theta + 2n\pi)}{3} = \cos \frac{2\theta}{3}$$

For this equality to hold, $4n\pi/3$ must be a multiple of 2π, and this first happens when $n = 3$. Therefore, we obtain the entire graph if we choose values of θ between $\theta = 0$ and $\theta = 0 + 2(3)\pi = 6\pi$. The graph is shown in Figure 13.

✎ NOW TRY EXERCISE **43**

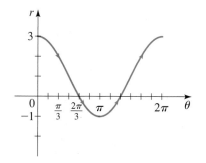

FIGURE 10

$\theta = \frac{2\pi}{3}$

$\theta = \frac{4\pi}{3}$

FIGURE 11 $r = 1 + 2 \cos \theta$

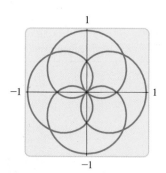

FIGURE 12 $r = \sin \theta + \sin^3(5\theta/2)$

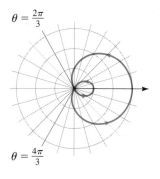

FIGURE 13 $r = \cos(2\theta/3)$

EXAMPLE 8 | A Family of Polar Equations

Graph the family of polar equations $r = 1 + c \sin \theta$ for $c = 3, 2.5, 2, 1.5, 1$. How does the shape of the graph change as c changes?

SOLUTION   Figure 14 shows computer-drawn graphs for the given values of c. When $c > 1$, the graph has an inner loop; the loop decreases in size as c decreases. When $c = 1$, the loop disappears, and the graph becomes a cardioid (see Example 4).

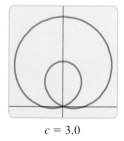
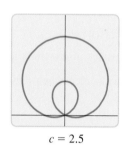

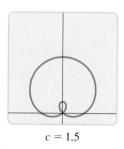

| $c = 3.0$ | $c = 2.5$ | $c = 2.0$ | $c = 1.5$ | $c = 1.0$ |

FIGURE 14   A family of limaçons $r = 1 + c \sin \theta$ in the viewing rectangle $[-2.5, 2.5]$ by $[-0.5, 4.5]$

✎ NOW TRY EXERCISE **47**

The box below gives a summary of some of the basic polar graphs used in calculus.

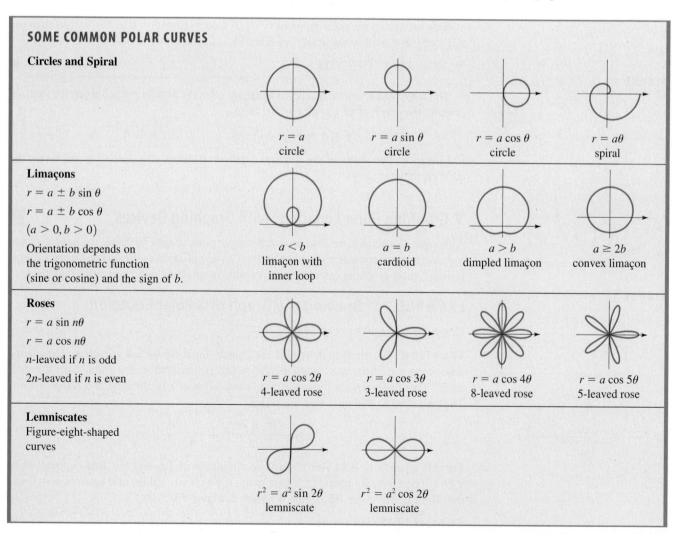

SOME COMMON POLAR CURVES

Circles and Spiral

| $r = a$ circle | $r = a \sin \theta$ circle | $r = a \cos \theta$ circle | $r = a\theta$ spiral |

Limaçons

$r = a \pm b \sin \theta$

$r = a \pm b \cos \theta$

$(a > 0, b > 0)$

Orientation depends on the trigonometric function (sine or cosine) and the sign of b.

| $a < b$ limaçon with inner loop | $a = b$ cardioid | $a > b$ dimpled limaçon | $a \geq 2b$ convex limaçon |

Roses

$r = a \sin n\theta$

$r = a \cos n\theta$

n-leaved if n is odd

$2n$-leaved if n is even

| $r = a \cos 2\theta$ 4-leaved rose | $r = a \cos 3\theta$ 3-leaved rose | $r = a \cos 4\theta$ 8-leaved rose | $r = a \cos 5\theta$ 5-leaved rose |

Lemniscates

Figure-eight-shaped curves

| $r^2 = a^2 \sin 2\theta$ lemniscate | $r^2 = a^2 \cos 2\theta$ lemniscate |

5.2 EXERCISES

CONCEPTS

1. To plot points in polar coordinates, we use a grid consisting of

_____ centered at the pole and _____ emanating from the pole.

2. (a) To graph a polar equation $r = f(\theta)$, we plot all the points (r, θ) that _____ the equation.

(b) The simplest polar equations are obtained by setting r or θ equal to a constant. The graph of the polar equation $r = 3$ is a _____ with radius _____ centered at the _____. The graph of the polar equation $\theta = \pi/4$ is a _____ passing through the _____ with slope _____. Graph these polar equations below.

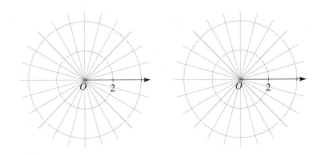

SKILLS

3–8 ■ Match the polar equation with the graphs labeled I–VI. Use the table on page 282 to help you.

3. $r = 3 \cos \theta$

4. $r = 3$

5. $r = 2 + 2 \sin \theta$

6. $r = 1 + 2 \cos \theta$

7. $r = \sin 3\theta$

8. $r = \sin 4\theta$

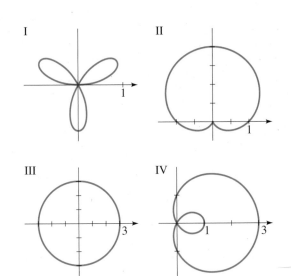

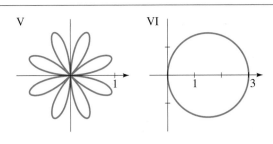

9–16 ■ Test the polar equation for symmetry with respect to the polar axis, the pole, and the line $\theta = \pi/2$.

9. $r = 2 - \sin \theta$

10. $r = 4 + 8 \cos \theta$

11. $r = 3 \sec \theta$

12. $r = 5 \cos \theta \csc \theta$

13. $r = \dfrac{4}{3 - 2 \sin \theta}$

14. $r = \dfrac{5}{1 + 3 \cos \theta}$

15. $r^2 = 4 \cos 2\theta$

16. $r^2 = 9 \sin \theta$

17–22 ■ Sketch a graph of the polar equation, and express the equation in rectangular coordinates.

17. $r = 2$

18. $r = -1$

19. $\theta = -\pi/2$

20. $\theta = 5\pi/6$

21. $r = 6 \sin \theta$

22. $r = \cos \theta$

23–42 ■ Sketch a graph of the polar equation.

23. $r = -2 \cos \theta$

24. $r = 2 \sin \theta + 2 \cos \theta$

25. $r = 2 - 2 \cos \theta$

26. $r = 1 + \sin \theta$

27. $r = -3(1 + \sin \theta)$

28. $r = \cos \theta - 1$

29. $r = \sin 2\theta$

30. $r = 2 \cos 3\theta$

31. $r = -\cos 5\theta$

32. $r = \sin 4\theta$

33. $r = \sqrt{3} - 2 \sin \theta$

34. $r = 2 + \sin \theta$

35. $r = \sqrt{3} + \cos \theta$

36. $r = 1 - 2 \cos \theta$

37. $r^2 = \cos 2\theta$

38. $r^2 = 4 \sin 2\theta$

39. $r = \theta, \quad \theta \geq 0$ (spiral)

40. $r\theta = 1, \quad \theta > 0$ (reciprocal spiral)

41. $r = 2 + \sec \theta$ (conchoid)

42. $r = \sin \theta \tan \theta$ (cissoid)

43–46 ■ Use a graphing device to graph the polar equation. Choose the domain of θ to make sure you produce the entire graph.

43. $r = \cos(\theta/2)$

44. $r = \sin(8\theta/5)$

45. $r = 1 + 2 \sin(\theta/2)$ (nephroid)

46. $r = \sqrt{1 - 0.8 \sin^2\theta}$ (hippopede)

47. Graph the family of polar equations $r = 1 + \sin n\theta$ for $n = 1, 2, 3, 4,$ and 5. How is the number of loops related to n?

48. Graph the family of polar equations $r = 1 + c \sin 2\theta$ for $c = 0.3, 0.6, 1, 1.5,$ and 2. How does the graph change as c increases?

49–52 ■ Match the polar equation with the graphs labeled I–IV. Give reasons for your answers.

49. $r = \sin(\theta/2)$

50. $r = 1/\sqrt{\theta}$

51. $r = \theta \sin \theta$

52. $r = 1 + 3 \cos(3\theta)$

I

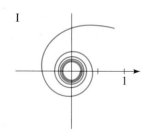

II

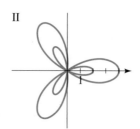

III

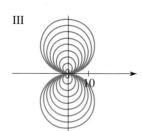

IV

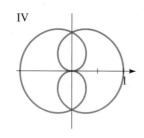

53–56 ■ Sketch a graph of the rectangular equation. [*Hint:* First convert the equation to polar coordinates.]

53. $(x^2 + y^2)^3 = 4x^2y^2$

54. $(x^2 + y^2)^3 = (x^2 - y^2)^2$

55. $(x^2 + y^2)^2 = x^2 - y^2$

56. $x^2 + y^2 = (x^2 + y^2 - x)^2$

57. Show that the graph of $r = a \cos \theta + b \sin \theta$ is a circle, and find its center and radius.

 58. (a) Graph the polar equation $r = \tan \theta \sec \theta$ in the viewing rectangle $[-3, 3]$ by $[-1, 9]$.
   (b) Note that your graph in part (a) looks like a parabola (see Section 1.6). Confirm this by converting the equation to rectangular coordinates.

APPLICATIONS

59. Orbit of a Satellite Scientists and engineers often use polar equations to model the motion of satellites in earth orbit. Let's consider a satellite whose orbit is modeled by the equation $r = 22500/(4 - \cos \theta)$, where r is the distance in miles between the satellite and the center of the earth and θ is the angle shown in the following figure.
   (a) On the same viewing screen, graph the circle $r = 3960$ (to represent the earth, which we will assume to be a sphere of radius 3960 mi) and the polar equation of the satellite's orbit. Describe the motion of the satellite as θ increases from 0 to 2π.

(b) For what angle θ is the satellite closest to the earth? Find the height of the satellite above the earth's surface for this value of θ.

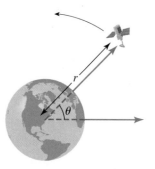

 60. An Unstable Orbit The orbit described in Exercise 59 is stable because the satellite traverses the same path over and over as θ increases. Suppose that a meteor strikes the satellite and changes its orbit to

$$r = \frac{22500\left(1 - \dfrac{\theta}{40}\right)}{4 - \cos \theta}$$

(a) On the same viewing screen, graph the circle $r = 3960$ and the new orbit equation, with θ increasing from 0 to 3π. Describe the new motion of the satellite.
   (b) Use the TRACE feature on your graphing calculator to find the value of θ at the moment the satellite crashes into the earth.

DISCOVERY ■ DISCUSSION ■ WRITING

 61. A Transformation of Polar Graphs How are the graphs of

$$r = 1 + \sin\left(\theta - \frac{\pi}{6}\right)$$

and

$$r = 1 + \sin\left(\theta - \frac{\pi}{3}\right)$$

related to the graph of $r = 1 + \sin \theta$? In general, how is the graph of $r = f(\theta - \alpha)$ related to the graph of $r = f(\theta)$?

62. Choosing a Convenient Coordinate System Compare the polar equation of the circle $r = 2$ with its equation in rectangular coordinates. In which coordinate system is the equation simpler? Do the same for the equation of the four-leaved rose $r = \sin 2\theta$. Which coordinate system would you choose to study these curves?

63. Choosing a Convenient Coordinate System Compare the rectangular equation of the line $y = 2$ with its polar equation. In which coordinate system is the equation simpler? Which coordinate system would you choose to study lines?

5.3 POLAR FORM OF COMPLEX NUMBERS; DE MOIVRE'S THEOREM

| Graphing Complex Numbers ▶ Polar Form of Complex Numbers ▶
De Moivre's Theorem ▶ nth Roots of Complex Numbers

Complex numbers are reviewed in Appendix D.1.

In this section we represent complex numbers in polar (or trigonometric) form. This enables us to find the nth roots of complex numbers. To describe the polar form of complex numbers, we must first learn to work with complex numbers graphically.

▼ Graphing Complex Numbers

To graph real numbers or sets of real numbers, we have been using the number line, which has just one dimension. Complex numbers, however, have two components: a real part and an imaginary part. This suggests that we need two axes to graph complex numbers: one for the real part and one for the imaginary part. We call these the **real axis** and the **imaginary axis**, respectively. The plane determined by these two axes is called the **complex plane**. To graph the complex number $a + bi$, we plot the ordered pair of numbers (a, b) in this plane, as indicated in Figure 1.

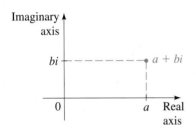

FIGURE 1

EXAMPLE 1 | Graphing Complex Numbers

Graph the complex numbers $z_1 = 2 + 3i$, $z_2 = 3 - 2i$, and $z_1 + z_2$.

SOLUTION We have $z_1 + z_2 = (2 + 3i) + (3 - 2i) = 5 + i$. The graph is shown in Figure 2.

✎ NOW TRY EXERCISE **19** ■

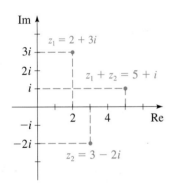

FIGURE 2

EXAMPLE 2 | Graphing Sets of Complex Numbers

Graph each set of complex numbers.

(a) $S = \{a + bi \mid a \geq 0\}$

(b) $T = \{a + bi \mid a < 1, b \geq 0\}$

SOLUTION

(a) S is the set of complex numbers whose real part is nonnegative. The graph is shown in Figure 3(a).

(b) T is the set of complex numbers for which the real part is less than 1 and the imaginary part is nonnegative. The graph is shown in Figure 3(b).

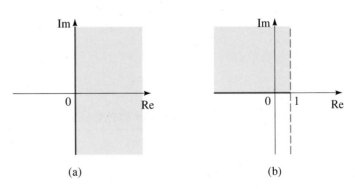

FIGURE 3   (a)   (b)

✎ NOW TRY EXERCISE **21** ■

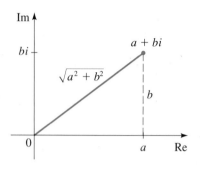

FIGURE 4

Recall that the absolute value of a real number can be thought of as its distance from the origin on the real number line (see Appendix A.1). We define absolute value for complex numbers in a similar fashion. Using the Pythagorean Theorem, we can see from Figure 4 that the distance between $a + bi$ and the origin in the complex plane is $\sqrt{a^2 + b^2}$. This leads to the following definition.

MODULUS OF A COMPLEX NUMBER

The **modulus** (or **absolute value**) of the complex number $z = a + bi$ is

$$|z| = \sqrt{a^2 + b^2}$$

The plural of modulus *is* moduli.

EXAMPLE 3 | Calculating the Modulus

Find the moduli of the complex numbers $3 + 4i$ and $8 - 5i$.

SOLUTION

$$|3 + 4i| = \sqrt{3^2 + 4^2} = \sqrt{25} = 5$$
$$|8 - 5i| = \sqrt{8^2 + (-5)^2} = \sqrt{89}$$

✎ NOW TRY EXERCISE **9**

EXAMPLE 4 | Absolute Value of Complex Numbers

Graph each set of complex numbers.

(a) $C = \{z \mid |z| = 1\}$ **(b)** $D = \{z \mid |z| \leq 1\}$

SOLUTION

(a) C is the set of complex numbers whose distance from the origin is 1. Thus, C is a circle of radius 1 with center at the origin, as shown in Figure 5.

(b) D is the set of complex numbers whose distance from the origin is less than or equal to 1. Thus, D is the disk that consists of all complex numbers on and inside the circle C of part (a), as shown in Figure 6.

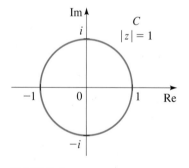

FIGURE 5

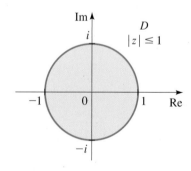

FIGURE 6

✎ NOW TRY EXERCISES **23** AND **25**

FIGURE 7

▼ Polar Form of Complex Numbers

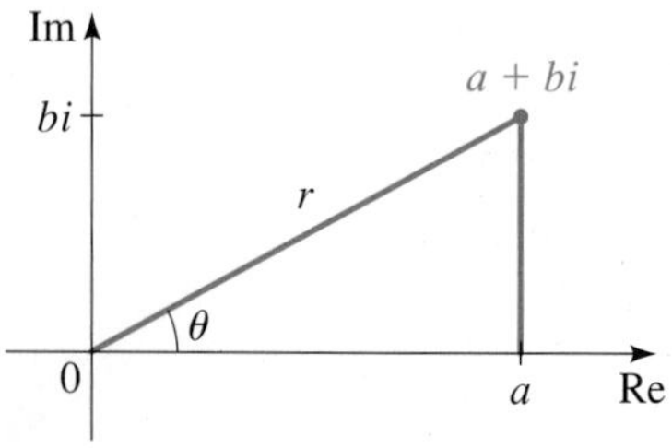

Let $z = a + bi$ be a complex number, and in the complex plane let's draw the line segment joining the origin to the point $a + bi$ (see Figure 7). The length of this line segment is $r = |z| = \sqrt{a^2 + b^2}$. If θ is an angle in standard position whose terminal side coincides with this line segment, then by the definitions of sine and cosine (see Section 3.2)

$$a = r \cos \theta \quad \text{and} \quad b = r \sin \theta$$

so $z = r \cos \theta + ir \sin \theta = r(\cos \theta + i \sin \theta)$. We have shown the following.

POLAR FORM OF COMPLEX NUMBERS

A complex number $z = a + bi$ has the **polar form** (or **trigonometric form**)

$$z = r(\cos \theta + i \sin \theta)$$

where $r = |z| = \sqrt{a^2 + b^2}$ and $\tan \theta = b/a$. The number r is the **modulus** of z, and θ is an **argument** of z.

The argument of z is not unique, but any two arguments of z differ by a multiple of 2π. When determining the argument, we must consider the quadrant in which z lies, as we see in the next example.

EXAMPLE 5 | Writing Complex Numbers in Polar Form

Write each complex number in polar form.

(a) $1 + i$ **(b)** $-1 + \sqrt{3}i$ **(c)** $-4\sqrt{3} - 4i$ **(d)** $3 + 4i$

SOLUTION These complex numbers are graphed in Figure 8, which helps us find their arguments.

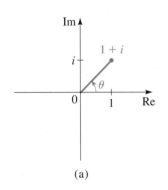

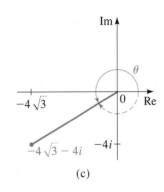

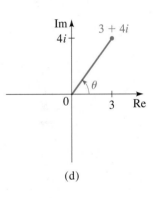

(a)   (b)   (c)   (d)

FIGURE 8

$\tan \theta = \frac{1}{1} = 1$

$\theta = \frac{\pi}{4}$

(a) An argument is $\theta = \pi/4$ and $r = \sqrt{1 + 1} = \sqrt{2}$. Thus

$$1 + i = \sqrt{2}\left(\cos \frac{\pi}{4} + i \sin \frac{\pi}{4}\right)$$

$\tan \theta = \frac{\sqrt{3}}{-1} = -\sqrt{3}$

$\theta = \frac{2\pi}{3}$

(b) An argument is $\theta = 2\pi/3$ and $r = \sqrt{1 + 3} = 2$. Thus

$$-1 + \sqrt{3}i = 2\left(\cos \frac{2\pi}{3} + i \sin \frac{2\pi}{3}\right)$$

$\tan \theta = \frac{-4}{-4\sqrt{3}} = \frac{1}{\sqrt{3}}$

$\theta = \frac{7\pi}{6}$

(c) An argument is $\theta = 7\pi/6$ (or we could use $\theta = -5\pi/6$), and $r = \sqrt{48 + 16} = 8$. Thus

$$-4\sqrt{3} - 4i = 8\left(\cos \frac{7\pi}{6} + i \sin \frac{7\pi}{6}\right)$$

$\tan \theta = \frac{4}{3}$

$\theta = \tan^{-1} \frac{4}{3}$

(d) An argument is $\theta = \tan^{-1} \frac{4}{3}$ and $r = \sqrt{3^2 + 4^2} = 5$. So

$$3 + 4i = 5[\cos(\tan^{-1} \tfrac{4}{3}) + i \sin(\tan^{-1} \tfrac{4}{3})]$$

✎ NOW TRY EXERCISES **29, 31,** AND **33** ■

The Addition Formulas for Sine and Cosine that we discussed in Section 4.2 greatly simplify the multiplication and division of complex numbers in polar form. The following theorem shows how.

MULTIPLICATION AND DIVISION OF COMPLEX NUMBERS

If the two complex numbers z_1 and z_2 have the polar forms

$$z_1 = r_1(\cos\theta_1 + i\sin\theta_1) \qquad \text{and} \qquad z_2 = r_2(\cos\theta_2 + i\sin\theta_2)$$

then

$$z_1 z_2 = r_1 r_2[\cos(\theta_1 + \theta_2) + i\sin(\theta_1 + \theta_2)] \qquad \text{Multiplication}$$

$$\frac{z_1}{z_2} = \frac{r_1}{r_2}[\cos(\theta_1 - \theta_2) + i\sin(\theta_1 - \theta_2)] \qquad (z_2 \neq 0) \qquad \text{Division}$$

This theorem says:

To multiply two complex numbers, multiply the moduli and add the arguments.

To divide two complex numbers, divide the moduli and subtract the arguments.

PROOF To prove the Multiplication Formula, we simply multiply the two complex numbers:

$$z_1 z_2 = r_1 r_2(\cos\theta_1 + i\sin\theta_1)(\cos\theta_2 + i\sin\theta_2)$$

$$= r_1 r_2[\cos\theta_1\cos\theta_2 - \sin\theta_1\sin\theta_2 + i(\sin\theta_1\cos\theta_2 + \cos\theta_1\sin\theta_2)]$$

$$= r_1 r_2[\cos(\theta_1 + \theta_2) + i\sin(\theta_1 + \theta_2)]$$

In the last step we used the Addition Formulas for Sine and Cosine.

The proof of the Division Formula is left as an exercise. ■

EXAMPLE 6 | Multiplying and Dividing Complex Numbers

Let

$$z_1 = 2\left(\cos\frac{\pi}{4} + i\sin\frac{\pi}{4}\right) \qquad \text{and} \qquad z_2 = 5\left(\cos\frac{\pi}{3} + i\sin\frac{\pi}{3}\right)$$

Find **(a)** $z_1 z_2$ and **(b)** z_1/z_2.

SOLUTION

(a) By the Multiplication Formula

$$z_1 z_2 = (2)(5)\left[\cos\left(\frac{\pi}{4} + \frac{\pi}{3}\right) + i\sin\left(\frac{\pi}{4} + \frac{\pi}{3}\right)\right]$$

$$= 10\left(\cos\frac{7\pi}{12} + i\sin\frac{7\pi}{12}\right)$$

To approximate the answer, we use a calculator in radian mode and get

$$z_1 z_2 \approx 10(-0.2588 + 0.9659i)$$

$$= -2.588 + 9.659i$$

(b) By the Division Formula

$$\frac{z_1}{z_2} = \frac{2}{5}\left[\cos\left(\frac{\pi}{4} - \frac{\pi}{3}\right) + i\sin\left(\frac{\pi}{4} - \frac{\pi}{3}\right)\right]$$

$$= \frac{2}{5}\left[\cos\left(-\frac{\pi}{12}\right) + i\sin\left(-\frac{\pi}{12}\right)\right]$$

$$= \frac{2}{5}\left(\cos\frac{\pi}{12} - i\sin\frac{\pi}{12}\right)$$

Using a calculator in radian mode, we get the approximate answer:

$$\frac{z_1}{z_2} \approx \tfrac{2}{5}(0.9659 - 0.2588i) = 0.3864 - 0.1035i$$

✎ NOW TRY EXERCISE **55**

▼ De Moivre's Theorem

Repeated use of the Multiplication Formula gives the following useful formula for raising a complex number to a power n for any positive integer n.

DE MOIVRE'S THEOREM

If $z = r(\cos\theta + i\sin\theta)$, then for any integer n

$$z^n = r^n(\cos n\theta + i\sin n\theta)$$

This theorem says: *To take the nth power of a complex number, we take the nth power of the modulus and multiply the argument by n.*

PROOF   By the Multiplication Formula

$$z^2 = zz = r^2[\cos(\theta + \theta) + i\sin(\theta + \theta)]$$

$$= r^2(\cos 2\theta + i\sin 2\theta)$$

Now we multiply z^2 by z to get

$$z^3 = z^2z = r^3[\cos(2\theta + \theta) + i\sin(2\theta + \theta)]$$

$$= r^3(\cos 3\theta + i\sin 3\theta)$$

Repeating this argument, we see that for any positive integer n

$$z^n = r^n(\cos n\theta + i\sin n\theta)$$

A similar argument using the Division Formula shows that this also holds for negative integers. ◼

EXAMPLE 7 | Finding a Power Using De Moivre's Theorem

Find $\left(\tfrac{1}{2} + \tfrac{1}{2}i\right)^{10}$.

SOLUTION   Since $\tfrac{1}{2} + \tfrac{1}{2}i = \tfrac{1}{2}(1 + i)$, it follows from Example 5(a) that

$$\frac{1}{2} + \frac{1}{2}i = \frac{\sqrt{2}}{2}\left(\cos\frac{\pi}{4} + i\sin\frac{\pi}{4}\right)$$

So by De Moivre's Theorem

$$\left(\frac{1}{2} + \frac{1}{2}i\right)^{10} = \left(\frac{\sqrt{2}}{2}\right)^{10}\left(\cos\frac{10\pi}{4} + i\sin\frac{10\pi}{4}\right)$$

$$= \frac{2^5}{2^{10}}\left(\cos\frac{5\pi}{2} + i\sin\frac{5\pi}{2}\right) = \frac{1}{32}i$$

✎ NOW TRY EXERCISE **69**   ∎

▼ *n*th Roots of Complex Numbers

An ***n*th root** of a complex number z is any complex number w such that $w^n = z$. De Moivre's Theorem gives us a method for calculating the *n*th roots of any complex number.

---

nth ROOTS OF COMPLEX NUMBERS

If $z = r(\cos\theta + i\sin\theta)$ and n is a positive integer, then z has the n distinct *n*th roots

$$w_k = r^{1/n}\left[\cos\left(\frac{\theta + 2k\pi}{n}\right) + i\sin\left(\frac{\theta + 2k\pi}{n}\right)\right]$$

for $k = 0, 1, 2, \ldots, n - 1$.

PROOF   To find the *n*th roots of z, we need to find a complex number w such that

$$w^n = z$$

Let's write z in polar form:

$$z = r(\cos\theta + i\sin\theta)$$

One *n*th root of z is

$$w = r^{1/n}\left(\cos\frac{\theta}{n} + i\sin\frac{\theta}{n}\right)$$

since by De Moivre's Theorem, $w^n = z$. But the argument θ of z can be replaced by $\theta + 2k\pi$ for any integer k. Since this expression gives a different value of w for $k = 0, 1, 2, \ldots, n - 1$, we have proved the formula in the theorem.   ∎

The following observations help us use the preceding formula.

FINDING THE nth ROOTS OF $z = r(\cos\theta + i\sin\theta)$

1. The modulus of each *n*th root is $r^{1/n}$.

2. The argument of the first root is θ/n.

3. We repeatedly add $2\pi/n$ to get the argument of each successive root.

These observations show that, when graphed, the *n*th roots of z are spaced equally on the circle of radius $r^{1/n}$.

EXAMPLE 8 | Finding Roots of a Complex Number

Find the six sixth roots of $z = -64$, and graph these roots in the complex plane.

SOLUTION In polar form, $z = 64(\cos \pi + i \sin \pi)$. Applying the formula for nth roots with $n = 6$, we get

$$w_k = 64^{1/6}\left[\cos\left(\frac{\pi + 2k\pi}{6}\right) + i \sin\left(\frac{\pi + 2k\pi}{6}\right)\right]$$

for $k = 0, 1, 2, 3, 4, 5$. Using $64^{1/6} = 2$, we find that the six sixth roots of -64 are

$$w_0 = 2\left(\cos\frac{\pi}{6} + i \sin\frac{\pi}{6}\right) = \sqrt{3} + i$$

$$w_1 = 2\left(\cos\frac{\pi}{2} + i \sin\frac{\pi}{2}\right) = 2i$$

$$w_2 = 2\left(\cos\frac{5\pi}{6} + i \sin\frac{5\pi}{6}\right) = -\sqrt{3} + i$$

$$w_3 = 2\left(\cos\frac{7\pi}{6} + i \sin\frac{7\pi}{6}\right) = -\sqrt{3} - i$$

$$w_4 = 2\left(\cos\frac{3\pi}{2} + i \sin\frac{3\pi}{2}\right) = -2i$$

$$w_5 = 2\left(\cos\frac{11\pi}{6} + i \sin\frac{11\pi}{6}\right) = \sqrt{3} - i$$

All these points lie on a circle of radius 2, as shown in Figure 9.

✎ NOW TRY EXERCISE **85**

We add $2\pi/6 = \pi/3$ to each argument to get the argument of the next root.

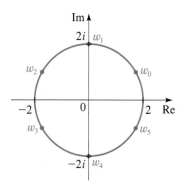

FIGURE 9 The six sixth roots of $z = -64$

When finding roots of complex numbers, we sometimes write the argument θ of the complex number in degrees. In this case the nth roots are obtained from the formula

$$w_k = r^{1/n}\left[\cos\left(\frac{\theta + 360°k}{n}\right) + i \sin\left(\frac{\theta + 360°k}{n}\right)\right]$$

for $k = 0, 1, 2, \ldots, n - 1$.

EXAMPLE 9 | Finding Cube Roots of a Complex Number

Find the three cube roots of $z = 2 + 2i$, and graph these roots in the complex plane.

SOLUTION First we write z in polar form using degrees. We have $r = \sqrt{2^2 + 2^2} = 2\sqrt{2}$ and $\theta = 45°$. Thus

$$z = 2\sqrt{2}(\cos 45° + i \sin 45°)$$

Applying the formula for nth roots (in degrees) with $n = 3$, we find that the cube roots of z are of the form

$$w_k = (2\sqrt{2})^{1/3}\left[\cos\left(\frac{45° + 360°k}{3}\right) + i \sin\left(\frac{45° + 360°k}{3}\right)\right]$$

where $k = 0, 1, 2$. Thus the three cube roots are

$w_0 = \sqrt{2}(\cos 15° + i \sin 15°) \approx 1.366 + 0.366i$ $\qquad (2\sqrt{2})^{1/3} = (2^{3/2})^{1/3} = 2^{1/2} = \sqrt{2}$

$w_1 = \sqrt{2}(\cos 135° + i \sin 135°) = -1 + i$

$w_2 = \sqrt{2}(\cos 255° + i \sin 255°) \approx -0.366 - 1.366i$

The three cube roots of z are graphed in Figure 10. These roots are spaced equally on a circle of radius $\sqrt{2}$.

We add $360°/3 = 120°$ to each argument to get the argument of the next root.

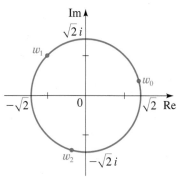

FIGURE 10 The three cube roots of $z = 2 + 2i$

✎ NOW TRY EXERCISE **81**

EXAMPLE 10 | Solving an Equation Using the *n*th Roots Formula

Solve the equation $z^6 + 64 = 0$.

SOLUTION This equation can be written as $z^6 = -64$. Thus the solutions are the sixth roots of -64, which we found in Example 8.

✎ NOW TRY EXERCISE **91** ∎

5.3 EXERCISES

CONCEPTS

1. A complex number $z = a + bi$ has two parts: a is the _____ part, and b is the _____ part. To graph $a + bi$, we graph the ordered pair (⬜ , ⬜) in the complex plane.

2. Let $z = a + bi$.

(a) The modulus of z is $r =$ _____, and an argument of z is an angle θ satisfying $\tan \theta =$ _____.

(b) We can express z in polar form as $z =$ _____, where r is the modulus of z and θ is the argument of z.

3. (a) The complex number $z = -1 + i$ in polar form is

$z =$ _____. The complex number

$z = 2\left(\cos\dfrac{\pi}{6} + i \sin\dfrac{\pi}{6} \right)$ in rectangular form is

$z =$ _____.

(b) The complex number graphed below can be expressed in rectangular form as _____ or in polar form as _____.

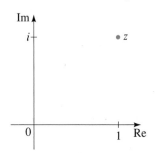

4. How many different *n*th roots does a nonzero complex number

have? _____. The number 16 has _____ fourth roots.

These roots are _____, _____, _____, and

_____. In the complex plane these roots all lie on a circle

of radius _____. Graph the roots on the following graph.

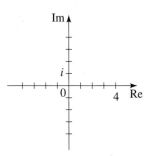

SKILLS

5–14 ■ Graph the complex number and find its modulus.

5. $4i$

6. $-3i$

7. -2

8. 6

9. $5 + 2i$

10. $7 - 3i$

11. $\sqrt{3} + i$

12. $-1 - \dfrac{\sqrt{3}}{3}i$

13. $\dfrac{3 + 4i}{5}$

14. $\dfrac{-\sqrt{2} + i\sqrt{2}}{2}$

15–16 ■ Sketch the complex number z, and also sketch $2z$, $-z$, and $\frac{1}{2}z$ on the same complex plane.

15. $z = 1 + i$

16. $z = -1 + i\sqrt{3}$

17–18 ■ Sketch the complex number z and its complex conjugate $\bar{z}$ on the same complex plane.

17. $z = 8 + 2i$

18. $z = -5 + 6i$

19–20 ■ Sketch z_1, z_2, $z_1 + z_2$, and z_1z_2 on the same complex plane.

19. $z_1 = 2 - i$, $z_2 = 2 + i$

20. $z_1 = -1 + i$, $z_2 = 2 - 3i$

21–28 ■ Sketch the set in the complex plane.

21. $\{z = a + bi \mid a \le 0, b \ge 0\}$

22. $\{z = a + bi \mid a > 1, b > 1\}$

23. $\{z \mid |z| = 3\}$

24. $\{z \mid |z| \ge 1\}$

25. $\{z \mid |z| < 2\}$

26. $\{z \mid 2 \le |z| \le 5\}$

27. $\{z = a + bi \mid a + b < 2\}$

28. $\{z = a + bi \mid a \ge b\}$

29–52 ■ Write the complex number in polar form with argument θ between 0 and 2π.

29. $1 + i$

30. $1 + \sqrt{3}\,i$

31. $\sqrt{2} - \sqrt{2}\,i$

32. $1 - i$

33. $2\sqrt{3} - 2i$

34. $-1 + i$

35. $-3i$

36. $-3 - 3\sqrt{3}\,i$

37. $5 + 5i$

38. 4

39. $4\sqrt{3} - 4i$

40. $8i$

41. -20

42. $\sqrt{3} + i$

43. $3 + 4i$

44. $i(2 - 2i)$

45. $3i(1 + i)$

46. $2(1 - i)$

47. $4(\sqrt{3} + i)$ **48.** $-3 - 3i$ **49.** $2 + i$

50. $3 + \sqrt{3}\,i$ **51.** $\sqrt{2} + \sqrt{2}\,i$ **52.** $-\pi i$

53–60 ■ Find the product $z_1 z_2$ and the quotient z_1/z_2. Express your answer in polar form.

53. $z_1 = \cos \pi + i \sin \pi, \quad z_2 = \cos \dfrac{\pi}{3} + i \sin \dfrac{\pi}{3}$

54. $z_1 = \cos \dfrac{\pi}{4} + i \sin \dfrac{\pi}{4}, \quad z_2 = \cos \dfrac{3\pi}{4} + i \sin \dfrac{3\pi}{4}$

55. $z_1 = 3\left(\cos \dfrac{\pi}{6} + i \sin \dfrac{\pi}{6} \right), \quad z_2 = 5\left(\cos \dfrac{4\pi}{3} + i \sin \dfrac{4\pi}{3} \right)$

56. $z_1 = 7\left(\cos \dfrac{9\pi}{8} + i \sin \dfrac{9\pi}{8} \right), \quad z_2 = 2\left(\cos \dfrac{\pi}{8} + i \sin \dfrac{\pi}{8} \right)$

57. $z_1 = 4(\cos 120° + i \sin 120°),$

     $z_2 = 2(\cos 30° + i \sin 30°)$

58. $z_1 = \sqrt{2}(\cos 75° + i \sin 75°),$

     $z_2 = 3\sqrt{2}(\cos 60° + i \sin 60°)$

59. $z_1 = 4(\cos 200° + i \sin 200°),$

     $z_2 = 25(\cos 150° + i \sin 150°)$

60. $z_1 = \frac{4}{5}(\cos 25° + i \sin 25°),$

     $z_2 = \frac{1}{5}(\cos 155° + i \sin 155°)$

61–68 ■ Write z_1 and z_2 in polar form, and then find the product $z_1 z_2$ and the quotients z_1/z_2 and $1/z_1$.

61. $z_1 = \sqrt{3} + i, \quad z_2 = 1 + \sqrt{3}\,i$

62. $z_1 = \sqrt{2} - \sqrt{2}\,i, \quad z_2 = 1 - i$

63. $z_1 = 2\sqrt{3} - 2i, \quad z_2 = -1 + i$

64. $z_1 = -\sqrt{2}\,i, \quad z_2 = -3 - 3\sqrt{3}\,i$

65. $z_1 = 5 + 5i, \quad z_2 = 4$ **66.** $z_1 = 4\sqrt{3} - 4i, \quad z_2 = 8i$

67. $z_1 = -20, \quad z_2 = \sqrt{3} + i$ **68.** $z_1 = 3 + 4i, \quad z_2 = 2 - 2i$

69–80 ■ Find the indicated power using De Moivre's Theorem.

69. $(1 + i)^{20}$ **70.** $(1 - \sqrt{3}\,i)^5$

71. $(2\sqrt{3} + 2i)^5$ **72.** $(1 - i)^8$

73. $\left(\dfrac{\sqrt{2}}{2} + \dfrac{\sqrt{2}}{2}i \right)^{12}$ **74.** $(\sqrt{3} - i)^{-10}$

75. $(2 - 2i)^8$ **76.** $\left(-\dfrac{1}{2} - \dfrac{\sqrt{3}}{2}i \right)^{15}$

77. $(-1 - i)^7$ **78.** $(3 + \sqrt{3}\,i)^4$

79. $(2\sqrt{3} + 2i)^{-5}$ **80.** $(1 - i)^{-8}$

81–90 ■ Find the indicated roots, and graph the roots in the complex plane.

81. The square roots of $4\sqrt{3} + 4i$

82. The cube roots of $4\sqrt{3} + 4i$

83. The fourth roots of $-81i$

84. The fifth roots of 32

85. The eighth roots of 1

86. The cube roots of $1 + i$

87. The cube roots of i

88. The fifth roots of i

89. The fourth roots of -1

90. The fifth roots of $-16 - 16\sqrt{3}i$

91–96 ■ Solve the equation.

91. $z^4 + 1 = 0$ **92.** $z^8 - i = 0$

93. $z^3 - 4\sqrt{3} - 4i = 0$ **94.** $z^6 - 1 = 0$

95. $z^3 + 1 = -i$ **96.** $z^3 - 1 = 0$

97. (a) Let $w = \cos \dfrac{2\pi}{n} + i \sin \dfrac{2\pi}{n}$ where n is a positive integer. Show that $1, w, w^2, w^3, \ldots, w^{n-1}$ are the n distinct nth roots of 1.

     (b) If $z \neq 0$ is any complex number and $s^n = z$, show that the n distinct nth roots of z are

$$s, \, sw, \, sw^2, \, sw^3, \ldots, \, sw^{n-1}$$

DISCOVERY ■ DISCUSSION ■ WRITING

98. Sums of Roots of Unity   Find the exact values of all three cube roots of 1 (see Exercise 97) and then add them. Do the same for the fourth, fifth, sixth, and eighth roots of 1. What do you think is the sum of the nth roots of 1 for any n?

99. Products of Roots of Unity   Find the product of the three cube roots of 1 (see Exercise 97). Do the same for the fourth, fifth, sixth, and eighth roots of 1. What do you think is the product of the nth roots of 1 for any n?

100. Complex Coefficients and the Quadratic Formula The quadratic formula works whether the coefficients of the equation are real or complex. Solve these equations using the quadratic formula and, if necessary, De Moivre's Theorem.

     (a) $z^2 + (1 + i)z + i = 0$

     (b) $z^2 - iz + 1 = 0$

     (c) $z^2 - (2 - i)z - \frac{1}{4}i = 0$

 **DISCOVERY PROJECT**   **Fractals**

In this project we use graphs of complex numbers to create fractal images. You can find the project at the book companion website: **www.stewartmath.com**

5.4 PLANE CURVES AND PARAMETRIC EQUATIONS

Plane Curves and Parametric Equations ▶ Eliminating the Parameter ▶
Finding Parametric Equations for a Curve ▶ Using Graphing Devices to Graph
Parametric Curves

So far, we have described a curve by giving an equation (in rectangular or polar coordinates) that the coordinates of all the points on the curve must satisfy. But not all curves in the plane can be described in this way. In this section we study parametric equations, which are a general method for describing any curve.

▼ Plane Curves and Parametric Equations

We can think of a curve as the path of a point moving in the plane; the x- and y-coordinates of the point are then functions of time. This idea leads to the following definition.

PLANE CURVES AND PARAMETRIC EQUATIONS

If f and g are functions defined on an interval I, then the set of points $(f(t), g(t))$ is a **plane curve**. The equations

$$x = f(t) \qquad y = g(t)$$

where $t \in I$, are **parametric equations** for the curve, with **parameter** t.

EXAMPLE 1 │ Sketching a Plane Curve

Sketch the curve defined by the parametric equations

$$x = t^2 - 3t \qquad y = t - 1$$

SOLUTION   For every value of t, we get a point on the curve. For example, if $t = 0$, then $x = 0$ and $y = -1$, so the corresponding point is $(0, -1)$. In Figure 1 we plot the points (x, y) determined by the values of t shown in the following table.

t	x	y
-2	10	-3
-1	4	-2
0	0	-1
1	-2	0
2	-2	1
3	0	2
4	4	3
5	10	4

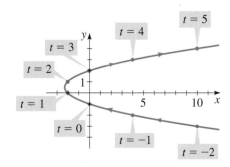

FIGURE 1

As t increases, a particle whose position is given by the parametric equations moves along the curve in the direction of the arrows.

✎ NOW TRY EXERCISE **3**   ■

MARIA GAETANA AGNESI (1718–1799) is famous for having written *Instituzioni Analitiche,* one of the first calculus textbooks.

Maria was born into a wealthy family in Milan, Italy, the oldest of 21 children. She was a child prodigy, mastering many languages at an early age, including Latin, Greek, and Hebrew. At the age of 20 she published a series of essays on philosophy and natural science. After Maria's mother died, she took on the task of educating her brothers. In 1748 Agnesi published her famous textbook, which she originally wrote as a text for tutoring her brothers. The book compiled and explained the mathematical knowledge of the day. It contains many carefully chosen examples, one of which is the curve now known as the "witch of Agnesi" (see Exercise 64, page 301). One review calls her book an "exposition by examples rather than by theory." The book gained Agnesi immediate recognition. Pope Benedict XIV appointed her to a position at the University of Bologna, writing, "we have had the idea that you should be awarded the well-known chair of mathematics, by which it comes of itself that you should not thank us but we you." This appointment was an extremely high honor for a woman, since very few women then were even allowed to attend university. Just two years later, Agnesi's father died, and she left mathematics completely. She became a nun and devoted the rest of her life and her wealth to caring for sick and dying women, herself dying in poverty at a poorhouse of which she had once been director.

If we replace t by $-t$ in Example 1, we obtain the parametric equations

$$x = t^2 + 3t \qquad y = -t - 1$$

The graph of these parametric equations (see Figure 2) is the same as the curve in Figure 1, but traced out in the opposite direction. On the other hand, if we replace t by $2t$ in Example 1, we obtain the parametric equations

$$x = 4t^2 - 6t \qquad y = 2t - 1$$

The graph of these parametric equations (see Figure 3) is again the same, but is traced out "twice as fast." Thus, *a parametrization contains more information than just the shape of the curve; it also indicates* how *the curve is being traced out.*

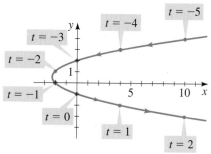

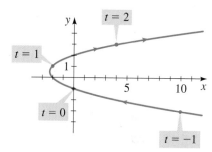

FIGURE 2 $x = t^2 + 3t, y = -t - 1$ **FIGURE 3** $x = 4t^2 - 6t, y = 2t - 1$

▼ Eliminating the Parameter

Often a curve given by parametric equations can also be represented by a single rectangular equation in x and y. The process of finding this equation is called *eliminating the parameter.* One way to do this is to solve for t in one equation, then substitute into the other.

EXAMPLE 2 | Eliminating the Parameter

Eliminate the parameter in the parametric equations of Example 1.

SOLUTION First we solve for t in the simpler equation, then we substitute into the other equation. From the equation $y = t - 1$, we get $t = y + 1$. Substituting into the equation for x, we get

$$x = t^2 - 3t = (y + 1)^2 - 3(y + 1) = y^2 - y - 2$$

Thus the curve in Example 1 has the rectangular equation $x = y^2 - y - 2$, so it is a parabola.

✎ NOW TRY EXERCISE **5** ∎

Eliminating the parameter often helps us identify the shape of a curve, as we see in the next two examples.

EXAMPLE 3 | Modeling Circular Motion

The following parametric equations model the position of a moving object at time t (in seconds):

$$x = \cos t \qquad y = \sin t \qquad t \geq 0$$

Describe and graph the path of the object.

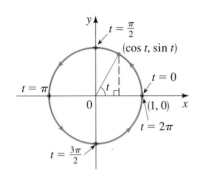

FIGURE 4

SOLUTION To identify the curve, we eliminate the parameter. Since $\cos^2 t + \sin^2 t = 1$ and since $x = \cos t$ and $y = \sin t$ for every point (x, y) on the curve, we have

$$x^2 + y^2 = (\cos t)^2 + (\sin t)^2 = 1$$

This means that all points on the curve satisfy the equation $x^2 + y^2 = 1$, so the graph is a circle of radius 1 centered at the origin. As t increases from 0 to 2π, the point given by the parametric equations starts at $(1, 0)$ and moves counterclockwise once around the circle, as shown in Figure 4. So the object completes one revolution around the circle in 2π seconds. Notice that the parameter t can be interpreted as the angle shown in the figure.

✎ **NOW TRY EXERCISE 25** ◼

EXAMPLE 4 | Sketching a Parametric Curve

Eliminate the parameter, and sketch the graph of the parametric equations

$$x = \sin t \qquad y = 2 - \cos^2 t$$

SOLUTION To eliminate the parameter, we first use the trigonometric identity $\cos^2 t = 1 - \sin^2 t$ to change the second equation:

$$y = 2 - \cos^2 t = 2 - (1 - \sin^2 t) = 1 + \sin^2 t$$

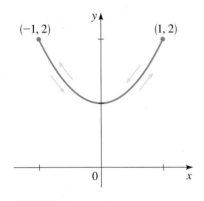

FIGURE 5

Now we can substitute $\sin t = x$ from the first equation to get

$$y = 1 + x^2$$

so the point (x, y) moves along the parabola $y = 1 + x^2$. However, since $-1 \le \sin t \le 1$, we have $-1 \le x \le 1$, so the parametric equations represent only the part of the parabola between $x = -1$ and $x = 1$. Since $\sin t$ is periodic, the point $(x, y) = (\sin t, 2 - \cos^2 t)$ moves back and forth infinitely often along the parabola between the points $(-1, 2)$ and $(1, 2)$, as shown in Figure 5.

✎ **NOW TRY EXERCISE 17** ◼

▼ Finding Parametric Equations for a Curve

It is often possible to find parametric equations for a curve by using some geometric properties that define the curve, as in the next two examples.

EXAMPLE 5 | Finding Parametric Equations for a Graph

Find parametric equations for the line of slope 3 that passes through the point $(2, 6)$.

SOLUTION Let's start at the point $(2, 6)$ and move up and to the right along this line. Because the line has slope 3, for every 1 unit we move to the right, we must move up 3 units. In other words, if we increase the x-coordinate by t units, we must correspondingly increase the y-coordinate by $3t$ units. This leads to the parametric equations

$$x = 2 + t \qquad y = 6 + 3t$$

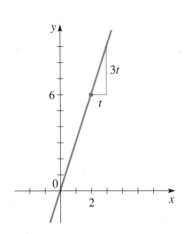

FIGURE 6

To confirm that these equations give the desired line, we eliminate the parameter. We solve for t in the first equation and substitute into the second to get

$$y = 6 + 3(x - 2) = 3x$$

Thus the slope-intercept form of the equation of this line is $y = 3x$, which is a line of slope 3 that does pass through $(2, 6)$ as required. The graph is shown in Figure 6.

✎ **NOW TRY EXERCISE 29** ◼

EXAMPLE 6 | Parametric Equations for the Cycloid

As a circle rolls along a straight line, the curve traced out by a fixed point P on the circumference of the circle is called a **cycloid** (see Figure 7). If the circle has radius a and rolls along the x-axis, with one position of the point P being at the origin, find parametric equations for the cycloid.

FIGURE 7

SOLUTION Figure 8 shows the circle and the point P after the circle has rolled through an angle θ (in radians). The distance $d(O, T)$ that the circle has rolled must be the same as the length of the arc PT, which, by the arc length formula, is $a\theta$ (see Section 3.1). This means that the center of the circle is $C(a\theta, a)$.

Let the coordinates of P be (x, y). Then from Figure 8 (which illustrates the case $0 < \theta < \pi/2$), we see that

$$x = d(O, T) - d(P, Q) = a\theta - a \sin \theta = a(\theta - \sin \theta)$$

$$y = d(T, C) - d(Q, C) = a - a \cos \theta = a(1 - \cos \theta)$$

so parametric equations for the cycloid are

$$x = a(\theta - \sin \theta) \qquad y = a(1 - \cos \theta)$$

✎ NOW TRY EXERCISE **57** ▪

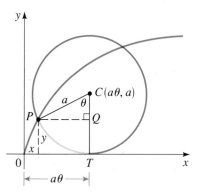

FIGURE 8

The cycloid has a number of interesting physical properties. It is the "curve of quickest descent" in the following sense. Let's choose two points P and Q that are not directly above each other, and join them with a wire. Suppose we allow a bead to slide down the wire under the influence of gravity (ignoring friction). Of all possible shapes into which the wire can be bent, the bead will slide from P to Q the fastest when the shape is half of an arch of an inverted cycloid (see Figure 9). The cycloid is also the "curve of equal descent" in the sense that no matter where we place a bead B on a cycloid-shaped wire, it takes the same time to slide to the bottom (see Figure 10). These rather surprising properties of the cycloid were proved (using calculus) in the 17th century by several mathematicians and physicists, including Johann Bernoulli, Blaise Pascal, and Christiaan Huygens.

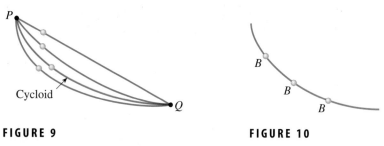

FIGURE 9   **FIGURE 10**

▼ Using Graphing Devices to Graph Parametric Curves

Most graphing calculators and computer graphing programs can be used to graph parametric equations. Such devices are particularly useful in sketching complicated curves like the one shown in Figure 11.

EXAMPLE 7 | Graphing Parametric Curves

Use a graphing device to draw the following parametric curves. Discuss their similarities and differences.

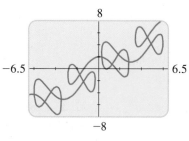

FIGURE 11 $x = t + 2 \sin 2t$, $y = t + 2 \cos 5t$

(a) $x = \sin 2t$
$\quad\; y = 2 \cos t$

(b) $x = \sin 3t$
$\quad\; y = 2 \cos t$

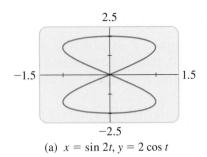

(a) $x = \sin 2t$, $y = 2 \cos t$

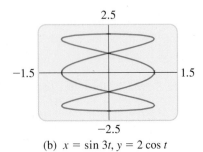

(b) $x = \sin 3t$, $y = 2 \cos t$

FIGURE 12

SOLUTION In both parts (a) and (b) the graph will lie inside the rectangle given by $-1 \le x \le 1$, $-2 \le y \le 2$, since both the sine and the cosine of any number will be between -1 and 1. Thus, we may use the viewing rectangle $[-1.5, 1.5]$ by $[-2.5, 2.5]$.

(a) Since $2 \cos t$ is periodic with period 2π (see Section 2.3) and since $\sin 2t$ has period π, letting t vary over the interval $0 \le t \le 2\pi$ gives us the complete graph, which is shown in Figure 12(a).

(b) Again, letting t take on values between 0 and 2π gives the complete graph shown in Figure 12(b).

Both graphs are *closed curves*, which means they form loops with the same starting and ending point; also, both graphs cross over themselves. However, the graph in Figure 12(a) has two loops, like a figure eight, whereas the graph in Figure 12(b) has three loops.

✎. NOW TRY EXERCISE **43**

The curves graphed in Example 7 are called Lissajous figures. A **Lissajous figure** is the graph of a pair of parametric equations of the form

$$x = A \sin \omega_1 t \qquad y = B \cos \omega_2 t$$

where A, B, ω_1, and ω_2 are real constants. Since $\sin \omega_1 t$ and $\cos \omega_2 t$ are both between -1 and 1, a Lissajous figure will lie inside the rectangle determined by $-A \le x \le A$, $-B \le y \le B$. This fact can be used to choose a viewing rectangle when graphing a Lissajous figure, as in Example 7.

Recall from Section 5.1 that rectangular coordinates (x, y) and polar coordinates (r, θ) are related by the equations $x = r \cos \theta$, $y = r \sin \theta$. Thus we can graph the polar equation $r = f(\theta)$ by changing it to parametric form as follows:

$$x = r \cos \theta = f(\theta) \cos \theta \qquad \text{Since } r = f(\theta)$$

$$y = r \sin \theta = f(\theta) \sin \theta$$

Replacing θ by the standard parametric variable t, we have the following result.

POLAR EQUATIONS IN PARAMETRIC FORM

The graph of the polar equation $r = f(\theta)$ is the same as the graph of the parametric equations

$$x = f(t) \cos t \qquad y = f(t) \sin t$$

EXAMPLE 8 | Parametric Form of a Polar Equation

Consider the polar equation $r = \theta$, $1 \le \theta \le 10\pi$.

(a) Express the equation in parametric form.

(b) Draw a graph of the parametric equations from part (a).

SOLUTION

(a) The given polar equation is equivalent to the parametric equations

$$x = t \cos t \qquad y = t \sin t$$

(b) Since $10\pi \approx 31.42$, we use the viewing rectangle $[-32, 32]$ by $[-32, 32]$, and we let t vary from 1 to 10π. The resulting graph shown in Figure 13 is a *spiral*.

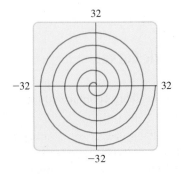

FIGURE 13 $x = t \cos t$, $y = t \sin t$

✎. NOW TRY EXERCISE **51**

5.4 EXERCISES

CONCEPTS

1. (a) The parametric equations $x = f(t)$ and $y = g(t)$ give the coordinates of a point $(x, y) = (f(t), g(t))$ for appropriate values of t. The variable t is called a _____.

(b) Suppose that the parametric equations $x = t$, $y = t^2$, $t \geq 0$, model the position of a moving object at time t. When $t = 0$, the object is at ([ ], [ ]), and when $t = 1$, the object is at ([ ], [ ]).

(c) If we eliminate the parameter in part (b), we get the equation $y =$ _____. We see from this equation that the path of the moving object is a _____.

2. (a) _True or false?_ The same curve can be described by parametric equations in many different ways.

(b) The parametric equations $x = 2t$, $y = (2t)^2$ model the position of a moving object at time t. When $t = 0$, the object is at ([ ], [ ]), and when $t = 1$, the object is at ([ ], [ ]).

(c) If we eliminate the parameter, we get the equation $y =$ _____, which is the same equation as in Exercise 1(b). So the objects in Exercises 1(b) and 2(b) move along the same _____ but traverse the path differently. Indicate the position of each object when $t = 0$ and when $t = 1$ on the following graph.

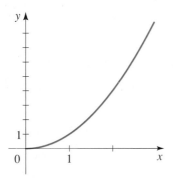

SKILLS

3–24 ■ A pair of parametric equations is given. **(a)** Sketch the curve represented by the parametric equations. **(b)** Find a rectangular-coordinate equation for the curve by eliminating the parameter.

3. $x = 2t$, $y = t + 6$

4. $x = 6t - 4$, $y = 3t$, $t \geq 0$

5. $x = t^2$, $y = t - 2$, $2 \leq t \leq 4$

6. $x = 2t + 1$, $y = \left(t + \frac{1}{2}\right)^2$

7. $x = \sqrt{t}$, $y = 1 - t$

8. $x = t^2$, $y = t^4 + 1$

9. $x = \dfrac{1}{t}$, $y = t + 1$ **10.** $x = t + 1$, $y = \dfrac{t}{t + 1}$

11. $x = 4t^2$, $y = 8t^3$ **12.** $x = |t|$, $y = |1 - |t||$

13. $x = 2 \sin t$, $y = 2 \cos t$, $0 \leq t \leq \pi$

14. $x = 2 \cos t$, $y = 3 \sin t$, $0 \leq t \leq 2\pi$

15. $x = \sin^2 t$, $y = \sin^4 t$ **16.** $x = \sin^2 t$, $y = \cos t$

17. $x = \cos t$, $y = \cos 2t$ **18.** $x = \cos 2t$, $y = \sin 2t$

19. $x = \sec t$, $y = \tan t$, $0 \leq t < \pi/2$

20. $x = \cot t$, $y = \csc t$, $0 < t < \pi$

21. $x = \tan t$, $y = \cot t$, $0 < t < \pi/2$

22. $x = \sec t$, $y = \tan^2 t$, $0 \leq t < \pi/2$

23. $x = \cos^2 t$, $y = \sin^2 t$

24. $x = \cos^3 t$, $y = \sin^3 t$, $0 \leq t \leq 2\pi$

25–28 ■ The position of an object in circular motion is modeled by the given parametric equations. Describe the path of the object by stating the radius of the circle, the position at time $t = 0$, the orientation of the motion (clockwise or counterclockwise), and the time t that it takes to complete one revolution around the circle.

25. $x = 3 \cos t$, $y = 3 \sin t$ **26.** $x = 2 \sin t$, $y = 2 \cos t$

27. $x = \sin 2t$, $y = \cos 2t$ **28.** $x = 4 \cos 3t$, $y = 4 \sin 3t$

29–34 ■ Find parametric equations for the line with the given properties.

29. Slope $\frac{1}{2}$, passing through $(4, -1)$

30. Slope -2, passing through $(-10, -20)$

31. Passing through $(6, 7)$ and $(7, 8)$

32. Passing through $(12, 7)$ and the origin

33. Find parametric equations for the circle $x^2 + y^2 = a^2$.

34. Find parametric equations for the ellipse
$$\frac{x^2}{a^2} + \frac{y^2}{b^2} = 1$$

35. Show by eliminating the parameter θ that the following parametric equations represent a hyperbola:
$$x = a \tan \theta \qquad y = b \sec \theta$$

36. Show that the following parametric equations represent a part of the hyperbola of Exercise 35:
$$x = a\sqrt{t} \qquad y = b\sqrt{t + 1}$$

37–40 ■ Sketch the curve given by the parametric equations.

37. $x = t \cos t$, $y = t \sin t$, $t \geq 0$

38. $x = \sin t$, $y = \sin 2t$

39. $x = \dfrac{3t}{1 + t^3}$, $y = \dfrac{3t^2}{1 + t^3}$

40. $x = \cot t$, $y = 2 \sin^2 t$, $0 < t < \pi$

41. If a projectile is fired with an initial speed of v_0 ft/s at an angle α above the horizontal, then its position after t seconds is given by the parametric equations
$$x = (v_0 \cos \alpha)t \qquad y = (v_0 \sin \alpha)t - 16t^2$$
(where x and y are measured in feet). Show that the path of the projectile is a parabola by eliminating the parameter t.

42. Referring to Exercise 41, suppose a gun fires a bullet into the air with an initial speed of 2048 ft/s at an angle of 30° to the horizontal.

  (a) After how many seconds will the bullet hit the ground?

  (b) How far from the gun will the bullet hit the ground?

  (c) What is the maximum height attained by the bullet?

 43–48 ■ Use a graphing device to draw the curve represented by the parametric equations.

 43. $x = \sin t, \quad y = 2 \cos 3t$

44. $x = 2 \sin t, \quad y = \cos 4t$

45. $x = 3 \sin 5t, \quad y = 5 \cos 3t$

46. $x = \sin 4t, \quad y = \cos 3t$

47. $x = \sin(\cos t), \quad y = \cos(t^{3/2}), \quad 0 \le t \le 2\pi$

48. $x = 2 \cos t + \cos 2t, \quad y = 2 \sin t - \sin 2t$

 49–52 ■ A polar equation is given. **(a)** Express the polar equation in parametric form. **(b)** Use a graphing device to graph the parametric equations you found in part (a).

49. $r = 2^{\theta/12}, \quad 0 \le \theta \le 4\pi$

50. $r = \sin \theta + 2 \cos \theta$

51. $r = \dfrac{4}{2 - \cos \theta}$

52. $r = 2^{\sin \theta}$

53–56 ■ Match the parametric equations with the graphs labeled I–IV. Give reasons for your answers.

53. $x = t^3 - 2t, \quad y = t^2 - t$

54. $x = \sin 3t, \quad y = \sin 4t$

55. $x = t + \sin 2t, \quad y = t + \sin 3t$

56. $x = \sin(t + \sin t), \quad y = \cos(t + \cos t)$

I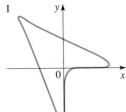

II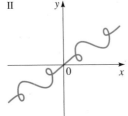

III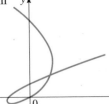

IV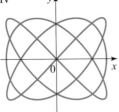

 57. (a) In Example 6 suppose the point P that traces out the curve lies not on the edge of the circle, but rather at a fixed point inside the rim, at a distance b from the center (with $b < a$). The curve traced out by P is called a **curtate cycloid** (or **trochoid**). Show that parametric equations for the curtate cycloid are

$$x = a\theta - b \sin \theta \qquad y = a - b \cos \theta$$

 (b) Sketch the graph using $a = 3$ and $b = 2$.

58. (a) In Exercise 57 if the point P lies *outside* the circle at a distance b from the center (with $b > a$), then the curve traced out by P is called a **prolate cycloid**. Show that parametric equations for the prolate cycloid are the same as the equations for the curtate cycloid.

 (b) Sketch the graph for the case in which $a = 1$ and $b = 2$.

59. A circle C of radius b rolls on the inside of a larger circle of radius a centered at the origin. Let P be a fixed point on the smaller circle, with initial position at the point $(a, 0)$ as shown in the figure. The curve traced out by P is called a **hypocycloid**.

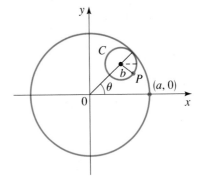

  (a) Show that parametric equations for the hypocycloid are

$$x = (a - b) \cos \theta + b \cos\left(\frac{a - b}{b}\theta\right)$$

$$y = (a - b) \sin \theta - b \sin\left(\frac{a - b}{b}\theta\right)$$

  (b) If $a = 4b$, the hypocycloid is called an **astroid**. Show that in this case the parametric equations can be reduced to

$$x = a \cos^3\theta \qquad y = a \sin^3\theta$$

  Sketch the curve. Eliminate the parameter to obtain an equation for the astroid in rectangular coordinates.

60. If the circle C of Exercise 59 rolls on the *outside* of the larger circle, the curve traced out by P is called an **epicycloid**. Find parametric equations for the epicycloid.

61. In the figure, the circle of radius a is stationary, and for every θ, the point P is the midpoint of the segment QR. The curve traced out by P for $0 < \theta < \pi$ is called the **longbow curve**. Find parametric equations for this curve.

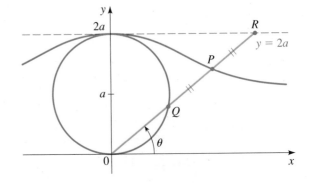

62. Two circles of radius a and b are centered at the origin, as shown in the figure. As the angle θ increases, the point P traces out a curve that lies between the circles.
   (a) Find parametric equations for the curve, using θ as the parameter.
    (b) Graph the curve using a graphing device, with $a = 3$ and $b = 2$.
   (c) Eliminate the parameter, and identify the curve.

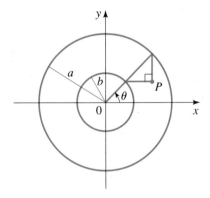

63. Two circles of radius a and b are centered at the origin, as shown in the figure.
   (a) Find parametric equations for the curve traced out by the point P, using the angle θ as the parameter. (Note that the line segment AB is always tangent to the larger circle.)
    (b) Graph the curve using a graphing device, with $a = 3$ and $b = 2$.

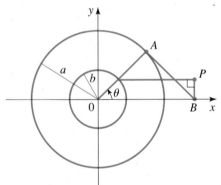

64. A curve, called a **witch of Agnesi**, consists of all points P determined as shown in the figure.
   (a) Show that parametric equations for this curve can be written as
   $$x = 2a \cot \theta \qquad y = 2a \sin^2 \theta$$
    (b) Graph the curve using a graphing device, with $a = 3$.

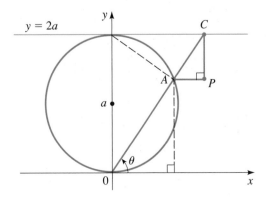

65. Eliminate the parameter θ in the parametric equations for the cycloid (Example 6) to obtain a rectangular coordinate equation for the section of the curve given by $0 \le \theta \le \pi$.

APPLICATIONS

66. **The Rotary Engine** The Mazda RX-8 uses an unconventional engine (invented by Felix Wankel in 1954) in which the pistons are replaced by a triangular rotor that turns in a special housing as shown in the figure. The vertices of the rotor maintain contact with the housing at all times, while the center of the triangle traces out a circle of radius r, turning the drive shaft. The shape of the housing is given by the parametric equations below (where R is the distance between the vertices and center of the rotor):
   $$x = r \cos 3\theta + R \cos \theta \qquad y = r \sin 3\theta + R \sin \theta$$
    (a) Suppose that the drive shaft has radius $r = 1$. Graph the curve given by the parametric equations for the following values of R: 0.5, 1, 3, 5.
   (b) Which of the four values of R given in part (a) seems to best model the engine housing illustrated in the figure?

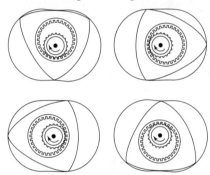

67. **Spiral Path of a Dog** A dog is tied to a circular tree trunk of radius 1 ft by a long leash. He has managed to wrap the entire leash around the tree while playing in the yard, and he finds himself at the point $(1, 0)$ in the figure. Seeing a squirrel, he runs around the tree counterclockwise, keeping the leash taut while chasing the intruder.
   (a) Show that parametric equations for the dog's path (called an **involute of a circle**) are
   $$x = \cos \theta + \theta \sin \theta \qquad y = \sin \theta - \theta \cos \theta$$
   [*Hint:* Note that the leash is always tangent to the tree, so OT is perpendicular to TD.]
    (b) Graph the path of the dog for $0 \le \theta \le 4\pi$.

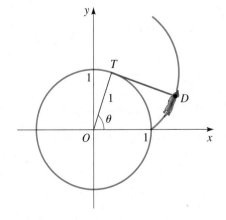

DISCOVERY ■ DISCUSSION ■ WRITING

68. More Information in Parametric Equations In this section we stated that parametric equations contain more information than just the shape of a curve. Write a short paragraph explaining this statement. Use the following example and your answers to parts (a) and (b) below in your explanation.

The position of a particle is given by the parametric equations

$$x = \sin t \qquad y = \cos t$$

where t represents time. We know that the shape of the path of the particle is a circle.

(a) How long does it take the particle to go once around the circle? Find parametric equations if the particle moves twice as fast around the circle.

(b) Does the particle travel clockwise or counterclockwise around the circle? Find parametric equations if the particle moves in the opposite direction around the circle.

69. Different Ways of Tracing Out a Curve The curves C, D, E, and F are defined parametrically as follows, where the parameter t takes on all real values unless otherwise stated:

$$C: \quad x = t, \quad y = t^2$$
$$D: \quad x = \sqrt{t}, \quad y = t, \quad t \geq 0$$
$$E: \quad x = \sin t, \quad y = \sin^2 t$$
$$F: \quad x = 3^t, \quad y = 3^{2t}$$

(a) Show that the points on all four of these curves satisfy the same rectangular coordinate equation.

(b) Draw the graph of each curve and explain how the curves differ from one another.

CHAPTER 5 | REVIEW

■ CONCEPT CHECK

1. Describe how polar coordinates represent the position of a point in the plane.

2. (a) What equations do you use to change from polar to rectangular coordinates?
   (b) What equations do you use to change from rectangular to polar coordinates?

3. How do you sketch the graph of a polar equation $r = f(\theta)$?

4. What type of curve has a polar equation of the given form?
   (a) $r = a \cos \theta$ or $r = a \sin \theta$
   (b) $r = a(1 \pm \cos \theta)$ or $r = a(1 \pm \sin \theta)$
   (c) $r = a \pm b \cos \theta$ or $r = a \pm b \sin \theta$
   (d) $r = a \cos n\theta$ or $r = a \sin n\theta$

5. How do you graph a complex number z? What is the polar form of a complex number z? What is the modulus of z? What is the argument of z?

6. (a) How do you multiply two complex numbers if they are given in polar form?
   (b) How do you divide two such numbers?

7. (a) State De Moivre's Theorem.
   (b) How do you find the nth roots of a complex number?

8. A curve is given by the parametric equations $x = f(t), y = g(t)$.
   (a) How do you sketch the curve?
   (b) How do you eliminate the parameter?

■ EXERCISES

1–6 ■ A point $P(r, \theta)$ is given in polar coordinates.
(a) Plot the point P. (b) Find rectangular coordinates for P.

1. $\left(12, \frac{\pi}{6}\right)$
2. $\left(8, -\frac{3\pi}{4}\right)$
3. $\left(-3, \frac{7\pi}{4}\right)$
4. $\left(-\sqrt{3}, \frac{2\pi}{3}\right)$
5. $\left(4\sqrt{3}, -\frac{5\pi}{3}\right)$
6. $\left(-6\sqrt{2}, -\frac{\pi}{4}\right)$

7–12 ■ A point $P(x, y)$ is given in rectangular coordinates.
(a) Plot the point P. (b) Find polar coordinates for P with $r \geq 0$.
(c) Find polar coordinates for P with $r \leq 0$.

7. $(8, 8)$
8. $(-\sqrt{2}, \sqrt{6})$
9. $(-6\sqrt{2}, -6\sqrt{2})$
10. $(3\sqrt{3}, 3)$
11. $(-3, \sqrt{3})$
12. $(4, -4)$

13–16 ■ (a) Convert the equation to polar coordinates and simplify. (b) Graph the equation.   [*Hint:* Use the form of the equation that you find easier to graph.]

13. $x + y = 4$
14. $xy = 1$
15. $x^2 + y^2 = 4x + 4y$
16. $(x^2 + y^2)^2 = 2xy$

17–24 ■ (a) Sketch the graph of the polar equation.
(b) Express the equation in rectangular coordinates.

17. $r = 3 + 3 \cos \theta$
18. $r = 3 \sin \theta$
19. $r = 2 \sin 2\theta$
20. $r = 4 \cos 3\theta$
21. $r^2 = \sec 2\theta$
22. $r^2 = 4 \sin 2\theta$
23. $r = \sin \theta + \cos \theta$
24. $r = \dfrac{4}{2 + \cos \theta}$

25–28 ■ Use a graphing device to graph the polar equation. Choose the domain of θ to make sure you produce the entire graph.

25. $r = \cos(\theta/3)$

26. $r = \sin(9\theta/4)$

27. $r = 1 + 4\cos(\theta/3)$

28. $r = \theta \sin \theta, \quad -6\pi \le \theta \le 6\pi$

29–34 ■ A complex number is given. **(a)** Graph the complex number in the complex plane. **(b)** Find the modulus and argument. **(c)** Write the number in polar form.

29. $4 + 4i$ **30.** $-10i$

31. $5 + 3i$ **32.** $1 + \sqrt{3}i$

33. $-1 + i$ **34.** -20

35–38 ■ Use De Moivre's Theorem to find the indicated power.

35. $(1 - \sqrt{3}i)^4$ **36.** $(1 + i)^8$

37. $(\sqrt{3} + i)^{-4}$ **38.** $\left(\dfrac{1}{2} + \dfrac{\sqrt{3}}{2}i\right)^{20}$

39–42 ■ Find the indicated roots.

39. The square roots of $-16i$

40. The cube roots of $4 + 4\sqrt{3}i$

41. The sixth roots of 1

42. The eighth roots of i

43–46 ■ A pair of parametric equations is given. **(a)** Sketch the curve represented by the parametric equations. **(b)** Find a rectangular-coordinate equation for the curve by eliminating the parameter.

43. $x = 1 - t^2, \quad y = 1 + t$

44. $x = t^2 - 1, \quad y = t^2 + 1$

45. $x = 1 + \cos t, \quad y = 1 - \sin t, \quad 0 \le t \le \pi/2$

46. $x = \dfrac{1}{t} + 2, \quad y = \dfrac{2}{t^2}, \quad 0 < t \le 2$

47–48 ■ Use a graphing device to draw the parametric curve.

47. $x = \cos 2t, \quad y = \sin 3t$

48. $x = \sin(t + \cos 2t), \quad y = \cos(t + \sin 3t)$

49. In the figure, the point P is the midpoint of the segment QR and $0 \le \theta < \pi/2$. Using θ as the parameter, find a parametric representation for the curve traced out by P.

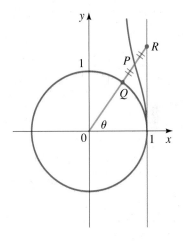

1. (a) Convert the point whose polar coordinates are $(8, 5\pi/4)$ to rectangular coordinates.

   (b) Find two polar coordinate representations for the rectangular coordinate point $(-6, 2\sqrt{3})$, one with $r > 0$ and one with $r < 0$ and both with $0 \le \theta < 2\pi$.

2. (a) Graph the polar equation $r = 8 \cos \theta$. What type of curve is this?

   (b) Convert the equation to rectangular coordinates.

3. Graph the polar equation $r = 3 + 6 \sin \theta$. What type of curve is this?

4. Let $z = 1 + \sqrt{3}i$.

   (a) Graph z in the complex plane.

   (b) Write z in polar form.

   (c) Find the complex number z^9.

5. Let $z_1 = 4\left(\cos \dfrac{7\pi}{12} + i \sin \dfrac{7\pi}{12}\right)$ and $z_2 = 2\left(\cos \dfrac{5\pi}{12} + i \sin \dfrac{5\pi}{12}\right)$.

   Find $z_1 z_2$ and $\dfrac{z_1}{z_2}$.

6. Find the cube roots of $27i$, and sketch these roots in the complex plane.

7. (a) Sketch the graph of the parametric curve

   $$x = 3 \sin t + 3 \qquad y = 2 \cos t \qquad (0 \le t \le \pi)$$

   (b) Eliminate the parameter t in part (a) to obtain an equation for this curve in rectangular coordinates.

8. Find parametric equations for the line of slope 2 that passes through the point $(3, 5)$.

Modeling motion is one of the most important ideas in both classical and modern physics. Much of Isaac Newton's work dealt with creating a mathematical model for how objects move and interact—this was the main reason for his invention of calculus. Albert Einstein developed his Special Theory of Relativity in the early 1900s to refine Newton's laws of motion.

In this section we use coordinate geometry to model the motion of a projectile, such as a ball thrown upward into the air, a bullet fired from a gun, or any other sort of missile. A similar model was created by Galileo, but we have the advantage of using our modern mathematical notation to make describing the model much easier than it was for Galileo!

▼ Parametric Equations for the Path of a Projectile

Suppose that we fire a projectile into the air from ground level, with an initial speed v_0 and at an angle θ upward from the ground. If there were no gravity (and no air resistance), the projectile would just keep moving indefinitely at the same speed and in the same direction. Since distance = speed × time, the projectile would travel a distance $v_0 t$, so its position at time t would therefore be given by the following parametric equations (assuming that the origin of our coordinate system is placed at the initial location of the projectile; see Figure 1):

$$x = (v_0 \cos \theta)t \qquad y = (v_0 \sin \theta)t \qquad \text{No gravity}$$

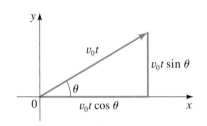

FIGURE 1

But, of course, we know that gravity will pull the projectile back to ground level. By using calculus, it can be shown that the effect of gravity can be accounted for by subtracting $\frac{1}{2}gt^2$ from the vertical position of the projectile. In this expression, g is the gravitational acceleration: $g \approx 32$ ft/s$^2 \approx 9.8$ m/s^2. Thus we have the following parametric equations for the path of the projectile:

$$x = (v_0 \cos \theta)t \qquad y = (v_0 \sin \theta)t - \tfrac{1}{2}gt^2 \qquad \text{With gravity}$$

EXAMPLE | The Path of a Cannonball

Find parametric equations that model the path of a cannonball fired into the air with an initial speed of 150.0 m/s at a 30° angle of elevation. Sketch the path of the cannonball.

SOLUTION Substituting the given initial speed and angle into the general parametric equations of the path of a projectile, we get

$$x = (150.0 \cos 30°)t \qquad y = (150.0 \sin 30°)t - \tfrac{1}{2}(9.8)t^2 \qquad \begin{array}{l}\text{Substitute}\\ v_0 = 150.0,\ \theta = 30°\end{array}$$

$$x = 129.9t \qquad y = 75.0t - 4.9t^2 \qquad \text{Simplify}$$

This path is graphed in Figure 2.

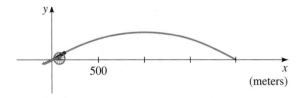

FIGURE 2 Path of a cannonball

305

▼ Range of a Projectile

How can we tell where and when the cannonball of the above example hits the ground? Since ground level corresponds to $y = 0$, we substitute this value for y and solve for t:

$$0 = 75.0t - 4.9t^2 \qquad \text{Set } y = 0$$

$$0 = t(75.0 - 4.9t) \qquad \text{Factor}$$

$$t = 0 \quad \text{or} \quad t = \frac{75.0}{4.9} \approx 15.3 \qquad \text{Solve for } t$$

The first solution, $t = 0$, is the time when the cannon was fired; the second solution means that the cannonball hits the ground after 15.3 s of flight. To see *where* this happens, we substitute this value into the equation for x, the horizontal location of the cannonball.

$$x = 129.9(15.3) \approx 1987.5 \text{ m}$$

The cannonball travels almost 2 km before hitting the ground.

Figure 3 shows the paths of several projectiles, all fired with the same initial speed but at different angles. From the graphs we see that if the firing angle is too high or too low, the projectile doesn't travel very far.

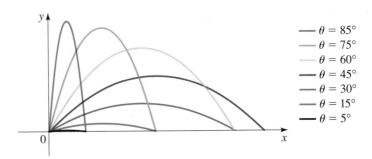

$$\theta = 85°$$
$$\theta = 75°$$
$$\theta = 60°$$
$$\theta = 45°$$
$$\theta = 30°$$
$$\theta = 15°$$
$$\theta = 5°$$

FIGURE 3 Paths of projectiles

Let's try to find the optimal firing angle—the angle that shoots the projectile as far as possible. We'll go through the same steps as we did in the preceding example, but we'll use the general parametric equations instead. First, we solve for the time when the projectile hits the ground by substituting $y = 0$:

$$0 = (v_0 \sin \theta)t - \tfrac{1}{2}gt^2 \qquad \text{Substitute } y = 0$$

$$0 = t(v_0 \sin \theta - \tfrac{1}{2}gt) \qquad \text{Factor}$$

$$0 = v_0 \sin \theta - \tfrac{1}{2}gt \qquad \text{Set second factor equal to 0}$$

$$t = \frac{2v_0 \sin \theta}{g} \qquad \text{Solve for } t$$

GALILEO GALILEI (1564–1642) was born in Pisa, Italy. He studied medicine but later abandoned this in favor of science and mathematics. At the age of 25, by dropping cannonballs of various sizes from the Leaning Tower of Pisa, he demonstrated that light objects fall at the same rate as heavier ones. This contradicted the then-accepted view of Aristotle that heavier objects fall more quickly. He also showed that the distance an object falls is proportional to the square of the time it has been falling, and from this he was able to prove that the path of a projectile is a parabola.

Galileo constructed the first telescope and, using it, discovered the moons of Jupiter. His advocacy of the Copernican view that the earth revolves around the sun (rather than being stationary) led to his being called before the Inquisition. By then an old man, he was forced to recant his views, but he is said to have muttered under his breath, "Nevertheless, it does move." Galileo revolutionized science by expressing scientific principles in the language of mathematics. He said, "The great book of nature is written in mathematical symbols."

Now we substitute this into the equation for x to see how far the projectile has traveled horizontally when it hits the ground:

$$x = (v_0 \cos \theta)t \qquad \text{Parametric equation for } x$$

$$= (v_0 \cos \theta)\left(\frac{2v_0 \sin \theta}{g}\right) \qquad \text{Substitute } t = (2v_0 \sin \theta)/g$$

$$= \frac{2v_0^2 \sin \theta \cos \theta}{g} \qquad \text{Simplify}$$

$$= \frac{v_0^2 \sin 2\theta}{g} \qquad \text{Use identity } \sin 2\theta = 2 \sin \theta \cos \theta$$

We want to choose θ so that x is as large as possible. The largest value that the sine of any angle can have is 1, the sine of 90°. Thus we want $2\theta = 90°$, or $\theta = 45°$. So to send the projectile as far as possible, it should be shot up at an angle of 45°. From the last equation in the preceding display, we can see that it will then travel a distance $x = v_0^2/g$.

PROBLEMS

1. **Trajectories Are Parabolas**   From the graphs in Figure 3 the paths of projectiles appear to be parabolas that open downward. Eliminate the parameter t from the general parametric equations to verify that these are indeed parabolas.

2. **Path of a Baseball**   Suppose a baseball is thrown at 30 ft/s at a 60° angle to the horizontal from a height of 4 ft above the ground.

   (a) Find parametric equations for the path of the baseball, and sketch its graph.

   (b) How far does the baseball travel, and when does it hit the ground?

3. **Path of a Rocket**   Suppose that a rocket is fired at an angle of 5° from the vertical with an initial speed of 1000 ft/s.

   (a) Find the length of time the rocket is in the air.

   (b) Find the greatest height it reaches.

   (c) Find the horizontal distance it has traveled when it hits the ground.

   (d) Graph the rocket's path.

4. **Firing a Missile**   The initial speed of a missile is 330 m/s.

   (a) At what angle should the missile be fired so that it hits a target 10 km away? (You should find that there are two possible angles.) Graph the missile paths for both angles.

   (b) For which angle is the target hit sooner?

5. **Maximum Height**   Show that the maximum height reached by a projectile as a function of its initial speed v_0 and its firing angle θ is

$$y = \frac{v_0^2 \sin^2 \theta}{2g}$$

6. **Shooting Into the Wind**   Suppose that a projectile is fired into a headwind that pushes it back so as to reduce its horizontal speed by a constant amount w. Find parametric equations for the path of the projectile.

7. **Shooting Into the Wind**   Using the parametric equations you derived in Problem 6, draw graphs of the path of a projectile with initial speed $v_0 = 32$ ft/s, fired into a headwind of $w = 24$ ft/s, for the angles $\theta = 5°$, 15°, 30°, 40°, 45°, 55°, 60°, and 75°. Is it still true that the greatest range is attained when firing at 45°? Draw some more graphs for different angles, and use these graphs to estimate the optimal firing angle.

8. **Simulating the Path of a Projectile** The path of a projectile can be simulated on a graphing calculator. On the TI-83, use the "Path" graph style to graph the general parametric equations for the path of a projectile, and watch as the circular cursor moves, simulating the motion of the projectile. Selecting the size of the **Tstep** determines the speed of the "projectile."

(a) Simulate the path of a projectile. Experiment with various values of θ. Use $v_0 = 10$ ft/s and **Tstep** $= 0.02$. Part (a) of the figure below shows one such path.

(b) Simulate the path of two projectiles, fired simultaneously, one at $\theta = 30°$ and the other at $\theta = 60°$. This can be done on the TI-83 using **Simul** mode ("simultaneous" mode). Use $v_0 = 10$ ft/s and **Tstep** $= 0.02$. See part (b) of the figure. Where do the projectiles land? Which lands first?

(c) Simulate the path of a ball thrown straight up ($\theta = 90°$). Experiment with values of v_0 between 5 and 20 ft/s. Use the "Animate" graph style and **Tstep** $= 0.02$. Simulate the path of two balls thrown simultaneously at different speeds. To better distinguish the two balls, place them at different x-coordinates (for example, $x = 1$ and $x = 2$). See part (c) of the figure. How does doubling v_0 change the maximum height the ball reaches?

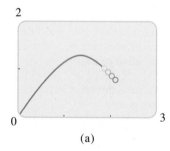

(a)

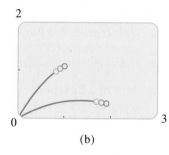

(b)

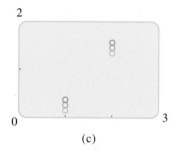

(c)

© James L. Amos/SuperStock

VECTORS IN TWO AND THREE DIMENSIONS

Many real-world quantities are described mathematically by just one number: their "size" or magnitude. For example, quantities such as mass, volume, distance, and temperature are described by their magnitude. But many other real-world quantities involve both magnitude *and* direction. Such quantities are described mathematically by vectors. For example, if you push a car with a certain force, the direction in which you push on the car is important; you get different results if you push the car forward, backward, or perhaps sideways. So force is a vector. The result of several forces acting on an object can be evaluated by using vectors. For example, we'll see how we can combine the vector forces of wind and water on the sails and hull of a sailboat to find the direction in which the boat will sail. Analyzing these vector forces helps sailors to sail against the wind by tacking. (See the Discovery Project *Sailing Against the Wind* referenced on page 327.)

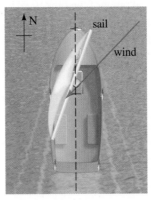

Vector forces

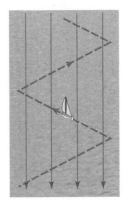

Tacking against the wind

6.1 VECTORS IN TWO DIMENSIONS

Geometric Description of Vectors ▶ Vectors in the Coordinate Plane
▶ Using Vectors to Model Velocity and Force

In applications of mathematics, certain quantities are determined completely by their magnitude—for example, length, mass, area, temperature, and energy. We speak of a length of 5 m or a mass of 3 kg; only one number is needed to describe each of these quantities. Such a quantity is called a **scalar**.

On the other hand, to describe the displacement of an object, two numbers are required: the *magnitude* and the *direction* of the displacement. To describe the velocity of a moving object, we must specify both the *speed* and the *direction* of travel. Quantities such as displacement, velocity, acceleration, and force that involve magnitude as well as direction are called *directed quantities*. One way to represent such quantities mathematically is through the use of *vectors*.

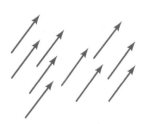

FIGURE 1

$\mathbf{u} = \overrightarrow{AB}$

▼ Geometric Description of Vectors

A **vector** in the plane is a line segment with an assigned direction. We sketch a vector as shown in Figure 1 with an arrow to specify the direction. We denote this vector by $\overrightarrow{AB}$. Point A is the **initial point**, and B is the **terminal point** of the vector $\overrightarrow{AB}$. The length of the line segment AB is called the **magnitude** or **length** of the vector and is denoted by $|\overrightarrow{AB}|$. We use boldface letters to denote vectors. Thus we write $\mathbf{u} = \overrightarrow{AB}$.

Two vectors are considered **equal** if they have equal magnitude and the same direction. Thus all the vectors in Figure 2 are equal. This definition of equality makes sense if we think of a vector as representing a displacement. Two such displacements are the same if they have equal magnitudes and the same direction. So the vectors in Figure 2 can be thought of as the *same* displacement applied to objects in different locations in the plane.

If the displacement $\mathbf{u} = \overrightarrow{AB}$ is followed by the displacement $\mathbf{v} = \overrightarrow{BC}$, then the resulting displacement is $\overrightarrow{AC}$ as shown in Figure 3. In other words, the single displacement represented by the vector $\overrightarrow{AC}$ has the same effect as the other two displacements together. We call the vector $\overrightarrow{AC}$ the **sum** of the vectors $\overrightarrow{AB}$ and $\overrightarrow{BC}$, and we write $\overrightarrow{AC} = \overrightarrow{AB} + \overrightarrow{BC}$. (The **zero vector**, denoted by $\mathbf{0}$, represents no displacement.) Thus to find the sum of any two vectors $\mathbf{u}$ and $\mathbf{v}$, we sketch vectors equal to $\mathbf{u}$ and $\mathbf{v}$ with the initial point of one at the terminal point of the other (see Figure 4(a)). If we draw $\mathbf{u}$ and $\mathbf{v}$ starting at the same point, then $\mathbf{u} + \mathbf{v}$ is the vector that is the diagonal of the parallelogram formed by $\mathbf{u}$ and $\mathbf{v}$ shown in Figure 4(b).

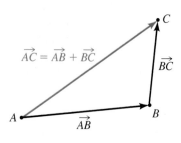

FIGURE 2

$\overrightarrow{AC} = \overrightarrow{AB} + \overrightarrow{BC}$

FIGURE 3

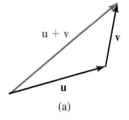

$\mathbf{u} + \mathbf{v}$

$\mathbf{v}$

$\mathbf{u}$

(a)

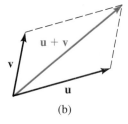

$\mathbf{u} + \mathbf{v}$

$\mathbf{v}$

$\mathbf{u}$

(b)

FIGURE 4 Addition of vectors

If a is a real number and $\mathbf{v}$ is a vector, we define a new vector $a\mathbf{v}$ as follows: The vector $a\mathbf{v}$ has magnitude $|a|\,|\mathbf{v}|$ and has the same direction as $\mathbf{v}$ if $a > 0$ and the opposite direction if $a < 0$. If $a = 0$, then $a\mathbf{v} = \mathbf{0}$, the zero vector. This process is called **multiplication of a vector by a scalar**. Multiplying a vector by a scalar has the effect of stretching or shrinking the vector. Figure 5 shows graphs of the vector $a\mathbf{v}$ for different values of a. We write the vector $(-1)\mathbf{v}$ as $-\mathbf{v}$. Thus $-\mathbf{v}$ is the vector with the same length as $\mathbf{v}$ but with the opposite direction.

The **difference** of two vectors **u** and **v** is defined by $\mathbf{u} - \mathbf{v} = \mathbf{u} + (-\mathbf{v})$. Figure 6 shows that the vector $\mathbf{u} - \mathbf{v}$ is the other diagonal of the parallelogram formed by **u** and **v**.

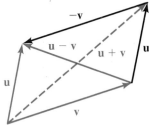

FIGURE 5 Multiplication of a vector by a scalar

FIGURE 6 Subtraction of vectors

▼ Vectors in the Coordinate Plane

So far, we've discussed vectors geometrically. By placing a vector in a coordinate plane, we can describe it analytically (that is, by using components). In Figure 7(a), to go from the initial point of the vector **v** to the terminal point, we move a units to the right and b units upward. We represent **v** as an ordered pair of real numbers.

$$\mathbf{v} = \langle a, b \rangle$$

Note the distinction between the vector *$\langle a, b \rangle$ and the* point *(a, b).*

where a is the **horizontal component** of **v** and b is the **vertical component** of **v**. Remember that a vector represents a magnitude and a direction, not a particular arrow in the plane. Thus the vector $\langle a, b \rangle$ has many different representations, depending on its initial point (see Figure 7(b)).

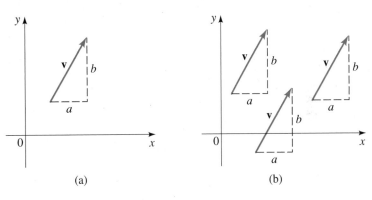

(a)　　　　(b)

FIGURE 7

Using Figure 8, we can state the relationship between a geometric representation of a vector and the analytic one as follows.

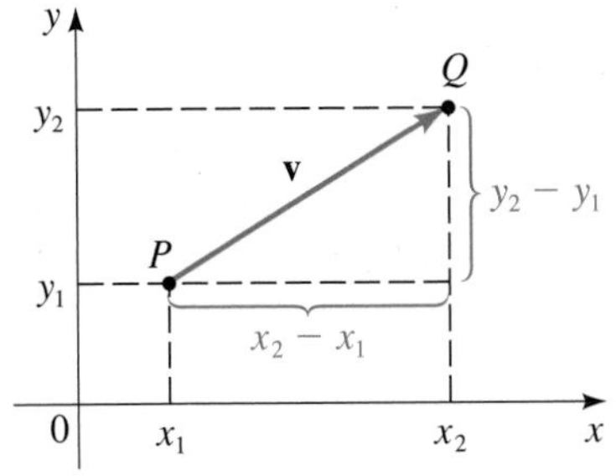

FIGURE 8

COMPONENT FORM OF A VECTOR

If a vector **v** is represented in the plane with initial point $P(x_1, y_1)$ and terminal point $Q(x_2, y_2)$, then

$$\mathbf{v} = \langle x_2 - x_1, y_2 - y_1 \rangle$$

EXAMPLE 1 | Describing Vectors in Component Form

(a) Find the component form of the vector **u** with initial point $(-2, 5)$ and terminal point $(3, 7)$.

(b) If the vector $\mathbf{v} = \langle 3, 7 \rangle$ is sketched with initial point $(2, 4)$, what is its terminal point?

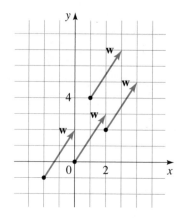

FIGURE 9

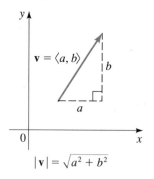

$$|\mathbf{v}| = \sqrt{a^2 + b^2}$$

FIGURE 10

(c) Sketch representations of the vector $\mathbf{w} = \langle 2, 3 \rangle$ with initial points at $(0, 0)$, $(2, 2)$, $(-2, -1)$, and $(1, 4)$.

SOLUTION

(a) The desired vector is

$$\mathbf{u} = \langle 3 - (-2), 7 - 5 \rangle = \langle 5, 2 \rangle$$

(b) Let the terminal point of $\mathbf{v}$ be (x, y). Then

$$\langle x - 2, y - 4 \rangle = \langle 3, 7 \rangle$$

So $x - 2 = 3$ and $y - 4 = 7$, or $x = 5$ and $y = 11$. The terminal point is $(5, 11)$.

(c) Representations of the vector $\mathbf{w}$ are sketched in Figure 9.

✎ NOW TRY EXERCISES **11**, **19**, AND **23** ■

We now give analytic definitions of the various operations on vectors that we have described geometrically. Let's start with equality of vectors. We've said that two vectors are equal if they have equal magnitude and the same direction. For the vectors $\mathbf{u} = \langle a_1, b_1 \rangle$ and $\mathbf{v} = \langle a_2, b_2 \rangle$, this means that $a_1 = a_2$ and $b_1 = b_2$. In other words, two vectors are **equal** if and only if their corresponding components are equal. Thus all the arrows in Figure 7(b) represent the same vector, as do all the arrows in Figure 9.

Applying the Pythagorean Theorem to the triangle in Figure 10, we obtain the following formula for the magnitude of a vector.

MAGNITUDE OF A VECTOR

The **magnitude** or **length** of a vector $\mathbf{v} = \langle a, b \rangle$ is

$$|\mathbf{v}| = \sqrt{a^2 + b^2}$$

EXAMPLE 2 | Magnitudes of Vectors

Find the magnitude of each vector.

(a) $\mathbf{u} = \langle 2, -3 \rangle$ (b) $\mathbf{v} = \langle 5, 0 \rangle$ (c) $\mathbf{w} = \left\langle \frac{3}{5}, \frac{4}{5} \right\rangle$

SOLUTION

(a) $|\mathbf{u}| = \sqrt{2^2 + (-3)^2} = \sqrt{13}$

(b) $|\mathbf{v}| = \sqrt{5^2 + 0^2} = \sqrt{25} = 5$

(c) $|\mathbf{w}| = \sqrt{\left(\frac{3}{5}\right)^2 + \left(\frac{4}{5}\right)^2} = \sqrt{\frac{9}{25} + \frac{16}{25}} = 1$

✎ NOW TRY EXERCISE **37** ■

The following definitions of addition, subtraction, and scalar multiplication of vectors correspond to the geometric descriptions given earlier. Figure 11 shows how the analytic definition of addition corresponds to the geometric one.

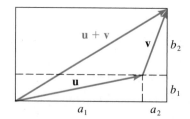

FIGURE 11

ALGEBRAIC OPERATIONS ON VECTORS

If $\mathbf{u} = \langle a_1, b_1 \rangle$ and $\mathbf{v} = \langle a_2, b_2 \rangle$, then

$$\mathbf{u} + \mathbf{v} = \langle a_1 + a_2, b_1 + b_2 \rangle$$

$$\mathbf{u} - \mathbf{v} = \langle a_1 - a_2, b_1 - b_2 \rangle$$

$$c\mathbf{u} = \langle ca_1, cb_1 \rangle, \qquad c \in \mathbb{R}$$

EXAMPLE 3 | Operations with Vectors

If $\mathbf{u} = \langle 2, -3 \rangle$ and $\mathbf{v} = \langle -1, 2 \rangle$, find $\mathbf{u} + \mathbf{v}, \mathbf{u} - \mathbf{v}, 2\mathbf{u}, -3\mathbf{v},$ and $2\mathbf{u} + 3\mathbf{v}$.

SOLUTION   By the definitions of the vector operations we have

$$\mathbf{u} + \mathbf{v} = \langle 2, -3 \rangle - \langle -1, 2 \rangle = \langle 1, -1 \rangle$$

$$\mathbf{u} - \mathbf{v} = \langle 2, -3 \rangle - \langle -1, 2 \rangle = \langle 3, -5 \rangle$$

$$2\mathbf{u} = 2\langle 2, -3 \rangle = \langle 4, -6 \rangle$$

$$-3\mathbf{v} = -3\langle -1, 2 \rangle = \langle 3, -6 \rangle$$

$$2\mathbf{u} + 3\mathbf{v} = 2\langle 2, -3 \rangle + 3\langle -1, 2 \rangle = \langle 4, -6 \rangle + \langle -3, 6 \rangle = \langle 1, 0 \rangle$$

NOW TRY EXERCISE 31

The following properties for vector operations can be easily proved from the definitions. The **zero vector** is the vector $\mathbf{0} = \langle 0, 0 \rangle$. It plays the same role for addition of vectors as the number 0 does for addition of real numbers.

PROPERTIES OF VECTORS

Vector addition	**Multiplication by a scalar**
$\mathbf{u} + \mathbf{v} = \mathbf{v} + \mathbf{u}$	$c(\mathbf{u} + \mathbf{v}) = c\mathbf{u} + c\mathbf{v}$
$\mathbf{u} + (\mathbf{v} + \mathbf{w}) = (\mathbf{u} + \mathbf{v}) + \mathbf{w}$	$(c + d)\mathbf{u} = c\mathbf{u} + d\mathbf{u}$
$\mathbf{u} + \mathbf{0} = \mathbf{u}$	$(cd)\mathbf{u} = c(d\mathbf{u}) = d(c\mathbf{u})$
$\mathbf{u} + (-\mathbf{u}) = \mathbf{0}$	$1\mathbf{u} = \mathbf{u}$
Length of a vector	$0\mathbf{u} = \mathbf{0}$
$|c\mathbf{u}| = |c||\mathbf{u}|$	$c\mathbf{0} = \mathbf{0}$

A vector of length 1 is called a **unit vector**. For instance, in Example 2(c) the vector $\mathbf{w} = \langle \frac{3}{5}, \frac{4}{5} \rangle$ is a unit vector. Two useful unit vectors are $\mathbf{i}$ and $\mathbf{j}$, defined by

$$\mathbf{i} = \langle 1, 0 \rangle \qquad \mathbf{j} = \langle 0, 1 \rangle$$

(See Figure 12.) These vectors are special because any vector can be expressed in terms of them. (See Figure 13.)

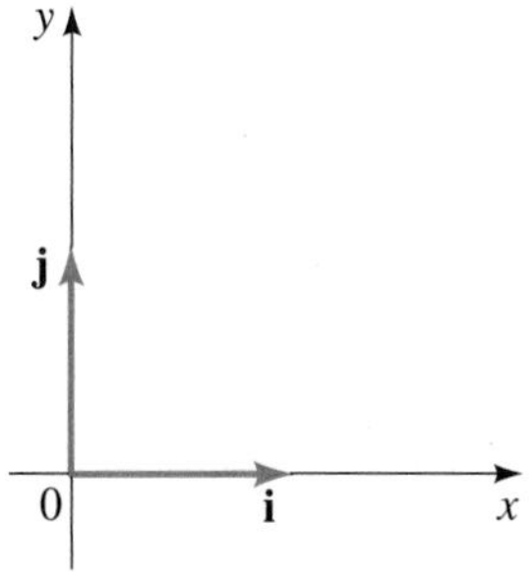

VECTORS IN TERMS OF i AND j

The vector $\mathbf{v} = \langle a, b \rangle$ can be expressed in terms of $\mathbf{i}$ and $\mathbf{j}$ by

$$\mathbf{v} = \langle a, b \rangle = a\mathbf{i} + b\mathbf{j}$$

EXAMPLE 4 | Vectors in Terms of i and j

(a) Write the vector $\mathbf{u} = \langle 5, -8 \rangle$ in terms of $\mathbf{i}$ and $\mathbf{j}$.

(b) If $\mathbf{u} = 3\mathbf{i} + 2\mathbf{j}$ and $\mathbf{v} = -\mathbf{i} + 6\mathbf{j}$, write $2\mathbf{u} + 5\mathbf{v}$ in terms of $\mathbf{i}$ and $\mathbf{j}$.

SOLUTION

(a) $\mathbf{u} = 5\mathbf{i} + (-8)\mathbf{j} = 5\mathbf{i} - 8\mathbf{j}$

FIGURE 12

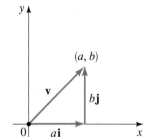

FIGURE 13

(b) The properties of addition and scalar multiplication of vectors show that we can manipulate vectors in the same way as algebraic expressions. Thus

$$2\mathbf{u} + 5\mathbf{v} = 2(3\mathbf{i} + 2\mathbf{j}) + 5(-\mathbf{i} + 6\mathbf{j})$$

$$= (6\mathbf{i} + 4\mathbf{j}) + (-5\mathbf{i} + 30\mathbf{j})$$

$$= \mathbf{i} + 34\mathbf{j}$$

✎ NOW TRY EXERCISES **27** AND **35** ■

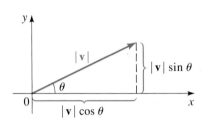

FIGURE 14

Let $\mathbf{v}$ be a vector in the plane with its initial point at the origin. The **direction** of $\mathbf{v}$ is θ, the smallest positive angle in standard position formed by the positive x-axis and $\mathbf{v}$ (see Figure 14). If we know the magnitude and direction of a vector, then Figure 14 shows that we can find the horizontal and vertical components of the vector.

HORIZONTAL AND VERTICAL COMPONENTS OF A VECTOR

Let $\mathbf{v}$ be a vector with magnitude $|\mathbf{v}|$ and direction θ.
Then $\mathbf{v} = \langle a, b \rangle = a\mathbf{i} + b\mathbf{j}$, where

$$a = |\mathbf{v}| \cos \theta \qquad \text{and} \qquad b = |\mathbf{v}| \sin \theta$$

Thus we can express $\mathbf{v}$ as

$$\mathbf{v} = |\mathbf{v}| \cos \theta\, \mathbf{i} + |\mathbf{v}| \sin \theta\, \mathbf{j}$$

EXAMPLE 5 | Components and Direction of a Vector

(a) A vector $\mathbf{v}$ has length 8 and direction $\pi/3$. Find the horizontal and vertical components, and write $\mathbf{v}$ in terms of $\mathbf{i}$ and $\mathbf{j}$.

(b) Find the direction of the vector $\mathbf{u} = -\sqrt{3}\mathbf{i} + \mathbf{j}$.

SOLUTION

(a) We have $\mathbf{v} = \langle a, b \rangle$, where the components are given by

$$a = 8 \cos \frac{\pi}{3} = 4 \qquad \text{and} \qquad b = 8 \sin \frac{\pi}{3} = 4\sqrt{3}$$

Thus $\mathbf{v} = \langle 4, 4\sqrt{3} \rangle = 4\mathbf{i} + 4\sqrt{3}\mathbf{j}$.

(b) From Figure 15 we see that the direction θ has the property that

$$\tan \theta = \frac{1}{-\sqrt{3}} = -\frac{\sqrt{3}}{3}$$

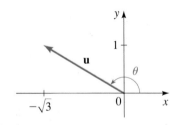

FIGURE 15

Thus the reference angle for θ is $\pi/6$. Since the terminal point of the vector $\mathbf{u}$ is in Quadrant II, it follows that $\theta = 5\pi/6$.

✎ NOW TRY EXERCISES **41** AND **51** ■

▼ Using Vectors to Model Velocity and Force

The **velocity** of a moving object is modeled by a vector whose direction is the direction of motion and whose magnitude is the speed. Figure 16 on the next page shows some vectors $\mathbf{u}$, representing the velocity of wind flowing in the direction N 30° E, and a vector $\mathbf{v}$, representing the velocity of an airplane flying through this wind at the point P. It's obvious from our experience that wind affects both the speed and the direction of an airplane.

The use of bearings (such as N 30° E) to describe directions is explained on page 210 in Section 3.6.

Figure 17 indicates that the true velocity of the plane (relative to the ground) is given by the vector **w** = **u** + **v**.

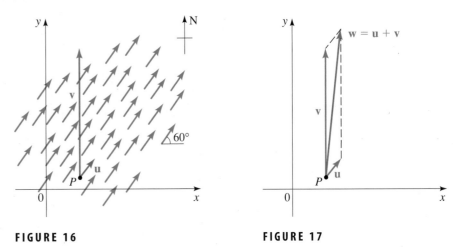

FIGURE 16   **FIGURE 17**

EXAMPLE 6 | The True Speed and Direction of an Airplane

An airplane heads due north at 300 mi/h. It experiences a 40 mi/h crosswind flowing in the direction N 30° E, as shown in Figure 16.

(a) Express the velocity **v** of the airplane relative to the air, and the velocity **u** of the wind, in component form.

(b) Find the true velocity of the airplane as a vector.

(c) Find the true speed and direction of the airplane.

SOLUTION

(a) The velocity of the airplane relative to the air is **v** = 0**i** + 300**j** = 300**j**. By the formulas for the components of a vector, we find that the velocity of the wind is

$$\mathbf{u} = (40 \cos 60°)\mathbf{i} + (40 \sin 60°)\mathbf{j}$$

$$= 20\mathbf{i} + 20\sqrt{3}\mathbf{j}$$

$$\approx 20\mathbf{i} + 34.64\mathbf{j}$$

(b) The true velocity of the airplane is given by the vector **w** = **u** + **v**:

$$\mathbf{w} = \mathbf{u} + \mathbf{v} = (20\mathbf{i} + 20\sqrt{3}\mathbf{j}) + (300\mathbf{j})$$

$$= 20\mathbf{i} + (20\sqrt{3} + 300)\mathbf{j}$$

$$\approx 20\mathbf{i} + 334.64\mathbf{j}$$

(c) The true speed of the airplane is given by the magnitude of **w**:

$$|\mathbf{w}| \approx \sqrt{(20)^2 + (334.64)^2} \approx 335.2 \text{ mi/h}$$

The direction of the airplane is the direction θ of the vector **w**. The angle θ has the property that $\tan \theta \approx 334.64/20 = 16.732$, so $\theta \approx 86.6°$. Thus the airplane is heading in the direction N 3.4° E.

✎ NOW TRY EXERCISE **59**   ■

EXAMPLE 7 | Calculating a Heading

A woman launches a boat from one shore of a straight river and wants to land at the point directly on the opposite shore. If the speed of the boat (relative to the water) is 10 mi/h and the river is flowing east at the rate of 5 mi/h, in what direction should she head the boat in order to arrive at the desired landing point?

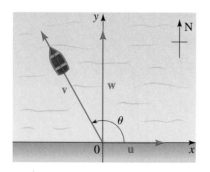

FIGURE 18

SOLUTION We choose a coordinate system with the origin at the initial position of the boat as shown in Figure 18. Let **u** and **v** represent the velocities of the river and the boat, respectively. Clearly, **u** = 5**i**, and since the speed of the boat is 10 mi/h, we have $|\mathbf{v}| = 10$, so

$$\mathbf{v} = (10 \cos \theta)\mathbf{i} + (10 \sin \theta)\mathbf{j}$$

where the angle θ is as shown in Figure 16. The true course of the boat is given by the vector **w** = **u** + **v**. We have

$$\mathbf{w} = \mathbf{u} + \mathbf{v} = 5\mathbf{i} + (10 \cos \theta)\mathbf{i} + (10 \sin \theta)\mathbf{j}$$
$$= (5 + 10 \cos \theta)\mathbf{i} + (10 \sin \theta)\mathbf{j}$$

Since the woman wants to land at a point directly across the river, her direction should have horizontal component 0. In other words, she should choose θ in such a way that

$$5 + 10 \cos \theta = 0$$
$$\cos \theta = -\tfrac{1}{2}$$
$$\theta = 120°$$

Thus she should head the boat in the direction $\theta = 120°$ (or N 30° W).

🖊 NOW TRY EXERCISE **57**

■

Force is also represented by a vector. Intuitively, we can think of force as describing a push or a pull on an object, for example, a horizontal push of a book across a table or the downward pull of the earth's gravity on a ball. Force is measured in pounds (or in newtons, in the metric system). For instance, a man weighing 200 lb exerts a force of 200 lb downward on the ground. If several forces are acting on an object, the **resultant force** experienced by the object is the vector sum of these forces.

EXAMPLE 8 | Resultant Force

Two forces $\mathbf{F}_1$ and $\mathbf{F}_2$ with magnitudes 10 and 20 lb, respectively, act on an object at a point P as shown in Figure 19. Find the resultant force acting at P.

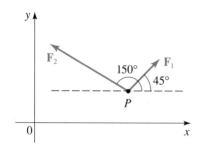

FIGURE 19

SOLUTION We write $\mathbf{F}_1$ and $\mathbf{F}_2$ in component form:

$$\mathbf{F}_1 = (10 \cos 45°)\mathbf{i} + (10 \sin 45°)\mathbf{j} = 10\frac{\sqrt{2}}{2}\mathbf{i} + 10\frac{\sqrt{2}}{2}\mathbf{j}$$
$$= 5\sqrt{2}\,\mathbf{i} + 5\sqrt{2}\,\mathbf{j}$$
$$\mathbf{F}_2 = (20 \cos 150°)\mathbf{i} + (20 \sin 150°)\mathbf{j} = -20\frac{\sqrt{3}}{2}\mathbf{i} + 20\left(\frac{1}{2}\right)\mathbf{j}$$
$$= -10\sqrt{3}\,\mathbf{i} + 10\mathbf{j}$$

So the resultant force **F** is

$$\mathbf{F} = \mathbf{F}_1 + \mathbf{F}_2$$
$$= (5\sqrt{2}\,\mathbf{i} + 5\sqrt{2}\,\mathbf{j}) + (-10\sqrt{3}\,\mathbf{i} + 10\mathbf{j})$$
$$= (5\sqrt{2} - 10\sqrt{3})\mathbf{i} + (5\sqrt{2} + 10)\mathbf{j}$$
$$\approx -10\mathbf{i} + 17\mathbf{j}$$

The resultant force **F** is shown in Figure 20.

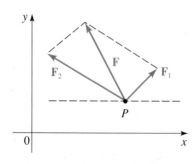

FIGURE 20

🖊 NOW TRY EXERCISE **67**

■

6.1 EXERCISES

CONCEPTS

1. (a) A vector in the plane is a line segment with an assigned direction. In Figure I below, the vector **u** has initial point

_____ and terminal point _____. Sketch the vectors 2**u** and **u** + **v**.

(b) A vector in a coordinate plane is expressed by using components. In Figure II below, the vector **u** has initial point (,) and terminal point (,). In component form we write **u** = ⟨ , ⟩, and **v** = ⟨ , ⟩. Then 2**u** = ⟨ , ⟩ and **u** + **v** = ⟨ , ⟩.

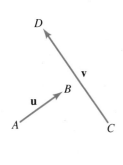
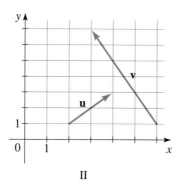

I   II

2. (a) The length of a vector **w** = ⟨a, b⟩ is |**w**| = _____, so the length of the vector **u** in Figure II is

|**u**| = _____.

(b) If we know the length |**w**| and direction θ of a vector **w**, then we can express the vector in component form as

w = ⟨ _____ , _____ ⟩.

SKILLS

3–8 ■ Sketch the vector indicated. (The vectors **u** and **v** are shown in the figure.)

3. 2**u**

4. −**v**

5. **u** + **v**

6. **u** − **v**

7. **v** − 2**u**

8. 2**u** + **v**

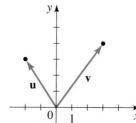

9–18 ■ Express the vector with initial point P and terminal point Q in component form.

9.
10.

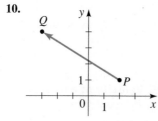

11.
12.

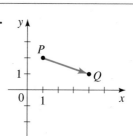
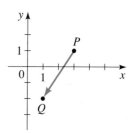

13. P(3, 2), Q(8, 9)
14. P(1, 1), Q(9, 9)

15. P(5, 3), Q(1, 0)
16. P(−1, 3), Q(−6, −1)

17. P(−1, −1), Q(−1, 1)

18. P(−8, −6), Q(−1, −1)

19–22 ■ Sketch the given vector with initial point (4, 3), and find the terminal point.

19. **u** = ⟨2, 4⟩
20. **u** = ⟨−1, 2⟩

21. **u** = ⟨4, −3⟩
22. **u** = ⟨−8, −1⟩

23–26 ■ Sketch representations of the given vector with initial points at (0, 0), (2, 3), and (−3, 5).

23. **u** = ⟨3, 5⟩
24. **u** = ⟨4, −6⟩

25. **u** = ⟨−7, 2⟩
26. **u** = ⟨0, −9⟩

27–30 ■ Write the given vector in terms of **i** and **j**.

27. **u** = ⟨1, 4⟩
28. **u** = ⟨−2, 10⟩

29. **u** = ⟨3, 0⟩
30. **u** = ⟨0, −5⟩

31–36 ■ Find 2**u**, −3**v**, **u** + **v**, and 3**u** − 4**v** for the given vectors **u** and **v**.

31. **u** = ⟨2, 7⟩, **v** = ⟨3, 1⟩

32. **u** = ⟨−2, 5⟩, **v** = ⟨2, −8⟩

33. **u** = ⟨0, −1⟩, **v** = ⟨−2, 0⟩

34. **u** = **i**, **v** = −2**j**

35. **u** = 2**i**, **v** = 3**i** − 2**j**   **36.** **u** = **i** + **j**, **v** = **i** − **j**

37–40 ■ Find |**u**|, |**v**|, |2**u**|, |½**v**|, |**u** + **v**|, |**u** − **v**|, and |**u**| − |**v**|.

37. **u** = 2**i** + **j**, **v** = 3**i** − 2**j**

38. **u** = −2**i** + 3**j**, **v** = **i** − 2**j**

39. **u** = ⟨10, −1⟩, **v** = ⟨−2, −2⟩

40. **u** = ⟨−6, 6⟩, **v** = ⟨−2, −1⟩

41–46 ■ Find the horizontal and vertical components of the vector with given length and direction, and write the vector in terms of the vectors **i** and **j**.

41. |**v**| = 40, θ = 30°
42. |**v**| = 50, θ = 120°

43. |**v**| = 1, θ = 225°
44. |**v**| = 800, θ = 125°

45. |**v**| = 4, θ = 10°
46. |**v**| = √3, θ = 300°

47–52 ■ Find the magnitude and direction (in degrees) of the vector.

47. $\mathbf{v} = \langle 3, 4 \rangle$

48. $\mathbf{v} = \left\langle -\dfrac{\sqrt{2}}{2}, -\dfrac{\sqrt{2}}{2} \right\rangle$

49. $\mathbf{v} = \langle -12, 5 \rangle$

50. $\mathbf{v} = \langle 40, 9 \rangle$

51. $\mathbf{v} = \mathbf{i} + \sqrt{3}\,\mathbf{j}$

52. $\mathbf{v} = \mathbf{i} + \mathbf{j}$

APPLICATIONS

53. Components of a Force A man pushes a lawn mower with a force of 30 lb exerted at an angle of 30° to the ground. Find the horizontal and vertical components of the force.

54. Components of a Velocity A jet is flying in a direction N 20° E with a speed of 500 mi/h. Find the north and east components of the velocity.

55. Velocity A river flows due south at 3 mi/h. A swimmer attempting to cross the river heads due east swimming at 2 mi/h relative to the water. Find the true velocity of the swimmer as a vector.

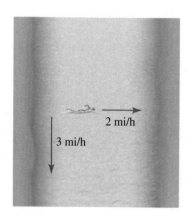

56. Velocity Suppose that in Exercise 55 the current is flowing at 1.2 mi/h due south. In what direction should the swimmer head in order to arrive at a landing point due east of his starting point?

57. Velocity The speed of an airplane is 300 mi/h relative to the air. The wind is blowing due north with a speed of 30 mi/h. In what direction should the airplane head in order to arrive at a point due west of its location?

58. Velocity A migrating salmon heads in the direction N 45° E, swimming at 5 mi/h relative to the water. The prevailing ocean currents flow due east at 3 mi/h. Find the true velocity of the fish as a vector.

59. True Velocity of a Jet A pilot heads his jet due east. The jet has a speed of 425 mi/h relative to the air. The wind is blowing due north with a speed of 40 mi/h.
(a) Express the velocity of the wind as a vector in component form.
(b) Express the velocity of the jet relative to the air as a vector in component form.
(c) Find the true velocity of the jet as a vector.
(d) Find the true speed and direction of the jet.

60. True Velocity of a Jet A jet is flying through a wind that is blowing with a speed of 55 mi/h in the direction N 30° E (see the figure). The jet has a speed of 765 mi/h relative to the air, and the pilot heads the jet in the direction N 45° E.
(a) Express the velocity of the wind as a vector in component form.
(b) Express the velocity of the jet relative to the air as a vector in component form.
(c) Find the true velocity of the jet as a vector.
(d) Find the true speed and direction of the jet.

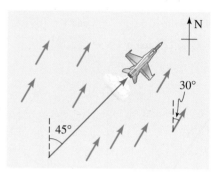

61. True Velocity of a Jet Find the true speed and direction of the jet in Exercise 60 if the pilot heads the plane in the direction N 30° W.

62. True Velocity of a Jet In what direction should the pilot in Exercise 60 head the plane for the true course to be due north?

63. Velocity of a Boat A straight river flows east at a speed of 10 mi/h. A boater starts at the south shore of the river and heads in a direction 60° from the shore (see the figure). The motorboat has a speed of 20 mi/h relative to the water.
(a) Express the velocity of the river as a vector in component form.
(b) Express the velocity of the motorboat relative to the water as a vector in component form.
(c) Find the true velocity of the motorboat.
(d) Find the true speed and direction of the motorboat.

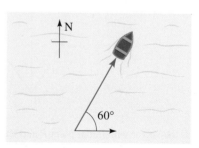

64. Velocity of a Boat The boater in Exercise 63 wants to arrive at a point on the north shore of the river directly opposite the starting point. In what direction should the boat be headed?

65. Velocity of a Boat A boat heads in the direction N 72° E. The speed of the boat relative to the water is 24 mi/h. The water is flowing directly south. It is observed that the true direction of the boat is directly east.

(a) Express the velocity of the boat relative to the water as a vector in component form.

(b) Find the speed of the water and the true speed of the boat.

66. Velocity A woman walks due west on the deck of an ocean liner at 2 mi/h. The ocean liner is moving due north at a speed of 25 mi/h. Find the speed and direction of the woman relative to the surface of the water.

67–72 ■ Equilibrium of Forces The forces $\mathbf{F}_1, \mathbf{F}_2, \ldots, \mathbf{F}_n$ acting at the same point P are said to be in equilibrium if the resultant force is zero, that is, if $\mathbf{F}_1 + \mathbf{F}_2 + \cdots + \mathbf{F}_n = 0$. Find (a) the resultant forces acting at P, and (b) the additional force required (if any) for the forces to be in equilibrium.

67. $\mathbf{F}_1 = \langle 2, 5 \rangle$, $\quad \mathbf{F}_2 = \langle 3, -8 \rangle$

68. $\mathbf{F}_1 = \langle 3, -7 \rangle$, $\quad \mathbf{F}_2 = \langle 4, -2 \rangle$, $\quad \mathbf{F}_3 = \langle -7, 9 \rangle$

69. $\mathbf{F}_1 = 4\mathbf{i} - \mathbf{j}$, $\quad \mathbf{F}_2 = 3\mathbf{i} - 7\mathbf{j}$, $\quad \mathbf{F}_3 = -8\mathbf{i} + 3\mathbf{j}$, $\mathbf{F}_4 = \mathbf{i} + \mathbf{j}$

70. $\mathbf{F}_1 = \mathbf{i} - \mathbf{j}$, $\quad \mathbf{F}_2 = \mathbf{i} + \mathbf{j}$, $\quad \mathbf{F}_3 = -2\mathbf{i} + \mathbf{j}$

71.

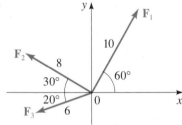

72.

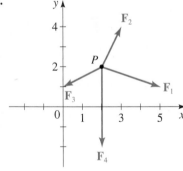

73. Equilibrium of Tensions A 100-lb weight hangs from a string as shown in the figure. Find the tensions $\mathbf{T}_1$ and $\mathbf{T}_2$ in the string.

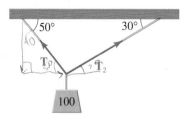

74. Equilibrium of Tensions The cranes in the figure are lifting an object that weighs 18,278 lb. Find the tensions $\mathbf{T}_1$ and $\mathbf{T}_2$.

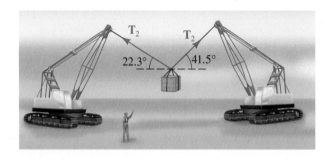

DISCOVERY ▪ DISCUSSION ▪ WRITING

75. Vectors That Form a Polygon Suppose that n vectors can be placed head to tail in the plane so that they form a polygon. (The figure shows the case of a hexagon.) Explain why the sum of these vectors is **0**.

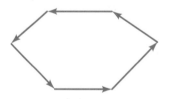

6.2 THE DOT PRODUCT

The Dot Product of Vectors ▶ The Component of **u** Along **v** ▶ The Projection of **u** Onto **v** ▶ Work

In this section we define an operation on vectors called the dot product. This concept is especially useful in calculus and in applications of vectors to physics and engineering.

▼ The Dot Product of Vectors

We begin by defining the dot product of two vectors.

DEFINITION OF THE DOT PRODUCT

If $\mathbf{u} = \langle a_1, b_1 \rangle$ and $\mathbf{v} = \langle a_2, b_2 \rangle$ are vectors, then their **dot product**, denoted by $\mathbf{u} \cdot \mathbf{v}$, is defined by

$$\mathbf{u} \cdot \mathbf{v} = a_1 a_2 + b_1 b_2$$

Thus to find the dot product of $\mathbf{u}$ and $\mathbf{v}$, we multiply corresponding components and add. The dot product is *not* a vector; it is a real number, or scalar.

EXAMPLE 1 | Calculating Dot Products

(a) If $\mathbf{u} = \langle 3, -2 \rangle$ and $\mathbf{v} = \langle 4, 5 \rangle$ then

$$\mathbf{u} \cdot \mathbf{v} = (3)(4) + (-2)(5) = 2$$

(b) If $\mathbf{u} = 2\mathbf{i} + \mathbf{j}$ and $\mathbf{v} = 5\mathbf{i} - 6\mathbf{j}$, then

$$\mathbf{u} \cdot \mathbf{v} = (2)(5) + (1)(-6) = 4$$

✎ NOW TRY EXERCISES **5(a)** AND **11(a)**

The proofs of the following properties of the dot product follow easily from the definition.

PROPERTIES OF THE DOT PRODUCT

1. $\mathbf{u} \cdot \mathbf{v} = \mathbf{v} \cdot \mathbf{u}$

2. $(a\mathbf{u}) \cdot \mathbf{v} = a(\mathbf{u} \cdot \mathbf{v}) = \mathbf{u} \cdot (a\mathbf{v})$

3. $(\mathbf{u} + \mathbf{v}) \cdot \mathbf{w} = \mathbf{u} \cdot \mathbf{w} + \mathbf{v} \cdot \mathbf{w}$

4. $|\mathbf{u}|^2 = \mathbf{u} \cdot \mathbf{u}$

PROOF We prove only the last property. The proofs of the others are left as exercises. Let $\mathbf{u} = \langle a, b \rangle$. Then

$$\mathbf{u} \cdot \mathbf{u} = \langle a, b \rangle \cdot \langle a, b \rangle = a^2 + b^2 = |\mathbf{u}|^2$$

Let $\mathbf{u}$ and $\mathbf{v}$ be vectors, and sketch them with initial points at the origin. We define the **angle θ between $\mathbf{u}$ and $\mathbf{v}$** to be the smaller of the angles formed by these representations of $\mathbf{u}$ and $\mathbf{v}$ (see Figure 1). Thus $0 \le \theta \le \pi$. The next theorem relates the angle between two vectors to their dot product.

THE DOT PRODUCT THEOREM

If θ is the angle between two nonzero vectors $\mathbf{u}$ and $\mathbf{v}$, then

$$\mathbf{u} \cdot \mathbf{v} = |\mathbf{u}||\mathbf{v}| \cos \theta$$

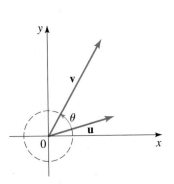

FIGURE 1

PROOF Applying the Law of Cosines to triangle AOB in Figure 2 gives

$$|\mathbf{u} - \mathbf{v}|^2 = |\mathbf{u}|^2 + |\mathbf{v}|^2 - 2|\mathbf{u}||\mathbf{v}| \cos \theta$$

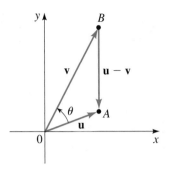

FIGURE 2

Using the properties of the dot product, we write the left-hand side as follows:

$$|\mathbf{u} - \mathbf{v}|^2 = (\mathbf{u} - \mathbf{v}) \cdot (\mathbf{u} - \mathbf{v})$$
$$= \mathbf{u} \cdot \mathbf{u} - \mathbf{u} \cdot \mathbf{v} - \mathbf{v} \cdot \mathbf{u} + \mathbf{v} \cdot \mathbf{v}$$
$$= |\mathbf{u}|^2 - 2(\mathbf{u} \cdot \mathbf{v}) + |\mathbf{v}|^2$$

Equating the right-hand sides of the displayed equations, we get

$$|\mathbf{u}|^2 - 2(\mathbf{u} \cdot \mathbf{v}) + |\mathbf{v}|^2 = |\mathbf{u}|^2 + |\mathbf{v}|^2 - 2|\mathbf{u}||\mathbf{v}|\cos\theta$$
$$-2(\mathbf{u} \cdot \mathbf{v}) = -2|\mathbf{u}||\mathbf{v}|\cos\theta$$
$$\mathbf{u} \cdot \mathbf{v} = |\mathbf{u}||\mathbf{v}|\cos\theta$$

This proves the theorem. ■

The Dot Product Theorem is useful because it allows us to find the angle between two vectors if we know the components of the vectors. The angle is obtained simply by solving the equation in the Dot Product Theorem for $\cos\theta$. We state this important result explicitly.

ANGLE BETWEEN TWO VECTORS

If θ is the angle between two nonzero vectors $\mathbf{u}$ and $\mathbf{v}$, then

$$\cos\theta = \frac{\mathbf{u} \cdot \mathbf{v}}{|\mathbf{u}||\mathbf{v}|}$$

EXAMPLE 2 | Finding the Angle Between Two Vectors

Find the angle between the vectors $\mathbf{u} = \langle 2, 5 \rangle$ and $\mathbf{v} = \langle 4, -3 \rangle$.

SOLUTION By the formula for the angle between two vectors we have

$$\cos\theta = \frac{\mathbf{u} \cdot \mathbf{v}}{|\mathbf{u}||\mathbf{v}|} = \frac{(2)(4) + (5)(-3)}{\sqrt{4 + 25}\sqrt{16 + 9}} = \frac{-7}{5\sqrt{29}}$$

Thus the angle between $\mathbf{u}$ and $\mathbf{v}$ is

$$\theta = \cos^{-1}\left(\frac{-7}{5\sqrt{29}}\right) \approx 105.1°$$

✎ NOW TRY EXERCISES **5(b)** AND **11(b)** ■

Two nonzero vectors $\mathbf{u}$ and $\mathbf{v}$ are called **perpendicular**, or **orthogonal**, if the angle between them is $\pi/2$. The following theorem shows that we can determine whether two vectors are perpendicular by finding their dot product.

ORTHOGONAL VECTORS

Two nonzero vectors $\mathbf{u}$ and $\mathbf{v}$ are perpendicular if and only if $\mathbf{u} \cdot \mathbf{v} = 0$.

PROOF If $\mathbf{u}$ and $\mathbf{v}$ are perpendicular, then the angle between them is $\pi/2$, so

$$\mathbf{u} \cdot \mathbf{v} = |\mathbf{u}||\mathbf{v}|\cos\frac{\pi}{2} = 0$$

Conversely, if $\mathbf{u} \cdot \mathbf{v} = 0$, then

$$|\mathbf{u}||\mathbf{v}|\cos\theta = 0$$

Since $\mathbf{u}$ and $\mathbf{v}$ are nonzero vectors, we conclude that $\cos\theta = 0$, so $\theta = \pi/2$. Thus $\mathbf{u}$ and $\mathbf{v}$ are orthogonal. ■

EXAMPLE 3 | Checking Vectors for Perpendicularity

Determine whether the vectors in each pair are perpendicular.

(a) $\mathbf{u} = \langle 3, 5 \rangle$ and $\mathbf{v} = \langle 2, -8 \rangle$ (b) $\mathbf{u} = \langle 2, 1 \rangle$ and $\mathbf{v} = \langle -1, 2 \rangle$

SOLUTION

(a) $\mathbf{u} \cdot \mathbf{v} = (3)(2) + (5)(-8) = -34 \neq 0$, so $\mathbf{u}$ and $\mathbf{v}$ are not perpendicular.

(b) $\mathbf{u} \cdot \mathbf{v} = (2)(-1) + (1)(2) = 0$, so $\mathbf{u}$ and $\mathbf{v}$ are perpendicular.

✎ NOW TRY EXERCISES **15** AND **17**   ■

▼ The Component of u Along v

The **component of u along v** (or the **component of u in the direction of v**) is defined to be

$$|\mathbf{u}| \cos \theta$$

Note that the component of **u** along **v** is a scalar, not a vector.

where θ is the angle between $\mathbf{u}$ and $\mathbf{v}$. Figure 3 gives a geometric interpretation of this concept. Intuitively, the component of $\mathbf{u}$ along $\mathbf{v}$ is the magnitude of the portion of $\mathbf{u}$ that points in the direction of $\mathbf{v}$. Notice that the component of $\mathbf{u}$ along $\mathbf{v}$ is negative if $\pi/2 < \theta \leq \pi$.

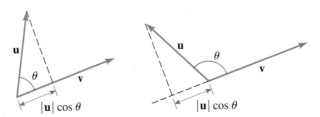

FIGURE 3

In analyzing forces in physics and engineering, it's often helpful to express a vector as a sum of two vectors lying in perpendicular directions. For example, suppose a car is parked on an inclined driveway as in Figure 4. The weight of the car is a vector $\mathbf{w}$ that points directly downward. We can write

$$\mathbf{w} = \mathbf{u} + \mathbf{v}$$

where $\mathbf{u}$ is parallel to the driveway and $\mathbf{v}$ is perpendicular to the driveway. The vector $\mathbf{u}$ is the force that tends to roll the car down the driveway, and $\mathbf{v}$ is the force experienced by the surface of the driveway. The magnitudes of these forces are the components of $\mathbf{w}$ along $\mathbf{u}$ and $\mathbf{v}$, respectively.

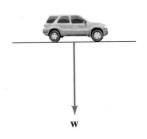

FIGURE 4

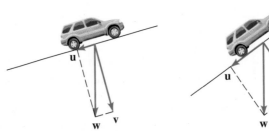

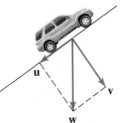

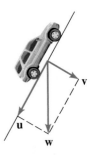

EXAMPLE 4 | Resolving a Force into Components

A car weighing 3000 lb is parked on a driveway that is inclined 15° to the horizontal, as shown in Figure 5.

(a) Find the magnitude of the force required to prevent the car from rolling down the driveway.

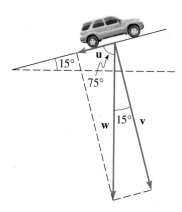

FIGURE 5

(b) Find the magnitude of the force experienced by the driveway due to the weight of the car.

SOLUTION   The car exerts a force **w** of 3000 lb directly downward. We resolve **w** into the sum of two vectors **u** and **v**, one parallel to the surface of the driveway and the other perpendicular to it, as shown in Figure 5.

(a) The magnitude of the part of the force **w** that causes the car to roll down the driveway is

$$|\mathbf{u}| = \text{component of } \mathbf{w} \text{ along } \mathbf{u} = 3000 \cos 75° \approx 776$$

Thus the force needed to prevent the car from rolling down the driveway is about 776 lb.

(b) The magnitude of the force exerted by the car on the driveway is

$$|\mathbf{v}| = \text{component of } \mathbf{w} \text{ along } \mathbf{v} = 3000 \cos 15° \approx 2898$$

The force experienced by the driveway is about 2898 lb.

✎. NOW TRY EXERCISE **49**   ◼

The component of **u** along **v** can be computed by using dot products:

$$|\mathbf{u}| \cos \theta = \frac{|\mathbf{v}||\mathbf{u}| \cos \theta}{|\mathbf{v}|} = \frac{\mathbf{u} \cdot \mathbf{v}}{|\mathbf{v}|}$$

We have shown the following.

CALCULATING COMPONENTS

The component of **u** along **v** is $\dfrac{\mathbf{u} \cdot \mathbf{v}}{|\mathbf{v}|}$.

EXAMPLE 5 | Finding Components

Let $\mathbf{u} = \langle 1, 4 \rangle$ and $\mathbf{v} = \langle -2, 1 \rangle$. Find the component of **u** along **v**.

SOLUTION   We have

$$\text{component of } \mathbf{u} \text{ along } \mathbf{v} = \frac{\mathbf{u} \cdot \mathbf{v}}{|\mathbf{v}|} = \frac{(1)(-2) + (4)(1)}{\sqrt{4 + 1}} = \frac{2}{\sqrt{5}}$$

✎. NOW TRY EXERCISE **25**   ◼

▼ The Projection of u Onto v

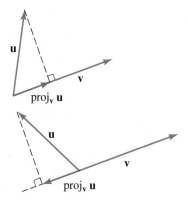

FIGURE 6

Figure 6 shows representations of the vectors **u** and **v**. The projection of **u** onto **v**, denoted by proj$_\mathbf{v}$ **u**, is the vector *parallel* to **v** and whose *length* is the component of **u** along **v** as shown in Figure 6. To find an expression for proj$_\mathbf{v}$ **u**, we first find a unit vector in the direction of **v** and then multiply it by the component of **u** along **v**:

$$\text{proj}_\mathbf{v} \mathbf{u} = (\text{component of } \mathbf{u} \text{ along } \mathbf{v})(\text{unit vector in direction of } \mathbf{v})$$

$$= \left(\frac{\mathbf{u} \cdot \mathbf{v}}{|\mathbf{v}|} \right) \frac{\mathbf{v}}{|\mathbf{v}|} = \left(\frac{\mathbf{u} \cdot \mathbf{v}}{|\mathbf{v}|^2} \right) \mathbf{v}$$

We often need to **resolve** a vector **u** into the sum of two vectors, one parallel to **v** and one orthogonal to **v**. That is, we want to write $\mathbf{u} = \mathbf{u}_1 + \mathbf{u}_2$, where $\mathbf{u}_1$ is parallel to **v** and $\mathbf{u}_2$ is orthogonal to **v**. In this case, $\mathbf{u}_1 = \text{proj}_\mathbf{v} \mathbf{u}$ and $\mathbf{u}_2 = \mathbf{u} - \text{proj}_\mathbf{v} \mathbf{u}$ (see Exercise 43).

CALCULATING PROJECTIONS

The **projection of u onto v** is the vector $\text{proj}_\mathbf{v}\,\mathbf{u}$ given by

$$\text{proj}_\mathbf{v}\,\mathbf{u} = \left(\frac{\mathbf{u} \cdot \mathbf{v}}{|\mathbf{v}|^2} \right)\mathbf{v}$$

If the vector **u** is **resolved** into $\mathbf{u}_1$ and $\mathbf{u}_2$, where $\mathbf{u}_1$ is parallel to **v** and $\mathbf{u}_2$ is orthogonal to **v**, then

$$\mathbf{u}_1 = \text{proj}_\mathbf{v}\,\mathbf{u} \qquad \text{and} \qquad \mathbf{u}_2 = \mathbf{u} - \text{proj}_\mathbf{v}\,\mathbf{u}$$

EXAMPLE 6 | Resolving a Vector into Orthogonal Vectors

Let $\mathbf{u} = \langle -2, 9 \rangle$ and $\mathbf{v} = \langle -1, 2 \rangle$.

(a) Find $\text{proj}_\mathbf{v}\,\mathbf{u}$.

(b) Resolve **u** into $\mathbf{u}_1$ and $\mathbf{u}_2$, where $\mathbf{u}_1$ is parallel to **v** and $\mathbf{u}_2$ is orthogonal to **v**.

SOLUTION

(a) By the formula for the projection of one vector onto another we have

$$\text{proj}_\mathbf{v}\,\mathbf{u} = \left(\frac{\mathbf{u} \cdot \mathbf{v}}{|\mathbf{v}|^2} \right)\mathbf{v} \qquad \text{\small Formula for projection}$$

$$= \left(\frac{\langle -2, 9 \rangle \cdot \langle -1, 2 \rangle}{(-1)^2 + 2^2} \right)\langle -1, 2 \rangle \qquad \text{\small Definition of \textbf{u} and \textbf{v}}$$

$$= 4\langle -1, 2 \rangle = \langle -4, 8 \rangle$$

(b) By the formula in the preceding box we have $\mathbf{u} = \mathbf{u}_1 + \mathbf{u}_2$, where

$$\mathbf{u}_1 = \text{proj}_\mathbf{v}\,\mathbf{u} = \langle -4, 8 \rangle \qquad \text{\small From part (a)}$$

$$\mathbf{u}_2 = \mathbf{u} - \text{proj}_\mathbf{v}\,\mathbf{u} = \langle -2, 9 \rangle - \langle -4, 8 \rangle = \langle 2, 1 \rangle$$

✎ NOW TRY EXERCISE **29**

▼ Work

One use of the dot product occurs in calculating work. In everyday use, the term *work* means the total amount of effort required to perform a task. In physics, *work* has a technical meaning that conforms to this intuitive meaning. If a constant force of magnitude F moves an object through a distance d along a straight line, then the **work** done is

$$W = Fd \qquad \text{or} \qquad \text{work} = \text{force} \times \text{distance}$$

If F is measured in pounds and d in feet, then the unit of work is a foot-pound (ft-lb). For example, how much work is done in lifting a 20-lb weight 6 ft off the ground? Since a force of 20 lb is required to lift this weight and since the weight moves through a distance of 6 ft, the amount of work done is

$$W = Fd = (20)(6) = 120 \text{ ft-lb}$$

This formula applies only when the force is directed along the direction of motion. In the general case, if the force **F** moves an object from P to Q, as in Figure 7, then only the component of the force in the direction of $\mathbf{D} = \overrightarrow{PQ}$ affects the object. Thus the effective magnitude of the force on the object is

$$\text{component of } \mathbf{F} \text{ along } \mathbf{D} = |\mathbf{F}| \cos \theta$$

So the work done is

$$W = \text{force} \times \text{distance} = (|\mathbf{F}| \cos \theta)|\mathbf{D}| = |\mathbf{F}||\mathbf{D}| \cos \theta = \mathbf{F} \cdot \mathbf{D}$$

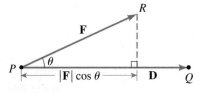

FIGURE 7

We have derived the following simple formula for calculating work.

WORK

The **work** W done by a force $\mathbf{F}$ in moving along a vector $\mathbf{D}$ is

$$W = \mathbf{F} \cdot \mathbf{D}$$

EXAMPLE 7 | Calculating Work

A force is given by the vector $\mathbf{F} = \langle 2, 3 \rangle$ and moves an object from the point $(1, 3)$ to the point $(5, 9)$. Find the work done.

SOLUTION The displacement vector is

$$\mathbf{D} = \langle 5 - 1, 9 - 3 \rangle = \langle 4, 6 \rangle$$

So the work done is

$$W = \mathbf{F} \cdot \mathbf{D} = \langle 2, 3 \rangle \cdot \langle 4, 6 \rangle = 26$$

If the unit of force is pounds and the distance is measured in feet, then the work done is 26 ft-lb.

✎ NOW TRY EXERCISE 35 ■

EXAMPLE 8 | Calculating Work

A man pulls a wagon horizontally by exerting a force of 20 lb on the handle. If the handle makes an angle of 60° with the horizontal, find the work done in moving the wagon 100 ft.

SOLUTION We choose a coordinate system with the origin at the initial position of the wagon (see Figure 8). That is, the wagon moves from the point $P(0, 0)$ to the point $Q(100, 0)$. The vector that represents this displacement is

$$\mathbf{D} = 100\,\mathbf{i}$$

The force on the handle can be written in terms of components (see Section 6.1) as

$$\mathbf{F} = (20 \cos 60°)\mathbf{i} + (20 \sin 60°)\mathbf{j} = 10\,\mathbf{i} + 10\sqrt{3}\,\mathbf{j}$$

Thus the work done is

$$W = \mathbf{F} \cdot \mathbf{D} = (10\,\mathbf{i} + 10\sqrt{3}\,\mathbf{j}) \cdot (100\,\mathbf{i}) = 1000 \text{ ft-lb}$$

✎ NOW TRY EXERCISE 47 ■

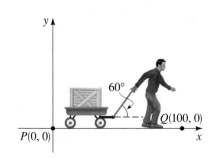

FIGURE 8

6.2 EXERCISES

CONCEPTS

1–2 ■ Let $\mathbf{a} = \langle a_1, a_2 \rangle$ and $\mathbf{b} = \langle b_1, b_2 \rangle$ be nonzero vectors in the plane, and let θ be the angle between them.

1. The dot product of $\mathbf{a}$ and $\mathbf{b}$ is defined by

$$\mathbf{a} \cdot \mathbf{b} = \underline{\hspace{4cm}}$$

The dot product of two vectors is a _____, not a vector.

2. The angle θ satisfies

$$\cos \theta = \frac{\rule{1.2cm}{0pt}}{\rule{1.2cm}{0pt}}$$

So if $\mathbf{a} \cdot \mathbf{b} = 0$, the vectors are _____.

3. (a) The component of **a** along **b** is the scalar $|\mathbf{a}|\cos\theta$ and can be expressed in terms of the dot product as

_____. Sketch this component in the figure below.

(b) The projection of **a** onto **b** is the vector

$\text{proj}_{\mathbf{b}}\,\mathbf{a} = $ _____. Sketch this projection in the figure below.

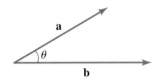

4. The work done by a force **F** in moving an object along a vector **D** is $W = $ _____.

SKILLS

5–14 ■ Find **(a)** $\mathbf{u}\cdot\mathbf{v}$ and **(b)** the angle between **u** and **v** to the nearest degree.

5. $\mathbf{u} = \langle 2, 0\rangle$, $\mathbf{v} = \langle 1, 1\rangle$

6. $\mathbf{u} = \mathbf{i} + \sqrt{3}\mathbf{j}$, $\mathbf{v} = -\sqrt{3}\mathbf{i} + \mathbf{j}$

7. $\mathbf{u} = \langle 2, 7\rangle$, $\mathbf{v} = \langle 3, 1\rangle$

8. $\mathbf{u} = \langle -6, 6\rangle$, $\mathbf{v} = \langle 1, -1\rangle$

9. $\mathbf{u} = \langle 3, -2\rangle$, $\mathbf{v} = \langle 1, 2\rangle$

10. $\mathbf{u} = 2\mathbf{i} + \mathbf{j}$, $\mathbf{v} = 3\mathbf{i} - 2\mathbf{j}$

11. $\mathbf{u} = -5\mathbf{j}$, $\mathbf{v} = -\mathbf{i} - \sqrt{3}\mathbf{j}$

12. $\mathbf{u} = \mathbf{i} + \mathbf{j}$, $\mathbf{v} = \mathbf{i} - \mathbf{j}$

13. $\mathbf{u} = \mathbf{i} + 3\mathbf{j}$, $\mathbf{v} = 4\mathbf{i} - \mathbf{j}$

14. $\mathbf{u} = 3\mathbf{i} + 4\mathbf{j}$, $\mathbf{v} = -2\mathbf{i} - \mathbf{j}$

15–20 ■ Determine whether the given vectors are perpendicular.

15. $\mathbf{u} = \langle 6, 4\rangle$, $\mathbf{v} = \langle -2, 3\rangle$ **16.** $\mathbf{u} = \langle 0, -5\rangle$, $\mathbf{v} = \langle 4, 0\rangle$

17. $\mathbf{u} = \langle -2, 6\rangle$, $\mathbf{v} = \langle 4, 2\rangle$ **18.** $\mathbf{u} = 2\mathbf{i}$, $\mathbf{v} = -7\mathbf{j}$

19. $\mathbf{u} = 2\mathbf{i} - 8\mathbf{j}$, $\mathbf{v} = -12\mathbf{i} - 3\mathbf{j}$

20. $\mathbf{u} = 4\mathbf{i}$, $\mathbf{v} = -\mathbf{i} + 3\mathbf{j}$

21–24 ■ Find the indicated quantity, assuming $\mathbf{u} = 2\mathbf{i} + \mathbf{j}$, $\mathbf{v} = \mathbf{i} - 3\mathbf{j}$, and $\mathbf{w} = 3\mathbf{i} + 4\mathbf{j}$.

21. $\mathbf{u}\cdot\mathbf{v} + \mathbf{u}\cdot\mathbf{w}$ **22.** $\mathbf{u}\cdot(\mathbf{v} + \mathbf{w})$

23. $(\mathbf{u} + \mathbf{v})\cdot(\mathbf{u} - \mathbf{v})$ **24.** $(\mathbf{u}\cdot\mathbf{v})(\mathbf{u}\cdot\mathbf{w})$

25–28 ■ Find the component of **u** along **v**.

25. $\mathbf{u} = \langle 4, 6\rangle$, $\mathbf{v} = \langle 3, -4\rangle$

26. $\mathbf{u} = \langle -3, 5\rangle$, $\mathbf{v} = \langle 1/\sqrt{2}, 1/\sqrt{2}\rangle$

27. $\mathbf{u} = 7\mathbf{i} - 24\mathbf{j}$, $\mathbf{v} = \mathbf{j}$

28. $\mathbf{u} = 7\mathbf{i}$, $\mathbf{v} = 8\mathbf{i} + 6\mathbf{j}$

29–34 ■ **(a)** Calculate $\text{proj}_{\mathbf{v}}\,\mathbf{u}$. **(b)** Resolve **u** into $\mathbf{u}_1$ and $\mathbf{u}_2$, where $\mathbf{u}_1$ is parallel to **v** and $\mathbf{u}_2$ is orthogonal to **v**.

29. $\mathbf{u} = \langle -2, 4\rangle$, $\mathbf{v} = \langle 1, 1\rangle$

30. $\mathbf{u} = \langle 7, -4\rangle$, $\mathbf{v} = \langle 2, 1\rangle$

31. $\mathbf{u} = \langle 1, 2\rangle$, $\mathbf{v} = \langle 1, -3\rangle$

32. $\mathbf{u} = \langle 11, 3\rangle$, $\mathbf{v} = \langle -3, -2\rangle$

33. $\mathbf{u} = \langle 2, 9\rangle$, $\mathbf{v} = \langle -3, 4\rangle$

34. $\mathbf{u} = \langle 1, 1\rangle$, $\mathbf{v} = \langle 2, -1\rangle$

35–38 ■ Find the work done by the force **F** in moving an object from P to Q.

35. $\mathbf{F} = 4\mathbf{i} - 5\mathbf{j}$; $P(0, 0)$, $Q(3, 8)$

36. $\mathbf{F} = 400\mathbf{i} + 50\mathbf{j}$; $P(-1, 1)$, $Q(200, 1)$

37. $\mathbf{F} = 10\mathbf{i} + 3\mathbf{j}$; $P(2, 3)$, $Q(6, -2)$

38. $\mathbf{F} = -4\mathbf{i} + 20\mathbf{j}$; $P(0, 10)$, $Q(5, 25)$

39–42 ■ Let **u**, **v**, and **w** be vectors, and let a be a scalar. Prove the given property.

39. $\mathbf{u}\cdot\mathbf{v} = \mathbf{v}\cdot\mathbf{u}$

40. $(a\mathbf{u})\cdot\mathbf{v} = a(\mathbf{u}\cdot\mathbf{v}) = \mathbf{u}\cdot(a\mathbf{v})$

41. $(\mathbf{u} + \mathbf{v})\cdot\mathbf{w} = \mathbf{u}\cdot\mathbf{w} + \mathbf{v}\cdot\mathbf{w}$

42. $(\mathbf{u} - \mathbf{v})\cdot(\mathbf{u} + \mathbf{v}) = |\mathbf{u}|^2 - |\mathbf{v}|^2$

43. Show that the vectors $\text{proj}_{\mathbf{v}}\,\mathbf{u}$ and $\mathbf{u} - \text{proj}_{\mathbf{v}}\,\mathbf{u}$ are orthogonal.

44. Evaluate $\mathbf{v}\cdot\text{proj}_{\mathbf{v}}\,\mathbf{u}$.

APPLICATIONS

45. Work   The force $\mathbf{F} = 4\mathbf{i} - 7\mathbf{j}$ moves an object 4 ft along the x-axis in the positive direction. Find the work done if the unit of force is the pound.

46. Work   A constant force $\mathbf{F} = \langle 2, 8\rangle$ moves an object along a straight line from the point $(2, 5)$ to the point $(11, 13)$. Find the work done if the distance is measured in feet and the force is measured in pounds.

47. Work   A lawn mower is pushed a distance of 200 ft along a horizontal path by a constant force of 50 lb. The handle of the lawn mower is held at an angle of 30° from the horizontal (see the figure). Find the work done.

48. Work   A car drives 500 ft on a road that is inclined 12° to the horizontal, as shown in the following figure. The car weighs 2500 lb. Thus gravity acts straight down on the car with a constant force $\mathbf{F} = -2500\mathbf{j}$. Find the work done by the car in overcoming gravity.

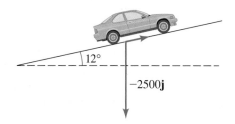

49. Force A car is on a driveway that is inclined 25° to the horizontal. If the car weighs 2755 lb, find the force required to keep it from rolling down the driveway.

50. Force A car is on a driveway that is inclined 10° to the horizontal. A force of 490 lb is required to keep the car from rolling down the driveway.
   (a) Find the weight of the car.
   (b) Find the force the car exerts against the driveway.

51. Force A package that weighs 200 lb is placed on an inclined plane. If a force of 80 lb is just sufficient to keep the package from sliding, find the angle of inclination of the plane. (Ignore the effects of friction.)

52. Force A cart weighing 40 lb is placed on a ramp inclined at 15° to the horizontal. The cart is held in place by a rope inclined at 60° to the horizontal, as shown in the figure. Find the force that the rope must exert on the cart to keep it from rolling down the ramp.

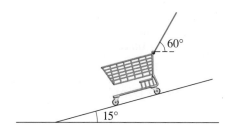

DISCOVERY ▪ DISCUSSION ▪ WRITING

53. Distance from a Point to a Line Let L be the line $2x + 4y = 8$ and let P be the point $(3, 4)$.
   (a) Show that the points $Q(0, 2)$ and $R(2, 1)$ lie on L.
   (b) Let $\mathbf{u} = \overrightarrow{QP}$ and $\mathbf{v} = \overrightarrow{QR}$, as shown in the figure. Find $\mathbf{w} = \text{proj}_{\mathbf{v}}\, \mathbf{u}$.
   (c) Sketch a graph that explains why $|\mathbf{u} - \mathbf{w}|$ is the distance from P to L. Find this distance.
   (d) Write a short paragraph describing the steps you would take to find the distance from a given point to a given line.

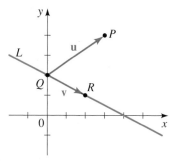

DISCOVERY PROJECT **Sailing Against the Wind**

In this project we study how sailors use the method of taking a zigzag path, or *tacking*, to sail against the wind. You can find the project at the book companion website:
www.stewartmath.com

6.3 THREE-DIMENSIONAL COORDINATE GEOMETRY

| The Three-Dimensional Rectangular Coordinate System ▶ Distance Formula in Three Dimensions ▶ The Equation of a Sphere

To locate a point in a plane, two numbers are necessary. We know that any point in the Cartesian plane can be represented as an ordered pair (a, b) of real numbers, where a is the x-coordinate and b is the y-coordinate. In three-dimensional space, a third dimension is added, so any point in space is represented by an ordered triple (a, b, c) of real numbers.

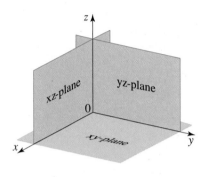

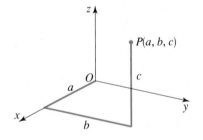

▼ The Three-Dimensional Rectangular Coordinate System

To represent points in space, we first choose a fixed point O (the origin) and three directed lines through O that are perpendicular to each other, called the **coordinate axes** and labeled the x-axis, y-axis, and z-axis. Usually we think of the x- and y-axes as being horizontal and the z-axis as being vertical, and we draw the orientation of the axes as in Figure 1.

The three coordinate axes determine the three **coordinate planes** illustrated in Figure 2(a). The xy-plane is the plane that contains the x- and y-axes; the yz-plane is the plane that contains the y- and z-axes; the xz-plane is the plane that contains the x- and z-axes. These three coordinate planes divide space into eight parts, called **octants**.

FIGURE 1 Coordinate axes

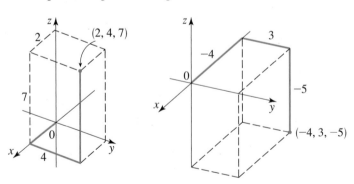

FIGURE 2   (a) Coordinate planes   (b) Coordinate "walls"

Because people often have difficulty visualizing diagrams of three-dimensional figures, you may find it helpful to do the following (see Figure 2(b)). Look at any bottom corner of a room and call the corner the origin. The wall on your left is in the xz-plane, the wall on your right is in the yz-plane, and the floor is in the xy-plane. The x-axis runs along the intersection of the floor and the left wall; the y-axis runs along the intersection of the floor and the right wall. The z-axis runs up from the floor toward the ceiling along the intersection of the two walls.

Now any point P in space can be located by a unique **ordered triple** of real numbers (a, b, c), as shown in Figure 3. The first number a is the x-coordinate of P, the second number b is the y-coordinate of P, and the third number c is the z-coordinate of P. The set of all ordered triples $\{(x, y, z) \mid x, y, z \in \mathbb{R}\}$ forms the **three-dimensional rectangular coordinate system**.

FIGURE 3   Point $P(a, b, c)$

EXAMPLE 1 | Plotting Points in Three Dimensions

Plot the points $(2, 4, 7)$ and $(-4, 3, -5)$.

SOLUTION   The points are plotted in Figure 4.

FIGURE 4

✎ NOW TRY EXERCISE **3(a)** ■

In two-dimensional geometry the graph of an equation involving x and y is a *curve* in the plane. In three-dimensional geometry an equation in x, y, and z represents a *surface* in space.

EXAMPLE 2 | Surfaces in Three-Dimensional Space

Describe and sketch the surfaces represented by the following equations:
(a) $z = 3$ **(b)** $y = 5$

SOLUTION

(a) The surface consists of the points $P(x, y, z)$ where the z-coordinate is 3. This is the horizontal plane that is parallel to the xy-plane and three units above it, as in Figure 5.

(b) The surface consists of the points $P(x, y, z)$ where the y-coordinate is 5. This is the vertical plane that is parallel to the xz-plane and five units to the right of it, as in Figure 6.

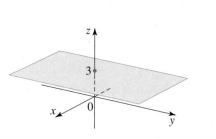

FIGURE 5 The plane $z = 3$

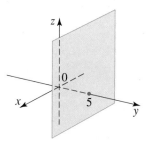

FIGURE 6 The plane $y = 5$

✎. NOW TRY EXERCISE **7**

▼ Distance Formula in Three Dimensions

The familiar formula for the distance between two points in a plane is easily extended to the following three-dimensional formula.

DISTANCE FORMULA IN THREE DIMENSIONS

The distance between the points $P(x_1, y_1, z_1)$ and $Q(x_2, y_2, z_2)$ is

$$d(P, Q) = \sqrt{(x_2 - x_1)^2 + (y_2 - y_1)^2 + (z_2 - z_1)^2}$$

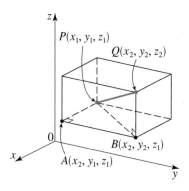

FIGURE 7

PROOF   To prove this formula, we construct a rectangular box as in Figure 7, where $P(x_1, y_1, z_1)$ and $Q(x_2, y_2, z_2)$ are diagonally opposite vertices and the faces of the box are parallel to the coordinate planes. If A and B are the vertices of the box that are indicated in the figure, then

$$d(P, A) = |x_2 - x_1| \qquad d(A, B) = |y_2 - y_1| \qquad d(Q, B) = |z_2 - z_1|$$

Triangles PAB and PBQ are right triangles, so by the Pythagorean Theorem we have

$$(d(P, Q))^2 = (d(P, B))^2 + (d(Q, B))^2$$
$$(d(P, B))^2 = (d(P, A))^2 + (d(A, B))^2$$

Combining these equations, we get

$$(d(P, Q))^2 = (d(P, A))^2 + (d(A, B))^2 + (d(Q, B))^2$$
$$= |x_2 - x_1|^2 + |y_2 - y_1|^2 + |z_2 - z_1|^2$$

Therefore

$$d(P, Q) = \sqrt{(x_2 - x_1)^2 + (y_2 - y_1)^2 + (z_2 - z_1)^2}$$

EXAMPLE 3 | Using the Distance Formula

Find the distance between the points $P(2, -1, 7)$ and $Q(1, -3, 5)$.

SOLUTION We use the Distance Formula:

$$d(P, Q) = \sqrt{(1 - 2)^2 + (-3 - (-1))^2 + (5 - 7)^2} = \sqrt{1 + 4 + 4} = 3$$

✎ NOW TRY EXERCISE **3(b)** ■

▼ The Equation of a Sphere

We can use the Distance Formula to find an equation for a sphere in a three-dimensional coordinate space.

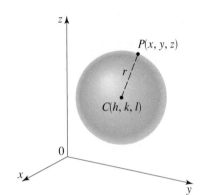

FIGURE 8 Sphere with radius r and center $C(h, k, l)$

> **EQUATION OF A SPHERE**
>
> An equation of a sphere with center $C(h, k, l)$ and radius r is
> $$(x - h)^2 + (y - k)^2 + (z - l)^2 = r^2$$

PROOF A sphere with radius r is the set of all points $P(x, y, z)$ whose distance from the center C is the constant r (see Figure 8). By the Distance Formula we have

$$[d(P, C)]^2 = (x - h)^2 + (y - k)^2 + (z - l)^2$$

Since the distance $d(P, C)$ is equal to r, we get the desired formula. ■

EXAMPLE 4 | Finding the Equation of a Sphere

Find an equation of a sphere with radius 5 and center $C(-2, 1, 3)$.

SOLUTION We use the general equation of a sphere, with $r = 5$, $h = -2$, $k = 1$, and $l = 3$:

$$(x + 2)^2 + (y - 1)^2 + (z - 3)^2 = 25$$

✎ NOW TRY EXERCISE **11** ■

EXAMPLE 5 | Finding the Center and Radius of a Sphere

Show that $x^2 + y^2 + z^2 + 4x - 6y + 2z + 6 = 0$ is the equation of a sphere, and find its center and radius.

SOLUTION We complete the squares in the x-, y-, and z-terms to rewrite the given equation in the form of an equation of a sphere:

$$x^2 + y^2 + z^2 + 4x - 6y + 2z + 6 = 0 \qquad \text{Given equation}$$
$$(x^2 + 4x + 4) + (y^2 - 6y + 9) + (z^2 + 2z + 1) = -6 + 4 + 9 + 1 \qquad \text{Complete squares}$$
$$(x + 2)^2 + (y - 3)^2 + (z + 1)^2 = 8 \qquad \text{Factor into squares}$$

Comparing this with the standard equation of a sphere, we can see that the center is $(-2, 3, -1)$ and the radius is $\sqrt{8} = 2\sqrt{2}$.

✎ NOW TRY EXERCISE **15** ■

The intersection of a sphere with a plane is called the **trace** of the sphere in a plane.

EXAMPLE 6 | Finding the Trace of a Sphere

Describe the trace of the sphere $(x - 2)^2 + (y - 4)^2 + (z - 5)^2 = 36$ in (a) the xy-plane and (b) the plane $z = 9$.

SOLUTION

(a) In the xy-plane the z-coordinate is 0. So the trace of the sphere in the xy-plane consists of all the points on the sphere whose z-coordinate is 0. We replace z by 0 in the equation of the sphere and get

$$(x - 2)^2 + (y - 4)^2 + (0 - 5)^2 = 36 \qquad \text{Replace } z \text{ by } 0$$
$$(x - 2)^2 + (y - 4)^2 + 25 = 36 \qquad \text{Calculate}$$
$$(x - 2)^2 + (y - 4)^2 = 11 \qquad \text{Subtract } 25$$

Thus the trace of the sphere is the circle

$$(x - 2)^2 + (y - 4)^2 = 11, \qquad z = 0$$

which is a circle of radius $\sqrt{11}$ that is in the xy-plane, centered at $(2, 4, 0)$ (see Figure 9(a)).

(b) The trace of the sphere in the plane $z = 9$ consists of all the points on the sphere whose z-coordinate is 9. So we replace z by 9 in the equation of the sphere and get

$$(x - 2)^2 + (y - 4)^2 + (9 - 5)^2 = 36 \qquad \text{Replace } z \text{ by } 0$$
$$(x - 2)^2 + (y - 4)^2 + 16 = 36 \qquad \text{Calculate}$$
$$(x - 2)^2 + (y - 4)^2 = 20 \qquad \text{Subtract } 16$$

Thus the trace of the sphere is the circle

$$(x - 2)^2 + (y - 4)^2 = 20, \qquad z = 9$$

which is a circle of radius $\sqrt{20}$ that is 9 units above the xy-plane, centered at $(2, 4, 9)$ (see Figure 9(b)).

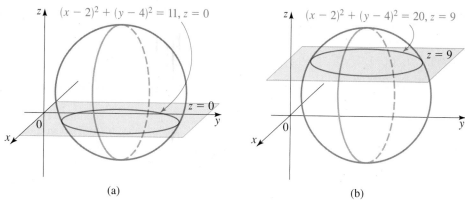

FIGURE 9 The trace of a sphere in the planes $z = 0$ and $z = 9$

✎ NOW TRY EXERCISE **19**

Courtesy of NASA

MATHEMATICS IN THE MODERN WORLD

Global Positioning System (GPS)

On a cold, foggy day in 1707 a British naval fleet was sailing home at a fast clip. The fleet's navigators didn't know it, but the fleet was only a few yards from the rocky shores of England. In the ensuing disaster the fleet was totally destroyed. This tragedy could have been avoided had the navigators known their positions. In those days latitude was determined by the position of the North Star (and this could only be done at night in good weather), and longitude by the position of the sun relative to where it would be in England *at that same time.* So navigation required an accurate method of telling time on ships. (The invention of the spring-loaded clock brought about the eventual solution.)

Since then, several different methods have been developed to determine position, and all rely heavily on mathematics (see LORAN, page 383). The latest method, called the Global Positioning System (GPS), uses triangulation. In this system, 24 satellites are strategically located above the surface of the earth. A handheld GPS device measures distance from a satellite, using the travel time of radio signals emitted from the satellite. Knowing the distances to three different satellites tells us that we are at the point of intersection of three different spheres. This uniquely determines our position.

6.3 EXERCISES

CONCEPTS

1–2 ■ Refer to the figure.

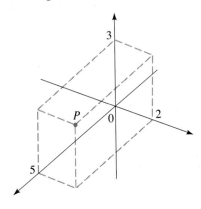

1. In a three-dimensional coordinate system the three mutually perpendicular axes are called the ___-axis, the ___-axis, and the ___-axis. Label the axes in the figure. The point P in the figure has coordinates (__ , __ , __). The equation of the plane passing through P and parallel to the xz-plane is

_____.

2. The distance between the point $P(x_1, y_1, z_1)$ and $Q(x_2, y_2, z_2)$ is given by the formula $d(P, Q) = $ _____.
The distance between the point P in the figure and the origin is _____. The equation of the sphere centered at P with radius 3 is _____.

SKILLS

3–6 ■ Two points P and Q are given. **(a)** Plot P and Q. **(b)** Find the distance between P and Q.

3. $P(3, 1, 0)$, $Q(-1, 2, -5)$

4. $P(5, 0, 10)$, $Q(3, -6, 7)$

5. $P(-2, -1, 0)$, $Q(-12, 3, 0)$

6. $P(5, -4, -6)$, $Q(8, -7, 4)$

7–10 ■ Describe and sketch the surface represented by the given equation.

7. $x = 4$

8. $y = -2$

9. $z = 8$

10. $y = -1$

11–14 ■ Find an equation of a sphere with the given radius r and center C.

11. $r = 5$; $C(2, -5, 3)$

12. $r = 3$; $C(-1, 4, -7)$

13. $r = \sqrt{6}$; $C(3, -1, 0)$

14. $r = \sqrt{11}$; $C(-10, 0, 1)$

15–18 ■ Show that the equation represents a sphere, and find its center and radius.

15. $x^2 + y^2 + z^2 - 10x + 2y + 8z = 9$

16. $x^2 + y^2 + z^2 + 4x - 6y + 2z = 10$

17. $x^2 + y^2 + z^2 = 12x + 2y$

18. $x^2 + y^2 + z^2 = 14y - 6z$

19. Describe the trace of the sphere

$$(x + 1)^2 + (y - 2)^2 + (z + 10)^2 = 100$$

in **(a)** the yz-plane and **(b)** the plane $x = 4$.

20. Describe the trace of the sphere

$$x^2 + (y - 4)^2 + (z - 3)^2 = 144$$

in **(a)** the xz-plane and in **(b)** the plane $z = -2$.

APPLICATIONS

21. Spherical Water Tank   A water tank is in the shape of a sphere of radius 5 feet. The tank is supported on a metal circle 4 feet below the center of the sphere, as shown in the figure. Find the radius of the metal circle.

22. A Spherical Buoy   A spherical buoy of radius 2 feet floats in a calm lake. Six inches of the buoy are submerged. Place a coordinate system with the origin at the center of the sphere.
(a) Find an equation of the sphere.
(b) Find an equation of the circle formed at the waterline of the buoy.

NOW TRY EXERCISES **3** AND **7**

DISCOVERY ▪ DISCUSSION ▪ WRITING

23. Visualizing a Set in Space Try to visualize the set of all points (x, y, z) in a coordinate space that are equidistant from the points $P(0, 0, 0)$ and $Q(0, 3, 0)$. Use the Distance Formula to find an equation for this surface, and observe that it is a plane.

24. Visualizing a Set in Space Try to visualize the set of all points (x, y, z) in a coordinate space that are twice as far from the points $Q(0, 3, 0)$ as from the point $P(0, 0, 0)$. Use the Distance Formula to show that the set is a sphere, and find its center and radius.

6.4 VECTORS IN THREE DIMENSIONS

Vectors in Space ▶ Combining Vectors in Space ▶ The Dot Product for Vectors in Space

Recall that vectors are used to indicate a quantity that has both magnitude and direction. In Section 6.1 we studied vectors in the coordinate plane, where the direction is restricted to two dimensions. Vectors in space have a direction that is in three-dimensional space. The properties that hold for vectors in the plane hold for vectors in space as well.

▼ Vectors in Space

Recall from Section 6.1 that a vector can be described geometrically by its initial point and terminal point. When we place a vector **a** in space with its initial point at the origin, we can describe it algebraically as an ordered triple:

$$\mathbf{a} = \langle a_1, a_2, a_3 \rangle$$

where a_1, a_2, and a_3 are the **components** of **a** (see Figure 1). Recall also that a vector has many different representations, depending on its initial point. The following definition gives the relationship between the algebraic and geometric representations of a vector.

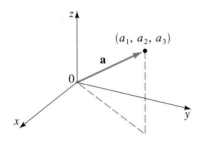

FIGURE 1 $\mathbf{a} = \langle a_1, a_2, a_3 \rangle$

COMPONENT FORM OF A VECTOR IN SPACE

If a vector **a** is represented in space with initial point $P(x_1, y_1, z_1)$ and terminal point $Q(x_2, y_2, z_2)$, then

$$\mathbf{a} = \langle x_2 - x_1, y_2 - y_1, z_2 - z_1 \rangle$$

EXAMPLE 1 | Describing Vectors in Component Form

(a) Find the components of the vector **a** with initial point $P(1, -4, 5)$ and terminal point $Q(3, 1, -1)$.

(b) If the vector $\mathbf{b} = \langle -2, 1, 3 \rangle$ has initial point $(2, 1, -1)$, what is its terminal point?

SOLUTION

(a) The desired vector is

$$\mathbf{a} = \langle 3 - 1, 1 - (-4), -1 - 5 \rangle = \langle 2, 5, -6 \rangle$$

See Figure 2.

(b) Let the terminal point of **b** be (x, y, z). Then

$$\mathbf{b} = \langle x - 2, y - 1, z - (-1) \rangle$$

Since $\mathbf{b} = \langle -2, 1, 3 \rangle$ we have $x - 2 = -2$, $y - 1 = 1$, and $z + 1 = 3$. So $x = 0$, $y = 2$, and $z = 2$, and the terminal point is $(0, 2, 2)$.

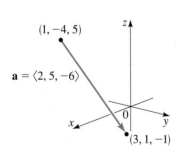

FIGURE 2 $\mathbf{a} = \langle 2, 5, -6 \rangle$

The following formula is a consequence of the Distance Formula, since the vector $\mathbf{a} = \langle a_1, a_2, a_3 \rangle$ in standard position has initial point $(0, 0, 0)$ and terminal point (a_1, a_2, a_3).

MAGNITUDE OF A VECTOR IN THREE DIMENSIONS

The magnitude of the vector $\mathbf{a} = \langle a_1, a_2, a_3 \rangle$ is

$$|\mathbf{a}| = \sqrt{a_1^2 + a_2^2 + a_3^2}$$

EXAMPLE 2 | Magnitude of Vectors in Three Dimensions

Find the magnitude of the given vector.

(a) $\mathbf{u} = \langle 3, 2, 5 \rangle$ **(b)** $\mathbf{v} = \langle 0, 3, -1 \rangle$ **(c)** $\mathbf{w} = \langle 0, 0, -1 \rangle$

SOLUTION

(a) $|\mathbf{u}| = \sqrt{3^2 + 2^2 + 5^2} = \sqrt{38}$

(b) $|\mathbf{v}| = \sqrt{0^2 + 3^2 + (-1)^2} = \sqrt{10}$

(c) $|\mathbf{w}| = \sqrt{0^2 + 0^2 + (-1)^2} = 1$

✎ NOW TRY EXERCISE **11** ∎

▼ Combining Vectors in Space

We now give definitions of the algebraic operations involving vectors in three dimensions.

ALGEBRAIC OPERATIONS ON VECTORS IN THREE DIMENSIONS

If $\mathbf{a} = \langle a_1, a_2, a_3 \rangle$, $\mathbf{b} = \langle b_1, b_2, b_3 \rangle$, and c is a scalar, then

$$\mathbf{a} + \mathbf{b} = \langle a_1 + b_1, a_2 + b_2, a_3 + b_3 \rangle$$

$$\mathbf{a} - \mathbf{b} = \langle a_1 - b_1, a_2 - b_2, a_3 - b_3 \rangle$$

$$c\mathbf{a} = \langle ca_1, ca_2, ca_3 \rangle$$

EXAMPLE 3 | Operations with Three-Dimensional Vectors

If $\mathbf{u} = \langle 1, -2, 4 \rangle$ and $\mathbf{v} = \langle 6, -1, 1 \rangle$ find $\mathbf{u} + \mathbf{v}$, $\mathbf{u} - \mathbf{v}$, and $5\mathbf{u} - 3\mathbf{v}$.

SOLUTION Using the definitions of algebraic operations we have

$$\mathbf{u} + \mathbf{v} = \langle 1 + 6, -2 - 1, 4 + 1 \rangle = \langle 7, -3, 5 \rangle$$

$$\mathbf{u} - \mathbf{v} = \langle 1 - 6, -2 - (-1), 4 - 1 \rangle = \langle -5, -1, 3 \rangle$$

$$5\mathbf{u} - 3\mathbf{v} = 5\langle 1, -2, 4 \rangle - 3\langle 6, -1, 1 \rangle = \langle 5, -10, 20 \rangle - \langle 18, -3, 3 \rangle = \langle -13, -7, 17 \rangle$$

✎ NOW TRY EXERCISE **15** ∎

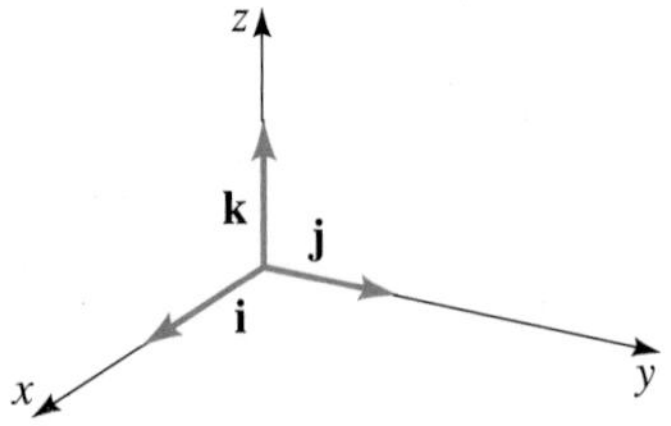

Recall that a unit vector is a vector of length 1. The vector $\mathbf{w}$ in Example 2(c) is an example of a unit vector. Some other unit vectors in three dimensions are

$$\mathbf{i} = \langle 1, 0, 0 \rangle \qquad \mathbf{j} = \langle 0, 1, 0 \rangle \qquad \mathbf{k} = \langle 0, 0, 1 \rangle$$

as shown in Figure 3. Any vector in three dimensions can be written in terms of these three vectors (see Figure 4).

FIGURE 3

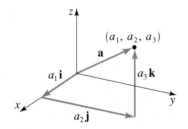

FIGURE 4

EXPRESSING VECTORS IN TERMS OF i, j, AND k

The vector $\mathbf{a} = \langle a_1, a_2, a_3 \rangle$ can be expressed in terms of $\mathbf{i}$, $\mathbf{j}$, and $\mathbf{k}$ by

$$\mathbf{a} = \langle a_1, a_2, a_3 \rangle = a_1\mathbf{i} + a_2\mathbf{j} + a_3\mathbf{k}$$

All the properties of vectors on page 313 in Section 6.1 hold for vectors in three dimensions as well. We use these properties in the next example.

EXAMPLE 4 | Vectors in Terms of i, j, and k

(a) Write the vector $\mathbf{u} = \langle 5, -3, 6 \rangle$ in terms of $\mathbf{i}$, $\mathbf{j}$, and $\mathbf{k}$.

(b) If $\mathbf{u} = \mathbf{i} + 2\mathbf{j} - 3\mathbf{k}$ and $\mathbf{v} = 4\mathbf{i} + 7\mathbf{k}$, express the vector $2\mathbf{u} + 3\mathbf{v}$ in terms of $\mathbf{i}$, $\mathbf{j}$, and $\mathbf{k}$.

SOLUTION

(a) $\mathbf{u} = 5\mathbf{i} + (-3)\mathbf{j} + 6\mathbf{k} = 5\mathbf{i} - 3\mathbf{j} + 6\mathbf{k}$

(b) We use the properties of vectors to get the following:

$$2\mathbf{u} + 3\mathbf{v} = 2(2\mathbf{i} + 2\mathbf{j} - 3\mathbf{k}) + 3(4\mathbf{i} + 7\mathbf{k})$$

$$= 4\mathbf{i} + 4\mathbf{j} - 6\mathbf{k} + 12\mathbf{i} + 21\mathbf{k}$$

$$= 16\mathbf{i} + 4\mathbf{j} + 15\mathbf{k}$$

✎ NOW TRY EXERCISE **19**

▼ The Dot Product for Vectors in Space

We define the dot product for vectors in three dimensions. All the properties of the dot product, including the Dot Product Theorem (page 320), hold for vectors in three dimensions.

DEFINITION OF THE DOT PRODUCT FOR VECTORS IN THREE DIMENSIONS

If $\mathbf{a} = \langle a_1, a_2, a_3 \rangle$ and $\mathbf{b} = \langle b_1, b_2, b_3 \rangle$ are vectors in three dimensions, then their **dot product** is defined by

$$\mathbf{a} \cdot \mathbf{b} = a_1 b_1 + a_2 b_2 + a_3 b_3$$

EXAMPLE 5 | Calculating Dot Products for Vectors in Three Dimensions

Find the given dot product.

(a) $\langle -1, 2, 3 \rangle \cdot \langle 6, 5, -1 \rangle$

(b) $(2\mathbf{i} - 3\mathbf{j} - \mathbf{k}) \cdot (-\mathbf{i} + 2\mathbf{j} + 8\mathbf{k})$

SOLUTION

(a) $\langle -1, 2, 3 \rangle \cdot \langle 6, 5, -1 \rangle = (-1)(6) + (2)(5) + (3)(-1) = 1$

(b) $(2\mathbf{i} - 3\mathbf{j} - \mathbf{k}) \cdot (-\mathbf{i} + 2\mathbf{j} + 8\mathbf{k}) = \langle 2, -3, -1 \rangle \cdot \langle -1, 2, 8 \rangle$

$$= (2)(-1) + (-3)(2) + (-1)(8) = -16$$

✎ NOW TRY EXERCISES **25** AND **27**

Recall that the cosine of the angle between two vectors can be calculated using the dot product (page 320). The same property holds for vectors in three dimensions. We restate this property here for emphasis.

ANGLE BETWEEN TWO VECTORS

Let **u** and **v** be vectors in space and θ be the angle between them. Then

$$\cos \theta = \frac{\mathbf{u} \cdot \mathbf{v}}{|\mathbf{u}| |\mathbf{v}|}$$

In particular, **u** and **v** are **perpendicular** (or **orthogonal**) if and only if $\mathbf{u} \cdot \mathbf{v} = 0$.

EXAMPLE 6 | Checking Vectors for Perpendicularity

Show that the vector $\mathbf{u} = 2\mathbf{i} + 2\mathbf{j} - \mathbf{k}$ is perpendicular to $5\mathbf{i} - 4\mathbf{j} + 2\mathbf{k}$.

SOLUTION We find the dot product.

$$(2\mathbf{i} + 2\mathbf{j} - \mathbf{k}) \cdot (5\mathbf{i} - 4\mathbf{j} + 2\mathbf{k}) = (2)(5) + (2)(-4) + (-1)(2) = 0$$

Since the dot product is 0, the vectors are perpendicular. See Figure 5.

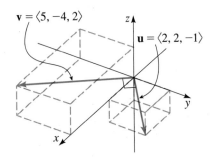

FIGURE 5 The vectors **u** and **v** are perpendicular.

✎ NOW TRY EXERCISE **29**

▼ Direction Angles of a Vector

The **direction angles** of a nonzero vector $\mathbf{a} = a_1 \mathbf{i} + a_2 \mathbf{j} + a_3 \mathbf{k}$ are the angles α, β, and γ in the interval $[0, \pi]$ that the vector **a** makes with the positive x-, y-, and z-axes (see Figure 6). The cosines of these angles, $\cos \alpha$, $\cos \beta$, and $\cos \gamma$, are called the **direction cosines** of the vector **a**. By using the formula for the angle between two vectors, we can find the direction cosines of **a**:

$$\cos \alpha = \frac{\mathbf{a} \cdot \mathbf{i}}{|\mathbf{a}| |\mathbf{i}|} = \frac{a_1}{|\mathbf{a}|} \qquad \cos \beta = \frac{\mathbf{a} \cdot \mathbf{j}}{|\mathbf{a}| |\mathbf{j}|} = \frac{a_2}{|\mathbf{a}|} \qquad \cos \gamma = \frac{\mathbf{a} \cdot \mathbf{k}}{|\mathbf{a}| |\mathbf{k}|} = \frac{a_3}{|\mathbf{a}|}$$

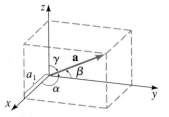

FIGURE 6 Direction angles of the vector **a**

DIRECTION ANGLES OF A VECTOR

If $\mathbf{a} = a_1\mathbf{i} + a_2\mathbf{j} + a_3\mathbf{k}$ is a nonzero vector in space, the direction angles α, β, and γ satisfy

$$\cos \alpha = \frac{a_1}{|\mathbf{a}|} \qquad \cos \beta = \frac{a_2}{|\mathbf{a}|} \qquad \cos \gamma = \frac{a_3}{|\mathbf{a}|}$$

In particular, if $|\mathbf{a}| = 1$ then the direction cosines of $\mathbf{a}$ are simply the components of $\mathbf{a}$.

EXAMPLE 7 | Finding the Direction Angles of a Vector

Find the direction angles of the vector $\mathbf{a} = \mathbf{i} + 2\mathbf{j} + 3\mathbf{k}$.

SOLUTION   The length of the vector $\mathbf{a}$ is $|\mathbf{a}| = \sqrt{1^2 + 2^2 + 3^2} = \sqrt{14}$. From the above box we get

$$\cos \alpha = \frac{1}{\sqrt{14}} \qquad \cos \beta = \frac{2}{\sqrt{14}} \qquad \cos \gamma = \frac{3}{\sqrt{14}}$$

Since the direction angles are in the interval $[0, \pi]$ and since $\cos^{-1}$ gives angles in that same interval, we get α, β, and γ by simply taking $\cos^{-1}$ of the above equations.

$$\alpha = \cos^{-1}\frac{1}{\sqrt{14}} \approx 74° \qquad \beta = \cos^{-1}\frac{2}{\sqrt{14}} \approx 58° \qquad \gamma = \cos^{-1}\frac{3}{\sqrt{14}} \approx 37°$$

✎ NOW TRY EXERCISE **37**   ∎

The direction angles of a vector uniquely determine its direction, but not its length. If we also know the length of the vector $\mathbf{a}$, the expressions for the direction cosines of $\mathbf{a}$ allow us to express the vector as

$$\mathbf{a} = \langle\, |\mathbf{a}|\cos \alpha, |\mathbf{a}|\cos \beta, |\mathbf{a}|\cos \gamma\,\rangle$$

From this we get

$$\mathbf{a} = |\mathbf{a}|\langle\cos \alpha, \cos \beta, \cos \gamma\rangle$$

$$\frac{\mathbf{a}}{|\mathbf{a}|} = \langle\cos \alpha, \cos \beta, \cos \gamma\rangle$$

Since $\mathbf{a}/|\mathbf{a}|$ is a unit vector we get the following.

PROPERTY OF DIRECTION COSINES

The direction angles α, β, and γ of a nonzero vector $\mathbf{a}$ in space satisfy the following equation:

$$\cos^2\alpha + \cos^2\beta + \cos^2\gamma = 1$$

This property indicates that if we know two of the direction cosines of a vector, we can find the third except for its sign.

EXAMPLE 8 | Finding the Direction Angles of a Vector

An angle θ is **acute** if $0 \le \theta < \pi/2$ and is **obtuse** if $\pi/2 < \theta \le \pi$.

A vector makes an angle $\alpha = \pi/3$ with the positive x-axis and an angle $\beta = 3\pi/4$ with the positive y-axis. Find the angle γ that the vector makes with the positive z-axis, given that γ is an obtuse angle.

SOLUTION By the property of the direction angles we have

$$\cos^2\alpha + \cos^2\beta + \cos^2\gamma = 1$$

$$\cos^2\frac{\pi}{3} + \cos^2\frac{3\pi}{4} + \cos^2\gamma = 1$$

$$\left(\frac{1}{2}\right)^2 + \left(-\frac{1}{\sqrt{2}}\right)^2 + \cos^2\gamma = 1$$

$$\cos^2\gamma = \frac{1}{4}$$

$$\cos\gamma = \frac{1}{2} \qquad \text{or} \qquad \cos\gamma = -\frac{1}{2}$$

$$\gamma = \frac{\pi}{3} \qquad \text{or} \qquad \gamma = \frac{2\pi}{3}$$

Since we require γ to be an obtuse angle, we conclude that $\gamma = 2\pi/3$.

✎ NOW TRY EXERCISE **41**

6.4 EXERCISES

CONCEPTS

1. A vector in three dimensions can be written in either of two forms: in coordinate form as $\mathbf{a} = \langle a_1, a_2, a_3 \rangle$ and in terms of

the _____ vectors $\mathbf{i}$, $\mathbf{j}$, and $\mathbf{k}$ as $\mathbf{a} =$ _____.

The magnitude of the vector $\mathbf{a}$ is $|\mathbf{a}| =$ _____.

So $\langle 4, -2, 4 \rangle = \boxed{}\mathbf{i} + \boxed{}\mathbf{j} + \boxed{}\mathbf{k}$ and

$7\mathbf{j} - 24\mathbf{k} = \langle \boxed{}, \boxed{}, \boxed{} \rangle$.

2. The angle θ between the vectors $\mathbf{u}$ and $\mathbf{v}$ satisfies

$\cos\theta = \dfrac{\boxed{}}{\boxed{}}$. So if $\mathbf{u}$ and $\mathbf{v}$ are perpendicular, then

$\mathbf{u} \cdot \mathbf{v} =$ _____. If $\mathbf{u} = \langle 4, 5, 6 \rangle$ and $\mathbf{v} = \langle 3, 0, -2 \rangle$ then

$\mathbf{u} \cdot \mathbf{v} =$ _____, so $\mathbf{u}$ and $\mathbf{v}$ are _____.

SKILLS

3–6 ■ Find the vector $\mathbf{v}$ with initial point P and terminal point Q.

3. $P(1, -1, 0)$, $Q(0, -2, 5)$ **4.** $P(1, 2, -1)$, $Q(3, -1, 2)$

5. $P(6, -1, 0)$, $Q(0, -3, 0)$ **6.** $P(1, -1, -1)$, $Q(0, 0, -1)$

7–10 ■ If the vector $\mathbf{v}$ has initial point P, what is its terminal point?

7. $\mathbf{v} = \langle 3, 4, -2 \rangle$, $P(2, 0, 1)$

8. $\mathbf{v} = \langle 0, 0, 1 \rangle$, $P(0, 1, -1)$

9. $\mathbf{v} = \langle -2, 0, 2 \rangle$, $P(3, 0, -3)$

10. $\mathbf{v} = \langle 23, -5, 12 \rangle$, $P(-6, 4, 2)$

11–14 ■ Find the magnitude of the given vector.

11. $\langle -2, 1, 2 \rangle$

12. $\langle 5, 0, -12 \rangle$

13. $\langle 3, 5, -4 \rangle$

14. $\langle 1, -6, 2\sqrt{2} \rangle$

15–18 ■ Find the vectors $\mathbf{u} + \mathbf{v}$, $\mathbf{u} - \mathbf{v}$, and $3\mathbf{u} - \frac{1}{2}\mathbf{v}$.

15. $\mathbf{u} = \langle 2, -7, 3 \rangle$, $\mathbf{v} = \langle 0, 4, -1 \rangle$

16. $\mathbf{u} = \langle 0, 1, -3 \rangle$, $\mathbf{v} = \langle 4, 2, 0 \rangle$

17. $\mathbf{u} = \mathbf{i} + \mathbf{j}$, $\mathbf{v} = -\mathbf{j} - 2\mathbf{k}$

18. $\mathbf{u} = \langle a, 2b, 3c \rangle$, $\mathbf{v} = \langle -4a, b, -2c \rangle$

19–22 ■ Express the given vector in terms of the unit vectors **i**, **j**, and **k**.

19. $\langle 12, 0, 2 \rangle$

20. $\langle 0, -3, 5 \rangle$

21. $\langle 3, -3, 0 \rangle$

22. $\langle -a, \frac{1}{3}a, 4 \rangle$

23–24 ■ Two vectors **u** and **v** are given. Express the vector $-2\mathbf{u} + 3\mathbf{v}$ **(a)** in component form $\langle a_1, a_2, a_3 \rangle$ and **(b)** in terms of the unit vectors **i**, **j**, and **k**.

23. $\mathbf{u} = \langle 0, -2, 1 \rangle$, $\mathbf{v} = \langle 1, -1, 0 \rangle$

24. $\mathbf{u} = \langle 3, 1, 0 \rangle$, $\mathbf{v} = \langle 3, 0, -5 \rangle$

25–28 ■ Two vectors **u** and **v** are given. Find their dot product $\mathbf{u} \cdot \mathbf{v}$.

25. $\mathbf{u} = \langle 2, 5, 0 \rangle$, $\mathbf{v} = \langle \frac{1}{2}, -1, 10 \rangle$

26. $\mathbf{u} = \langle -3, 0, 4 \rangle$, $\mathbf{v} = \langle 2, 4, \frac{1}{2} \rangle$

27. $\mathbf{u} = 6\mathbf{i} - 4\mathbf{j} - 2\mathbf{k}$, $\mathbf{v} = \frac{5}{6}\mathbf{i} + \frac{3}{2}\mathbf{j} - \mathbf{k}$

28. $\mathbf{u} = 3\mathbf{j} - 2\mathbf{k}$, $\mathbf{v} = \frac{5}{6}\mathbf{i} - \frac{5}{3}\mathbf{j}$

29–32 ■ Determine whether or not the given vectors are perpendicular.

29. $\langle 4, -2, -4 \rangle$, $\langle 1, -2, 2 \rangle$

30. $4\mathbf{j} - \mathbf{k}$, $\mathbf{i} + 2\mathbf{j} + 9\mathbf{k}$

31. $\langle 0.3, 1.2, -0.9 \rangle$, $\langle 10, -5, 10 \rangle$

32. $\langle x, -2x, 3x \rangle$, $\langle 5, 7, 3 \rangle$

33–36 ■ Two vectors **u** and **v** are given. Find the angle (expressed in degrees) between **u** and **v**.

33. $\mathbf{u} = \langle 2, -2, -1 \rangle$, $\mathbf{v} = \langle 1, 2, 2 \rangle$

34. $\mathbf{u} = \langle 4, 0, 2 \rangle$, $\mathbf{v} = \langle 2, -1, 0 \rangle$

35. $\mathbf{u} = \mathbf{j} + \mathbf{k}$, $\mathbf{v} = \mathbf{i} + 2\mathbf{j} - 3\mathbf{k}$

36. $\mathbf{u} = \mathbf{i} + 2\mathbf{j} - 2\mathbf{k}$, $\mathbf{v} = 4\mathbf{i} - 3\mathbf{k}$

37–40 ■ Find the direction angles of the given vector, rounded to the nearest degree.

37. $3\mathbf{i} + 4\mathbf{j} + 5\mathbf{k}$

38. $\mathbf{i} - 2\mathbf{j} - \mathbf{k}$

39. $\langle 2, 3, -6 \rangle$

40. $\langle 2, -1, 2 \rangle$

41–44 ■ Two direction angles of a vector are given. Find the third direction angle, given that it is either obtuse or acute as indicated. (In Exercises 43 and 44, round your answers to the nearest degree.)

41. $\alpha = \dfrac{\pi}{3}$, $\gamma = \dfrac{2\pi}{3}$; β is acute

42. $\beta = \dfrac{2\pi}{3}$, $\gamma = \dfrac{\pi}{4}$; α is acute

43. $\alpha = 60°$, $\beta = 50°$; γ is obtuse

44. $\alpha = 75°$, $\gamma = 15°$

45–46 ■ Explain why it is impossible for a vector to have the given direction angles.

45. $\alpha = 20°$, $\beta = 45°$

46. $\alpha = 150°$, $\gamma = 25°$

APPLICATIONS

47. Resultant of Four Forces An object located at the origin in a three-dimensional coordinate system is held in equilibrium by four forces. One has magnitude 7 lb and points in the direction of the positive x-axis, so it is represented by the vector 7**i**. The second has magnitude 24 lb and points in the direction of the positive y-axis. The third has magnitude 25 lb and points in the direction of the *negative* z-axis.
  (a) Use the fact that the four forces are in equilibrium (that is, their sum is **0**) to find the fourth force. Express it in terms of the unit vectors **i**, **j**, and **k**.
  (b) What is the magnitude of the fourth force?

48. Central Angle of a Tetrahedron A *tetrahedron* is a solid with four triangular faces, four vertices, and six edges, as shown in the figure. In a *regular* tetrahedron, the edges are all of the same length. Consider the tetrahedron with vertices $A(1, 0, 0)$, $B(0, 1, 0)$, $C(0, 0, 1)$, and $D(1, 1, 1)$.
  (a) Show that the tetrahedron is regular.
  (b) The center of the tetrahedron is the point $E(\frac{1}{2}, \frac{1}{2}, \frac{1}{2})$ (the "average" of the vertices). Find the angle between the vectors that join the center to any two of the vertices (for instance, $\angle AEB$). This angle is called the *central angle* of the tetrahedron.

NOTE: In a molecule of methane (CH_4) the four hydrogen atoms form the vertices of a regular tetrahedron with the carbon atom at the center. In this case chemists refer to the central angle as the *bond angle*. In the figure, the tetrahedron in the exercise is shown, with the vertices labeled H for hydrogen, and the center labeled C for carbon.

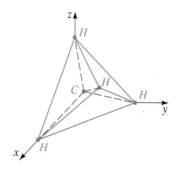

DISCOVERY ■ DISCUSSION ■ WRITING

49. Parallel Vectors Two nonzero vectors are *parallel* if they point in the same direction or in opposite directions. This means that if two vectors are parallel, one must be a scalar multiple of the other. Determine whether the given vectors **u** and **v** are parallel. If they are, express **v** as a scalar multiple of **u**.
  (a) $\mathbf{u} = \langle 3, -2, 4 \rangle$, $\mathbf{v} = \langle -6, 4, -8 \rangle$
  (b) $\mathbf{u} = \langle -9, -6, 12 \rangle$, $\mathbf{v} = \langle 12, 8, -16 \rangle$
  (c) $\mathbf{u} = \mathbf{i} + \mathbf{j} + \mathbf{k}$, $\mathbf{v} = 2\mathbf{i} + 2\mathbf{j} - 2\mathbf{k}$

50. Unit Vectors A *unit vector* is a vector of magnitude 1. Multiplying a vector by a scalar changes its magnitude but not its direction.
  (a) If a vector **v** has magnitude m, what scalar multiple of **v** has magnitude 1 (i.e., is a unit vector)?

(b) Multiply each of the following vectors by an appropriate scalar to change them into unit vectors:

$$\langle 1, -2, 2 \rangle \quad \langle -6, 8, -10 \rangle \quad \langle 6, 5, 9 \rangle$$

51. Vector Equation of a Sphere Let $\mathbf{a} = \langle 2, 2, 2 \rangle$, $\mathbf{b} = \langle -2, -2, 0 \rangle$, and $\mathbf{r} = \langle x, y, z \rangle$.
  (a) Show that the vector equation $(\mathbf{r} - \mathbf{a}) \cdot (\mathbf{r} - \mathbf{b}) = 0$ represents a sphere, by expanding the dot product and simplifying the resulting algebraic equation.
  (b) Find the center and radius of the sphere.

(c) Interpret the result of part (a) geometrically, using the fact that the dot product of two vectors is 0 only if the vectors are perpendicular. [*Hint:* Draw a diagram showing the endpoints of the vectors **a**, **b**, and **r**, noting that the endpoints of **a** and **b** are the endpoints of a diameter and the endpoint of **r** is an arbitrary point on the sphere.]

(d) Using your observations from part (a), find a vector equation for the sphere in which the points $(0, 1, 3)$ and $(2, -1, 4)$ form the endpoints of a diameter. Simplify the vector equation to obtain an algebraic equation for the sphere. What are its center and radius?

6.5 THE CROSS PRODUCT

| The Cross Product ▶ Properties of the Cross Product ▶ Area of a Parallelogram ▶ Volume of a Parallelepiped

In this section we define an operation on vectors that allows us to find a vector which is perpendicular to two given vectors.

▼ The Cross Product

Given two vectors $\mathbf{a} = \langle a_1, a_2, a_3 \rangle$ and $\mathbf{b} = \langle b_1, b_2, b_3 \rangle$ we often need to find a vector **c** perpendicular to both **a** and **b**. If we write $\mathbf{c} = \langle c_1, c_2, c_3 \rangle$ then $\mathbf{a} \cdot \mathbf{c} = 0$ and $\mathbf{b} \cdot \mathbf{c} = 0$, so

$$a_1 c_1 + a_2 c_2 + a_3 c_3 = 0$$

$$b_1 c_1 + b_2 c_2 + b_3 c_3 = 0$$

You can check that one of the solutions of this system of equations is the vector $\mathbf{c} = \langle a_2 b_3 - a_3 b_2, a_3 b_1 - a_1 b_3, a_1 b_2 - a_2 b_1 \rangle$. This vector is called the *cross product* of **a** and **b** and is denoted by $\mathbf{a} \times \mathbf{b}$.

THE CROSS PRODUCT

If $\mathbf{a} = \langle a_1, a_2, a_3 \rangle$ and $\mathbf{b} = \langle b_1, b_2, b_3 \rangle$ are three-dimensional vectors, then the **cross product** of **a** and **b** is the vector

$$\mathbf{a} \times \mathbf{b} = \langle a_2 b_3 - a_3 b_2, \ a_3 b_1 - a_1 b_3, \ a_1 b_2 - a_2 b_1 \rangle$$

The *cross product* $\mathbf{a} \times \mathbf{b}$ of two vectors **a** and **b**, unlike the dot product, is a vector (not a scalar). For this reason it is also called the *vector product*. Note that $\mathbf{a} \times \mathbf{b}$ is defined only when **a** and **b** are vectors in *three dimensions*.

To help us remember the definition of the cross product, we use the notation of determinants. A **determinant of order two** is defined by

$$\begin{vmatrix} a & b \\ c & d \end{vmatrix} = ad - bc$$

For example,

$$\begin{vmatrix} 2 & 1 \\ -6 & 4 \end{vmatrix} = 2(4) - 1(-6) = 14$$

A **determinant of order three** is defined in terms of second-order determinants as

$$\begin{vmatrix} a_1 & a_2 & a_3 \\ b_1 & b_2 & b_3 \\ c_1 & c_2 & c_3 \end{vmatrix} = a_1 \begin{vmatrix} b_2 & b_3 \\ c_2 & c_3 \end{vmatrix} - a_2 \begin{vmatrix} b_1 & b_3 \\ c_1 & c_3 \end{vmatrix} + a_3 \begin{vmatrix} b_1 & b_2 \\ c_1 & c_2 \end{vmatrix}$$

Observe that each term on the right side of the above equation involves a number a_i in the first row of the determinant, and a_i is multiplied by the second-order determinant obtained from the left side by deleting the row and column in which a_i appears. Notice also the minus sign in the second term. For example,

$$\begin{vmatrix} 1 & 2 & -1 \\ 3 & 0 & 1 \\ -5 & 4 & 2 \end{vmatrix} = 1 \begin{vmatrix} 0 & 1 \\ 4 & 2 \end{vmatrix} - 2 \begin{vmatrix} 3 & 1 \\ -5 & 2 \end{vmatrix} + (-1) \begin{vmatrix} 3 & 0 \\ -5 & 4 \end{vmatrix}$$

$$= 1(0 - 4) - 2(6 - (-5)) + (-1)(12 - 0) = -38$$

We can write the definition of the cross product using determinants as

$$\begin{vmatrix} \mathbf{i} & \mathbf{j} & \mathbf{k} \\ a_1 & a_2 & a_3 \\ b_1 & b_2 & b_3 \end{vmatrix} = \begin{vmatrix} a_2 & a_3 \\ b_2 & b_3 \end{vmatrix} \mathbf{i} - \begin{vmatrix} a_1 & a_3 \\ b_1 & b_3 \end{vmatrix} \mathbf{j} + \begin{vmatrix} a_1 & a_2 \\ b_1 & b_2 \end{vmatrix} \mathbf{k}$$

$$= (a_2 b_3 - a_3 b_2)\mathbf{i} + (a_3 b_1 - a_1 b_3)\mathbf{j} + (a_1 b_2 - a_2 b_1)\mathbf{k}$$

Although the first row of the above determinant consists of vectors, we expand it as if it were an ordinary determinant of order 3. The symbolic formula given by the above determinant is probably the easiest way to remember and compute cross products.

EXAMPLE 1 | Finding a Cross Product

If $\mathbf{a} = \langle 0, -1, 3 \rangle$ and $\mathbf{b} = \langle 2, 0, -1 \rangle$, find $\mathbf{a} \times \mathbf{b}$.

SOLUTION   We use the formula above to find the cross product of $\mathbf{a}$ and $\mathbf{b}$:

$$\mathbf{a} \times \mathbf{b} = \begin{vmatrix} \mathbf{i} & \mathbf{j} & \mathbf{k} \\ 0 & -1 & 3 \\ 2 & 0 & -1 \end{vmatrix}$$

$$= \begin{vmatrix} -1 & 3 \\ 0 & -1 \end{vmatrix} \mathbf{i} - \begin{vmatrix} 0 & 3 \\ 2 & -1 \end{vmatrix} \mathbf{j} + \begin{vmatrix} 0 & -1 \\ 2 & 0 \end{vmatrix} \mathbf{k}$$

$$= (1 - 0)\mathbf{i} - (0 - 6)\mathbf{j} + (0 - (-2))\mathbf{k}$$

$$= \mathbf{i} + 6\mathbf{j} + 2\mathbf{k}$$

So the desired vector is $\mathbf{i} + 6\mathbf{j} + 2\mathbf{k}$.

✎ NOW TRY EXERCISE **3**

WILLIAM ROWAN HAMILTON
(1805–1865) was an Irish mathematician and physicist. He was raised by his uncle (a linguist) who noticed that Hamilton had a remarkable ability to learn languages. When he was five years old, he could read Latin, Greek, and Hebrew. At age eight he added French and Italian, and by age ten he had mastered Arabic and Sanskrit.

Hamilton was also a calculating prodigy and competed in contests of mental arithmetic. He entered Trinity College in Dublin, Ireland, where he studied science; he was appointed Professor of Astronomy there while still an undergraduate.

Hamilton made many contributions to mathematics and physics, but he is best known for his invention of quaternions. Hamilton knew that we can multiply vectors in the plane by considering them as complex numbers. He was looking for a similar multiplication for points in space. After thinking about this problem for over 20 years he discovered the solution in a flash of insight while walking near Brougham Bridge in Dublin: He realized that a fourth dimension is needed to make the multiplication work. He carved the formula for his quaternions into the bridge, where it still stands. Later, the American mathematician Josiah Willard Gibbs extracted the dot product and cross product of vectors from the properties of quaternion multiplication. Quaternions are used today in computer graphics because of their ability to easily describe special rotations.

▼ Properties of the Cross Product

One of the most important properties of the cross product is the following theorem.

CROSS PRODUCT THEOREM

The vector $\mathbf{a} \times \mathbf{b}$ is orthogonal (perpendicular) to both $\mathbf{a}$ and $\mathbf{b}$.

PROOF To show that $\mathbf{a} \times \mathbf{b}$ is orthogonal to $\mathbf{a}$, we compute their dot product and show that it is 0:

$$(\mathbf{a} \times \mathbf{b}) \cdot \mathbf{a} = \begin{vmatrix} a_2 & a_3 \\ b_2 & b_3 \end{vmatrix} a_1 - \begin{vmatrix} a_1 & a_3 \\ b_1 & b_3 \end{vmatrix} a_2 + \begin{vmatrix} a_1 & a_2 \\ b_1 & b_2 \end{vmatrix} a_3$$

$$= a_1(a_2 b_3 - a_3 b_2) - a_2(a_1 b_3 - a_3 b_1) + a_3(a_1 b_2 - a_2 b_1)$$

$$= a_1 a_2 b_3 - a_1 a_3 b_2 - a_1 a_2 b_3 + a_2 a_3 b_1 + a_1 a_3 b_2 - a_2 a_3 b_1$$

$$= 0$$

A similar computation shows that $(\mathbf{a} \times \mathbf{b}) \cdot \mathbf{b} = 0$. Therefore, $\mathbf{a} \times \mathbf{b}$ is orthogonal to $\mathbf{a}$ and to $\mathbf{b}$. ∎

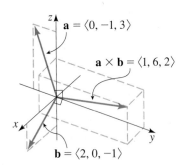

FIGURE 1 The vector $\mathbf{a} \times \mathbf{b}$ is perpendicular to $\mathbf{a}$ and $\mathbf{b}$.

EXAMPLE 2 | Finding an Orthogonal Vector

If $\mathbf{a} = -\mathbf{j} + 3\mathbf{k}$ and $\mathbf{b} = 2\mathbf{i} - \mathbf{k}$, find a unit vector that is orthogonal to the plane containing the vectors $\mathbf{a}$ and $\mathbf{b}$.

SOLUTION By the Cross Product Theorem the vector $\mathbf{a} \times \mathbf{b}$ is orthogonal to the plane containing the vectors $\mathbf{a}$ and $\mathbf{b}$. (See Figure 1.) In Example 1 we found $\mathbf{a} \times \mathbf{b} = \mathbf{i} + 6\mathbf{j} + 2\mathbf{k}$. To obtain an orthogonal unit vector, we multiply $\mathbf{a} \times \mathbf{b}$ by the scalar $1/|\mathbf{a} \times \mathbf{b}|$:

$$\frac{\mathbf{a} \times \mathbf{b}}{|\mathbf{a} \times \mathbf{b}|} = \frac{\mathbf{i} + 6\mathbf{j} + 2\mathbf{k}}{\sqrt{1^2 + 6^2 + 2^2}} = \frac{\mathbf{i} + 6\mathbf{j} + 2\mathbf{k}}{\sqrt{41}}$$

So the desired vector is $\dfrac{1}{\sqrt{41}}(\mathbf{i} + 6\mathbf{j} + 2\mathbf{k})$.

✎ NOW TRY EXERCISE **9**

EXAMPLE 3 | Finding a Vector Perpendicular to a Plane

Find a vector perpendicular to the plane that passes through the points $P(1, 4, 6)$, $Q(-2, 5, -1)$, and $R(1, -1, 1)$.

SOLUTION By the Cross Product Theorem, the vector $\overrightarrow{PQ} \times \overrightarrow{PR}$ is perpendicular to both $\overrightarrow{PQ}$ and $\overrightarrow{PR}$, and is therefore perpendicular to the plane through P, Q, and R. We know that

$$\overrightarrow{PQ} = (-2 - 1)\mathbf{i} + (5 - 4)\mathbf{j} + (-1 - 6)\mathbf{k} = -3\mathbf{i} + \mathbf{j} - 7\mathbf{k}$$

$$\overrightarrow{PR} = (1 - 1)\mathbf{i} + ((-1) - 4)\mathbf{j} + (1 - 6)\mathbf{k} = -5\mathbf{j} - 5\mathbf{k}$$

We compute the cross product of these vectors:

$$\overrightarrow{PQ} \times \overrightarrow{PR} = \begin{vmatrix} \mathbf{i} & \mathbf{j} & \mathbf{k} \\ -3 & 1 & -7 \\ 0 & -5 & -5 \end{vmatrix}$$

$$= (-5 - 35)\mathbf{i} - (15 - 0)\mathbf{j} + (15 - 0)\mathbf{k} = -40\mathbf{i} - 15\mathbf{j} + 15\mathbf{k}$$

So the vector $\langle -40, -15, 15 \rangle$ is perpendicular to the given plane. Notice that any nonzero scalar multiple of this vector, such as $\langle -8, -3, 3 \rangle$, is also perpendicular to the plane.

✎ NOW TRY EXERCISE **17**

If $\mathbf{a}$ and $\mathbf{b}$ are represented by directed line segments with the same initial point (as in Figure 2), then the Cross Product Theorem says that the cross product $\mathbf{a} \times \mathbf{b}$ points in a

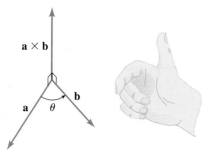

FIGURE 2 Right-hand rule

direction perpendicular to the plane through **a** and **b**. It turns out that the direction of **a** × **b** is given by the *right-hand rule:* If the fingers of your right hand curl in the direction of a rotation (through an angle less than 180°) from **a** to **b**, then your thumb points in the direction of **a** × **b** (as in Figure 2). You can check that the vector **a** × **b** in Figure 1 satisfies the right-hand rule.

Now that we know the direction of the vector **a** × **b**, the remaining thing we need is the length $|\mathbf{a} \times \mathbf{b}|$.

LENGTH OF THE CROSS PRODUCT

If θ is the angle between **a** and **b** (so $0 \le \theta \le \pi$), then

$$|\mathbf{a} \times \mathbf{b}| = |\mathbf{a}||\mathbf{b}|\sin\theta$$

In particular, two nonzero vectors **a** and **b** are parallel if and only if

$$\mathbf{a} \times \mathbf{b} = \mathbf{0}$$

PROOF We apply the definitions of the cross product and length of a vector. You can verify the algebra in the first step by expanding the right-hand sides of the first and second lines, and then comparing the results.

$$
\begin{aligned}
|\mathbf{a} \times \mathbf{b}|^2 &= (a_2 b_3 - a_3 b_2)^2 + (a_3 b_1 - a_1 b_3)^2 + (a_1 b_2 - a_2 b_1)^2 && \text{Definitions} \\
&= (a_1^2 + a_2^2 + a_3^2)(b_1^2 + b_2^2 + b_3^2) - (a_1 b_1 + a_2 b_2 + a_3 b_3)^2 && \text{Verify algebra} \\
&= |\mathbf{a}|^2 |\mathbf{b}|^2 - (\mathbf{a} \cdot \mathbf{b})^2 && \text{Definitions} \\
&= |\mathbf{a}|^2 |\mathbf{b}|^2 - |\mathbf{a}|^2 |\mathbf{b}|^2 \cos^2\theta && \begin{array}{l}\text{Property of}\\\text{Dot Product}\end{array} \\
&= |\mathbf{a}|^2 |\mathbf{b}|^2 (1 - \cos^2\theta) && \text{Factor} \\
&= |\mathbf{a}|^2 |\mathbf{b}|^2 \sin^2\theta && \begin{array}{l}\text{Pythagorean}\\\text{Identity}\end{array}
\end{aligned}
$$

The result follows by taking square roots and observing that $\sqrt{\sin^2\theta} = \sin\theta$ because $\sin\theta \ge 0$ when $0 \le \theta \le \pi$. ∎

We have now completely determined the vector **a** × **b** geometrically. The vector **a** × **b** is perpendicular to both **a** and **b**, and its orientation is determined by the right-hand rule. The length of **a** × **b** is $|\mathbf{a}||\mathbf{b}|\sin\theta$.

▼ Area of a Parallelogram

We can use the cross product to find the area of a parallelogram. If **a** and **b** are represented by directed line segments with the same initial point, then they determine a parallelogram with base $|\mathbf{a}|$, altitude $|\mathbf{b}|\sin\theta$, and area

$$A = |\mathbf{a}|(|\mathbf{b}|\sin\theta) = |\mathbf{a} \times \mathbf{b}|$$

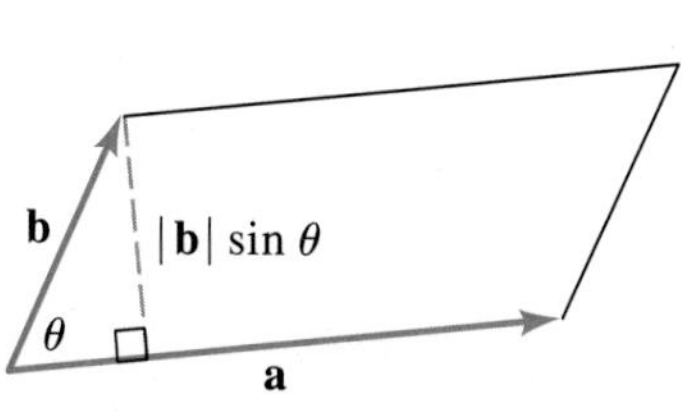

FIGURE 3 Parallelogram determined by **a** and **b**

(See Figure 3.) Thus we have the following way of interpreting the magnitude of a cross product.

AREA OF A PARALLELOGRAM

The length of the cross product **a** × **b** is the area of the parallelogram determined by **a** and **b**.

EXAMPLE 4 | Finding the Area of a Triangle

Find the area of the triangle with vertices $P(1, 4, 6)$, $Q(-2, 5, -1)$, and $R(1, -1, 1)$.

SOLUTION In Example 3 we computed that $\overrightarrow{PQ} \times \overrightarrow{PR} = \langle -40, -15, 15 \rangle$. The area of the parallelogram with adjacent sides PQ and PR is the length of this cross product:

$$|\overrightarrow{PQ} \times \overrightarrow{PR}| = \sqrt{(-40)^2 + (-15)^2 + 15^2} = 5\sqrt{82}$$

The area A of the triangle PQR is half the area of this parallelogram, that is, $\frac{5}{2}\sqrt{82}$.

✎ NOW TRY EXERCISES **21** AND **25** ∎

▼ Volume of a Parallelepiped

The product $\mathbf{a} \cdot (\mathbf{b} \times \mathbf{c})$ is called the **scalar triple product** of the vectors $\mathbf{a}$, $\mathbf{b}$, and $\mathbf{c}$. You can check that the scalar triple product can be written as the following determinant:

$$\mathbf{a} \cdot (\mathbf{b} \times \mathbf{c}) = \begin{vmatrix} a_1 & a_2 & a_3 \\ b_1 & b_2 & b_3 \\ c_1 & c_2 & c_3 \end{vmatrix}$$

The geometric significance of the scalar triple product can be seen by considering the parallelepiped* determined by the vectors $\mathbf{a}$, $\mathbf{b}$, and $\mathbf{c}$ (see Figure 4). The area of the base parallelogram is $A = |\mathbf{b} \times \mathbf{c}|$. If θ is the angle between $\mathbf{a}$ and $\mathbf{b} \times \mathbf{c}$, then the height h of the parallelepiped is $h = |\mathbf{a}||\cos \theta|$. (We must use $|\cos \theta|$ instead of $\cos \theta$ in case $\theta > \pi/2$.) Therefore, the volume of the parallelepiped is

$$V = Ah = |\mathbf{b} \times \mathbf{c}||\mathbf{a}||\cos \theta| = |\mathbf{a} \cdot (\mathbf{b} \times \mathbf{c})|$$

The last equality follows from the Dot Product Theorem on page 320.

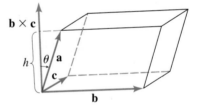

FIGURE 4 Parallelepiped determined by $\mathbf{a}$, $\mathbf{b}$, and $\mathbf{c}$

We have proved the following formula.

VOLUME OF A PARALLELEPIPED

The volume of the parallelepiped determined by the vectors $\mathbf{a}$, $\mathbf{b}$, and $\mathbf{c}$ is the magnitude of their scalar triple product:

$$V = |\mathbf{a} \cdot (\mathbf{b} \times \mathbf{c})|$$

In particular, if the volume of the parallelepiped is 0, then the vectors $\mathbf{a}$, $\mathbf{b}$, and $\mathbf{c}$ are coplanar.

EXAMPLE 5 | Coplanar Vectors

Use the scalar triple product to show that the vectors $\mathbf{a} = \langle 1, 4, -7 \rangle$, $\mathbf{b} = \langle 2, -1, 4 \rangle$, and $\mathbf{c} = \langle 0, -9, 18 \rangle$ are coplanar, that is, lie in the same plane.

---
*The word *parallelepiped* is derived from Greek roots which together mean, roughly, "parallel faces." While the word is often pronounced "par-al-lel-uh-PIE-ped," the more etymologically correct pronunciation is "par-al-lel-EP-uh-ped."

SOLUTION We compute the scalar triple product:

$$\mathbf{a} \cdot (\mathbf{b} \times \mathbf{c}) = \begin{vmatrix} 1 & 4 & -7 \\ 2 & -1 & 4 \\ 0 & -9 & 18 \end{vmatrix}$$

$$= 1 \begin{vmatrix} -1 & 4 \\ -9 & 18 \end{vmatrix} - 4 \begin{vmatrix} 2 & 4 \\ 0 & 18 \end{vmatrix} + (-7) \begin{vmatrix} 2 & -1 \\ 0 & -9 \end{vmatrix}$$

$$= 1(18) - 4(36) - 7(-18) = 0$$

So the volume of the parallelepiped is 0, and hence the vectors $\mathbf{a}$, $\mathbf{b}$, and $\mathbf{c}$ are coplanar.

✎ NOW TRY EXERCISE **29** ■

6.5 EXERCISES

CONCEPTS

1. The cross product of the vectors $\mathbf{a} = \langle a_1, a_2, a_3 \rangle$ and $\mathbf{b} = \langle b_1, b_2, b_3 \rangle$ is the vector

$$\mathbf{a} \times \mathbf{b} = \begin{vmatrix} \mathbf{i} & \mathbf{j} & \mathbf{k} \\ \square & \square & \square \\ \square & \square & \square \end{vmatrix}$$

$$= \underline{\hspace{1cm}} \mathbf{i} + \underline{\hspace{1cm}} \mathbf{j} + \underline{\hspace{1cm}} \mathbf{k}$$

So the cross product of $\mathbf{a} = \langle 1, 0, 1 \rangle$ and $\mathbf{b} = \langle 2, 3, 0 \rangle$

is $\mathbf{a} \times \mathbf{b} = \underline{\hspace{2cm}}$.

2. The cross product of two vectors $\mathbf{a}$ and $\mathbf{b}$ is _____ to $\mathbf{a}$ and to $\mathbf{b}$. Thus if both vectors $\mathbf{a}$ and $\mathbf{b}$ lie in a plane, the vector $\mathbf{a} \times \mathbf{b}$ is _____ to the plane.

SKILLS

3–8 ■ For the given vectors $\mathbf{a}$ and $\mathbf{b}$, find the cross product $\mathbf{a} \times \mathbf{b}$.

✎ **3.** $\mathbf{a} = \langle 1, 0, -3 \rangle$, $\mathbf{b} = \langle 2, 3, 0 \rangle$

4. $\mathbf{a} = \langle 0, -4, 1 \rangle$, $\mathbf{b} = \langle 1, 1, -2 \rangle$

5. $\mathbf{a} = \langle 6, -2, 8 \rangle$, $\mathbf{b} = \langle -9, 3, -12 \rangle$

6. $\mathbf{a} = \langle -2, 3, 4 \rangle$, $\mathbf{b} = \langle \frac{1}{6}, -\frac{1}{4}, -\frac{1}{3} \rangle$

7. $\mathbf{a} = \mathbf{i} + \mathbf{j} + \mathbf{k}$, $\mathbf{b} = 3\mathbf{i} - 4\mathbf{k}$

8. $\mathbf{a} = 3\mathbf{i} - \mathbf{j}$, $\mathbf{b} = -3\mathbf{j} + \mathbf{k}$

9–12 ■ Two vectors $\mathbf{a}$ and $\mathbf{b}$ are given. **(a)** Find a vector perpendicular to both $\mathbf{a}$ and $\mathbf{b}$. **(b)** Find a unit vector perpendicular to both $\mathbf{a}$ and $\mathbf{b}$.

✎ **9.** $\mathbf{a} = \langle 1, 1, -1 \rangle$, $\mathbf{b} = \langle -1, 1, -1 \rangle$

10. $\mathbf{a} = \langle 2, 5, 3 \rangle$, $\mathbf{b} = \langle 3, -2, -1 \rangle$

11. $\mathbf{a} = \frac{1}{2}\mathbf{i} - \mathbf{j} + \frac{2}{3}\mathbf{k}$, $\mathbf{b} = 6\mathbf{i} - 12\mathbf{j} - 6\mathbf{k}$

12. $\mathbf{a} = 3\mathbf{j} + 5\mathbf{k}$, $\mathbf{b} = -\mathbf{i} + 2\mathbf{k}$

13–16 ■ The lengths of two vectors $\mathbf{a}$ and $\mathbf{b}$ and the angle θ between them are given. Find the length of their cross product, $|\mathbf{a} \times \mathbf{b}|$.

13. $|\mathbf{a}| = 6$, $|\mathbf{b}| = \frac{1}{2}$, $\theta = 60°$

14. $|\mathbf{a}| = 4$, $|\mathbf{b}| = 5$, $\theta = 30°$

15. $|\mathbf{a}| = 10$, $|\mathbf{b}| = 10$, $\theta = 90°$

16. $|\mathbf{a}| = 0.12$, $|\mathbf{b}| = 1.25$, $\theta = 75°$

17–20 ■ Find a vector that is perpendicular to the plane passing through the three given points.

✎ **17.** $P(0, 1, 0)$, $Q(1, 2, -1)$, $R(-2, 1, 0)$

18. $P(3, 4, 5)$, $Q(1, 2, 3)$, $R(4, 7, 6)$

19. $P(1, 1, -5)$, $Q(2, 2, 0)$, $R(0, 0, 0)$

20. $P(3, 0, 0)$, $Q(0, 2, -5)$, $R(-2, 0, 6)$

21–24 ■ Find the area of the parallelogram determined by the given vectors.

✎ **21.** $\mathbf{u} = \langle 3, 2, 1 \rangle$, $\mathbf{v} = \langle 1, 2, 3 \rangle$

22. $\mathbf{u} = \langle 0, -3, 2 \rangle$, $\mathbf{v} = \langle 5, -6, 0 \rangle$

23. $\mathbf{u} = 2\mathbf{i} - \mathbf{j} + 4\mathbf{k}$, $\mathbf{v} = \frac{1}{2}\mathbf{i} + 2\mathbf{j} - \frac{3}{2}\mathbf{k}$

24. $\mathbf{u} = \mathbf{i} - \mathbf{j} + \mathbf{k}$, $\mathbf{v} = \mathbf{i} + \mathbf{j} - \mathbf{k}$

25–28 ■ Find the area of $\triangle PQR$.

✎ **25.** $P(1, 0, 1)$, $Q(0, 1, 0)$, $R(2, 3, 4)$

26. $P(2, 1, 0)$, $Q(0, 0, -1)$, $R(-4, 2, 0)$

27. $P(6, 0, 0)$, $Q(0, -6, 0)$, $R(0, 0, -6)$

28. $P(3, -2, 6)$, $Q(-1, -4, -6)$, $R(3, 4, 6)$

29–34 ■ Three vectors **a**, **b**, and **c** are given. **(a)** Find their scalar triple product $\mathbf{a} \cdot (\mathbf{b} \times \mathbf{c})$. **(b)** Are the vectors coplanar? If not, find the volume of the parallelepiped that they determine.

29. $\mathbf{a} = \langle 1, 2, 3 \rangle$, $\quad \mathbf{b} = \langle -3, 2, 1 \rangle$, $\quad \mathbf{c} = \langle 0, 8, 10 \rangle$

30. $\mathbf{a} = \langle 3, 0, -4 \rangle$, $\quad \mathbf{b} = \langle 1, 1, 1 \rangle$, $\quad \mathbf{c} = \langle 7, 4, 0 \rangle$

31. $\mathbf{a} = \langle 2, 3, -2 \rangle$, $\quad \mathbf{b} = \langle -1, 4, 0 \rangle$, $\quad \mathbf{c} = \langle 3, -1, 3 \rangle$

32. $\mathbf{a} = \langle 1, -1, 0 \rangle$, $\quad \mathbf{b} = \langle -1, 0, 1 \rangle$, $\quad \mathbf{c} = \langle 0, -1, 1 \rangle$

33. $\mathbf{a} = \mathbf{i} - \mathbf{j} + \mathbf{k}$, $\quad \mathbf{b} = -\mathbf{j} + \mathbf{k}$, $\quad \mathbf{c} = \mathbf{i} + \mathbf{j} + \mathbf{k}$

34. $\mathbf{a} = 2\mathbf{i} - 2\mathbf{j} - 3\mathbf{k}$, $\quad \mathbf{b} = 3\mathbf{i} - \mathbf{j} - \mathbf{k}$, $\quad \mathbf{c} = 6\mathbf{i}$

APPLICATIONS

35. Volume of a Fish Tank A fish tank in an avant-garde restaurant is in the shape of a parallelepiped with a rectangular base that is 300 cm long and 120 cm wide. The front and back faces are vertical, but the left and right faces are slanted at 30° from the vertical and measure 120 cm by 150 cm. (See the figure.)

(a) Let **a**, **b**, and **c** be the three vectors shown in the figure. Find $\mathbf{a} \cdot (\mathbf{b} \times \mathbf{c})$.   [*Hint:* Recall that $\mathbf{u} \cdot \mathbf{v} = |\mathbf{u}| |\mathbf{v}| \cos \theta$ and $|\mathbf{u} \times \mathbf{v}| = |\mathbf{u}| |\mathbf{v}| \sin \theta$.]

(b) What is the capacity of the tank in liters? [*Note:* 1 L = 1000 cm³.]

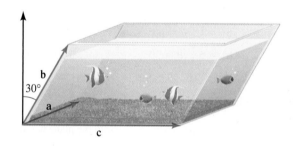

36. Rubik's Tetrahedron Rubik's Cube, a puzzle craze of the 1980s that remains popular to this day, inspired many similar puzzles. The one illustrated in the figure is called Rubik's

Tetrahedron; it is in the shape of a regular tetrahedron, with each edge $\sqrt{2}$ inches long. The volume of a regular tetrahedron is one-sixth the volume of the parallelepiped determined by any three edges that meet at a corner.

(a) Use the triple product to find the volume of Rubik's Tetrahedron.   [*Hint:* See Exercise 48 in Section 6.4, which gives the corners of a tetrahedron that has the same shape and size as Rubik's Tetrahedron.]

(b) Construct six identical regular tetrahedra using modeling clay. Experiment to see how they can be put together to create a parallelepiped that is determined by three edges of one of the tetrahedra (thus confirming the above statement about the volume of a regular tetrahedron).

DISCOVERY ■ DISCUSSION ■ WRITING

37. Order of Operations in the Triple Product Given three vectors **u**, **v**, and **w**, their scalar triple product can be performed in six different orders:

$$\mathbf{u} \cdot (\mathbf{v} \times \mathbf{w}), \quad \mathbf{u} \cdot (\mathbf{w} \times \mathbf{v}), \quad \mathbf{v} \cdot (\mathbf{u} \times \mathbf{w}),$$

$$\mathbf{v} \cdot (\mathbf{w} \times \mathbf{u}), \quad \mathbf{w} \cdot (\mathbf{u} \times \mathbf{v}), \quad \mathbf{w} \cdot (\mathbf{v} \times \mathbf{u})$$

(a) Calculate each of these six triple products for the vectors:

$$\mathbf{u} = \langle 0, 1, 1 \rangle, \quad \mathbf{v} = \langle 1, 0, 1 \rangle, \quad \mathbf{w} = \langle 1, 1, 0 \rangle$$

(b) On the basis of your observations in part (a), make a conjecture about the relationships between these six triple products.

(c) Prove the conjecture you made in part (b).

6.6 EQUATIONS OF LINES AND PLANES

| Equations of Lines ▶ Equations of Planes

In this section we find equations for lines and planes in a three-dimensional coordinate space. We use vectors to help us find such equations.

▼ Equations of Lines

The **position vector** of a point (a_1, a_2, a_3) is the vector $\langle a_1, a_2, a_3 \rangle$; that is, it is the vector from the origin to the point.

A line L in three-dimensional space is determined when we know a point $P_0(x_0, y_0, z_0)$ on L and the direction of L. In three dimensions the direction of a line is described by a vector **v** parallel to L. If we let $\mathbf{r}_0$ be the position vector of P_0 (that is, the vector $\overrightarrow{OP_0}$), then for all real numbers t, the terminal points P of the position vectors $\mathbf{r}_0 + t\mathbf{v}$ trace out a line

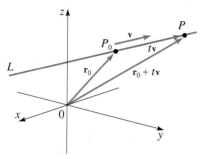

FIGURE 1

parallel to **v** and passing through P_0 (see Figure 1). Each value of the parameter t gives a point P on L. So the line L is given by the position vector **r**, where

$$\mathbf{r} = \mathbf{r}_0 + t\mathbf{v}$$

for $t \in \mathbb{R}$. This is the **vector equation of a line**.

Let's write the vector **v** in component form $\mathbf{v} = \langle a, b, c \rangle$ and let $\mathbf{r}_0 = \langle x_0, y_0, z_0 \rangle$ and $\mathbf{r} = \langle x, y, z \rangle$. Then the vector equation of the line becomes

$$\langle x, y, z \rangle = \langle x_0, y_0, z_0 \rangle + t \langle a, b, c \rangle$$
$$= \langle x_0 + ta, y_0 + tb, z_0 + tc \rangle$$

Since two vectors are equal if and only if their corresponding components are equal, we have the following result.

PARAMETRIC EQUATIONS FOR A LINE

A line passing through the point $P(x_0, y_0, z_0)$ and parallel to the vector $\mathbf{v} = \langle a, b, c \rangle$ is described by the parametric equations

$$x = x_0 + at$$
$$y = y_0 + bt$$
$$z = z_0 + ct$$

where t is any real number.

EXAMPLE 1 | Equations of a Line

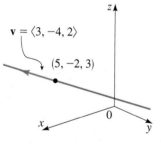

FIGURE 2 Line through $(5, -2, 3)$ with direction $\mathbf{v} = \langle 3, -4, 2 \rangle$

Find parametric equations for the line that passes through the point $(5, -2, 3)$ and is parallel to the vector $\mathbf{v} = \langle 3, -4, 2 \rangle$.

SOLUTION   We use the above formula to find the parametric equations:

$$x = 5 + 3t$$
$$y = -2 - 4t$$
$$z = 3 + 2t$$

where t is any real number. (See Figure 2.)

✎ NOW TRY EXERCISE **3**

EXAMPLE 2 | Equations of a Line

Find parametric equations for the line that passes through the points $(-1, 2, 6)$ and $(2, -3, -7)$.

SOLUTION   We first find a vector determined by the two points:

$$\mathbf{v} = \langle 2 - (-1), -3 - 2, -7 - 6 \rangle = \langle 3, -5, -13 \rangle$$

Now we use **v** and the point $(-1, 2, 6)$ to find the parametric equations:

$$x = -1 + 3t$$
$$y = 2 - 5t$$
$$z = 6 - 13t$$

where t is any real number. A graph of the line is shown in Figure 3.

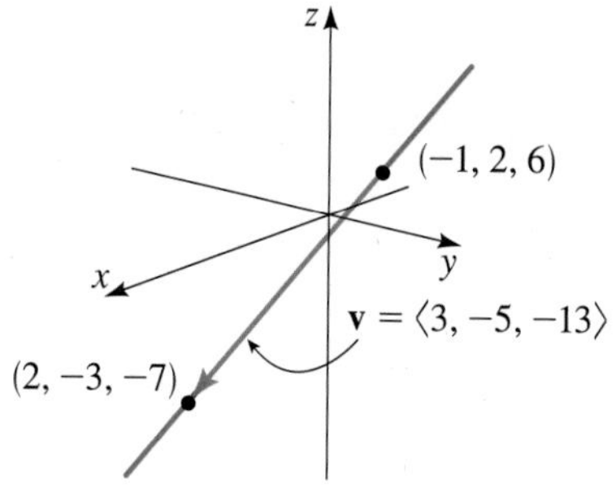

FIGURE 3 Line through $(-1, 2, 6)$ and $(2, -3, -7)$

✎ NOW TRY EXERCISE **9**

In Example 2 we used the point $(-1, 2, 6)$ to get the parametric equations of the line. We could instead use the point $(2, -3, -7)$. The resulting parametric equations would look different but would still describe the same line (see Exercise 37).

▼ Equations of Planes

Although a line in space is determined by a point and a direction, the "direction" of a plane cannot be described by a vector in the plane. In fact, different vectors in a plane can have different directions. But a vector perpendicular to a plane *does* completely specify the direction of the plane. Thus a plane in space is determined by a point $P_0(x_0, y_0, z_0)$ in the plane and a vector $\mathbf{n}$ that is orthogonal to the plane. This orthogonal vector $\mathbf{n}$ is called a **normal vector**. To determine whether a point $P(x, y, z)$ is in the plane, we check whether the vector $\overrightarrow{P_0P}$ with initial point P_0 and terminal point P is orthogonal to the normal vector. Let $\mathbf{r}_0$ and $\mathbf{r}$ be the position vectors of P_0 and P, respectively. Then the vector $\overrightarrow{P_0P}$ is represented by $\mathbf{r} - \mathbf{r}_0$ (see Figure 4). So the plane is described by the tips of the vectors $\mathbf{r}$ satisfying

$$\mathbf{n} \cdot (\mathbf{r} - \mathbf{r}_0) = 0$$

This is the **vector equation of the plane**.

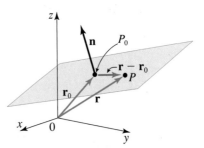

FIGURE 4

Let's write the normal vector $\mathbf{n}$ in component form $\mathbf{n} = \langle a, b, c \rangle$ and let $\mathbf{r}_0 = \langle x_0, y_0, z_0 \rangle$ and $\mathbf{r} = \langle x, y, z \rangle$. Then the vector equation of the plane becomes

$$\langle a, b, c \rangle \cdot \langle x - x_0, y - y_0, z - z_0 \rangle = 0$$

Performing the dot product, we arrive at the following equation of the plane in the variables x, y, and z.

EQUATION OF A PLANE

The plane containing the point $P(x_0, y_0, z_0)$ and having the normal vector $\mathbf{n} = \langle a, b, c \rangle$ is described by the equation

$$a(x - x_0) + b(y - y_0) + c(z - z_0) = 0$$

EXAMPLE 3 | Finding an Equation for a Plane

A plane has normal vector $\mathbf{n} = \langle 4, -6, 3 \rangle$ and passes through the point $P(3, -1, -2)$.

(a) Find an equation of the plane.

(b) Find the intercepts, and sketch a graph of the plane.

SOLUTION

(a) By the above formula for the equation of a plane we have

$$4(x - 3) - 6(y - (-1)) + 3(z - (-2)) = 0 \qquad \text{Formula}$$

$$4x - 12 - 6y - 6 + 3z + 6 = 0 \qquad \text{Expand}$$

$$4x - 6y + 3z = 12 \qquad \text{Simplify}$$

Thus an equation of the plane is $4x - 6y + 3z = 32$.

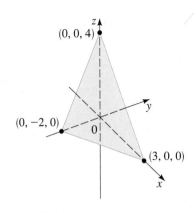

FIGURE 5 The plane
$4x - 6y + 3z = 32$

Notice that in Figure 5 the axes have been rotated so that we get a better view.

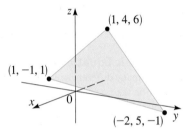

FIGURE 6 A plane through three points

(b) To find the x-intercept, we set $y = 0$ and $z = 0$ in the equation of the plane and solve for x. Similarly, we find the y- and z- intercepts.

x-intercept: Setting $y = 0$, $z = 0$, we get $x = 3$.

y-intercept: Setting $x = 0$, $z = 0$, we get $y = -2$.

z-intercept: Setting $x = 0$, $y = 0$, we get $z = 4$.

So the graph of the plane intersects the coordinate axes at the points $(3, 0, 0)$, $(0, -2, 0)$, and $(0, 0, 4)$. This enables us to sketch the portion of the plane shown in Figure 5.

NOW TRY EXERCISE **15**

EXAMPLE 4 | Finding an Equation for a Plane

Find an equation of the plane that passes through the points $P(1, 4, 6)$, $Q(-2, 5, -1)$, and $R(1, -1, 1)$.

SOLUTION   The vector $\mathbf{n} = \overrightarrow{PQ} \times \overrightarrow{PR}$ is perpendicular to both $\overrightarrow{PQ}$ and $\overrightarrow{PR}$ and is therefore perpendicular to the plane through P, Q, and R. In Example 3 of Section 6.5 we found $\overrightarrow{PQ} \times \overrightarrow{PR} = \langle -40, -15, 15 \rangle$. Using the formula for an equation of a plane, we have

$$-40(x - 1) - 15(y - 4) + 15(z - 6) = 0 \qquad \text{Formula}$$
$$-40x + 40 - 15y + 60 + 15z - 90 = 0 \qquad \text{Expand}$$
$$-40x - 15y + 15z = -10 \qquad \text{Simplify}$$
$$-8x - 3y + 3z = -2 \qquad \text{Divide by 5}$$

So an equation of the plane is $-8x - 3y + 3z = -2$. A graph of this plane is shown in Figure 6.

NOW TRY EXERCISE **21**

In Example 4 we used the point P to obtain the equation of the plane. You can check that using Q or R gives the same equation.

6.6 EXERCISES

CONCEPTS

1. A line in space is described algebraically by using

_____ equations. The line that passes through the point $P(x_0, y_0, z_0)$ and is parallel to the vector $\mathbf{v} = \langle a, b, c \rangle$ is described by the equations $x = $ _____,

$y = $ _____, $z = $ _____.

2. The plane containing the point $P(x_0, y_0, z_0)$ and having the normal vector $\mathbf{n} = \langle a, b, c \rangle$ is described algebraically by the equation _____.

SKILLS

3–8 ■ Find parametric equations for the line that passes through the point P and is parallel to the vector $\mathbf{v}$.

3. $P(1, 0, -2)$, $\mathbf{v} = \langle 3, 2, -3 \rangle$

4. $P(0, -5, 3)$, $\mathbf{v} = \langle 2, 0, -4 \rangle$

5. $P(3, 2, 1)$, $\mathbf{v} = \langle 0, -4, 2 \rangle$

6. $P(0, 0, 0)$, $\mathbf{v} = \langle -4, 3, 5 \rangle$

7. $P(1, 0, -2)$, $\mathbf{v} = 2\mathbf{i} - 5\mathbf{k}$

8. $P(1, 1, 1)$, $\mathbf{v} = \mathbf{i} - \mathbf{j} + \mathbf{k}$

9–14 ■ Find parametric equations for the line that passes through the points P and Q.

9. $P(1, -3, 2)$, $Q(2, 1, -1)$ **10.** $P(2, -1, -2)$, $Q(0, 1, -3)$

11. $P(1, 1, 0)$, $Q(0, 2, 2)$ **12.** $P(3, 3, 3)$, $Q(7, 0, 0)$

13. $P(3, 7, -5)$, $Q(7, 3, -5)$

14. $P(12, 16, 18)$, $Q(12, -6, 0)$

15–20 ■ A plane has normal vector $\mathbf{n}$ and passes through the point P. **(a)** Find an equation for the plane. **(b)** Find the intercepts and sketch a graph of the plane.

15. $\mathbf{n} = \langle 1, 1, -1 \rangle$, $P(0, 2, -3)$

16. $\mathbf{n} = \langle 3, 2, 0 \rangle$, $P(1, 2, 7)$

17. $\mathbf{n} = \langle 3, 0, -\frac{1}{2} \rangle$, $P(2, 4, 8)$

18. $\mathbf{n} = \langle -\frac{2}{3}, -\frac{1}{3}, 1 \rangle$, $P(-6, 0, -3)$

19. $\mathbf{n} = 3\mathbf{i} - \mathbf{j} + 2\mathbf{k}$, $P(0, 2, -3)$

20. $\mathbf{n} = \mathbf{i} + 4\mathbf{j}$, $P(1, 0, -9)$

21–26 ■ Find an equation of the plane that passes through the points P, Q, and R.

21. $P(6, -2, 1)$, $Q(5, -3, -1)$, $R(7, 0, 0)$

22. $P(3, 4, 5)$, $Q(1, 2, 3)$, $R(4, 7, 6)$

23. $P(3, \frac{1}{3}, -5)$, $Q(4, \frac{2}{3}, -3)$, $R(2, 0, 1)$

24. $P(\frac{3}{2}, 4, -2)$, $Q(-\frac{1}{2}, 2, 0)$, $R(-\frac{1}{2}, 0, 2)$

25. $P(6, 1, 1)$, $Q(3, 2, 0)$, $R(0, 0, 0)$

26. $P(2, 0, 0)$, $Q(0, 2, -2)$, $R(0, 0, 4)$

27–30 ■ A description of a line is given. Find parametric equations for the line.

27. The line crosses the z-axis where $z = 4$ and crosses the xy-plane where $x = 2$ and $y = 5$.

28. The line crosses the x-axis where $x = -2$ and crosses the z-axis where $z = 10$.

29. The line perpendicular to the xz-plane that contains the point $(2, -1, 5)$.

30. The line parallel to the y-axis that crosses the xz-plane where $x = -3$ and $z = 2$.

31–34 ■ A description of a plane is given. Find an equation for the plane.

31. The plane that crosses the x-axis where $x = 1$, the y-axis where $y = 3$, and the z-axis where $z = 4$.

32. The plane that crosses the x-axis where $x = -2$, the y-axis where $y = -1$, and the z-axis where $z = 3$.

33. The plane that is parallel to the plane $x - 2y + 4z = 6$ and contains the origin.

34. The plane that contains the line $x = 1 - t$, $y = 2 + t$, $z = -3t$ and the point $P(2, 0, -6)$. [*Hint:* A vector from any point on the line to P will lie in the plane.]

DISCOVERY ■ DISCUSSION ■ WRITING

35. Intersection of a Line and a Plane A line has parametric equations

$$x = 2 + t, \qquad y = 3t, \qquad z = 5 - t$$

and a plane has equation $5x - 2y - 2z = 1$.

(a) For what value of t does the corresponding point on the line intersect the plane?

(b) At what point do the line and the plane intersect?

36. Lines and Planes A line is parallel to the vector $\mathbf{v}$, and a plane has normal vector $\mathbf{n}$.

(a) If the line is perpendicular to the plane, what is the relationship between $\mathbf{v}$ and $\mathbf{n}$ (parallel or perpendicular)?

(b) If the line is parallel to the plane (that is, the line and the plane do not intersect), what is the relationship between $\mathbf{v}$ and $\mathbf{n}$ (parallel or perpendicular)?

(c) Parametric equations for two lines are given. Which line is parallel to the plane $x - y + 4z = 6$? Which line is perpendicular to this plane?

Line 1: $x = 2t$, $y = 3 - 2t$, $z = 4 + 8t$

Line 2: $x = -2t$, $y = 5 + 2t$, $z = 3 + t$

37. Same Line: Different Parametric Equations Every line can be described by infinitely many different sets of parametric equations, since *any* point on the line and *any* vector parallel to the line can be used to construct the equations. But how can we tell whether two sets of parametric equations represent the same line? Consider the following two sets of parametric equations:

Line 1: $x = 1 - t$, $y = 3t$, $z = -6 + 5t$

Line 2: $x = -1 + 2t$, $y = 6 - 6t$, $z = 4 - 10t$

(a) Find two points that lie on Line 1 by setting $t = 0$ and $t = 1$ in its parametric equations. Then show that these points also lie on Line 2 by finding two values of the parameter that give these points when substituted into the parametric equations for Line 2.

(b) Show that the following two lines are not the same by finding a point on Line 3 and then showing that it does not lie on Line 4.

Line 3: $x = 4t$, $y = 3 - 6t$, $z = -5 + 2t$

Line 4: $x = 8 - 2t$, $y = -9 + 3t$, $z = 6 - t$

CHAPTER 6 | REVIEW

■ CONCEPT CHECK

1. (a) What is the difference between a scalar and a vector?

(b) Draw a diagram to show how to add two vectors.

(c) Draw a diagram to show how to subtract two vectors.

(d) Draw a diagram to show how to multiply a vector by the scalars $2, \frac{1}{2}, -2$, and $-\frac{1}{2}$.

2. If $\mathbf{u} = \langle u_1, u_2 \rangle$ and $\mathbf{v} = \langle v_1, v_2 \rangle$ are vectors in two dimensions and c is a scalar, write expressions for $\mathbf{u} + \mathbf{v}$, $\mathbf{u} - \mathbf{v}$, and $c\mathbf{u}$.

3. If $\mathbf{u} = \langle u_1, u_2 \rangle$ and $\mathbf{v} = \langle v_1, v_2, v_3 \rangle$ are vectors in two and three dimensions, respectively, write expressions for their magnitudes $|\mathbf{u}|$ and $|\mathbf{v}|$.

4. (a) If $\mathbf{u} = \langle u_1, u_2 \rangle$, write $\mathbf{u}$ in terms of $\mathbf{i}$ and $\mathbf{j}$.

(b) If $\mathbf{v} = \langle v_1, v_2, v_3 \rangle$, write $\mathbf{v}$ in terms of $\mathbf{i}$, $\mathbf{j}$, and $\mathbf{k}$.

5. Write the components of the vector $\mathbf{u} = \langle u_1, u_2 \rangle$ in terms of its magnitude $|\mathbf{u}|$ and direction θ.

6. Express the dot product $\mathbf{u} \cdot \mathbf{v}$ in terms of the components of the vectors.

   (a) $\mathbf{u} = \langle u_1, u_2 \rangle, \quad \mathbf{v} = \langle v_1, v_2 \rangle$

   (b) $\mathbf{u} = \langle u_1, u_2, u_3 \rangle, \quad \mathbf{v} = \langle v_1, v_2, v_3 \rangle$

7. (a) How do you use the dot product to find the angle between two vectors?

   (b) How do you use the dot product to determine whether two vectors are perpendicular?

8. What is the component of $\mathbf{u}$ along $\mathbf{v}$, and how do you calculate it?

9. What is the projection of $\mathbf{u}$ onto $\mathbf{v}$, and how do you calculate it?

10. How much work is done by the force $\mathbf{F}$ in moving an object along a displacement $\mathbf{D}$?

11. How do you find the distance between two points $P(x_1, y_1, z_1)$ and $Q(x_2, y_2, z_2)$ in three-dimensional space?

12. What is the equation of the sphere with center $C(a, b, c)$ and radius r?

13. (a) How do you calculate the cross product $\mathbf{a} \times \mathbf{b}$ if you know the components of $\mathbf{a}$ and $\mathbf{b}$?

   (b) How do you calculate $\mathbf{a} \times \mathbf{b}$ if you know the lengths of $\mathbf{a}$ and $\mathbf{b}$ and the angle between them?

   (c) What is the angle between $\mathbf{a} \times \mathbf{b}$ and each of $\mathbf{a}$ and $\mathbf{b}$?

14. (a) How do you find the area of the parallelogram determined by $\mathbf{a}$ and $\mathbf{b}$?

   (b) How do you find the volume of the parallelepiped determined by $\mathbf{a}$, $\mathbf{b}$, and $\mathbf{c}$?

15. Write parametric equations for the line that contains the point $P(x_0, y_0, z_0)$ and that is parallel to the vector $\mathbf{v} = \langle a, b, c \rangle$.

16. Write an equation for the plane that contains the point $P(x_0, y_0, z_0)$ and has normal vector $\mathbf{n} = \langle a, b, c \rangle$.

17. How do you find parametric equations for the line that contains the points $P(x_1, y_1, z_1)$ and $Q(x_2, y_2, z_2)$?

18. How do you find an equation for the plane that contains the points $P(x_1, y_1, z_1)$, $Q(x_2, y_2, z_2)$, and $R(x_3, y_3, z_3)$?

■ EXERCISES

Exercises 1–24 deal with vectors in two dimensions.

1–4 ■ Find $|\mathbf{u}|$, $\mathbf{u} + \mathbf{v}$, $\mathbf{u} - \mathbf{v}$, $2\mathbf{u}$, and $3\mathbf{u} - 2\mathbf{v}$.

1. $\mathbf{u} = \langle -2, 3 \rangle, \quad \mathbf{v} = \langle 8, 1 \rangle$

2. $\mathbf{u} = \langle 5, -2 \rangle, \quad \mathbf{v} = \langle -3, 0 \rangle$

3. $\mathbf{u} = 2\mathbf{i} + \mathbf{j}, \quad \mathbf{v} = \mathbf{i} - 2\mathbf{j}$

4. $\mathbf{u} = 3\mathbf{j}, \quad \mathbf{v} = -\mathbf{i} + 2\mathbf{j}$

5. Find the vector with initial point $P(0, 3)$ and terminal point $Q(3, -1)$.

6. If the vector $5\mathbf{i} - 8\mathbf{j}$ is placed in the plane with its initial point at $P(5, 6)$, find its terminal point.

7–8 ■ Find the length and direction of the given vector.

7. $\mathbf{u} = \langle -2, 2\sqrt{3} \rangle$ **8.** $\mathbf{v} = 2\mathbf{i} - 5\mathbf{j}$

9–10 ■ The length $|\mathbf{u}|$ and direction θ of a vector $\mathbf{u}$ are given. Express $\mathbf{u}$ in component form.

9. $|\mathbf{u}| = 20, \quad \theta = 60°$ **10.** $|\mathbf{u}| = 13.5, \quad \theta = 125°$

11. Two tugboats are pulling a barge as shown in the figure. One pulls with a force of 2.0×10^4 lb in the direction N 50° E, and the other pulls with a force of 3.4×10^4 lb in the direction S 75° E.

   (a) Find the resultant force on the barge as a vector.

   (b) Find the magnitude and direction of the resultant force.

12. An airplane heads N 60° E at a speed of 600 mi/h relative to the air. A wind begins to blow in the direction N 30° W at 50 mi/h. (See the figure.)

   (a) Find the velocity of the airplane as a vector.

   (b) Find the true speed and direction of the airplane.

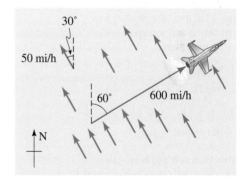

13–16 ■ Find the vectors $|\mathbf{u}|$, $\mathbf{u} \cdot \mathbf{u}$, and $\mathbf{u} \cdot \mathbf{v}$.

13. $\mathbf{u} = \langle 4, -3 \rangle, \quad \mathbf{v} = \langle 9, -8 \rangle$

14. $\mathbf{u} = \langle 5, 12 \rangle, \quad \mathbf{v} = \langle 10, -4 \rangle$

15. $\mathbf{u} = -2\mathbf{i} + 2\mathbf{j}, \quad \mathbf{v} = \mathbf{i} + \mathbf{j}$

16. $\mathbf{u} = 10\mathbf{j}, \quad \mathbf{v} = 5\mathbf{i} - 3\mathbf{j}$

17–20 ■ Are $\mathbf{u}$ and $\mathbf{v}$ orthogonal? If not, find the angle between them.

17. $\mathbf{u} = \langle -4, 2 \rangle, \quad \mathbf{v} = \langle 3, 6 \rangle$

18. $\mathbf{u} = \langle 5, 3 \rangle, \quad \mathbf{v} = \langle -2, 6 \rangle$

19. $\mathbf{u} = 2\mathbf{i} + \mathbf{j}, \quad \mathbf{v} = \mathbf{i} + 3\mathbf{j}$

20. $\mathbf{u} = \mathbf{i} - \mathbf{j}, \quad \mathbf{v} = \mathbf{i} + \mathbf{j}$

21–24 ■ Two vectors **u** and **v** are given.
  (a) Find the component of **u** along **v**.
  (b) Find $\text{proj}_\mathbf{v}\,\mathbf{u}$.
  (c) Resolve **u** into the vectors $\mathbf{u}_1$ and $\mathbf{u}_2$, where $\mathbf{u}_1$ is parallel to **v** and $\mathbf{u}_2$ is perpendicular to **v**.

21. $\mathbf{u} = \langle 3, 1 \rangle, \quad \mathbf{v} = \langle 6, -1 \rangle$

22. $\mathbf{u} = \langle -8, 6 \rangle, \quad \mathbf{v} = \langle 20, 20 \rangle$

23. $\mathbf{u} = \mathbf{i} + 2\mathbf{j}, \quad \mathbf{v} = 4\mathbf{i} - 9\mathbf{j}$

24. $\mathbf{u} = 2\mathbf{i} + 4\mathbf{j}, \quad \mathbf{v} = 10\mathbf{j}$

Exercises 25–54 deal with three-dimensional coordinate geometry.

25–26 ■ Plot the given points, and find the distance between them.

25. $P(1, 0, 2), \quad Q(3, -2, 3)$ **26.** $P(0, 2, 4), \quad Q(1, 3, 0)$

27–28 ■ Find an equation of the sphere with the given radius r and center C.

27. $r = 6, \quad C(0, 0, 0)$ **28.** $r = 2, \quad C(1, -2, 4)$

29–30 ■ Show that the equation represents a sphere, and find its center and radius.

29. $x^2 + y^2 + z^2 - 2x - 6y + 4z = 2$

30. $x^2 + y^2 + z^2 = 4y + 4z$

31–32 ■ Find $|\mathbf{u}|$, $\mathbf{u} + \mathbf{v}$, $\mathbf{u} - \mathbf{v}$, and $\frac{3}{4}\mathbf{u} - 2\mathbf{v}$.

31. $\mathbf{u} = \langle 4, -2, 4 \rangle, \quad \mathbf{v} = \langle 2, 3, -1 \rangle$

32. $\mathbf{u} = 6\mathbf{i} - 8\mathbf{k}, \quad \mathbf{v} = \mathbf{i} - \mathbf{j} + \mathbf{k}$

33–36 ■ Two vectors **u** and **v** are given.
  (a) Find their dot product $\mathbf{u} \cdot \mathbf{v}$.
  (b) Are **u** and **v** perpendicular? If not, find the angle between them.

33. $\mathbf{u} = \langle 3, -2, 4 \rangle, \quad \mathbf{v} = \langle 3, 1, -2 \rangle$

34. $\mathbf{u} = \langle 2, -6, 5 \rangle, \quad \mathbf{v} = \langle 1, -\frac{1}{2}, -1 \rangle$

35. $\mathbf{u} = 2\mathbf{i} - \mathbf{j} + 4\mathbf{k}, \quad \mathbf{v} = 3\mathbf{i} + 2\mathbf{j} - \mathbf{k}$

36. $\mathbf{u} = \mathbf{j} - \mathbf{k}, \quad \mathbf{v} = \mathbf{i} + \mathbf{j}$

37–40 ■ Two vectors **a** and **b** are given.
  (a) Find their cross product $\mathbf{a} \times \mathbf{b}$.
  (b) Find a unit vector **u** that is perpendicular to both **a** and **b**.

37. $\mathbf{a} = \langle 1, 1, 3 \rangle, \quad \mathbf{b} = \langle 5, 0, -2 \rangle$

38. $\mathbf{a} = \langle 2, 3, 0 \rangle, \quad \mathbf{b} = \langle 0, 4, -1 \rangle$

39. $\mathbf{a} = \mathbf{i} - \mathbf{j}, \quad \mathbf{b} = 2\mathbf{j} - \mathbf{k}$

40. $\mathbf{a} = \mathbf{i} + \mathbf{j} - \mathbf{k}, \quad \mathbf{b} = \mathbf{i} - \mathbf{j} + \mathbf{k}$

41. Find the area of the triangle with vertices $P(2, 1, 1)$, $Q(0, 0, 3)$, and $R(-2, 4, 0)$.

42. Find the area of the parallelogram determined by the vectors $\mathbf{a} = \langle 4, 1, 1 \rangle$ and $\mathbf{b} = \langle -1, 2, 2 \rangle$.

43. Find the volume of the parallelepiped determined by the vectors $\mathbf{a} = 2\mathbf{i} - \mathbf{j}$, $\mathbf{b} = 2\mathbf{j} + \mathbf{k}$, and $\mathbf{c} = 3\mathbf{i} + \mathbf{j} - \mathbf{k}$.

44. A parallelepiped has one vertex at the origin; the three edges that have the origin as one endpoint extend to the points $P(0, 2, 2)$, $Q(3, 1, -1)$, and $R(1, 4, 1)$. Find the volume of the parallelepiped.

45–46 ■ Find parametric equations for the line that passes through P and is parallel to **v**.

45. $P(2, 0, -6), \quad \mathbf{v} = \langle 3, 1, 0 \rangle$

46. $P(5, 2, 8), \quad \mathbf{v} = 2\mathbf{i} - \mathbf{j} + 5\mathbf{k}$

47–48 ■ Find parametric equations for the line that passes through the points P and Q.

47. $P(6, -2, -3), \quad Q(4, 1, -2)$

48. $P(1, 0, 0), \quad Q(3, -4, 2)$

49–50 ■ Find an equation for the plane with normal vector **n** and passing through the point P.

49. $\mathbf{n} = \langle 2, 3, -5 \rangle, \quad P(2, 1, 1)$

50. $\mathbf{n} = -\mathbf{i} - 2\mathbf{j} + 7\mathbf{k}, \quad P(-2, 5, 2)$

51–52 ■ Find an equation of the plane that passes through the points P, Q, and R.

51. $P(1, 1, 1), \quad Q(3, -4, 2), \quad R(6, -1, 0)$

52. $P(4, 0, 0), \quad Q(0, -3, 0), \quad R(0, 0, -5)$

53. Find parametric equations for the line that crosses the x-axis where $x = 2$ and the z-axis where $z = -4$.

54. Find an equation of the plane that contains the line $x = 2 + 2t, y = 4t, z = -6$ and the point $P(5, 3, 0)$.

1. Let $\mathbf{u}$ be the vector with initial point $P(3, -1)$ and terminal point $Q(-3, 9)$.
   (a) Graph $\mathbf{u}$ in the coordinate plane.
   (b) Express $\mathbf{u}$ in terms of $\mathbf{i}$ and $\mathbf{j}$.
   (c) Find the length of $\mathbf{u}$.

2. Let $\mathbf{u} = \langle 1, 3 \rangle$ and $\mathbf{v} = \langle -6, 2 \rangle$.
   (a) Find $\mathbf{u} - 3\mathbf{v}$.
   (b) Find $|\mathbf{u} + \mathbf{v}|$.
   (c) Find $\mathbf{u} \cdot \mathbf{v}$.
   (d) Are $\mathbf{u}$ and $\mathbf{v}$ perpendicular?

3. Let $\mathbf{u} = \langle -4\sqrt{3}, 4 \rangle$.
   (a) Graph $\mathbf{u}$ in the coordinate plane, with initial point $(0, 0)$.
   (b) Find the length and direction of $\mathbf{u}$.

4. A river is flowing due east at 8 mi/h. A man heads his motorboat in the direction N 30° E in the river. The speed of the motorboat relative to the water is 12 mi/h.
   (a) Express the true velocity of the motorboat as a vector.
   (b) Find the true speed and direction of the motorboat.

5. Let $\mathbf{u} = 3\mathbf{i} + 2\mathbf{j}$ and $\mathbf{v} = 5\mathbf{i} - \mathbf{j}$.
   (a) Find the angle between $\mathbf{u}$ and $\mathbf{v}$.
   (b) Find the component of $\mathbf{u}$ along $\mathbf{v}$.
   (c) Find $\text{proj}_{\mathbf{v}}\, \mathbf{u}$.

6. Find the work done by the force $\mathbf{F} = 3\mathbf{i} - 5\mathbf{j}$ in moving an object from the point $(2, 2)$ to the point $(7, -13)$.

7. Let $P(4, 3, -1)$ and $Q(6, -1, 3)$ be two points in three-dimensional space.
   (a) Find the distance between P and Q.
   (b) Find an equation for the sphere whose center is P and for which the segment $\overrightarrow{PQ}$ is a radius of the sphere.
   (c) The vector $\mathbf{u}$ has initial point P and terminal point Q. Express $\mathbf{u}$ both in component form and using the vectors $\mathbf{i}$, $\mathbf{j}$, and $\mathbf{k}$.

8. Calculate the given quantity if

$$\mathbf{a} = \mathbf{i} + \mathbf{j} - 2\mathbf{k} \qquad \mathbf{b} = 3\mathbf{i} - 2\mathbf{j} + \mathbf{k} \qquad \mathbf{c} = \mathbf{j} - 5\mathbf{k}$$

   (a) $2\mathbf{a} + 3\mathbf{b}$ (b) $|\mathbf{a}|$
   (c) $\mathbf{a} \cdot \mathbf{b}$ (d) $\mathbf{a} \times \mathbf{b}$
   (e) $|\mathbf{b} \times \mathbf{c}|$ (f) $\mathbf{a} \cdot (\mathbf{b} \times \mathbf{c})$
   (g) The angle between $\mathbf{a}$ and $\mathbf{b}$ (rounded to the nearest degree)

9. Find two unit vectors that are perpendicular to both $\mathbf{j} + 2\mathbf{k}$ and $\mathbf{i} - 2\mathbf{j} + 3\mathbf{k}$.

10. (a) Find a vector perpendicular to the plane that contains the points $P(1, 0, 0)$, $Q(2, 0, -1)$, and $R(1, 4, 3)$.
    (b) Find an equation for the plane that contains P, Q, and R.
    (c) Find the area of triangle PQR.

11. Find parametric equations for the line that contains the points $P(2, -4, 7)$ and $Q(0, -3, 5)$.

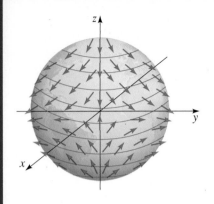

FIGURE 1 Wind represented by a vector field

To model the gravitational force near the earth or the flow of wind on a surface of the earth, we use vectors. For example, at each point on the surface of the earth air flows with a certain speed and direction. We represent the air currents by vectors. If we graph many of these vectors we get a "picture" or a graph of the flow of the air. (See Figure 1.)

▼ Vector Fields in the Plane

A **vector field** in the coordinate plane is a function that assigns a vector to each point in the plane (or to each point in some subset of the plane). For example,

$$\mathbf{F}(x, y) = x\mathbf{i} + y\mathbf{j}$$

is a vector field that assigns the vector $x\mathbf{i} + y\mathbf{j}$ to the point (x, y). We graph this vector field in the next example.

EXAMPLE 1 │ Graphing a Vector Field in the Plane

Graph the vector field $\mathbf{F}(x, y) = x\mathbf{i} + y\mathbf{j}$. What does the graph indicate?

SOLUTION   The table gives the vector field at several points. In Figure 2 we sketch the vectors in the table together with several other vectors in the vector field.

(x, y)	$\mathbf{F} = x\mathbf{i} + y\mathbf{j}$
$(1, 3)$	$\mathbf{i} + 3\mathbf{j}$
$(3, 3)$	$3\mathbf{i} + 3\mathbf{j}$
$(-4, 6)$	$-4\mathbf{i} + 6\mathbf{j}$
$(-6, -1)$	$-6\mathbf{i} - \mathbf{j}$
$(6, -6)$	$6\mathbf{i} - 6\mathbf{j}$

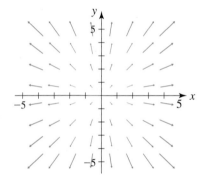

FIGURE 2

We see from the graph that the vectors in the field point away from the origin, and the farther from the origin, the greater the magnitude of the vector. ■

EXAMPLE 2 │ Graphing a Vector Field in the Plane

A potter's wheel has a radius of 5 inches. The velocity of each point on the wheel is given by the vector field $\mathbf{F}(x, y) = -y\mathbf{i} + x\mathbf{j}$. What does the graph indicate?

SOLUTION   The table gives the vector field at several points. In Figure 3 we sketch the vectors in the table.

(x, y)	$\mathbf{F}(x, y)$	(x, y)	$\mathbf{F}(x, y)$
$(1, 0)$	$\langle 0, 1 \rangle$	$(-1, 0)$	$\langle 0, -1 \rangle$
$(2, 2)$	$\langle -2, 2 \rangle$	$(-2, -2)$	$\langle 2, -2 \rangle$
$(3, 0)$	$\langle 0, 3 \rangle$	$(-3, 0)$	$\langle 0, -3 \rangle$
$(0, 1)$	$\langle -1, 0 \rangle$	$(0, -1)$	$\langle 1, 0 \rangle$
$(-2, 2)$	$\langle -2, -2 \rangle$	$(2, -2)$	$\langle 2, 2 \rangle$
$(0, 3)$	$\langle -3, 0 \rangle$	$(0, -3)$	$\langle 3, 0 \rangle$

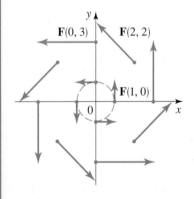

FIGURE 3

We see from the graph that the wheel is rotating counterclockwise and that the points at the edge of the wheel have a higher velocity than do the points near the center of the wheel. ■

Graphing vector fields requires graphing a lot of vectors. Some graphing calculators and computer programs are capable of graphing vector fields. You can also find many Internet sites that have applets for graphing vector fields. The vector field in Example 2 is graphed with a computer program in Figure 4. Notice how the computer scales the lengths of the vectors so that they are not too long yet are proportional to their true lengths.

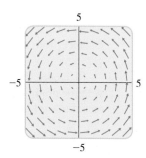

FIGURE 4

▼ Vector Fields in Space

A **vector field** in three-dimensional space is a function that assigns a vector to each point in space (or to each point in some subset of space). For example,

$$\mathbf{F}(x, y, z) = 2x\mathbf{i} - y\mathbf{j} + z^2\mathbf{k}$$

is a vector field that assigns the vector $2x\mathbf{i} - y\mathbf{j} + z^3\mathbf{k}$ to the point (x, y, z). In general, it is difficult to draw a vector field in space by hand, since we must draw many vectors with the proper perspective. The vector field in the next example is particularly simple, so we'll sketch it by hand.

EXAMPLE 3 │ Graphing a Vector Field in Space

Graph the vector field $\mathbf{F}(x, y, z) = z\mathbf{k}$. What does the graph indicate?

SOLUTION A graph is shown in Figure 5. Notice that all vectors are vertical and point upward above the xy-plane and downward below it. The magnitude of each vector increases with the distance from the xy-plane.

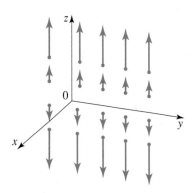

FIGURE 5

The gravitational pull of the earth in the space surrounding it is mathematically modeled by a vector field. According to Newton's Law of Gravitation, the gravitational force **F** is directed toward the center of the earth and is inversely proportional to the distance from the center of the earth. The magnitude of the force is

$$F = G\frac{Mm}{r^2}$$

where M is the mass of the earth, m is mass of an object in proximity to the earth, r is the distance from the object to the center of the earth, and G is the universal gravitational constant.

To model the gravitational force, let's place a three-dimensional coordinate system with the origin at the center of the earth. The gravitational force at the point (x, y, z) is directed toward the origin. A unit vector pointing toward the origin is

$$\mathbf{u} = -\frac{x\mathbf{i} + y\mathbf{j} + z\mathbf{k}}{\sqrt{x^2 + y^2 + z^2}}$$

To obtain the gravitational vector field, we multiply this unit vector by the appropriate magnitude, namely, GMm/r^2. Since the distance r from the point (x, y, z) to the origin is $r = \sqrt{x^2 + y^2 + z^2}$, it follows that $r^2 = x^2 + y^2 + z^2$. So we can express the gravitational vector field as

$$\mathbf{F}(x, y, z) = -GMm\frac{x\mathbf{i} + y\mathbf{j} + z\mathbf{k}}{(x^2 + y^2 + z^2)^{3/2}}$$

Some of the vectors in the gravitational field **F** are pictured in Figure 6.

FIGURE 6 The gravitational field

PROBLEMS

1–6 ■ Sketch the vector field **F** by drawing a diagram as in Figure 3.

1. $\mathbf{F}(x, y) = \frac{1}{2}\mathbf{i} + \frac{1}{2}\mathbf{j}$

2. $\mathbf{F}(x, y) = \mathbf{i} + x\mathbf{j}$

3. $\mathbf{F}(x, y) = y\mathbf{i} + \frac{1}{2}\mathbf{j}$

4. $\mathbf{F}(x, y) = (x - y)\mathbf{i} + x\mathbf{j}$

5. $\mathbf{F}(x, y) = \dfrac{y\mathbf{i} + x\mathbf{j}}{\sqrt{x^2 + y^2}}$

6. $\mathbf{F}(x, y) = \dfrac{y\mathbf{i} - x\mathbf{j}}{\sqrt{x^2 + y^2}}$

7–10 ■ Sketch the vector field **F** by drawing a diagram as in Figure 5.

7. $\mathbf{F}(x, y, z) = \mathbf{j}$

8. $\mathbf{F}(x, y, z) = \mathbf{j} - \mathbf{k}$

9. $\mathbf{F}(x, y, z) = z\mathbf{j}$

10. $\mathbf{F}(x, y, z) = y\mathbf{k}$

11–14 ■ Match the vector field **F** with the graphs labeled I–IV.

11. $\mathbf{F}(x, y) = \langle y, x \rangle$ **12.** $\mathbf{F}(x, y) = \langle 1, \sin y \rangle$

13. $\mathbf{F}(x, y) = \langle x - 2, x + 1 \rangle$ **14.** $\mathbf{F}(x, y) = \langle y, 1/x \rangle$

I

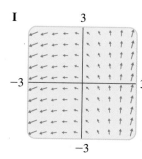

II

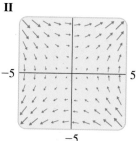

III

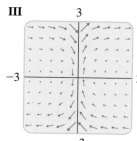

IV

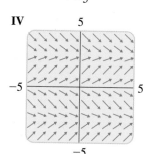

15–18 ■ Match the vector field **F** with the graphs labeled I–IV.

15. $\mathbf{F}(x, y, z) = \mathbf{i} + 2\mathbf{j} + 3\mathbf{k}$ **16.** $\mathbf{F}(x, y, z) = \mathbf{i} + 2\mathbf{j} + z\mathbf{k}$

17. $\mathbf{F}(x, y, z) = x\mathbf{i} + y\mathbf{j} + 3\mathbf{k}$ **18.** $\mathbf{F}(x, y, z) = x\mathbf{i} + y\mathbf{j} + z\mathbf{k}$

I

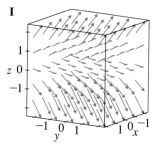

II

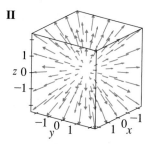

III

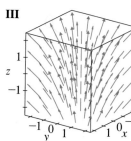

IV

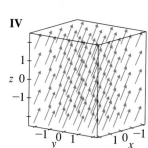

19. Flow Lines in a Current   The current in a turbulent bay is described by the velocity vector field

$$\mathbf{F}(x, y) = (x + y)\mathbf{i} + (x - y)\mathbf{j}$$

A graph of the vector field **F** is shown. If a small toy boat is put in this bay, we can tell from the graph of the vector field what path the boat would follow. Such paths are called **flow lines** (or **streamlines**) of the vector field. A streamline starting at $(1, -3)$ is shown in blue in the figure. Sketch streamlines starting at the given point.

(a) $(1, 4)$ **(b)** $(-2, 1)$ **(c)** $(-1, -2)$

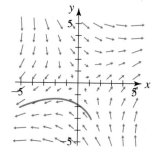

CONIC SECTIONS

Conic sections are the curves we get when we make a straight cut in a cone, as shown in the figure. For example, if a cone is cut horizontally, the cross section is a circle. So a circle is a conic section. Other ways of cutting a cone produce parabolas, ellipses, and hyperbolas.

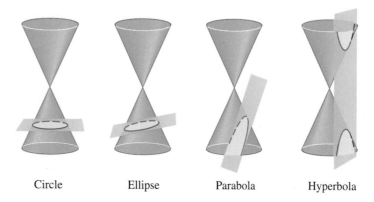

Circle   Ellipse   Parabola   Hyperbola

Our goal in this chapter is to find equations whose graphs are the conic sections. We already know from Section 1.1 that the graph of the equation $x^2 + y^2 = r^2$ is a circle. We will find equations for each of the other conic sections by analyzing their *geometric* properties.

The conic sections have interesting properties that make them useful for many real-world applications. For instance, a reflecting surface with parabolic cross-sections concentrates light at a single point. This property of a parabola is used in the construction of solar power plants, like the one in California pictured above.

7.1 PARABOLAS

▶ Geometric Definition of a Parabola ▶ Equations and Graphs of Parabolas
▶ Applications

▼ Geometric Definition of a Parabola

The graph of the equation

$$y = ax^2 + bx + c$$

is a U-shaped curve called a *parabola* that opens either upward or downward, depending on whether the sign of a is positive or negative.

In this section we study parabolas from a geometric rather than an algebraic point of view. We begin with the geometric definition of a parabola and show how this leads to the algebraic formula that we are already familiar with.

> ### GEOMETRIC DEFINITION OF A PARABOLA
>
> A **parabola** is the set of points in the plane that are equidistant from a fixed point F (called the **focus**) and a fixed line l (called the **directrix**).

This definition is illustrated in Figure 1. The **vertex** V of the parabola lies halfway between the focus and the directrix, and the **axis of symmetry** is the line that runs through the focus perpendicular to the directrix.

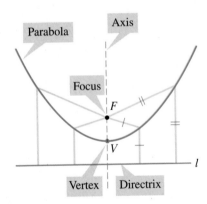

FIGURE 1

In this section we restrict our attention to parabolas that are situated with the vertex at the origin and that have a vertical or horizontal axis of symmetry. (Parabolas in more general positions will be considered in Sections 7.4 and 7.5.) If the focus of such a parabola is the point $F(0, p)$, then the axis of symmetry must be vertical, and the directrix has the equation $y = -p$. Figure 2 illustrates the case $p > 0$.

If $P(x, y)$ is any point on the parabola, then the distance from P to the focus F (using the Distance Formula) is

$$\sqrt{x^2 + (y - p)^2}$$

The distance from P to the directrix is

$$|y - (-p)| = |y + p|$$

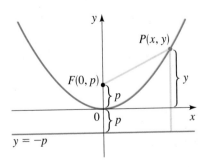

FIGURE 2

By the definition of a parabola these two distances must be equal:

$$\sqrt{x^2 + (y-p)^2} = |y+p|$$

$$x^2 + (y-p)^2 = |y+p|^2 = (y+p)^2 \qquad \text{Square both sides}$$

$$x^2 + y^2 - 2py + p^2 = y^2 + 2py + p^2 \qquad \text{Expand}$$

$$x^2 - 2py = 2py \qquad \text{Simplify}$$

$$x^2 = 4py$$

If $p > 0$, then the parabola opens upward; but if $p < 0$, it opens downward. When x is replaced by $-x$, the equation remains unchanged, so the graph is symmetric about the y-axis.

▼ Equations and Graphs of Parabolas

The following box summarizes what we have just proved about the equation and features of a parabola with a vertical axis.

PARABOLA WITH VERTICAL AXIS

The graph of the equation

$$x^2 = 4py$$

is a parabola with the following properties.

VERTEX	$V(0, 0)$
FOCUS	$F(0, p)$
DIRECTRIX	$y = -p$

The parabola opens upward if $p > 0$ or downward if $p < 0$.

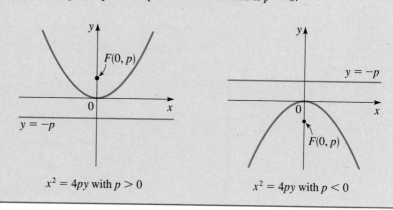

$x^2 = 4py$ with $p > 0$ $x^2 = 4py$ with $p < 0$

EXAMPLE 1 | Finding the Equation of a Parabola

Find an equation for the parabola with vertex $V(0, 0)$ and focus $F(0, 2)$, and sketch its graph.

SOLUTION   Since the focus is $F(0, 2)$, we conclude that $p = 2$ (so the directrix is $y = -2$). Thus the equation of the parabola is

$$x^2 = 4(2)y \qquad x^2 = 4py \text{ with } p = 2$$

$$x^2 = 8y$$

Since $p = 2 > 0$, the parabola opens upward. See Figure 3.

✎. NOW TRY EXERCISES **29** AND **41**

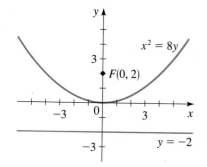

FIGURE 3

EXAMPLE 2 | Finding the Focus and Directrix of a Parabola from Its Equation

Find the focus and directrix of the parabola $y = -x^2$, and sketch the graph.

SOLUTION To find the focus and directrix, we put the given equation in the standard form $x^2 = -y$. Comparing this to the general equation $x^2 = 4py$, we see that $4p = -1$, so $p = -\frac{1}{4}$. Thus the focus is $F\left(0, -\frac{1}{4}\right)$, and the directrix is $y = \frac{1}{4}$. The graph of the parabola, together with the focus and the directrix, is shown in Figure 4(a). We can also draw the graph using a graphing calculator as shown in Figure 4(b).

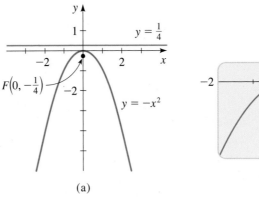

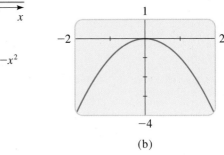

FIGURE 4
(a)   (b)

🔍 NOW TRY EXERCISE **11**

Reflecting the graph in Figure 2 about the diagonal line $y = x$ has the effect of interchanging the roles of x and y. This results in a parabola with horizontal axis. By the same method as before, we can prove the following properties.

PARABOLA WITH HORIZONTAL AXIS

The graph of the equation

$$y^2 = 4px$$

is a parabola with the following properties.

VERTEX	$V(0, 0)$
FOCUS	$F(p, 0)$
DIRECTRIX	$x = -p$

The parabola opens to the right if $p > 0$ or to the left if $p < 0$.

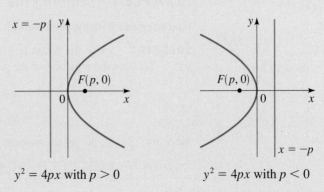

$y^2 = 4px$ with $p > 0$ $y^2 = 4px$ with $p < 0$

EXAMPLE 3 | A Parabola with Horizontal Axis

A parabola has the equation $6x + y^2 = 0$.

(a) Find the focus and directrix of the parabola, and sketch the graph.

(b) Use a graphing calculator to draw the graph.

SOLUTION

(a) To find the focus and directrix, we put the given equation in the standard form $y^2 = -6x$. Comparing this to the general equation $y^2 = 4px$, we see that $4p = -6$, so $p = -\frac{3}{2}$. Thus the focus is $F\left(-\frac{3}{2}, 0\right)$ and the directrix is $x = \frac{3}{2}$. Since $p < 0$, the parabola opens to the left. The graph of the parabola, together with the focus and the directrix, is shown in Figure 5(a) below.

(b) To draw the graph using a graphing calculator, we need to solve for y.

$$6x + y^2 = 0$$

$$y^2 = -6x \qquad \text{Subtract } 6x$$

$$y = \pm\sqrt{-6x} \qquad \text{Take square roots}$$

To obtain the graph of the parabola, we graph both functions

$$y = \sqrt{-6x} \qquad \text{and} \qquad y = -\sqrt{-6x}$$

as shown in Figure 5(b).

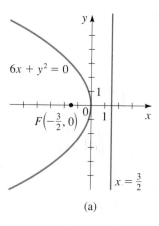

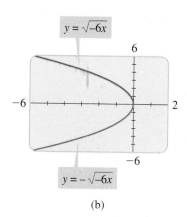

FIGURE 5
(a)   (b)

✎ NOW TRY EXERCISE **13**

The equation $y^2 = 4px$, does not define y as a function of x (see page 44). So to use a graphing calculator to graph a parabola with a horizontal axis, we must first solve for y. This leads to two functions: $y = \sqrt{4px}$ and $y = -\sqrt{4px}$. We need to graph both functions to get the complete graph of the parabola. For example, in Figure 5(b) we had to graph both $y = \sqrt{-6x}$ and $y = -\sqrt{-6x}$ to graph the parabola $y^2 = -6x$.

We can use the coordinates of the focus to estimate the "width" of a parabola when sketching its graph. The line segment that runs through the focus perpendicular to the axis, with endpoints on the parabola, is called the **latus rectum**, and its length is the **focal diameter** of the parabola. From Figure 6 we can see that the distance from an endpoint Q of the latus rectum to the directrix is $|2p|$. Thus the distance from Q to the focus must be $|2p|$ as well (by the definition of a parabola), so the focal diameter is $|4p|$. In the next example we use the focal diameter to determine the "width" of a parabola when graphing it.

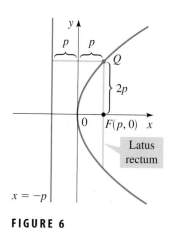

FIGURE 6

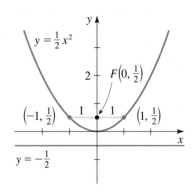

FIGURE 7

EXAMPLE 4 | The Focal Diameter of a Parabola

Find the focus, directrix, and focal diameter of the parabola $y = \frac{1}{2}x^2$, and sketch its graph.

SOLUTION   We first put the equation in the form $x^2 = 4py$.

$$y = \tfrac{1}{2}x^2$$

$$x^2 = 2y \qquad \text{Multiply by 2, switch sides}$$

From this equation we see that $4p = 2$, so the focal diameter is 2. Solving for p gives $p = \frac{1}{2}$, so the focus is $\left(0, \frac{1}{2}\right)$ and the directrix is $y = -\frac{1}{2}$. Since the focal diameter is 2, the latus rectum extends 1 unit to the left and 1 unit to the right of the focus. The graph is sketched in Figure 7.

✎ NOW TRY EXERCISE **15**

In the next example we graph a family of parabolas, to show how changing the distance between the focus and the vertex affects the "width" of a parabola.

EXAMPLE 5 | A Family of Parabolas

(a) Find equations for the parabolas with vertex at the origin and foci $F_1\left(0, \frac{1}{8}\right)$, $F_2\left(0, \frac{1}{2}\right)$, $F_3(0, 1)$, and $F_4(0, 4)$.

(b) Draw the graphs of the parabolas in part (a). What do you conclude?

SOLUTION

(a) Since the foci are on the positive y-axis, the parabolas open upward and have equations of the form $x^2 = 4py$. This leads to the following equations.

Focus	p	Equation $x^2 = 4py$	Form of the equation for graphing calculator
$F_1\left(0, \frac{1}{8}\right)$	$p = \frac{1}{8}$	$x^2 = \frac{1}{2}y$	$y = 2x^2$
$F_2\left(0, \frac{1}{2}\right)$	$p = \frac{1}{2}$	$x^2 = 2y$	$y = 0.5x^2$
$F_3(0, 1)$	$p = 1$	$x^2 = 4y$	$y = 0.25x^2$
$F_4(0, 4)$	$p = 4$	$x^2 = 16y$	$y = 0.0625x^2$

(b) The graphs are drawn in Figure 8. We see that the closer the focus is to the vertex, the narrower the parabola.

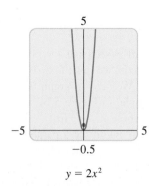

$y = 2x^2$

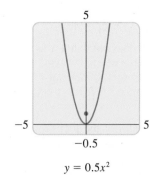

$y = 0.5x^2$

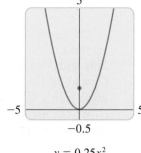

$y = 0.25x^2$

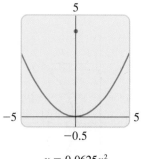

$y = 0.0625x^2$

FIGURE 8 A family of parabolas

✎ NOW TRY EXERCISE **51**

▼ Applications

Parabolas have an important property that makes them useful as reflectors for lamps and telescopes. Light from a source placed at the focus of a surface with parabolic cross section will be reflected in such a way that it travels parallel to the axis of the parabola (see Figure 9). Thus, a parabolic mirror reflects the light into a beam of parallel rays. Conversely, light approaching the reflector in rays parallel to its axis of symmetry is concentrated to the focus. This *reflection property*, which can be proved by using calculus, is used in the construction of reflecting telescopes.

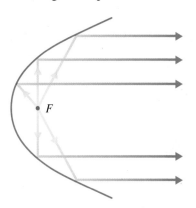

FIGURE 9 Parabolic reflector

EXAMPLE 6 | Finding the Focal Point of a Searchlight Reflector

A searchlight has a parabolic reflector that forms a "bowl," which is 12 in. wide from rim to rim and 8 in. deep, as shown in Figure 10. If the filament of the light bulb is located at the focus, how far from the vertex of the reflector is it?

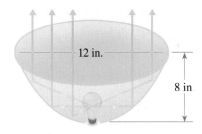

12 in.

8 in

FIGURE 10 A parabolic reflector

ARCHIMEDES (287–212 B.C.) was the greatest mathematician of the ancient world. He was born in Syracuse, a Greek colony on Sicily, a generation after Euclid (see page 229). One of his many discoveries is the Law of the Lever. He famously said, "Give me a place to stand and a fulcrum for my lever, and I can lift the earth."

Renowned as a mechanical genius for his many engineering inventions, he designed pulleys for lifting heavy ships and the spiral screw for transporting water to higher levels. He is said to have used parabolic mirrors to concentrate the rays of the sun to set fire to Roman ships attacking Syracuse.

King Hieron II of Syracuse once suspected a goldsmith of keeping part of the gold intended for the king's crown and replacing it with an equal amount of silver. The king asked Archimedes for advice. While in deep thought at a public bath, Archimedes discovered the solution to the king's problem when he noticed that his body's volume was the same as the volume of water it displaced from the tub. Using this insight he was able to measure the volume of each crown, and so determine which was the denser, all-gold crown. As the story is told, he ran home naked, shouting "Eureka, eureka!" ("I have found it, I have found it!") This incident attests to his enormous powers of concentration.

In spite of his engineering prowess, Archimedes was most proud of his mathematical discoveries. These include the formulas for the volume of a sphere, $\left(V = \frac{4}{3}\pi r^3\right)$ and the surface area of a sphere $\left(S = 4\pi r^2\right)$ and a careful analysis of the properties of parabolas and other conics.

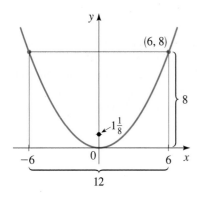

FIGURE 11

SOLUTION We introduce a coordinate system and place a parabolic cross section of the reflector so that its vertex is at the origin and its axis is vertical (see Figure 11). Then the equation of this parabola has the form $x^2 = 4py$. From Figure 11 we see that the point $(6, 8)$ lies on the parabola. We use this to find p.

$$6^2 = 4p(8) \qquad \text{The point } (6, 8) \text{ satisfies the equation } x^2 = 4py$$

$$36 = 32p$$

$$p = \tfrac{9}{8}$$

The focus is $F\left(0, \tfrac{9}{8}\right)$, so the distance between the vertex and the focus is $\tfrac{9}{8} = 1\tfrac{1}{8}$ in. Because the filament is positioned at the focus, it is located $1\tfrac{1}{8}$ in. from the vertex of the reflector.

✎ NOW TRY EXERCISE **53** ∎

7.1 EXERCISES

CONCEPTS

1. A parabola is the set of all points in the plane that are equidistant from a fixed point called the _____ and a fixed line called the _____ of the parabola.

2. The graph of the equation $x^2 = 4py$ is a parabola with focus $F(\_\_, \_\_)$ and directrix $y = $ _____. So the graph of $x^2 = 12y$ is a parabola with focus $F(\_\_, \_\_)$ and directrix $y = $ _____.

3. The graph of the equation $y^2 = 4px$ is a parabola with focus $F(\_\_, \_\_)$ and directrix $x = $ _____. So the graph of $y^2 = 12x$ is a parabola with focus $F(\_\_, \_\_)$ and directrix $x = $ _____.

4. Label the focus, directrix, and vertex on the graphs given for the parabolas in Exercises 2 and 3.
(a) $x^2 = 12y$ **(b)** $y^2 = 12x$

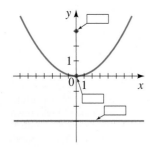
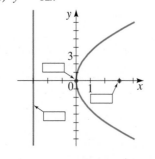

SKILLS

5–10 ■ Match the equation with the graphs labeled I–VI. Give reasons for your answers.

5. $y^2 = 2x$ **6.** $y^2 = -\tfrac{1}{4}x$

7. $x^2 = -6y$ **8.** $2x^2 = y$

9. $y^2 - 8x = 0$ **10.** $12y + x^2 = 0$

I 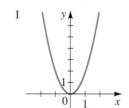   II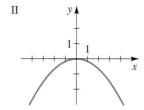

III 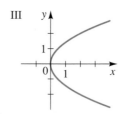   IV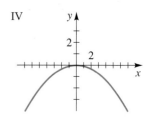

V 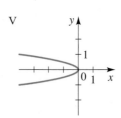   VI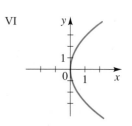

11–22 ■ Find the focus, directrix, and focal diameter of the parabola, and sketch its graph.

11. $x^2 = 9y$ **12.** $x^2 = y$

13. $y^2 = 4x$ **14.** $y^2 = 3x$

15. $y = 5x^2$ **16.** $y = -2x^2$

17. $x = -8y^2$ **18.** $x = \tfrac{1}{2}y^2$

19. $x^2 + 6y = 0$ **20.** $x - 7y^2 = 0$

21. $5x + 3y^2 = 0$ **22.** $8x^2 + 12y = 0$

23–28 ■ Use a graphing device to graph the parabola.

23. $x^2 = 16y$ **24.** $x^2 = -8y$

25. $y^2 = -\frac{1}{3}x$ **26.** $8y^2 = x$

27. $4x + y^2 = 0$ **28.** $x - 2y^2 = 0$

29–40 ■ Find an equation for the parabola that has its vertex at the origin and satisfies the given condition(s).

29. Focus: $F(0, 2)$ **30.** Focus: $F\left(0, -\frac{1}{2}\right)$

31. Focus: $F(-8, 0)$ **32.** Focus: $F(5, 0)$

33. Directrix: $x = 2$ **34.** Directrix: $y = 6$

35. Directrix: $y = -10$ **36.** Directrix: $x = -\frac{1}{8}$

37. Focus on the positive x-axis, 2 units away from the directrix

38. Directrix has y-intercept 6

39. Opens upward with focus 5 units from the vertex

40. Focal diameter 8 and focus on the negative y-axis

41–50 ■ Find an equation of the parabola whose graph is shown.

41.

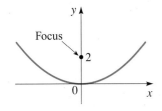

42.

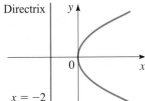

43.

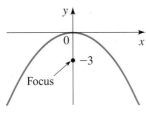

44.

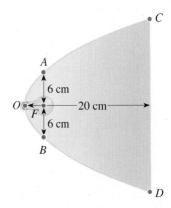

45.

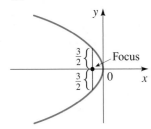

46.

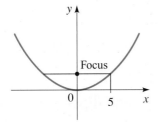

47.

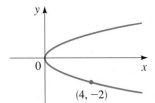

48.

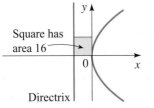

49.

50.

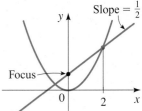

51. (a) Find equations for the family of parabolas with vertex at the origin and with directrixes $y = \frac{1}{2}$, $y = 1$, $y = 4$, and $y = 8$.

    (b) Draw the graphs. What do you conclude?

52. (a) Find equations for the family of parabolas with vertex at the origin, focus on the positive y-axis, and with focal diameters 1, 2, 4, and 8.

    (b) Draw the graphs. What do you conclude?

APPLICATIONS

53. Parabolic Reflector A lamp with a parabolic reflector is shown in the figure. The bulb is placed at the focus, and the focal diameter is 12 cm.

    (a) Find an equation of the parabola.

    (b) Find the diameter $d(C, D)$ of the opening, 20 cm from the vertex.

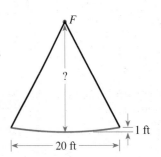

54. Satellite Dish A reflector for a satellite dish is parabolic in cross section, with the receiver at the focus F. The reflector is 1 ft deep and 20 ft wide from rim to rim (see the figure). How far is the receiver from the vertex of the parabolic reflector?

55. Suspension Bridge In a suspension bridge the shape of the suspension cables is parabolic. The bridge shown in the figure has towers that are 600 m apart, and the lowest point of the suspension cables is 150 m below the top of the towers. Find the equation of the parabolic part of the cables, placing the origin of the coordinate system at the vertex. [*Note*: This equation is used to find the length of cable needed in the construction of the bridge.]

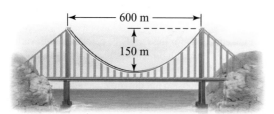

56. Reflecting Telescope The Hale telescope at the Mount Palomar Observatory has a 200-in. mirror, as shown in the figure. The mirror is constructed in a parabolic shape that collects light from the stars and focuses it at the **prime focus**, that is, the focus of the parabola. The mirror is 3.79 in. deep at its center. Find the **focal length** of this parabolic mirror, that is, the distance from the vertex to the focus.

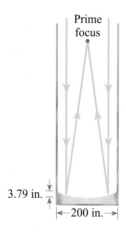

DISCOVERY ■ DISCUSSION ■ WRITING

57. Parabolas in the Real World Several examples of the uses of parabolas are given in the text. Find other situations in real life in which parabolas occur. Consult a scientific encyclopedia in the reference section of your library, or search the Internet.

58. Light Cone from a Flashlight A flashlight is held to form a lighted area on the ground, as shown in the figure. Is it possible to angle the flashlight in such a way that the boundary of the lighted area is a parabola? Explain your answer.

DISCOVERY PROJECT | **Rolling Down a Ramp**

In this project we investigate the process of modeling the motion of falling objects using a calculator-based motion detector. You can find the project at the book companion website: **www.stewartmath.com**

7.2 ELLIPSES

Geometric Definition of an Ellipse ▶ Equations and Graphs of Ellipses ▶ Eccentricity of an Ellipse

▼ Geometric Definition of an Ellipse

An ellipse is an oval curve that looks like an elongated circle. More precisely, we have the following definition.

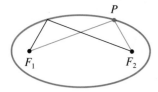

FIGURE 1

GEOMETRIC DEFINITION OF AN ELLIPSE

An **ellipse** is the set of all points in the plane the sum of whose distances from two fixed points F_1 and F_2 is a constant. (See Figure 1.) These two fixed points are the **foci** (plural of **focus**) of the ellipse.

The geometric definition suggests a simple method for drawing an ellipse. Place a sheet of paper on a drawing board, and insert thumbtacks at the two points that are to be the foci of the ellipse. Attach the ends of a string to the tacks, as shown in Figure 2(a). With the point of a pencil, hold the string taut. Then carefully move the pencil around the foci, keeping the string taut at all times. The pencil will trace out an ellipse, because the sum of the distances from the point of the pencil to the foci will always equal the length of the string, which is constant.

If the string is only slightly longer than the distance between the foci, then the ellipse that is traced out will be elongated in shape, as in Figure 2(a), but if the foci are close together relative to the length of the string, the ellipse will be almost circular, as shown in Figure 2(b).

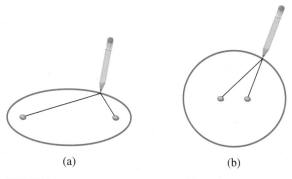

(a)   (b)

FIGURE 2

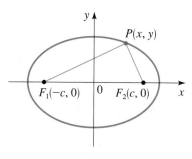

FIGURE 3

To obtain the simplest equation for an ellipse, we place the foci on the x-axis at $F_1(-c, 0)$ and $F_2(c, 0)$ so that the origin is halfway between them (see Figure 3).

For later convenience we let the sum of the distances from a point on the ellipse to the foci be $2a$. Then if $P(x, y)$ is any point on the ellipse, we have

$$d(P, F_1) + d(P, F_2) = 2a$$

So from the Distance Formula we have

$$\sqrt{(x + c)^2 + y^2} + \sqrt{(x - c)^2 + y^2} = 2a$$

or

$$\sqrt{(x - c)^2 + y^2} = 2a - \sqrt{(x + c)^2 + y^2}$$

Squaring each side and expanding, we get

$$x^2 - 2cx + c^2 + y^2 = 4a^2 - 4a\sqrt{(x + c)^2 + y^2} + (x^2 + 2cx + c^2 + y^2)$$

which simplifies to

$$4a\sqrt{(x + c)^2 + y^2} = 4a^2 + 4cx$$

Dividing each side by 4 and squaring again, we get

$$a^2[(x + c)^2 + y^2] = (a^2 + cx)^2$$

$$a^2x^2 + 2a^2cx + a^2c^2 + a^2y^2 = a^4 + 2a^2cx + c^2x^2$$

$$(a^2 - c^2)x^2 + a^2y^2 = a^2(a^2 - c^2)$$

Since the sum of the distances from P to the foci must be larger than the distance between the foci, we have that $2a > 2c$, or $a > c$. Thus $a^2 - c^2 > 0$, and we can divide each side of the preceding equation by $a^2(a^2 - c^2)$ to get

$$\frac{x^2}{a^2} + \frac{y^2}{a^2 - c^2} = 1$$

For convenience let $b^2 = a^2 - c^2$ (with $b > 0$). Since $b^2 < a^2$, it follows that $b < a$. The preceding equation then becomes

$$\frac{x^2}{a^2} + \frac{y^2}{b^2} = 1 \qquad \text{with } a > b$$

This is the equation of the ellipse. To graph it, we need to know the x- and y-intercepts. Setting $y = 0$, we get

$$\frac{x^2}{a^2} = 1$$

so $x^2 = a^2$, or $x = \pm a$. Thus, the ellipse crosses the x-axis at $(a, 0)$ and $(-a, 0)$, as in Figure 4. These points are called the **vertices** of the ellipse, and the segment that joins them is called the **major axis**. Its length is $2a$.

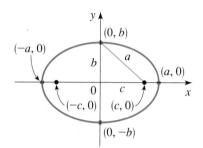

FIGURE 4
$\dfrac{x^2}{a^2} + \dfrac{y^2}{b^2} = 1$ with $a > b$

Similarly, if we set $x = 0$, we get $y = \pm b$, so the ellipse crosses the y-axis at $(0, b)$ and $(0, -b)$. The segment that joins these points is called the **minor axis**, and it has length $2b$. Note that $2a > 2b$, so the major axis is longer than the minor axis. The origin is the **center** of the ellipse.

If the foci of the ellipse are placed on the y-axis at $(0, \pm c)$ rather than on the x-axis, then the roles of x and y are reversed in the preceding discussion, and we get a vertical ellipse.

▼ Equations and Graphs of Ellipses

The following box summarizes what we have just proved about the equation and features of an ellipse centered at the origin.

In the standard equation for an ellipse, a^2 is the *larger* denominator and b^2 is the *smaller*. To find c^2, we subtract: larger denominator minus smaller denominator.

ELLIPSE WITH CENTER AT THE ORIGIN

The graph of each of the following equations is an ellipse with center at the origin and having the given properties.

EQUATION	$\dfrac{x^2}{a^2} + \dfrac{y^2}{b^2} = 1$	$\dfrac{x^2}{b^2} + \dfrac{y^2}{a^2} = 1$
	$a > b > 0$	$a > b > 0$
VERTICES	$(\pm a, 0)$	$(0, \pm a)$
MAJOR AXIS	Horizontal, length $2a$	Vertical, length $2a$
MINOR AXIS	Vertical, length $2b$	Horizontal, length $2b$
FOCI	$(\pm c, 0)$, $c^2 = a^2 - b^2$	$(0, \pm c)$, $c^2 = a^2 - b^2$
GRAPH		

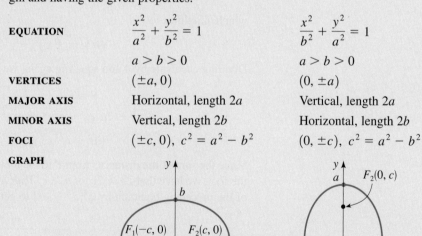

EXAMPLE 1 | Sketching an Ellipse

An ellipse has the equation

$$\frac{x^2}{9} + \frac{y^2}{4} = 1$$

(a) Find the foci, the vertices, and the lengths of the major and minor axes, and sketch the graph.

(b) Draw the graph using a graphing calculator.

SOLUTION

(a) Since the denominator of x^2 is larger, the ellipse has a horizontal major axis. This gives $a^2 = 9$ and $b^2 = 4$, so $c^2 = a^2 - b^2 = 9 - 4 = 5$. Thus $a = 3$, $b = 2$, and $c = \sqrt{5}$.

FOCI	$(\pm\sqrt{5}, 0)$
VERTICES	$(\pm 3, 0)$
LENGTH OF MAJOR AXIS	6
LENGTH OF MINOR AXIS	4

The graph is shown in Figure 5(a).

(b) To draw the graph using a graphing calculator, we need to solve for y.

$$\frac{x^2}{9} + \frac{y^2}{4} = 1$$

$$\frac{y^2}{4} = 1 - \frac{x^2}{9} \qquad \text{Subtract } \frac{x^2}{9}$$

$$y^2 = 4\left(1 - \frac{x^2}{9}\right) \qquad \text{Multiply by 4}$$

$$y = \pm 2\sqrt{1 - \frac{x^2}{9}} \qquad \text{Take square roots}$$

The orbits of the planets are ellipses, with the sun at one focus.

To obtain the graph of the ellipse, we graph both functions

$$y = 2\sqrt{1 - x^2/9} \qquad \text{and} \qquad y = -2\sqrt{1 - x^2/9}$$

as shown in Figure 5(b).

Note that the equation of an ellipse does not define y as a function of x (see page 44). That's why we need to graph two functions to graph an ellipse.

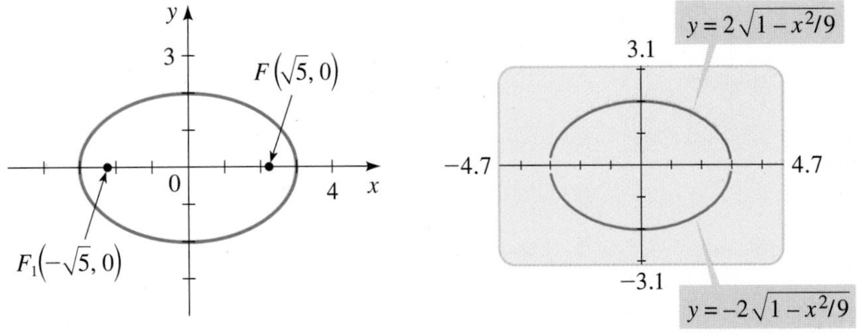

FIGURE 5
$$\frac{x^2}{9} + \frac{y^2}{4} = 1$$

(a)

(b)

NOW TRY EXERCISE **9**

EXAMPLE 2 | Finding the Foci of an Ellipse

Find the foci of the ellipse $16x^2 + 9y^2 = 144$, and sketch its graph.

SOLUTION   First we put the equation in standard form. Dividing by 144, we get

$$\frac{x^2}{9} + \frac{y^2}{16} = 1$$

Since $16 > 9$, this is an ellipse with its foci on the y-axis and with $a = 4$ and $b = 3$. We have

$$c^2 = a^2 - b^2 = 16 - 9 = 7$$
$$c = \sqrt{7}$$

Thus the foci are $(0, \pm\sqrt{7})$. The graph is shown in Figure 6(a).

We can also draw the graph using a graphing calculator as shown in Figure 6(b).

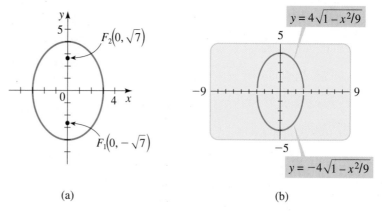

FIGURE 6
$16x^2 + 9y^2 = 144$

(a)   (b)

✎ NOW TRY EXERCISE **11**

EXAMPLE 3 | Finding the Equation of an Ellipse

The vertices of an ellipse are $(\pm 4, 0)$, and the foci are $(\pm 2, 0)$. Find its equation, and sketch the graph.

SOLUTION   Since the vertices are $(\pm 4, 0)$, we have $a = 4$ and the major axis is horizontal. The foci are $(\pm 2, 0)$, so $c = 2$. To write the equation, we need to find b. Since $c^2 = a^2 - b^2$, we have

$$2^2 = 4^2 - b^2$$
$$b^2 = 16 - 4 = 12$$

Thus the equation of the ellipse is

$$\frac{x^2}{16} + \frac{y^2}{12} = 1$$

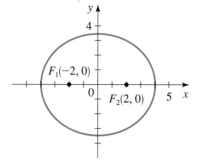

FIGURE 7
$\dfrac{x^2}{16} + \dfrac{y^2}{12} = 1$

The graph is shown in Figure 7.

✎ NOW TRY EXERCISES **25** AND **33**

▼ Eccentricity of an Ellipse

We saw earlier in this section (Figure 2) that if $2a$ is only slightly greater than $2c$, the ellipse is long and thin, whereas if $2a$ is much greater than $2c$, the ellipse is almost circular. We measure the deviation of an ellipse from being circular by the ratio of a and c.

DEFINITION OF ECCENTRICITY

For the ellipse $\dfrac{x^2}{a^2} + \dfrac{y^2}{b^2} = 1$ or $\dfrac{x^2}{b^2} + \dfrac{y^2}{a^2} = 1$ (with $a > b > 0$), the **eccentricity** e is the number

$$e = \frac{c}{a}$$

where $c = \sqrt{a^2 - b^2}$. The eccentricity of every ellipse satisfies $0 < e < 1$.

Thus if e is close to 1, then c is almost equal to a, and the ellipse is elongated in shape, but if e is close to 0, then the ellipse is close to a circle in shape. The eccentricity is a measure of how "stretched" the ellipse is.

In Figure 8 we show a number of ellipses to demonstrate the effect of varying the eccentricity e.

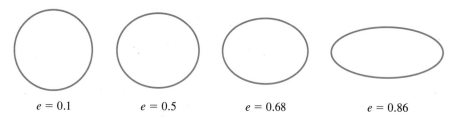

$e = 0.1$ $e = 0.5$ $e = 0.68$ $e = 0.86$

FIGURE 8 Ellipses with various eccentricities

EXAMPLE 4 | Finding the Equation of an Ellipse from Its Eccentricity and Foci

Find the equation of the ellipse with foci $(0, \pm 8)$ and eccentricity $e = \frac{4}{5}$, and sketch its graph.

SOLUTION We are given $e = \frac{4}{5}$ and $c = 8$. Thus

$$\frac{4}{5} = \frac{8}{a} \qquad \text{Eccentricity } e = \frac{c}{a}$$

$$4a = 40 \qquad \text{Cross-multiply}$$

$$a = 10$$

To find b, we use the fact that $c^2 = a^2 - b^2$.

$$8^2 = 10^2 - b^2$$

$$b^2 = 10^2 - 8^2 = 36$$

$$b = 6$$

Thus the equation of the ellipse is

$$\frac{x^2}{36} + \frac{y^2}{100} = 1$$

Because the foci are on the y-axis, the ellipse is oriented vertically. To sketch the ellipse, we find the intercepts: The x-intercepts are ± 6, and the y-intercepts are ± 10. The graph is sketched in Figure 9.

✎ NOW TRY EXERCISE **43**

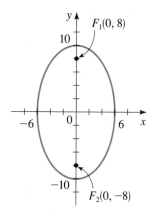

FIGURE 9

$$\frac{x^2}{36} + \frac{y^2}{100} = 1$$

Eccentricities of the Orbits of the Planets

The orbits of the planets are ellipses with the sun at one focus. For most planets these ellipses have very small eccentricity, so they are nearly circular. However, Mercury and Pluto, the innermost and outermost known planets, have visibly elliptical orbits.

Planet	Eccentricity
Mercury	0.206
Venus	0.007
Earth	0.017
Mars	0.093
Jupiter	0.048
Saturn	0.056
Uranus	0.046
Neptune	0.010
Pluto	0.248

Gravitational attraction causes the planets to move in elliptical orbits around the sun with the sun at one focus. This remarkable property was first observed by Johannes Kepler and was later deduced by Isaac Newton from his inverse square Law of Gravity, using calculus. The orbits of the planets have different eccentricities, but most are nearly circular (see the margin).

Ellipses, like parabolas, have an interesting *reflection property* that leads to a number of practical applications. If a light source is placed at one focus of a reflecting surface with elliptical cross sections, then all the light will be reflected off the surface to the other focus, as shown in Figure 10. This principle, which works for sound waves as well as for light, is used in *lithotripsy*, a treatment for kidney stones. The patient is placed in a tub of water with elliptical cross sections in such a way that the kidney stone is accurately located at one focus. High-intensity sound waves generated at the other focus are reflected to the stone and destroy it with minimal damage to surrounding tissue. The patient is spared the trauma of surgery and recovers within days instead of weeks.

The reflection property of ellipses is also used in the construction of *whispering galleries*. Sound coming from one focus bounces off the walls and ceiling of an elliptical room and passes through the other focus. In these rooms even quiet whispers spoken at one focus can be heard clearly at the other. Famous whispering galleries include the National Statuary Hall of the U.S. Capitol in Washington, D.C. (see page 412), and the Mormon Tabernacle in Salt Lake City, Utah.

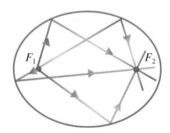

FIGURE 10

7.2 EXERCISES

CONCEPTS

1. An ellipse is the set of all points in the plane for which the _____ of the distances from two fixed points F_1 and F_2 is constant. The points F_1 and F_2 are called the _____ of the ellipse.

2. The graph of the equation $\dfrac{x^2}{a^2} + \dfrac{y^2}{b^2} = 1$ with $a > b > 0$ is an ellipse with vertices (___, ___) and (___, ___) and foci $(\pm c, 0)$, where $c =$ _____. So the graph of $\dfrac{x^2}{5^2} + \dfrac{y^2}{4^2} = 1$ is an ellipse with vertices (___, ___) and (___, ___) and foci (___, ___) and (___, ___).

3. The graph of the equation $\dfrac{x^2}{b^2} + \dfrac{y^2}{a^2} = 1$ with $a > b > 0$ is an ellipse with vertices (___, ___) and (___, ___) and foci $(0, \pm c)$, where $c =$ _____. So the graph of $\dfrac{x^2}{4^2} + \dfrac{y^2}{5^2} = 1$

is an ellipse with vertices (___, ___) and (___, ___) and foci (___, ___) and (___, ___).

4. Label the vertices and foci on the graphs given for the ellipses in Exercises 2 and 3.

(a) $\dfrac{x^2}{5^2} + \dfrac{y^2}{4^2} = 1$

(b) $\dfrac{x^2}{4^2} + \dfrac{y^2}{5^2} = 1$

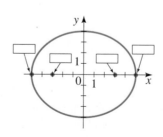

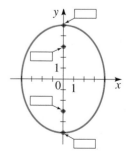

SKILLS

5–8 ■ Match the equation with the graphs labeled I–IV. Give reasons for your answers.

5. $\dfrac{x^2}{16} + \dfrac{y^2}{4} = 1$

6. $x^2 + \dfrac{y^2}{9} = 1$

7. $4x^2 + y^2 = 4$

8. $16x^2 + 25y^2 = 400$

I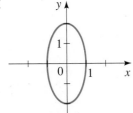

II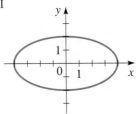

III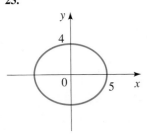

IV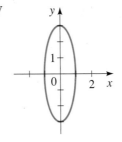

9–22 ■ Find the vertices, foci, and eccentricity of the ellipse. Determine the lengths of the major and minor axes, and sketch the graph.

9. $\dfrac{x^2}{25} + \dfrac{y^2}{9} = 1$

10. $\dfrac{x^2}{16} + \dfrac{y^2}{25} = 1$

11. $9x^2 + 4y^2 = 36$

12. $4x^2 + 25y^2 = 100$

13. $x^2 + 4y^2 = 16$

14. $4x^2 + y^2 = 16$

15. $2x^2 + y^2 = 3$

16. $5x^2 + 6y^2 = 30$

17. $x^2 + 4y^2 = 1$

18. $9x^2 + 4y^2 = 1$

19. $\frac{1}{2}x^2 + \frac{1}{8}y^2 = \frac{1}{4}$

20. $x^2 = 4 - 2y^2$

21. $y^2 = 1 - 2x^2$

22. $20x^2 + 4y^2 = 5$

23–28 ■ Find an equation for the ellipse whose graph is shown.

23.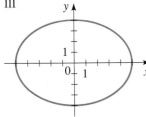

24.

25.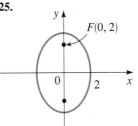

26.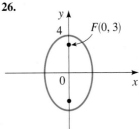

27.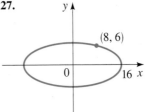

28.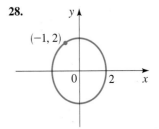

29–32 ■ Use a graphing device to graph the ellipse.

29. $\dfrac{x^2}{25} + \dfrac{y^2}{20} = 1$

30. $x^2 + \dfrac{y^2}{12} = 1$

31. $6x^2 + y^2 = 36$

32. $x^2 + 2y^2 = 8$

33–44 ■ Find an equation for the ellipse that satisfies the given conditions.

33. Foci: $(\pm 4, 0)$, vertices: $(\pm 5, 0)$

34. Foci: $(0, \pm 3)$, vertices: $(0, \pm 5)$

35. Length of major axis: 4, length of minor axis: 2, foci on y-axis

36. Length of major axis: 6, length of minor axis: 4, foci on x-axis

37. Foci: $(0, \pm 2)$, length of minor axis: 6

38. Foci: $(\pm 5, 0)$, length of major axis: 12

39. Endpoints of major axis: $(\pm 10, 0)$, distance between foci: 6

40. Endpoints of minor axis: $(0, \pm 3)$, distance between foci: 8

41. Length of major axis: 10, foci on x-axis, ellipse passes through the point $(\sqrt{5}, 2)$

42. Eccentricity: $\frac{1}{9}$, foci: $(0, \pm 2)$

43. Eccentricity: 0.8, foci: $(\pm 1.5, 0)$

44. Eccentricity: $\sqrt{3}/2$, foci on y-axis, length of major axis: 4

45–47 ■ Find the intersection points of the pair of ellipses. Sketch the graphs of each pair of equations on the same coordinate axes, and label the points of intersection.

45. $\begin{cases} 4x^2 + y^2 = 4 \\ 4x^2 + 9y^2 = 36 \end{cases}$

46. $\begin{cases} \dfrac{x^2}{16} + \dfrac{y^2}{9} = 1 \\ \dfrac{x^2}{9} + \dfrac{y^2}{16} = 1 \end{cases}$

47. $\begin{cases} 100x^2 + 25y^2 = 100 \\ x^2 + \dfrac{y^2}{9} = 1 \end{cases}$

48. The **ancillary circle** of an ellipse is the circle with radius equal to half the length of the minor axis and center the same as the ellipse (see the figure). The ancillary circle is thus the largest circle that can fit within an ellipse.

  (a) Find an equation for the ancillary circle of the ellipse $x^2 + 4y^2 = 16$.

  (b) For the ellipse and ancillary circle of part (a), show that if (s, t) is a point on the ancillary circle, then $(2s, t)$ is a point on the ellipse.

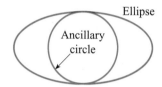

49. (a) Use a graphing device to sketch the top half (the portion in the first and second quadrants) of the family of ellipses $x^2 + ky^2 = 100$ for $k = 4, 10, 25$, and 50.

  (b) What do the members of this family of ellipses have in common? How do they differ?

50. If $k > 0$, the following equation represents an ellipse:

$$\frac{x^2}{k} + \frac{y^2}{4 + k} = 1$$

Show that all the ellipses represented by this equation have the same foci, no matter what the value of k.

APPLICATIONS

51. Perihelion and Aphelion The planets move around the sun in elliptical orbits with the sun at one focus. The point in the orbit at which the planet is closest to the sun is called **perihelion**, and the point at which it is farthest is called **aphelion**. These points are the vertices of the orbit. The earth's distance from the sun is 147,000,000 km at perihelion and 153,000,000 km at aphelion. Find an equation for the earth's orbit. (Place the origin at the center of the orbit with the sun on the x-axis.)

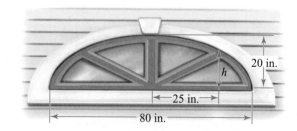

52. The Orbit of Pluto With an eccentricity of 0.25, Pluto's orbit is the most eccentric in the solar system. The length of the minor axis of its orbit is approximately 10,000,000,000 km. Find the distance between Pluto and the sun at perihelion and at aphelion. (See Exercise 51.)

53. Lunar Orbit For an object in an elliptical orbit around the moon, the points in the orbit that are closest to and farthest from the center of the moon are called **perilune** and **apolune**, respectively. These are the vertices of the orbit. The center of the moon is at one focus of the orbit. The *Apollo 11* spacecraft was placed in a lunar orbit with perilune at 68 mi and apolune at 195 mi above the surface of the moon. Assuming that the moon is a sphere of radius 1075 mi, find an equation for the orbit of *Apollo 11*. (Place the coordinate axes so that the origin is at the center of the orbit and the foci are located on the x-axis.)

54. Plywood Ellipse A carpenter wishes to construct an elliptical table top from a sheet of plywood, 4 ft by 8 ft. He will trace out the ellipse using the "thumbtack and string" method illustrated in Figures 2 and 3. What length of string should he use, and how far apart should the tacks be located, if the ellipse is to be the largest possible that can be cut out of the plywood sheet?

55. Sunburst Window A "sunburst" window above a doorway is constructed in the shape of the top half of an ellipse, as shown in the figure. The window is 20 in. tall at its highest point and 80 in. wide at the bottom. Find the height of the window 25 in. from the center of the base.

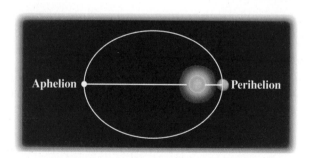

DISCOVERY ▪ DISCUSSION ▪ WRITING

56. Drawing an Ellipse on a Blackboard Try drawing an ellipse as accurately as possible on a blackboard. How would a piece of string and two friends help this process?

57. Light Cone from a Flashlight A flashlight shines on a wall, as shown in the figure. What is the shape of the boundary of the lighted area? Explain your answer.

58. How Wide Is an Ellipse at Its Foci? A *latus rectum* for an ellipse is a line segment perpendicular to the major axis at a focus, with endpoints on the ellipse, as shown in the figure at the top of the next column. Show that the length of a latus rectum is $2b^2/a$ for the ellipse

$$\frac{x^2}{a^2} + \frac{y^2}{b^2} = 1 \quad \text{with } a > b$$

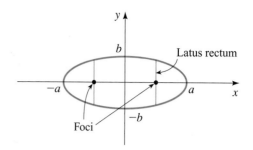

59. Is It an Ellipse? A piece of paper is wrapped around a cylindrical bottle, and then a compass is used to draw a circle on the paper, as shown in the figure. When the paper is laid flat, is the shape drawn on the paper an ellipse? (You don't need to prove your answer, but you might want to do the experiment and see what you get.)

7.3 HYPERBOLAS

❘ Geometric Definition of a Hyperbola ▶ Equations and Graphs of Hyperbolas

▼ Geometric Definition of a Hyperbola

Although ellipses and hyperbolas have completely different shapes, their definitions and equations are similar. Instead of using the *sum* of distances from two fixed foci, as in the case of an ellipse, we use the *difference* to define a hyperbola.

> **GEOMETRIC DEFINITION OF A HYPERBOLA**
>
> A **hyperbola** is the set of all points in the plane, the difference of whose distances from two fixed points F_1 and F_2 is a constant. (See Figure 1.) These two fixed points are the **foci** of the hyperbola.

As in the case of the ellipse, we get the simplest equation for the hyperbola by placing the foci on the x-axis at $(\pm c, 0)$, as shown in Figure 1. By definition, if $P(x, y)$ lies on the hyperbola, then either $d(P, F_1) - d(P, F_2)$ or $d(P, F_2) - d(P, F_1)$ must equal some positive constant, which we call $2a$. Thus we have

$$d(P, F_1) - d(P, F_2) = \pm 2a$$

or

$$\sqrt{(x + c)^2 + y^2} - \sqrt{(x - c)^2 + y^2} = \pm 2a$$

FIGURE 1 P is on the hyperbola if $|d(P, F_1) - d(P, F_2)| = 2a$.

Proceeding as we did in the case of the ellipse (Section 7.2), we simplify this to

$$(c^2 - a^2)x^2 - a^2y^2 = a^2(c^2 - a^2)$$

From triangle PF_1F_2 in Figure 1 we see that $\left| d(P, F_1) - d(P, F_2) \right| < 2c$. It follows that $2a < 2c$, or $a < c$. Thus $c^2 - a^2 > 0$, so we can set $b^2 = c^2 - a^2$. We then simplify the last displayed equation to get

$$\frac{x^2}{a^2} - \frac{y^2}{b^2} = 1$$

This is the *equation of the hyperbola*. If we replace x by $-x$ or y by $-y$ in this equation, it remains unchanged, so the hyperbola is symmetric about both the x- and y-axes and about the origin. The x-intercepts are $\pm a$, and the points $(a, 0)$ and $(-a, 0)$ are the **vertices** of the hyperbola. There is no y-intercept, because setting $x = 0$ in the equation of the hyperbola leads to $-y^2 = b^2$, which has no real solution. Furthermore, the equation of the hyperbola implies that

$$\frac{x^2}{a^2} = \frac{y^2}{b^2} + 1 \geq 1$$

so $x^2/a^2 \geq 1$; thus $x^2 \geq a^2$, and hence $x \geq a$ or $x \leq -a$. This means that the hyperbola consists of two parts, called its **branches**. The segment joining the two vertices on the separate branches is the **transverse axis** of the hyperbola, and the origin is called its **center**.

If we place the foci of the hyperbola on the y-axis rather than on the x-axis, this has the effect of reversing the roles of x and y in the derivation of the equation of the hyperbola. This leads to a hyperbola with a vertical transverse axis.

▼ Equations and Graphs of Hyperbolas

The main properties of hyperbolas are listed in the following box.

HYPERBOLA WITH CENTER AT THE ORIGIN

The graph of each of the following equations is a hyperbola with center at the origin and having the given properties.

EQUATION	$\dfrac{x^2}{a^2} - \dfrac{y^2}{b^2} = 1 \quad (a > 0, b > 0)$	$\dfrac{y^2}{a^2} - \dfrac{x^2}{b^2} = 1 \quad (a > 0, b > 0)$
VERTICES	$(\pm a, 0)$	$(0, \pm a)$
TRANSVERSE AXIS	Horizontal, length $2a$	Vertical, length $2a$
ASYMPTOTES	$y = \pm \dfrac{b}{a}x$	$y = \pm \dfrac{a}{b}x$
FOCI	$(\pm c, 0), \quad c^2 = a^2 + b^2$	$(0, \pm c), \quad c^2 = a^2 + b^2$
GRAPH	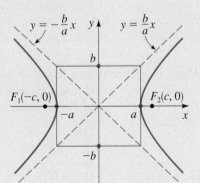	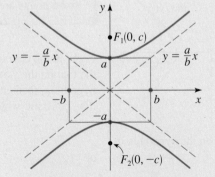

The *asymptotes* mentioned in this box are lines that the hyperbola approaches for large values of x and y. To find the asymptotes in the first case in the box, we solve the equation for y to get

$$y = \pm \frac{b}{a} \sqrt{x^2 - a^2}$$

$$= \pm \frac{b}{a} x \sqrt{1 - \frac{a^2}{x^2}}$$

As x gets large, a^2/x^2 gets closer to zero. In other words, as $x \to \infty$, we have $a^2/x^2 \to 0$. So for large x the value of y can be approximated as $y = \pm (b/a)x$. This shows that these lines are asymptotes of the hyperbola.

Asymptotes are an essential aid for graphing a hyperbola; they help us to determine its shape. A convenient way to find the asymptotes, for a hyperbola with horizontal transverse axis, is to first plot the points $(a, 0)$, $(-a, 0)$, $(0, b)$, and $(0, -b)$. Then sketch horizontal and vertical segments through these points to construct a rectangle, as shown in Figure 2(a). We call this rectangle the **central box** of the hyperbola. The slopes of the diagonals of the central box are $\pm b/a$, so by extending them, we obtain the asymptotes $y = \pm (b/a)x$, as sketched in Figure 2(b). Finally, we plot the vertices and use the asymptotes as a guide in sketching the hyperbola shown in Figure 2(c). (A similar procedure applies to graphing a hyperbola that has a vertical transverse axis.)

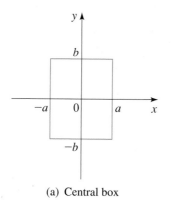

(a) Central box

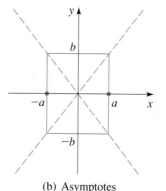

(b) Asymptotes

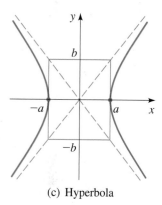

(c) Hyperbola

FIGURE 2 Steps in graphing the hyperbola $\dfrac{x^2}{a^2} - \dfrac{y^2}{b^2} = 1$

HOW TO SKETCH A HYPERBOLA

1. **Sketch the Central Box.** This is the rectangle centered at the origin, with sides parallel to the axes, that crosses one axis at $\pm a$, the other at $\pm b$.

2. **Sketch the Asymptotes.** These are the lines obtained by extending the diagonals of the central box.

3. **Plot the Vertices.** These are the two x-intercepts or the two y-intercepts.

4. **Sketch the Hyperbola.** Start at a vertex, and sketch a branch of the hyperbola, approaching the asymptotes. Sketch the other branch in the same way.

EXAMPLE 1 | A Hyperbola with Horizontal Transverse Axis

A hyperbola has the equation

$$9x^2 - 16y^2 = 144$$

(a) Find the vertices, foci, and asymptotes, and sketch the graph.

(b) Draw the graph using a graphing calculator.

SOLUTION

(a) First we divide both sides of the equation by 144 to put it into standard form:

$$\frac{x^2}{16} - \frac{y^2}{9} = 1$$

Because the x^2-term is positive, the hyperbola has a horizontal transverse axis; its vertices and foci are on the x-axis. Since $a^2 = 16$ and $b^2 = 9$, we get $a = 4$, $b = 3$, and $c = \sqrt{16 + 9} = 5$. Thus we have

VERTICES	$(\pm 4, 0)$
FOCI	$(\pm 5, 0)$
ASYMPTOTES	$y = \pm \frac{3}{4} x$

After sketching the central box and asymptotes, we complete the sketch of the hyperbola as in Figure 3(a).

Note that the equation of a hyperbola does not define y as a function of x (see page 44). That's why we need to graph two functions to graph a hyperbola.

(b) To draw the graph using a graphing calculator, we need to solve for y.

$$9x^2 - 16y^2 = 144$$

$$-16y^2 = -9x^2 + 144 \qquad \text{Subtract } 9x^2$$

$$y^2 = 9\left(\frac{x^2}{16} - 1\right) \qquad \text{Divide by } -16 \text{ and factor } 9$$

$$y = \pm 3 \sqrt{\frac{x^2}{16} - 1} \qquad \text{Take square roots}$$

To obtain the graph of the hyperbola, we graph the functions

$$y = 3\sqrt{(x^2/16) - 1} \qquad \text{and} \qquad y = -3\sqrt{(x^2/16) - 1}$$

as shown in Figure 3(b).

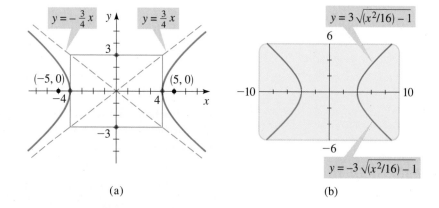

FIGURE 3
$9x^2 - 16y^2 = 144$

(a)   (b)

✎ NOW TRY EXERCISE **9**

EXAMPLE 2 | A Hyperbola with Vertical Transverse Axis

Find the vertices, foci, and asymptotes of the hyperbola, and sketch its graph.

$$x^2 - 9y^2 + 9 = 0$$

SOLUTION We begin by writing the equation in the standard form for a hyperbola.

$$x^2 - 9y^2 = -9$$

$$y^2 - \frac{x^2}{9} = 1 \qquad \text{Divide by } -9$$

Because the y^2-term is positive, the hyperbola has a vertical transverse axis; its foci and vertices are on the y-axis. Since $a^2 = 1$ and $b^2 = 9$, we get $a = 1$, $b = 3$, and $c = \sqrt{1 + 9} = \sqrt{10}$. Thus we have

VERTICES	$(0, \pm 1)$
FOCI	$(0, \pm\sqrt{10})$
ASYMPTOTES	$y = \pm\frac{1}{3}x$

We sketch the central box and asymptotes, then complete the graph, as shown in Figure 4(a). We can also draw the graph using a graphing calculator, as shown in Figure 4(b).

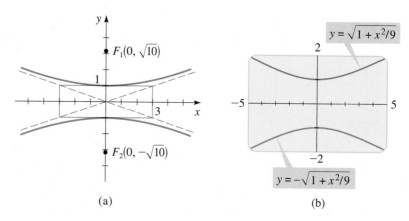

(a) (b)

FIGURE 4
$x^2 - 9y^2 + 9 = 0$

NOW TRY EXERCISE 17

Paths of Comets

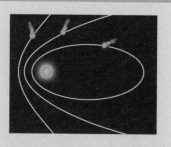

The path of a comet is an ellipse, a parabola, or a hyperbola with the sun at a focus. This fact can be proved by using calculus and Newton's laws of motion.* If the path is a parabola or a hyperbola, the comet will never return. If the path is an ellipse, it can be determined precisely when and where the comet can be seen again. Halley's comet has an elliptical path and returns every 75 years; it was last seen in 1987. The brightest comet of the 20th century was comet Hale-Bopp, seen in 1997. Its orbit is a very eccentric ellipse; it is expected to return to the inner solar system around the year 4377.

*James Stewart, *Calculus*, 7th ed. (Belmont, CA: Brooks/Cole, 2012), pages 868 and 872.

EXAMPLE 3 | Finding the Equation of a Hyperbola from Its Vertices and Foci

Find the equation of the hyperbola with vertices $(\pm 3, 0)$ and foci $(\pm 4, 0)$. Sketch the graph.

SOLUTION Since the vertices are on the x-axis, the hyperbola has a horizontal transverse axis. Its equation is of the form

$$\frac{x^2}{3^2} - \frac{y^2}{b^2} = 1$$

We have $a = 3$ and $c = 4$. To find b, we use the relation $a^2 + b^2 = c^2$:

$$3^2 + b^2 = 4^2$$

$$b^2 = 4^2 - 3^2 = 7$$

$$b = \sqrt{7}$$

Thus the equation of the hyperbola is

$$\frac{x^2}{9} - \frac{y^2}{7} = 1$$

The graph is shown in Figure 5.

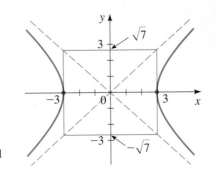

FIGURE 5
$$\frac{x^2}{9} - \frac{y^2}{7} = 1$$

✎ NOW TRY EXERCISES **21** AND **31**

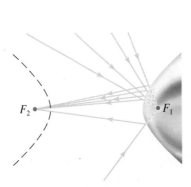

FIGURE 6
$$\frac{y^2}{4} - x^2 = 1$$

EXAMPLE 4 | Finding the Equation of a Hyperbola from Its Vertices and Asymptotes

Find the equation and the foci of the hyperbola with vertices $(0, \pm 2)$ and asymptotes $y = \pm 2x$. Sketch the graph.

SOLUTION Since the vertices are on the y-axis, the hyperbola has a vertical transverse axis with $a = 2$. From the asymptote equation we see that $a/b = 2$. Since $a = 2$, we get $2/b = 2$, so $b = 1$. Thus the equation of the hyperbola is

$$\frac{y^2}{4} - x^2 = 1$$

To find the foci, we calculate $c^2 = a^2 + b^2 = 2^2 + 1^2 = 5$, so $c = \sqrt{5}$. Thus the foci are $(0, \pm\sqrt{5})$. The graph is shown in Figure 6.

✎ NOW TRY EXERCISES **25** AND **35**

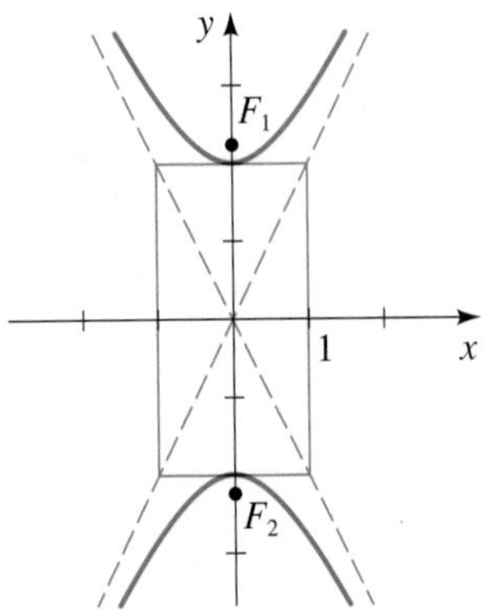

Like parabolas and ellipses, hyperbolas have an interesting *reflection property*. Light aimed at one focus of a hyperbolic mirror is reflected toward the other focus, as shown in Figure 7. This property is used in the construction of Cassegrain-type telescopes. A hyperbolic mirror is placed in the telescope tube so that light reflected from the primary parabolic reflector is aimed at one focus of the hyperbolic mirror. The light is then refocused at a more accessible point below the primary reflector (Figure 8).

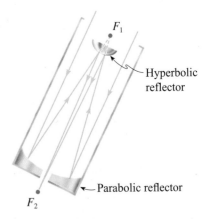

FIGURE 7 Reflection property of hyperbolas

FIGURE 8 Cassegrain-type telescope

The LORAN (LOng RAnge Navigation) system was used until the early 1990s; it has now been superseded by the GPS system (see page 331). In the LORAN system, hyperbolas are used onboard a ship to determine its location. In Figure 9, radio stations at A and B transmit signals simultaneously for reception by the ship at P. The onboard computer converts the time difference in reception of these signals into a distance difference $d(P, A) - d(P, B)$. From the definition of a hyperbola this locates the ship on one branch of a hyperbola with foci at A and B (sketched in black in the figure). The same procedure is carried out with two other radio stations at C and D, and this locates the ship on a second hyperbola (shown in red in the figure). (In practice, only three stations are needed because one station can be used as a focus for both hyperbolas.) The coordinates of the intersection point of these two hyperbolas, which can be calculated precisely by the computer, give the location of P.

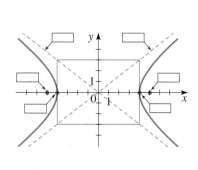

FIGURE 9 LORAN system for finding the location of a ship

7.3 EXERCISES

CONCEPTS

1. A hyperbola is the set of all points in the plane for which the _____ of the distances from two fixed points F_1 and F_2 is constant. The points F_1 and F_2 are called the _____ of the hyperbola.

2. The graph of the equation $\dfrac{x^2}{a^2} - \dfrac{y^2}{b^2} = 1$ with $a > 0, b > 0$

   is a hyperbola with vertices (___, ___) and (___, ___) and foci $(\pm c, 0)$, where $c = $ _____. So the graph of

   $\dfrac{x^2}{4^2} - \dfrac{y^2}{3^2} = 1$ is a hyperbola with vertices (___, ___) and

   (___, ___) and foci (___, ___) and (___, ___).

3. The graph of the equation $\dfrac{y^2}{a^2} - \dfrac{x^2}{b^2} = 1$ with $a > 0, b > 0$

   is a hyperbola with vertices (___, ___) and (___, ___) and foci $(0, \pm c)$, where $c = $ _____. So the graph of

   $\dfrac{y^2}{4^2} - \dfrac{x^2}{3^2} = 1$ is a hyperbola with vertices (___, ___) and

   (___, ___) and foci (___, ___) and (___, ___).

4. Label the vertices, foci, and asymptotes on the graphs given for the hyperbolas in Exercises 2 and 3.

   (a) $\dfrac{x^2}{4^2} - \dfrac{y^2}{3^2} = 1$ **(b)** $\dfrac{y^2}{4^2} - \dfrac{x^2}{3^2} = 1$

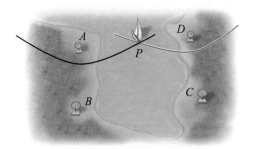

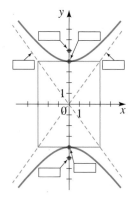

SKILLS

5–8 ■ Match the equation with the graphs labeled I–IV. Give reasons for your answers.

5. $\dfrac{x^2}{4} - y^2 = 1$

6. $y^2 - \dfrac{x^2}{9} = 1$

7. $16y^2 - x^2 = 144$

8. $9x^2 - 25y^2 = 225$

I

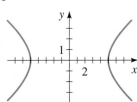

II

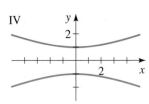

III

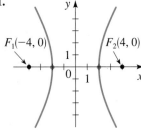

IV

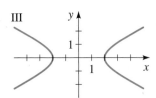

9–20 ■ Find the vertices, foci, and asymptotes of the hyperbola, and sketch its graph.

9. $\dfrac{x^2}{4} - \dfrac{y^2}{16} = 1$

10. $\dfrac{y^2}{9} - \dfrac{x^2}{16} = 1$

11. $y^2 - \dfrac{x^2}{25} = 1$

12. $\dfrac{x^2}{2} - y^2 = 1$

13. $x^2 - y^2 = 1$

14. $9x^2 - 4y^2 = 36$

15. $25y^2 - 9x^2 = 225$

16. $x^2 - y^2 + 4 = 0$

17. $x^2 - 4y^2 - 8 = 0$

18. $x^2 - 2y^2 = 3$

19. $4y^2 - x^2 = 1$

20. $9x^2 - 16y^2 = 1$

21–26 ■ Find the equation for the hyperbola whose graph is shown.

21.

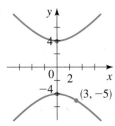

22.

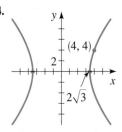

23.

24.

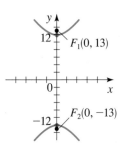

25.

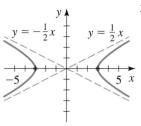

26.

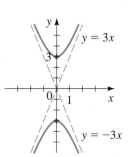

27–30 ■ Use a graphing device to graph the hyperbola.

27. $x^2 - 2y^2 = 8$

28. $3y^2 - 4x^2 = 24$

29. $\dfrac{y^2}{2} - \dfrac{x^2}{6} = 1$

30. $\dfrac{x^2}{100} - \dfrac{y^2}{64} = 1$

31–42 ■ Find an equation for the hyperbola that satisfies the given conditions.

31. Foci: $(\pm 5, 0)$, vertices: $(\pm 3, 0)$

32. Foci: $(0, \pm 10)$, vertices: $(0, \pm 8)$

33. Foci: $(0, \pm 2)$, vertices: $(0, \pm 1)$

34. Foci: $(\pm 6, 0)$, vertices: $(\pm 2, 0)$

35. Vertices: $(\pm 1, 0)$, asymptotes: $y = \pm 5x$

36. Vertices: $(0, \pm 6)$, asymptotes: $y = \pm \frac{1}{3}x$

37. Foci: $(0, \pm 8)$, asymptotes: $y = \pm \frac{1}{2}x$

38. Vertices: $(0, \pm 6)$, hyperbola passes through $(-5, 9)$

39. Asymptotes: $y = \pm x$, hyperbola passes through $(5, 3)$

40. Foci: $(\pm 3, 0)$, hyperbola passes through $(4, 1)$

41. Foci: $(\pm 5, 0)$, length of transverse axis: 6

42. Foci: $(0, \pm 1)$, length of transverse axis: 1

43. **(a)** Show that the asymptotes of the hyperbola $x^2 - y^2 = 5$ are perpendicular to each other.

(b) Find an equation for the hyperbola with foci $(\pm c, 0)$ and with asymptotes perpendicular to each other.

44. The hyperbolas

$$\frac{x^2}{a^2} - \frac{y^2}{b^2} = 1 \quad \text{and} \quad \frac{x^2}{a^2} - \frac{y^2}{b^2} = -1$$

are said to be **conjugate** to each other.

(a) Show that the hyperbolas

$$x^2 - 4y^2 + 16 = 0 \quad \text{and} \quad 4y^2 - x^2 + 16 = 0$$

are conjugate to each other, and sketch their graphs on the same coordinate axes.

(b) What do the hyperbolas of part (a) have in common?

(c) Show that any pair of conjugate hyperbolas have the relationship you discovered in part (b).

45. In the derivation of the equation of the hyperbola at the beginning of this section, we said that the equation

$$\sqrt{(x + c)^2 + y^2} - \sqrt{(x - c)^2 + y^2} = \pm 2a$$

simplifies to

$$(c^2 - a^2)x^2 - a^2y^2 = a^2(c^2 - a^2)$$

Supply the steps needed to show this.

46. **(a)** For the hyperbola

$$\frac{x^2}{9} - \frac{y^2}{16} = 1$$

determine the values of a, b, and c, and find the coordinates of the foci F_1 and F_2.

(b) Show that the point $P(5, \frac{16}{3})$ lies on this hyperbola.

(c) Find $d(P, F_1)$ and $d(P, F_2)$.

(d) Verify that the difference between $d(P, F_1)$ and $d(P, F_2)$ is $2a$.

47. Hyperbolas are called **confocal** if they have the same foci.

(a) Show that the hyperbolas

$$\frac{y^2}{k} - \frac{x^2}{16 - k} = 1 \quad \text{with } 0 < k < 16$$

are confocal.

 (b) Use a graphing device to draw the top branches of the family of hyperbolas in part (a) for $k = 1, 4, 8$, and 12. How does the shape of the graph change as k increases?

APPLICATIONS

48. **Navigation** In the figure, the LORAN stations at A and B are 500 mi apart, and the ship at P receives station A's signal 2640 microseconds (μs) before it receives the signal from station B.

(a) Assuming that radio signals travel at 980 ft/μs, find $d(P, A) - d(P, B)$.

(b) Find an equation for the branch of the hyperbola indicated in red in the figure. (Use miles as the unit of distance.)

(c) If A is due north of B and if P is due east of A, how far is P from A?

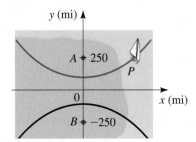

49. **Comet Trajectories** Some comets, such as Halley's comet, are a permanent part of the solar system, traveling in elliptical orbits around the sun. Other comets pass through the solar system only once, following a hyperbolic path with the sun at a focus. The figure at the top of the next column shows the path of such a comet. Find an equation for the path, assuming that the closest the comet comes to the sun is 2×10^9 mi and that the path the comet was taking before it neared the solar system is at a right angle to the path it continues on after leaving the solar system.

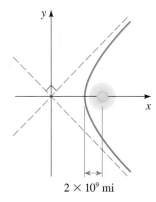

2 × 10⁹ mi

50. **Ripples in Pool** Two stones are dropped simultaneously into a calm pool of water. The crests of the resulting waves form equally spaced concentric circles, as shown in the figures. The waves interact with each other to create certain interference patterns.

(a) Explain why the red dots lie on an ellipse.

(b) Explain why the blue dots lie on a hyperbola.

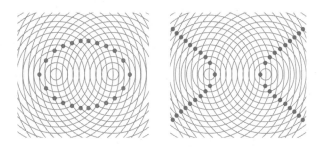

DISCOVERY ▪ DISCUSSION ▸ WRITING

51. **Hyperbolas in the Real World** Several examples of the uses of hyperbolas are given in the text. Find other situations in real life in which hyperbolas occur. Consult a scientific encyclopedia in the reference section of your library, or search the Internet.

52. **Light from a Lamp** The light from a lamp forms a lighted area on a wall, as shown in the figure. Why is the boundary of this lighted area a hyperbola? How can one hold a flashlight so that its beam forms a hyperbola on the ground?

7.4 SHIFTED CONICS

In the preceding sections we studied parabolas with vertices at the origin and ellipses and hyperbolas with centers at the origin. We restricted ourselves to these cases because these equations have the simplest form. In this section we consider conics whose vertices and centers are not necessarily at the origin, and we determine how this affects their equations.

▼ Shifting Graphs of Equations

In Section 1.6 we studied transformations of functions that have the effect of shifting their graphs. In general, for any equation in x and y, if we replace x by $x - h$ or by $x + h$, the graph of the new equation is simply the old graph shifted horizontally; if y is replaced by $y - k$ or by $y + k$, the graph is shifted vertically. The following box gives the details.

SHIFTING GRAPHS OF EQUATIONS

If h and k are positive real numbers, then replacing x by $x - h$ or by $x + h$ and replacing y by $y - k$ or by $y + k$ has the following effect(s) on the graph of any equation in x and y.

Replacement	How the graph is shifted
1. x replaced by $x - h$	Right h units
2. x replaced by $x + h$	Left h units
3. y replaced by $y - k$	Upward k units
4. y replaced by $y + k$	Downward k units

▼ Shifted Ellipses

Let's apply horizontal and vertical shifting to the ellipse with equation

$$\frac{x^2}{a^2} + \frac{y^2}{b^2} = 1$$

whose graph is shown in Figure 1. If we shift it so that its center is at the point (h, k) instead of at the origin, then its equation becomes

$$\frac{(x - h)^2}{a^2} + \frac{(y - k)^2}{b^2} = 1$$

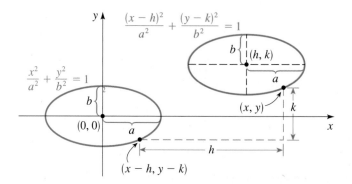

FIGURE 1 Shifted ellipse

EXAMPLE 1 | Sketching the Graph of a Shifted Ellipse

Sketch a graph of the ellipse

$$\frac{(x+1)^2}{4} + \frac{(y-2)^2}{9} = 1$$

and determine the coordinates of the foci.

SOLUTION The ellipse

$$\frac{(x+1)^2}{4} + \frac{(y-2)^2}{9} = 1 \qquad \text{Shifted ellipse}$$

is shifted so that its center is at $(-1, 2)$. It is obtained from the ellipse

$$\frac{x^2}{4} + \frac{y^2}{9} = 1 \qquad \text{Ellipse with center at origin}$$

by shifting it left 1 unit and upward 2 units. The endpoints of the minor and major axes of the ellipse with center at the origin are $(2, 0)$, $(-2, 0)$, $(0, 3)$, $(0, -3)$. We apply the required shifts to these points to obtain the corresponding points on the shifted ellipse:

$$(2, 0) \rightarrow (2-1, 0+2) = (1, 2)$$
$$(-2, 0) \rightarrow (-2-1, 0+2) = (-3, 2)$$
$$(0, 3) \rightarrow (0-1, 3+2) = (-1, 5)$$
$$(0, -3) \rightarrow (0-1, -3+2) = (-1, -1)$$

This helps us sketch the graph in Figure 2.

To find the foci of the shifted ellipse, we first find the foci of the ellipse with center at the origin. Since $a^2 = 9$ and $b^2 = 4$, we have $c^2 = 9 - 4 = 5$, so $c = \sqrt{5}$. So the foci are $(0, \pm\sqrt{5})$. Shifting left 1 unit and upward 2 units, we get

$$(0, \sqrt{5}) \rightarrow (0-1, \sqrt{5}+2) = (-1, 2+\sqrt{5})$$
$$(0, -\sqrt{5}) \rightarrow (0-1, -\sqrt{5}+2) = (-1, 2-\sqrt{5})$$

Thus the foci of the shifted ellipse are

$$(-1, 2+\sqrt{5}) \qquad \text{and} \qquad (-1, 2-\sqrt{5})$$

NOW TRY EXERCISE 7

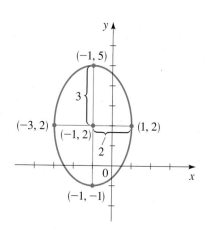

FIGURE 2
$$\frac{(x+1)^2}{4} + \frac{(y-2)^2}{9} = 1$$

▼ Shifted Parabolas

Applying shifts to parabolas leads to the equations and graphs shown in Figure 3.

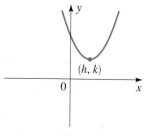

(a) $(x-h)^2 = 4p(y-k)$
$p > 0$

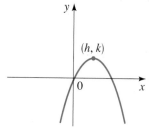

(b) $(x-h)^2 = 4p(y-k)$
$p < 0$

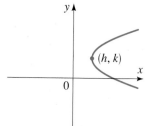

(c) $(y-k)^2 = 4p(x-h)$
$p > 0$

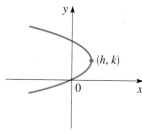

(d) $(y-k)^2 = 4p(x-h)$
$p < 0$

FIGURE 3 Shifted parabolas

EXAMPLE 2 | Graphing a Shifted Parabola

Determine the vertex, focus, and directrix, and sketch a graph of the parabola.

$$x^2 - 4x = 8y - 28$$

SOLUTION We complete the square in x to put this equation into one of the forms in Figure 3.

$$x^2 - 4x + 4 = 8y - 28 + 4 \qquad \text{Add 4 to complete the square}$$

$$(x - 2)^2 = 8y - 24$$

$$(x - 2)^2 = 8(y - 3) \qquad \text{Shifted parabola}$$

This parabola opens upward with vertex at $(2, 3)$. It is obtained from the parabola

$$x^2 = 8y \qquad \text{Parabola with vertex at origin}$$

by shifting right 2 units and upward 3 units. Since $4p = 8$, we have $p = 2$, so the focus is 2 units above the vertex and the directrix is 2 units below the vertex. Thus the focus is $(2, 5)$, and the directrix is $y = 1$. The graph is shown in Figure 4.

◣ NOW TRY EXERCISES **9** AND **23** ◼

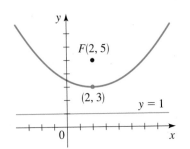

FIGURE 4
$x^2 - 4x = 8y - 28$

▼ Shifted Hyperbolas

Applying shifts to hyperbolas leads to the equations and graphs shown in Figure 5.

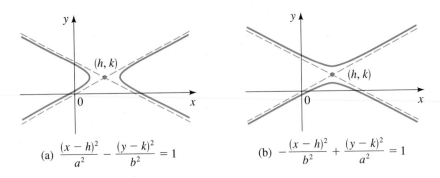

(a) $\dfrac{(x - h)^2}{a^2} - \dfrac{(y - k)^2}{b^2} = 1$ (b) $-\dfrac{(x - h)^2}{b^2} + \dfrac{(y - k)^2}{a^2} = 1$

FIGURE 5 Shifted hyperbolas

EXAMPLE 3 | Graphing a Shifted Hyperbola

A shifted conic has the equation

$$9x^2 - 72x - 16y^2 - 32y = 16$$

(a) Complete the square in x and y to show that the equation represents a hyperbola.

(b) Find the center, vertices, foci, and asymptotes of the hyperbola, and sketch its graph.

(c) Draw the graph using a graphing calculator.

SOLUTION

(a) We complete the squares in both x and y:

$$9(x^2 - 8x \qquad) - 16(y^2 + 2y \qquad) = 16 \qquad \text{Group terms and factor}$$

$$9(x^2 - 8x + 16) - 16(y^2 + 2y + 1) = 16 + 9 \cdot 16 - 16 \cdot 1 \qquad \text{Complete the squares}$$

$$9(x - 4)^2 - 16(y + 1)^2 = 144 \qquad \text{Divide this by 144}$$

$$\frac{(x - 4)^2}{16} - \frac{(y + 1)^2}{9} = 1 \qquad \text{Shifted hyperbola}$$

Comparing this to Figure 5(a), we see that this is the equation of a shifted hyperbola.

(b) The shifted hyperbola has center $(4, -1)$ and a horizontal transverse axis.

$$\text{CENTER} \quad (4, -1)$$

Its graph will have the same shape as the unshifted hyperbola

$$\frac{x^2}{16} - \frac{y^2}{9} = 1 \qquad \text{Hyperbola with center at origin}$$

Since $a^2 = 16$ and $b^2 = 9$, we have $a = 4$, $b = 3$, and $c = \sqrt{a^2 + b^2} = \sqrt{16 + 9} = 5$. Thus the foci lie 5 units to the left and to the right of the center, and the vertices lie 4 units to either side of the center.

$$\text{FOCI} \qquad (-1, -1) \quad \text{and} \quad (9, -1)$$

$$\text{VERTICES} \quad (0, -1) \quad \text{and} \quad (8, -1)$$

The asymptotes of the unshifted hyperbola are $y = \pm\frac{3}{4}x$, so the asymptotes of the shifted hyperbola are found as follows.

$$\text{ASYMPTOTES} \quad y + 1 = \pm\tfrac{3}{4}(x - 4)$$

$$y + 1 = \pm\tfrac{3}{4}x \mp 3$$

$$y = \tfrac{3}{4}x - 4 \qquad \text{and} \qquad y = -\tfrac{3}{4}x + 2$$

To help us sketch the hyperbola, we draw the central box; it extends 4 units left and right from the center and 3 units upward and downward from the center. We then draw the asymptotes and complete the graph of the shifted hyperbola as shown in Figure 6(a).

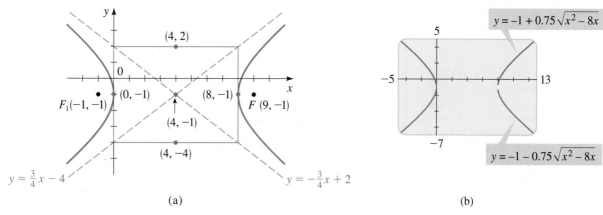

(a)   (b)

FIGURE 6 $9x^2 - 72x - 16y^2 - 32y = 16$

(c) To draw the graph using a graphing calculator, we need to solve for y. The given equation is a quadratic equation in y, so we use the Quadratic Formula to solve for y. Writing the equation in the form

$$16y^2 + 32y - 9x^2 + 72x + 16 = 0$$

we get

Note that the equation of a hyperbola does not define y as a function of x (see page 44). That's why we need to graph two functions to graph a hyperbola.

$$y = \frac{-32 \pm \sqrt{32^2 - 4(16)(-9x^2 + 72x + 16)}}{2(16)} \qquad \text{Quadratic Formula}$$

$$= \frac{-32 \pm \sqrt{576x^2 - 4608x}}{32} \qquad \text{Expand}$$

$$= \frac{-32 \pm 24\sqrt{x^2 - 8x}}{32} \qquad \text{Factor 576 from under the radical}$$

$$= -1 \pm \tfrac{3}{4}\sqrt{x^2 - 8x} \qquad \text{Simplify}$$

JOHANNES KEPLER (1571–1630) was the first to give a correct description of the motion of the planets. The cosmology of his time postulated complicated systems of circles moving on circles to describe these motions. Kepler sought a simpler and more harmonious description. As the official astronomer at the imperial court in Prague, he studied the astronomical observations of the Danish astronomer Tycho Brahe, whose data were the most accurate available at the time. After numerous attempts to find a theory, Kepler made the momentous discovery that the orbits of the planets are elliptical. His three great laws of planetary motion are

1. The orbit of each planet is an ellipse with the sun at one focus.

2. The line segment that joins the sun to a planet sweeps out equal areas in equal time (see the figure).

3. The square of the period of revolution of a planet is proportional to the cube of the length of the major axis of its orbit.

His formulation of these laws is perhaps the most impressive deduction from empirical data in the history of science.

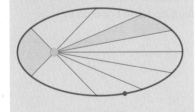

To obtain the graph of the hyperbola, we graph the functions

$$y = -1 + 0.75\sqrt{x^2 - 8x}$$

and

$$y = -1 - 0.75\sqrt{x^2 - 8x}$$

as shown in Figure 6(b).

✎ NOW TRY EXERCISES **13** AND **25** ∎

▼ The General Equation of a Shifted Conic

If we expand and simplify the equations of any of the shifted conics illustrated in Figures 1, 3, and 5, then we will always obtain an equation of the form

$$Ax^2 + Cy^2 + Dx + Ey + F = 0$$

where A and C are not both 0. Conversely, if we begin with an equation of this form, then we can complete the square in x and y to see which type of conic section the equation represents. In some cases the graph of the equation turns out to be just a pair of lines or a single point, or there might be no graph at all. These cases are called **degenerate conics**. If the equation is not degenerate, then we can tell whether it represents a parabola, an ellipse, or a hyperbola simply by examining the signs of A and C, as described in the box below.

GENERAL EQUATION OF A SHIFTED CONIC

The graph of the equation

$$Ax^2 + Cy^2 + Dx + Ey + F = 0$$

where A and C are not both 0, is a conic or a degenerate conic. In the nondegenerate cases the graph is

1. a parabola if A or C is 0,

2. an ellipse if A and C have the same sign (or a circle if $A = C$),

3. a hyperbola if A and C have opposite signs.

EXAMPLE 4 | An Equation That Leads to a Degenerate Conic

Sketch the graph of the equation

$$9x^2 - y^2 + 18x + 6y = 0$$

SOLUTION Because the coefficients of x^2 and y^2 are of opposite sign, this equation looks as if it should represent a hyperbola (like the equation of Example 3). To see whether this is in fact the case, we complete the squares:

$$9(x^2 + 2x \quad) - (y^2 - 6y \quad) = 0 \qquad \text{Group terms and factor 9}$$

$$9(x^2 + 2x + 1) - (y^2 - 6y + 9) = 0 + 9 \cdot 1 - 9 \qquad \text{Complete the squares}$$

$$9(x + 1)^2 - (y - 3)^2 = 0 \qquad \text{Factor}$$

$$(x + 1)^2 - \frac{(y - 3)^2}{9} = 0 \qquad \text{Divide by 9}$$

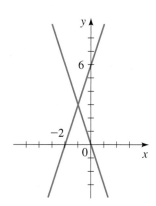

FIGURE 7

$9x^2 - y^2 + 18x + 6y = 0$

For this to fit the form of the equation of a hyperbola, we would need a nonzero constant to the right of the equal sign. In fact, further analysis shows that this is the equation of a pair of intersecting lines:

$$(y - 3)^2 = 9(x + 1)^2$$

$$y - 3 = \pm 3(x + 1) \quad \text{Take square roots}$$

$$y = 3(x + 1) + 3 \quad \text{or} \quad y = -3(x + 1) + 3$$

$$y = 3x + 6 \qquad\qquad\qquad y = -3x$$

These lines are graphed in Figure 7.

✎. NOW TRY EXERCISE **31**

Because the equation in Example 4 looked at first glance like the equation of a hyperbola but, in fact, turned out to represent simply a pair of lines, we refer to its graph as a **degenerate hyperbola**. Degenerate ellipses and parabolas can also arise when we complete the square(s) in an equation that seems to represent a conic. For example, the equation

$$4x^2 + y^2 - 8x + 2y + 6 = 0$$

looks as if it should represent an ellipse, because the coefficients of x^2 and y^2 have the same sign. But completing the squares leads to

$$(x - 1)^2 + \frac{(y + 1)^2}{4} = -\frac{1}{4}$$

which has no solution at all (since the sum of two squares cannot be negative). This equation is therefore degenerate.

7.4 EXERCISES

CONCEPTS

1. Suppose we want to graph an equation in x and y.

(a) If we replace x by $x - 3$, the graph of the equation is

shifted to the _____ by 3 units. If we replace x by

$x + 3$, the graph of the equation is shifted to the

_____ by 3 units.

(b) If we replace y by $y - 1$, the graph of the equation is

shifted _____ by 1 unit. If we replace y by $y + 1$, the

graph of the equation is shifted _____ by 1 unit.

2. The graphs of $x^2 = 12y$ and $(x - 3)^2 = 12(y - 1)$ are given. Label the focus, directrix, and vertex on each parabola.

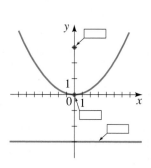

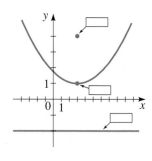

3. The graphs of $\dfrac{x^2}{5^2} + \dfrac{y^2}{4^2} = 1$ and $\dfrac{(x - 3)^2}{5^2} + \dfrac{(y - 1)^2}{4^2} = 1$

are given. Label the vertices and foci on each ellipse.

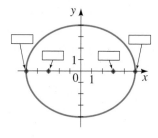

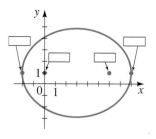

4. The graphs of $\dfrac{x^2}{4^2} - \dfrac{y^2}{3^2} = 1$ and $\dfrac{(x-3)^2}{4^2} - \dfrac{(y-1)^2}{3^2} = 1$

are given. Label the vertices, foci, and asymptotes on each hyperbola.

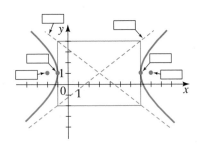

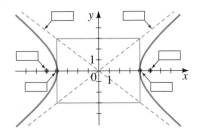

SKILLS

5–8 ■ Find the center, foci, and vertices of the ellipse, and determine the lengths of the major and minor axes. Then sketch the graph.

5. $\dfrac{(x-2)^2}{9} + \dfrac{(y-1)^2}{4} = 1$ **6.** $\dfrac{(x-3)^2}{16} + (y+3)^2 = 1$

7. $\dfrac{x^2}{9} + \dfrac{(y+5)^2}{25} = 1$ **8.** $\dfrac{(x+2)^2}{4} + y^2 = 1$

9–12 ■ Find the vertex, focus, and directrix of the parabola. Then sketch the graph.

9. $(x-3)^2 = 8(y+1)$ **10.** $(y+5)^2 = -6x + 12$

11. $-4\left(x + \tfrac{1}{2}\right)^2 = y$ **12.** $y^2 = 16x - 8$

13–16 ■ Find the center, foci, vertices, and asymptotes of the hyperbola. Then sketch the graph.

13. $\dfrac{(x+1)^2}{9} - \dfrac{(y-3)^2}{16} = 1$ **14.** $(x-8)^2 - (y+6)^2 = 1$

15. $y^2 - \dfrac{(x+1)^2}{4} = 1$ **16.** $\dfrac{(y-1)^2}{25} - (x+3)^2 = 1$

17–22 ■ Find an equation for the conic whose graph is shown.

17.

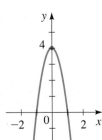

18.

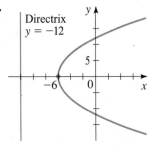

19.

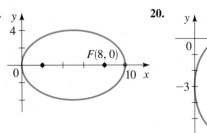

20.

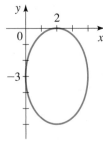

21.

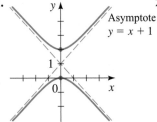

22.

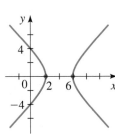

23–34 ■ Complete the square to determine whether the equation represents an ellipse, a parabola, a hyperbola, or a degenerate conic. If the graph is an ellipse, find the center, foci, vertices, and lengths of the major and minor axes. If it is a parabola, find the vertex, focus, and directrix. If it is a hyperbola, find the center, foci, vertices, and asymptotes. Then sketch the graph of the equation. If the equation has no graph, explain why.

23. $y^2 = 4(x + 2y)$

24. $9x^2 - 36x + 4y^2 = 0$

25. $x^2 - 4y^2 - 2x + 16y = 20$

26. $x^2 + 6x + 12y + 9 = 0$

27. $4x^2 + 25y^2 - 24x + 250y + 561 = 0$

28. $2x^2 + y^2 = 2y + 1$

29. $16x^2 - 9y^2 - 96x + 288 = 0$

30. $4x^2 - 4x - 8y + 9 = 0$

31. $x^2 + 16 = 4(y^2 + 2x)$

32. $x^2 - y^2 = 10(x - y) + 1$

33. $3x^2 + 4y^2 - 6x - 24y + 39 = 0$

34. $x^2 + 4y^2 + 20x - 40y + 300 = 0$

35–38 ■ Use a graphing device to graph the conic.

35. $2x^2 - 4x + y + 5 = 0$

36. $4x^2 + 9y^2 - 36y = 0$

37. $9x^2 + 36 = y^2 + 36x + 6y$

38. $x^2 - 4y^2 + 4x + 8y = 0$

39. Determine what the value of F must be if the graph of the equation

$$4x^2 + y^2 + 4(x - 2y) + F = 0$$

is **(a)** an ellipse, **(b)** a single point, or **(c)** the empty set.

40. Find an equation for the ellipse that shares a vertex and a focus with the parabola $x^2 + y = 100$ and has its other focus at the origin.

41. This exercise deals with **confocal parabolas**, that is, families of parabolas that have the same focus.

(a) Draw graphs of the family of parabolas

$$x^2 = 4p(y + p)$$

for $p = -2, -\frac{3}{2}, -1, -\frac{1}{2}, \frac{1}{2}, 1, \frac{3}{2}, 2$.

(b) Show that each parabola in this family has its focus at the origin.

(c) Describe the effect on the graph of moving the vertex closer to the origin.

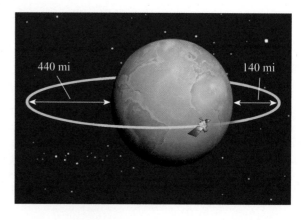

APPLICATIONS

42. Path of a Cannonball A cannon fires a cannonball as shown in the figure. The path of the cannonball is a parabola with vertex at the highest point of the path. If the cannonball lands 1600 ft from the cannon and the highest point it reaches is 3200 ft above the ground, find an equation for the path of the cannonball. Place the origin at the location of the cannon.

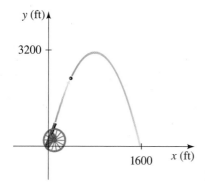

43. Orbit of a Satellite A satellite is in an elliptical orbit around the earth with the center of the earth at one focus, as shown in the figure at the top of the right-hand column. The height of the satellite above the earth varies between 140 mi and 440 mi. Assume that the earth is a sphere with radius 3960 mi. Find an equation for the path of the satellite with the origin at the center of the earth.

DISCOVERY ▪ DISCUSSION ▪ WRITING

44. A Family of Confocal Conics Conics that share a focus are called **confocal**. Consider the family of conics that have a focus at $(0, 1)$ and a vertex at the origin, as shown in the figure.

(a) Find equations of two different ellipses that have these properties.

(b) Find equations of two different hyperbolas that have these properties.

(c) Explain why only one parabola satisfies these properties. Find its equation.

(d) Sketch the conics you found in parts (a), (b), and (c) on the same coordinate axes (for the hyperbolas, sketch the top branches only).

(e) How are the ellipses and hyperbolas related to the parabola?

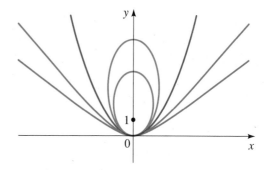

7.5 ROTATION OF AXES

| Rotation of Axes ▶ General Equation of a Conic ▶ The Discriminant

In Section 7.4 we studied conics with equations of the form

$$Ax^2 + Cy^2 + Dx + Ey + F = 0$$

We saw that the graph is always an ellipse, parabola, or hyperbola with horizontal or vertical axes (except in the degenerate cases). In this section we study the most general second-degree equation

$$Ax^2 + Bxy + Cy^2 + Dx + Ey + F = 0$$

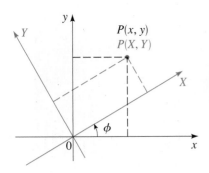

FIGURE 1

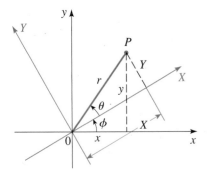

FIGURE 2

We will see that the graph of an equation of this form is also a conic. In fact, by rotating the coordinate axes through an appropriate angle, we can eliminate the term Bxy and then use our knowledge of conic sections to analyze the graph.

▼ Rotation of Axes

In Figure 1 the x- and y-axes have been rotated through an acute angle ϕ about the origin to produce a new pair of axes, which we call the X- and Y-axes. A point P that has coordinates (x, y) in the old system has coordinates (X, Y) in the new system. If we let r denote the distance of P from the origin and let θ be the angle that the segment OP makes with the new X-axis, then we can see from Figure 2 (by considering the two right triangles in the figure) that

$$X = r \cos \theta \qquad\qquad Y = r \sin \theta$$
$$x = r \cos(\theta + \phi) \qquad y = r \sin(\theta + \phi)$$

Using the Addition Formula for Cosine, we see that

$$x = r \cos(\theta + \phi)$$
$$= r(\cos \theta \cos \phi - \sin \theta \sin \phi)$$
$$= (r \cos \theta) \cos \phi - (r \sin \theta) \sin \phi$$
$$= X \cos \phi - Y \sin \phi$$

Similarly, we can apply the Addition Formula for Sine to the expression for y to obtain $y = X \sin \phi + Y \cos \phi$. By treating these equations for x and y as a system of linear equations in the variables X and Y (see Exercise 35), we obtain expressions for X and Y in terms of x and y, as detailed in the following box.

ROTATION OF AXES FORMULAS

Suppose the x- and y-axes in a coordinate plane are rotated through the acute angle ϕ to produce the X- and Y-axes, as shown in Figure 1. Then the coordinates (x, y) and (X, Y) of a point in the xy- and the XY-planes are related as follows:

$$x = X \cos \phi - Y \sin \phi \qquad\qquad X = x \cos \phi + y \sin \phi$$
$$y = X \sin \phi + Y \cos \phi \qquad\qquad Y = -x \sin \phi + y \cos \phi$$

EXAMPLE 1 | Rotation of Axes

If the coordinate axes are rotated through 30°, find the XY-coordinates of the point with xy-coordinates $(2, -4)$.

SOLUTION Using the Rotation of Axes Formulas with $x = 2$, $y = -4$, and $\phi = 30°$, we get

$$X = 2 \cos 30° + (-4) \sin 30° = 2\left(\frac{\sqrt{3}}{2}\right) - 4\left(\frac{1}{2}\right) = \sqrt{3} - 2$$

$$Y = -2 \sin 30° + (-4) \cos 30° = -2\left(\frac{1}{2}\right) - 4\left(\frac{\sqrt{3}}{2}\right) = -1 - 2\sqrt{3}$$

The XY-coordinates are $(-2 + \sqrt{3}, -1 - 2\sqrt{3})$.

✎. NOW TRY EXERCISE **3**

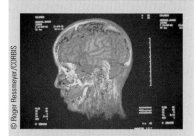

Looking Inside Your Head
How would you like to look inside your head? The idea isn't particularly appealing to most of us, but doctors often need to do just that. If they can look without invasive surgery, all the better. An X-ray doesn't really give a look inside, it simply gives a "graph" of the density of tissue the X-rays must pass through. So an X-ray is a "flattened" view in one direction. Suppose you get an X-ray view from many different directions. Can these "graphs" be used to reconstruct the three-dimensional inside view? This is a purely mathematical problem and was solved by mathematicians a long time ago. However, reconstructing the inside view requires thousands of tedious computations. Today, mathematics and high-speed computers make it possible to "look inside" by a process called computer-aided tomography (or CAT scan). Mathematicians continue to search for better ways of using mathematics to reconstruct images. One of the latest techniques, called magnetic resonance imaging (MRI), combines molecular biology and mathematics for a clear "look inside."

EXAMPLE 2 | Rotating a Hyperbola

Rotate the coordinate axes through $45°$ to show that the graph of the equation $xy = 2$ is a hyperbola.

SOLUTION   We use the Rotation of Axes Formulas with $\phi = 45°$ to obtain

$$x = X \cos 45° - Y \sin 45° = \frac{X}{\sqrt{2}} - \frac{Y}{\sqrt{2}}$$

$$y = X \sin 45° + Y \cos 45° = \frac{X}{\sqrt{2}} + \frac{Y}{\sqrt{2}}$$

Substituting these expressions into the original equation gives

$$\left(\frac{X}{\sqrt{2}} - \frac{Y}{\sqrt{2}}\right)\left(\frac{X}{\sqrt{2}} + \frac{Y}{\sqrt{2}}\right) = 2$$

$$\frac{X^2}{2} - \frac{Y^2}{2} = 2$$

$$\frac{X^2}{4} - \frac{Y^2}{4} = 1$$

We recognize this as a hyperbola with vertices $(\pm 2, 0)$ in the XY-coordinate system. Its asymptotes are $Y = \pm X$, which correspond to the coordinate axes in the xy-system (see Figure 3).

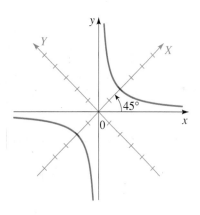

FIGURE 3
$xy = 2$

↘. NOW TRY EXERCISE **11**

▼ General Equation of a Conic

The method of Example 2 can be used to transform any equation of the form

$$Ax^2 + Bxy + Cy^2 + Dx + Ey + F = 0$$

into an equation in X and Y that doesn't contain an XY-term by choosing an appropriate angle of rotation. To find the angle that works, we rotate the axes through an angle ϕ and substitute for x and y using the Rotation of Axes Formulas:

$$A(X \cos \phi - Y \sin \phi)^2 + B(X \cos \phi - Y \sin \phi)(X \sin \phi + Y \cos \phi)$$
$$+ C(X \sin \phi + Y \cos \phi)^2 + D(X \cos \phi - Y \sin \phi)$$
$$+ E(X \sin \phi + Y \cos \phi) + F = 0$$

If we expand this and collect like terms, we obtain an equation of the form

$$A'X^2 + B'XY + C'Y^2 + D'X + E'Y + F' = 0$$

where

$$A' = A\cos^2\phi + B\sin\phi\cos\phi + C\sin^2\phi$$
$$B' = 2(C - A)\sin\phi\cos\phi + B(\cos^2\phi - \sin^2\phi)$$
$$C' = A\sin^2\phi - B\sin\phi\cos\phi + C\cos^2\phi$$
$$D' = D\cos\phi + E\sin\phi$$
$$E' = -D\sin\phi + E\cos\phi$$
$$F' = F$$

To eliminate the XY-term, we would like to choose ϕ so that $B' = 0$, that is,

Double-Angle Formulas

$\sin 2\phi = 2\sin\phi\cos\phi$

$\cos 2\phi = \cos^2\phi - \sin^2\phi$

$$2(C - A)\sin\phi\cos\phi + B(\cos^2\phi - \sin^2\phi) = 0 \qquad \text{Double-Angle}$$
$$(C - A)\sin 2\phi + B\cos 2\phi = 0 \qquad \text{Formulas for Sine and Cosine}$$

$$B\cos 2\phi = (A - C)\sin 2\phi$$

$$\cot 2\phi = \frac{A - C}{B} \qquad \text{Divide by } B\sin 2\phi$$

The preceding calculation proves the following theorem.

SIMPLIFYING THE GENERAL CONIC EQUATION

To eliminate the xy-term in the general conic equation

$$Ax^2 + Bxy + Cy^2 + Dx + Ey + F = 0$$

rotate the axes through the acute angle ϕ that satisfies

$$\cot 2\phi = \frac{A - C}{B}$$

EXAMPLE 3 | Eliminating the xy-Term

Use a rotation of axes to eliminate the xy-term in the equation

$$6\sqrt{3}x^2 + 6xy + 4\sqrt{3}y^2 = 21\sqrt{3}$$

Identify and sketch the curve.

SOLUTION To eliminate the xy-term, we rotate the axes through an angle ϕ that satisfies

$$\cot 2\phi = \frac{A - C}{B} = \frac{6\sqrt{3} - 4\sqrt{3}}{6} = \frac{\sqrt{3}}{3}$$

Thus $2\phi = 60°$ and hence $\phi = 30°$. With this value of ϕ, we get

$$x = X\left(\frac{\sqrt{3}}{2}\right) - Y\left(\frac{1}{2}\right) \qquad \text{Rotation of Axes Formulas}$$

$$y = X\left(\frac{1}{2}\right) + Y\left(\frac{\sqrt{3}}{2}\right) \qquad \cos\phi = \frac{\sqrt{3}}{2}, \sin\phi = \frac{1}{2}$$

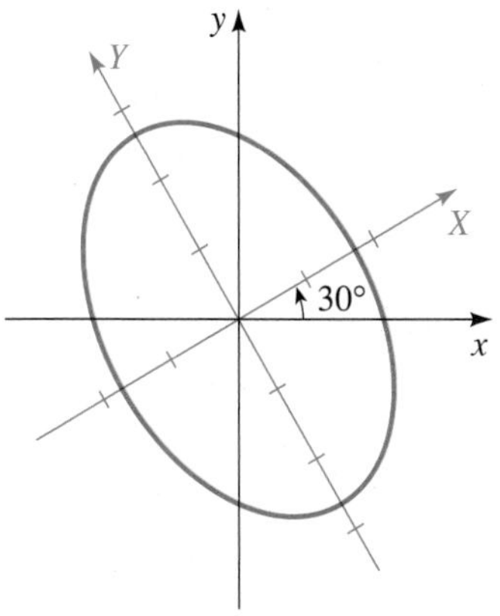

Substituting these values for x and y into the given equation leads to

$$6\sqrt{3}\left(\frac{X\sqrt{3}}{2} - \frac{Y}{2}\right)^2 + 6\left(\frac{X\sqrt{3}}{2} - \frac{Y}{2}\right)\left(\frac{X}{2} + \frac{Y\sqrt{3}}{2}\right) + 4\sqrt{3}\left(\frac{X}{2} + \frac{Y\sqrt{3}}{2}\right)^2 = 21\sqrt{3}$$

Expanding and collecting like terms, we get

$$7\sqrt{3}X^2 + 3\sqrt{3}Y^2 = 21\sqrt{3}$$

$$\frac{X^2}{3} + \frac{Y^2}{7} = 1 \qquad \text{Divide by } 21\sqrt{3}$$

This is the equation of an ellipse in the XY-coordinate system. The foci lie on the Y-axis. Because $a^2 = 7$ and $b^2 = 3$, the length of the major axis is $2\sqrt{7}$, and the length of the minor axis is $2\sqrt{3}$. The ellipse is sketched in Figure 4.

✎ NOW TRY EXERCISE **17**   ■

FIGURE 4
$6\sqrt{3}x^2 + 6xy + 4\sqrt{3}y^2 = 21\sqrt{3}$

In the preceding example we were able to determine ϕ without difficulty, since we remembered that $\cot 60° = \sqrt{3}/3$. In general, finding ϕ is not quite so easy. The next example illustrates how the following Half-Angle Formulas, which are valid for $0 < \phi < \pi/2$, are useful in determining ϕ (see Section 4.3).

$$\cos\phi = \sqrt{\frac{1 + \cos 2\phi}{2}} \qquad \sin\phi = \sqrt{\frac{1 - \cos 2\phi}{2}}$$

EXAMPLE 4 | Graphing a Rotated Conic

A conic has the equation

$$64x^2 + 96xy + 36y^2 - 15x + 20y - 25 = 0$$

(a) Use a rotation of axes to eliminate the xy-term.

(b) Identify and sketch the graph.

(c) Draw the graph using a graphing calculator.

SOLUTION

(a) To eliminate the xy-term, we rotate the axes through an angle ϕ that satisfies

$$\cot 2\phi = \frac{A - C}{B} = \frac{64 - 36}{96} = \frac{7}{24}$$

In Figure 5 we sketch a triangle with $\cot 2\phi = \frac{7}{24}$. We see that

$$\cos 2\phi = \frac{7}{25}$$

so, using the Half-Angle Formulas, we get

$$\cos\phi = \sqrt{\frac{1 + \frac{7}{25}}{2}} = \sqrt{\frac{16}{25}} = \frac{4}{5}$$

$$\sin\phi = \sqrt{\frac{1 - \frac{7}{25}}{2}} = \sqrt{\frac{9}{25}} = \frac{3}{5}$$

The Rotation of Axes Formulas then give

$$x = \tfrac{4}{5}X - \tfrac{3}{5}Y \qquad \text{and} \qquad y = \tfrac{3}{5}X + \tfrac{4}{5}Y$$

Substituting into the given equation, we have

$$64\left(\tfrac{4}{5}X - \tfrac{3}{5}Y\right)^2 + 96\left(\tfrac{4}{5}X - \tfrac{3}{5}Y\right)\left(\tfrac{3}{5}X + \tfrac{4}{5}Y\right)$$

$$+ 36\left(\tfrac{3}{5}X + \tfrac{4}{5}Y\right)^2 - 15\left(\tfrac{4}{5}X - \tfrac{3}{5}Y\right) + 20\left(\tfrac{3}{5}X + \tfrac{4}{5}Y\right) - 25 = 0$$

FIGURE 5

Expanding and collecting like terms, we get

$$100X^2 + 25Y - 25 = 0$$

$$-4X^2 = Y - 1 \qquad \text{Simplify}$$

$$X^2 = -\tfrac{1}{4}(Y - 1) \qquad \text{Divide by 4}$$

(b) We recognize this as the equation of a parabola that opens along the negative Y-axis and has vertex $(0, 1)$ in XY-coordinates. Since $4p = -\tfrac{1}{4}$, we have $p = -\tfrac{1}{16}$, so the focus is $\left(0, \tfrac{15}{16}\right)$ and the directrix is $Y = \tfrac{17}{16}$. Using

$$\phi = \cos^{-1}\tfrac{4}{5} \approx 37°$$

we sketch the graph in Figure 6(a).

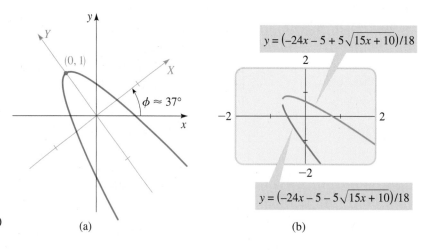

FIGURE 6

$64x^2 + 96xy + 36y^2 - 15x + 20y - 25 = 0$

(a)   (b)

(c) To draw the graph using a graphing calculator, we need to solve for y. The given equation is a quadratic equation in y, so we can use the Quadratic Formula to solve for y. Writing the equation in the form

$$36y^2 + (96x + 20)y + (64x^2 - 15x - 25) = 0$$

we get

$$y = \frac{-(96x + 20) \pm \sqrt{(96x + 20)^2 - 4(36)(64x^2 - 15x - 25)}}{2(36)} \qquad \text{Quadratic Formula}$$

$$= \frac{-(96x + 20) \pm \sqrt{6000x + 4000}}{72} \qquad \text{Expand}$$

$$= \frac{-96x - 20 \pm 20\sqrt{15x + 10}}{72} \qquad \text{Simplify}$$

$$= \frac{-24x - 5 \pm 5\sqrt{15x + 10}}{18} \qquad \text{Simplify}$$

To obtain the graph of the parabola, we graph the functions

$$y = \left(-24x - 5 + 5\sqrt{15x + 10}\right)/18 \qquad \text{and} \qquad y = \left(-24x - 5 - 5\sqrt{15x + 10}\right)/18$$

as shown in Figure 6(b).

✎ NOW TRY EXERCISE **23**

▼ The Discriminant

In Examples 3 and 4 we were able to identify the type of conic by rotating the axes. The next theorem gives rules for identifying the type of conic directly from the equation, without rotating axes.

IDENTIFYING CONICS BY THE DISCRIMINANT

The graph of the equation

$$Ax^2 + Bxy + Cy^2 + Dx + Ey + F = 0$$

is either a conic or a degenerate conic. In the nondegenerate cases, the graph is

1. a parabola if $B^2 - 4AC = 0$

2. an ellipse if $B^2 - 4AC < 0$

3. a hyperbola if $B^2 - 4AC > 0$

The quantity $B^2 - 4AC$ is called the **discriminant** of the equation.

PROOF   If we rotate the axes through an angle ϕ, we get an equation of the form

$$A'X^2 + B'XY + C'Y^2 + D'X + E'Y + F' = 0$$

where A', B', C', ... are given by the formulas on page 396. A straightforward calculation shows that

$$(B')^2 - 4A'C' = B^2 - 4AC$$

Thus the expression $B^2 - 4AC$ remains unchanged for any rotation. In particular, if we choose a rotation that eliminates the xy-term ($B' = 0$), we get

$$A'X^2 + C'Y^2 + D'X + E'Y + F' = 0$$

In this case, $B^2 - 4AC = -4A'C'$. So $B^2 - 4AC = 0$ if either A' or C' is zero; $B^2 - 4AC < 0$ if A' and C' have the same sign; and $B^2 - 4AC > 0$ if A' and C' have opposite signs. According to the box on page 390, these cases correspond to the graph of the last displayed equation being a parabola, an ellipse, or a hyperbola, respectively. ∎

In the proof we indicated that the discriminant is unchanged by any rotation; for this reason, the discriminant is said to be **invariant** under rotation.

EXAMPLE 5 | Identifying a Conic by the Discriminant

A conic has the equation

$$3x^2 + 5xy - 2y^2 + x - y + 4 = 0$$

(a) Use the discriminant to identify the conic.

(b) Confirm your answer to part (a) by graphing the conic with a graphing calculator.

SOLUTION

(a) Since $A = 3$, $B = 5$, and $C = -2$, the discriminant is

$$B^2 - 4AC = 5^2 - 4(3)(-2) = 49 > 0$$

So the conic is a hyperbola.

(b) Using the Quadratic Formula, we solve for y to get

$$y = \frac{5x - 1 \pm \sqrt{49x^2 - 2x + 33}}{4}$$

We graph these functions in Figure 7. The graph confirms that this is a hyperbola.

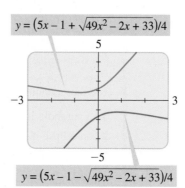

$$y = (5x - 1 + \sqrt{49x^2 - 2x + 33})/4$$

$$y = (5x - 1 - \sqrt{49x^2 - 2x + 33})/4$$

FIGURE 7

✎ NOW TRY EXERCISE **29**

7.5 EXERCISES

CONCEPTS

1. Suppose the x- and y-axes are rotated through an acute angle ϕ to produce the new X- and Y-axes. A point P in the plane can be described by its xy-coordinates (x, y) or its XY-coordinates (X, Y). These coordinates are related by the following formulas.

$x = $ _____ $X = $ _____

$y = $ _____ $Y = $ _____

2. Consider the equation

$$Ax^2 + Bxy + Cy^2 + Dx + Ey + F = 0$$

(a) In general, the graph of this equation is a _____.

(b) To eliminate the xy-term from this equation, we rotate the axes through an angle ϕ that satisfies

$\cot 2\phi = $ _____.

(c) The discriminant of this equation is _____.

If the discriminant is 0, the graph is a _____;

if it is negative, the graph is _____; and

if it is positive, the graph is _____.

SKILLS

3–8 ■ Determine the XY-coordinates of the given point if the co-ordinate axes are rotated through the indicated angle.

3. $(1, 1)$, $\phi = 45°$ **4.** $(-2, 1)$, $\phi = 30°$

5. $(3, -\sqrt{3})$, $\phi = 60°$ **6.** $(2, 0)$, $\phi = 15°$

7. $(0, 2)$, $\phi = 55°$ **8.** $(\sqrt{2}, 4\sqrt{2})$, $\phi = 45°$

9–14 ■ Determine the equation of the given conic in XY-coordinates when the coordinate axes are rotated through the indicated angle.

9. $x^2 - 3y^2 = 4$, $\phi = 60°$

10. $y = (x - 1)^2$, $\phi = 45°$

11. $x^2 - y^2 = 2y$, $\phi = \cos^{-1}\frac{3}{5}$

12. $x^2 + 2y^2 = 16$, $\phi = \sin^{-1}\frac{3}{5}$

13. $x^2 + 2\sqrt{3}xy - y^2 = 4$, $\phi = 30°$

14. $xy = x + y$, $\phi = \pi/4$

15–28 ■ **(a)** Use the discriminant to determine whether the graph of the equation is a parabola, an ellipse, or a hyperbola. **(b)** Use a rotation of axes to eliminate the xy-term. **(c)** Sketch the graph.

15. $xy = 8$

16. $xy + 4 = 0$

17. $x^2 + 2\sqrt{3}xy - y^2 + 2 = 0$

18. $13x^2 + 6\sqrt{3}xy + 7y^2 = 16$

19. $11x^2 - 24xy + 4y^2 + 20 = 0$

20. $21x^2 + 10\sqrt{3}xy + 31y^2 = 144$

21. $\sqrt{3}x^2 + 3xy = 3$

22. $153x^2 + 192xy + 97y^2 = 225$

23. $x^2 + 2xy + y^2 + x - y = 0$

24. $25x^2 - 120xy + 144y^2 - 156x - 65y = 0$

25. $2\sqrt{3}x^2 - 6xy + \sqrt{3}x + 3y = 0$

26. $9x^2 - 24xy + 16y^2 = 100(x - y - 1)$

27. $52x^2 + 72xy + 73y^2 = 40x - 30y + 75$

28. $(7x + 24y)^2 = 600x - 175y + 25$

29–32 ■ (a) Use the discriminant to identify the conic.
(b) Confirm your answer by graphing the conic using a graphing device.

29. $2x^2 - 4xy + 2y^2 - 5x - 5 = 0$

30. $x^2 - 2xy + 3y^2 = 8$

31. $6x^2 + 10xy + 3y^2 - 6y = 36$

32. $9x^2 - 6xy + y^2 + 6x - 2y = 0$

33. (a) Use rotation of axes to show that the following equation represents a hyperbola.

$$7x^2 + 48xy - 7y^2 - 200x - 150y + 600 = 0$$

(b) Find the XY- and xy-coordinates of the center, vertices, and foci.

(c) Find the equations of the asymptotes in XY- and xy-coordinates.

34. (a) Use rotation of axes to show that the following equation represents a parabola.

$$2\sqrt{2}(x + y)^2 = 7x + 9y$$

(b) Find the XY- and xy-coordinates of the vertex and focus.

(c) Find the equation of the directrix in XY- and xy-coordinates.

35. Solve the equations

$$x = X \cos \phi - Y \sin \phi$$

$$y = X \sin \phi + Y \cos \phi$$

for X and Y in terms of x and y.   [*Hint:* To begin, multiply the first equation by $\cos \phi$ and the second by $\sin \phi$, and then add the two equations to solve for X.]

36. Show that the graph of the equation

$$\sqrt{x} + \sqrt{y} = 1$$

is part of a parabola by rotating the axes through an angle of 45°.   [*Hint:* First convert the equation to one that does not involve radicals.]

DISCOVERY ■ DISCUSSION ■ WRITING

37. Matrix Form of Rotation of Axes Formulas

Let Z, Z', and R be the matrices

$$Z = \begin{bmatrix} x \\ y \end{bmatrix} \qquad Z' = \begin{bmatrix} X \\ Y \end{bmatrix}$$

$$R = \begin{bmatrix} \cos \phi & -\sin \phi \\ \sin \phi & \cos \phi \end{bmatrix}$$

Show that the Rotation of Axes Formulas can be written as

$$Z = RZ' \qquad \text{and} \qquad Z' = R^{-1}Z$$

38. Algebraic Invariants   A quantity is invariant under rotation if it does not change when the axes are rotated. It was stated in the text that for the general equation of a conic the quantity $B^2 - 4AC$ is invariant under rotation.

(a) Use the formulas for A', B', and C' on page 396 to prove that the quantity $B^2 - 4AC$ is invariant under rotation; that is, show that

$$B^2 - 4AC = B'^2 - 4A'C'$$

(b) Prove that $A + C$ is invariant under rotation.

(c) Is the quantity F invariant under rotation?

39. Geometric Invariants   Do you expect that the distance between two points is invariant under rotation? Prove your answer by comparing the distance $d(P, Q)$ and $d(P', Q')$ where P' and Q' are the images of P and Q under a rotation of axes.

 **DISCOVERY PROJECT**   **Computer Graphics II**

In this project we investigate how matrices are used to rotate images on a computer screen. You can find the project at the book companion website: **www.stewartmath.com**

7.6 POLAR EQUATIONS OF CONICS

| A Unified Geometric Description of Conics ▶ Polar Equations of Conics

▼ A Unified Geometric Description of Conics

Earlier in this chapter, we defined a parabola in terms of a focus and directrix, but we defined the ellipse and hyperbola in terms of two foci. In this section we give a more unified treatment of all three types of conics in terms of a focus and directrix. If we place the focus at the origin, then a conic section has a simple polar equation. Moreover, in polar form, rotation of conics becomes a simple matter. Polar equations of ellipses are crucial in the derivation of Kepler's Laws (see page 390).

EQUIVALENT DESCRIPTION OF CONICS

Let F be a fixed point (the **focus**), ℓ a fixed line (the **directrix**), and let e be a fixed positive number (the **eccentricity**). The set of all points P such that the ratio of the distance from P to F to the distance from P to ℓ is the constant e is a conic. That is, the set of all points P such that

$$\frac{d(P, F)}{d(P, \ell)} = e$$

is a conic. The conic is a parabola if $e = 1$, an ellipse if $e < 1$, or a hyperbola if $e > 1$.

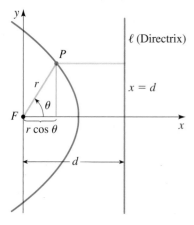

FIGURE 1

PROOF If $e = 1$, then $d(P, F) = d(P, \ell)$, and so the given condition becomes the definition of a parabola as given in Section 7.1.

Now, suppose $e \neq 1$. Let's place the focus F at the origin and the directrix parallel to the y-axis and d units to the right. In this case the directrix has equation $x = d$ and is perpendicular to the polar axis. If the point P has polar coordinates (r, θ), we see from Figure 1 that $d(P, F) = r$ and $d(P, \ell) = d - r\cos\theta$. Thus the condition $d(P, F)/d(P, \ell) = e$, or $d(P, F) = e \cdot d(P, \ell)$, becomes

$$r = e(d - r\cos\theta)$$

If we square both sides of this polar equation and convert to rectangular coordinates, we get

$$x^2 + y^2 = e^2(d - x)^2$$

$$(1 - e^2)x^2 + 2de^2x + y^2 = e^2d^2 \qquad \text{Expand and simplify}$$

$$\left(x + \frac{e^2d}{1 - e^2}\right)^2 + \frac{y^2}{1 - e^2} = \frac{e^2d^2}{(1 - e^2)^2} \qquad \begin{array}{l}\text{Divide by } 1 - e^2 \text{ and complete} \\ \text{the square}\end{array}$$

If $e < 1$, then dividing both sides of this equation by $e^2d^2/(1 - e^2)^2$ gives an equation of the form

$$\frac{(x - h)^2}{a^2} + \frac{y^2}{b^2} = 1$$

where

$$h = \frac{-e^2d}{1 - e^2} \qquad a^2 = \frac{e^2d^2}{(1 - e^2)^2} \qquad b^2 = \frac{e^2d^2}{1 - e^2}$$

This is the equation of an ellipse with center $(h, 0)$. In Section 7.2 we found that the foci of an ellipse are a distance c from the center, where $c^2 = a^2 - b^2$. In our case

$$c^2 = a^2 - b^2 = \frac{e^4d^2}{(1 - e^2)^2}$$

Thus $c = e^2d/(1 - e^2) = -h$, which confirms that the focus defined in the theorem (namely the origin) is the same as the focus defined in Section 7.2. It also follows that

$$e = \frac{c}{a}$$

If $e > 1$, a similar proof shows that the conic is a hyperbola with $e = c/a$, where $c^2 = a^2 + b^2$. ∎

▼ Polar Equations of Conics

In the proof we saw that the polar equation of the conic in Figure 1 is $r = e(d - r \cos \theta)$. Solving for r, we get

$$r = \frac{ed}{1 + e \cos \theta}$$

If the directrix is chosen to be to the *left* of the focus $(x = -d)$, then we get the equation $r = ed/(1 - e \cos \theta)$. If the directrix is *parallel* to the polar axis $(y = d$ or $y = -d)$, then we get $\sin \theta$ instead of $\cos \theta$ in the equation. These observations are summarized in the following box and in Figure 2.

POLAR EQUATIONS OF CONICS

A polar equation of the form

$$r = \frac{ed}{1 \pm e \cos \theta} \qquad \text{or} \qquad r = \frac{ed}{1 \pm e \sin \theta}$$

represents a conic with one focus at the origin and with eccentricity e. The conic is

1. a parabola if $e = 1$

2. an ellipse if $0 < e < 1$

3. a hyperbola if $e > 1$

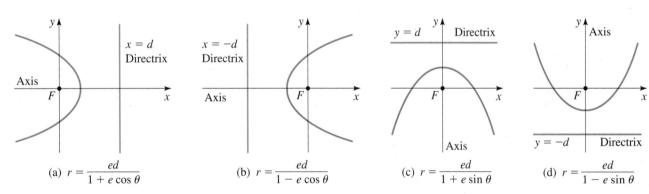

(a) $r = \dfrac{ed}{1 + e \cos \theta}$ (b) $r = \dfrac{ed}{1 - e \cos \theta}$ (c) $r = \dfrac{ed}{1 + e \sin \theta}$ (d) $r = \dfrac{ed}{1 - e \sin \theta}$

FIGURE 2   The form of the polar equation of a conic indicates the location of the directrix.

To graph the polar equation of a conic, we first determine the location of the directrix from the form of the equation. The four cases that arise are shown in Figure 2. (The figure shows only the parts of the graphs that are close to the focus at the origin. The shape of the rest of the graph depends on whether the equation represents a parabola, an ellipse, or a hyperbola.) The axis of a conic is perpendicular to the directrix—specifically we have the following:

1. For a parabola the axis of symmetry is perpendicular to the directrix.

2. For an ellipse the major axis is perpendicular to the directrix.

3. For a hyperbola the transverse axis is perpendicular to the directrix.

EXAMPLE 1 | Finding a Polar Equation for a Conic

Find a polar equation for the parabola that has its focus at the origin and whose directrix is the line $y = -6$.

SOLUTION Using $e = 1$ and $d = 6$ and using part (d) of Figure 2, we see that the polar equation of the parabola is

$$r = \frac{6}{1 - \sin \theta}$$

✎ NOW TRY EXERCISE **3**

To graph a polar conic, it is helpful to plot the points for which $\theta = 0, \pi/2, \pi,$ and $3\pi/2$. Using these points and a knowledge of the type of conic (which we obtain from the eccentricity), we can easily get a rough idea of the shape and location of the graph.

EXAMPLE 2 | Identifying and Sketching a Conic

A conic is given by the polar equation

$$r = \frac{10}{3 - 2\cos \theta}$$

(a) Show that the conic is an ellipse, and sketch the graph.

(b) Find the center of the ellipse and the lengths of the major and minor axes.

SOLUTION

(a) Dividing the numerator and denominator by 3, we have

$$r = \frac{\frac{10}{3}}{1 - \frac{2}{3}\cos \theta}$$

Since $e = \frac{2}{3} < 1$, the equation represents an ellipse. For a rough graph we plot the points for which $\theta = 0, \pi/2, \pi, 3\pi/2$ (see Figure 3).

θ	r
0	10
$\frac{\pi}{2}$	$\frac{10}{3}$
π	2
$\frac{3\pi}{2}$	$\frac{10}{3}$

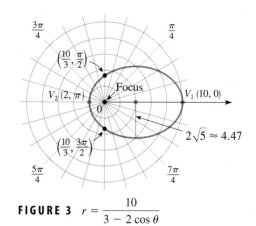

FIGURE 3 $\quad r = \dfrac{10}{3 - 2\cos \theta}$

(b) Comparing the equation to those in Figure 2, we see that the major axis is horizontal. Thus the endpoints of the major axis are $V_1(10, 0)$ and $V_2(2, \pi)$. So the center of the ellipse is at $C(4, 0)$, the midpoint of $V_1 V_2$.

The distance between the vertices V_1 and V_2 is 12; thus the length of the major axis is $2a = 12$, so $a = 6$. To determine the length of the minor axis, we need to find b. From page 402 we have $c = ae = 6\left(\frac{2}{3}\right) = 4$, so

$$b^2 = a^2 - c^2 = 6^2 - 4^2 = 20$$

Thus $b = \sqrt{20} = 2\sqrt{5} \approx 4.47$, and the length of the minor axis is $2b = 4\sqrt{5} \approx 8.94$.

✎ NOW TRY EXERCISES **17** AND **21**

EXAMPLE 3 | Identifying and Sketching a Conic

A conic is given by the polar equation

$$r = \frac{12}{2 + 4 \sin \theta}$$

(a) Show that the conic is a hyperbola and sketch the graph.

(b) Find the center of the hyperbola and sketch the asymptotes.

SOLUTION

(a) Dividing the numerator and denominator by 2, we have

$$r = \frac{6}{1 + 2 \sin \theta}$$

Since $e = 2 > 1$, the equation represents a hyperbola. For a rough graph we plot the points for which $\theta = 0, \pi/2, \pi, 3\pi/2$ (see Figure 4).

(b) Comparing the equation to those in Figure 2, we see that the transverse axis is vertical. Thus the endpoints of the transverse axis (the vertices of the hyperbola) are $V_1(2, \pi/2)$ and $V_2(-6, 3\pi/2) = V_2(6, \pi/2)$. So the center of the hyperbola is $C(4, \pi/2)$, the midpoint of $V_1 V_2$.

To sketch the asymptotes, we need to find a and b. The distance between V_1 and V_2 is 4; thus the length of the transverse axis is $2a = 4$, so $a = 2$. To find b, we first find c. From page 402 we have $c = ae = 2 \cdot 2 = 4$, so

$$b^2 = c^2 - a^2 = 4^2 - 2^2 = 12$$

Thus $b = \sqrt{12} = 2\sqrt{3} \approx 3.46$. Knowing a and b allows us to sketch the central box, from which we obtain the asymptotes shown in Figure 4.

θ	r
0	6
$\frac{\pi}{2}$	2
π	6
$\frac{3\pi}{2}$	-6

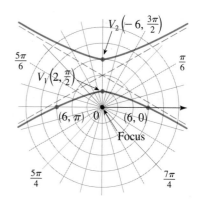

FIGURE 4 $r = \dfrac{12}{2 + 4 \sin \theta}$

✎ NOW TRY EXERCISE **25**

When we rotate conic sections, it is much more convenient to use polar equations than Cartesian equations. We use the fact that the graph of $r = f(\theta - \alpha)$ is the graph of $r = f(\theta)$ rotated counterclockwise about the origin through an angle α (see Exercise 61 in Section 5.2).

EXAMPLE 4 | Rotating an Ellipse

Suppose the ellipse of Example 2 is rotated through an angle $\pi/4$ about the origin. Find a polar equation for the resulting ellipse, and draw its graph.

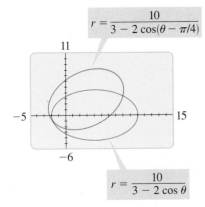

$$r = \frac{10}{3 - 2\cos(\theta - \pi/4)}$$

11

-5 15

-6

$$r = \frac{10}{3 - 2\cos\theta}$$

FIGURE 5

SOLUTION We get the equation of the rotated ellipse by replacing θ with $\theta - \pi/4$ in the equation given in Example 2. So the new equation is

$$r = \frac{10}{3 - 2\cos(\theta - \pi/4)}$$

We use this equation to graph the rotated ellipse in Figure 5. Notice that the ellipse has been rotated about the focus at the origin.

✎ NOW TRY EXERCISE **37**   ■

In Figure 6 we use a computer to sketch a number of conics to demonstrate the effect of varying the eccentricity e. Notice that when e is close to 0, the ellipse is nearly circular, and it becomes more elongated as e increases. When $e = 1$, of course, the conic is a parabola. As e increases beyond 1, the conic is an ever steeper hyperbola.

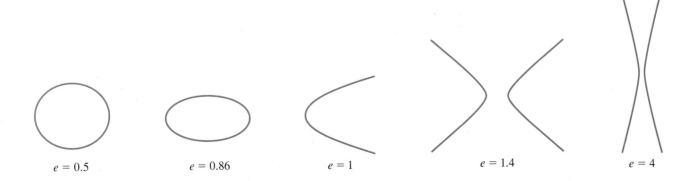

$e = 0.5$ $e = 0.86$ $e = 1$ $e = 1.4$ $e = 4$

FIGURE 6

7.6 EXERCISES

CONCEPTS

1. All conics can be described geometrically using a fixed point F called the _____ and a fixed line ℓ called the

_____. For a fixed positive number e the set of all points P satisfying

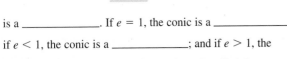 $= e$

is a _____. If $e = 1$, the conic is a _____;

if $e < 1$, the conic is a _____; and if $e > 1$, the

conic is a _____. The number e is called the

_____ of the conic.

2. The polar equation of a conic with eccentricity e has one of the following forms:

$r =$ _____   or $r =$ _____

SKILLS

3–10 ■ Write a polar equation of a conic that has its focus at the origin and satisfies the given conditions.

✎ **3.** Ellipse, eccentricity $\frac{2}{3}$, directrix $x = 3$

4. Hyperbola, eccentricity $\frac{4}{3}$, directrix $x = -3$

5. Parabola, directrix $y = 2$

6. Ellipse, eccentricity $\frac{1}{2}$, directrix $y = -4$

7. Hyperbola, eccentricity 4, directrix $r = 5\sec\theta$

8. Ellipse, eccentricity 0.6, directrix $r = 2\csc\theta$

9. Parabola, vertex at $(5, \pi/2)$

10. Ellipse, eccentricity 0.4, vertex at $(2, 0)$

11–16 ■ Match the polar equations with the graphs labeled I–VI. Give reasons for your answer.

11. $r = \dfrac{6}{1 + \cos \theta}$

12. $r = \dfrac{2}{2 - \cos \theta}$

13. $r = \dfrac{3}{1 - 2 \sin \theta}$

14. $r = \dfrac{5}{3 - 3 \sin \theta}$

15. $r = \dfrac{12}{3 + 2 \sin \theta}$

16. $r = \dfrac{12}{2 + 3 \cos \theta}$

I
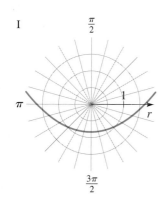

II
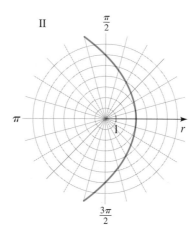

III
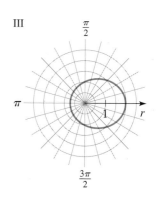

IV
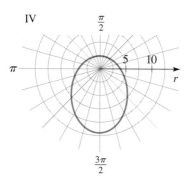

V
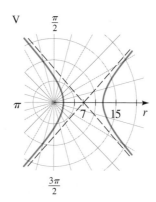

VI
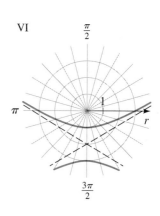

17–20 ■ A polar equation of a conic is given. **(a)** Show that the conic is a parabola and sketch its graph. **(b)** Find the vertex and directrix and indicate them on the graph.

17. $r = \dfrac{4}{1 - \sin \theta}$

18. $r = \dfrac{3}{2 + 2 \sin \theta}$

19. $r = \dfrac{5}{3 + 3 \cos \theta}$

20. $r = \dfrac{2}{5 - 5 \cos \theta}$

21–24 ■ A polar equation of a conic is given. **(a)** Show that the conic is an ellipse, and sketch its graph. **(b)** Find the vertices and directrix, and indicate them on the graph. **(c)** Find the center of the ellipse and the lengths of the major and minor axes.

21. $r = \dfrac{4}{2 - \cos \theta}$

22. $r = \dfrac{6}{3 - 2 \sin \theta}$

23. $r = \dfrac{12}{4 + 3 \sin \theta}$

24. $r = \dfrac{18}{4 + 3 \cos \theta}$

25–28 ■ A polar equation of a conic is given. **(a)** Show that the conic is a hyperbola, and sketch its graph. **(b)** Find the vertices and directrix, and indicate them on the graph. **(c)** Find the center of the hyperbola, and sketch the asymptotes.

25. $r = \dfrac{8}{1 + 2 \cos \theta}$

26. $r = \dfrac{10}{1 - 4 \sin \theta}$

27. $r = \dfrac{20}{2 - 3 \sin \theta}$

28. $r = \dfrac{6}{2 + 7 \cos \theta}$

29–36 ■ **(a)** Find the eccentricity and identify the conic. **(b)** Sketch the conic and label the vertices.

29. $r = \dfrac{4}{1 + 3 \cos \theta}$

30. $r = \dfrac{8}{3 + 3 \cos \theta}$

31. $r = \dfrac{2}{1 - \cos \theta}$

32. $r = \dfrac{10}{3 - 2 \sin \theta}$

33. $r = \dfrac{6}{2 + \sin \theta}$

34. $r = \dfrac{5}{2 - 3 \sin \theta}$

35. $r = \dfrac{7}{2 - 5 \sin \theta}$

36. $r = \dfrac{8}{3 + \cos \theta}$

37–40 ■ A polar equation of a conic is given. (a) Find the eccentricity and the directrix of the conic. (b) If this conic is rotated about the origin through the given angle θ, write the resulting equation. (c) Draw graphs of the original conic and the rotated conic on the same screen.

 37. $r = \dfrac{1}{4 - 3\cos\theta}; \quad \theta = \dfrac{\pi}{3}$

38. $r = \dfrac{2}{5 - 3\sin\theta}; \quad \theta = \dfrac{2\pi}{3}$

39. $r = \dfrac{2}{1 + \sin\theta}; \quad \theta = -\dfrac{\pi}{4}$

40. $r = \dfrac{9}{2 + 2\cos\theta}; \quad \theta = -\dfrac{5\pi}{6}$

 41. Graph the conics $r = e/(1 - e\cos\theta)$ with $e = 0.4, 0.6, 0.8$, and 1.0 on a common screen. How does the value of e affect the shape of the curve?

 42. (a) Graph the conics

$$r = \frac{ed}{(1 + e\sin\theta)}$$

for $e = 1$ and various values of d. How does the value of d affect the shape of the conic?

(b) Graph these conics for $d = 1$ and various values of e. How does the value of e affect the shape of the conic?

APPLICATIONS

43. Orbit of the Earth The polar equation of an ellipse can be expressed in terms of its eccentricity e and the length a of its major axis.

(a) Show that the polar equation of an ellipse with directrix $x = -d$ can be written in the form

$$r = \frac{a(1 - e^2)}{1 - e\cos\theta}$$

[*Hint:* Use the relation $a^2 = e^2d^2/(1 - e^2)^2$ given in the proof on page 402.]

(b) Find an approximate polar equation for the elliptical orbit of the earth around the sun (at one focus) given that the eccentricity is about 0.017 and the length of the major axis is about 2.99×10^8 km.

44. Perihelion and Aphelion The planets move around the sun in elliptical orbits with the sun at one focus. The positions of a planet that are closest to, and farthest from, the sun are called its **perihelion** and **aphelion**, respectively.

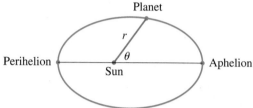

(a) Use Exercise 43(a) to show that the perihelion distance from a planet to the sun is $a(1 - e)$ and the aphelion distance is $a(1 + e)$.

(b) Use the data of Exercise 43(b) to find the distances from the earth to the sun at perihelion and at aphelion.

45. Orbit of Pluto The distance from Pluto to the sun is 4.43×10^9 km at perihelion and 7.37×10^9 km at aphelion. Use Exercise 44 to find the eccentricity of Pluto's orbit.

DISCOVERY ▪ DISCUSSION ▪ WRITING

46. Distance to a Focus When we found polar equations for the conics, we placed one focus at the pole. It's easy to find the distance from that focus to any point on the conic. Explain how the polar equation gives us this distance.

47. Polar Equations of Orbits When a satellite orbits the earth, its path is an ellipse with one focus at the center of the earth. Why do scientists use polar (rather than rectangular) coordinates to track the position of satellites? [*Hint:* Your answer to Exercise 46 is relevant here.]

CHAPTER 7 | REVIEW

■ CONCEPT CHECK

1. (a) Give the geometric definition of a parabola. What are the focus and directrix of the parabola?

(b) Sketch the parabola $x^2 = 4py$ for the case $p > 0$. Identify on your diagram the vertex, focus, and directrix. What happens if $p < 0$?

(c) Sketch the parabola $y^2 = 4px$, together with its vertex, focus, and directrix, for the case $p > 0$. What happens if $p < 0$?

2. (a) Give the geometric definition of an ellipse. What are the foci of the ellipse?

(b) For the ellipse with equation

$$\frac{x^2}{a^2} + \frac{y^2}{b^2} = 1$$

where $a > b > 0$, what are the coordinates of the vertices and the foci? What are the major and minor axes? Illustrate with a graph.

(c) Give an expression for the eccentricity of the ellipse in part (b).

(d) State the equation of an ellipse with foci on the y-axis.

3. (a) Give the geometric definition of a hyperbola. What are the foci of the hyperbola?
   (b) For the hyperbola with equation

   $$\frac{x^2}{a^2} - \frac{y^2}{b^2} = 1$$

   what are the coordinates of the vertices and foci? What are the equations of the asymptotes? What is the transverse axis? Illustrate with a graph.
   (c) State the equation of a hyperbola with foci on the y-axis.
   (d) What steps would you take to sketch a hyperbola with a given equation?

4. Suppose h and k are positive numbers. What is the effect on the graph of an equation in x and y if
   (a) x is replaced by $x - h$? By $x + h$?
   (b) y is replaced by $y - k$? By $y + k$?

5. How can you tell whether the following nondegenerate conic is a parabola, an ellipse, or a hyperbola?

   $$Ax^2 + Cy^2 + Dx + Ey + F = 0$$

6. Suppose the x- and y-axes are rotated through an acute angle ϕ to produce the X- and Y-axes. Write equations that relate the coordinates (x, y) and (X, Y) of a point in the xy-plane and XY-plane, respectively.

7. (a) How do you eliminate the xy-term in this equation?

   $$Ax^2 + Bxy + Cy^2 + Dx + Ey + F = 0$$

   (b) What is the discriminant of the conic in part (a)? How can you use the discriminant to determine whether the conic is a parabola, an ellipse, or a hyperbola?

8. (a) Write polar equations that represent a conic with eccentricity e.
   (b) For what values of e is the conic an ellipse? A hyperbola? A parabola?

■ EXERCISES

1–8 ■ Find the vertex, focus, and directrix of the parabola, and sketch the graph.

1. $y^2 = 4x$
2. $x = \frac{1}{12}y^2$
3. $x^2 + 8y = 0$
4. $2x - y^2 = 0$
5. $x - y^2 + 4y - 2 = 0$
6. $2x^2 + 6x + 5y + 10 = 0$
7. $\frac{1}{2}x^2 + 2x = 2y + 4$
8. $x^2 = 3(x + y)$

9–16 ■ Find the center, vertices, foci, and the lengths of the major and minor axes of the ellipse, and sketch the graph.

9. $\dfrac{x^2}{9} + \dfrac{y^2}{25} = 1$
10. $\dfrac{x^2}{49} + \dfrac{y^2}{9} = 1$
11. $x^2 + 4y^2 = 16$
12. $9x^2 + 4y^2 = 1$
13. $\dfrac{(x-3)^2}{9} + \dfrac{y^2}{16} = 1$
14. $\dfrac{(x-2)^2}{25} + \dfrac{(y+3)^2}{16} = 1$
15. $4x^2 + 9y^2 = 36y$
16. $2x^2 + y^2 = 2 + 4(x - y)$

17–24 ■ Find the center, vertices, foci, and asymptotes of the hyperbola, and sketch the graph.

17. $-\dfrac{x^2}{9} + \dfrac{y^2}{16} = 1$
18. $\dfrac{x^2}{49} - \dfrac{y^2}{32} = 1$
19. $x^2 - 2y^2 = 16$
20. $x^2 - 4y^2 + 16 = 0$
21. $\dfrac{(x+4)^2}{16} - \dfrac{y^2}{16} = 1$
22. $\dfrac{(x-2)^2}{8} - \dfrac{(y+2)^2}{8} = 1$
23. $9y^2 + 18y = x^2 + 6x + 18$
24. $y^2 = x^2 + 6y$

25–30 ■ Find an equation for the conic whose graph is shown.

25.
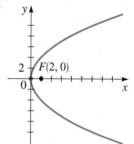

26.
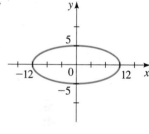

27.
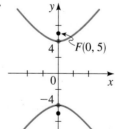

28.
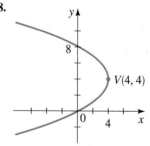

29.
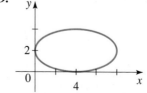

30.
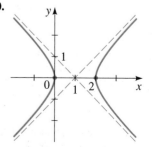

31–42 ■ Determine the type of curve represented by the equation. Find the foci and vertices (if any), and sketch the graph.

31. $\dfrac{x^2}{12} + y = 1$

32. $\dfrac{x^2}{12} + \dfrac{y^2}{144} = \dfrac{y}{12}$

33. $x^2 - y^2 + 144 = 0$

34. $x^2 + 6x = 9y^2$

35. $4x^2 + y^2 = 8(x + y)$

36. $3x^2 - 6(x + y) = 10$

37. $x = y^2 - 16y$

38. $2x^2 + 4 = 4x + y^2$

39. $2x^2 - 12x + y^2 + 6y + 26 = 0$

40. $36x^2 - 4y^2 - 36x - 8y = 31$

41. $9x^2 + 8y^2 - 15x + 8y + 27 = 0$

42. $x^2 + 4y^2 = 4x + 8$

43–50 ■ Find an equation for the conic section with the given properties.

43. The parabola with focus $F(0, 1)$ and directrix $y = -1$

44. The ellipse with center $C(0, 4)$, foci $F_1(0, 0)$ and $F_2(0, 8)$, and major axis of length 10

45. The hyperbola with vertices $V(0, \pm2)$ and asymptotes $y = \pm\frac{1}{2}x$

46. The hyperbola with center $C(2, 4)$, foci $F_1(2, 1)$ and $F_2(2, 7)$, and vertices $V_1(2, 6)$ and $V_2(2, 2)$

47. The ellipse with foci $F_1(1, 1)$ and $F_2(1, 3)$, and with one vertex on the x-axis

48. The parabola with vertex $V(5, 5)$ and directrix the y-axis

49. The ellipse with vertices $V_1(7, 12)$ and $V_2(7, -8)$, and passing through the point $P(1, 8)$

50. The parabola with vertex $V(-1, 0)$ and horizontal axis of symmetry, and crossing the y-axis at $y = 2$

51. The path of the earth around the sun is an ellipse with the sun at one focus. The ellipse has major axis 186,000,000 mi and eccentricity 0.017. Find the distance between the earth and the sun when the earth is **(a)** closest to the sun and **(b)** farthest from the sun.

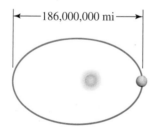

52. A ship is located 40 mi from a straight shoreline. LORAN stations A and B are located on the shoreline, 300 mi apart. From the LORAN signals, the captain determines that his ship is

80 mi closer to A than to B. Find the location of the ship. (Place A and B on the y-axis with the x-axis halfway between them. Find the x- and y-coordinates of the ship.)

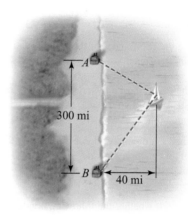

53. (a) Draw graphs of the following family of ellipses for $k = 1, 2, 4,$ and 8.

$$\frac{x^2}{16 + k^2} + \frac{y^2}{k^2} = 1$$

(b) Prove that all the ellipses in part (a) have the same foci.

54. (a) Draw graphs of the following family of parabolas for $k = \frac{1}{2}, 1, 2,$ and 4.

$$y = kx^2$$

(b) Find the foci of the parabolas in part (a).

(c) How does the location of the focus change as k increases?

55–58 ■ An equation of a conic is given. **(a)** Use the discriminant to determine whether the graph of the equation is a parabola, an ellipse, or a hyperbola. **(b)** Use a rotation of axes to eliminate the xy-term. **(c)** Sketch the graph.

55. $x^2 + 4xy + y^2 = 1$

56. $5x^2 - 6xy + 5y^2 - 8x + 8y - 8 = 0$

57. $7x^2 - 6\sqrt{3}\,xy + 13y^2 - 4\sqrt{3}\,x - 4y = 0$

58. $9x^2 + 24xy + 16y^2 = 25$

59–62 ■ Use a graphing device to graph the conic. Identify the type of conic from the graph.

59. $5x^2 + 3y^2 = 60$ **60.** $9x^2 - 12y^2 + 36 = 0$

61. $6x + y^2 - 12y = 30$ **62.** $52x^2 - 72xy + 73y^2 = 100$

63–66 ■ A polar equation of a conic is given. **(a)** Find the eccentricity and identify the conic. **(b)** Sketch the conic and label the vertices.

63. $r = \dfrac{1}{1 - \cos\theta}$ **64.** $r = \dfrac{2}{3 + 2\sin\theta}$

65. $r = \dfrac{4}{1 + 2\sin\theta}$ **66.** $r = \dfrac{12}{1 - 4\cos\theta}$

1. Find the focus and directrix of the parabola $x^2 = -12y$, and sketch its graph.

2. Find the vertices, foci, and the lengths of the major and minor axes for the ellipse
   $\dfrac{x^2}{16} + \dfrac{y^2}{4} = 1$. Then sketch its graph.

3. Find the vertices, foci, and asymptotes of the hyperbola $\dfrac{y^2}{9} - \dfrac{x^2}{16} = 1$. Then sketch its graph.

4–6 ■ Find an equation for the conic whose graph is shown.

4.

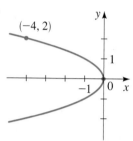

5.

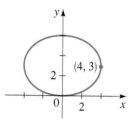

6.

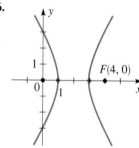

7–9 ■ Sketch the graph of the equation.

7. $16x^2 + 36y^2 - 96x + 36y + 9 = 0$

8. $9x^2 - 8y^2 + 36x + 64y = 164$

9. $2x + y^2 + 8y + 8 = 0$

10. Find an equation for the hyperbola with foci $(0, \pm 5)$ and with asymptotes $y = \pm\frac{3}{4}x$.

11. Find an equation for the parabola with focus $(2, 4)$ and directrix the x-axis.

6 in.

3 in.

12. A parabolic reflector for a car headlight forms a bowl shape that is 6 in. wide at its opening and 3 in. deep, as shown in the figure at the left. How far from the vertex should the filament of the bulb be placed if it is to be located at the focus?

13. (a) Use the discriminant to determine whether the graph of this equation is a parabola, an ellipse, or a hyperbola:
   $$5x^2 + 4xy + 2y^2 = 18$$
   (b) Use rotation of axes to eliminate the xy-term in the equation.
   (c) Sketch the graph of the equation.
   (d) Find the coordinates of the vertices of this conic (in the xy-coordinate system).

14. (a) Find the polar equation of the conic that has a focus at the origin, eccentricity $e = \frac{1}{2}$, and directrix $x = 2$. Sketch the graph.
   (b) What type of conic is represented by the following equation? Sketch its graph.
   $$r = \frac{3}{2 - \sin\theta}$$

Many buildings employ conic sections in their design. Architects have various reasons for using these curves, ranging from structural stability to simple beauty. But how can a huge parabola, ellipse, or hyperbola be accurately constructed in concrete and steel? In this *Focus on Modeling,* we will see how the geometric properties of the conics can be used to construct these shapes.

▼ Conics in Buildings

In ancient times architecture was part of mathematics, so architects had to be mathematicians. Many of the structures they built—pyramids, temples, amphitheaters, and irrigation projects—still stand. In modern times architects employ even more sophisticated mathematical principles. The photographs below show some structures that employ conic sections in their design.

Roman Amphitheater in
Alexandria, Egypt (circle)
© Nick Wheeler/CORBIS

Ceiling of Statuary Hall in the
U.S. Capitol (ellipse)
Architect of the Capitol

Roof of the Skydome in
Toronto, Canada (parabola)
Walter Schmid/© Stone/Getty Images

Roof of Washington Dulles Airport
(hyperbola and parabola)
© Richard T. Nowitz/CORBIS

McDonnell Planetarium,
St. Louis, MO (hyperbola)
VisionsofAmerica/Joe Sohm

Attic in La Pedrera,
Barcelona, Spain (parabola)
© O. Alamany & E. Vincens/CORBIS

Architects have different reasons for using conics in their designs. For example, the Spanish architect Antoni Gaudí used parabolas in the attic of La Pedrera (see photo above). He reasoned that since a rope suspended between two points with an equally distributed load (as in a suspension bridge) has the shape of a parabola, an inverted parabola would provide the best support for a flat roof.

▼ Constructing Conics

The equations of the conics are helpful in manufacturing small objects, because a computer-controlled cutting tool can accurately trace a curve given by an equation. But in a building project, how can we construct a portion of a parabola, ellipse, or hyperbola that spans the ceiling or walls of a building? The geometric properties of the conics provide practical ways of constructing them. For example, if you were building a circular tower, you would choose a center point, then make sure that the walls of the tower were a fixed distance from that point. Elliptical walls can be constructed using a string anchored at two points, as shown in Figure 1.

To construct a parabola, we can use the apparatus shown in Figure 2. A piece of string of length a is anchored at F and A. The T-square, also of length a, slides along the straight bar L. A pencil at P holds the string taut against the T-square. As the T-square slides to the right the pencil traces out a curve.

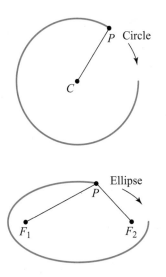

FIGURE 1 Constructing a circle and an ellipse

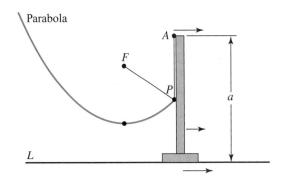

FIGURE 2 Constructing a parabola

From the figure we see that

$$d(F, P) + d(P, A) = a \qquad \text{The string is of length } a$$

$$d(L, P) + d(P, A) = a \qquad \text{The T-square is of length } a$$

It follows that $d(F, P) + d(P, A) = d(L, P) + d(P, A)$. Subtracting $d(P, A)$ from each side, we get

$$d(F, P) = d(L, P)$$

The last equation says that the distance from F to P is equal to the distance from P to the line L. Thus, the curve is a parabola with focus F and directrix L.

In building projects it is easier to construct a straight line than a curve. So in some buildings, such as in the Kobe Tower (see Problem 4), a curved surface is produced by using many straight lines. We can also produce a curve using straight lines, such as the parabola shown in Figure 3.

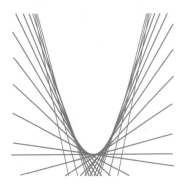

FIGURE 3 Tangent lines to a parabola

Each line is **tangent** to the parabola; that is, the line meets the parabola at exactly one point and does not cross the parabola. The line tangent to the parabola $y = x^2$ at the point (a, a^2) is

$$y = 2ax - a^2$$

You are asked to show this in Problem 6. The parabola is called the **envelope** of all such lines.

PROBLEMS

1. **Conics in Architecture** The photographs on page 412 show six examples of buildings that contain conic sections. Search the Internet to find other examples of structures that employ parabolas, ellipses, or hyperbolas in their design. Find at least one example for each type of conic.

2. **Constructing a Hyperbola** In this problem we construct a hyperbola. The wooden bar in the figure can pivot at F_1. A string that is shorter than the bar is anchored at F_2 and at A, the other end of the bar. A pencil at P holds the string taut against the bar as it moves counterclockwise around F_1.

   (a) Show that the curve traced out by the pencil is one branch of a hyperbola with foci at F_1 and F_2.

   (b) How should the apparatus be reconfigured to draw the other branch of the hyperbola?

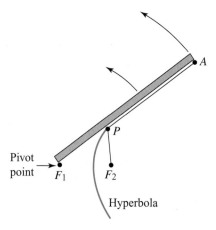

3. **A Parabola in a Rectangle** The following method can be used to construct a parabola that fits in a given rectangle. The parabola will be approximated by many short line segments.

   First, draw a rectangle. Divide the rectangle in half by a vertical line segment, and label the top endpoint V. Next, divide the length and width of each half rectangle into an equal number of parts to form grid lines, as shown in the figure below. Draw lines from V to the endpoints of horizontal grid line 1, and mark the points where these lines cross the vertical grid lines labeled 1. Next, draw lines from V to the endpoints of horizontal grid line 2, and mark the points where these lines cross the vertical grid lines labeled 2. Continue in this way until you have used all the horizontal grid lines. Now use line segments to connect the points you have marked to obtain an approximation to the desired parabola. Apply this procedure to draw a parabola that fits into a 6 ft by 10 ft rectangle on a lawn.

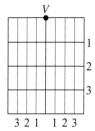
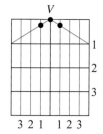
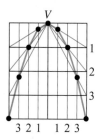

4. **Hyperbolas from Straight Lines** In this problem we construct hyperbolic shapes using straight lines. Punch equally spaced holes into the edges of two large plastic lids. Connect corresponding holes with strings of equal lengths as shown in the figure on the next page. Holding the strings taut, twist one lid against the other. An imaginary surface passing through the strings has hyperbolic cross sections. (An architectural example of this is the

Martin Mette/Shutterstock.com

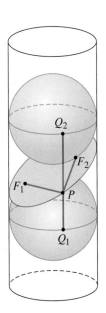

Kobe Tower in Japan, shown in the photograph.) What happens to the vertices of the hyperbolic cross sections as the lids are twisted more?

5. Tangent Lines to a Parabola   In this problem we show that the line tangent to the parabola $y = x^2$ at the point (a, a^2) has the equation $y = 2ax - a^2$.

(a) Let m be the slope of the tangent line at (a, a^2). Show that the equation of the tangent line is $y - a^2 = m(x - a)$.

(b) Use the fact that the tangent line intersects the parabola at only one point to show that (a, a^2) is the only solution of the system.

$$\begin{cases} y - a^2 = m(x - a) \\ y = x^2 \end{cases}$$

(c) Eliminate y from the system in part (b) to get a quadratic equation in x. Show that the discriminant of this quadratic is $(m - 2a)^2$. Since the system in part (b) has exactly one solution, the discriminant must equal 0. Find m.

(d) Substitute the value for m you found in part (c) into the equation in part (a), and simpify to get the equation of the tangent line.

6. A Cut Cylinder   In this problem we prove that when a cylinder is cut by a plane, an ellipse is formed. An architectural example of this is the Tycho Brahe Planetarium in Copenhagen (see the photograph). In the figure, a cylinder is cut by a plane, resulting in the red curve. Two spheres with the same radius as the cylinder slide inside the cylinder so that they just touch the plane at F_1 and F_2. Choose an arbitrary point P on the curve, and let Q_1 and Q_2 be the two points on the cylinder where a vertical line through P touches the "equator" of each sphere.

(a) Show that $PF_1 = PQ_1$ and $PF_2 = PQ_2$.   [*Hint:* Use the fact that all tangents to a sphere from a given point outside the sphere are of the same length.]

(b) Explain why $PQ_1 + PQ_2$ is the same for all points P on the curve.

(c) Show that $PF_1 + PF_2$ is the same for all points P on the curve.

(d) Conclude that the curve is an ellipse with foci F_1 and F_2.

© Bob Krist /CORBIS

George Marks/Retrofile/Getty Images

EXPONENTIAL AND LOGARITHMIC FUNCTIONS

In this chapter we study a class of functions called *exponential functions*. These are functions, like $f(x) = 2^x$, where the independent variable is in the exponent. Exponential functions are used in modeling many real-world phenomena, such as the growth of a population or the growth of an investment that earns compound interest. Once an exponential model is obtained, we can use the model to predict population size or calculate the amount of an investment for any future date. To find out *when* a population will reach a certain level, we use the inverse functions of exponential functions, called *logarithmic functions*. So if we have an exponential model for population growth, we can answer questions like: When will my city be as crowded as the New York City street pictured above?

8.1 EXPONENTIAL FUNCTIONS

Exponential Functions ▶ Graphs of Exponential Functions ▶ Compound Interest

In this chapter we study a new class of functions called *exponential functions*. For example,

$$f(x) = 2^x$$

is an exponential function (with base 2). Notice how quickly the values of this function increase:

$$f(3) = 2^3 = 8$$

$$f(10) = 2^{10} = 1024$$

$$f(30) = 2^{30} = 1{,}073{,}741{,}824$$

Compare this with the function $g(x) = x^2$, where $g(30) = 30^2 = 900$. The point is that when the variable is in the exponent, even a small change in the variable can cause a dramatic change in the value of the function.

▼ Exponential Functions

To study exponential functions, we must first define what we mean by the exponential expression a^x when x is any real number. In Appendix A.2 we define a^x for $a > 0$ and x a rational number, but we have not yet defined irrational powers. So what is meant by $5^{\sqrt{3}}$ or 2^π? To define a^x when x is irrational, we approximate x by rational numbers.

For example, since

$$\sqrt{3} \approx 1.73205\ldots$$

is an irrational number, we successively approximate $a^{\sqrt{3}}$ by the following rational powers:

$$a^{1.7}, a^{1.73}, a^{1.732}, a^{1.7320}, a^{1.73205}, \ldots$$

Intuitively, we can see that these rational powers of a are getting closer and closer to $a^{\sqrt{3}}$. It can be shown by using advanced mathematics that there is exactly one number that these powers approach. We define $a^{\sqrt{3}}$ to be this number.

For example, using a calculator, we find

$$5^{\sqrt{3}} \approx 5^{1.732}$$

$$\approx 16.2411\ldots$$

The more decimal places of $\sqrt{3}$ we use in our calculation, the better our approximation of $5^{\sqrt{3}}$.

The Laws of Exponents are listed in Appendix A.2 on page 496.

It can be proved that the *Laws of Exponents are still true when the exponents are real numbers.*

EXPONENTIAL FUNCTIONS

The **exponential function with base a** is defined for all real numbers x by

$$f(x) = a^x$$

where $a > 0$ and $a \neq 1$.

We assume that $a \neq 1$ because the function $f(x) = 1^x = 1$ is just a constant function. Here are some examples of exponential functions:

$$f(x) = 2^x \qquad g(x) = 3^x \qquad h(x) = 10^x$$

Base 2   Base 3   Base 10

EXAMPLE 1 | Evaluating Exponential Functions

Let $f(x) = 3^x$, and evaluate the following:

(a) $f(2)$ (b) $f\left(-\frac{2}{3}\right)$

(c) $f(\pi)$ (d) $f(\sqrt{2})$

SOLUTION   We use a calculator to obtain the values of f.

	Calculator keystrokes	**Output**
(a) $f(2) = 3^2 = 9$	3 ∧ 2 ENTER	9
(b) $f\left(-\frac{2}{3}\right) = 3^{-2/3} \approx 0.4807$	3 ∧ ((−) 2 ÷ 3) ENTER	0.4807498
(c) $f(\pi) = 3^\pi \approx 31.544$	3 ∧ π ENTER	31.5442807
(d) $f(\sqrt{2}) = 3^{\sqrt{2}} \approx 4.7288$	3 ∧ √ 2 ENTER	4.7288043

✎. NOW TRY EXERCISE **5**

▼ Graphs of Exponential Functions

We first graph exponential functions by plotting points. We will see that the graphs of such functions have an easily recognizable shape.

EXAMPLE 2 | Graphing Exponential Functions by Plotting Points

Draw the graph of each function.

(a) $f(x) = 3^x$ (b) $g(x) = \left(\frac{1}{3}\right)^x$

SOLUTION   We calculate values of $f(x)$ and $g(x)$ and plot points to sketch the graphs in Figure 1.

x	$f(x) = 3^x$	$g(x) = \left(\frac{1}{3}\right)^x$
−3	$\frac{1}{27}$	27
−2	$\frac{1}{9}$	9
−1	$\frac{1}{3}$	3
0	1	1
1	3	$\frac{1}{3}$
2	9	$\frac{1}{9}$
3	27	$\frac{1}{27}$

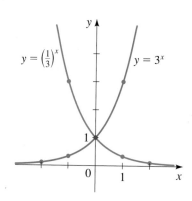

FIGURE 1

Notice that

$$g(x) = \left(\frac{1}{3}\right)^x = \frac{1}{3^x} = 3^{-x} = f(-x)$$

Reflecting graphs is explained in Section 1.6.

so we could have obtained the graph of g from the graph of f by reflecting in the y-axis.

✎. NOW TRY EXERCISE **15**

To see just how quickly $f(x) = 2^x$ increases, let's perform the following thought experiment. Suppose we start with a piece of paper that is a thousandth of an inch thick, and we fold it in half 50 times. Each time we fold the paper, the thickness of the paper stack doubles, so the thickness of the resulting stack would be $2^{50}/1000$ inches. How thick do you think that is? It works out to be more than 17 million miles!

Figure 2 shows the graphs of the family of exponential functions $f(x) = a^x$ for various values of the base a. All of these graphs pass through the point $(0, 1)$ because $a^0 = 1$ for $a \neq 0$. You can see from Figure 2 that there are two kinds of exponential functions: If $0 < a < 1$, the exponential function decreases rapidly. If $a > 1$, the function increases rapidly (see the margin note).

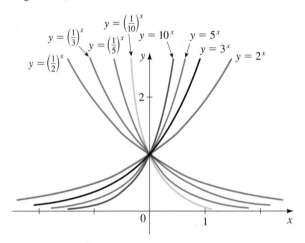

FIGURE 2 A family of exponential functions

The x-axis is a horizontal asymptote for the exponential function $f(x) = a^x$. This is because when $a > 1$, we have $a^x \to 0$ as $x \to -\infty$, and when $0 < a < 1$, we have $a^x \to 0$ as $x \to \infty$ (see Figure 2). Also, $a^x > 0$ for all $x \in \mathbb{R}$, so the function $f(x) = a^x$ has domain $\mathbb{R}$ and range $(0, \infty)$. These observations are summarized in the following box.

GRAPHS OF EXPONENTIAL FUNCTIONS

The exponential function

$$f(x) = a^x \qquad (a > 0, a \neq 1)$$

has domain $\mathbb{R}$ and range $(0, \infty)$. The line $y = 0$ (the x-axis) is a horizontal asymptote of f. The graph of f has one of the following shapes.

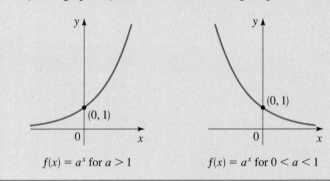

$f(x) = a^x$ for $a > 1$ $f(x) = a^x$ for $0 < a < 1$

EXAMPLE 3 | Identifying Graphs of Exponential Functions

Find the exponential function $f(x) = a^x$ whose graph is given.

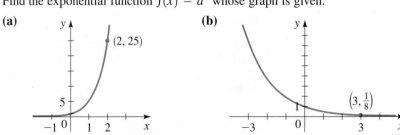

SOLUTION

(a) Since $f(2) = a^2 = 25$, we see that the base is $a = 5$. So $f(x) = 5^x$.

(b) Since $f(3) = a^3 = \frac{1}{8}$, we see that the base is $a = \frac{1}{2}$. So $f(x) = \left(\frac{1}{2}\right)^x$.

✎ NOW TRY EXERCISE **19**

In the next example we see how to graph certain functions, not by plotting points, but by taking the basic graphs of the exponential functions in Figure 2 and applying the shifting and reflecting transformations of Section 1.6.

EXAMPLE 4 | Transformations of Exponential Functions

Use the graph of $f(x) = 2^x$ to sketch the graph of each function.

(a) $g(x) = 1 + 2^x$ **(b)** $h(x) = -2^x$ **(c)** $k(x) = 2^{x-1}$

SOLUTION

Shifting and reflecting of graphs is explained in Section 1.6.

(a) To obtain the graph of $g(x) = 1 + 2^x$, we start with the graph of $f(x) = 2^x$ and shift it upward 1 unit. Notice from Figure 3(a) that the line $y = 1$ is now a horizontal asymptote.

(b) Again we start with the graph of $f(x) = 2^x$, but here we reflect in the x-axis to get the graph of $h(x) = -2^x$ shown in Figure 3(b).

(c) This time we start with the graph of $f(x) = 2^x$ and shift it to the right by 1 unit to get the graph of $k(x) = 2^{x-1}$ shown in Figure 3(c).

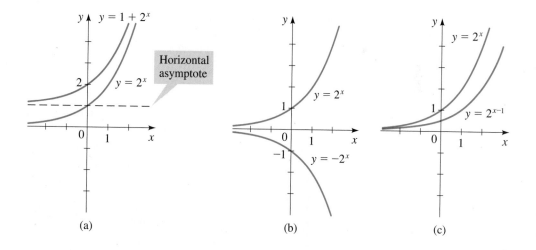

FIGURE 3

(a) (b) (c)

✎ NOW TRY EXERCISES **25, 27,** AND **31**

EXAMPLE 5 | Comparing Exponential and Power Functions

Compare the rates of growth of the exponential function $f(x) = 2^x$ and the power function $g(x) = x^2$ by drawing the graphs of both functions in the following viewing rectangles.

(a) $[0, 3]$ by $[0, 8]$

(b) $[0, 6]$ by $[0, 25]$

(c) $[0, 20]$ by $[0, 1000]$

SOLUTION

(a) Figure 4(a) shows that the graph of $g(x) = x^2$ catches up with, and becomes higher than, the graph of $f(x) = 2^x$ at $x = 2$.

(b) The larger viewing rectangle in Figure 4(b) shows that the graph of $f(x) = 2^x$ overtakes that of $g(x) = x^2$ when $x = 4$.

(c) Figure 4(c) gives a more global view and shows that when x is large, $f(x) = 2^x$ is much larger than $g(x) = x^2$.

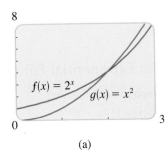

(a)

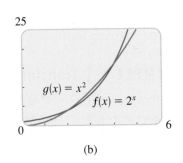

(b)

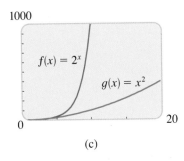

(c)

FIGURE 4

✎ NOW TRY EXERCISE **41**   ■

▼ Compound Interest

Exponential functions occur in calculating compound interest. If an amount of money P, called the **principal**, is invested at an interest rate i per time period, then after one time period the interest is Pi, and the amount A of money is

$$A = P + Pi = P(1 + i)$$

If the interest is reinvested, then the new principal is $P(1 + i)$, and the amount after another time period is $A = P(1 + i)(1 + i) = P(1 + i)^2$. Similarly, after a third time period the amount is $A = P(1 + i)^3$. In general, after k periods the amount is

$$A = P(1 + i)^k$$

Notice that this is an exponential function with base $1 + i$.

If the annual interest rate is r and if interest is compounded n times per year, then in each time period the interest rate is $i = r/n$, and there are nt time periods in t years. This leads to the following formula for the amount after t years.

COMPOUND INTEREST

Compound interest is calculated by the formula

$$A(t) = P\left(1 + \frac{r}{n}\right)^{nt}$$

where $A(t)$ = amount after t years

P = principal

r = interest rate per year

n = number of times interest is compounded per year

t = number of years

---

r is often referred to as the nominal annual interest rate.

EXAMPLE 6 │ Calculating Compound Interest

A sum of $1000 is invested at an interest rate of 12% per year. Find the amounts in the account after 3 years if interest is compounded annually, semiannually, quarterly, monthly, and daily.

SOLUTION We use the compound interest formula with $P = \$1000$, $r = 0.12$, and $t = 3$.

Compounding	n	Amount after 3 years	
Annual	1	$1000\left(1 + \dfrac{0.12}{1}\right)^{1(3)}$	$= \$1404.93$
Semiannual	2	$1000\left(1 + \dfrac{0.12}{2}\right)^{2(3)}$	$= \$1418.52$
Quarterly	4	$1000\left(1 + \dfrac{0.12}{4}\right)^{4(3)}$	$= \$1425.76$
Monthly	12	$1000\left(1 + \dfrac{0.12}{12}\right)^{12(3)}$	$= \$1430.77$
Daily	365	$1000\left(1 + \dfrac{0.12}{365}\right)^{365(3)}$	$= \$1433.24$

✎ NOW TRY EXERCISE **51** ■

If an investment earns compound interest, then the **annual percentage yield** (APY) is the *simple* interest rate that yields the same amount at the end of one year.

EXAMPLE 7 | Calculating the Annual Percentage Yield

Find the annual percentage yield for an investment that earns interest at a rate of 6% per year, compounded daily.

SOLUTION After one year, a principal P will grow to the amount

$$A = P\left(1 + \frac{0.06}{365}\right)^{365} = P(1.06183)$$

The formula for simple interest is

$$A = P(1 + r)$$

Comparing, we see that $1 + r = 1.06183$, so $r = 0.06183$. Thus, the annual percentage yield is 6.183%.

✎ NOW TRY EXERCISE **57** ■

8.1 EXERCISES

CONCEPTS

1. The function $f(x) = 5^x$ is an exponential function with base

_____; $f(-2) =$ _____, $f(0) =$ _____,

$f(2) =$ _____, and $f(6) =$ _____.

2. Match the exponential function with its graph.

(a) $f(x) = 2^x$

(b) $f(x) = 2^{-x}$

(c) $f(x) = -2^x$

(d) $f(x) = -2^{-x}$

I
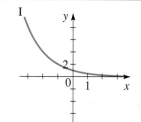

II
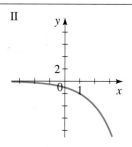

III
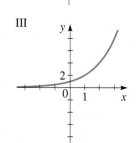

IV
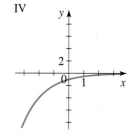

3. (a) To obtain the graph of $g(x) = 2^x - 1$, we start with the graph of $f(x) = 2^x$ and shift it _____ (upward/downward) 1 unit.

(b) To obtain the graph of $h(x) = 2^{x-1}$, we start with the graph of $f(x) = 2^x$ and shift it to the _____ (left/right) 1 unit.

4. In the formula $A(t) = P\left(1 + \frac{r}{n}\right)^{nt}$ for compound interest the letters P, r, n, and t stand for _____ , _____ , _____ , and _____ , respectively, and $A(t)$ stands for _____ . So if $100 is invested at an interest rate of 6% compounded quarterly, then the amount after 2 years is _____ .

SKILLS

5–10 ■ Use a calculator to evaluate the function at the indicated values. Round your answers to three decimals.

5. $f(x) = 4^x$; $f(0.5), f(\sqrt{2}), f(\pi), f\left(\frac{1}{3}\right)$

6. $f(x) = 3^{x+1}$; $f(-1.5), f(\sqrt{3}), f(-\pi), f\left(-\frac{5}{4}\right)$

7. $g(x) = \left(\frac{2}{3}\right)^{x-1}$; $g(1.3), g(\sqrt{5}), g(2\pi), g\left(-\frac{1}{2}\right)$

8. $g(x) = \left(\frac{3}{4}\right)^{2x}$; $g(0.7), g(\sqrt{7}/2), g(1/\pi), g\left(\frac{2}{3}\right)$

9–14 ■ Sketch the graph of the function by making a table of values. Use a calculator if necessary.

9. $f(x) = 2^x$

10. $g(x) = 8^x$

11. $f(x) = \left(\frac{1}{3}\right)^x$

12. $h(x) = (1.1)^x$

13. $g(x) = 3(1.3)^x$

14. $h(x) = 2\left(\frac{1}{4}\right)^x$

15–18 ■ Graph both functions on one set of axes.

15. $f(x) = 2^x$ and $g(x) = 2^{-x}$

16. $f(x) = 3^{-x}$ and $g(x) = \left(\frac{1}{3}\right)^x$

17. $f(x) = 4^x$ and $g(x) = 7^x$

18. $f(x) = \left(\frac{2}{3}\right)^x$ and $g(x) = \left(\frac{4}{3}\right)^x$

19–22 ■ Find the exponential function $f(x) = a^x$ whose graph is given.

19.

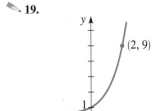

20.

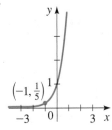

21.

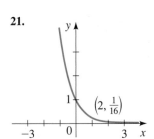

22.

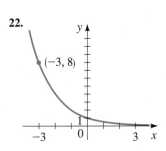

23–24 ■ Match the exponential function with one of the graphs labeled I or II.

23. $f(x) = 5^{x+1}$

24. $f(x) = 5^x + 1$

I

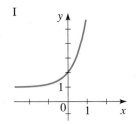

II

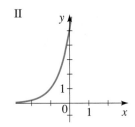

25–36 ■ Graph the function, not by plotting points, but by starting from the graphs in Figure 2. State the domain, range, and asymptote.

25. $f(x) = -3^x$

26. $f(x) = 10^{-x}$

27. $g(x) = 2^x - 3$

28. $g(x) = 2^{x-3}$

29. $h(x) = 4 + \left(\frac{1}{2}\right)^x$

30. $h(x) = 6 - 3^x$

31. $f(x) = 10^{x+3}$

32. $f(x) = -\left(\frac{1}{5}\right)^x$

33. $y = 5^{-x} + 1$

34. $g(x) = 1 - 3^{-x}$

35. $y = 3 - 10^{x-1}$

36. $h(x) = 2^{x-4} + 1$

37. (a) Sketch the graphs of $f(x) = 2^x$ and $g(x) = 3(2^x)$.
(b) How are the graphs related?

38. (a) Sketch the graphs of $f(x) = 9^{x/2}$ and $g(x) = 3^x$.
(b) Use the Laws of Exponents to explain the relationship between these graphs.

39. Compare the functions $f(x) = x^3$ and $g(x) = 3^x$ by evaluating both of them for $x = 0, 1, 2, 3, 4, 5, 6, 7, 8, 9, 10, 15,$ and 20. Then draw the graphs of f and g on the same set of axes.

40. If $f(x) = 10^x$, show that $\dfrac{f(x+h) - f(x)}{h} = 10^x\left(\dfrac{10^h - 1}{h}\right)$.

41. (a) Compare the rates of growth of the functions $f(x) = 2^x$ and $g(x) = x^5$ by drawing the graphs of both functions in the following viewing rectangles.
  (i) $[0, 5]$ by $[0, 20]$
  (ii) $[0, 25]$ by $[0, 10^7]$
  (iii) $[0, 50]$ by $[0, 10^8]$
(b) Find the solutions of the equation $2^x = x^5$, rounded to one decimal place.

42. (a) Compare the rates of growth of the functions $f(x) = 3^x$ and $g(x) = x^4$ by drawing the graphs of both functions in the following viewing rectangles:
  (i) $[-4, 4]$ by $[0, 20]$
  (ii) $[0, 10]$ by $[0, 5000]$
  (iii) $[0, 20]$ by $[0, 10^5]$
  (b) Find the solutions of the equation $3^x = x^4$, rounded to two decimal places.

 43–44 ▪ Draw graphs of the given family of functions for $c = 0.25, 0.5, 1, 2, 4$. How are the graphs related?

43. $f(x) = c2^x$ 　　　　　**44.** $f(x) = 2^{cx}$

 45–46 ▪ Find, rounded to two decimal places, **(a)** the intervals on which the function is increasing or decreasing and **(b)** the range of the function.

45. $y = 10^{x - x^2}$ 　　　　**46.** $y = x2^x$

APPLICATIONS

47. Bacteria Growth A bacteria culture contains 1500 bacteria initially and doubles every hour.
  (a) Find a function that models the number of bacteria after t hours.
  (b) Find the number of bacteria after 24 hours.

48. Mouse Population A certain breed of mouse was introduced onto a small island with an initial population of 320 mice, and scientists estimate that the mouse population is doubling every year.
  (a) Find a function that models the number of mice after t years.
  (b) Estimate the mouse population after 8 years.

49–50 ▪ **Compound Interest** An investment of $5000 is deposited into an account in which interest is compounded monthly. Complete the table by filling in the amounts to which the investment grows at the indicated times or interest rates.

49. $r = 4\%$ 　　　　　　　　**50.** $t = 5$ years

Time (years)	Amount
1	
2	
3	
4	
5	
6	

Rate per year	Amount
1%	
2%	
3%	
4%	
5%	
6%	

51. Compound Interest If $10,000 is invested at an interest rate of 3% per year, compounded semiannually, find the value of the investment after the given number of years.
  (a) 5 years 　　**(b)** 10 years 　　**(c)** 15 years

52. Compound Interest If $2500 is invested at an interest rate of 2.5% per year, compounded daily, find the value of the investment after the given number of years.
  (a) 2 years 　　**(b)** 3 years 　　**(c)** 6 years

53. Compound Interest If $500 is invested at an interest rate of 3.75% per year, compounded quarterly, find the value of the investment after the given number of years.
  (a) 1 year 　　**(b)** 2 years 　　**(c)** 10 years

54. Compound Interest If $4000 is borrowed at a rate of 5.75% interest per year, compounded quarterly, find the amount due at the end of the given number of years.
  (a) 4 years 　　**(b)** 6 years 　　**(c)** 8 years

55–56 ▪ **Present Value** The **present value** of a sum of money is the amount that must be invested now, at a given rate of interest, to produce the desired sum at a later date.

55. Find the present value of $10,000 if interest is paid at a rate of 9% per year, compounded semiannually, for 3 years.

56. Find the present value of $100,000 if interest is paid at a rate of 8% per year, compounded monthly, for 5 years.

57. Annual Percentage Yield Find the annual percentage yield for an investment that earns 8% per year, compounded monthly.

58. Annual Percentage Yield Find the annual percentage yield for an investment that earns $5\frac{1}{2}\%$ per year, compounded quarterly.

DISCOVERY ▪ DISCUSSION ▪ WRITING

59. Growth of an Exponential Function Suppose you are offered a job that lasts one month, and you are to be very well paid. Which of the following methods of payment is more profitable for you?
  (a) One million dollars at the end of the month
  (b) Two cents on the first day of the month, 4 cents on the second day, 8 cents on the third day, and, in general, 2^n cents on the nth day

60. The Height of the Graph of an Exponential Function Your mathematics instructor asks you to sketch a graph of the exponential function

$$f(x) = 2^x$$

for x between 0 and 40, using a scale of 10 units to one inch. What are the dimensions of the sheet of paper you will need to sketch this graph?

 **DISCOVERY PROJECT** 　**Exponential Explosion**

In this project we explore an example about collecting pennies that helps us experience how exponential growth works. You can find the project at the book companion website:
www.stewartmath.com

8.2 THE NATURAL EXPONENTIAL FUNCTION

The Number *e* ▶ The Natural Exponential Function ▶ Continuously Compounded Interest

Any positive number can be used as a base for an exponential function. In this section we study the special base *e*, which is convenient for applications involving calculus.

▼ The Number *e*

The number *e* is defined as the value that $(1 + 1/n)^n$ approaches as *n* becomes large. (In calculus this idea is made more precise through the concept of a limit.) The table shows the values of the expression $(1 + 1/n)^n$ for increasingly large values of *n*.

The **Gateway Arch** in St. Louis, Missouri, is shaped in the form of the graph of a combination of exponential functions (*not* a parabola, as it might first appear). Specifically, it is a **catenary**, which is the graph of an equation of the form

$$y = a(e^{bx} + e^{-bx})$$

(see Exercise 17). This shape was chosen because it is optimal for distributing the internal structural forces of the arch. Chains and cables suspended between two points (for example, the stretches of cable between pairs of telephone poles) hang in the shape of a catenary.

n	$\left(1 + \dfrac{1}{n}\right)^n$
1	2.00000
5	2.48832
10	2.59374
100	2.70481
1000	2.71692
10,000	2.71815
100,000	2.71827
1,000,000	2.71828

It appears that, correct to five decimal places, $e \approx 2.71828$; in fact, the approximate value to 20 decimal places is

$$e \approx 2.71828182845904523536$$

It can be shown that *e* is an irrational number, so we cannot write its exact value in decimal form.

▼ The Natural Exponential Function

The number *e* is the base for the natural exponential function. Why use such a strange base for an exponential function? It might seem at first that a base such as 10 is easier to work with. We will see, however, that in certain applications the number *e* is the best possible base. In this section we study how *e* occurs in the description of compound interest.

The notation *e* was chosen by Leonhard Euler (see page 541), probably because it is the first letter of the word *exponential*.

> **THE NATURAL EXPONENTIAL FUNCTION**
>
> The **natural exponential function** is the exponential function
>
> $$f(x) = e^x$$
>
> with base *e*. It is often referred to as *the* exponential function.

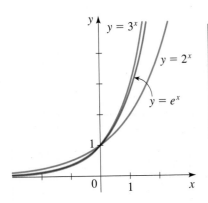

FIGURE 1 Graph of the natural exponential function

Since $2 < e < 3$, the graph of the natural exponential function lies between the graphs of $y = 2^x$ and $y = 3^x$, as shown in Figure 1.

Scientific calculators have a special key for the function $f(x) = e^x$. We use this key in the next example.

EXAMPLE 1 | Evaluating the Exponential Function

Evaluate each expression rounded to five decimal places.

(a) e^3 **(b)** $2e^{-0.53}$ **(c)** $e^{4.8}$

SOLUTION   We use the $\boxed{e^x}$ key on a calculator to evaluate the exponential function.

(a) $e^3 \approx 20.08554$ **(b)** $2e^{-0.53} \approx 1.17721$ **(c)** $e^{4.8} \approx 121.51042$

✎ NOW TRY EXERCISE **3**   ■

EXAMPLE 2 | Transformations of the Exponential Function

Sketch the graph of each function.

(a) $f(x) = e^{-x}$ **(b)** $g(x) = 3e^{0.5x}$

SOLUTION

(a) We start with the graph of $y = e^x$ and reflect in the y-axis to obtain the graph of $y = e^{-x}$ as in Figure 2.

(b) We calculate several values, plot the resulting points, then connect the points with a smooth curve. The graph is shown in Figure 3.

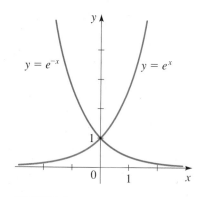

$y = e^{-x}$ $y = e^x$

FIGURE 2

x	$f(x) = 3e^{0.5x}$
-3	0.67
-2	1.10
-1	1.82
0	3.00
1	4.95
2	8.15
3	13.45

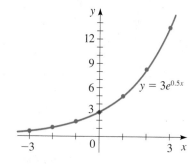

$y = 3e^{0.5x}$

FIGURE 3

✎ NOW TRY EXERCISES **5** AND **7**   ■

EXAMPLE 3 | An Exponential Model for the Spread of a Virus

An infectious disease begins to spread in a small city of population 10,000. After t days, the number of people who have succumbed to the virus is modeled by the function

$$v(t) = \frac{10{,}000}{5 + 1245e^{-0.97t}}$$

(a) How many infected people are there initially (at time $t = 0$)?

(b) Find the number of infected people after one day, two days, and five days.

(c) Graph the function v, and describe its behavior.

SOLUTION

(a) Since $v(0) = 10{,}000/(5 + 1245e^0) = 10{,}000/1250 = 8$, we conclude that 8 people initially have the disease.

(b) Using a calculator, we evaluate $v(1)$, $v(2)$, and $v(5)$ and then round off to obtain the following values.

Days	Infected people
1	21
2	54
5	678

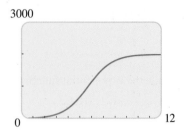

FIGURE 4

$$v(t) = \frac{10,000}{5 + 1245e^{-0.97t}}$$

(c) From the graph in Figure 4 we see that the number of infected people first rises slowly, then rises quickly between day 3 and day 8, and then levels off when about 2000 people are infected.

✎ NOW TRY EXERCISE **25** ∎

The graph in Figure 4 is called a *logistic curve* or a *logistic growth model*. Curves like it occur frequently in the study of population growth. (See Exercises 25–28.)

▼ Continuously Compounded Interest

In Example 6 of Section 8.1 we saw that the interest paid increases as the number of compounding periods n increases. Let's see what happens as n increases indefinitely. If we let $m = n/r$, then

$$A(t) = P\left(1 + \frac{r}{n}\right)^{nt} = P\left[\left(1 + \frac{r}{n}\right)^{n/r}\right]^{rt} = P\left[\left(1 + \frac{1}{m}\right)^{m}\right]^{rt}$$

Recall that as m becomes large, the quantity $(1 + 1/m)^m$ approaches the number e. Thus, the amount approaches $A = Pe^{rt}$. This expression gives the amount when the interest is compounded at "every instant."

CONTINUOUSLY COMPOUNDED INTEREST

Continuously compounded interest is calculated by the formula

$$A(t) = Pe^{rt}$$

where $A(t)$ = amount after t years

 P = principal

 r = interest rate per year

 t = number of years

EXAMPLE 4 | Calculating Continuously Compounded Interest

Find the amount after 3 years if $1000 is invested at an interest rate of 12% per year, compounded continuously.

SOLUTION We use the formula for continuously compounded interest with P = $1000, r = 0.12, and t = 3 to get

$$A(3) = 1000e^{(0.12)3} = 1000e^{0.36} = \$1433.33$$

Compare this amount with the amounts in Example 6 of Section 8.1.

✎ NOW TRY EXERCISE **31** ∎

8.2 EXERCISES

CONCEPTS

1. The function $f(x) = e^x$ is called the ——— exponential function. The number e is approximately equal to ———.

2. In the formula $A(t) = Pe^{rt}$ for continuously compound interest, the letters P, r, and t stand for ———, ———, and ———, respectively, and $A(t)$ stands for ———. So if $100 is invested at an interest rate of 6% compounded continuously, then the amount after 2 years is ———.

SKILLS

3–4 ■ Use a calculator to evaluate the function at the indicated values. Round your answers to three decimals.

3. $h(x) = e^x$; $h(3), h(0.23), h(1), h(-2)$

4. $h(x) = e^{-2x}$; $h(1), h(\sqrt{2}), h(-3), h(\frac{1}{2})$

5–6 ■ Complete the table of values, rounded to two decimal places, and sketch a graph of the function.

5.

x	$f(x) = 3e^x$
-2	
-1	
-0.5	
0	
0.5	
1	
2	

6.

x	$f(x) = 2e^{-0.5x}$
-3	
-2	
-1	
0	
1	
2	
3	

7–14 ■ Graph the function, not by plotting points, but by starting from the graph of $y = e^x$ in Figure 1. State the domain, range, and asymptote.

7. $f(x) = -e^x$

8. $y = 1 - e^x$

9. $y = e^{-x} - 1$

10. $f(x) = -e^{-x}$

11. $f(x) = e^{x-2}$

12. $y = e^{x-3} + 4$

13. $h(x) = e^{x+1} - 3$

14. $g(x) = -e^{x-1} - 2$

15. The *hyperbolic cosine function* is defined by

$$\cosh(x) = \frac{e^x + e^{-x}}{2}$$

(a) Sketch the graphs of the functions $y = \frac{1}{2}e^x$ and $y = \frac{1}{2}e^{-x}$ on the same axes, and use graphical addition (see Section 1.7) to sketch the graph of $y = \cosh(x)$.

(b) Use the definition to show that $\cosh(-x) = \cosh(x)$.

16. The *hyperbolic sine function* is defined by

$$\sinh(x) = \frac{e^x - e^{-x}}{2}$$

(a) Sketch the graph of this function using graphical addition as in Exercise 15.

(b) Use the definition to show that $\sinh(-x) = -\sinh(x)$

17. **(a)** Draw the graphs of the family of functions

$$f(x) = \frac{a}{2}(e^{x/a} + e^{-x/a})$$

for $a = 0.5, 1, 1.5,$ and 2.

(b) How does a larger value of a affect the graph?

18–19 ■ Find the local maximum and minimum values of the function and the value of x at which each occurs. State each answer correct to two decimal places.

18. $g(x) = x^x$ $(x > 0)$

19. $g(x) = e^x + e^{-3x}$

APPLICATIONS

20. Medical Drugs   When a certain medical drug is administered to a patient, the number of milligrams remaining in the patient's bloodstream after t hours is modeled by

$$D(t) = 50e^{-0.2t}$$

How many milligrams of the drug remain in the patient's bloodstream after 3 hours?

21. Radioactive Decay   A radioactive substance decays in such a way that the amount of mass remaining after t days is given by the function

$$m(t) = 13e^{-0.015t}$$

where $m(t)$ is measured in kilograms.
(a) Find the mass at time $t = 0$.
(b) How much of the mass remains after 45 days?

22. Radioactive Decay   Doctors use radioactive iodine as a tracer in diagnosing certain thyroid gland disorders. This type of iodine decays in such a way that the mass remaining after t days is given by the function

$$m(t) = 6e^{-0.087t}$$

where $m(t)$ is measured in grams.
(a) Find the mass at time $t = 0$.
(b) How much of the mass remains after 20 days?

23. Sky Diving   A sky diver jumps from a reasonable height above the ground. The air resistance she experiences is proportional to her velocity, and the constant of proportionality is 0.2. It can be shown that the downward velocity of the sky diver at time t is given by

$$v(t) = 80(1 - e^{-0.2t})$$

where t is measured in seconds and $v(t)$ is measured in feet per second (ft/s).
(a) Find the initial velocity of the sky diver.
(b) Find the velocity after 5 s and after 10 s.
(c) Draw a graph of the velocity function $v(t)$.
(d) The maximum velocity of a falling object with wind resistance is called its *terminal velocity*. From the graph in part (c) find the terminal velocity of this sky diver.

$v(t) = 80(1 - e^{-0.2t})$

24. Mixtures and Concentrations   A 50-gallon barrel is filled completely with pure water. Salt water with a concentration of 0.3 lb/gal is then pumped into the barrel, and the resulting mixture overflows at the same rate. The amount of salt in the barrel at time t is given by

$$Q(t) = 15(1 - e^{-0.04t})$$

where t is measured in minutes and $Q(t)$ is measured in pounds.
(a) How much salt is in the barrel after 5 min?
(b) How much salt is in the barrel after 10 min?
(c) Draw a graph of the function $Q(t)$.

(d) Use the graph in part (c) to determine the value that the amount of salt in the barrel approaches as t becomes large. Is this what you would expect?

$$Q(t) = 15(1 - e^{-0.04t})$$

25. Logistic Growth   Animal populations are not capable of unrestricted growth because of limited habitat and food supplies. Under such conditions the population follows a *logistic growth model*:

$$P(t) = \frac{d}{1 + ke^{-ct}}$$

where c, d, and k are positive constants. For a certain fish population in a small pond $d = 1200$, $k = 11$, $c = 0.2$, and t is measured in years. The fish were introduced into the pond at time $t = 0$.
(a) How many fish were originally put in the pond?
(b) Find the population after 10, 20, and 30 years.
(c) Evaluate $P(t)$ for large values of t. What value does the population approach as $t \to \infty$? Does the graph shown confirm your calculations?

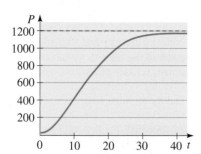

26. Bird Population   The population of a certain species of bird is limited by the type of habitat required for nesting. The population behaves according to the logistic growth model

$$n(t) = \frac{5600}{0.5 + 27.5e^{-0.044t}}$$

where t is measured in years.
(a) Find the initial bird population.
(b) Draw a graph of the function $n(t)$.
(c) What size does the population approach as time goes on?

27. World Population   The relative growth rate of world population has been decreasing steadily in recent years. On the basis of this, some population models predict that world population will eventually stabilize at a level that the planet can support. One such logistic model is

$$P(t) = \frac{73.2}{6.1 + 5.9e^{-0.02t}}$$

where $t = 0$ is the year 2000 and population is measured in billions.
(a) What world population does this model predict for the year 2200? For 2300?
(b) Sketch a graph of the function P for the years 2000 to 2500.
(c) According to this model, what size does the world population seem to approach as time goes on?

28. Tree Diameter   For a certain type of tree the diameter D (in feet) depends on the tree's age t (in years) according to the logistic growth model

$$D(t) = \frac{5.4}{1 + 2.9e^{-0.01t}}$$

Find the diameter of a 20-year-old tree.

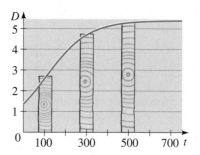

29–30 ■ Compound Interest   An investment of $7,000 is deposited into an account in which interest is compounded continuously. Complete the table by filling in the amounts to which the investment grows at the indicated times or interest rates.

29. $r = 3\%$

Time (years)	Amount
1	
2	
3	
4	
5	
6	

30. $t = 10$ years

Rate per year	Amount
1%	
2%	
3%	
4%	
5%	
6%	

31. Compound Interest   If $2000 is invested at an interest rate of 3.5% per year, compounded continuously, find the value of the investment after the given number of years.
(a) 2 years   **(b)** 4 years   **(c)** 12 years

32. Compound Interest   If $3500 is invested at an interest rate of 6.25% per year, compounded continuously, find the value of the investment after the given number of years.
(a) 3 years   **(b)** 6 years   **(c)** 9 years

33. Compound Interest   If $600 is invested at an interest rate of 2.5% per year, find the amount of the investment at the end of 10 years for the following compounding methods.
(a) Annually   **(b)** Semiannually
(c) Quarterly   **(d)** Continuously

34. Compound Interest   If $8000 is invested in an account for which interest is compounded continuously, find the amount of the investment at the end of 12 years for the following interest rates.
(a) 2%   **(b)** 3%   **(c)** 4.5%   **(d)** 7%

35. Compound Interest Which of the given interest rates and compounding periods would provide the best investment?
(a) $2\frac{1}{2}\%$ per year, compounded semiannually
(b) $2\frac{1}{4}\%$ per year, compounded monthly
(c) 2% per year, compounded continuously

36. Compound Interest Which of the given interest rates and compounding periods would provide the better investment?
(a) $5\frac{1}{8}\%$ per year, compounded semiannually
(b) 5% per year, compounded continuously

37. Investment A sum of $5000 is invested at an interest rate of 9% per year, compounded continuously.
(a) Find the value $A(t)$ of the investment after t years.

(b) Draw a graph of $A(t)$.
(c) Use the graph of $A(t)$ to determine when this investment will amount to $25,000.

___

DISCOVERY ▪ DISCUSSION ▪ WRITING

 38. The Definition of e Illustrate the definition of the number e by graphing the curve $y = (1 + 1/x)^x$ and the line $y = e$ on the same screen, using the viewing rectangle $[0, 40]$ by $[0, 4]$.

8.3 LOGARITHMIC FUNCTIONS

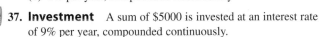 Logarithmic Functions ▶ Graphs of Logarithmic Functions ▶ Common Logarithms ▶ Natural Logarithms

In this section we study the inverses of exponential functions.

▼ Logarithmic Functions

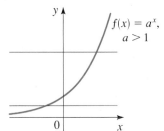

FIGURE 1 $f(x) = a^x$ is one-to-one.

Every exponential function $f(x) = a^x$, with $a > 0$ and $a \neq 1$, is a one-to-one function by the Horizontal Line Test (see Figure 1 for the case $a > 1$) and therefore has an inverse function. The inverse function f^{-1} is called the *logarithmic function with base a* and is denoted by $\log_a$. Recall from Section 1.8 that f^{-1} is defined by

$$f^{-1}(x) = y \quad \Leftrightarrow \quad f(y) = x$$

This leads to the following definition of the logarithmic function.

We read $\log_a x = y$ as "log base a of x is y."

___

DEFINITION OF THE LOGARITHMIC FUNCTION

Let a be a positive number with $a \neq 1$. The **logarithmic function with base a**, denoted by $\log_a$, is defined by

$$\log_a x = y \quad \Leftrightarrow \quad a^y = x$$

So $\log_a x$ is the *exponent* to which the base a must be raised to give x.

___

By tradition the name of the logarithmic function is $\log_a$, not just a single letter. Also, we usually omit the parentheses in the function notation and write

$$\log_a(x) = \log_a x$$

When we use the definition of logarithms to switch back and forth between the **logarithmic form** $\log_a x = y$ and the **exponential form** $a^y = x$, it is helpful to notice that, in both forms, the base is the same:

Logarithmic form	Exponential form
Exponent	Exponent
$\log_a x = y$	$a^y = x$
Base	Base

EXAMPLE 1 | Logarithmic and Exponential Forms

The logarithmic and exponential forms are equivalent equations: If one is true, then so is the other. So we can switch from one form to the other as in the following illustrations.

Logarithmic form	Exponential form
$\log_{10} 100{,}000 = 5$	$10^5 = 100{,}000$
$\log_2 8 = 3$	$2^3 = 8$
$\log_2\left(\frac{1}{8}\right) = -3$	$2^{-3} = \frac{1}{8}$
$\log_5 s = r$	$5^r = s$

✎ NOW TRY EXERCISE **5**

It is important to understand that $\log_a x$ is an *exponent*. For example, the numbers in the right column of the table in the margin are the logarithms (base 10) of the numbers in the left column. This is the case for all bases, as the following example illustrates.

x	$\log_{10} x$
10^4	4
10^3	3
10^2	2
10	1
1	0
10^{-1}	-1
10^{-2}	-2
10^{-3}	-3
10^{-4}	-4

EXAMPLE 2 | Evaluating Logarithms

(a) $\log_{10} 1000 = 3$ because $10^3 = 1000$

(b) $\log_2 32 = 5$ because $2^5 = 32$

(c) $\log_{10} 0.1 = -1$ because $10^{-1} = 0.1$

(d) $\log_{16} 4 = \frac{1}{2}$ because $16^{1/2} = 4$

✎ NOW TRY EXERCISES **7** AND **9**

When we apply the Inverse Function Property described on page 79 to $f(x) = a^x$ and $f^{-1}(x) = \log_a x$, we get

Inverse Function Property:

$$f^{-1}(f(x)) = x$$
$$f(f^{-1}(x)) = x$$

$$\log_a(a^x) = x, \qquad x \in \mathbb{R}$$
$$a^{\log_a x} = x, \qquad x > 0$$

We list these and other properties of logarithms discussed in this section.

PROPERTIES OF LOGARITHMS

Property	Reason
1. $\log_a 1 = 0$	We must raise a to the power 0 to get 1.
2. $\log_a a = 1$	We must raise a to the power 1 to get a.
3. $\log_a a^x = x$	We must raise a to the power x to get a^x.
4. $a^{\log_a x} = x$	$\log_a x$ is the power to which a must be raised to get x.

EXAMPLE 3 | Applying Properties of Logarithms

We illustrate the properties of logarithms when the base is 5.

$$\log_5 1 = 0 \quad \text{Property 1} \qquad\qquad \log_5 5 = 1 \quad \text{Property 2}$$

$$\log_5 5^8 = 8 \quad \text{Property 3} \qquad\qquad 5^{\log_5 12} = 12 \quad \text{Property 4}$$

✎ NOW TRY EXERCISES **19** AND **25**

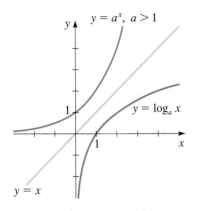

FIGURE 2 Graph of the logarithmic function $f(x) = \log_a x$

▼ Graphs of Logarithmic Functions

Recall that if a one-to-one function f has domain A and range B, then its inverse function f^{-1} has domain B and range A. Since the exponential function $f(x) = a^x$ with $a \neq 1$ has domain $\mathbb{R}$ and range $(0, \infty)$, we conclude that its inverse function, $f^{-1}(x) = \log_a x$, has domain $(0, \infty)$ and range $\mathbb{R}$.

The graph of $f^{-1}(x) = \log_a x$ is obtained by reflecting the graph of $f(x) = a^x$ in the line $y = x$. Figure 2 shows the case $a > 1$. The fact that $y = a^x$ (for $a > 1$) is a very rapidly increasing function for $x > 0$ implies that $y = \log_a x$ is a very slowly increasing function for $x > 1$ (see Exercise 92).

Since $\log_a 1 = 0$, the x-intercept of the function $y = \log_a x$ is 1. The y-axis is a vertical asymptote of $y = \log_a x$ because $\log_a x \to -\infty$ as $x \to 0^+$.

EXAMPLE 4 | Graphing a Logarithmic Function by Plotting Points

Sketch the graph of $f(x) = \log_2 x$.

SOLUTION To make a table of values, we choose the x-values to be powers of 2 so that we can easily find their logarithms. We plot these points and connect them with a smooth curve as in Figure 3.

x	$\log_2 x$
2^3	3
2^2	2
2	1
1	0
2^{-1}	-1
2^{-2}	-2
2^{-3}	-3
2^{-4}	-4

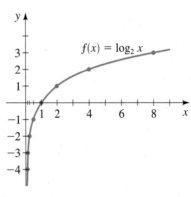

FIGURE 3

✎ NOW TRY EXERCISE **41**

Figure 4 shows the graphs of the family of logarithmic functions with bases 2, 3, 5, and 10. These graphs are drawn by reflecting the graphs of $y = 2^x$, $y = 3^x$, $y = 5^x$, and $y = 10^x$ (see Figure 2 in Section 8.1) in the line $y = x$. We can also plot points as an aid to sketching these graphs, as illustrated in Example 4.

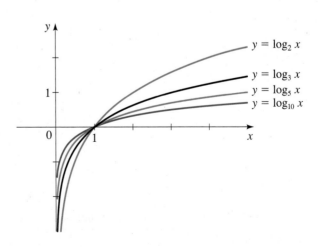

FIGURE 4 A family of logarithmic functions

In the next two examples we graph logarithmic functions by starting with the basic graphs in Figure 4 and using the transformations of Section 1.6.

EXAMPLE 5 | Reflecting Graphs of Logarithmic Functions

Sketch the graph of each function.

(a) $g(x) = -\log_2 x$

(b) $h(x) = \log_2(-x)$

SOLUTION

(a) We start with the graph of $f(x) = \log_2 x$ and reflect in the x-axis to get the graph of $g(x) = -\log_2 x$ in Figure 5(a).

(b) We start with the graph of $f(x) = \log_2 x$ and reflect in the y-axis to get the graph of $h(x) = \log_2(-x)$ in Figure 5(b).

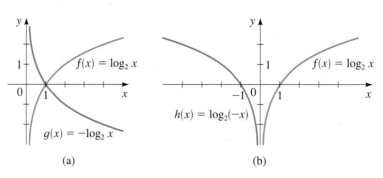

FIGURE 5

✎ NOW TRY EXERCISE **55**

EXAMPLE 6 | Shifting Graphs of Logarithmic Functions

Find the domain of each function, and sketch the graph.

(a) $g(x) = 2 + \log_5 x$

(b) $h(x) = \log_{10}(x - 3)$

SOLUTION

(a) The graph of g is obtained from the graph of $f(x) = \log_5 x$ (Figure 4) by shifting upward 2 units (see Figure 6). The domain of f is $(0, \infty)$.

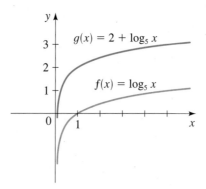

FIGURE 6

(b) The graph of h is obtained from the graph of $f(x) = \log_{10} x$ (Figure 4) by shifting to the right 3 units (see Figure 7). The line $x = 3$ is a vertical asymptote. Since $\log_{10} x$ is defined only when $x > 0$, the domain of $h(x) = \log_{10}(x - 3)$ is

$$\{x \mid x - 3 > 0\} = \{x \mid x > 3\} = (3, \infty)$$

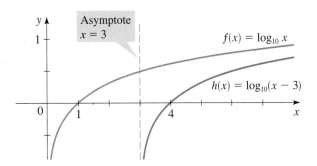

FIGURE 7

✎. NOW TRY EXERCISES **53** AND **57**

JOHN NAPIER (1550–1617) was a Scottish landowner for whom mathematics was a hobby. We know him today because of his key invention: logarithms, which he published in 1614 under the title *A Description of the Marvelous Rule of Logarithms*. In Napier's time, logarithms were used exclusively for simplifying complicated calculations. For example, to multiply two large numbers, we would write them as powers of 10. The exponents are simply the logarithms of the numbers. For instance,

$$4532 \times 57783$$
$$\approx 10^{3.65629} \times 10^{4.76180}$$
$$= 10^{8.41809}$$
$$\approx 261{,}872{,}564$$

The idea is that multiplying powers of 10 is easy (we simply add their exponents). Napier produced extensive tables giving the logarithms (or exponents) of numbers. Since the advent of calculators and computers, logarithms are no longer used for this purpose. The logarithmic functions, however, have found many applications, some of which are described in this chapter.

Napier wrote on many topics. One of his most colorful works is a book entitled *A Plaine Discovery of the Whole Revelation of Saint John*, in which he predicted that the world would end in the year 1700.

▼ Common Logarithms

We now study logarithms with base 10.

> **COMMON LOGARITHM**
>
> The logarithm with base 10 is called the **common logarithm** and is denoted by omitting the base:
>
> $$\log x = \log_{10} x$$

From the definition of logarithms we can easily find that

$$\log 10 = 1 \quad \text{and} \quad \log 100 = 2$$

But how do we find log 50? We need to find the exponent y such that $10^y = 50$. Clearly, 1 is too small and 2 is too large. So

$$1 < \log 50 < 2$$

To get a better approximation, we can experiment to find a power of 10 closer to 50. Fortunately, scientific calculators are equipped with a [LOG] key that directly gives values of common logarithms.

EXAMPLE 7 | Evaluating Common Logarithms

Use a calculator to find appropriate values of $f(x) = \log x$ and use the values to sketch the graph.

SOLUTION We make a table of values, using a calculator to evaluate the function at those values of x that are not powers of 10. We plot those points and connect them by a smooth curve as in Figure 8.

x	$\log x$
0.01	−2
0.1	−1
0.5	−0.301
1	0
4	0.602
5	0.699
10	1

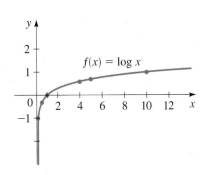

FIGURE 8

✎. NOW TRY EXERCISE **43**

Human response to sound and light intensity is logarithmic.

We study the decibel scale in more detail in Section 8.6.

Scientists model human response to stimuli (such as sound, light, or pressure) using logarithmic functions. For example, the intensity of a sound must be increased manyfold before we "feel" that the loudness has simply doubled. The psychologist Gustav Fechner formulated the law as

$$S = k \log\left(\frac{I}{I_0}\right)$$

where S is the subjective intensity of the stimulus, I is the physical intensity of the stimulus, I_0 stands for the threshold physical intensity, and k is a constant that is different for each sensory stimulus.

EXAMPLE 8 | Common Logarithms and Sound

The perception of the loudness B (in decibels, dB) of a sound with physical intensity I (in W/m^2) is given by

$$B = 10 \log\left(\frac{I}{I_0}\right)$$

where I_0 is the physical intensity of a barely audible sound. Find the decibel level (loudness) of a sound whose physical intensity I is 100 times that of I_0.

SOLUTION We find the decibel level B by using the fact that $I = 100I_0$.

$$B = 10 \log\left(\frac{I}{I_0}\right) \qquad \text{Definition of } B$$

$$= 10 \log\left(\frac{100I_0}{I_0}\right) \qquad I = 100I_0$$

$$= 10 \log 100 \qquad \text{Cancel } I_0$$

$$= 10 \cdot 2 = 20 \qquad \text{Definition of log}$$

The loudness of the sound is 20 dB.

✎ NOW TRY EXERCISE **87**

▼ Natural Logarithms

The notation ln is an abbreviation for the Latin name *logarithmus naturalis*.

Of all possible bases a for logarithms, it turns out that the most convenient choice for the purposes of calculus is the number e, which we defined in Section 8.2.

NATURAL LOGARITHM

The logarithm with base e is called the **natural logarithm** and is denoted by **ln**:

$$\ln x = \log_e x$$

The natural logarithmic function $y = \ln x$ is the inverse function of the natural exponential function $y = e^x$. Both functions are graphed in Figure 9. By the definition of inverse functions we have

$$\ln x = y \quad \Leftrightarrow \quad e^y = x$$

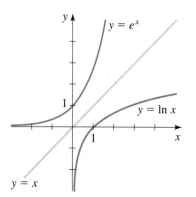

FIGURE 9 Graph of the natural logarithmic function

If we substitute $a = e$ and write "ln" for "log$_e$" in the properties of logarithms mentioned earlier, we obtain the following properties of natural logarithms.

> **PROPERTIES OF NATURAL LOGARITHMS**
>
> | Property | Reason |
> |---|---|
> | **1.** $\ln 1 = 0$ | We must raise e to the power 0 to get 1. |
> | **2.** $\ln e = 1$ | We must raise e to the power 1 to get e. |
> | **3.** $\ln e^x = x$ | We must raise e to the power x to get e^x. |
> | **4.** $e^{\ln x} = x$ | $\ln x$ is the power to which e must be raised to get x. |

Calculators are equipped with an $\boxed{\text{LN}}$ key that directly gives the values of natural logarithms.

EXAMPLE 9 | Evaluating the Natural Logarithm Function

(a) $\ln e^8 = 8$ Definition of natural logarithm

(b) $\ln\left(\dfrac{1}{e^2}\right) = \ln e^{-2} = -2$ Definition of natural logarithm

(c) $\ln 5 \approx 1.609$ Use $\boxed{\text{LN}}$ key on calculator

✎ NOW TRY EXERCISE **39**

EXAMPLE 10 | Finding the Domain of a Logarithmic Function

Find the domain of the function $f(x) = \ln(4 - x^2)$.

SOLUTION As with any logarithmic function, $\ln x$ is defined when $x > 0$. Thus, the domain of f is

$$\{x \mid 4 - x^2 > 0\} = \{x \mid x^2 < 4\} = \{x \mid |x| < 2\}$$
$$= \{x \mid -2 < x < 2\} = (-2, 2)$$

✎ NOW TRY EXERCISE **63**

EXAMPLE 11 | Drawing the Graph of a Logarithmic Function

Draw the graph of the function $y = x \ln(4 - x^2)$, and use it to find the asymptotes and local maximum and minimum values.

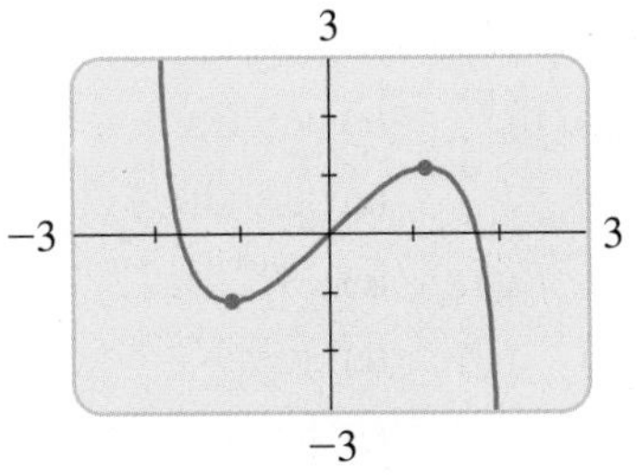

SOLUTION As in Example 10 the domain of this function is the interval $(-2, 2)$, so we choose the viewing rectangle $[-3, 3]$ by $[-3, 3]$. The graph is shown in Figure 10, and from it we see that the lines $x = -2$ and $x = 2$ are vertical asymptotes.

The function has a local maximum point to the right of $x = 1$ and a local minimum point to the left of $x = -1$. By zooming in and tracing along the graph with the cursor, we find that the local maximum value is approximately 1.13 and this occurs when $x \approx 1.15$. Similarly (or by noticing that the function is odd), we find that the local minimum value is about -1.13, and it occurs when $x \approx -1.15$.

✎ NOW TRY EXERCISE **69**

FIGURE 10
$y = x \ln(4 - x^2)$

8.3 EXERCISES

CONCEPTS

1. log x is the exponent to which the base 10 must be raised to get _____. So we can complete the following table for log x.

x	10^3	10^2	10^1	10^0	10^{-1}	10^{-2}	10^{-3}	$10^{1/2}$
log x								

2. The function $f(x) = \log_9 x$ is the logarithm function with base _____. So $f(9) =$ _____, $f(1) =$ _____, $f(\frac{1}{9}) =$ _____, $f(81) =$ _____, and $f(3) =$ _____.

3. **(a)** $5^3 = 125$, so $\log_{\square} \square = \square$

(b) $\log_5 25 = 2$, so $\square^{\square} = \square$

4. Match the logarithmic function with its graph.
(a) $f(x) = \log_2 x$ **(b)** $f(x) = \log_2(-x)$
(c) $f(x) = -\log_2 x$ **(d)** $f(x) = -\log_2(-x)$

I
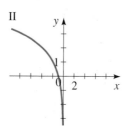

II
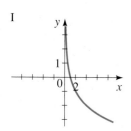

III
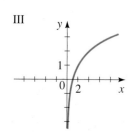

IV
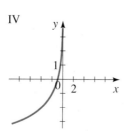

SKILLS

5–6 ■ Complete the table by finding the appropriate logarithmic or exponential form of the equation, as in Example 1.

5.

Logarithmic form	Exponential form
$\log_8 8 = 1$	
$\log_8 64 = 2$	
	$8^{2/3} = 4$
	$8^3 = 512$
$\log_8\left(\frac{1}{8}\right) = -1$	
	$8^{-2} = \frac{1}{64}$

6.

Logarithmic form	Exponential form
	$4^3 = 64$
$\log_4 2 = \frac{1}{2}$	
	$4^{3/2} = 8$
$\log_4\left(\frac{1}{16}\right) = -2$	
$\log_4\left(\frac{1}{2}\right) = -\frac{1}{2}$	
	$4^{-5/2} = \frac{1}{32}$

7–12 ■ Express the equation in exponential form.

7. **(a)** $\log_5 25 = 2$ **(b)** $\log_5 1 = 0$

8. **(a)** $\log_{10} 0.1 = -1$ **(b)** $\log_8 512 = 3$

9. **(a)** $\log_8 2 = \frac{1}{3}$ **(b)** $\log_2\left(\frac{1}{8}\right) = -3$

10. **(a)** $\log_3 81 = 4$ **(b)** $\log_8 4 = \frac{2}{3}$

11. **(a)** $\ln 5 = x$ **(b)** $\ln y = 5$

12. **(a)** $\ln(x + 1) = 2$ **(b)** $\ln(x - 1) = 4$

13–18 ■ Express the equation in logarithmic form.

13. **(a)** $5^3 = 125$ **(b)** $10^{-4} = 0.0001$

14. **(a)** $10^3 = 1000$ **(b)** $81^{1/2} = 9$

15. **(a)** $8^{-1} = \frac{1}{8}$ **(b)** $2^{-3} = \frac{1}{8}$

16. **(a)** $4^{-3/2} = 0.125$ **(b)** $7^3 = 343$

17. **(a)** $e^x = 2$ **(b)** $e^3 = y$

18. **(a)** $e^{x+1} = 0.5$ **(b)** $e^{0.5x} = t$

19–28 ■ Evaluate the expression.

19. **(a)** $\log_3 3$ **(b)** $\log_3 1$ **(c)** $\log_3 3^2$

20. **(a)** $\log_5 5^4$ **(b)** $\log_4 64$ **(c)** $\log_3 9$

21. **(a)** $\log_6 36$ **(b)** $\log_9 81$ **(c)** $\log_7 7^{10}$

22. **(a)** $\log_2 32$ **(b)** $\log_8 8^{17}$ **(c)** $\log_6 1$

23. **(a)** $\log_3\left(\frac{1}{27}\right)$ **(b)** $\log_{10} \sqrt{10}$ **(c)** $\log_5 0.2$

24. **(a)** $\log_5 125$ **(b)** $\log_{49} 7$ **(c)** $\log_9 \sqrt{3}$

25. **(a)** $2^{\log_2 37}$ **(b)** $3^{\log_3 8}$ **(c)** $e^{\ln\sqrt{5}}$

26. **(a)** $e^{\ln \pi}$ **(b)** $10^{\log 5}$ **(c)** $10^{\log 87}$

27. **(a)** $\log_8 0.25$ **(b)** $\ln e^4$ **(c)** $\ln(1/e)$

28. **(a)** $\log_4 \sqrt{2}$ **(b)** $\log_4\left(\frac{1}{2}\right)$ **(c)** $\log_4 8$

29–36 ■ Use the definition of the logarithmic function to find x.

29. **(a)** $\log_2 x = 5$ **(b)** $\log_2 16 = x$

30. **(a)** $\log_5 x = 4$ **(b)** $\log_{10} 0.1 = x$

31. **(a)** $\log_3 243 = x$ **(b)** $\log_3 x = 3$

32. **(a)** $\log_4 2 = x$ **(b)** $\log_4 x = 2$

33. **(a)** $\log_{10} x = 2$ **(b)** $\log_5 x = 2$

34. (a) $\log_x 1000 = 3$ **(b)** $\log_x 25 = 2$

35. (a) $\log_x 16 = 4$ **(b)** $\log_x 8 = \frac{3}{2}$

36. (a) $\log_x 6 = \frac{1}{2}$ **(b)** $\log_x 3 = \frac{1}{3}$

37–40 ■ Use a calculator to evaluate the expression, correct to four decimal places.

37. (a) $\log 2$ **(b)** $\log 35.2$ **(c)** $\log\left(\frac{2}{3}\right)$

38. (a) $\log 50$ **(b)** $\log \sqrt{2}$ **(c)** $\log(3\sqrt{2})$

 39. (a) $\ln 5$ **(b)** $\ln 25.3$ **(c)** $\ln(1 + \sqrt{3})$

40. (a) $\ln 27$ **(b)** $\ln 7.39$ **(c)** $\ln 54.6$

41–44 ■ Sketch the graph of the function by plotting points.

41. $f(x) = \log_3 x$ **42.** $g(x) = \log_4 x$

43. $f(x) = 2\log x$ **44.** $g(x) = 1 + \log x$

45–48 ■ Find the function of the form $y = \log_a x$ whose graph is given.

45.
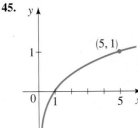

46.
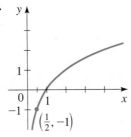

47.
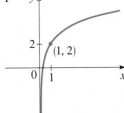

48.
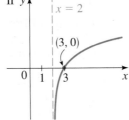

49–50 ■ Match the logarithmic function with one of the graphs labeled I or II.

49. $f(x) = 2 + \ln x$ **50.** $f(x) = \ln(x - 2)$

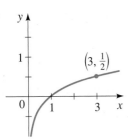

51. Draw the graph of $y = 4^x$, then use it to draw the graph of $y = \log_4 x$.

52. Draw the graph of $y = 3^x$, then use it to draw the graph of $y = \log_3 x$.

53–62 ■ Graph the function, not by plotting points, but by starting from the graphs in Figures 4 and 9. State the domain, range, and asymptote.

53. $f(x) = \log_2(x - 4)$ **54.** $f(x) = -\log_{10} x$

55. $g(x) = \log_5(-x)$ **56.** $g(x) = \ln(x + 2)$

57. $y = 2 + \log_3 x$ **58.** $y = \log_3(x - 1) - 2$

59. $y = 1 - \log_{10} x$ **60.** $y = 1 + \ln(-x)$

61. $y = |\ln x|$ **62.** $y = \ln|x|$

63–68 ■ Find the domain of the function.

63. $f(x) = \log_{10}(x + 3)$ **64.** $f(x) = \log_5(8 - 2x)$

65. $g(x) = \log_3(x^2 - 1)$ **66.** $g(x) = \ln(x - x^2)$

67. $h(x) = \ln x + \ln(2 - x)$

68. $h(x) = \sqrt{x - 2} - \log_5(10 - x)$

 69–74 ■ Draw the graph of the function in a suitable viewing rectangle, and use it to find the domain, the asymptotes, and the local maximum and minimum values.

69. $y = \log_{10}(1 - x^2)$ **70.** $y = \ln(x^2 - x)$

71. $y = x + \ln x$ **72.** $y = x(\ln x)^2$

73. $y = \dfrac{\ln x}{x}$ **74.** $y = x\log_{10}(x + 10)$

75-78 ■ Find the functions $f \circ g$ and $g \circ f$ and their domains.

75. $f(x) = 2^x$, $g(x) = x + 1$

76. $f(x) = 3^x$, $g(x) = x^2 + 1$

77. $f(x) = \log_2 x$, $g(x) = x - 2$

78. $f(x) = \log x$, $g(x) = x^2$

 79. Compare the rates of growth of the functions $f(x) = \ln x$ and $g(x) = \sqrt{x}$ by drawing their graphs on a common screen using the viewing rectangle $[-1, 30]$ by $[-1, 6]$.

 80. (a) By drawing the graphs of the functions

$$f(x) = 1 + \ln(1 + x) \qquad \text{and} \qquad g(x) = \sqrt{x}$$

in a suitable viewing rectangle, show that even when a logarithmic function starts out higher than a root function, it is ultimately overtaken by the root function.

(b) Find, correct to two decimal places, the solutions of the equation $\sqrt{x} = 1 + \ln(1 + x)$.

81–82 ■ A family of functions is given. **(a)** Draw graphs of the family for $c = 1, 2, 3,$ and 4. **(b)** How are the graphs in part (a) related?

81. $f(x) = \log(cx)$ **82.** $f(x) = c\log x$

83–84 ■ A function $f(x)$ is given. **(a)** Find the domain of the function f. **(b)** Find the inverse function of f.

83. $f(x) = \log_2(\log_{10} x)$ **84.** $f(x) = \ln(\ln(\ln x))$

85. (a) Find the inverse of the function $f(x) = \dfrac{2^x}{1 + 2^x}$.

(b) What is the domain of the inverse function?

APPLICATIONS

86. Absorption of Light A spectrophotometer measures the concentration of a sample dissolved in water by shining a light through it and recording the amount of light that emerges. In other words, if we know the amount of light that is absorbed, we can calculate the concentration of the sample. For a certain substance the concentration (in moles per liter) is found by using the formula

$$C = -2500 \ln\left(\frac{I}{I_0}\right)$$

where I_0 is the intensity of the incident light and I is the intensity of light that emerges. Find the concentration of the substance if the intensity I is 70% of I_0.

87. Carbon Dating The age of an ancient artifact can be determined by the amount of radioactive carbon-14 remaining in it. If D_0 is the original amount of carbon-14 and D is the amount remaining, then the artifact's age A (in years) is given by

$$A = -8267 \ln\left(\frac{D}{D_0}\right)$$

Find the age of an object if the amount D of carbon-14 that remains in the object is 73% of the original amount D_0.

88. Bacteria Colony A certain strain of bacteria divides every three hours. If a colony is started with 50 bacteria, then the time t (in hours) required for the colony to grow to N bacteria is given by

$$t = 3\frac{\log(N/50)}{\log 2}$$

Find the time required for the colony to grow to a million bacteria.

89. Investment The time required to double the amount of an investment at an interest rate r compounded continuously is given by

$$t = \frac{\ln 2}{r}$$

Find the time required to double an investment at 6%, 7%, and 8%.

90. Charging a Battery The rate at which a battery charges is slower the closer the battery is to its maximum charge C_0. The time (in hours) required to charge a fully discharged battery to a charge C is given by

$$t = -k \ln\left(1 - \frac{C}{C_0}\right)$$

where k is a positive constant that depends on the battery. For a certain battery, $k = 0.25$. If this battery is fully discharged, how long will it take to charge to 90% of its maximum charge C_0?

91. Difficulty of a Task The difficulty in "acquiring a target" (such as using your mouse to click on an icon on your computer screen) depends on the distance to the target and the size of the target. According to Fitts's Law, the index of difficulty (ID) is given by

$$ID = \frac{\log(2A/W)}{\log 2}$$

where W is the width of the target and A is the distance to the center of the target. Compare the difficulty of clicking on an icon that is 5 mm wide to clicking on one that is 10 mm wide. In each case, assume that the mouse is 100 mm from the icon.

DISCOVERY ▪ DISCUSSION ▪ WRITING

92. The Height of the Graph of a Logarithmic Function Suppose that the graph of $y = 2^x$ is drawn on a coordinate plane where the unit of measurement is an inch.
  (a) Show that at a distance 2 ft to the right of the origin the height of the graph is about 265 mi.
  (b) If the graph of $y = \log_2 x$ is drawn on the same set of axes, how far to the right of the origin do we have to go before the height of the curve reaches 2 ft?

93. The Googolplex A **googol** is 10^{100}, and a **googolplex** is 10^{googol}. Find

$$\log(\log(\text{googol})) \quad \text{and} \quad \log(\log(\log(\text{googolplex})))$$

94. Comparing Logarithms Which is larger, $\log_4 17$ or $\log_5 24$? Explain your reasoning.

95. The Number of Digits in an Integer Compare $\log 1000$ to the number of digits in 1000. Do the same for 10,000. How many digits does any number between 1000 and 10,000 have? Between what two values must the common logarithm of such a number lie? Use your observations to explain why the number of digits in any positive integer x is $[\![\log x]\!] + 1$. (The symbol $[\![n]\!]$ is the greatest integer function defined in Section 1.4.) How many digits does the number 2^{100} have?

8.4 LAWS OF LOGARITHMS

| Laws of Logarithms ▶ Expanding and Combining Logarithmic Expressions
| ▶ Change of Base Formula

In this section we study properties of logarithms. These properties give logarithmic functions a wide range of applications, as we will see in Section 8.6.

▼ Laws of Logarithms

Since logarithms are exponents, the Laws of Exponents give rise to the Laws of Logarithms.

LAWS OF LOGARITHMS

Let a be a positive number, with $a \neq 1$. Let A, B, and C be any real numbers with $A > 0$ and $B > 0$.

Law	Description
1. $\log_a(AB) = \log_a A + \log_a B$	The logarithm of a product of numbers is the sum of the logarithms of the numbers.
2. $\log_a\left(\dfrac{A}{B}\right) = \log_a A - \log_a B$	The logarithm of a quotient of numbers is the difference of the logarithms of the numbers.
3. $\log_a(A^C) = C \log_a A$	The logarithm of a power of a number is the exponent times the logarithm of the number.

PROOF We make use of the property $\log_a a^x = x$ from Section 8.3.

Law 1 Let $\log_a A = u$ and $\log_a B = v$. When written in exponential form, these equations become

$$a^u = A \qquad \text{and} \qquad a^v = B$$

Thus

$$\log_a(AB) = \log_a(a^u a^v) = \log_a(a^{u+v})$$
$$= u + v = \log_a A + \log_a B$$

Law 2 Using Law 1, we have

$$\log_a A = \log_a\left[\left(\frac{A}{B}\right)B\right] = \log_a\left(\frac{A}{B}\right) + \log_a B$$

so

$$\log_a\left(\frac{A}{B}\right) = \log_a A - \log_a B$$

Law 3 Let $\log_a A = u$. Then $a^u = A$, so

$$\log_a(A^C) = \log_a(a^u)^C = \log_a(a^{uC}) = uC = C \log_a A \qquad ■$$

EXAMPLE 1 | Using the Laws of Logarithms to Evaluate Expressions

Evaluate each expression.

(a) $\log_4 2 + \log_4 32$

(b) $\log_2 80 - \log_2 5$

(c) $-\frac{1}{3} \log 8$

SOLUTION

(a) $\log_4 2 + \log_4 32 = \log_4(2 \cdot 32)$ Law 1

                $= \log_4 64 = 3$ Because $64 = 4^3$

(b) $\log_2 80 - \log_2 5 = \log_2\left(\frac{80}{5}\right)$ Law 2

                $= \log_2 16 = 4$ Because $16 = 2^4$

(c) $-\frac{1}{3} \log 8 = \log 8^{-1/3}$ Law 3

           $= \log\left(\frac{1}{2}\right)$ Property of negative exponents

           ≈ -0.301 Calculator

✎ NOW TRY EXERCISES **7**, **9**, AND **11** ■

▼ Expanding and Combining Logarithmic Expressions

The Laws of Logarithms allow us to write the logarithm of a product or a quotient as the sum or difference of logarithms. This process, called *expanding* a logarithmic expression, is illustrated in the next example.

EXAMPLE 2 | Expanding Logarithmic Expressions

Use the Laws of Logarithms to expand each expression.

(a) $\log_2(6x)$ **(b)** $\log_5(x^3 y^6)$ **(c)** $\ln\left(\dfrac{ab}{\sqrt[3]{c}}\right)$

SOLUTION

(a) $\log_2(6x) = \log_2 6 + \log_2 x$ Law 1

(b) $\log_5(x^3 y^6) = \log_5 x^3 + \log_5 y^6$ Law 1

              $= 3 \log_5 x + 6 \log_5 y$ Law 3

(c) $\ln\left(\dfrac{ab}{\sqrt[3]{c}}\right) = \ln(ab) - \ln\sqrt[3]{c}$ Law 2

           $= \ln a + \ln b - \ln c^{1/3}$ Law 1

           $= \ln a + \ln b - \frac{1}{3} \ln c$ Law 3

✎ NOW TRY EXERCISES **19**, **21**, AND **33** ■

The Laws of Logarithms also allow us to reverse the process of expanding that was done in Example 2. That is, we can write sums and differences of logarithms as a single logarithm. This process, called *combining* logarithmic expressions, is illustrated in the next example.

EXAMPLE 3 | Combining Logarithmic Expressions

Combine $3 \log x + \frac{1}{2} \log(x + 1)$ into a single logarithm.

SOLUTION

$$3 \log x + \tfrac{1}{2} \log(x + 1) = \log x^3 + \log(x + 1)^{1/2} \quad \text{Law 3}$$

$$= \log(x^3(x + 1)^{1/2}) \quad \text{Law 1}$$

✎ NOW TRY EXERCISE **47** ■

EXAMPLE 4 | Combining Logarithmic Expressions

Combine $3 \ln s + \frac{1}{2} \ln t - 4 \ln(t^2 + 1)$ into a single logarithm.

SOLUTION

$$3 \ln s + \tfrac{1}{2} \ln t - 4 \ln(t^2 + 1) = \ln s^3 + \ln t^{1/2} - \ln(t^2 + 1)^4 \qquad \text{Law 3}$$

$$= \ln(s^3 t^{1/2}) - \ln(t^2 + 1)^4 \qquad \text{Law 1}$$

$$= \ln\left(\frac{s^3 \sqrt{t}}{(t^2 + 1)^4} \right) \qquad \text{Law 2}$$

✎. NOW TRY EXERCISE **49**   ■

Warning   Although the Laws of Logarithms tell us how to compute the logarithm of a product or a quotient, *there is no corresponding rule for the logarithm of a sum or a difference*. For instance,

$$\log_a(x + y) \;\cancel{=}\; \log_a x + \log_a y$$

In fact, we know that the right side is equal to $\log_a(xy)$. Also, don't improperly simplify quotients or powers of logarithms. For instance,

$$\frac{\log 6}{\log 2} \;\cancel{=}\; \log\left(\frac{6}{2} \right) \qquad \text{and} \qquad (\log_2 x)^3 \;\cancel{=}\; 3 \log_2 x$$

Logarithmic functions are used to model a variety of situations involving human behavior. One such behavior is how quickly we forget things we have learned. For example, if you learn algebra at a certain performance level (say, 90% on a test) and then don't use algebra for a while, how much will you retain after a week, a month, or a year? Hermann Ebbinghaus (1850–1909) studied this phenomenon and formulated the law described in the next example.

EXAMPLE 5 | The Law of Forgetting

If a task is learned at a performance level P_0, then after a time interval t the performance level P satisfies

$$\log P = \log P_0 - c \log(t + 1)$$

where c is a constant that depends on the type of task and t is measured in months.

(a) Solve for P.

(b) If your score on a history test is 90, what score would you expect to get on a similar test after two months? After a year? (Assume that $c = 0.2$.)

SOLUTION

Forgetting what we've learned depends on how long ago we learned it.

(a) We first combine the right-hand side.

$$\log P = \log P_0 - c \log(t + 1) \qquad \text{Given equation}$$

$$\log P = \log P_0 - \log(t + 1)^c \qquad \text{Law 3}$$

$$\log P = \log \frac{P_0}{(t + 1)^c} \qquad \text{Law 2}$$

$$P = \frac{P_0}{(t + 1)^c} \qquad \text{Because log is one-to-one}$$

(b) Here $P_0 = 90$, $c = 0.2$, and t is measured in months.

In two months: $t = 2$ and $P = \dfrac{90}{(2 + 1)^{0.2}} \approx 72$

In one year: $t = 12$ and $P = \dfrac{90}{(12 + 1)^{0.2}} \approx 54$

Your expected scores after two months and one year are 72 and 54, respectively.

✎. NOW TRY EXERCISE **69**   ■

▼ Change of Base Formula

For some purposes we find it useful to change from logarithms in one base to logarithms in another base. Suppose we are given $\log_a x$ and want to find $\log_b x$. Let

$$y = \log_b x$$

We write this in exponential form and take the logarithm, with base a, of each side.

$$b^y = x \qquad \text{Exponential form}$$

$$\log_a(b^y) = \log_a x \qquad \text{Take } \log_a \text{ of each side}$$

$$y \log_a b = \log_a x \qquad \text{Law 3}$$

$$y = \frac{\log_a x}{\log_a b} \qquad \text{Divide by } \log_a b$$

This proves the following formula.

We may write the Change of Base Formula as

$$\log_b x = \left(\frac{1}{\log_a b}\right)\log_a x$$

So $\log_b x$ is just a constant multiple of $\log_a x$; the constant is $\dfrac{1}{\log_a b}$.

CHANGE OF BASE FORMULA

$$\log_b x = \frac{\log_a x}{\log_a b}$$

In particular, if we put $x = a$, then $\log_a a = 1$, and this formula becomes

$$\log_b a = \frac{1}{\log_a b}$$

We can now evaluate a logarithm to *any* base by using the Change of Base Formula to express the logarithm in terms of common logarithms or natural logarithms and then using a calculator.

EXAMPLE 6 | Evaluating Logarithms with the Change of Base Formula

Use the Change of Base Formula and common or natural logarithms to evaluate each logarithm, correct to five decimal places.

(a) $\log_8 5$ **(b)** $\log_9 20$

SOLUTION

(a) We use the Change of Base Formula with $b = 8$ and $a = 10$:

$$\log_8 5 = \frac{\log_{10} 5}{\log_{10} 8} \approx 0.77398$$

(b) We use the Change of Base Formula with $b = 9$ and $a = e$:

$$\log_9 20 = \frac{\ln 20}{\ln 9} \approx 1.36342$$

✎ NOW TRY EXERCISES **55** AND **57**

EXAMPLE 7 | Using the Change of Base Formula to Graph a Logarithmic Function

Use a graphing calculator to graph $f(x) = \log_6 x$.

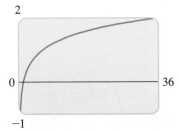

FIGURE 1 $f(x) = \log_6 x = \dfrac{\ln x}{\ln 6}$

SOLUTION Calculators don't have a key for $\log_6$, so we use the Change of Base Formula to write

$$f(x) = \log_6 x = \frac{\ln x}{\ln 6}$$

Since calculators do have an $\boxed{\text{LN}}$ key, we can enter this new form of the function and graph it. The graph is shown in Figure 1.

✎ NOW TRY EXERCISE **63**   ■

8.4 EXERCISES

CONCEPTS

1. The logarithm of a product of two numbers is the same as the _____ of the logarithms of these numbers. So $\log_5(25 \cdot 125) =$ _____ $+$ _____.

2. The logarithm of a quotient of two numbers is the same as the _____ of the logarithms of these numbers. So $\log_5\left(\frac{25}{125}\right) =$ _____ $-$ _____.

3. The logarithm of a number raised to a power is the same as the power _____ the logarithm of the number. So $\log_5(25^{10}) =$ _____ $\cdot$ _____.

4. (a) We can expand $\log\left(\dfrac{x^2 y}{z}\right)$ to get _____.
   (b) We can combine $2\log x + \log y - \log z$ to get _____.

5. Most calculators can find logarithms with base _____ and base _____. To find logarithms with different bases, we use the _____ Formula. To find $\log_7 12$, we write

$$\log_7 12 = \frac{\log \blacksquare}{\log \blacksquare} = \text{_____}$$

6. *True or false?* We get the same answer if we do the calculation in Exercise 5 using ln in place of log.

SKILLS

7–18 ■ Evaluate the expression.

7. $\log_3 \sqrt{27}$

8. $\log_2 160 - \log_2 5$

9. $\log 4 + \log 25$

10. $\log \dfrac{1}{\sqrt{1000}}$

11. $\log_4 192 - \log_4 3$

12. $\log_{12} 9 + \log_{12} 16$

13. $\log_2 6 - \log_2 15 + \log_2 20$

14. $\log_3 100 - \log_3 18 - \log_3 50$

15. $\log_4 16^{100}$

16. $\log_2 8^{33}$

17. $\log(\log 10^{10,000})$

18. $\ln(\ln e^{e^{200}})$

19–44 ■ Use the Laws of Logarithms to expand the expression.

19. $\log_2(2x)$

20. $\log_3(5y)$

21. $\log_2(x(x-1))$

22. $\log_5 \dfrac{x}{2}$

23. $\log 6^{10}$

24. $\ln \sqrt{z}$

25. $\log_2(AB^2)$

26. $\log_6 \sqrt[4]{17}$

27. $\log_3(x\sqrt{y})$

28. $\log_2(xy)^{10}$

29. $\log_5 \sqrt[3]{x^2 + 1}$

30. $\log_a\left(\dfrac{x^2}{yz^3}\right)$

31. $\ln \sqrt{ab}$

32. $\ln \sqrt[3]{3r^2 s}$

33. $\log\left(\dfrac{x^3 y^4}{z^6}\right)$

34. $\log\left(\dfrac{a^2}{b^4 \sqrt{c}}\right)$

35. $\log_2\left(\dfrac{x(x^2+1)}{\sqrt{x^2-1}}\right)$

36. $\log_5 \sqrt{\dfrac{x-1}{x+1}}$

37. $\ln\left(x\sqrt{\dfrac{y}{z}}\right)$

38. $\ln \dfrac{3x^2}{(x+1)^{10}}$

39. $\log \sqrt[4]{x^2 + y^2}$

40. $\log\left(\dfrac{x}{\sqrt[3]{1-x}}\right)$

41. $\log \sqrt{\dfrac{x^2+4}{(x^2+1)(x^3-7)^2}}$

42. $\log \sqrt{x\sqrt{y\sqrt{z}}}$

43. $\ln\left(\dfrac{x^3\sqrt{x-1}}{3x+4}\right)$

44. $\log\left(\dfrac{10^x}{x(x^2+1)(x^4+2)}\right)$

45–54 ■ Use the Laws of Logarithms to combine the expression.

45. $\log_3 5 + 5\log_3 2$

46. $\log 12 + \frac{1}{2}\log 7 - \log 2$

47. $\log_2 A + \log_2 B - 2\log_2 C$

48. $\log_5(x^2 - 1) - \log_5(x - 1)$

49. $4\log x - \frac{1}{3}\log(x^2 + 1) + 2\log(x - 1)$

50. $\ln(a + b) + \ln(a - b) - 2\ln c$

51. $\ln 5 + 2\ln x + 3\ln(x^2 + 5)$

52. $2(\log_5 x + 2\log_5 y - 3\log_5 z)$

53. $\frac{1}{3}\log(x + 2)^3 + \frac{1}{2}[\log x^4 - \log(x^2 - x - 6)^2]$

54. $\log_a b + c \log_a d - r \log_a s$

55–62 ■ Use the Change of Base Formula and a calculator to evaluate the logarithm, rounded to six decimal places. Use either natural or common logarithms.

 55. $\log_2 5$ **56.** $\log_5 2$

 57. $\log_3 16$ **58.** $\log_6 92$

59. $\log_7 2.61$ **60.** $\log_6 532$

61. $\log_4 125$ **62.** $\log_{12} 2.5$

63. Use the Change of Base Formula to show that

$$\log_3 x = \frac{\ln x}{\ln 3}$$

Then use this fact to draw the graph of the function $f(x) = \log_3 x$.

64. Draw graphs of the family of functions $y = \log_a x$ for $a = 2, e, 5,$ and 10 on the same screen, using the viewing rectangle $[0, 5]$ by $[-3, 3]$. How are these graphs related?

65. Use the Change of Base Formula to show that

$$\log e = \frac{1}{\ln 10}$$

66. Simplify: $(\log_2 5)(\log_5 7)$

67. Show that $-\ln(x - \sqrt{x^2 - 1}) = \ln(x + \sqrt{x^2 - 1})$.

APPLICATIONS

68. Forgetting Use the Law of Forgetting (Example 5) to estimate a student's score on a biology test two years after he got a score of 80 on a test covering the same material. Assume that $c = 0.3$ and t is measured in months.

69. Wealth Distribution Vilfredo Pareto (1848–1923) observed that most of the wealth of a country is owned by a few members of the population. **Pareto's Principle** is

$$\log P = \log c - k \log W$$

where W is the wealth level (how much money a person has) and P is the number of people in the population having that much money.

(a) Solve the equation for P.

(b) Assume that $k = 2.1$, $c = 8000$, and W is measured in millions of dollars. Use part (a) to find the number of people who have \$2 million or more. How many people have \$10 million or more?

70. Biodiversity Some biologists model the number of species S in a fixed area A (such as an island) by the species-area relationship

$$\log S = \log c + k \log A$$

where c and k are positive constants that depend on the type of species and habitat.

(a) Solve the equation for S.

(b) Use part (a) to show that if $k = 3$, then doubling the area increases the number of species eightfold.

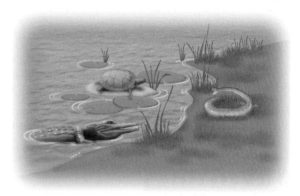

71. Magnitude of Stars The magnitude M of a star is a measure of how bright a star appears to the human eye. It is defined by

$$M = -2.5 \log\left(\frac{B}{B_0}\right)$$

where B is the actual brightness of the star and B_0 is a constant.

(a) Expand the right-hand side of the equation.

(b) Use part (a) to show that the brighter a star, the less its magnitude.

(c) Betelgeuse is about 100 times brighter than Albiero. Use part (a) to show that Betelgeuse is 5 magnitudes less bright than Albiero.

DISCOVERY ■ DISCUSSION ■ WRITING

72. True or False? Discuss each equation and determine whether it is true for all possible values of the variables. (Ignore values of the variables for which any term is undefined.)

(a) $\log\left(\dfrac{x}{y}\right) = \dfrac{\log x}{\log y}$

(b) $\log_2(x - y) = \log_2 x - \log_2 y$

(c) $\log_5\left(\dfrac{a}{b^2}\right) = \log_5 a - 2 \log_5 b$

(d) $\log 2^z = z \log 2$

(e) $(\log P)(\log Q) = \log P + \log Q$

(f) $\dfrac{\log a}{\log b} = \log a - \log b$

(g) $(\log_2 7)^x = x \log_2 7$

(h) $\log_a a^a = a$

(i) $\log(x - y) = \dfrac{\log x}{\log y}$

(j) $-\ln\left(\dfrac{1}{A}\right) = \ln A$

73. Find the Error What is wrong with the following argument?

$$\log 0.1 < 2 \log 0.1$$
$$= \log(0.1)^2$$
$$= \log 0.01$$
$$\log 0.1 < \log 0.01$$
$$0.1 < 0.01$$

74. Shifting, Shrinking, and Stretching Graphs of Functions Let $f(x) = x^2$. Show that $f(2x) = 4f(x)$, and explain how this shows that shrinking the graph of f horizontally has the same effect as stretching it vertically. Then use the identities $e^{2+x} = e^2 e^x$ and $\ln(2x) = \ln 2 + \ln x$ to show that for $g(x) = e^x$ a horizontal shift is the same as a vertical stretch and for $h(x) = \ln x$ a horizontal shrinking is the same as a vertical shift.

8.5 EXPONENTIAL AND LOGARITHMIC EQUATIONS

| Exponential Equations ▶ Logarithmic Equations ▶ Compound Interest

In this section we solve equations that involve exponential or logarithmic functions. The techniques that we develop here will be used in the next section for solving applied problems.

▼ Exponential Equations

An *exponential equation* is one in which the variable occurs in the exponent. For example,

$$2^x = 7$$

The variable x presents a difficulty because it is in the exponent. To deal with this difficulty, we take the logarithm of each side and then use the Laws of Logarithms to "bring down x" from the exponent.

$2^x = 7$	Given equation
$\ln 2^x = \ln 7$	Take ln of each side
$x \ln 2 = \ln 7$	Law 3 (bring down exponent)
$x = \dfrac{\ln 7}{\ln 2}$	Solve for x
≈ 2.807	Calculator

Recall that Law 3 of the Laws of Logarithms says that $\log_a A^C = C \log_a A$.

The method that we used to solve $2^x = 7$ is typical of how we solve exponential equations in general.

GUIDELINES FOR SOLVING EXPONENTIAL EQUATIONS

1. Isolate the exponential expression on one side of the equation.

2. Take the logarithm of each side, then use the Laws of Logarithms to "bring down the exponent."

3. Solve for the variable.

EXAMPLE 1 | Solving an Exponential Equation

Find the solution of the equation $3^{x+2} = 7$, rounded to six decimal places.

SOLUTION We take the common logarithm of each side and use Law 3.

$3^{x+2} = 7$	Given equation
$\log(3^{x+2}) = \log 7$	Take log of each side
$(x + 2)\log 3 = \log 7$	Law 3 (bring down exponent)
$x + 2 = \dfrac{\log 7}{\log 3}$	Divide by log 3
$x = \dfrac{\log 7}{\log 3} - 2$	Subtract 2
≈ -0.228756	Calculator

We could have used natural logarithms instead of common logarithms. In fact, using the same steps, we get

$$x = \frac{\ln 7}{\ln 3} - 2 \approx -0.228756$$

CHECK YOUR ANSWER

Substituting $x = -0.228756$ into the original equation and using a calculator, we get

$$3^{(-0.228756)+2} \approx 7 \quad ✔$$

✎ NOW TRY EXERCISE **7**

EXAMPLE 2 | Solving an Exponential Equation

Solve the equation $8e^{2x} = 20$.

SOLUTION We first divide by 8 to isolate the exponential term on one side of the equation.

$8e^{2x} = 20$	Given equation
$e^{2x} = \frac{20}{8}$	Divide by 8
$\ln e^{2x} = \ln 2.5$	Take ln of each side
$2x = \ln 2.5$	Property of ln
$x = \dfrac{\ln 2.5}{2}$	Divide by 2
≈ 0.458	Calculator

CHECK YOUR ANSWER

Substituting $x = 0.458$ into the original equation and using a calculator, we get

$$8e^{2(0.458)} \approx 20 \quad ✔$$

✎ NOW TRY EXERCISE **9**

EXAMPLE 3 | Solving an Exponential Equation Algebraically and Graphically

Solve the equation $e^{3-2x} = 4$ algebraically and graphically.

SOLUTION 1: Algebraic

Since the base of the exponential term is e, we use natural logarithms to solve this equation.

$e^{3-2x} = 4$	Given equation
$\ln(e^{3-2x}) = \ln 4$	Take ln of each side
$3 - 2x = \ln 4$	Property of ln
$-2x = -3 + \ln 4$	Subtract 3
$x = \frac{1}{2}(3 - \ln 4) \approx 0.807$	Multiply by $-\frac{1}{2}$

You should check that this answer satisfies the original equation.

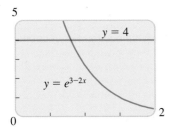

FIGURE 1

If we let $w = e^x$, we get the quadratic equation

$$w^2 - w - 6 = 0$$

which factors as

$$(w - 3)(w + 2) = 0$$

SOLUTION 2: Graphical

We graph the equations $y = e^{3-2x}$ and $y = 4$ in the same viewing rectangle as in Figure 1. The solutions occur where the graphs intersect. Zooming in on the point of intersection of the two graphs, we see that $x \approx 0.81$.

✎ **NOW TRY EXERCISE 11** ■

EXAMPLE 4 | An Exponential Equation of Quadratic Type

Solve the equation $e^{2x} - e^x - 6 = 0$.

SOLUTION   To isolate the exponential term, we factor.

$e^{2x} - e^x - 6 = 0$	Given equation
$(e^x)^2 - e^x - 6 = 0$	Law of Exponents
$(e^x - 3)(e^x + 2) = 0$	Factor (a quadratic in e^x)
$e^x - 3 = 0 \quad$ or $\quad e^x + 2 = 0$	Zero-Product Property
$e^x = 3 \qquad\qquad e^x = -2$	

The equation $e^x = 3$ leads to $x = \ln 3$. But the equation $e^x = -2$ has no solution because $e^x > 0$ for all x. Thus, $x = \ln 3 \approx 1.0986$ is the only solution. You should check that this answer satisfies the original equation.

✎ **NOW TRY EXERCISE 29** ■

EXAMPLE 5 | Solving an Exponential Equation

Solve the equation $3xe^x + x^2e^x = 0$.

SOLUTION   First we factor the left side of the equation.

$3xe^x + x^2e^x = 0$	Given equation
$x(3 + x)e^x = 0$	Factor out common factors
$x(3 + x) = 0$	Divide by e^x (because $e^x \neq 0$)
$x = 0 \quad$ or $\quad 3 + x = 0$	Zero-Product Property

Thus the solutions are $x = 0$ and $x = -3$.

✎ **NOW TRY EXERCISE 33** ■

CHECK YOUR ANSWER

$x = 0$:
$$3(0)e^0 + 0^2e^0 = 0 \quad ✔$$

$x = -3$:
$$3(-3)e^{-3} + (-3)^2e^{-3}$$
$$= -9e^{-3} + 9e^{-3} = 0 \quad ✔$$

Radiocarbon Dating is a method archeologists use to determine the age of ancient objects. The carbon dioxide in the atmosphere always contains a fixed fraction of radioactive carbon, carbon-14 (^{14}C), with a half-life of about 5730 years. Plants absorb carbon dioxide from the atmosphere, which then makes its way to animals through the food chain. Thus, all living creatures contain the same fixed proportions of ^{14}C to nonradioactive ^{12}C as the atmosphere.

After an organism dies, it stops assimilating ^{14}C, and the amount of ^{14}C in it begins to decay exponentially. We can then determine the time elapsed since the death of the organism by measuring the amount of ^{14}C left in it.

For example, if a donkey bone contains 73% as much ^{14}C as a living donkey and it died t years ago, then by the formula for radioactive decay (Section 8.6),

$$0.73 = (1.00)e^{-(t \ln 2)/5730}$$

We solve this exponential equation to find $t \approx 2600$, so the bone is about 2600 years old.

▼ Logarithmic Equations

A *logarithmic equation* is one in which a logarithm of the variable occurs. For example,

$$\log_2(x + 2) = 5$$

To solve for x, we write the equation in exponential form.

$$x + 2 = 2^5 \qquad \text{Exponential form}$$

$$x = 32 - 2 = 30 \qquad \text{Solve for } x$$

Another way of looking at the first step is to raise the base, 2, to each side of the equation.

$$2^{\log_2(x+2)} = 2^5 \qquad \text{Raise 2 to each side}$$

$$x + 2 = 2^5 \qquad \text{Property of logarithms}$$

$$x = 32 - 2 = 30 \qquad \text{Solve for } x$$

The method used to solve this simple problem is typical. We summarize the steps as follows.

GUIDELINES FOR SOLVING LOGARITHMIC EQUATIONS

1. Isolate the logarithmic term on one side of the equation; you might first need to combine the logarithmic terms.

2. Write the equation in exponential form (or raise the base to each side of the equation).

3. Solve for the variable.

EXAMPLE 6 | Solving Logarithmic Equations

Solve each equation for x.

(a) $\ln x = 8$ **(b)** $\log_2(25 - x) = 3$

SOLUTION

(a)

$$\ln x = 8 \qquad \text{Given equation}$$

$$x = e^8 \qquad \text{Exponential form}$$

Therefore, $x = e^8 \approx 2981$.

We can also solve this problem another way:

$$\ln x = 8 \qquad \text{Given equation}$$

$$e^{\ln x} = e^8 \qquad \text{Raise } e \text{ to each side}$$

$$x = e^8 \qquad \text{Property of ln}$$

(b) The first step is to rewrite the equation in exponential form.

$$\log_2(25 - x) = 3 \qquad \text{Given equation}$$

$$25 - x = 2^3 \qquad \text{Exponential form (or raise 2 to each side)}$$

$$25 - x = 8$$

$$x = 25 - 8 = 17$$

CHECK YOUR ANSWER

If $x = 17$, we get

$$\log_2(25 - 17) = \log_2 8 = 3 \quad ✔$$

🔖 NOW TRY EXERCISES **37** AND **41**

EXAMPLE 7 | Solving a Logarithmic Equation

Solve the equation $4 + 3 \log(2x) = 16$.

SOLUTION   We first isolate the logarithmic term. This allows us to write the equation in exponential form.

$$
\begin{array}{lll}
4 + 3 \log(2x) = 16 & \quad & \text{Given equation} \\
3 \log(2x) = 12 & \quad & \text{Subtract 4} \\
\log(2x) = 4 & \quad & \text{Divide by 3} \\
2x = 10^4 & \quad & \text{Exponential form (or raise 10 to each side)} \\
x = 5000 & \quad & \text{Divide by 2}
\end{array}
$$

CHECK YOUR ANSWER

If $x = 5000$, we get

$$
\begin{aligned}
4 + 3 \log 2(5000) &= 4 + 3 \log 10{,}000 \\
&= 4 + 3(4) \\
&= 16 \quad ✔
\end{aligned}
$$

✎ NOW TRY EXERCISE **43**

EXAMPLE 8 | Solving a Logarithmic Equation Algebraically and Graphically

Solve the equation $\log(x + 2) + \log(x - 1) = 1$ algebraically and graphically.

SOLUTION 1: Algebraic

We first combine the logarithmic terms, using the Laws of Logarithms.

$$
\begin{array}{lll}
\log[(x + 2)(x - 1)] = 1 & \quad & \text{Law 1} \\
(x + 2)(x - 1) = 10 & \quad & \text{Exponential form (or raise 10 to each side)} \\
x^2 + x - 2 = 10 & \quad & \text{Expand left side} \\
x^2 + x - 12 = 0 & \quad & \text{Subtract 10} \\
(x + 4)(x - 3) = 0 & \quad & \text{Factor} \\
x = -4 \quad \text{or} \quad x = 3
\end{array}
$$

We check these potential solutions in the original equation and find that $x = -4$ is not a solution (because logarithms of negative numbers are undefined), but $x = 3$ is a solution. (See *Check Your Answers*.)

SOLUTION 2: Graphical

We first move all terms to one side of the equation:

$$\log(x + 2) + \log(x - 1) - 1 = 0$$

Then we graph

$$y = \log(x + 2) + \log(x - 1) - 1$$

as in Figure 2. The solutions are the x-intercepts of the graph. Thus, the only solution is $x \approx 3$.

✎ NOW TRY EXERCISE **49**

CHECK YOUR ANSWERS

$x = -4$:

$$
\begin{aligned}
&\log(-4 + 2) + \log(-4 - 1) \\
&\quad = \log(-2) + \log(-5) \\
&\qquad\qquad\qquad \text{undefined} \quad ✗
\end{aligned}
$$

$x = 3$:

$$
\begin{aligned}
&\log(3 + 2) + \log(3 - 1) \\
&\quad = \log 5 + \log 2 = \log(5 \cdot 2) \\
&\quad = \log 10 = 1 \quad ✔
\end{aligned}
$$

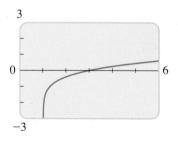

FIGURE 2

In Example 9 it's not possible to isolate *x* algebraically, so we must solve the equation graphically.

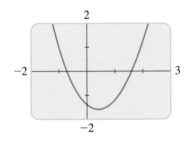

FIGURE 3

EXAMPLE 9 │ Solving a Logarithmic Equation Graphically

Solve the equation $x^2 = 2 \ln(x + 2)$.

SOLUTION   We first move all terms to one side of the equation

$$x^2 - 2 \ln(x + 2) = 0$$

Then we graph

$$y = x^2 - 2 \ln(x + 2)$$

as in Figure 3. The solutions are the *x*-intercepts of the graph. Zooming in on the *x*-intercepts, we see that there are two solutions:

$$x \approx -0.71 \quad \text{and} \quad x \approx 1.60$$

✎ NOW TRY EXERCISE 59   ■

Logarithmic equations are used in determining the amount of light that reaches various depths in a lake. (This information helps biologists to determine the types of life a lake can support.) As light passes through water (or other transparent materials such as glass or plastic), some of the light is absorbed. It's easy to see that the murkier the water, the more light is absorbed. The exact relationship between light absorption and the distance light travels in a material is described in the next example.

EXAMPLE 10 │ Transparency of a Lake

The intensity of light in a lake diminishes with depth.

If I_0 and I denote the intensity of light before and after going through a material and *x* is the distance (in feet) the light travels in the material, then according to the **Beer-Lambert Law**,

$$-\frac{1}{k} \ln\left(\frac{I}{I_0}\right) = x$$

where *k* is a constant depending on the type of material.

(a) Solve the equation for *I*.

(b) For a certain lake $k = 0.025$, and the light intensity is $I_0 = 14$ lumens (lm). Find the light intensity at a depth of 20 ft.

SOLUTION

(a) We first isolate the logarithmic term.

$$-\frac{1}{k} \ln\left(\frac{I}{I_0}\right) = x \qquad \text{Given equation}$$

$$\ln\left(\frac{I}{I_0}\right) = -kx \qquad \text{Multiply by } -k$$

$$\frac{I}{I_0} = e^{-kx} \qquad \text{Exponential form}$$

$$I = I_0 e^{-kx} \qquad \text{Multiply by } I_0$$

(b) We find *I* using the formula from part (a).

$$I = I_0 e^{-kx} \qquad \text{From part (a)}$$

$$= 14e^{(-0.025)(20)} \qquad I_0 = 14, \ k = 0.025, \ x = 20$$

$$\approx 8.49 \qquad \text{Calculator}$$

The light intensity at a depth of 20 ft is about 8.5 lm.

✎ NOW TRY EXERCISE 85   ■

▼ Compound Interest

Recall the formulas for interest that we found in Section 8.1. If a principal P is invested at an interest rate r for a period of t years, then the amount A of the investment is given by

$$A = P(1 + r)$$ Simple interest (for one year)

$$A(t) = P\left(1 + \frac{r}{n}\right)^{nt}$$ Interest compounded n times per year

$$A(t) = Pe^{rt}$$ Interest compounded continuously

We can use logarithms to determine the time it takes for the principal to increase to a given amount.

EXAMPLE 11 | Finding the Term for an Investment to Double

A sum of $5000 is invested at an interest rate of 5% per year. Find the time required for the money to double if the interest is compounded according to the following method.

(a) Semiannually   **(b)** Continuously

SOLUTION

(a) We use the formula for compound interest with $P =$ \$5000, $A(t) =$ \$10,000, $r = 0.05$, and $n = 2$ and solve the resulting exponential equation for t.

$$5000\left(1 + \frac{0.05}{2}\right)^{2t} = 10{,}000 \qquad P\left(1 + \frac{r}{n}\right)^{nt} = A$$

$$(1.025)^{2t} = 2 \qquad \text{Divide by 5000}$$

$$\log 1.025^{2t} = \log 2 \qquad \text{Take log of each side}$$

$$2t \log 1.025 = \log 2 \qquad \text{Law 3 (bring down the exponent)}$$

$$t = \frac{\log 2}{2 \log 1.025} \qquad \text{Divide by } 2 \log 1.025$$

$$t \approx 14.04 \qquad \text{Calculator}$$

The money will double in 14.04 years.

(b) We use the formula for continuously compounded interest with $P =$ \$5000, $A(t) =$ \$10,000, and $r = 0.05$ and solve the resulting exponential equation for t.

$$5000e^{0.05t} = 10{,}000 \qquad Pe^{rt} = A$$

$$e^{0.05t} = 2 \qquad \text{Divide by 5000}$$

$$\ln e^{0.05t} = \ln 2 \qquad \text{Take ln of each side}$$

$$0.05t = \ln 2 \qquad \text{Property of ln}$$

$$t = \frac{\ln 2}{0.05} \qquad \text{Divide by 0.05}$$

$$t \approx 13.86 \qquad \text{Calculator}$$

The money will double in 13.86 years.

✎ NOW TRY EXERCISE **75**   ■

EXAMPLE 12 | Time Required to Grow an Investment

A sum of $1000 is invested at an interest rate of 4% per year. Find the time required for the amount to grow to $4000 if interest is compounded continuously.

SOLUTION We use the formula for continuously compounded interest with $P = \$1000$, $A(t) = \$4000$, and $r = 0.04$ and solve the resulting exponential equation for t.

$$1000e^{0.04t} = 4000 \qquad Pe^{rt} = A$$

$$e^{0.04t} = 4 \qquad \text{Divide by 1000}$$

$$0.04t = \ln 4 \qquad \text{Take ln of each side}$$

$$t = \frac{\ln 4}{0.04} \qquad \text{Divide by 0.04}$$

$$t \approx 34.66 \qquad \text{Calculator}$$

The amount will be \$4000 in about 34 years and 8 months.

✎ NOW TRY EXERCISE **77**

8.5 EXERCISES

CONCEPTS

1. Let's solve the exponential equation $2e^x = 50$.

    (a) First, we isolate e^x to get the equivalent equation _____.

    (b) Next, we take ln of each side to get the equivalent equation

    _____.

    (c) Now we use a calculator to find $x =$ _____.

2. Let's solve the logarithmic equation
$\log 3 + \log(x - 2) = \log x$.

    (a) First, we combine the logarithms to get the equivalent

    equation _____.

    (b) Next, we write each side in exponential form to get the

    equivalent equation _____.

    (c) Now we find $x =$ _____.

SKILLS

3–28 ■ Find the solution of the exponential equation, rounded to four decimal places.

3. $10^x = 25$

4. $10^{-x} = 4$

5. $e^{-2x} = 7$

6. $e^{3x} = 12$

7. $2^{1-x} = 3$

8. $3^{2x-1} = 5$

9. $3e^x = 10$

10. $2e^{12x} = 17$

11. $e^{1-4x} = 2$

12. $4(1 + 10^{5x}) = 9$

13. $4 + 3^{5x} = 8$

14. $2^{3x} = 34$

15. $8^{0.4x} = 5$

16. $3^{x/14} = 0.1$

17. $5^{-x/100} = 2$

18. $e^{3-5x} = 16$

19. $e^{2x+1} = 200$

20. $\left(\frac{1}{4}\right)^x = 75$

21. $5^x = 4^{x+1}$

22. $10^{1-x} = 6^x$

23. $2^{3x+1} = 3^{x-2}$

24. $7^{x/2} = 5^{1-x}$

25. $\dfrac{50}{1 + e^{-x}} = 4$

26. $\dfrac{10}{1 + e^{-x}} = 2$

27. $100(1.04)^{2t} = 300$

28. $(1.00625)^{12t} = 2$

29–36 ■ Solve the equation.

29. $e^{2x} - 3e^x + 2 = 0$

30. $e^{2x} - e^x - 6 = 0$

31. $e^{4x} + 4e^{2x} - 21 = 0$

32. $e^x - 12e^{-x} - 1 = 0$

33. $x^2 2^x - 2^x = 0$

34. $x^2 10^x - x10^x = 2(10^x)$

35. $4x^3 e^{-3x} - 3x^4 e^{-3x} = 0$

36. $x^2 e^x + xe^x - e^x = 0$

37–54 ■ Solve the logarithmic equation for x.

37. $\ln x = 10$

38. $\ln(2 + x) = 1$

39. $\log x = -2$

40. $\log(x - 4) = 3$

41. $\log(3x + 5) = 2$

42. $\log_3(2 - x) = 3$

43. $4 - \log(3 - x) = 3$

44. $\log_2(x^2 - x - 2) = 2$

45. $\log_2 3 + \log_2 x = \log_2 5 + \log_2(x - 2)$

46. $2 \log x = \log 2 + \log(3x - 4)$

47. $\log x + \log(x - 1) = \log(4x)$

48. $\log_5 x + \log_5(x + 1) = \log_5 20$

49. $\log_5(x + 1) - \log_5(x - 1) = 2$

50. $\log_3(x + 15) - \log_3(x - 1) = 2$

51. $\log_2 x + \log_2(x - 3) = 2$

52. $\log x + \log(x - 3) = 1$

53. $\log_9(x - 5) + \log_9(x + 3) = 1$

54. $\ln(x - 1) + \ln(x + 2) = 1$

55. For what value of x is the following true?

$$\log(x + 3) = \log x + \log 3$$

56. For what value of x is it true that $(\log x)^3 = 3 \log x$?

57. Solve for x: $\quad 2^{2/\log_5 x} = \frac{1}{16}$

58. Solve for x: $\quad \log_2(\log_3 x) = 4$

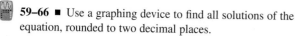

59–66 ■ Use a graphing device to find all solutions of the equation, rounded to two decimal places.

59. $\ln x = 3 - x$ **60.** $\log x = x^2 - 2$

61. $x^3 - x = \log(x + 1)$ **62.** $x = \ln(4 - x^2)$

63. $e^x = -x$ **64.** $2^{-x} = x - 1$

65. $4^{-x} = \sqrt{x}$ **66.** $e^{x^2} - 2 = x^3 - x$

67–70 ■ Solve the inequality.

67. $\log(x - 2) + \log(9 - x) < 1$

68. $3 \le \log_2 x \le 4$

69. $2 < 10^x < 5$ **70.** $x^2 e^x - 2e^x < 0$

71-74 ■ Find the inverse function of f.

71. $f(x) = 2^{2x}$ **72.** $f(x) = 3^{x+1}$

73. $f(x) = \log_2(x - 1)$ **74.** $f(x) = \log 3x$

APPLICATIONS

75. Compound Interest   A man invests $5000 in an account that pays 8.5% interest per year, compounded quarterly.
(a) Find the amount after 3 years.
(b) How long will it take for the investment to double?

76. Compound Interest   A woman invests $6500 in an account that pays 6% interest per year, compounded continuously.
(a) What is the amount after 2 years?
(b) How long will it take for the amount to be $8000?

77. Compound Interest   Find the time required for an investment of $5000 to grow to $8000 at an interest rate of 7.5% per year, compounded quarterly.

78. Compound Interest   Nancy wants to invest $4000 in saving certificates that bear an interest rate of 9.75% per year, compounded semiannually. How long a time period should she choose to save an amount of $5000?

79. Doubling an Investment   How long will it take for an investment of $1000 to double in value if the interest rate is 8.5% per year, compounded continuously?

80. Interest Rate   A sum of $1000 was invested for 4 years, and the interest was compounded semiannually. If this sum amounted to $1435.77 in the given time, what was the interest rate?

81. Radioactive Decay   A 15-g sample of radioactive iodine decays in such a way that the mass remaining after t days is given by $m(t) = 15e^{-0.087t}$, where $m(t)$ is measured in grams. After how many days is there only 5 g remaining?

82. Sky Diving   The velocity of a sky diver t seconds after jumping is given by $v(t) = 80(1 - e^{-0.2t})$. After how many seconds is the velocity 70 ft/s?

83. Fish Population   A small lake is stocked with a certain species of fish. The fish population is modeled by the function

$$P = \frac{10}{1 + 4e^{-0.8t}}$$

where P is the number of fish in thousands and t is measured in years since the lake was stocked.
(a) Find the fish population after 3 years.
(b) After how many years will the fish population reach 5000 fish?

84. Transparency of a Lake   Environmental scientists measure the intensity of light at various depths in a lake to find the "transparency" of the water. Certain levels of transparency are required for the biodiversity of the submerged macrophyte population. In a certain lake the intensity of light at depth x is given by

$$I = 10e^{-0.008x}$$

where I is measured in lumens and x in feet.
(a) Find the intensity I at a depth of 30 ft.
(b) At what depth has the light intensity dropped to $I = 5$?

85. Atmospheric Pressure   Atmospheric pressure P (in kilopascals, kPa) at altitude h (in kilometers, km) is governed by the formula

$$\ln\left(\frac{P}{P_0}\right) = -\frac{h}{k}$$

where $k = 7$ and $P_0 = 100$ kPa are constants.
(a) Solve the equation for P.
(b) Use part (a) to find the pressure P at an altitude of 4 km.

86. Cooling an Engine   Suppose you're driving your car on a cold winter day (20°F outside) and the engine overheats (at about 220°F). When you park, the engine begins to cool down. The temperature T of the engine t minutes after you park satisfies the equation

$$\ln\left(\frac{T - 20}{200}\right) = -0.11t$$

(a) Solve the equation for T.
(b) Use part (a) to find the temperature of the engine after 20 min ($t = 20$).

87. Electric Circuits   An electric circuit contains a battery that produces a voltage of 60 volts (V), a resistor with a resistance of 13 ohms (Ω), and an inductor with an inductance of 5 henrys (H), as shown in the figure. Using calculus, it can be shown that the current $I = I(t)$ (in amperes, A) t seconds after the switch is closed is $I = \frac{60}{13}(1 - e^{-13t/5})$.
(a) Use this equation to express the time t as a function of the current I.
(b) After how many seconds is the current 2 A?

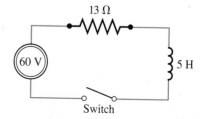

88. Learning Curve A *learning curve* is a graph of a function $P(t)$ that measures the performance of someone learning a skill as a function of the training time t. At first, the rate of learning is rapid. Then, as performance increases and approaches a maximal value M, the rate of learning decreases. It has been found that the function

$$P(t) = M - Ce^{-kt}$$

where k and C are positive constants and $C < M$ is a reasonable model for learning.

(a) Express the learning time t as a function of the performance level P.

(b) For a pole-vaulter in training, the learning curve is given by

$$P(t) = 20 - 14e^{-0.024t}$$

where $P(t)$ is the height he is able to pole-vault after t months. After how many months of training is he able to vault 12 ft?

 (c) Draw a graph of the learning curve in part (b).

DISCOVERY ▪ DISCUSSION ▪ WRITING

89. Estimating a Solution Without actually solving the equation, find two whole numbers between which the solution of $9^x = 20$ must lie. Do the same for $9^x = 100$. Explain how you reached your conclusions.

90. A Surprising Equation Take logarithms to show that the equation

$$x^{1/\log x} = 5$$

has no solution. For what values of k does the equation

$$x^{1/\log x} = k$$

have a solution? What does this tell us about the graph of the function $f(x) = x^{1/\log x}$? Confirm your answer using a graphing device.

91. Disguised Equations Each of these equations can be transformed into an equation of linear or quadratic type by applying the hint. Solve each equation.

(a) $(x - 1)^{\log(x-1)} = 100(x - 1)$ [Take log of each side.]

(b) $\log_2 x + \log_4 x + \log_8 x = 11$ [Change all logs to base 2.]

(c) $4^x - 2^{x+1} = 3$ [Write as a quadratic in 2^x.]

8.6 MODELING WITH EXPONENTIAL AND LOGARITHMIC FUNCTIONS

Exponential Growth (Doubling Time) ▶ Exponential Growth (Relative Growth Rate) ▶ Radioactive Decay ▶ Newton's Law of Cooling ▶ Logarithmic Scales

Many processes that occur in nature, such as population growth, radioactive decay, heat diffusion, and numerous others, can be modeled by using exponential functions. Logarithmic functions are used in models for the loudness of sounds, the intensity of earthquakes, and many other phenomena. In this section we study exponential and logarithmic models.

▼ Exponential Growth (Doubling Time)

Suppose we start with a single bacterium, which divides every hour. After one hour we have 2 bacteria, after two hours we have 2^2 or 4 bacteria, after three hours we have 2^3 or 8 bacteria, and so on (see Figure 1). We see that we can model the bacteria population after t hours by $f(t) = 2^t$.

FIGURE 1 Bacteria population

0   1   2   3   4   5   6

If we start with 10 of these bacteria, then the population is modeled by $f(t) = 10 \cdot 2^t$. A slower-growing strain of bacteria doubles every 3 hours; in this case the population is modeled by $f(t) = 10 \cdot 2^{t/3}$. In general, we have the following.

EXPONENTIAL GROWTH (DOUBLING TIME)

If the intial size of a population is n_0 and the doubling time is a, then the size of the population at time t is

$$n(t) = n_0 2^{t/a}$$

where a and t are measured in the same time units (minutes, hours, days, years, and so on).

EXAMPLE 1 | Bacteria Population

Under ideal conditions a certain bacteria population doubles every three hours. Initially there are 1000 bacteria in a colony.

(a) Find a model for the bacteria population after t hours.

(b) How many bacteria are in the colony after 15 hours?

(c) When will the bacteria count reach 100,000?

SOLUTION

(a) The population at time t is modeled by

$$n(t) = 1000 \cdot 2^{t/3}$$

where t is measured in hours.

(b) After 15 hours the number of bacteria is

$$n(15) = 1000 \cdot 2^{15/3} = 32,000$$

(c) We set $n(t) = 100,000$ in the model that we found in part (a) and solve the resulting exponential equation for t.

$$100,000 = 1000 \cdot 2^{t/3} \qquad n(t) = 1000 \cdot 2^{t/3}$$

$$100 = 2^{t/3} \qquad \text{Divide by 1000}$$

$$\log 100 = \log 2^{t/3} \qquad \text{Take log of each side}$$

$$2 = \frac{t}{3} \log 2 \qquad \text{Properties of log}$$

$$t = \frac{6}{\log 2} \approx 19.93 \qquad \text{Solve for } t$$

The bacteria level reaches 100,000 in about 20 hours.

✎ NOW TRY EXERCISE 1   ■

EXAMPLE 2 | Rabbit Population

A certain breed of rabbit was introduced onto a small island 8 months ago. The current rabbit population on the island is estimated to be 4100 and doubling every 3 months.

(a) What was the initial size of the rabbit population?

(b) Estimate the population one year after the rabbits were introduced to the island.

(c) Sketch a graph of the rabbit population.

SOLUTION

(a) The doubling time is $a = 3$, so the population at time t is

$$n(t) = n_0 2^{t/3} \qquad \text{Model}$$

where n_0 is the initial population. Since the population is 4100 when t is 8 months, we have

$$n(8) = n_0 2^{8/3} \qquad \text{From model}$$

$$4100 = n_0 2^{8/3} \qquad \text{Because } n(8) = 4100$$

$$n_0 = \frac{4100}{2^{8/3}} \qquad \text{Divide by } 2^{8/3} \text{ and switch sides}$$

$$n_0 \approx 645 \qquad \text{Calculate}$$

Thus we estimate that 645 rabbits were introduced onto the island.

(b) From part (a) we know that the initial population is $n_0 = 645$, so we can model the population after t months by

$$n(t) = 645 \cdot 2^{t/3} \qquad \text{Model}$$

After one year $t = 12$, so

$$n(12) = 645 \cdot 2^{12/3} \approx 10{,}320$$

So after one year there would be about 10,000 rabbits.

(c) We first note that the domain is $t \geq 0$. The graph is shown in Figure 2.

✎ NOW TRY EXERCISE **3** ∎

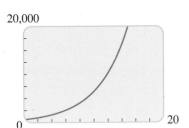

20,000

0   20

FIGURE 2 $n(t) = 645 \cdot 2^{t/3}$

▼ Exponential Growth (Relative Growth Rate)

We have used an exponential function with base 2 to model population growth (in terms of the doubling time). We could also model the same population with an exponential function with base 3 (in terms of the tripling time). In fact, we can find an exponential model with any base. If we use the base e, we get the following model of a population in terms of the **relative growth rate** r: the rate of population growth expressed as a proportion of the population at any time. For instance, if $r = 0.02$, then at any time t the growth rate is 2% of the population at time t.

EXPONENTIAL GROWTH (RELATIVE GROWTH RATE)

A population that experiences **exponential growth** increases according to the model

$$n(t) = n_0 e^{rt}$$

where $n(t) = $ population at time t

  $n_0 = $ initial size of the population

  $r = $ relative rate of growth (expressed as a proportion of the population)

  $t = $ time

Notice that the formula for population growth is the same as that for continuously compounded interest. In fact, the same principle is at work in both cases: The growth of a population (or an investment) per time period is proportional to the size of the population (or the amount of the investment). A population of 1,000,000 will increase more in one year than a population of 1000; in exactly the same way, an investment of $1,000,000 will increase more in one year than an investment of $1000.

In the following examples we assume that the populations grow exponentially.

EXAMPLE 3 | Predicting the Size of a Population

The initial bacterium count in a culture is 500. A biologist later makes a sample count of bacteria in the culture and finds that the relative rate of growth is 40% per hour.

(a) Find a function that models the number of bacteria after t hours.

(b) What is the estimated count after 10 hours?

(c) When will the bacteria count reach 80,000?

(d) Sketch the graph of the function $n(t)$.

SOLUTION

(a) We use the exponential growth model with $n_0 = 500$ and $r = 0.4$ to get

$$n(t) = 500e^{0.4t}$$

where t is measured in hours.

(b) Using the function in part (a), we find that the bacterium count after 10 hours is

$$n(10) = 500e^{0.4(10)} = 500e^4 \approx 27,300$$

(c) We set $n(t) = 80,000$ and solve the resulting exponential equation for t:

$$80,000 = 500 \cdot e^{0.4t} \qquad n(t) = 500 \cdot e^{0.4t}$$

$$160 = e^{0.4t} \qquad \text{Divide by 500}$$

$$\ln 160 = 0.4t \qquad \text{Take ln of each side}$$

$$t = \frac{\ln 160}{0.4} \approx 12.68 \qquad \text{Solve for } t$$

The bacteria level reaches 80,000 in about 12.7 hours.

(d) The graph is shown in Figure 3.

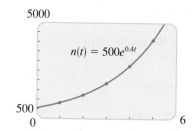

5000

$n(t) = 500e^{0.4t}$

500

0   6

FIGURE 3

NOW TRY EXERCISE **5**

EXAMPLE 4 | Comparing Different Rates of Population Growth

In 2000 the population of the world was 6.1 billion, and the relative rate of growth was 1.4% per year. It is claimed that a rate of 1.0% per year would make a significant difference in the total population in just a few decades. Test this claim by estimating the population of the world in the year 2050 using a relative rate of growth of (a) 1.4% per year and (b) 1.0% per year.

Graph the population functions for the next 100 years for the two relative growth rates in the same viewing rectangle.

SOLUTION

(a) By the exponential growth model we have

$$n(t) = 6.1e^{0.014t}$$

where $n(t)$ is measured in billions and t is measured in years since 2000. Because the year 2050 is 50 years after 2000, we find

$$n(50) = 6.1e^{0.014(50)} = 6.1e^{0.7} \approx 12.3$$

The estimated population in the year 2050 is about 12.3 billion.

The relative growth of world population has been declining over the past few decades—from 2% in 1995 to 1.3% in 2006.

Standing Room Only

The population of the world was about 6.1 billion in 2000 and was increasing at 1.4% per year. Assuming that each person occupies an average of 4 ft² of the surface of the earth, the exponential model for population growth projects that by the year 2801 there will be standing room only! (The total land surface area of the world is about 1.8×10^{15} ft².)

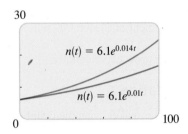

FIGURE 4

(b) We use the function

$$n(t) = 6.1e^{0.010t}$$

and find

$$n(50) = 6.1e^{0.010(50)} = 6.1e^{0.50} \approx 10.1$$

The estimated population in the year 2050 is about 10.1 billion.

The graphs in Figure 4 show that a small change in the relative rate of growth will, over time, make a large difference in population size.

✎ NOW TRY EXERCISE **7**

EXAMPLE 5 | Expressing a Model in Terms of e

A culture starts with 10,000 bacteria, and the number doubles every 40 minutes.

(a) Find a function $n(t) = n_0 2^{t/a}$ that models the number of bacteria after t minutes.

(b) Find a function $n(t) = n_0 e^{rt}$ that models the number of bacteria after t minutes.

(c) Sketch a graph of the number of bacteria at time t.

SOLUTION

(a) The initial population is $n_0 = 10,000$. The doubling time is $a = 40$ min $= 2/3$ h. Since $1/a = 3/2 = 1.5$, the model is

$$n(t) = 10,000 \cdot 2^{1.5t}$$

(b) The initial population is $n_0 = 10,000$. We need to find the relative growth rate r. Since there are 20,000 bacteria when $t = 2/3$ h, we have

$$20,000 = 10,000e^{r(2/3)} \qquad n(t) = 10,000e^{rt}$$

$$2 = e^{r(2/3)} \qquad \text{Divide by 10,000}$$

$$\ln 2 = \ln e^{r(2/3)} \qquad \text{Take ln of each side}$$

$$\ln 2 = r(2/3) \qquad \text{Property of ln}$$

$$r = \frac{3 \ln 2}{2} \approx 1.0397 \qquad \text{Solve for } r$$

Now that we know the relative growth rate r, we can find the model:

$$n(t) = 10,000e^{1.0397t}$$

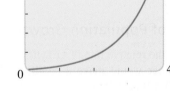

FIGURE 5
Graphs of $y = 10,000 \cdot 2^{1.5t}$
and $y = 10,000e^{1.0397t}$

(c) We can graph the model in part (a) or the one in part (b). The graphs are identical. See Figure 5.

✎ NOW TRY EXERCISE **9**

▼ Radioactive Decay

Radioactive substances decay by spontaneously emitting radiation. The rate of decay is proportional to the mass of the substance. This is analogous to population growth except that the mass *decreases*. Physicists express the rate of decay in terms of **half-life**. For example, the half-life of radium-226 is 1600 years, so a 100-g sample decays to 50 g (or $\frac{1}{2} \times 100$ g) in 1600 years, then to 25 g (or $\frac{1}{2} \times \frac{1}{2} \times 100$ g) in 3200 years, and so on. In

The half-lives of **radioactive elements** vary from very long to very short. Here are some examples.

Element	Half-life
Thorium-232	14.5 billion years
Uranium-235	4.5 billion years
Thorium-230	80,000 years
Plutonium-239	24,360 years
Carbon-14	5,730 years
Radium-226	1,600 years
Cesium-137	30 years
Strontium-90	28 years
Polonium-210	140 days
Thorium-234	25 days
Iodine-135	8 days
Radon-222	3.8 days
Lead-211	3.6 minutes
Krypton-91	10 seconds

general, for a radioactive substance with mass m_0 and half-life h, the amount remaining at time t is modeled by

$$m(t) = m_0 2^{-t/h}$$

where h and t are measured in the same time units (minutes, hours, days, years, and so on).

To express this model in the form $m(t) = m_0 e^{rt}$, we need to find the relative decay rate r. Since h is the half-life, we have

$$m(t) = m_0 e^{-rt} \qquad \text{Model}$$

$$\frac{m_0}{2} = m_0 e^{-rh} \qquad h \text{ is the half-life}$$

$$\frac{1}{2} = e^{-rh} \qquad \text{Divide by } m_0$$

$$\ln \frac{1}{2} = -rh \qquad \text{Take ln of each side}$$

$$r = \frac{\ln 2}{h} \qquad \text{Solve for } r$$

This last equation allows us to find the rate r from the half-life h.

RADIOACTIVE DECAY MODEL

If m_0 is the initial mass of a radioactive substance with half-life h, then the mass remaining at time t is modeled by the function

$$m(t) = m_0 e^{-rt}$$

where $r = \dfrac{\ln 2}{h}$.

EXAMPLE 6 | Radioactive Decay

Polonium-210 (^{210}Po) has a half-life of 140 days. Suppose a sample of this substance has a mass of 300 mg.

(a) Find a function $m(t) = m_0 2^{-t/h}$ that models the mass remaining after t days.

(b) Find a function $m(t) = m_0 e^{-rt}$ that models the mass remaining after t days.

(c) Find the mass remaining after one year.

(d) How long will it take for the sample to decay to a mass of 200 mg?

(e) Draw a graph of the sample mass as a function of time.

SOLUTION

(a) We have $m_0 = 300$ and $h = 140$, so the amount remaining after t days is

$$m(t) = 300 \cdot 2^{-t/140}$$

(b) We have $m_0 = 300$ and $r = \ln 2/140 \approx -0.00495$, so the amount remaining after t days is

$$m(t) = 300 \cdot e^{-0.00495t}$$

In parts (c) and (d) we can also use the model found in part (a). Check that the result is the same using either model.

(c) We use the function we found in part (a) with $t = 365$ (one year).

$$m(365) = 300 e^{-0.00495(365)} \approx 49.256$$

Thus, approximately 49 mg of ^{210}Po remains after one year.

Radioactive Waste

Harmful radioactive isotopes are produced whenever a nuclear reaction occurs, whether as the result of an atomic bomb test, a nuclear accident such as the one at Chernobyl in 1986, or the uneventful production of electricity at a nuclear power plant.

One radioactive material that is produced in atomic bombs is the isotope strontium-90 (^{90}Sr), with a half-life of 28 years. This is deposited like calcium in human bone tissue, where it can cause leukemia and other cancers. However, in the decades since atmospheric testing of nuclear weapons was halted, ^{90}Sr levels in the environment have fallen to a level that no longer poses a threat to health.

Nuclear power plants produce radioactive plutonium-239 (^{239}Pu), which has a half-life of 24,360 years. Because of its long half-life, ^{239}Pu could pose a threat to the environment for thousands of years. So great care must be taken to dispose of it properly. The difficulty of ensuring the safety of the disposed radioactive waste is one reason that nuclear power plants remain controversial.

(d) We use the function that we found in part (a) with $m(t) = 200$ and solve the resulting exponential equation for t.

$$300e^{-0.00495t} = 200 \qquad \text{\small $m(t) = m_0 e^{-rt}$}$$

$$e^{-0.00495t} = \tfrac{2}{3} \qquad \text{\small Divided by 300}$$

$$\ln e^{-0.00495t} = \ln \tfrac{2}{3} \qquad \text{\small Take ln of each side}$$

$$-0.00495t = \ln \tfrac{2}{3} \qquad \text{\small Property of ln}$$

$$t = -\frac{\ln \tfrac{2}{3}}{0.00495} \qquad \text{\small Solve for } t$$

$$t \approx 81.9 \qquad \text{\small Calculator}$$

The time required for the sample to decay to 200 mg is about 82 days.

(e) We can graph the model in part (a) or the one in part (b). The graphs are identical. See Figure 6.

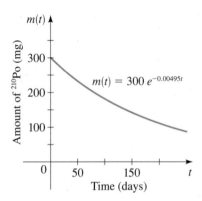

FIGURE 6

✎ NOW TRY EXERCISE **17**

▼ Newton's Law of Cooling

Newton's Law of Cooling states that the rate at which an object cools is proportional to the temperature difference between the object and its surroundings, provided that the temperature difference is not too large. By using calculus, the following model can be deduced from this law.

NEWTON'S LAW OF COOLING

If D_0 is the initial temperature difference between an object and its surroundings, and if its surroundings have temperature T_s, then the temperature of the object at time t is modeled by the function

$$T(t) = T_s + D_0 e^{-kt}$$

where k is a positive constant that depends on the type of object.

EXAMPLE 7 | Newton's Law of Cooling

A cup of coffee has a temperature of 200°F and is placed in a room that has a temperature of 70°F. After 10 min the temperature of the coffee is 150°F.

(a) Find a function that models the temperature of the coffee at time t.

(b) Find the temperature of the coffee after 15 min.

(c) When will the coffee have cooled to 100°F?

(d) Illustrate by drawing a graph of the temperature function.

SOLUTION

(a) The temperature of the room is $T_s = 70°F$, and the initial temperature difference is

$$D_0 = 200 - 70 = 130°F$$

So by Newton's Law of Cooling, the temperature after t minutes is modeled by the function

$$T(t) = 70 + 130e^{-kt}$$

We need to find the constant k associated with this cup of coffee. To do this, we use the fact that when $t = 10$, the temperature is $T(10) = 150$. So we have

$$70 + 130e^{-10k} = 150 \qquad T_s + D_0e^{-kt} = T(t)$$

$$130e^{-10k} = 80 \qquad \text{Subtract 70}$$

$$e^{-10k} = \tfrac{8}{13} \qquad \text{Divide by 130}$$

$$-10k = \ln \tfrac{8}{13} \qquad \text{Take ln of each side}$$

$$k = -\tfrac{1}{10} \ln \tfrac{8}{13} \qquad \text{Solve for } k$$

$$k \approx 0.04855 \qquad \text{Calculator}$$

Substituting this value of k into the expression for $T(t)$, we get

$$T(t) = 70 + 130e^{-0.04855t}$$

(b) We use the function that we found in part (a) with $t = 15$.

$$T(15) = 70 + 130e^{-0.04855(15)} \approx 133°F$$

(c) We use the function that we found in part (a) with $T(t) = 100$ and solve the resulting exponential equation for t.

$$70 + 130e^{-0.04855t} = 100 \qquad T_s + D_0e^{-kt} = T(t)$$

$$130e^{-0.04855t} = 30 \qquad \text{Subtract 70}$$

$$e^{-0.04855t} = \tfrac{3}{13} \qquad \text{Divide by 130}$$

$$-0.04855t = \ln \tfrac{3}{13} \qquad \text{Take ln of each side}$$

$$t = \frac{\ln \tfrac{3}{13}}{-0.04855} \qquad \text{Solve for } t$$

$$t \approx 30.2 \qquad \text{Calculator}$$

The coffee will have cooled to 100°F after about half an hour.

(d) The graph of the temperature function is sketched in Figure 7. Notice that the line $t = 70$ is a horizontal asymptote. (Why?)

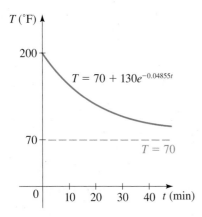

FIGURE 7 Temperature of coffee after t minutes

🖋 NOW TRY EXERCISE **25**

▼ Logarithmic Scales

When a physical quantity varies over a very large range, it is often convenient to take its logarithm in order to have a more manageable set of numbers. We discuss three such situations: the pH scale, which measures acidity; the Richter scale, which measures the intensity of earthquakes; and the decibel scale, which measures the loudness of sounds. Other quantities that are measured on logarithmic scales are light intensity, information capacity, and radiation.

The pH Scale Chemists measured the acidity of a solution by giving its hydrogen ion concentration until Søren Peter Lauritz Sørensen, in 1909, proposed a more convenient measure. He defined

$$pH = -\log[H^+]$$

where $[H^+]$ is the concentration of hydrogen ions measured in moles per liter (M). He did this to avoid very small numbers and negative exponents. For instance,

$$\text{if} \quad [H^+] = 10^{-4} \text{ M}, \quad \text{then} \quad pH = -\log_{10}(10^{-4}) = -(-4) = 4$$

Solutions with a pH of 7 are defined as *neutral*, those with pH < 7 are *acidic*, and those with pH > 7 are *basic*. Notice that when the pH increases by one unit, $[H^+]$ decreases by a factor of 10.

EXAMPLE 8 | pH Scale and Hydrogen Ion Concentration

(a) The hydrogen ion concentration of a sample of human blood was measured to be $[H^+] = 3.16 \times 10^{-8}$ M. Find the pH and classify the blood as acidic or basic.

(b) The most acidic rainfall ever measured occurred in Scotland in 1974; its pH was 2.4. Find the hydrogen ion concentration.

SOLUTION

(a) A calculator gives

$$pH = -\log[H^+] = -\log(3.16 \times 10^{-8}) \approx 7.5$$

Since this is greater than 7, the blood is basic.

(b) To find the hydrogen ion concentration, we need to solve for $[H^+]$ in the logarithmic equation

$$\log[H^+] = -pH$$

So we write it in exponential form.

$$[H^+] = 10^{-pH}$$

In this case pH $= 2.4$, so

$$[H^+] = 10^{-2.4} \approx 4.0 \times 10^{-3} \text{ M}$$

✎ NOW TRY EXERCISE **29**

The Richter Scale In 1935 the American geologist Charles Richter (1900–1984) defined the magnitude M of an earthquake to be

$$M = \log \frac{I}{S}$$

where I is the intensity of the earthquake (measured by the amplitude of a seismograph reading taken 100 km from the epicenter of the earthquake) and S is the intensity of a "standard" earthquake (whose amplitude is 1 micron $= 10^{-4}$ cm). The magnitude of a standard earthquake is

$$M = \log \frac{S}{S} = \log 1 = 0$$

pH for Some Common Substances

Substance	pH
Milk of magnesia	10.5
Seawater	8.0–8.4
Human blood	7.3–7.5
Crackers	7.0–8.5
Hominy	6.9–7.9
Cow's milk	6.4–6.8
Spinach	5.1–5.7
Tomatoes	4.1–4.4
Oranges	3.0–4.0
Apples	2.9–3.3
Limes	1.3–2.0
Battery acid	1.0

Largest Earthquakes

Location	Date	Magnitude
Chile	1960	9.5
Alaska	1964	9.2
Japan	2011	9.1
Sumatra	2004	9.1
Alaska	1957	9.1
Kamchatka	1952	9.0
Chile	2010	8.8
Ecuador	1906	8.8
Alaska	1965	8.7
Sumatra	2005	8.7
Tibet	1950	8.6
Kamchatka	1923	8.5
Indonesia	1938	8.5

Richter studied many earthquakes that occurred between 1900 and 1950. The largest had magnitude 8.9 on the Richter scale, and the smallest had magnitude 0. This corresponds to a ratio of intensities of 800,000,000, so the Richter scale provides more manageable numbers to work with. For instance, an earthquake of magnitude 6 is ten times stronger than an earthquake of magnitude 5.

EXAMPLE 9 | Magnitude of Earthquakes

The 1906 earthquake in San Francisco had an estimated magnitude of 8.3 on the Richter scale. In the same year a powerful earthquake occurred on the Colombia-Ecuador border that was four times as intense. What was the magnitude of the Colombia-Ecuador earthquake on the Richter scale?

SOLUTION   If I is the intensity of the San Francisco earthquake, then from the definition of magnitude we have

$$M = \log \frac{I}{S} = 8.3$$

The intensity of the Colombia-Ecuador earthquake was $4I$, so its magnitude was

$$M = \log \frac{4I}{S} = \log 4 + \log \frac{I}{S} = \log 4 + 8.3 \approx 8.9$$

✎ NOW TRY EXERCISE **35**   ■

EXAMPLE 10 | Intensity of Earthquakes

The 1989 Loma Prieta earthquake that shook San Francisco had a magnitude of 7.1 on the Richter scale. How many times more intense was the 1906 earthquake (see Example 9) than the 1989 event?

SOLUTION   If I_1 and I_2 are the intensities of the 1906 and 1989 earthquakes, then we are required to find I_1/I_2. To relate this to the definition of magnitude, we divide the numerator and denominator by S.

$$\log \frac{I_1}{I_2} = \log \frac{I_1/S}{I_2/S} \qquad \text{Divide numerator and denominator by } S$$

$$= \log \frac{I_1}{S} - \log \frac{I_2}{S} \qquad \text{Law 2 of logarithms}$$

$$= 8.3 - 7.1 = 1.2 \qquad \text{Definition of earthquake magnitude}$$

Therefore,

$$\frac{I_1}{I_2} = 10^{\log(I_1/I_2)} = 10^{1.2} \approx 16$$

The 1906 earthquake was about 16 times as intense as the 1989 earthquake.

✎ NOW TRY EXERCISE **37**   ■

The Decibel Scale   The ear is sensitive to an extremely wide range of sound intensities. We take as a reference intensity $I_0 = 10^{-12}$ W/m² (watts per square meter) at a frequency of 1000 hertz, which measures a sound that is just barely audible (the threshold of hearing). The psychological sensation of loudness varies with the logarithm of the intensity (the Weber-Fechner Law), so the **intensity level** B, measured in decibels (dB), is defined as

$$B = 10 \log \frac{I}{I_0}$$

© Roger Ressmeyer/CORBIS

The intensity level of the barely audible reference sound is

$$B = 10 \log \frac{I_0}{I_0} = 10 \log 1 = 0 \text{ dB}$$

The **intensity levels of sounds** that we can hear vary from very loud to very soft. Here are some examples of the decibel levels of commonly heard sounds.

Source of sound	B (dB)
Jet takeoff	140
Jackhammer	130
Rock concert	120
Subway	100
Heavy traffic	80
Ordinary traffic	70
Normal conversation	50
Whisper	30
Rustling leaves	10–20
Threshold of hearing	0

EXAMPLE 11 | Sound Intensity of a Jet Takeoff

Find the decibel intensity level of a jet engine during takeoff if the intensity was measured at 100 W/m^2.

SOLUTION   From the definition of intensity level we see that

$$B = 10 \log \frac{I}{I_0} = 10 \log \frac{10^2}{10^{-12}} = 10 \log 10^{14} = 140 \text{ dB}$$

Thus, the intensity level is 140 dB.

✎ NOW TRY EXERCISE **41**   ▪

The table in the margin lists decibel intensity levels for some common sounds ranging from the threshold of human hearing to the jet takeoff of Example 11. The threshold of pain is about 120 dB.

8.6 EXERCISES

APPLICATIONS

1–16 ■ These exercises use the population growth model.

✎ **1. Bacteria Culture**   A certain culture of the bacterium *Streptococcus A* initially has 10 bacteria and is observed to double every 1.5 hours.
  (a) Find an exponential model $n(t) = n_0 2^{t/a}$ for the number of bacteria in the culture after t hours.
  (b) Estimate the number of bacteria after 35 hours.
  (c) When will the bacteria count reach 10,000?

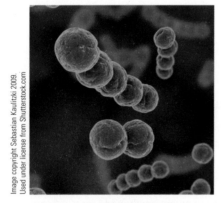

Streptococcus A
(12,000 × magnification)

2. Bacteria Culture   A certain culture of the bacterium *Rhodobacter sphaeroides* initially has 25 bacteria and is observed to double every 5 hours.
  (a) Find an exponential model $n(t) = n_0 2^{t/a}$ for the number of bacteria in the culture after t hours.
  (b) Estimate the number of bacteria after 18 hours.

  (c) After how many hours will the bacteria count reach 1 million?

✎ **3. Squirrel Population**   A grey squirrel population was introduced in a certain county of Great Britain 30 years ago. Biologists observe that the population doubles every 6 years, and now the population is 100,000.
  (a) What was the initial size of the squirrel population?
  (b) Estimate the squirrel population 10 years from now.
  (c) Sketch a graph of the squirrel population.

4. Bird Population   A certain species of bird was introduced in a certain county 25 years ago. Biologists observe that the population doubles every 10 years, and now the population is 13,000.
  (a) What was the initial size of the bird population?
  (b) Estimate the bird population 5 years from now.
  (c) Sketch a graph of the bird population.

✎ **5. Fox Population**   The fox population in a certain region has a relative growth rate of 8% per year. It is estimated that the population in 2005 was 18,000.
  (a) Find a function $n(t) = n_0 e^{rt}$ that models the population t years after 2005.
  (b) Use the function from part (a) to estimate the fox population in the year 2013.
  (c) Sketch a graph of the fox population function for the years 2005–2013.

6. Fish Population   The population of a certain species of fish has a relative growth rate of 1.2% per year. It is estimated that the population in 2000 was 12 million.
  (a) Find an exponential model $n(t) = n_0 e^{rt}$ for the population t years after 2000.
  (b) Estimate the fish population in the year 2005.
  (c) Sketch a graph of the fish population.

7. **Population of a Country** The population of a country has a relative growth rate of 3% per year. The government is trying to reduce the growth rate to 2%. The population in 1995 was approximately 110 million. Find the projected population for the year 2020 for the following conditions.
   (a) The relative growth rate remains at 3% per year.
   (b) The relative growth rate is reduced to 2% per year.

8. **Bacteria Culture** It is observed that a certain bacteria culture has a relative growth rate of 12% per hour, but in the presence of an antibiotic the relative growth rate is reduced to 5% per hour. The initial number of bacteria in the culture is 22. Find the projected population after 24 hours for the following conditions.
   (a) No antibiotic is present, so the relative growth rate is 12%.
   (b) An antibiotic is present in the culture, so the relative growth rate is reduced to 5%.

9. **Population of a City** The population of a certain city was 112,000 in 2006, and the observed doubling time for the population is 18 years.
   (a) Find an exponential model $n(t) = n_0 2^{t/a}$ for the population t years after 2006.
   (b) Find an exponential model $n(t) = n_0 e^{rt}$ for the population t years after 2006.
   (c) Sketch a graph of the population at time t.
   (d) Estimate when the population will reach 500,000.

10. **Bat Population** The bat population in a certain Midwestern county was 350,000 in 2009, and the observed doubling time for the population is 25 years.
    (a) Find an exponential model $n(t) = n_0 2^{t/a}$ for the population t years after 2006.
    (b) Find an exponential model $n(t) = n_0 e^{rt}$ for the population t years after 2006.
    (c) Sketch a graph of the population at time t.
    (d) Estimate when the population will reach 2 million.

11. **Deer Population** The graph shows the deer population in a Pennsylvania county between 2003 and 2007. Assume that the population grows exponentially.
    (a) What was the deer population in 2003?
    (b) Find a function that models the deer population t years after 2003.
    (c) What is the projected deer population in 2011?
    (d) In what year will the deer population reach 100,000?

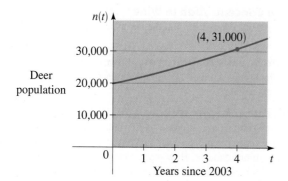

12. **Frog Population** Some bullfrogs were introduced into a small pond. The graph shows the bullfrog population for the next few years. Assume that the population grows exponentially.
    (a) What was the initial bullfrog population?

(b) Find a function that models the bullfrog population t years since the bullfrogs were put into the pond.
(c) What is the projected bullfrog population after 15 years?
(d) Estimate how long it takes the population to reach 75,000.

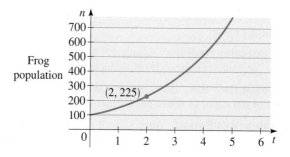

13. **Bacteria Culture** A culture starts with 8600 bacteria. After one hour the count is 10,000.
    (a) Find a function that models the number of bacteria $n(t)$ after t hours.
    (b) Find the number of bacteria after 2 hours.
    (c) After how many hours will the number of bacteria double?

14. **Bacteria Culture** The count in a culture of bacteria was 400 after 2 hours and 25,600 after 6 hours.
    (a) What is the relative rate of growth of the bacteria population? Express your answer as a percentage.
    (b) What was the initial size of the culture?
    (c) Find a function that models the number of bacteria $n(t)$ after t hours.
    (d) Find the number of bacteria after 4.5 hours.
    (e) When will the number of bacteria be 50,000?

15. **Population of California** The population of California was 29.76 million in 1990 and 33.87 million in 2000. Assume that the population grows exponentially.
    (a) Find a function that models the population t years after 1990.
    (b) Find the time required for the population to double.
    (c) Use the function from part (a) to predict the population of California in the year 2010. Look up California's actual population in 2010, and compare.

16. **World Population** The population of the world was 5.7 billion in 1995, and the observed relative growth rate was 2% per year.
    (a) By what year will the population have doubled?
    (b) By what year will the population have tripled?

17–24 ■ These exercises use the radioactive decay model.

17. **Radioactive Radium** The half-life of radium-226 is 1600 years. Suppose we have a 22-mg sample.
    (a) Find a function $m(t) = m_0 2^{-t/h}$ that models the mass remaining after t years.
    (b) Find a function $m(t) = m_0 e^{-rt}$ that models the mass remaining after t years.
    (c) How much of the sample will remain after 4000 years?
    (d) After how long will only 18 mg of the sample remain?

18. **Radioactive Cesium** The half-life of cesium-137 is 30 years. Suppose we have a 10-g sample.
    (a) Find a function $m(t) = m_0 2^{-t/h}$ that models the mass remaining after t years.

(b) Find a function $m(t) = m_0 e^{-rt}$ that models the mass remaining after t years.

(c) How much of the sample will remain after 80 years?

(d) After how long will only 2 g of the sample remain?

19. Radioactive Strontium The half-life of strontium-90 is 28 years. How long will it take a 50-mg sample to decay to a mass of 32 mg?

20. Radioactive Radium Radium-221 has a half-life of 30 s. How long will it take for 95% of a sample to decay?

21. Finding Half-life If 250 mg of a radioactive element decays to 200 mg in 48 hours, find the half-life of the element.

22. Radioactive Radon After 3 days a sample of radon-222 has decayed to 58% of its original amount.

(a) What is the half-life of radon-222?

(b) How long will it take the sample to decay to 20% of its original amount?

23. Carbon-14 Dating A wooden artifact from an ancient tomb contains 65% of the carbon-14 that is present in living trees. How long ago was the artifact made? (The half-life of carbon-14 is 5730 years.)

24. Carbon-14 Dating The burial cloth of an Egyptian mummy is estimated to contain 59% of the carbon-14 it contained originally. How long ago was the mummy buried? (The half-life of carbon-14 is 5730 years.)

25–28 ■ These exercises use Newton's Law of Cooling.

25. Cooling Soup A hot bowl of soup is served at a dinner party. It starts to cool according to Newton's Law of Cooling, so its temperature at time t is given by

$$T(t) = 65 + 145e^{-0.05t}$$

where t is measured in minutes and T is measured in °F.

(a) What is the initial temperature of the soup?

(b) What is the temperature after 10 min?

(c) After how long will the temperature be 100°F?

26. Time of Death Newton's Law of Cooling is used in homicide investigations to determine the time of death. The normal body temperature is 98.6°F. Immediately following death, the body begins to cool. It has been determined experimentally that the constant in Newton's Law of Cooling is approximately $k = 0.1947$, assuming that time is measured in hours. Suppose that the temperature of the surroundings is 60°F.

(a) Find a function $T(t)$ that models the temperature t hours after death.

(b) If the temperature of the body is now 72°F, how long ago was the time of death?

27. Cooling Turkey A roasted turkey is taken from an oven when its temperature has reached 185°F and is placed on a table in a room where the temperature is 75°F.

(a) If the temperature of the turkey is 150°F after half an hour, what is its temperature after 45 min?

(b) When will the turkey cool to 100°F?

 28. Boiling Water A kettle full of water is brought to a boil in a room with temperature 20°C. After 15 min the temperature of the water has decreased from 100°C to 75°C. Find the temperature after another 10 min. Illustrate by graphing the temperature function.

29–43 ■ These exercises deal with logarithmic scales.

29. Finding pH The hydrogen ion concentration of a sample of each substance is given. Calculate the pH of the substance.

(a) Lemon juice: $[H^+] = 5.0 \times 10^{-3}$ M

(b) Tomato juice: $[H^+] = 3.2 \times 10^{-4}$ M

(c) Seawater: $[H^+] = 5.0 \times 10^{-9}$ M

30. Finding pH An unknown substance has a hydrogen ion concentration of $[H^+] = 3.1 \times 10^{-8}$ M. Find the pH and classify the substance as acidic or basic.

31. Ion Concentration The pH reading of a sample of each substance is given. Calculate the hydrogen ion concentration of the substance.

(a) Vinegar: pH = 3.0

(b) Milk: pH = 6.5

32. Ion Concentration The pH reading of a glass of liquid is given. Find the hydrogen ion concentration of the liquid.

(a) Beer: pH = 4.6

(b) Water: pH = 7.3

33. Finding pH The hydrogen ion concentrations in cheeses range from 4.0×10^{-7} M to 1.6×10^{-5} M. Find the corresponding range of pH readings.

34. Ion Concentration in Wine The pH readings for wines vary from 2.8 to 3.8. Find the corresponding range of hydrogen ion concentrations.

35. Earthquake Magnitudes If one earthquake is 20 times as intense as another, how much larger is its magnitude on the Richter scale?

36. Earthquake Magnitudes The 1906 earthquake in San Francisco had a magnitude of 8.3 on the Richter scale. At the same time in Japan an earthquake with magnitude 4.9 caused only minor damage. How many times more intense was the San Francisco earthquake than the Japanese earthquake?

37. Earthquake Magnitudes The Alaska earthquake of 1964 had a magnitude of 8.6 on the Richter scale. How many times more intense was this than the 1906 San Francisco earthquake? (See Exercise 36.)

38. **Earthquake Magnitudes** The Northridge, California, earthquake of 1994 had a magnitude of 6.8 on the Richter scale. A year later, a 7.2-magnitude earthquake struck Kobe, Japan. How many times more intense was the Kobe earthquake than the Northridge earthquake?

39. **Earthquake Magnitudes** The 1985 Mexico City earthquake had a magnitude of 8.1 on the Richter scale. The 1976 earthquake in Tangshan, China, was 1.26 times as intense. What was the magnitude of the Tangshan earthquake?

40. **Subway Noise** The intensity of the sound of a subway train was measured at 98 dB. Find the intensity in W/m^2.

41. **Traffic Noise** The intensity of the sound of traffic at a busy intersection was measured at 2.0×10^{-5} W/m^2. Find the intensity level in decibels.

42. **Comparing Decibel Levels** The noise from a power mower was measured at 106 dB. The noise level at a rock concert was measured at 120 dB. Find the ratio of the intensity of the rock music to that of the power mower.

43. **Inverse Square Law for Sound** A law of physics states that the intensity of sound is inversely proportional to the square of the distance d from the source: $I = k/d^2$.
(a) Use this model and the equation

$$B = 10 \log \frac{I}{I_0}$$

(described in this section) to show that the decibel levels B_1 and B_2 at distances d_1 and d_2 from a sound source are related by the equation

$$B_2 = B_1 + 20 \log \frac{d_1}{d_2}$$

(b) The intensity level at a rock concert is 120 dB at a distance 2 m from the speakers. Find the intensity level at a distance of 10 m.

8.7 DAMPED HARMONIC MOTION

| Damped Harmonic Motion

In Section 2.6 we learned about modeling harmonic motion using sine and cosine functions. For instance, we saw that a mass suspended on a spring from the ceiling and then set into motion will move up and down, with its vertical displacement given by a function of the form $y = k \cos \omega t$. Of course, we know that in real life this motion won't continue forever. Air resistance and friction in the spring will cause the amplitude of the motion to decline and eventually become imperceptible. Motion of this kind is called *damped harmonic motion*.

▼ Damped Harmonic Motion

Using the laws of physics and a branch of mathematics called Differential Equations, it can be shown that damped harmonic motion can be described as in the following box.

> **DAMPED HARMONIC MOTION**
>
> If the equation describing the displacement y of an object at time t is
>
> $$y = ke^{-ct} \sin \omega t \quad \text{or} \quad y = ke^{-ct} \cos \omega t \quad (c > 0)$$
>
> then the object is in **damped harmonic motion**. The constant c is the **damping constant**, k is the initial amplitude, and $2\pi/\omega$ is the period.*

Damped harmonic motion is simply harmonic motion for which the amplitude is governed by the function $a(t) = ke^{-ct}$. Figure 1 shows the difference between harmonic motion and damped harmonic motion.

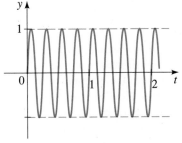

(a) Harmonic motion: $y = \sin 8\pi t$

(b) Damped harmonic motion:
   $y = e^{-t} \sin 8\pi t$

FIGURE 1

*In the case of damped harmonic motion the term *quasi-period* is often used instead of *period* because the motion is not actually periodic—it diminishes with time. However, we will continue to use the term *period* to avoid confusion.

EXAMPLE 1 | Modeling Damped Harmonic Motion

Two mass-spring systems are experiencing damped harmonic motion, both at 0.5 cycles per second and both with an initial maximum displacement of 10 cm. The first has a damping constant of 0.5, and the second has a damping constant of 0.1.

(a) Find functions of the form $g(t) = ke^{-ct} \cos \omega t$ to model the motion in each case.

(b) Graph the two functions you found in part (a). How do they differ?

SOLUTION

Hz is the abbreviation for hertz. One hertz is one cycle per second.

(a) At time $t = 0$ the displacement is 10 cm. Thus, $g(0) = ke^{-c \cdot 0} \cos(\omega \cdot 0) = k$, so $k = 10$. Also, the frequency is $f = 0.5$ Hz, and since $\omega = 2\pi f$ (see page 147), we get $\omega = 2\pi(0.5) = \pi$. Using the given damping constants, we find that the motions of the two springs are given by the functions

$$g_1(t) = 10e^{-0.5t} \cos \pi t \qquad \text{and} \qquad g_2(t) = 10e^{-0.1t} \cos \pi t$$

(b) The functions g_1 and g_2 are graphed in Figure 2. From the graphs we see that in the first case (where the damping constant is larger) the motion dies down quickly, whereas in the second case, perceptible motion continues much longer.

FIGURE 2

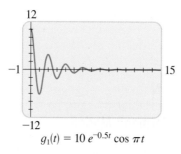

$g_1(t) = 10\, e^{-0.5t} \cos \pi t$

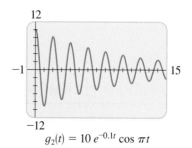

$g_2(t) = 10\, e^{-0.1t} \cos \pi t$

✎ NOW TRY EXERCISE **3**

As the preceding example indicates, the larger the damping constant c, the quicker the oscillation dies down. When a guitar string is plucked and then allowed to vibrate freely, a point on that string undergoes damped harmonic motion. We hear the damping of the motion as the sound produced by the vibration of the string fades. How fast the damping of the string occurs (as measured by the size of the constant c) is a property of the size of the string and the material it is made of. Another example of damped harmonic motion is the motion that a shock absorber on a car undergoes when the car hits a bump in the road. In this case the shock absorber is engineered to damp the motion as quickly as possible (large c) and to have the frequency as small as possible (small ω). On the other hand, the sound produced by a tuba player playing a note is undamped as long as the player can maintain the loudness of the note. The electromagnetic waves that produce light move in simple harmonic motion that is not damped.

EXAMPLE 2 | A Vibrating Violin String

The G-string on a violin is pulled a distance of 0.5 cm above its rest position, then released and allowed to vibrate. The damping constant c for this string is determined to be 1.4. Suppose that the note produced is a pure G (frequency = 200 Hz). Find an equation that describes the motion of the point at which the string was plucked.

SOLUTION   Let P be the point at which the string was plucked. We will find a function $f(t)$ that gives the distance at time t of the point P from its original rest position. Since the maximum displacement occurs at $t = 0$, we find an equation in the form

$$y = ke^{-ct} \cos \omega t$$

From this equation we see that $f(0) = k$. But we know that the original displacement of the string is 0.5 cm. Thus, $k = 0.5$. Since the frequency of the vibration is 200, we have $\omega = 2\pi f = 2\pi(200) = 400\pi$. Finally, since we know that the damping constant is 1.4, we get

$$f(t) = 0.5e^{-1.4t} \cos 400\pi t$$

✎. NOW TRY EXERCISE **15**

EXAMPLE 3 | Modeling Damped Harmonic Motion from Its Graph

The motion of a mass suspended from the ceiling by a spring is graphed in Figure 3. Find an equation to model this damped harmonic motion.

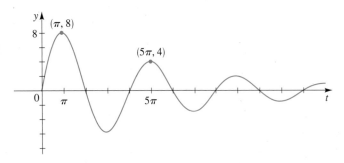

FIGURE 3

SOLUTION This is a damped sine function, so we are looking for a function of the form $y = ke^{-ct} \sin \omega t$. First, we find the value of ω. One complete cycle of the motion occurs between $t = 0$ and $t = 4\pi$, so the period is $p = 4\pi$. Thus $\omega = 2\pi/p = 2\pi/4\pi = 0.5$.

Now we determine the value of c. When $t = \pi$, the displacement is $y = 8$, and when $t = 5\pi$, the displacement is 4. Thus, we have

$$8 = ke^{-c\pi} \sin(0.5\pi) \qquad \text{Set } t = \pi \text{ and } \omega = 0.5$$

$$8 = ke^{-c\pi} \qquad \qquad \sin(0.5\pi) = 1$$

and

$$4 = ke^{-c5\pi} \sin(2.5\pi) \qquad \text{Set } t = 5\pi \text{ and } \omega = 0.5$$

$$4 = ke^{-c5\pi} \qquad \qquad \sin(2.5\pi) = 1$$

Dividing these two equations we get

$$\frac{8}{4} = \frac{ke^{-c\pi}}{ke^{-5c\pi}} \qquad \text{Divide}$$

$$2 = e^{5c\pi - c\pi} = e^{4c\pi} \qquad \text{Simplify}$$

$$\ln 2 = 4c\pi \qquad \text{Take natural logarithms}$$

$$c = \frac{\ln 2}{4\pi} \approx 0.05516 \qquad \text{Solve for } c$$

Finally, we must find k. Using the fact that when $t = \pi$ we have $y = 8$, we get

$$8 = ke^{-0.05516\pi} \sin(0.5\pi) \qquad \text{Set } t = \pi, c = 0.05516, \text{ and } \omega = 0.5$$

$$k = \frac{8}{e^{-0.05516\pi} \sin(0.5\pi)} \approx 9.51 \qquad \text{Solve for } k$$

Thus, the function that models this damped harmonic motion is $y = 9.51e^{-0.05516t} \sin(0.5t)$.

✎. NOW TRY EXERCISE **11**

EXAMPLE 4 | Ripples on a Pond

A stone is dropped in a calm lake, causing waves to form. The up-and-down motion of a point on the surface of the water is modeled by damped harmonic motion. At some time the amplitude of the wave is measured, and 20 s later it is found that the amplitude has dropped to $\frac{1}{10}$ of this value. Find the damping constant c.

SOLUTION The amplitude is governed by the coefficient ke^{-ct} in the equations for damped harmonic motion. Thus, the amplitude at time t is ke^{-ct}, and 20 s later, it is $ke^{-c(t+20)}$. So, because the later value is $\frac{1}{10}$ the earlier value, we have

$$ke^{-c(t+20)} = \tfrac{1}{10}ke^{-ct}$$

We now solve this equation for c. Canceling k and using the Laws of Exponents, we get

$$e^{-ct} \cdot e^{-20c} = \tfrac{1}{10}e^{-ct}$$

$$e^{-20c} = \tfrac{1}{10} \qquad \text{Cancel } e^{-ct}$$

$$e^{20c} = 10 \qquad \text{Take reciprocals}$$

Taking the natural logarithm of each side gives

$$20c = \ln(10)$$

$$c = \tfrac{1}{20}\ln(10) \approx \tfrac{1}{20}(2.30) \approx 0.12$$

Thus, the damping constant is $c \approx 0.12$.

✎ NOW TRY EXERCISE **17** ■

8.7 EXERCISES

CONCEPTS

1–2 ■ An object is in damped harmonic motion with initial amplitude k, period $2\pi/\omega$, and damping constant c. Find an equation that models the displacement y at time t under the given condition.

1. $y = 0$ at time $t = 0$: $y =$ _____.

2. $y = k$ at time $t = 0$: $y =$ _____.

SKILLS

3–6 ■ An initial amplitude k, damping constant c, and frequency f or period p are given. (Recall from Section 2.6 that frequency and period are related by the equation $f = 1/p$.)
(a) Find a function of the form $y = ke^{-ct} \cos \omega t$ that models damped harmonic motion.
(b) Graph the function.

3. $k = 2$, $c = 1.5$, $f = 3$

4. $k = 15$, $c = 0.25$, $f = 0.6$

5. $k = 100$, $c = 0.05$, $p = 4$

6. $k = 0.75$, $c = 3$, $p = 3\pi$

7–10 ■ An initial amplitude k, damping constant c, and frequency f or period p are given. (Recall from Section 2.6 that period and frequency are related by the equation $p = 1/f$.)
(a) Find a function of the form $y = ke^{-ct} \sin \omega t$ that models damped harmonic motion.
(b) Graph the function.

7. $k = 7$, $c = 10$, $p = \pi/6$

8. $k = 1$, $c = 1$, $p = 1$

9. $k = 0.3$, $c = 0.2$, $f = 20$

10. $k = 12$, $c = 0.01$, $f = 8$

11–14 ■ Find a function of the form $y = ke^{-ct} \sin \omega t$ or $y = ke^{-ct} \cos \omega t$ that models the damped harmonic motion whose graph is shown.

11.

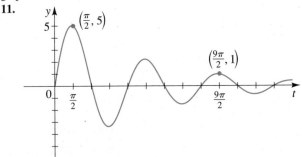

12.

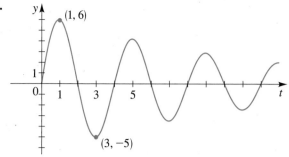

13.

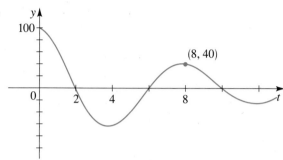

14.

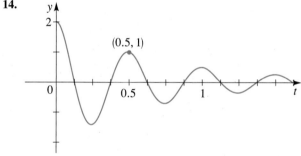

APPLICATIONS

15. Motion of a Building   A strong gust of wind strikes a tall building, causing it to sway back and forth in damped harmonic motion. The frequency of the oscillation is 0.5 cycle per second, and the damping constant is $c = 0.9$. Find an equation

that describes the motion of the building. (Assume that $k = 1$, and take $t = 0$ to be the instant when the gust of wind strikes the building.)

16. Shock Absorber   When a car hits a certain bump on the road, a shock absorber on the car is compressed a distance of 6 in., then released (see the figure). The shock absorber vibrates in damped harmonic motion with a frequency of 2 cycles per second. The damping constant for this particular shock absorber is 2.8.
  (a) Find an equation that describes the displacement of the shock absorber from its rest position as a function of time. Take $t = 0$ to be the instant that the shock absorber is released.
  (b) How long does it take for the amplitude of the vibration to decrease to 0.5 in?

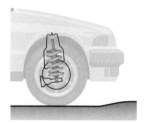 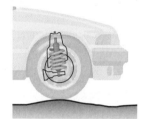

17. Tuning Fork   A tuning fork is struck and oscillates in damped harmonic motion. The amplitude of the motion is measured, and 3 s later it is found that the amplitude has dropped to $\frac{1}{4}$ of this value. Find the damping constant c for this tuning fork.

18. Guitar String   A guitar string is plucked at point P a distance of 3 cm above its rest position. It is then released and vibrates in damped harmonic motion with a frequency of 165 cycles per second. After 2 s, it is observed that the amplitude of the vibration at point P is 0.6 cm.
  (a) Find the damping constant c.
  (b) Find an equation that describes the position of point P above its rest position as a function of time. Take $t = 0$ to be the instant that the string is released.

CHAPTER 8 | REVIEW

■ CONCEPT CHECK

1. (a) Write an equation that defines the exponential function with base a.
  (b) What is the domain of this function?
  (c) What is the range of this function?
  (d) Sketch the general shape of the graph of the exponential function for each case.
      (i) $a > 1$ **(ii)** $0 < a < 1$

2. If x is large, which function grows faster, $y = 2^x$ or $y = x^2$?

3. (a) How is the number e defined?
  (b) What is the natural exponential function?

4. (a) How is the logarithmic function $y = \log_a x$ defined?
  (b) What is the domain of this function?
  (c) What is the range of this function?
  (d) Sketch the general shape of the graph of the function $y = \log_a x$ if $a > 1$.
  (e) What is the natural logarithm?
  (f) What is the common logarithm?

5. State the three Laws of Logarithms.

6. State the Change of Base Formula.

7. (a) How do you solve an exponential equation?
  (b) How do you solve a logarithmic equation?

8. Suppose an amount P is invested at an interest rate r and A is the amount after t years.

    (a) Write an expression for A if the interest is compounded n times per year.

    (b) Write an expression for A if the interest is compounded continuously.

9. The initial size of a population is n_0 and the population grows exponentially.

    (a) Write an expression for the population in terms of the doubling time a.

    (b) Write an expression for the population in terms of the relative growth rate r.

10. **(a)** What is the half-life of a radioactive substance?

    (b) If a radioactive substance has initial mass m_0 and half-life h, write an expression for the mass $m(t)$ remaining at time t.

11. What does Newton's Law of Cooling say?

12. What do the pH scale, the Richter scale, and the decibel scale have in common? What do they measure?

13. What are the equations used to model damped harmonic motion? What is the damping constant?

■ EXERCISES

1–4 ■ Use a calculator to find the indicated values of the exponential function, correct to three decimal places.

1. $f(x) = 5^x$; $f(-1.5), f(\sqrt{2}), f(2.5)$

2. $f(x) = 3 \cdot 2^x$; $f(-2.2), f(\sqrt{7}), f(5.5)$

3. $g(x) = 4 \cdot \left(\frac{2}{3}\right)^{x-2}$; $g(-0.7), g(e), g(\pi)$

4. $g(x) = \frac{7}{4}e^{x+1}$; $g(-2), g(\sqrt{3}), g(3.6)$

5–16 ■ Sketch the graph of the function. State the domain, range, and asymptote.

5. $f(x) = 2^{-x+1}$ **6.** $f(x) = 3^{x-2}$

7. $g(x) = 3 + 2^x$ **8.** $g(x) = 5^{-x} - 5$

9. $f(x) = \log_3(x - 1)$ **10.** $g(x) = \log(-x)$

11. $f(x) = 2 - \log_2 x$ **12.** $f(x) = 3 + \log_5(x + 4)$

13. $F(x) = e^x - 1$ **14.** $G(x) = \frac{1}{2}e^{x-1}$

15. $g(x) = 2 \ln x$ **16.** $g(x) = \ln(x^2)$

17–20 ■ Find the domain of the function.

17. $f(x) = 10^{x^2} + \log(1 - 2x)$

18. $g(x) = \log(2 + x - x^2)$

19. $h(x) = \ln(x^2 - 4)$ **20.** $k(x) = \ln|x|$

21–24 ■ Write the equation in exponential form.

21. $\log_2 1024 = 10$ **22.** $\log_6 37 = x$

23. $\log x = y$ **24.** $\ln c = 17$

25–28 ■ Write the equation in logarithmic form.

25. $2^6 = 64$ **26.** $49^{-1/2} = \frac{1}{7}$

27. $10^x = 74$ **28.** $e^k = m$

29–44 ■ Evaluate the expression without using a calculator.

29. $\log_2 128$ **30.** $\log_8 1$

31. $10^{\log 45}$ **32.** $\log 0.000001$

33. $\ln(e^6)$ **34.** $\log_4 8$

35. $\log_3\left(\frac{1}{27}\right)$ **36.** $2^{\log_2 13}$

37. $\log_5 \sqrt{5}$ **38.** $e^{2\ln 7}$

39. $\log 25 + \log 4$ **40.** $\log_3 \sqrt{243}$

41. $\log_2 16^{23}$ **42.** $\log_5 250 - \log_5 2$

43. $\log_8 6 - \log_8 3 + \log_8 2$ **44.** $\log \log 10^{100}$

45–50 ■ Expand the logarithmic expression.

45. $\log(AB^2C^3)$ **46.** $\log_2(x\sqrt{x^2 + 1})$

47. $\ln\sqrt{\dfrac{x^2 - 1}{x^2 + 1}}$ **48.** $\log\left(\dfrac{4x^3}{y^2(x-1)^5}\right)$

49. $\log_5\left(\dfrac{x^2(1 - 5x)^{3/2}}{\sqrt{x^3 - x}}\right)$ **50.** $\ln\left(\dfrac{\sqrt[3]{x^4 + 12}}{(x+16)\sqrt{x-3}}\right)$

51–56 ■ Combine into a single logarithm.

51. $\log 6 + 4 \log 2$ **52.** $\log x + \log(x^2 y) + 3 \log y$

53. $\frac{3}{2}\log_2(x - y) - 2\log_2(x^2 + y^2)$

54. $\log_5 2 + \log_5(x + 1) - \frac{1}{3}\log_5(3x + 7)$

55. $\log(x - 2) + \log(x + 2) - \frac{1}{2}\log(x^2 + 4)$

56. $\frac{1}{2}[\ln(x - 4) + 5\ln(x^2 + 4x)]$

57–68 ■ Solve the equation. Find the exact solution if possible; otherwise, use a calculator to approximate to two decimals.

57. $3^{2x-7} = 27$ **58.** $5^{4-x} = \frac{1}{125}$

59. $2^{3x-5} = 7$ **60.** $10^{6-3x} = 18$

61. $4^{1-x} = 3^{2x+5}$ **62.** $e^{3x/4} = 10$

63. $x^2 e^{2x} + 2xe^{2x} = 8e^{2x}$ **64.** $3^{2x} - 3^x - 6 = 0$

65. $\log_2(1 - x) = 4$

66. $\log x + \log(x + 1) = \log 12$

67. $\log_8(x + 5) - \log_8(x - 2) = 1$

68. $\ln(2x - 3) + 1 = 0$

69–72 ■ Use a calculator to find the solution of the equation, rounded to six decimal places.

69. $5^{-2x/3} = 0.63$ **70.** $2^{3x-5} = 7$

71. $5^{2x+1} = 3^{4x-1}$ **72.** $e^{-15k} = 10{,}000$

 73–76 ■ Draw a graph of the function and use it to determine the asymptotes and the local maximum and minimum values.

73. $y = e^{x/(x+2)}$ **74.** $y = 10^x - 5^x$

75. $y = \log(x^3 - x)$ **76.** $y = 2x^2 - \ln x$

 77–78 ■ Find the solutions of the equation, rounded to two decimal places.

77. $3 \log x = 6 - 2x$ **78.** $4 - x^2 = e^{-2x}$

 79–80 ■ Solve the inequality graphically.

79. $\ln x > x - 2$ **80.** $e^x < 4x^2$

 81. Use a graph of $f(x) = e^x - 3e^{-x} - 4x$ to find, approximately, the intervals on which f is increasing and on which f is decreasing.

82. Find an equation of the line shown in the figure.

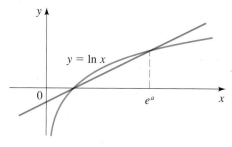

83–86 ■ Use the Change of Base Formula to evaluate the logarithm, rounded to six decimal places.

83. $\log_4 15$ **84.** $\log_7(\tfrac{3}{4})$

85. $\log_9 0.28$ **86.** $\log_{100} 250$

87. Which is larger, $\log_4 258$ or $\log_5 620$?

88. Find the inverse of the function $f(x) = 2^{3^x}$, and state its domain and range.

89. If $12{,}000 is invested at an interest rate of 10% per year, find the amount of the investment at the end of 3 years for each compounding method.
 (a) Semiannually   **(b)** Monthly
 (c) Daily   **(d)** Continuously

90. A sum of $5000 is invested at an interest rate of $8\frac{1}{2}$% per year, compounded semiannually.
 (a) Find the amount of the investment after $1\frac{1}{2}$ years.
 (b) After what period of time will the investment amount to $7000?
 (c) If interest were compounded continuously instead of semiannually, how long would it take for the amount to grow to $7000?

91. A money market account pays 5.2% annual interest, compounded daily. If $100{,}000 is invested in this account, how long will it take for the account to accumulate $10{,}000 in interest?

92. A retirement savings plan pays 4.5% interest, compounded continuously. How long will it take for an investment in this plan to double?

93–94 ■ Determine the annual percentage yield (APY) for the given nominal annual interest rate and compounding frequency.

93. 4.25%; daily   **94.** 3.2%; monthly

95. The stray-cat population in a small town grows exponentially. In 1999 the town had 30 stray cats, and the relative growth rate was 15% per year.
 (a) Find a function that models the stray-cat population $n(t)$ after t years.
 (b) Find the projected population after 4 years.
 (c) Find the number of years required for the stray-cat population to reach 500.

96. A culture contains 10,000 bacteria initially. After an hour the bacteria count is 25,000.
 (a) Find the doubling period.
 (b) Find the number of bacteria after 3 hours.

97. Uranium-234 has a half-life of 2.7×10^5 years.
 (a) Find the amount remaining from a 10-mg sample after a thousand years.
 (b) How long will it take this sample to decompose until its mass is 7 mg?

98. A sample of bismuth-210 decayed to 33% of its original mass after 8 days.
 (a) Find the half-life of this element.
 (b) Find the mass remaining after 12 days.

99. The half-life of radium-226 is 1590 years.
 (a) If a sample has a mass of 150 mg, find a function that models the mass that remains after t years.
 (b) Find the mass that will remain after 1000 years.
 (c) After how many years will only 50 mg remain?

100. The half-life of palladium-100 is 4 days. After 20 days a sample has been reduced to a mass of 0.375 g.
 (a) What was the initial mass of the sample?
 (b) Find a function that models the mass remaining after t days.
 (c) What is the mass after 3 days?
 (d) After how many days will only 0.15 g remain?

101. The graph shows the population of a rare species of bird, where t represents years since 1999 and $n(t)$ is measured in thousands.
 (a) Find a function that models the bird population at time t in the form $n(t) = n_0 e^{rt}$.
 (b) What is the bird population expected to be in the year 2010?

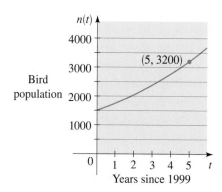

102. A car engine runs at a temperature of 190°F. When the engine is turned off, it cools according to Newton's Law of Cooling with constant $k = 0.0341$, where the time is measured in minutes. Find the time needed for the engine to cool to 90°F if the surrounding temperature is 60°F.

103. The hydrogen ion concentration of fresh egg whites was measured as

$$[H^+] = 1.3 \times 10^{-8} \, M$$

Find the pH, and classify the substance as acidic or basic.

104. The pH of lime juice is 1.9. Find the hydrogen ion concentration.

105. If one earthquake has magnitude 6.5 on the Richter scale, what is the magnitude of another quake that is 35 times as intense?

106. The drilling of a jackhammer was measured at 132 dB. The sound of whispering was measured at 28 dB. Find the ratio of the intensity of the drilling to that of the whispering.

107. The top floor of a building undergoes damped harmonic motion after a sudden brief earthquake. At time $t = 0$ the displacement is at a maximum, 16 cm from the normal position. The damping constant is $c = 0.72$ and the building vibrates at 1.4 cycles per second.
(a) Find a function of the form $y = ke^{-ct} \cos \omega t$ to model the motion.
(b) Graph the function you found in part (a).
(c) What is the displacement at time $t = 10$ s?

1. Sketch the graph of each function, and state its domain, range, and asymptote. Show the x- and y-intercepts on the graph.

    (a) $f(x) = 2^{-x} + 4$ (b) $g(x) = \log_3(x + 3)$

2. (a) Write the equation $6^{2x} = 25$ in logarithmic form.

    (b) Write the equation $\ln A = 3$ in exponential form.

3. Find the exact value of each expression.

    (a) $10^{\log 36}$ (b) $\ln e^3$

    (c) $\log_3 \sqrt{27}$ (d) $\log_2 80 - \log_2 10$

    (e) $\log_8 4$ (f) $\log_6 4 + \log_6 9$

4. Use the Laws of Logarithms to expand the expression:

$$\log \sqrt[3]{\frac{x + 2}{x^4(x^2 + 4)}}$$

5. Combine into a single logarithm: $\ln x - 2\ln(x^2 + 1) + \frac{1}{2}\ln(3 - x^4)$

6. Find the solution of the equation, correct to two decimal places.

    (a) $2^{x-1} = 10$ (b) $5\ln(3 - x) = 4$

    (c) $10^{x+3} = 6^{2x}$ (d) $\log_2(x + 2) + \log_2(x - 1) = 2$

7. The initial size of a culture of bacteria is 1000. After one hour the bacteria count is 8000.

    (a) Find a function that models the population after t hours.

    (b) Find the population after 1.5 hours.

    (c) When will the population reach 15,000?

    (d) Sketch the graph of the population function.

8. Suppose that $12,000 is invested in a savings account paying 5.6% interest per year.

    (a) Write the formula for the amount in the account after t years if interest is compounded monthly.

    (b) Find the amount in the account after 3 years if interest is compounded daily.

    (c) How long will it take for the amount in the account to grow to $20,000 if interest is compounded semiannually?

9. The half-life of krypton-91 (^{91}Kr) is 10 seconds. At time $t = 0$ a heavy canister contains 3 g of this radioactive gas.

    (a) Find a function that models the amount $A(t)$ of ^{91}Kr remaining in the canister after t seconds.

    (b) How much ^{91}Kr remains after one minute?

    (c) When will the amount of ^{91}Kr remaining be reduced to 1 μg (1 microgram, or 10^{-6} g)?

10. An earthquake measuring 6.4 on the Richter scale struck Japan in July 2007, causing extensive damage. Earlier that year, a minor earthquake measuring 3.1 on the Richter scale was felt in parts of Pennsylvania. How many times more intense was the Japanese earthquake than the Pennsylvania earthquake?

11. An object is moving up and down in damped harmonic motion. Its displacement at time $t = 0$ is 16 in; this is its maximum displacement. The damping constant is $c = 0.1$, and the frequency is 12 Hz.

    (a) Find a function that models this motion.

     (b) Graph the function.

Fitting Exponential and Power Curves to Data

In a previous *Focus on Modeling* (page 159) we learned that the shape of a scatter plot helps us to choose the type of curve to use in modeling data. The first plot in Figure 1 fairly screams for a line to be fitted through it, and the second one points to a cubic polynomial. For the third plot it is tempting to fit a second-degree polynomial. But what if an exponential curve fits better? How do we decide this? In this section we learn how to fit exponential and power curves to data and how to decide which type of curve fits the data better. We also learn that for scatter plots like those in the last two plots in Figure 1, the data can be modeled by logarithmic or logistic functions.

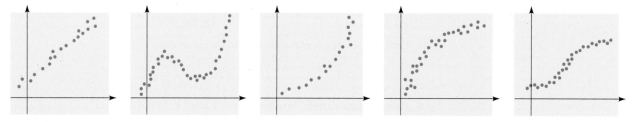

FIGURE 1

▼ Modeling with Exponential Functions

If a scatter plot shows that the data increase rapidly, we might want to model the data using an *exponential model*, that is, a function of the form

$$f(x) = Ce^{kx}$$

where C and k are constants. In the first example we model world population by an exponential model. Recall from Section 8.6 that population tends to increase exponentially.

TABLE 1
World population

Year (t)	World population (P in millions)
1900	1650
1910	1750
1920	1860
1930	2070
1940	2300
1950	2520
1960	3020
1970	3700
1980	4450
1990	5300
2000	6060

EXAMPLE 1 | An Exponential Model for World Population

Table 1 gives the population of the world in the 20th century.

(a) Draw a scatter plot, and note that a linear model is not appropriate.

(b) Find an exponential function that models population growth.

(c) Draw a graph of the function that you found together with the scatter plot. How well does the model fit the data?

(d) Use the model that you found to predict world population in the year 2020.

SOLUTION

(a) The scatter plot is shown in Figure 2. The plotted points do not appear to lie along a straight line, so a linear model is not appropriate.

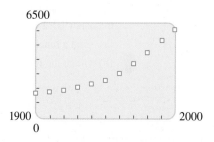

FIGURE 2 Scatter plot of world population

(b) Using a graphing calculator and the `ExpReg` command (see Figure 3(a)), we get the exponential model

$$P(t) = (0.0082543) \cdot (1.0137186)^t$$

This is a model of the form $y = Cb^t$. To convert this to the form $y = Ce^{kt}$, we use the properties of exponentials and logarithms as follows:

$$1.0137186^t = e^{\ln 1.0137186^t} \qquad A = e^{\ln A}$$

$$= e^{t \ln 1.0137186} \qquad \ln A^B = B \ln A$$

$$= e^{0.013625t} \qquad \ln 1.0137186 \approx 0.013625$$

Thus, we can write the model as

$$P(t) = 0.0082543 e^{0.013625t}$$

(c) From the graph in Figure 3(b) we see that the model appears to fit the data fairly well. The period of relatively slow population growth is explained by the depression of the 1930s and the two world wars.

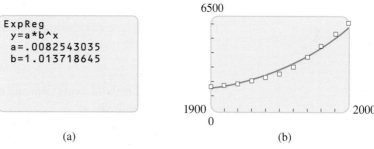

(a)   (b)

FIGURE 3 Exponential model for world population

(d) The model predicts that the world population in 2020 will be

$$P(2020) = 0.0082543 e^{(0.013625)(2020)}$$

$$\approx 7,405,400,000 \qquad \blacksquare$$

▼ Modeling with Power Functions

If the scatter plot of the data we are studying resembles the graph of $y = ax^2$, $y = ax^{1.32}$, or some other power function, then we seek a *power model*, that is, a function of the form

$$f(x) = ax^n$$

where a is a positive constant and n is any real number.

In the next example we seek a power model for some astronomical data. In astronomy, distance in the solar system is often measured in astronomical units. An *astronomical unit* (AU) is the mean distance from the earth to the sun. The *period* of a planet is the time it takes the planet to make a complete revolution around the sun (measured in earth years). In this example we derive the remarkable relationship, first discovered by Johannes Kepler (see page 390), between the mean distance of a planet from the sun and its period.

EXAMPLE 2 | A Power Model for Planetary Periods

Table 2 gives the mean distance d of each planet from the sun in astronomical units and its period T in years.

The population of the world increases exponentially.

TABLE 2
Distances and periods of the planets

Planet	d	T
Mercury	0.387	0.241
Venus	0.723	0.615
Earth	1.000	1.000
Mars	1.523	1.881
Jupiter	5.203	11.861
Saturn	9.541	29.457
Uranus	19.190	84.008
Neptune	30.086	164.784
Pluto	39.507	248.350

(a) Sketch a scatter plot. Is a linear model appropriate?

(b) Find a power function that models the data.

(c) Draw a graph of the function you found and the scatter plot on the same graph. How well does the model fit the data?

(d) Use the model that you found to calculate the period of an asteroid whose mean distance from the sun is 5 AU.

SOLUTION

(a) The scatter plot shown in Figure 4 indicates that the plotted points do not lie along a straight line, so a linear model is not appropriate.

FIGURE 4 Scatter plot of planetary data

(b) Using a graphing calculator and the `PwrReg` command (see Figure 5(a)), we get the power model

$$T = 1.000396d^{1.49966}$$

If we round both the coefficient and the exponent to three significant figures, we can write the model as

$$T = d^{1.5}$$

This is the relationship discovered by Kepler (see page 390). Sir Isaac Newton later used his Law of Gravity to derive this relationship theoretically, thereby providing strong scientific evidence that the Law of Gravity must be true.

(c) The graph is shown in Figure 5(b). The model appears to fit the data very well.

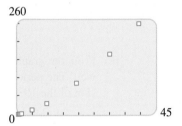

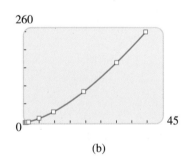

FIGURE 5 Power model for planetary data

(a)   (b)

(d) In this case $d = 5$ AU, so our model gives

$$T = 1.00039 \cdot 5^{1.49966} \approx 11.22$$

The period of the asteroid is about 11.2 years.

▼ Linearizing Data

We have used the shape of a scatter plot to decide which type of model to use: linear, exponential, or power. This works well if the data points lie on a straight line. But it's difficult to distinguish a scatter plot that is exponential from one that requires a power model. So to help decide which model to use, we can *linearize* the data, that is, apply a function that "straightens" the scatter plot. The inverse of the linearizing function is then an appropriate model. We now describe how to linearize data that can be modeled by exponential or power functions.

▶ Linearizing exponential data

If we suspect that the data points (x, y) lie on an exponential curve $y = Ce^{kx}$, then the points

$$(x, \ln y)$$

should lie on a straight line. We can see this from the following calculations:

$$\ln y = \ln Ce^{kx} \qquad \text{Assume that } y = Ce^{kx} \text{ and take } \ln$$

$$= \ln e^{kx} + \ln C \qquad \text{Property of } \ln$$

$$= kx + \ln C \qquad \text{Property of } \ln$$

To see that $\ln y$ is a linear function of x, let $Y = \ln y$ and $A = \ln C$; then

$$Y = kx + A$$

We apply this technique to the world population data (t, P) to obtain the points $(t, \ln P)$ in Table 3. The scatter plot of $(t, \ln P)$ in Figure 6, called a **semi-log plot**, shows that the linearized data lie approximately on a straight line, so an exponential model should be appropriate.

TABLE 3

World population data

t	Population P (in millions)	$\ln P$
1900	1650	21.224
1910	1750	21.283
1920	1860	21.344
1930	2070	21.451
1940	2300	21.556
1950	2520	21.648
1960	3020	21.829
1970	3700	22.032
1980	4450	22.216
1990	5300	22.391
2000	6060	22.525

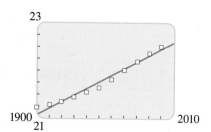

FIGURE 6 Semi-log plot of data in Table 3

▶ Linearizing power data

If we suspect that the data points (x, y) lie on a power curve $y = ax^n$, then the points

$$(\ln x, \ln y)$$

should be on a straight line. We can see this from the following calculations:

$$\ln y = \ln ax^n \qquad \text{Assume that } y = ax^n \text{ and take } \ln$$

$$= \ln a + \ln x^n \qquad \text{Property of } \ln$$

$$= \ln a + n \ln x \qquad \text{Property of } \ln$$

To see that $\ln y$ is a linear function of $\ln x$, let $Y = \ln y$, $X = \ln x$, and $A = \ln a$; then

$$Y = nX + A$$

We apply this technique to the planetary data (d, T) in Table 2 to obtain the points $(\ln d, \ln T)$ in Table 4. The scatter plot of $(\ln d, \ln T)$ in Figure 7, called a **log-log plot**, shows that the data lie on a straight line, so a power model seems appropriate.

TABLE 4

Log-log table

$\ln d$	$\ln T$
−0.94933	−1.4230
−0.32435	−0.48613
0	0
0.42068	0.6318
1.6492	2.4733
2.2556	3.3829
2.9544	4.4309
3.4041	5.1046
3.6765	5.5148

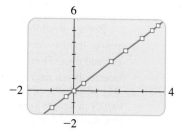

FIGURE 7 Log-log plot of data in Table 4

▼ An Exponential or Power Model?

Suppose that a scatter plot of the data points (x, y) shows a rapid increase. Should we use an exponential function or a power function to model the data? To help us decide, we draw two scatter plots: one for the points $(x, \ln y)$ and the other for the points $(\ln x, \ln y)$. If the first scatter plot appears to lie along a line, then an exponential model is appropriate. If the second plot appears to lie along a line, then a power model is appropriate.

EXAMPLE 3 | An Exponential or Power Model?

Data points (x, y) are shown in Table 5.

(a) Draw a scatter plot of the data.

(b) Draw scatter plots of $(x, \ln y)$ and $(\ln x, \ln y)$.

(c) Is an exponential function or a power function appropriate for modeling this data?

(d) Find an appropriate function to model the data.

SOLUTION

(a) The scatter plot of the data is shown in Figure 8.

TABLE 5

x	y
1	2
2	6
3	14
4	22
5	34
6	46
7	64
8	80
9	102
10	130

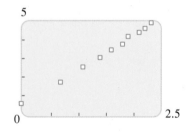

FIGURE 8

(b) We use the values from Table 6 to graph the scatter plots in Figures 9 and 10.

TABLE 6

x	ln x	ln y
1	0	0.7
2	0.7	1.8
3	1.1	2.6
4	1.4	3.1
5	1.6	3.5
6	1.8	3.8
7	1.9	4.2
8	2.1	4.4
9	2.2	4.6
10	2.3	4.9

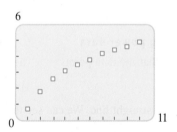

FIGURE 9 Semi-log plot   **FIGURE 10** Log-log plot

(c) The scatter plot of $(x, \ln y)$ in Figure 9 does not appear to be linear, so an exponential model is not appropriate. On the other hand, the scatter plot of $(\ln x, \ln y)$ in Figure 10 is very nearly linear, so a power model is appropriate.

(d) Using the PwrReg command on a graphing calculator, we find that the power function that best fits the data point is

$$y = 1.85x^{1.82}$$

The graph of this function and the original data points are shown in Figure 11. ■

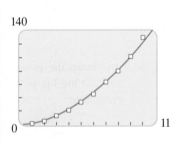

FIGURE 11

Before graphing calculators and statistical software became common, exponential and power models for data were often constructed by first finding a linear model for the linearized data. Then the model for the actual data was found by taking exponentials. For instance, if we find that $\ln y = A \ln x + B$, then by taking exponentials we get the model $y = e^B \cdot e^{A \ln x}$, or $y = Cx^A$ (where $C = e^B$). Special graphing paper called "log paper" or "log-log paper" was used to facilitate this process.

▼ Modeling with Logistic Functions

A logistic growth model is a function of the form

$$f(t) = \frac{c}{1 + ae^{-bt}}$$

where a, b, and c are positive constants. Logistic functions are used to model populations where the growth is constrained by available resources. (See Exercises 25–28 of Section 8.2.)

EXAMPLE 4 | Stocking a Pond with Catfish

TABLE 7

Week	Catfish
0	1000
15	1500
30	3300
45	4400
60	6100
75	6900
90	7100
105	7800
120	7900

Much of the fish that is sold in supermarkets today is raised on commercial fish farms, not caught in the wild. A pond on one such farm is initially stocked with 1000 catfish, and the fish population is then sampled at 15-week intervals to estimate its size. The population data are given in Table 7.

(a) Find an appropriate model for the data.

(b) Make a scatter plot of the data and graph the model that you found in part (a) on the scatter plot.

(c) How does the model predict that the fish population will change with time?

SOLUTION

(a) Since the catfish population is restricted by its habitat (the pond), a logistic model is appropriate. Using the `Logistic` command on a calculator (see Figure 12(a)), we find the following model for the catfish population $P(t)$:

$$P(t) = \frac{7925}{1 + 7.7e^{-0.052t}}$$

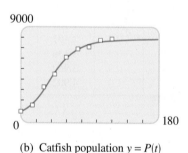

```
Logistic
 y=c/(1+ae^(-bx))
 a=7.69477503
 b=.0523020764
 c=7924.540299
```

FIGURE 12   (a)   (b) Catfish population $y = P(t)$

(b) The scatter plot and the logistic curve are shown in Figure 12(b).

(c) From the graph of P in Figure 12(b) we see that the catfish population increases rapidly until about $t = 80$ weeks. Then growth slows down, and at about $t = 120$ weeks the population levels off and remains more or less constant at slightly over 7900. ■

The behavior that is exhibited by the catfish population in Example 4 is typical of logistic growth. After a rapid growth phase, the population approaches a constant level called the **carrying capacity** of the environment. This occurs because as $t \to \infty$, we have $e^{-bt} \to 0$ (see Section 8.2), and so

$$P(t) = \frac{c}{1 + ae^{-bt}} \quad \longrightarrow \quad \frac{c}{1 + 0} = c$$

Thus, the carrying capacity is c.

PROBLEMS

1. U.S. Population The U.S. Constitution requires a census every 10 years. The census data for 1790–2000 are given in the table.

(a) Make a scatter plot of the data.

(b) Use a calculator to find an exponential model for the data.

(c) Use your model to predict the population at the 2010 census.

(d) Use your model to estimate the population in 1965.

(e) Compare your answers from parts (c) and (d) to the values in the table. Do you think an exponential model is appropriate for these data?

Year	Population (in millions)	Year	Population (in millions)	Year	Population (in millions)
1790	3.9	1870	38.6	1950	151.3
1800	5.3	1880	50.2	1960	179.3
1810	7.2	1890	63.0	1970	203.3
1820	9.6	1900	76.2	1980	226.5
1830	12.9	1910	92.2	1990	248.7
1840	17.1	1920	106.0	2000	281.4
1850	23.2	1930	123.2		
1860	31.4	1940	132.2		

Time (s)	Distance (m)
0.1	0.048
0.2	0.197
0.3	0.441
0.4	0.882
0.5	1.227
0.6	1.765
0.7	2.401
0.8	3.136
0.9	3.969
1.0	4.902

2. A Falling Ball In a physics experiment a lead ball is dropped from a height of 5 m. The students record the distance the ball has fallen every one-tenth of a second. (This can be done by using a camera and a strobe light.)

(a) Make a scatter plot of the data.

(b) Use a calculator to find a power model.

(c) Use your model to predict how far a dropped ball would fall in 3 s.

3. Health-Care Expenditures The U.S. health-care expenditures for 1970–2001 are given in the table below, and a scatter plot of the data is shown in the figure.

(a) Does the scatter plot shown suggest an exponential model?

(b) Make a table of the values $(t, \ln E)$ and a scatter plot. Does the scatter plot appear to be linear?

(c) Find the regression line for the data in part (b).

(d) Use the results of part (c) to find an exponential model for the growth of health-care expenditures.

(e) Use your model to predict the total health-care expenditures in 2009.

Year	Health-care expenditures (in billions of dollars)
1970	74.3
1980	251.1
1985	434.5
1987	506.2
1990	696.6
1992	820.3
1994	937.2
1996	1039.4
1998	1150.0
2000	1310.0
2001	1424.5

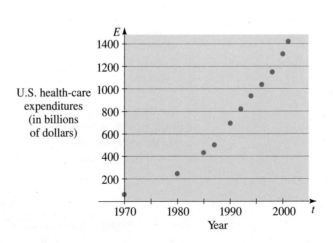

Time (h)	Amount of ^{131}I (g)
0	4.80
8	4.66
16	4.51
24	4.39
32	4.29
40	4.14
48	4.04

4. Half-Life of Radioactive Iodine   A student is trying to determine the half-life of radioactive iodine-131. He measures the amount of iodine-131 in a sample solution every 8 hours. His data are shown in the table in the margin.

(a) Make a scatter plot of the data.

(b) Use a calculator to find an exponential model.

(c) Use your model to find the half-life of iodine-131.

5. The Beer-Lambert Law   As sunlight passes through the waters of lakes and oceans, the light is absorbed, and the deeper it penetrates, the more its intensity diminishes. The light intensity I at depth x is given by the Beer-Lambert Law:

$$I = I_0 e^{-kx}$$

where I_0 is the light intensity at the surface and k is a constant that depends on the murkiness of the water (see page 452). A biologist uses a photometer to investigate light penetration in a northern lake, obtaining the data in the table.

(a) Use a graphing calculator to find an exponential function of the form given by the Beer-Lambert Law to model these data. What is the light intensity I_0 at the surface on this day, and what is the "murkiness" constant k for this lake?   [*Hint:* If your calculator gives you a function of the form $I = ab^x$, convert this to the form you want using the identities $b^x = e^{\ln(b^x)} = e^{x \ln b}$. See Example 1(b).]

(b) Make a scatter plot of the data, and graph the function that you found in part (a) on your scatter plot.

(c) If the light intensity drops below 0.15 lumen (lm), a certain species of algae can't survive because photosynthesis is impossible. Use your model from part (a) to determine the depth below which there is insufficient light to support this algae.

Light intensity decreases exponentially with depth.

Depth (ft)	Light intensity (lm)	Depth (ft)	Light intensity (lm)
5	13.0	25	1.8
10	7.6	30	1.1
15	4.5	35	0.5
20	2.7	40	0.3

6. Experimenting with "Forgetting" Curves   Every one of us is all too familiar with the phenomenon of forgetting. Facts that we clearly understood at the time we first learned them sometimes fade from our memory by the time the final exam rolls around. Psychologists have proposed several ways to model this process. One such model is Ebbinghaus' Law of Forgetting, described on page 443. Other models use exponential or logarithmic functions. To develop her own model, a psychologist performs an experiment on a group of volunteers by asking them to memorize a list of 100 related words. She then tests how many of these words they can recall after various periods of time. The average results for the group are shown in the table.

(a) Use a graphing calculator to find a *power* function of the form $y = at^b$ that models the average number of words y that the volunteers remember after t hours. Then find an *exponential* function of the form $y = ab^t$ to model the data.

(b) Make a scatter plot of the data, and graph both the functions that you found in part (a) on your scatter plot.

(c) Which of the two functions seems to provide the better model?

Time	Words recalled
15 min	64.3
1 h	45.1
8 h	37.3
1 day	32.8
2 days	26.9
3 days	25.6
5 days	22.9

The number of different bat species in a cave is related to the size of the cave by a power function.

7. **Modeling the Species-Area Relation**   The table gives the areas of several caves in central Mexico and the number of bat species that live in each cave.*

   (a) Find a power function that models the data.

   (b) Draw a graph of the function you found in part (a) and a scatter plot of the data on the same graph. Does the model fit the data well?

   (c) The cave called El Sapo near Puebla, Mexico, has a surface area of $A = 205$ m^2. Use the model to estimate the number of bat species you would expect to find in that cave.

Cave	Area (m^2)	Number of species
La Escondida	18	1
El Escorpion	19	1
El Tigre	58	1
Mision Imposible	60	2
San Martin	128	5
El Arenal	187	4
La Ciudad	344	6
Virgen	511	7

8. **Auto Exhaust Emissions**   A study by the U.S. Office of Science and Technology in 1972 estimated the cost of reducing automobile emissions by certain percentages. Find an exponential model that captures the "diminishing returns" trend of these data shown in the table below.

Reduction in emissions (%)	Cost per car ($)
50	45
55	55
60	62
65	70
70	80
75	90
80	100
85	200
90	375
95	600

9. **Exponential or Power Model?**   Data points (x, y) are shown in the table.

   (a) Draw a scatter plot of the data.

   (b) Draw scatter plots of $(x, \ln y)$ and $(\ln x, \ln y)$.

   (c) Which is more appropriate for modeling this data: an exponential function or a power function?

   (d) Find an appropriate function to model the data.

x	y
2	0.08
4	0.12
6	0.18
8	0.25
10	0.36
12	0.52
14	0.73
16	1.06

*A. K. Brunet and R. A. Medallin, "The Species-Area Relationship in Bat Assemblages of Tropical Caves." *Journal of Mammalogy, 82*(4):1114–1122, 2001.

x	y
10	29
20	82
30	151
40	235
50	330
60	430
70	546
80	669
90	797

10. Exponential or Power Model?   Data points (x, y) are shown in the table in the margin.

(a) Draw a scatter plot of the data.

(b) Draw scatter plots of $(x, \ln y)$ and $(\ln x, \ln y)$.

(c) Which is more appropriate for modeling this data: an exponential function or a power function?

(d) Find an appropriate function to model the data.

11. Logistic Population Growth   The table and scatter plot give the population of black flies in a closed laboratory container over an 18-day period.

(a) Use the Logistic command on your calculator to find a logistic model for these data.

(b) Use the model to estimate the time when there were 400 flies in the container.

Time (days)	Number of flies
0	10
2	25
4	66
6	144
8	262
10	374
12	446
16	492
18	498

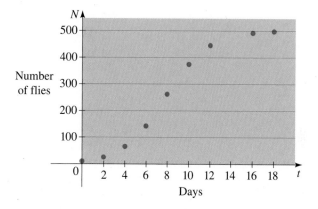

12. Logarithmic Models   A **logarithmic model** is a function of the form

$$y = a + b \ln x$$

Many relationships between variables in the real world can be modeled by this type of function. The table and the scatter plot show the coal production (in metric tons) from a small mine in northern British Columbia.

(a) Use the LnReg command on your calculator to find a logarithmic model for these production figures.

(b) Use the model to predict coal production from this mine in 2010.

Year	Metric tons of coal
1950	882
1960	889
1970	894
1980	899
1990	905
2000	909

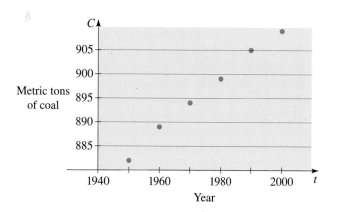

A.1 REAL NUMBERS

| Properties of Real Numbers ▶ The Real Line ▶ Sets and Intervals ▶
Absolute Value and Distance

The different types of real numbers were invented to meet specific needs. For example, natural numbers are needed for counting, negative numbers for describing debt or below-zero temperatures, rational numbers for concepts like "half a gallon of milk," and irrational numbers for measuring certain distances, like the diagonal of a square.

Let's recall the types of numbers that make up the real number system. We start with the **natural numbers**:

$$1, 2, 3, 4, \ldots$$

The **integers** consist of the natural numbers together with their negatives and 0:

$$\ldots, -3, -2, -1, 0, 1, 2, 3, 4, \ldots$$

We construct the **rational numbers** by taking ratios of integers. Thus, any rational number r can be expressed as

$$r = \frac{m}{n} \qquad \text{where } m \text{ and } n \text{ are integers and } n \neq 0$$

Examples are: $\qquad \frac{1}{2} \qquad -\frac{3}{7} \qquad 46 = \frac{46}{1} \qquad 0.17 = \frac{17}{100}$

(Recall that division by 0 is always ruled out, so expressions like $\frac{3}{0}$ and $\frac{0}{0}$ are undefined.) There are also real numbers, such as $\sqrt{2}$, that cannot be expressed as a ratio of integers and are therefore called **irrational numbers**. It can be shown, with varying degrees of difficulty, that these numbers are also irrational:

$$\sqrt{3} \qquad \sqrt{5} \qquad \sqrt[3]{2} \qquad \pi \qquad \frac{3}{\pi^2}$$

The set of all real numbers is usually denoted by the symbol $\mathbb{R}$. When we use the word *number* without qualification, we will mean "real number." Figure 1 is a diagram of the types of real numbers that we work with in this book.

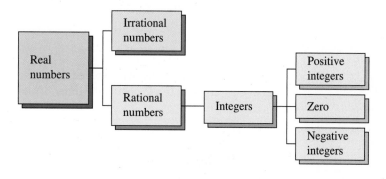

FIGURE 1 The real number system

Every real number has a decimal representation. If the number is rational, then its corresponding decimal is repeating. For example,

$$\frac{1}{2} = 0.5000 \ldots = 0.5\overline{0} \qquad\qquad \frac{2}{3} = 0.66666 \ldots = 0.\overline{6}$$

$$\frac{157}{495} = 0.3171717 \ldots = 0.3\overline{17} \qquad\qquad \frac{9}{7} = 1.285714285714 \ldots = 1.\overline{285714}$$

(The bar indicates that the sequence of digits repeats forever.) If the number is irrational, the decimal representation is nonrepeating:

$$\sqrt{2} = 1.414213562373095 \ldots \qquad\qquad \pi = 3.141592653589793 \ldots$$

A repeating decimal such as

$$x = 3.5474747\ldots$$

is a rational number. To convert it to a ratio of two integers, we write

$$1000x = 3547.47474747\ldots$$
$$10x = 35.47474747\ldots$$
$$990x = 3512.0$$

Thus, $x = \frac{3512}{990}$. (The idea is to multiply x by appropriate powers of 10, and then subtract to eliminate the repeating part.)

If we stop the decimal expansion of any number at a certain place, we get an approximation to the number. For instance, we can write

$$\pi \approx 3.14159265$$

where the symbol $\approx$ is read "is approximately equal to." The more decimal places we retain, the better the approximation we get.

▼ Properties of Real Numbers

In combining real numbers using the familiar operations of addition and multiplication, we use the following properties of real numbers.

PROPERTIES OF REAL NUMBERS

Property	Example
Commutative Properties	
$a + b = b + a$	$7 + 3 = 3 + 7$
$ab = ba$	$3 \cdot 5 = 5 \cdot 3$
Associative Properties	
$(a + b) + c = a + (b + c)$	$(2 + 4) + 7 = 2 + (4 + 7)$
$(ab)c = a(bc)$	$(3 \cdot 7) \cdot 5 = 3 \cdot (7 \cdot 5)$
Distributive Property	
$a(b + c) = ab + ac$	$2 \cdot (3 + 5) = 2 \cdot 3 + 2 \cdot 5$
$(b + c)a = ab + ac$	$(3 + 5) \cdot 2 = 2 \cdot 3 + 2 \cdot 5$

EXAMPLE 1 | Using the Properties of the Real Numbers

Let x, y, z, and w be real numbers.

(a) $(x + y)(2zw) = (2zw)(x + y)$ Commutative Property for multiplication

(b) $(x + y)(z + w) = (x + y)z + (x + y)w$ Distributive Property (with $a = x + y$)

$$= (zx + zy) + (wx + wy)$$ Distributive Property

$$= zx + zy + wx + wy$$ Associative Property of addition

In the last step we removed the parentheses because, according to the Associative Property, the order of addition doesn't matter. ∎

▼ The Real Line

The real numbers can be represented by points on a line, as shown in Figure 2. The positive direction (toward the right) is indicated by an arrow. We choose an arbitrary reference point O, called the **origin**, which corresponds to the real number 0. Given any convenient unit of measurement, each positive number x is represented by the point on the line a distance of x units to the right of the origin, and each negative number $-x$ is represented by the point x units to the left of the origin. Thus, every real number is represented by a point on the line, and every point P on the line corresponds to exactly one real number. The number associated with the point P is called the coordinate of P, and the line is then called a **coordinate line**, or a **real number line**, or simply a **real line**. Often we identify the point with its coordinate and think of a number as being a point on the real line.

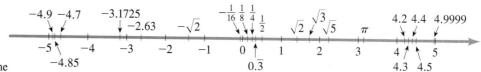

FIGURE 2 The real line

The real numbers are *ordered*. We say that *a* **is less than** *b* and write $a < b$ if $b - a$ is a positive number. Geometrically, this means that *a* lies to the left of *b* on the number line. (Equivalently, we can say that *b* **is greater than** *a* and write $b > a$.) The symbol $a \le b$ (or $b \ge a$) means that either $a < b$ or $a = b$ and is read "*a* is less than or equal to *b*." For instance, the following are true inequalities (see Figure 3):

$$7 < 7.4 < 7.5 \qquad -\pi < -3 \qquad \sqrt{2} < 2 \qquad 2 \le 2$$

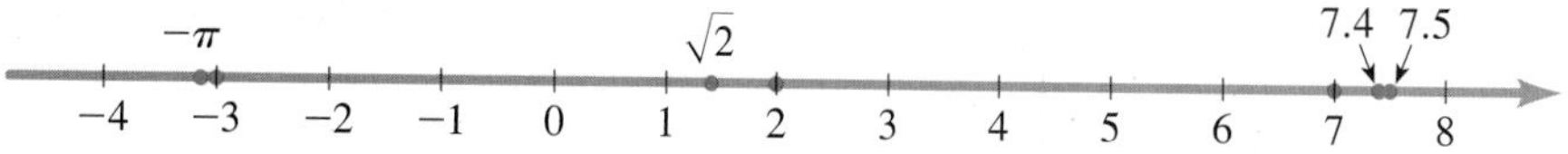

FIGURE 3

▼ Sets and Intervals

In the discussion that follows, we need to use set notation. A **set** is a collection of objects, and these objects are called the **elements** of the set. If *S* is a set, the notation $a \in S$ means that *a* is an element of *S*, and $b \notin S$ means that *b* is not an element of *S*. For example, if *Z* represents the set of integers, then $-3 \in Z$ but $\pi \notin Z$.

Some sets can be described by listing their elements within braces. For instance, the set *A* that consists of all positive integers less than 7 can be written as

$$A = \{1, 2, 3, 4, 5, 6\}$$

We could also write *A* in **set-builder notation** as

$$A = \{x \mid x \text{ is an integer and } 0 < x < 7\}$$

which is read "*A* is the set of all *x* such that *x* is an integer and $0 < x < 7$."

If *S* and *T* are sets, then their **union** $S \cup T$ is the set that consists of all elements that are in *S* or *T* (or in both). The **intersection** of *S* and *T* is the set $S \cap T$ consisting of all elements that are in both *S* and *T*. In other words, $S \cap T$ is the common part of *S* and *T*. The **empty set**, denoted by $\varnothing$, is the set that contains no element.

Certain sets of real numbers, called **intervals**, occur frequently in calculus and correspond geometrically to line segments. For example, if $a < b$, then the **open interval** from *a* to *b* consists of all numbers between *a* and *b* and is denoted by the symbol (a, b). Using set-builder notation, we can write

$$(a, b) = \{x \mid a < x < b\}$$

The **closed interval** from *a* to *b* is the set

$$[a, b] = \{x \mid a \le x \le b\}$$

We also need to consider infinite intervals, such as

$$(a, \infty) = \{x \mid x > a\}$$

The following table lists the nine possible types of intervals. When these intervals are discussed, we will always assume that $a < b$.

Notation	Set description	Graph
(a, b)	$\{x \mid a < x < b\}$	
$[a, b]$	$\{x \mid a \leqslant x \leqslant b\}$	
$[a, b)$	$\{x \mid a \leqslant x < b\}$	
$(a, b]$	$\{x \mid a < x \leqslant b\}$	
(a, ∞)	$\{x \mid a < x\}$	
$[a, \infty)$	$\{x \mid a \leqslant x\}$	
$(-\infty, b)$	$\{x \mid x < b\}$	
$(-\infty, b]$	$\{x \mid x \leqslant b\}$	
$(-\infty, \infty)$	$\mathbb{R}$ (set of all real numbers)	

EXAMPLE 2 | Finding Unions and Intersections of Intervals

Graph each set.

(a) $(1, 3) \cap [2, 7]$ **(b)** $(1, 3) \cup [2, 7]$

SOLUTION

(a) The intersection of two intervals consists of the numbers that are in both intervals. Therefore

$$(1, 3) \cap [2, 7] = \{x \mid 1 < x < 3 \text{ and } 2 \leqslant x \leqslant 7\}$$
$$= \{x \mid 2 \leqslant x < 3\}$$
$$= [2, 3)$$

This set is illustrated in Figure 4.

(b) The union of two intervals consists of the numbers that are in either one interval or the other (or both). Therefore

$$(1, 3) \cup [2, 7] = \{x \mid 1 < x < 3 \text{ or } 2 \leqslant x \leqslant 7\}$$
$$= \{x \mid 1 < x \leqslant 7\}$$
$$= (1, 7]$$

This set is illustrated in Figure 5.

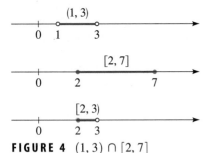

FIGURE 4 $(1, 3) \cap [2, 7]$

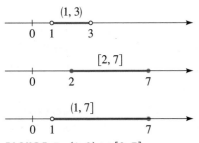

FIGURE 5 $(1, 3) \cup [2, 7]$

▼ Absolute Value and Distance

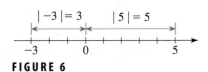

FIGURE 6

The **absolute value** of a number a, denoted by $|a|$, is the distance from a to 0 on the real number line (see Figure 6). Distance is always positive or zero, so we have $|a| \geqslant 0$ for every number a. Remembering that $-a$ is positive when a is negative, we have the following definition.

DEFINITION OF ABSOLUTE VALUE

If a is a real number, then the **absolute value** of a is

$$|a| = \begin{cases} a & \text{if } a \geq 0 \\ -a & \text{if } a < 0 \end{cases}$$

EXAMPLE 3 | Evaluating Absolute Values of Numbers

(a) $|3| = 3$

(b) $|-3| = -(-3) = 3$

(c) $|0| = 0$

(d) $|\sqrt{2} - 1| = \sqrt{2} - 1$ (since $\sqrt{2} > 1 \Rightarrow \sqrt{2} - 1 > 0$)

(e) $|3 - \pi| = -(3 - \pi) = \pi - 3$ (since $\pi > 3 \Rightarrow 3 - \pi < 0$)

When working with absolute values, we use the following properties.

PROPERTIES OF ABSOLUTE VALUE

Property	Example
1. $|a| \geq 0$	$|-3| = 3 \geq 0$
2. $|a| = |-a|$	$|5| = |-5|$
3. $|ab| = |a||b|$	$|-2 \cdot 5| = |-2||5|$
4. $\left|\dfrac{a}{b}\right| = \dfrac{|a|}{|b|}$	$\left|\dfrac{12}{-3}\right| = \dfrac{|12|}{|-3|}$

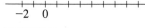

FIGURE 7

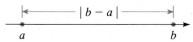

FIGURE 8 Length of a line segment $= |b - a|$

What is the distance on the real line between the numbers -2 and 11? From Figure 7 we see that the distance is 13. We arrive at this by finding either $|11 - (-2)| = 13$ or $|(-2) - 11| = 13$. From this observation we make the following definition (see Figure 8).

DISTANCE BETWEEN POINTS ON THE REAL LINE

If a and b are real numbers, then the **distance** between the points a and b on the real line is

$$d(a, b) = |b - a|$$

From Property 2 of absolute values it follows that

$$|b - a| = |-(a - b)| = |a - b|$$

This confirms that, as we would expect, the distance from a to b is the same as the distance from b to a.

EXAMPLE 4 | Distance between Points on the Real Line

The distance between the numbers -8 and 2 is

FIGURE 9

$$d(a, b) = |-8 - 2| = |-10| = 10$$

We can check this calculation geometrically, as shown in Figure 9.

A.1 EXERCISES

1–6 ■ State the property of real numbers being used.

1. $3x + 4y = 4y + 3x$

2. $c(a + b) = (a + b)c$

3. $(x + 2y) + 3z = x + (2y + 3z)$

4. $2(A + B) = 2A + 2B$

5. $(5x + 1)3 = 15x + 3$

6. $(x + a)(x + b) = (x + a)x + (x + a)b$

7–12 ■ Use properties of real numbers to write the expression without parentheses.

7. $3(x + y)$

8. $(a - b)8$

9. $4(2m)$

10. $\frac{4}{3}(-6y)$

11. $-\frac{5}{2}(2x - 4y)$

12. $(3a)(b + c - 2d)$

13–16 ■ State whether each inequality is true or false.

13. (a) $-6 < -10$

(b) $\sqrt{2} > 1.41$

14. (a) $\dfrac{10}{11} < \dfrac{12}{13}$

(b) $-\dfrac{1}{2} < -1$

15. (a) $-\pi > -3$

(b) $8 \leqslant 9$

16. (a) $1.1 > 1.\overline{1}$

(b) $8 \leqslant 8$

17–18 ■ Write each statement in terms of inequalities.

17. (a) x is positive
(b) t is less than 4
(c) a is greater than or equal to π
(d) x is less than $\frac{1}{3}$ and is greater than -5
(e) The distance from p to 3 is at most 5

18. (a) y is negative
(b) z is greater than 1
(c) b is at most 8
(d) w is positive and is less than or equal to 17
(e) y is at least 2 units from π

19–22 ■ Find the indicated set if $A = \{x \mid x \geqslant -2\}$, $B = \{x \mid x < 4\}$, and $C = \{x \mid -1 < x \leqslant 5\}$.

19. (a) $A \cup B$

(b) $A \cap B$

20. (a) $A \cup C$

(b) $A \cap C$

21. (a) $B \cup C$

(b) $B \cap C$

22. (a) $A \cup B \cup C$

(b) $A \cap B \cap C$

23–28 ■ Express the interval in terms of inequalities, and then graph the interval.

23. $(-3, 0)$

24. $(2, 8]$

25. $[2, 8)$

26. $\left[-6, -\frac{1}{2}\right]$

27. $[2, \infty)$

28. $(-\infty, 1)$

29–34 ■ Express the inequality in interval notation, and then graph the corresponding interval.

29. $x \leqslant 1$

30. $1 \leqslant x \leqslant 2$

31. $-2 < x \leqslant 1$

32. $x \geqslant -5$

33. $x > -1$

34. $-5 < x < 2$

35–40 ■ Graph the set.

35. $(-2, 0) \cup (-1, 1)$

36. $(-2, 0) \cap (-1, 1)$

37. $[-4, 6] \cap [0, 8)$

38. $[-4, 6) \cup [0, 8)$

39. $(-\infty, -4) \cup (4, \infty)$

40. $(-\infty, 6] \cap (2, 10)$

41–46 ■ Evaluate each expression.

41. (a) $|100|$

(b) $|-73|$

42. (a) $|\sqrt{5} - 5|$

(b) $|10 - \pi|$

43. (a) $\left||-6| - |-4|\right|$

(b) $\dfrac{-1}{|-1|}$

44. (a) $\left|2 - |-12|\right|$

(b) $-1 - \left|1 - |-1|\right|$

45. (a) $|(-2) \cdot 6|$

(b) $\left|\left(-\frac{1}{3}\right)(-15)\right|$

46. (a) $\left|\dfrac{-6}{24}\right|$

(b) $\left|\dfrac{7 - 12}{12 - 7}\right|$

47–48 ■ Find the distance between the given numbers.

47. (a) 2 and 17
(b) -3 and 21
(c) $\frac{11}{8}$ and $-\frac{3}{10}$

48. (a) $\frac{7}{15}$ and $-\frac{1}{21}$
(b) -38 and -57
(c) -2.6 and -1.8

A.2 EXPONENTS AND RADICALS

> Integer Exponents ▶ Radicals ▶ Rational Exponents ▶ Rationalizing the Denominator

In this section we give meaning to expressions such as $a^{m/n}$ in which the exponent m/n is a rational number. To do this, we need to recall some facts about integer exponents, radicals, and nth roots.

▼ Integer Exponents

A product of identical numbers is usually written in exponential notation. For example, $5 \cdot 5 \cdot 5$ is written as 5^3. In general, we have the following definition.

EXPONENTIAL NOTATION

If a is any real number and n is a positive integer, then the **nth power** of a is

$$a^n = \underbrace{a \cdot a \cdot \cdots \cdot a}_{n \text{ factors}}$$

The number a is called the **base** and n is called the **exponent**.

EXAMPLE 1 | Exponential Notation

Note the distinction between $(-3)^4$ and -3^4. In $(-3)^4$ the exponent applies to -3, but in -3^4 the exponent applies only to 3.

(a) $\left(\frac{1}{2}\right)^5 = \left(\frac{1}{2}\right)\left(\frac{1}{2}\right)\left(\frac{1}{2}\right)\left(\frac{1}{2}\right)\left(\frac{1}{2}\right) = \frac{1}{32}$

(b) $(-3)^4 = (-3) \cdot (-3) \cdot (-3) \cdot (-3) = 81$

(c) $-3^4 = -(3 \cdot 3 \cdot 3 \cdot 3) = -81$

ZERO AND NEGATIVE EXPONENTS

If $a \neq 0$ is any real number and n is a positive integer, then

$$a^0 = 1 \qquad \text{and} \qquad a^{-n} = \frac{1}{a^n}$$

EXAMPLE 2 | Zero and Negative Exponents

(a) $\left(\frac{4}{7}\right)^0 = 1$

(b) $x^{-1} = \frac{1}{x^1} = \frac{1}{x}$

(c) $(-2)^{-3} = \frac{1}{(-2)^3} = \frac{1}{-8} = -\frac{1}{8}$

Familiarity with the following rules is essential for our work with exponents and bases. In the table the bases a and b are real numbers, and the exponents m and n are integers.

LAWS OF EXPONENTS

Law	Example
1. $a^m a^n = a^{m+n}$	$3^2 \cdot 3^5 = 3^{2+5} = 3^7$
2. $\dfrac{a^m}{a^n} = a^{m-n}$	$\dfrac{3^5}{3^2} = 3^{5-2} = 3^3$
3. $(a^m)^n = a^{mn}$	$(3^2)^5 = 3^{2 \cdot 5} = 3^{10}$
4. $(ab)^n = a^n b^n$	$(3 \cdot 4)^2 = 3^2 \cdot 4^2$
5. $\left(\dfrac{a}{b}\right)^n = \dfrac{a^n}{b^n}$	$\left(\dfrac{3}{4}\right)^2 = \dfrac{3^2}{4^2}$
6. $\left(\dfrac{a}{b}\right)^{-n} = \left(\dfrac{b}{a}\right)^n$	$\left(\dfrac{3}{4}\right)^{-2} = \left(\dfrac{4}{3}\right)^2$
7. $\dfrac{a^{-n}}{b^{-m}} = \dfrac{b^m}{a^n}$	$\dfrac{3^{-2}}{4^{-5}} = \dfrac{4^5}{3^2}$

EXAMPLE 3 | Simplifying Expressions with Exponents

Simplify: **(a)** $(2a^3 b^2)(3ab^4)^3$ **(b)** $\left(\dfrac{x}{y}\right)^3\left(\dfrac{y^2 x}{z}\right)^4$

SOLUTION

(a)
$$
\begin{aligned}
(2a^3 b^2)(3ab^4)^3 &= (2a^3 b^2)[3^3 a^3 (b^4)^3] && \text{Law 4}\\
&= (2a^3 b^2)(27 a^3 b^{12}) && \text{Law 3}\\
&= (2)(27) a^3 a^3 b^2 b^{12} && \text{Group factors with the same base}\\
&= 54 a^6 b^{14} && \text{Law 1}
\end{aligned}
$$

(b)
$$
\begin{aligned}
\left(\frac{x}{y}\right)^3\left(\frac{y^2 x}{z}\right)^4 &= \frac{x^3}{y^3}\,\frac{(y^2)^4 x^4}{z^4} && \text{Laws 5 and 4}\\[2mm]
&= \frac{x^3}{y^3}\,\frac{y^8 x^4}{z^4} && \text{Law 3}\\[2mm]
&= (x^3 x^4)\left(\frac{y^8}{y^3}\right)\frac{1}{z^4} && \text{Group factors with the same base}\\[2mm]
&= \frac{x^7 y^5}{z^4} && \text{Laws 1 and 2}
\end{aligned}
$$

EXAMPLE 4 | Simplifying Expressions with Negative Exponents

Eliminate negative exponents and simplify each expression.

(a) $\dfrac{6st^{-4}}{2s^{-2}t^2}$ **(b)** $\left(\dfrac{y}{3z^2}\right)^{-2}$

SOLUTION

(a) We use Law 7, which allows us to move a number raised to a power from the numerator to the denominator (or vice versa) by changing the sign of the exponent.

$$\frac{6st^{-4}}{2s^{-2}t^2} = \frac{6ss^2}{2t^4t^2} \qquad \text{Law 7}$$

$$= \frac{3s^3}{t^6} \qquad \text{Law 1}$$

(b) We use Law 6, which allows us to change the sign of the exponent of a fraction by inverting the fraction.

$$\left(\frac{y}{3z^2}\right)^{-2} = \left(\frac{3z^2}{y}\right)^2 \qquad \text{Law 6}$$

$$= \frac{9z^4}{y^2} \qquad \text{Laws 5 and 4}$$

■

▼ Radicals

We know what 2^n means whenever n is an integer. To give meaning to a power, such as $2^{4/5}$, whose exponent is a rational number, we need to discuss radicals.

The symbol $\sqrt{}$ means "the positive square root of." Thus

$$\boxed{\sqrt{a} = b \qquad \text{means} \qquad b^2 = a \quad \text{and} \quad b \geq 0}$$

It is true that the number 9 has two square roots, 3 and -3, but the notation $\sqrt{9}$ is reserved for the *positive* square root of 9 (sometimes called the *principal square root* of 9). If we want the negative root, we *must* write $-\sqrt{9}$, which is -3.

Since $a = b^2 \geq 0$, the symbol $\sqrt{a}$ makes sense only when $a \geq 0$. For instance,

$$\sqrt{9} = 3 \qquad \text{because} \qquad 3^2 = 9 \quad \text{and} \quad 3 \geq 0$$

Square roots are special cases of nth roots. The nth root of x is the number that, when raised to the nth power, gives x.

DEFINITION OF nth ROOT

If n is any positive integer, then the **principal nth root** of a is defined as follows:

$$\sqrt[n]{a} = b \qquad \text{means} \qquad b^n = a$$

If n is even, we must have $a \geq 0$ and $b \geq 0$.

Thus

$$\sqrt[4]{81} = 3 \qquad \text{because} \qquad 3^4 = 81 \quad \text{and} \quad 3 \geq 0$$
$$\sqrt[3]{-8} = -2 \qquad \text{because} \qquad (-2)^3 = -8$$

But $\sqrt{-8}$, $\sqrt[4]{-8}$, and $\sqrt[6]{-8}$ are not defined. (For instance, $\sqrt{-8}$ is not defined because the square of every real number is nonnegative.)

Notice that

$$\sqrt{4^2} = \sqrt{16} = 4 \qquad \text{but} \qquad \sqrt{(-4)^2} = \sqrt{16} = 4 = |-4|$$

Thus, the equation $\sqrt{a^2} = a$ is not always true; it is true only when $a \geq 0$. However, we can always write $\sqrt{a^2} = |a|$. This last equation is true not only for square roots, but for any even root. This and other rules used in working with nth roots are listed in the following box. In each property we assume that all the given roots exist.

PROPERTIES OF nth ROOTS

Property	Example
1. $\sqrt[n]{ab} = \sqrt[n]{a}\,\sqrt[n]{b}$	$\sqrt[3]{-8 \cdot 27} = \sqrt[3]{-8}\,\sqrt[3]{27} = (-2)(3) = -6$
2. $\sqrt[n]{\dfrac{a}{b}} = \dfrac{\sqrt[n]{a}}{\sqrt[n]{b}}$	$\sqrt[4]{\dfrac{16}{81}} = \dfrac{\sqrt[4]{16}}{\sqrt[4]{81}} = \dfrac{2}{3}$
3. $\sqrt[m]{\sqrt[n]{a}} = \sqrt[mn]{a}$	$\sqrt{\sqrt[3]{729}} = \sqrt[6]{729} = 3$
4. $\sqrt[n]{a^n} = a$ if n is odd	$\sqrt[3]{(-5)^3} = -5,\ \ \sqrt[5]{2^5} = 2$
5. $\sqrt[n]{a^n} = |a|$ if n is even	$\sqrt[4]{(-3)^4} = |-3| = 3$

EXAMPLE 5 | Simplifying Expressions Involving nth Roots

(a) $\sqrt[3]{x^4} = \sqrt[3]{x^3 x}$ — Factor out the largest cube

$\qquad = \sqrt[3]{x^3}\,\sqrt[3]{x}$ — Property 1

$\qquad = x\sqrt[3]{x}$ — Property 4

(b) $\sqrt[4]{81x^8 y^4} = \sqrt[4]{81}\,\sqrt[4]{x^8}\,\sqrt[4]{y^4}$ — Property 1

$\qquad\qquad = 3\sqrt[4]{(x^2)^4}\,|y|$ — Property 5

$\qquad\qquad = 3x^2\,|y|$ — Property 5

■

It is frequently useful to combine like radicals in an expression such as $2\sqrt{3} + 5\sqrt{3}$. This can be done by using the Distributive Property. Thus

$$2\sqrt{3} + 5\sqrt{3} = (2 + 5)\sqrt{3} = 7\sqrt{3}$$

The next example further illustrates this process.

⊘ Avoid making the following error:

$$\sqrt{a + b} \ \times\ \sqrt{a} + \sqrt{b}$$

For instance, if we let $a = 9$ and $b = 16$, then we see the error:

$$\sqrt{9 + 16} \overset{?}{=} \sqrt{9} + \sqrt{16}$$

$$\sqrt{25} \overset{?}{=} 3 + 4$$

$$5 \overset{?}{=} 7 \quad \text{Wrong!}$$

EXAMPLE 6 | Combining Radicals

(a) $\sqrt{32} + \sqrt{200} = \sqrt{16 \cdot 2} + \sqrt{100 \cdot 2}$ — Factor out the largest squares

$\qquad\qquad = \sqrt{16}\,\sqrt{2} + \sqrt{100}\,\sqrt{2}$ — Property 1

$\qquad\qquad = 4\sqrt{2} + 10\sqrt{2} = 14\sqrt{2}$ — Distributive Property

(b) If $b > 0$, then

$$\sqrt{25b} - \sqrt{b^3} = \sqrt{25}\,\sqrt{b} - \sqrt{b^2}\,\sqrt{b} \qquad \text{Property 1}$$

$$= 5\sqrt{b} - b\sqrt{b} \qquad \text{Property 5, } b > 0$$

$$= (5 - b)\sqrt{b} \qquad \text{Distributive Property}$$

■

▼ Rational Exponents

To define what is meant by a *rational exponent* or, equivalently, a *fractional exponent* such as $a^{1/3}$, we need to use radicals. In order to give meaning to the symbol $a^{1/n}$ in a way that is consistent with the Laws of Exponents, we would have to have

$$(a^{1/n})^n = a^{(1/n)n} = a^1 = a$$

So, by the definition of nth root,

$$a^{1/n} = \sqrt[n]{a}$$

In general, we define rational exponents as follows.

DEFINITION OF RATIONAL EXPONENTS

For any rational exponent m/n in lowest terms, where m and n are integers and $n > 0$, we define

$$a^{m/n} = \left(\sqrt[n]{a}\right)^m$$

or equivalently

$$a^{m/n} = \sqrt[n]{a^m}$$

If n is even, then we require that $a \geq 0$.

With this definition it can be proved that *the Laws of Exponents also hold for rational exponents*.

EXAMPLE 7 | Using the Definition of Rational Exponents

(a) $4^{1/2} = \sqrt{4} = 2$

(b) $8^{2/3} = \left(\sqrt[3]{8}\right)^2 = 2^2 = 4$ Alternative solution: $8^{2/3} = \sqrt[3]{8^2} = \sqrt[3]{64} = 4$

(c) $(125)^{-1/3} = \dfrac{1}{125^{1/3}} = \dfrac{1}{\sqrt[3]{125}} = \dfrac{1}{5}$

■

EXAMPLE 8 | Using the Laws of Exponents with Rational Exponents

(a) $a^{1/3}a^{7/3} = a^{8/3}$ Law 1

(b) $\dfrac{a^{2/5}a^{7/5}}{a^{3/5}} = a^{2/5+7/5-3/5} = a^{6/5}$ Laws 1 and 2

(c) $\left(\dfrac{2x^{3/4}}{y^{1/3}}\right)^3 \left(\dfrac{y^4}{x^{-1/2}}\right) = \dfrac{2^3(x^{3/4})^3}{(y^{1/3})^3} \cdot (y^4 x^{1/2})$ Laws 5, 4, and 7

$\qquad\qquad\qquad = \dfrac{8x^{9/4}}{y} \cdot y^4 x^{1/2}$ Law 3

$\qquad\qquad\qquad = 8x^{11/4}y^3$ Laws 1 and 2

■

EXAMPLE 9 | Simplifying by Writing Radicals as Rational Exponents

(a) $(2\sqrt{x})(3\sqrt[3]{x}) = (2x^{1/2})(3x^{1/3})$ Definition of rational exponents

$= 6x^{1/2+1/3} = 6x^{5/6}$ Law 1

(b) $\sqrt{x\sqrt{x}} = (xx^{1/2})^{1/2}$ Definition of rational exponents

$= (x^{3/2})^{1/2}$ Law 1

$= x^{3/4}$ Law 3

▼ Rationalizing the Denominator

It is often useful to eliminate the radical in a denominator by multiplying both numerator and denominator by an appropriate expression. This procedure is called **rationalizing the denominator**. If the denominator is of the form $\sqrt{a}$, we multiply numerator and denominator by $\sqrt{a}$. In doing so we multiply the given quantity by 1, so we do not change its value. For instance,

$$\frac{1}{\sqrt{a}} = \frac{1}{\sqrt{a}} \cdot 1 = \frac{1}{\sqrt{a}} \cdot \frac{\sqrt{a}}{\sqrt{a}} = \frac{\sqrt{a}}{a}$$

Note that the denominator in the last fraction contains no radical. In general, if the denominator is of the form $\sqrt[n]{a^m}$ with $m < n$, then multiplying numerator and denominator by $\sqrt[n]{a^{n-m}}$ will rationalize the denominator, because (for $a > 0$)

$$\sqrt[n]{a^m}\,\sqrt[n]{a^{n-m}} = \sqrt[n]{a^{m+n-m}} = \sqrt[n]{a^n} = a$$

EXAMPLE 10 | Rationalizing Denominators

(a) $\dfrac{2}{\sqrt{3}} = \dfrac{2}{\sqrt{3}} \cdot \dfrac{\sqrt{3}}{\sqrt{3}} = \dfrac{2\sqrt{3}}{3}$

(b) $\dfrac{1}{\sqrt[3]{x^2}} = \dfrac{1}{\sqrt[3]{x^2}}\,\dfrac{\sqrt[3]{x}}{\sqrt[3]{x}} = \dfrac{\sqrt[3]{x}}{\sqrt[3]{x^3}} = \dfrac{\sqrt[3]{x}}{x}$

A.2 EXERCISES

1–6 ■ Write each radical expression using exponents, and each exponential expression using radicals.

1. $\dfrac{1}{\sqrt{17}}$ **2.** $\sqrt[5]{7^3}$ **3.** $4^{2/3}$

4. $\sqrt[3]{b^5}$ **5.** $a^{3/5}$ **6.** $w^{-3/2}$

7–14 ■ Evaluate each number.

7. (a) $(-2)^4$ **(b)** -2^4 **(c)** $(-2)^0$

8. (a) $\left(\frac{1}{2}\right)^4 4^{-2}$ **(b)** $\left(\frac{1}{4}\right)^{-2}$ **(c)** $\left(\frac{1}{4}\right)^0 2^{-1}$

9. (a) $2^4 5^{-2}$ **(b)** $\dfrac{10^7}{10^4}$ **(c)** $(2^3 \cdot 2^2)^2$

10. (a) $\sqrt{64}$ **(b)** $\sqrt[3]{-64}$ **(c)** $\sqrt[5]{-32}$

11. (a) $\sqrt{\frac{4}{9}}$ **(b)** $\sqrt[4]{256}$ **(c)** $\sqrt[6]{\frac{1}{64}}$

12. (a) $\sqrt{7}\sqrt{28}$ **(b)** $\dfrac{\sqrt{48}}{\sqrt{3}}$ **(c)** $\sqrt[4]{24}\sqrt[4]{54}$

13. (a) $\left(\frac{4}{9}\right)^{-1/2}$ **(b)** $(-32)^{2/5}$ **(c)** $(-125)^{-1/3}$

14. (a) $1024^{-0.1}$ **(b)** $\left(-\frac{27}{8}\right)^{2/3}$ **(c)** $\left(\frac{25}{64}\right)^{3/2}$

15–18 ■ Evaluate the expression using $x = 3$, $y = 4$, and $z = -1$.

15. $\sqrt{x^2 + y^2}$ **16.** $\sqrt[4]{x^3 + 14y + 2z}$

17. $(9x)^{2/3} + (2y)^{2/3} + z^{2/3}$ **18.** $(xy)^{2z}$

19–22 ■ Simplify the expression.

19. $\sqrt[3]{108} - \sqrt[3]{32}$

20. $\sqrt{8} + \sqrt{50}$

21. $\sqrt{245} - \sqrt{125}$

22. $\sqrt[3]{54} - \sqrt[3]{16}$

23–40 ■ Simplify the expression and eliminate any negative exponent(s).

23. $a^9 a^{-5}$

24. $(3y^2)(4y^5)$

25. $(12x^2 y^4)\left(\frac{1}{2}x^5 y\right)$

26. $(6y)^3$

27. $\dfrac{x^9(2x)^4}{x^3}$

28. $\dfrac{a^{-3} b^4}{a^{-5} b^5}$

29. $b^4\left(\frac{1}{3}b^2\right)(12b^{-8})$

30. $(2s^3 t^{-1})\left(\frac{1}{4}s^6\right)(16t^4)$

31. $(rs)^3(2s)^{-2}(4r)^4$

32. $(2u^2 v^3)^3(3u^3 v)^{-2}$

33. $\dfrac{(6y^3)^4}{2y^5}$

34. $\dfrac{(2x^3)^2(3x^4)}{(x^3)^4}$

35. $\dfrac{(x^2 y^3)^4(xy^4)^{-3}}{x^2 y}$

36. $\left(\dfrac{c^4 d^3}{cd^2}\right)\left(\dfrac{d^2}{c^3}\right)^3$

37. $\dfrac{(xy^2 z^3)^4}{(x^3 y^2 z)^3}$

38. $\left(\dfrac{xy^{-2} z^{-3}}{x^2 y^3 z^{-4}}\right)^{-3}$

39. $\left(\dfrac{q^{-1} rs^{-2}}{r^{-5} sq^{-8}}\right)^{-1}$

40. $(3ab^2 c)\left(\dfrac{2a^2 b}{c^3}\right)^{-2}$

41–56 ■ Simplify the expression and eliminate any negative exponent(s). Assume that all letters denote positive numbers.

41. $x^{2/3} x^{1/5}$

42. $(-2a^{3/4})(5a^{3/2})$

43. $(4b)^{1/2}(8b^{2/5})$

44. $(8x^6)^{-2/3}$

45. $(c^2 d^3)^{-1/3}$

46. $(4x^6 y^8)^{3/2}$

47. $(y^{3/4})^{2/3}$

48. $(a^{2/5})^{-3/4}$

49. $(2x^4 y^{-4/5})^3(8y^2)^{2/3}$

50. $(x^{-5} y^3 z^{10})^{-3/5}$

51. $\left(\dfrac{x^6 y}{y^4}\right)^{5/2}$

52. $\left(\dfrac{-2x^{1/3}}{y^{1/2} z^{1/6}}\right)^4$

53. $\left(\dfrac{3a^{-2}}{4b^{-1/3}}\right)^{-1}$

54. $\dfrac{(y^{10} z^{-5})^{1/5}}{(y^{-2} z^3)^{1/3}}$

55. $\dfrac{(9st)^{3/2}}{(27s^3 t^{-4})^{2/3}}$

56. $\left(\dfrac{a^2 b^{-3}}{x^{-1} y^2}\right)^3\left(\dfrac{x^{-2} b^{-1}}{a^{3/2} y^{1/3}}\right)$

57–64 ■ Simplify the expression. Assume the letters denote any real numbers.

57. $\sqrt[4]{x^4}$

58. $\sqrt[3]{x^3 y^6}$

59. $\sqrt[3]{x^3 y}$

60. $\sqrt{x^4 y^4}$

61. $\sqrt[5]{a^6 b^7}$

62. $\sqrt[3]{a^2 b}\,\sqrt[3]{a^4 b}$

63. $\sqrt[3]{\sqrt{64x^6}}$

64. $\sqrt[4]{x^4 y^2 z^2}$

65–68 ■ Rationalize the denominator.

65. (a) $\dfrac{1}{\sqrt{6}}$ (b) $\sqrt{\dfrac{x}{3y}}$ (c) $\sqrt{\dfrac{3}{20}}$

66. (a) $\sqrt{\dfrac{x^5}{2}}$ (b) $\sqrt{\dfrac{2}{3}}$ (c) $\sqrt{\dfrac{1}{2x^3 y^5}}$

67. (a) $\dfrac{1}{\sqrt[3]{x}}$ (b) $\dfrac{1}{\sqrt[5]{x^2}}$ (c) $\dfrac{1}{\sqrt[7]{x^3}}$

68. (a) $\dfrac{1}{\sqrt[3]{x^2}}$ (b) $\dfrac{1}{\sqrt[4]{x^3}}$ (c) $\dfrac{1}{\sqrt[3]{x^4}}$

A.3 ALGEBRAIC EXPRESSIONS

Combining Algebraic Expressions ▶ Factoring ▶ Fractional Expressions ▶ Multiplying and Dividing Fractional Expressions ▶ Adding and Subtracting Fractional Expressions ▶ Compound Fractions ▶ Rationalizing the Denominator or the Numerator

Algebraic expressions such as

$$2x^2 - 3x + 4 \qquad ax + b$$

$$\frac{y-1}{y^2+2} \qquad \frac{cx^2 y + dy^2 z}{\sqrt{x^2 + y^2 + z^2}}$$

are obtained by starting with variables such as x, y, and z and constants such as 2, -3, a, b, c, and d, and combining them using addition, subtraction, multiplication, division, and roots. A **variable** is a letter that can represent any number in a given set of numbers, whereas a **constant** represents a fixed (or specific) number. The **domain** of a variable is the set of numbers that the variable is permitted to have. For instance, in the expression $\sqrt{x}$ the domain of x is $\{x \mid x \geqslant 0\}$, whereas in the expression $2/(x-3)$ the domain of x is $\{x \mid x \neq 3\}$.

Monomials:

$$5, \quad 6x, \quad ax^2, \quad -10ab$$

Binomials:

$$2 + 3x, \quad 6x^2 - 5, \quad ax + b$$

Trinomials:

$$x^2 + x + 1, \quad ax + by + cz,$$
$$x^5 + 2x^3 + 3x$$

The simplest types of algebraic expressions use only addition, subtraction, and multiplication. Such expressions are called **polynomials**. The general form of a polynomial of degree n (where n is a nonnegative integer) in the variable x is

$$a_n x^n + a_{n-1} x^{n-1} + \cdots + a_1 x + a_0$$

where $a_0, a_1 \ldots, a_n$ are constants and $a_n \neq 0$. The **degree** of a polynomial is the highest power of the variable. Any polynomial is a sum of **terms** of the form ax^k, called **monomials**, where a is a constant and k is a nonnegative integer. A **binomial** is a sum of two monomials, a **trinomial** is the sum of three monomials, and so on. Thus, $2x^2 - 3x + 4$, $ax + b$, and $x^4 + 2x^3$ are polynomials of degree 2, 1, and 4, respectively; the first is a trinomial, the other two are binomials.

▼ Combining Algebraic Expressions

We **add** and **subtract** polynomials using the properties of real numbers that were discussed in Appendix A.1. The idea is to combine **like terms** (that is, terms with the same variables raised to the same powers) using the Distributive Property. For instance,

$$5x^7 + 3x^7 = (5 + 3)x^7 = 8x^7$$

Distributive Property

$$ac + bc = (a + b)c$$

EXAMPLE 1 | Adding and Subtracting Polynomials

(a) Find the sum $(x^3 - 6x^2 + 2x + 4) + (x^3 + 5x^2 - 7x)$.

(b) Find the difference $(x^3 - 6x^2 + 2x + 4) - (x^3 + 5x^2 - 7x)$.

SOLUTION

(a) $(x^3 - 6x^2 + 2x + 4) + (x^3 + 5x^2 - 7x)$

$$= (x^3 + x^3) + (-6x^2 + 5x^2) + (2x - 7x) + 4 \qquad \text{Group like terms}$$

$$= 2x^3 - x^2 - 5x + 4 \qquad \text{Combine like terms}$$

(b) $(x^3 - 6x^2 + 2x + 4) - (x^3 + 5x^2 - 7x)$

$$= x^3 - 6x^2 + 2x + 4 - x^3 - 5x^2 + 7x \qquad \text{Distributive Property}$$

$$= (x^3 - x^3) + (-6x^2 - 5x^2) + (2x + 7x) + 4 \qquad \text{Group like terms}$$

$$= -11x^2 + 9x + 4 \qquad \text{Combine like terms}$$

To find the **product** of polynomials or other algebraic expressions, we need to use the Distributive Property repeatedly. In particular, using it three times on the product of two binomials, we get

$$(a + b)(c + d) = a(c + d) + b(c + d) = ac + ad + bc + bd$$

This says that we multiply the two factors by multiplying each term in one factor by each term in the other factor and adding these products. Schematically we have

The acronym **FOIL** helps us remember that the product of two binomials is the sum of the products of the **F**irst terms, the **O**uter terms, the **I**nner terms, and the **L**ast terms.

$$(a + b)(c + d) = ac + ad + bc + bd$$
$$\qquad\qquad\quad \uparrow \quad \uparrow \quad \uparrow \quad \uparrow$$
$$\qquad\qquad\quad \text{F} \quad \text{O} \quad \text{I} \quad \text{L}$$

In general, we can multiply two algebraic expressions by using the Distributive Property and the Laws of Exponents.

EXAMPLE 2 | Multiplying Algebraic Expressions

(a) $(2x + 1)(3x - 5) = 6x^2 - 10x + 3x - 5$ Distributive Property

$\qquad\qquad\qquad\qquad = 6x^2 - 7x - 5$ Combine like terms

(b) $(1 + \sqrt{x})(2 - 3\sqrt{x}) = 2 - 3\sqrt{x} + 2\sqrt{x} - 3(\sqrt{x})^2$ Distributive Property

$\qquad\qquad\qquad\qquad\quad = 2 - \sqrt{x} - 3x$ Combine like terms

Certain types of products occur so frequently that you should memorize them. You can verify the following formulas by performing the multiplications.

SPECIAL PRODUCT FORMULAS

1. $(A - B)(A + B) = A^2 - B^2$

2. $(A + B)^2 = A^2 + 2AB + B^2$

3. $(A - B)^2 = A^2 - 2AB + B^2$

4. $(A + B)^3 = A^3 + 3A^2B + 3AB^2 + B^3$

5. $(A - B)^3 = A^3 - 3A^2B + 3AB^2 - B^3$

The key idea in using these formulas (or any other formula in algebra) is the **Principle of Substitution**: We may substitute any algebraic expression for any letter in a formula. For example, to find $(x^2 + y^3)^2$ we use Product Formula 2:

$$(A + B)^2 = A^2 + 2AB + B^2$$

We substitute x^2 for A and y^3 for B to get

$$(x^2 + y^3)^2 = (x^2)^2 + 2(x^2)(y^3) + (y^3)^2$$

This type of substitution is valid because every algebraic expression (in this case, x^2 or y^3) represents a number.

EXAMPLE 3 | Using the Special Product Formulas

Use the Special Product Formulas to find each product.

(a) $(3x + 5)^2$ $\qquad\qquad\qquad$ **(b)** $(2x - \sqrt{y})(2x + \sqrt{y})$

SOLUTION

(a) Product Formula 2, with $A = 3x$ and $B = 5$, gives

$$(3x + 5)^2 = (3x)^2 + 2(3x)(5) + 5^2 = 9x^2 + 30x + 25$$

(b) Using Product Formula 1 with $A = 2x$ and $B = \sqrt{y}$, we have

$$(2x - \sqrt{y})(2x + \sqrt{y}) = (2x)^2 - (\sqrt{y})^2 = 4x^2 - y$$

▼ Factoring

We used the Distributive Property to expand algebraic expressions. We sometimes need to reverse this process (again using the Distributive Property) by **factoring** an expression as a product of simpler ones. For example, we can write

▬▬ FACTORING ➡

$$x^2 - 4 = (x - 2)(x + 2)$$

⬅ EXPANDING ▬▬

We say that $x - 2$ and $x + 2$ are **factors** of $x^2 - 4$. The easiest type of factoring occurs when the terms have a common factor.

EXAMPLE 4 | Factoring Out Common Factors

Factor each expression.

(a) $3x^2 - 6x$

(b) $8x^4y^2 + 6x^3y^3 - 2xy^4$

SOLUTION

(a) The greatest common factor of the terms $3x^2$ and $-6x$ is $3x$, so we have

$$3x^2 - 6x = 3x(x - 2)$$

(b) The greatest common factor of the three terms in the polynomial is $2xy^2$, so

$$8x^4y^2 + 6x^3y^3 - 2xy^4 = (2xy^2)(4x^3) + (2xy^2)(3x^2y) + (2xy^2)(-y^2)$$
$$= 2xy^2(4x^3 + 3x^2y - y^2)$$

■

To factor a **quadratic** $x^2 + bx + c$, we look for factors of the form $x + r$ and $x + s$:

$$x^2 + bx + c = (x + r)(x + s) = x^2 + (r + s)x + rs$$

So we need to find numbers r and s so that $r + s = b$ and $rs = c$.

EXAMPLE 5 | Factoring $x^2 + bx + c$ by Trial and Error

Factor: $x^2 + 7x + 12$

SOLUTION

We need to find two integers whose product is 12 and whose sum is 7. By trial and error we find that the two integers are 3 and 4. Thus, the factorization is

$$x^2 + 7x + 12 = (x + 3)(x + 4)$$

■

To factor a quadratic $ax^2 + bx + c$ with $a \neq 1$, we look for factors of the form $px + r$ and $qx + s$:

$$ax^2 + bx + c = (px + r)(qx + s) = pqx^2 + (ps + qr)x + rs$$

Therefore, we try to find numbers p, q, r, and s such that $pq = a$, $rs = c$, $ps + qr = b$. If these numbers are all integers, then we will have a limited number of possibilities to try for p, q, r, and s.

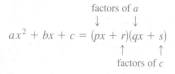

factors of a
↓   ↓
$ax^2 + bx + c = (px + r)(qx + s)$
↑   ↑
factors of c

EXAMPLE 6 | Factoring $ax^2 + bx + c$ by Trial and Error

Factor: $6x^2 + 7x - 5$

SOLUTION

We can factor 6 as $6 \cdot 1$ or $3 \cdot 2$, and -5 as $-5 \cdot 1$ or $5 \cdot (-1)$. By trying these possibilities, we arrive at the factorization

$$6x^2 + 7x - 5 = (3x + 5)(2x - 1)$$

CHECK YOUR ANSWER

Multiplying gives

$$(3x + 5)(2x - 1) = 6x^2 + 7x - 5$$

✔

Some special algebraic expressions can be factored using the following formulas. The first three are simply the Special Product Formulas written backward.

FACTORING FORMULAS

Formula	Name
1. $A^2 - B^2 = (A - B)(A + B)$	Difference of squares
2. $A^2 + 2AB + B^2 = (A + B)^2$	Perfect square
3. $A^2 - 2AB + B^2 = (A - B)^2$	Perfect square
4. $A^3 - B^3 = (A - B)(A^2 + AB + B^2)$	Difference of cubes
5. $A^3 + B^3 = (A + B)(A^2 - AB + B^2)$	Sum of cubes

EXAMPLE 7 | Factoring Differences of Squares

Factor each polynomial.

(a) $4x^2 - 25$ **(b)** $(x + y)^2 - z^2$

SOLUTION

(a) Using the formula for a difference of squares with $A = 2x$ and $B = 5$, we have

$$4x^2 - 25 = (2x)^2 - 5^2$$
$$= (2x - 5)(2x + 5)$$

(b) We use the formula for the difference of squares with $A = x + y$ and $B = z$.

$$(x + y)^2 - z^2 = (x + y - z)(x + y + z)$$

A trinomial is a perfect square if it is of the form

$$A^2 + 2AB + B^2 \quad \text{or} \quad A^2 - 2AB + B^2$$

So, we **recognize a perfect square** if the middle term ($2AB$ or $-2AB$) is plus or minus twice the product of the square roots of the outer two terms.

EXAMPLE 8 | Recognizing Perfect Squares

Factor each trinomial: **(a)** $x^2 + 6x + 9$ **(b)** $4x^2 - 4xy + y^2$

SOLUTION

(a) Here $A = x$ and $B = 3$, so $2AB = 2 \cdot x \cdot 3 = 6x$. Since the middle term is $6x$, the trinomial is a perfect square. By Formula 2 we have

$$x^2 + 6x + 9 = (x + 3)^2$$

(b) Here $A = 2x$ and $B = y$, so $2AB = 2 \cdot 2x \cdot y = 4xy$. Since the middle term is $-4xy$, the trinomial is a perfect square. By Formula 2 we have

$$4x^2 - 4xy + y^2 = (2x - y)^2$$

▼ Fractional Expressions

A quotient of two algebraic expressions is called a **fractional expression**. We assume that all fractions are defined; that is, *we deal only with values of the variables such that the denominators are not zero.*

To **simplify fractional expressions**, we factor both numerator and denominator and use the following property of fractions:

$$\frac{AC}{BC} = \frac{A}{B}$$

This says that we can **cancel** common factors from numerator and denominator.

EXAMPLE 9 | Simplifying Fractional Expressions by Cancellation

Simplify: $\dfrac{x^2 - 1}{x^2 + x - 2}$

SOLUTION

$$\frac{x^2 - 1}{x^2 + x - 2} = \frac{(x - 1)(x + 1)}{(x - 1)(x + 2)} \qquad \text{Factor}$$

$$= \frac{x + 1}{x + 2} \qquad \text{Cancel common factors}$$

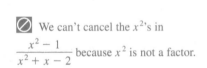 We can't cancel the x^2's in $\dfrac{x^2 - 1}{x^2 + x - 2}$ because x^2 is not a factor.

▼ Multiplying and Dividing Fractional Expressions

To **multiply fractional expressions**, we use the following property of fractions:

$$\frac{A}{B} \cdot \frac{C}{D} = \frac{AC}{BD}$$

This says that to multiply two fractions we multiply their numerators and multiply their denominators.

EXAMPLE 10 | Multiplying Fractional Expressions

Perform the indicated multiplication and simplify: $\dfrac{x^2 + 2x - 3}{x^2 + 8x + 16} \cdot \dfrac{3x + 12}{x - 1}$

SOLUTION

We first factor.

$$\frac{x^2 + 2x - 3}{x^2 + 8x + 16} \cdot \frac{3x + 12}{x - 1} = \frac{(x - 1)(x + 3)}{(x + 4)^2} \cdot \frac{3(x + 4)}{x - 1} \qquad \text{Factor}$$

$$= \frac{3(x + 3)}{x + 4} \qquad \text{Cancel common factors}$$

Diophantus lived in Alexandria about 250 A.D. His book *Arithmetica* is considered the first book on algebra. In it he gives methods for finding integer solutions of algebraic equations. *Arithmetica* was read and studied for more than a thousand years. Fermat (see page 241) made some of his most important discoveries while studying this book. Diophantus' major contribution is the use of symbols to stand for the unknowns in a problem. Although his symbolism is not as simple as what we use today, it was a major advance over writing everything in words. In Diophantus' notation the equation

$$x^5 - 7x^2 + 8x - 5 = 24$$

is written

$$\Delta K^\gamma \alpha \varsigma \eta \wedge \Delta^\gamma \zeta \mathring{M} \varepsilon \iota^\sigma \kappa \delta$$

Our modern algebraic notation did not come into common use until the 17th century.

To **divide fractional expressions**, we use the following property of fractions:

$$\frac{A}{B} \div \frac{C}{D} = \frac{A}{B} \cdot \frac{D}{C}$$

This says that to divide a fraction by another fraction we invert the divisor and multiply.

EXAMPLE 11 | Dividing Fractional Expressions

Perform the indicated division and simplify: $\dfrac{x-4}{x^2-4} \div \dfrac{x^2-3x-4}{x^2+5x+6}$

SOLUTION

$$\frac{x-4}{x^2-4} \div \frac{x^2-3x-4}{x^2+5x+6} = \frac{x-4}{x^2-4} \cdot \frac{x^2+5x+6}{x^2-3x-4} \qquad \text{Invert and multiply}$$

$$= \frac{(x-4)(x+2)(x+3)}{(x-2)(x+2)(x-4)(x+1)} \qquad \text{Factor}$$

$$= \frac{x+3}{(x-2)(x+1)} \qquad \text{Cancel common factors}$$

▼ Adding and Subtracting Fractional Expressions

⊘ Avoid making the following error:

$$\frac{A}{B+C} \;\;✗\;\; \frac{A}{B} + \frac{A}{C}$$

For instance, if we let $A = 2$, $B = 1$, and $C = 1$, then we see the error:

$$\frac{2}{1+1} \overset{?}{=} \frac{2}{1} + \frac{2}{1}$$

$$\frac{2}{2} \overset{?}{=} 2 + 2$$

$$1 \overset{?}{=} 4 \quad \text{Wrong!}$$

To **add or subtract fractional expressions**, we first find a common denominator and then use the following property of fractions:

$$\frac{A}{C} + \frac{B}{C} = \frac{A+B}{C}$$

Although any common denominator will work, it is best to use the **least common denominator** (LCD). The LCD is found by factoring each denominator and taking the product of the distinct factors, using the highest power that appears in any of the factors.

EXAMPLE 12 | Subtracting Fractional Expressions

Simplify: $\dfrac{1}{x^2-1} - \dfrac{2}{(x+1)^2}$

SOLUTION

The LCD of $x^2 - 1 = (x-1)(x+1)$ and $(x+1)^2$ is $(x-1)(x+1)^2$, so we have

$$\frac{1}{x^2-1} - \frac{2}{(x+1)^2} = \frac{1}{(x-1)(x+1)} - \frac{2}{(x+1)^2} \qquad \text{Factor}$$

$$= \frac{(x+1) - 2(x-1)}{(x-1)(x+1)^2} \qquad \text{Combine fractions using LCD}$$

$$= \frac{x+1-2x+2}{(x-1)(x+1)^2} \qquad \text{Distributive Property}$$

$$= \frac{3-x}{(x-1)(x+1)^2} \qquad \text{Combine terms in numerator}$$

▼ Compound Fractions

A **compound fraction** is a fraction in which the numerator, the denominator, or both, are themselves fractional expressions.

EXAMPLE 13 | Simplifying a Compound Fraction

Simplify: $\dfrac{\dfrac{x}{y} + 1}{1 - \dfrac{y}{x}}$

SOLUTION 1

We combine the terms in the numerator into a single fraction. We do the same in the denominator. Then we invert and multiply.

$$\frac{\dfrac{x}{y} + 1}{1 - \dfrac{y}{x}} = \frac{\dfrac{x + y}{y}}{\dfrac{x - y}{x}} = \frac{x + y}{y} \cdot \frac{x}{x - y}$$

$$= \frac{x(x + y)}{y(x - y)}$$

SOLUTION 2

We find the LCD of all the fractions in the expression, then multiply numerator and denominator by it. In this example the LCD of all the fractions is xy. Thus

$$\frac{\dfrac{x}{y} + 1}{1 - \dfrac{y}{x}} = \frac{\dfrac{x}{y} + 1}{1 - \dfrac{y}{x}} \cdot \frac{xy}{xy} \qquad \text{Multiply numerator and denominator by } xy$$

$$= \frac{x^2 + xy}{xy - y^2} \qquad \text{Simplify}$$

$$= \frac{x(x + y)}{y(x - y)} \qquad \text{Factor}$$

▼ Rationalizing the Denominator or the Numerator

If a fraction has a denominator of the form $A + B\sqrt{C}$, we may rationalize the denominator by multiplying numerator and denominator by the **conjugate radical** $A - B\sqrt{C}$. This is effective because, by Product Formula 1 on page 503, the product of the denominator and its conjugate radical does not contain a radical:

$$\left(A + B\sqrt{C}\right)\left(A - B\sqrt{C}\right) = A^2 - B^2C$$

EXAMPLE 14 | Rationalizing the Denominator

Rationalize the denominator: $\dfrac{1}{1 + \sqrt{2}}$

SOLUTION

We multiply both the numerator and the denominator by the conjugate radical of $1 + \sqrt{2}$, which is $1 - \sqrt{2}$.

Product Formula 1
$$(a + b)(a - b) = a^2 - b^2$$

$$\frac{1}{1 + \sqrt{2}} = \frac{1}{1 + \sqrt{2}} \cdot \frac{1 - \sqrt{2}}{1 - \sqrt{2}}$$
Multiply numerator and denominator by the conjugate radical

$$= \frac{1 - \sqrt{2}}{1^2 - (\sqrt{2})^2}$$
Product Formula 1

$$= \frac{1 - \sqrt{2}}{1 - 2} = \frac{1 - \sqrt{2}}{-1} = \sqrt{2} - 1$$

EXAMPLE 15 | Rationalizing the Numerator

Rationalize the numerator: $\dfrac{\sqrt{4 + h} - 2}{h}$

SOLUTION

We multiply numerator and denominator by the conjugate radical $\sqrt{4 + h} + 2$.

$$\frac{\sqrt{4 + h} - 2}{h} = \frac{\sqrt{4 + h} - 2}{h} \cdot \frac{\sqrt{4 + h} + 2}{\sqrt{4 + h} + 2}$$
Multiply numerator and denominator by the conjugate radical

$$= \frac{(\sqrt{4 + h})^2 - 2^2}{h(\sqrt{4 + h} + 2)}$$
Product Formula 1

$$= \frac{4 + h - 4}{h(\sqrt{4 + h} + 2)}$$

$$= \frac{h}{h(\sqrt{4 + h} + 2)} = \frac{1}{\sqrt{4 + h} + 2}$$
Property 5 of fractions (cancel common factors)

A.3 EXERCISES

1–26 ■ Perform the indicated operations and simplify.

1. $(3x^2 + x + 1) + (2x^2 - 3x - 5)$

2. $(3x^2 + x + 1) - (2x^2 - 3x - 5)$

3. $8(2x + 5) - 7(x - 9)$

4. $4(x^2 - 3x + 5) - 3(x^2 - 2x + 1)$

5. $2(2 - 5t) + t^2(t - 1) - (t^4 - 1)$

6. $5(3t - 4) - (t^2 + 2) - 2t(t - 3)$

7. $\sqrt{x}(x - \sqrt{x})$

8. $x^{3/2}(\sqrt{x} - 1/\sqrt{x})$

9. $(x + 2y)(3x - y)$

10. $(4x - 3y)(2x + 5y)$

11. $(1 - 2y)^2$

12. $(3x + 4)^2$

13. $(2x^2 + 3y^2)^2$

14. $\left(c + \dfrac{1}{c}\right)^2$

15. $(2x - 5)(x^2 - x + 1)$

16. $(1 + 2x)(x^2 - 3x + 1)$

17. $(x^2 - a^2)(x^2 + a^2)$

18. $(x^{1/2} + y^{1/2})(x^{1/2} - y^{1/2})$

19. $\left(\sqrt{a} - \dfrac{1}{b}\right)\left(\sqrt{a} + \dfrac{1}{b}\right)$

20. $(1 - 2y)^3$

21. $(x^2 + x - 2)(x^3 - x + 1)$

22. $(1 + x + x^2)(1 - x + x^2)$

23. $(1 + x^{4/3})(1 - x^{2/3})$

24. $(1 - b)^2(1 + b)^2$

25. $(3x^2y + 7xy^2)(x^2y^3 - 2y^2)$

26. $(x^4y - y^5)(x^2 + xy + y^2)$

27–60 ■ Factor the expression completely.

27. $12x^3 + 18x$

28. $30x^3 + 15x^4$

29. $6y^4 - 15y^3$

30. $5ab - 8abc$

31. $x^2 - 2x - 8$

32. $x^2 - 14x + 48$

33. $y^2 - 8y + 15$

34. $z^2 + 6z - 16$

35. $2x^2 + 5x + 3$

36. $2x^2 + 7x - 4$

37. $9x^2 - 36x - 45$

38. $8x^2 + 10x + 3$

39. $6x^2 - 5x - 6$

40. $6 + 5t - 6t^2$

41. $4t^2 - 12t + 9$

42. $4x^2 + 4xy + y^2$

43. $r^2 - 6rs + 9s^2$

44. $25s^2 - 10st + t^2$

45. $x^2 - 36$

46. $4x^2 - 25$

47. $49 - 4y^2$

48. $4t^2 - 9s^2$

49. $(a + b)^2 - (a - b)^2$

50. $\left(1 + \dfrac{1}{x}\right)^2 - \left(1 - \dfrac{1}{x}\right)^2$

51. $x^2(x^2 - 1) - 9(x^2 - 1)$

52. $(a^2 - 1)b^2 - 4(a^2 - 1)$

53. $t^3 + 1$

54. $x^3 - 27$

55. $x^3 + 2x^2 + x$

56. $3x^3 - 27x$

57. $(x - 1)(x + 2)^2 - (x - 1)^2(x + 2)$

58. $(x + 1)^3 x - 2(x + 1)^2 x^2 + x^3(x + 1)$

59. $y^4(y + 2)^3 + y^5(y + 2)^4$

60. $n(x - y) + (n - 1)(y - x)$

61–100 ■ Simplify the expression.

61. $\dfrac{x - 2}{x^2 - 4}$

62. $\dfrac{x^2 - x - 2}{x^2 - 1}$

63. $\dfrac{x^2 + 6x + 8}{x^2 + 5x + 4}$

64. $\dfrac{x^2 - x - 12}{x^2 + 5x + 6}$

65. $\dfrac{y^2 + y}{y^2 - 1}$

66. $\dfrac{y^2 - 3y - 18}{2y^2 + 5y + 3}$

67. $\dfrac{2x^3 - x^2 - 6x}{2x^2 - 7x + 6}$

68. $\dfrac{1 - x^2}{x^3 - 1}$

69. $\dfrac{t - 3}{t^2 + 9} \cdot \dfrac{t + 3}{t^2 - 9}$

70. $\dfrac{x^2 - x - 6}{x^2 + 2x} \cdot \dfrac{x^3 + x^2}{x^2 - 2x - 3}$

71. $\dfrac{x^2 + 7x + 12}{x^2 + 3x + 2} \cdot \dfrac{x^2 + 5x + 6}{x^2 + 6x + 9}$

72. $\dfrac{x^2 + 2xy + y^2}{x^2 - y^2} \cdot \dfrac{2x^2 - xy - y^2}{x^2 - xy - 2y^2}$

73. $\dfrac{2x^2 + 3x + 1}{x^2 + 2x - 15} \div \dfrac{x^2 + 6x + 5}{2x^2 - 7x + 3}$

74. $\dfrac{4y^2 - 9}{2y^2 + 9y - 18} \div \dfrac{2y^2 + y - 3}{y^2 + 5y - 6}$

75. $\dfrac{x/y}{z}$

76. $\dfrac{x}{y/z}$

77. $\dfrac{1}{x + 5} + \dfrac{2}{x - 3}$

78. $\dfrac{1}{x + 1} + \dfrac{1}{x - 1}$

79. $\dfrac{1}{x + 1} - \dfrac{1}{x + 2}$

80. $\dfrac{x}{x - 4} - \dfrac{3}{x + 6}$

81. $\dfrac{x}{(x + 1)^2} + \dfrac{2}{x + 1}$

82. $\dfrac{5}{2x - 3} - \dfrac{3}{(2x - 3)^2}$

83. $u + 1 + \dfrac{u}{u + 1}$

84. $\dfrac{2}{a^2} - \dfrac{3}{ab} + \dfrac{4}{b^2}$

85. $\dfrac{1}{x^2} + \dfrac{1}{x^2 + x}$

86. $\dfrac{1}{x} + \dfrac{1}{x^2} + \dfrac{1}{x^3}$

87. $\dfrac{2}{x + 3} - \dfrac{1}{x^2 + 7x + 12}$

88. $\dfrac{x}{x^2 - 4} + \dfrac{1}{x - 2}$

89. $\dfrac{1}{x + 3} + \dfrac{1}{x^2 - 9}$

90. $\dfrac{x}{x^2 + x - 2} - \dfrac{2}{x^2 - 5x + 4}$

91. $\dfrac{2}{x} + \dfrac{3}{x - 1} - \dfrac{4}{x^2 - x}$

92. $\dfrac{x}{x^2 - x - 6} - \dfrac{1}{x + 2} - \dfrac{2}{x - 3}$

93. $\dfrac{1}{x^2 + 3x + 2} - \dfrac{1}{x^2 - 2x - 3}$

94. $\dfrac{1}{x + 1} - \dfrac{2}{(x + 1)^2} + \dfrac{3}{x^2 - 1}$

95. $\dfrac{\dfrac{x}{y} - \dfrac{y}{x}}{\dfrac{1}{x^2} - \dfrac{1}{y^2}}$

96. $x - \dfrac{y}{\dfrac{x}{y} + \dfrac{y}{x}}$

97. $\dfrac{1 + \dfrac{1}{c - 1}}{1 - \dfrac{1}{c - 1}}$

98. $1 + \dfrac{1}{1 + \dfrac{1}{1 + x}}$

99. $\dfrac{x^{-2} - y^{-2}}{x^{-1} + y^{-1}}$

100. $\dfrac{\dfrac{1}{a + h} - \dfrac{1}{a}}{h}$

101–104 ■ Rationalize the denominator.

101. $\dfrac{2}{3 + \sqrt{5}}$

102. $\dfrac{1}{\sqrt{x} + 1}$

103. $\dfrac{2}{\sqrt{2} + \sqrt{7}}$

104. $\dfrac{y}{\sqrt{3} + \sqrt{y}}$

105–110 ■ Rationalize the numerator.

105. $\dfrac{1 - \sqrt{5}}{3}$

106. $\dfrac{\sqrt{3} + \sqrt{5}}{2}$

107. $\dfrac{\sqrt{r} + \sqrt{2}}{5}$

108. $\dfrac{\sqrt{x} - \sqrt{x + h}}{h\sqrt{x}\sqrt{x + h}}$

109. $\sqrt{x^2 + 1} - x$

110. $\sqrt{x + 1} - \sqrt{x}$

A.4 EQUATIONS

| Linear Equations ▶ Quadratic Equations ▶ Other Equations

An equation is a statement that two mathematical expressions are equal. For example,

$$3 + 5 = 8$$

is an equation. Most equations that we study in algebra contain variables, which are symbols (usually letters) that stand for numbers. In the equation

$$4x + 7 = 19$$

the letter x is the variable. We think of x as the "unknown" in the equation, and our goal is to find the value of x that makes the equation true. The values of the unknown that make the equation true are called the **solutions** or **roots** of the equation, and the process of finding the solutions is called **solving the equation**.

Two equations with exactly the same solutions are called **equivalent equations**. To solve an equation, we try to find a simpler, equivalent equation in which the variable stands alone on one side of the "equal" sign. Here are the properties that we use to solve an equation. (In these properties, A, B, and C stand for any algebraic expressions, and the symbol $\Leftrightarrow$ means "is equivalent to.")

PROPERTIES OF EQUALITY

Property	Description
1. $A = B \Leftrightarrow A + C = B + C$	Adding the same quantity to both sides of an equation gives an equivalent equation.
2. $A = B \Leftrightarrow CA = CB$ $(C \neq 0)$	Multiplying both sides of an equation by the same nonzero quantity gives an equivalent equation.

These properties require that you *perform the same operation on both sides of an equation* when solving it. Thus, if we say "*add* -7" when solving an equation, that is just a short way of saying "*add* -7 to each side of the equation."

▼ Linear Equations

The simplest type of equation is a **linear equation**, or first-degree equation, which is an equation in which each term is either a constant or a nonzero multiple of the variable. This means that it is equivalent to an equation of the form $ax + b = 0$. Here a and b represent real numbers with $a \neq 0$, and x is the unknown variable that we are solving for. The equation in the following example is linear.

EXAMPLE 1 | Solving a Linear Equation

Solve the equation $7x - 4 = 3x + 8$.

SOLUTION

We solve this by changing it to an equivalent equation with all terms that have the variable x on one side and all constant terms on the other.

$$7x - 4 = 3x + 8$$

$$7x = 3x + 12 \qquad \text{Add } 4$$

$$4x = 12 \qquad \text{Subtract } 3x$$

$$x = 3 \qquad \text{Multiply by } \tfrac{1}{4}$$

CHECK YOUR ANSWER

$x = 3$: $\qquad$ LHS $= 7(3) - 4$ $\qquad$ RHS $= 3(3) + 8$

$\qquad\qquad\qquad\qquad = 17 \qquad\qquad\qquad\qquad = 17$

LHS = RHS ✔

LHS stands for "left-hand side" and RHS stands for "right-hand side."

Many formulas in the sciences involve several variables, and it is often necessary to express one of the variables in terms of the others, as in the next example.

EXAMPLE 2 | Solving for One Variable in Terms of Others

The surface area A of the closed rectangular box shown in Figure 1 can be calculated from the length l, the width w, and the height h according to the formula

$$A = 2lw + 2wh + 2lh$$

Solve for w in terms of the other variables in this equation.

SOLUTION

Although this equation involves more than one variable, we solve it as usual by isolating w on one side, treating the other variables as we would numbers.

$$A = (2lw + 2wh) + 2lh \qquad \text{Collect terms involving } w$$

$$A - 2lh = 2lw + 2wh \qquad \text{Subtract } 2lh$$

$$A - 2lh = (2l + 2h)w \qquad \text{Factor } w \text{ from RHS}$$

$$\frac{A - 2lh}{2l + 2h} = w \qquad \text{Divide by } 2l + 2h$$

The solution is $w = \dfrac{A - 2lh}{2l + 2h}$.

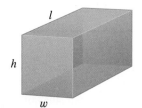

FIGURE 1 A closed rectangular box

▼ Quadratic Equations

Linear equations are first-degree equations of the form $ax + b = 0$. Quadratic equations are second-degree equations; they contain an additional term involving the square of the variable.

QUADRATIC EQUATIONS

A **quadratic equation** is an equation of the form

$$ax^2 + bx + c = 0$$

where a, b, and c are real numbers with $a \neq 0$.

Some quadratic equations can be solved by factoring and using the following basic property of real numbers.

ZERO-PRODUCT PROPERTY

$$AB = 0 \quad \text{if and only if} \quad A = 0 \quad \text{or} \quad B = 0$$

This means that if we can factor the left-hand side of a quadratic (or other) equation, then we can solve it by setting each factor equal to 0 in turn. This method works only when the right-hand side of the equation is 0.

EXAMPLE 3 | Solving a Quadratic Equation by Factoring

Solve the equation $x^2 + 5x = 24$.

SOLUTION

We must first rewrite the equation so that the right-hand side is 0.

$$x^2 + 5x = 24$$

$$x^2 + 5x - 24 = 0 \qquad \text{Subtract 24}$$

$$(x - 3)(x + 8) = 0 \qquad \text{Factor}$$

$$x - 3 = 0 \quad \text{or} \quad x + 8 = 0 \qquad \text{Zero-Product Property}$$

$$x = 3 \qquad\qquad x = -8 \qquad \text{Solve}$$

The solutions are $x = 3$ and $x = -8$.

■

CHECK YOUR ANSWERS

$x = 3$:

$(3)^2 + 5(3) = 9 + 15 = 24$ ✔

$x = -8$:

$(-8)^2 + 5(-8) = 64 - 40 = 24$ ✔

A quadratic equation of the form $x^2 - c = 0$, where c is a positive constant, factors as $(x - \sqrt{c})(x + \sqrt{c}) = 0$, and so the solutions are $x = \sqrt{c}$ and $x = -\sqrt{c}$. We often abbreviate this as $x = \pm\sqrt{c}$.

SOLVING A SIMPLE QUADRATIC EQUATION

The solutions of the equation $x^2 = c$ are $x = \sqrt{c}$ and $x = -\sqrt{c}$.

EXAMPLE 4 | Solving Simple Quadratics

Solve each equation: (a) $x^2 = 5$ (b) $(x - 4)^2 = 5$

SOLUTION

(a) From the principle in the preceding box, we get $x = \pm\sqrt{5}$.

(b) We can take the square root of each side of this equation as well.

$$(x - 4)^2 = 5$$

$$x - 4 = \pm\sqrt{5} \qquad \text{Take the square root}$$

$$x = 4 \pm \sqrt{5} \qquad \text{Add 4}$$

The solutions are $x = 4 + \sqrt{5}$ and $x = 4 - \sqrt{5}$.

■

See page 505 for how to recognize when a quadratic expression is a perfect square.

As we saw in Example 4, if a quadratic equation is of the form $(x \pm a)^2 = c$, then we can solve it by taking the square root of each side. In an equation of this form the left-hand side is a *perfect square*: the square of a linear expression in x. So, if a quadratic equation does not factor readily, then we can solve it using the technique of **completing the square**. This means that we add a constant to an expression to make it a perfect square. For example, to make $x^2 - 6x$ a perfect square we must add 9, since $x^2 - 6x + 9 = (x - 3)^2$.

Completing the Square

Area of blue region is

$$x^2 + 2\left(\frac{b}{2}\right)x = x^2 + bx$$

Add a small square of area $(b/2)^2$ to "complete" the square.

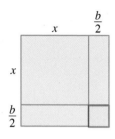

COMPLETING THE SQUARE

To make $x^2 + bx$ a perfect square, add $\left(\dfrac{b}{2}\right)^2$, the square of half the coefficient of x. This gives the perfect square

$$x^2 + bx + \left(\frac{b}{2}\right)^2 = \left(x + \frac{b}{2}\right)^2$$

EXAMPLE 5 | Solving Quadratic Equations by Completing the Square

Solve each equation.

(a) $x^2 - 8x + 13 = 0$ **(b)** $3x^2 - 12x + 6 = 0$

SOLUTION

(a) $x^2 - 8x + 13 = 0$

$x^2 - 8x \quad\quad = -13$	Subtract 13
$x^2 - 8x + 16 = -13 + 16$	Complete the square: add $\left(\dfrac{-8}{2}\right)^2 = 16$
$(x - 4)^2 = 3$	Perfect square
$x - 4 = \pm\sqrt{3}$	Take square root
$x = 4 \pm \sqrt{3}$	Add 4

⊘ When completing the square, make sure the coefficient of x^2 is 1. If it isn't, you must factor this coefficient from both terms that contain x:

$$ax^2 + bx = a\left(x^2 + \frac{b}{a}x \qquad\right)$$

Then complete the square inside the parentheses. Remember that the term added inside the parentheses is multiplied by a.

(b) After subtracting 6 from each side of the equation, we must factor the coefficient of x^2 (the 3) from the left side to put the equation in the correct form for completing the square.

$3x^2 - 12x + 6 = 0$	
$3x^2 - 12x \quad\quad = -6$	Subtract 6
$3(x^2 - 4x \quad) = -6$	Factor 3 from LHS

Now we complete the square by adding $(-2)^2 = 4$ inside the parentheses. Since everything inside the parentheses is multiplied by 3, this means that we are actually adding $3 \cdot 4 = 12$ to the left side of the equation. Thus, we must add 12 to the right side as well.

$3(x^2 - 4x + 4) = -6 + 3 \cdot 4$	Complete the square: add 4
$3(x - 2)^2 = 6$	Perfect square
$(x - 2)^2 = 2$	Divide by 3
$x - 2 = \pm\sqrt{2}$	Take square root
$x = 2 \pm \sqrt{2}$	Add 2

We can use the technique of completing the square to derive a formula for the roots of the general quadratic equation $ax^2 + bx + c = 0$. First, we divide each side of the equation by a and move the constant to the right side, giving

$$x^2 + \frac{b}{a}x = -\frac{c}{a}$$

We now complete the square by adding $(b/2a)^2$ to each side of the equation:

$$x^2 + \frac{b}{a}x + \left(\frac{b}{2a}\right)^2 = -\frac{c}{a} + \left(\frac{b}{2a}\right)^2 \qquad \text{Complete the square:}$$
$$\qquad\qquad\qquad\qquad\qquad\qquad\qquad \text{Add } \left(\frac{b}{2a}\right)^2$$

$$\left(x + \frac{b}{2a}\right)^2 = \frac{-4ac + b^2}{4a^2} \qquad \text{Perfect square}$$

$$x + \frac{b}{2a} = \pm\sqrt{\frac{-4ac + b^2}{4a^2}} = \pm\frac{\sqrt{b^2 - 4ac}}{2a} \qquad \text{Take square root}$$

$$x = \frac{-b \pm \sqrt{b^2 - 4ac}}{2a} \qquad \text{Subtract } \frac{b}{2a}$$

This is the quadratic formula.

THE QUADRATIC FORMULA

The roots of the quadratic equation $ax^2 + bx + c = 0$, where $a \neq 0$, are

$$x = \frac{-b \pm \sqrt{b^2 - 4ac}}{2a}$$

The quadratic formula could be used to solve the equations in Example 5. You should carry out the details of these calculations.

EXAMPLE 6 | Using the Quadratic Formula

Find all solutions of each equation.

(a) $3x^2 - 5x - 1 = 0$ **(b)** $4x^2 + 12x + 9 = 0$ **(c)** $x^2 + 2x + 2 = 0$

SOLUTION

(a) Using the quadratic formula with $a = 3$, $b = -5$, and $c = -1$, we get

$$x = \frac{-(-5) \pm \sqrt{(-5)^2 - 4(3)(-1)}}{2(3)} = \frac{5 \pm \sqrt{37}}{6}$$

If approximations are desired, we use a calculator and obtain

$$x = \frac{5 + \sqrt{37}}{6} \approx 1.8471 \qquad \text{and} \qquad x = \frac{5 - \sqrt{37}}{6} \approx -0.1805$$

(b) Using the quadratic formula with $a = 4$, $b = 12$, and $c = 9$ gives

$$x = \frac{-12 \pm \sqrt{(12)^2 - 4 \cdot 4 \cdot 9}}{2 \cdot 4} = \frac{-12 \pm 0}{8} = -\frac{3}{2}$$

This equation has only one solution, $x = -\frac{3}{2}$.

Library of Congress

François Viète (1540–1603) had a successful political career before taking up mathematics late in life. He became one of the most famous French mathematicians of the 16th century. Viète introduced a new level of abstraction in algebra by using letters to stand for *known* quantities in an equation. Before Viète's time, each equation had to be solved on its own. For instance, the quadratic equations

$$3x^2 + 2x + 8 = 0$$

and $\quad 5x^2 - 6x + 4 = 0$

had to be solved separately for the unknown x. Viète's idea was to consider all quadratic equations at once by writing

$$ax^2 + bx + c = 0$$

where a, b, and c are known quantities. Thus, he made it possible to write a *formula* (in this case, the quadratic formula) involving a, b, and c that can be used to solve all such equations in one fell swoop. Viète's mathematical genius proved valuable during a war between France and Spain. To communicate with their troops, the Spaniards used a complicated code which Viète managed to decipher. Unaware of Viète's accomplishment, King Philip II of Spain protested to the Pope, claiming that the French were using witchcraft to read his messages.

(c) Using the quadratic formula with $a = 1$, $b = 2$, and $c = 2$ gives

$$x = \frac{-2 \pm \sqrt{2^2 - 4 \cdot 2}}{2} = \frac{-2 \pm \sqrt{-4}}{2} = \frac{-2 \pm 2\sqrt{-1}}{2} = -1 \pm \sqrt{-1}$$

Since the square of any real number is nonnegative, $\sqrt{-1}$ is undefined in the real number system. The equation has no real solution.

In Appendix D.1 we study the complex number system, in which the square roots of negative numbers do exist. The equation in Example 6(c) does have solutions in the complex number system.

The quantity $b^2 - 4ac$ that appears under the square root sign in the quadratic formula is called the *discriminant* of the equation $ax^2 + bx + c = 0$ and is given the symbol D. If $D < 0$, then $\sqrt{b^2 - 4ac}$ is undefined, and the quadratic equation has no real solution, as in Example 6(c). If $D = 0$, then the equation has only one real solution, as in Example 6(b). Finally, if $D > 0$, then the equation has two distinct real solutions, as in Example 6(a). The following box summarizes these observations.

THE DISCRIMINANT

The **discriminant** of the general quadratic equation $ax^2 + bx + c = 0$ $(a \neq 0)$ is $D = b^2 - 4ac$.

1. If $D > 0$, then the equation has two distinct real solutions.

2. If $D = 0$, then the equation has exactly one real solution.

3. If $D < 0$, then the equation has no real solution.

EXAMPLE 7 | Using the Discriminant

Use the discriminant to determine how many real solutions each equation has.

(a) $x^2 + 4x - 1 = 0$ **(b)** $4x^2 - 12x + 9 = 0$ **(c)** $\frac{1}{3}x^2 - 2x + 4 = 0$

SOLUTION

(a) The discriminant is $D = 4^2 - 4(1)(-1) = 20 > 0$, so the equation has two distinct real solutions.

(b) The discriminant is $D = (-12)^2 - 4 \cdot 4 \cdot 9 = 0$, so the equation has exactly one real solution.

(c) The discriminant is $D = (-2)^2 - 4\left(\frac{1}{3}\right)4 = -\frac{4}{3} < 0$, so the equation has no real solution.

▼ Other Equations

So far we have learned how to solve linear and quadratic equations. Now we study other types of equations, including those that involve fractional expressions, radicals, and higher powers.

EXAMPLE 8 | An Equation Involving Fractional Expressions

Solve the equation $\dfrac{3}{x} + \dfrac{5}{x+2} = 2$.

SOLUTION

To simplify the equation, we multiply each side by the common denominator.

$$\left(\frac{3}{x} + \frac{5}{x+2}\right)x(x+2) = 2x(x+2) \qquad \text{Multiply by LCD } x(x+2)$$

$$3(x+2) + 5x = 2x^2 + 4x \qquad \text{Expand}$$

$$8x + 6 = 2x^2 + 4x \qquad \text{Expand LHS}$$

$$0 = 2x^2 - 4x - 6 \qquad \text{Subtract } 8x + 6$$

$$0 = x^2 - 2x - 3 \qquad \text{Divide both sides by 2}$$

$$0 = (x-3)(x+1) \qquad \text{Factor}$$

$$x - 3 = 0 \quad \text{or} \quad x + 1 = 0 \qquad \text{Zero-Product Property}$$

$$x = 3 \qquad\qquad x = -1 \qquad \text{Solve}$$

We must check our answers because multiplying by an expression that contains the variable can introduce extraneous solutions (see the *Warning* on the next page). From *Check Your Answers* we see that the solutions are $x = 3$ and -1.

CHECK YOUR ANSWERS

$x = 3$:

$\text{LHS} = \dfrac{3}{3} + \dfrac{5}{3+2}$

$\qquad = 1 + 1 = 2$

$\text{RHS} = 2$

$\text{LHS} = \text{RHS} \quad ✔$

$x = -1$:

$\text{LHS} = \dfrac{3}{-1} + \dfrac{5}{-1+2}$

$\qquad = -3 + 5 = 2$

$\text{RHS} = 2$

$\text{LHS} = \text{RHS} \quad ✔$

When you solve an equation that involves radicals, you must be especially careful to check your final answers. The next example demonstrates why.

EXAMPLE 9 | An Equation Involving a Radical

Solve the equation $2x = 1 - \sqrt{2-x}$.

SOLUTION

To eliminate the square root, we first isolate it on one side of the equal sign, then square.

$$2x - 1 = -\sqrt{2-x} \qquad \text{Subtract 1}$$

$$(2x-1)^2 = 2 - x \qquad \text{Square each side}$$

$$4x^2 - 4x + 1 = 2 - x \qquad \text{Expand LHS}$$

$$4x^2 - 3x - 1 = 0 \qquad \text{Add } -2 + x$$

$$(4x+1)(x-1) = 0 \qquad \text{Factor}$$

$$4x + 1 = 0 \quad \text{or} \quad x - 1 = 0 \qquad \text{Zero-Product Property}$$

$$x = -\tfrac{1}{4} \qquad\qquad x = 1 \qquad \text{Solve}$$

The values $x = -\tfrac{1}{4}$ and $x = 1$ are only potential solutions. We must check them to see if they satisfy the original equation. From *Check Your Answers* we see that $x = -\tfrac{1}{4}$ is a solution but $x = 1$ is not. The only solution is

$$x = -\tfrac{1}{4}$$

CHECK YOUR ANSWERS

$x = -\tfrac{1}{4}$:

$\text{LHS} = 2\left(-\tfrac{1}{4}\right) = -\tfrac{1}{2}$

$\text{RHS} = 1 - \sqrt{2 - \left(-\tfrac{1}{4}\right)}$

$\qquad = 1 - \sqrt{\tfrac{9}{4}}$

$\qquad = 1 - \tfrac{3}{2} = -\tfrac{1}{2}$

$\text{LHS} = \text{RHS} \quad ✔$

$x = 1$:

$\text{LHS} = 2(1) = 2$

$\text{RHS} = 1 - \sqrt{2-1}$

$\qquad = 1 - 1 = 0$

$\text{LHS} \ne \text{RHS} \quad ✗$

When we solve an equation, we may end up with one or more **extraneous solutions**, that is, potential solutions that do not satisfy the original equation. In Example 3 the value $x = 1$ is an extraneous solution. Extraneous solutions may be introduced when we square each side of an equation because the operation of squaring can turn a false equation into a true one. For example, $-1 \neq 1$, but $(-1)^2 = 1^2$. Thus, the squared equation may be true for more values of the variable than the original equation. That is why you must always check your answers to make sure that each satisfies the original equation.

An equation of the form $aw^2 + bw + c = 0$, where w is an algebraic expression, is an equation of **quadratic type**. We solve equations of quadratic type by substituting for the algebraic expression, as we see in the next two examples.

EXAMPLE 10 | A Fourth-Degree Equation of Quadratic Type

Find all solutions of the equation $x^4 - 8x^2 + 8 = 0$.

SOLUTION

If we set $w = x^2$, then we get a quadratic equation in the new variable w:

$$(x^2)^2 - 8x^2 + 8 = 0 \qquad \text{Write } x^4 \text{ as } (x^2)^2$$

$$w^2 - 8w + 8 = 0 \qquad \text{Let } w = x^2$$

$$w = \frac{-(-8) \pm \sqrt{(-8)^2 - 4 \cdot 8}}{2} = 4 \pm 2\sqrt{2} \qquad \text{Quadratic formula}$$

$$x^2 = 4 \pm 2\sqrt{2} \qquad w = x^2$$

$$x = \pm\sqrt{4 \pm 2\sqrt{2}} \qquad \text{Take square roots}$$

So, there are four solutions:

$$\sqrt{4 + 2\sqrt{2}}, \qquad \sqrt{4 - 2\sqrt{2}}, \qquad -\sqrt{4 + 2\sqrt{2}}, \qquad -\sqrt{4 - 2\sqrt{2}}$$

Using a calculator, we obtain the approximations $x \approx 2.61, 1.08, -2.61, -1.08$.

EXAMPLE 11 | An Equation Involving Fractional Powers

Find all solutions of the equation $x^{1/3} + x^{1/6} - 2 = 0$.

SOLUTION

This equation is of quadratic type because if we let $w = x^{1/6}$, then $w^2 = x^{1/3}$.

$$x^{1/3} + x^{1/6} - 2 = 0$$

$$w^2 + w - 2 = 0 \qquad \text{Let } w = x^{1/6}$$

$$(w - 1)(w + 2) = 0 \qquad \text{Factor}$$

$$w - 1 = 0 \quad \text{or} \quad w + 2 = 0 \qquad \text{Zero-Product Property}$$

$$w = 1 \qquad\qquad w = -2 \qquad \text{Solve}$$

$$x^{1/6} = 1 \qquad\qquad x^{1/6} = -2 \qquad w = x^{1/6}$$

$$x = 1^6 = 1 \qquad x = (-2)^6 = 64 \qquad \text{Take the 6th power}$$

From *Check Your Answers* we see that $x = 1$ is a solution but $x = 64$ is not. The only solution is $x = 1$.

$x = 1$:

$\text{LHS} = 1^{1/3} + 1^{1/6} - 2 = 0$

$\text{RHS} = 0$

$\text{LHS} = \text{RHS}$

$x = 64$:

$\text{LHS} = 64^{1/3} + 64^{1/6} - 2$

$= 4 + 2 - 2 = 4$

$\text{RHS} = 0$

$\text{LHS} \neq \text{RHS}$

EXAMPLE 12 | An Equation Involving Absolute Value

Solve the equation $|2x - 5| = 3$.

SOLUTION

By the definition of absolute value, $|2x - 5| = 3$ is equivalent to

$$2x - 5 = 3 \quad \text{or} \quad 2x - 5 = -3$$
$$2x = 8 \qquad\qquad 2x = 2$$
$$x = 4 \qquad\qquad x = 1$$

The solutions are $x = 1$, $x = 4$.

A.4 EXERCISES

1–4 ■ Determine whether the given value is a solution of the equation.

1. $3x + 7 = 5x - 1$
   (a) $x = 4$ (b) $x = \frac{3}{2}$

2. $\dfrac{1}{x} - \dfrac{1}{x + 3} = \dfrac{1}{6}$
   (a) $x = -3$ (b) $x = 3$

3. $1 - [2 - (3 - x)] = 4x - (6 + x)$
   (a) $x = 2$ (b) $x = 22$

4. $ax - 2b = 0 \quad (a \neq 0, b \neq 0)$
   (a) $x = 0$ (b) $x = \dfrac{2b}{a}$

5–16 ■ Solve the linear equation.

5. $3x - 5 = 7$

6. $4x + 12 = 28$

7. $x - 3 = 2x + 6$

8. $4x + 7 = 9x - 13$

9. $-7w = 15 - 2w$

10. $5t - 13 = 12 - 5t$

11. $\frac{1}{2}y - 2 = \frac{1}{3}y$

12. $\dfrac{z}{5} = \dfrac{3}{10}z + 7$

13. $2(1 - x) = 3(1 + 2x) + 5$

14. $5(x + 3) + 9 = -2(x - 2) - 1$

15. $4\left(y - \frac{1}{2}\right) - y = 6(5 - y)$ **16.** $\frac{2}{3}y + \frac{1}{2}(y - 3) = \dfrac{y + 1}{4}$

17–22 ■ Solve the equation by factoring.

17. $x^2 + 2x - 8 = 0$

18. $x^2 + 6x + 8 = 0$

19. $2y^2 + 7y + 3 = 0$

20. $4w^2 = 4w + 3$

21. $6x^2 + 5x = 4$

22. $x^2 = 5(x + 100)$

23–28 ■ Solve the equation by completing the square.

23. $x^2 + 2x - 2 = 0$

24. $x^2 - 6x - 9 = 0$

25. $x^2 + x - \frac{3}{4} = 0$

26. $x^2 + 22x + 21 = 0$

27. $2x^2 + 8x + 1 = 0$

28. $3x^2 - 6x - 1 = 0$

29–38 ■ Find all real solutions of the quadratic equation. Use any method.

29. $x^2 + 12x - 27 = 0$

30. $8x^2 - 6x - 9 = 0$

31. $3x^2 + 6x - 5 = 0$

32. $x^2 - 6x + 1 = 0$

33. $2y^2 - y - \frac{1}{2} = 0$

34. $\theta^2 - \frac{3}{2}\theta + \frac{9}{16} = 0$

35. $10y^2 - 16y + 5 = 0$

36. $25x^2 + 70x + 49 = 0$

37. $3x^2 + 2x + 2 = 0$

38. $5x^2 - 7x + 5 = 0$

39–62 ■ Find all real solutions of the equation.

39. $\dfrac{1}{x - 1} + \dfrac{1}{x + 2} = \dfrac{5}{4}$ **40.** $\dfrac{10}{x} - \dfrac{12}{x - 3} + 4 = 0$

41. $\dfrac{x^2}{x + 100} = 50$ **42.** $\dfrac{1}{x - 1} - \dfrac{2}{x^2} = 0$

43. $\dfrac{x+5}{x-2} = \dfrac{5}{x+2} + \dfrac{28}{x^2-4}$

44. $\dfrac{x}{2x+7} - \dfrac{x+1}{x+3} = 1$

45. $\sqrt{2x+1} + 1 = x$

46. $\sqrt{5-x} + 1 = x - 2$

47. $2x + \sqrt{x+1} = 8$

48. $\sqrt{\sqrt{x-5}+x} = 5$

49. $x^4 - 13x^2 + 40 = 0$

50. $x^4 - 5x^2 + 4 = 0$

51. $2x^4 + 4x^2 + 1 = 0$

52. $x^6 - 2x^3 - 3 = 0$

53. $x^{4/3} - 5x^{2/3} + 6 = 0$

54. $\sqrt{x} - 3\sqrt[4]{x} - 4 = 0$

55. $4(x+1)^{1/2} - 5(x+1)^{3/2} + (x+1)^{5/2} = 0$

56. $x^{1/2} + 3x^{-1/2} = 10x^{-3/2}$

57. $x^{1/2} - 3x^{1/3} = 3x^{1/6} - 9$

58. $x - 5\sqrt{x} + 6 = 0$

59. $|2x| = 3$

60. $|3x+5| = 1$

61. $|x-4| = 0.01$

62. $|x-6| = -1$

63–74 ■ Solve the equation for the indicated variable.

63. $PV = nRT$; for R

64. $F = G\dfrac{mM}{r^2}$; for m

65. $\dfrac{1}{R} = \dfrac{1}{R_1} + \dfrac{1}{R_2}$; for R_1

66. $P = 2l + 2w$; for w

67. $\dfrac{ax+b}{cx+d} = 2$; for x

68. $a - 2[b - 3(c - x)] = 6$; for x

69. $a^2x + (a-1) = (a+1)x$; for x

70. $\dfrac{a+1}{b} = \dfrac{a-1}{b} + \dfrac{b+1}{a}$; for a

71. $V = \tfrac{1}{3}\pi r^2 h$; for r

72. $F = G\dfrac{mM}{r^2}$; for r

73. $a^2 + b^2 = c^2$; for b

74. $A = P\left(1 + \dfrac{i}{100}\right)^2$; for i

75–78 ■ Use the discriminant to determine the number of real solutions of the equation. Do not solve the equation.

75. $x^2 - 6x + 1 = 0$

76. $3x^2 = 6x - 9$

77. $x^2 + 2.20x + 1.21 = 0$

78. $x^2 + rx - s = 0$ $(s > 0)$

79–80 ■ Suppose an object is dropped from a height h_0 above the ground. Then its height after t seconds is given by $h = -16t^2 + h_0$, where h is measured in feet. Use this information to solve the problem.

79. If a ball is dropped from 288 ft above the ground, how long does it take to reach ground level?

80. A ball is dropped from the top of a building 96 ft tall.
(a) How long will it take to fall half the distance to ground level?
(b) How long will it take to fall to ground level?

81. The fish population in a certain lake rises and falls according to the formula

$$F = 1000(30 + 17t - t^2)$$

Here F is the number of fish at time t, where t is measured in years since January 1, 1992, when the fish population was first estimated.
(a) On what date will the fish population again be the same as on January 1, 1992?
(b) By what date will all the fish in the lake have died?

82. If an imaginary line segment is drawn between the centers of the earth and the moon, then the net gravitational force F acting on an object situated on this line segment is

$$F = \dfrac{-K}{x^2} + \dfrac{0.012K}{(239 - x)^2}$$

where $K > 0$ is a constant and x is the distance of the object from the center of the earth, measured in thousands of miles. How far from the center of the earth is the "dead spot" where no net gravitational force acts upon the object? (Express your answer to the nearest thousand miles.)

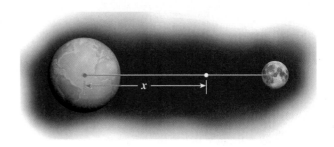

A.5 INEQUALITIES

| Linear Inequalities ▶ Absolute Value Inequalites

x	$4x + 7 \leqslant 19$
1	$11 \leqslant 19$ ✔
2	$15 \leqslant 19$ ✔
3	$19 \leqslant 19$ ✔
4	$23 \leqslant 19$ ✗
5	$27 \leqslant 19$ ✗

Some problems in algebra lead to **inequalities** instead of equations. An inequality looks just like an equation, except that in the place of the equal sign is one of the symbols, $<$, $>$, $\leqslant$, or $\geqslant$. Here is an example of an inequality:

$$4x + 7 \leqslant 19$$

The table in the margin shows that some numbers satisfy the inequality and some numbers don't.

To **solve** an inequality that contains a variable means to find all values of the variable that make the inequality true. Unlike an equation, an inequality generally has infinitely many solutions, which form an interval or a union of intervals on the real line. The following illustration shows how an inequality differs from its corresponding equation:

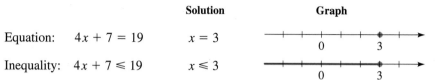

		Solution	Graph
Equation:	$4x + 7 = 19$	$x = 3$	
Inequality:	$4x + 7 \leq 19$	$x \leq 3$	

To solve inequalities, we use the following rules to isolate the variable on one side of the inequality sign. These rules tell us when two inequalities are *equivalent* (the symbol ⇔ means "is equivalent to"). In these rules the symbols A, B, and C stand for real numbers or algebraic expressions. Here we state the rules for inequalities involving the symbol ≤, but they apply to all four inequality symbols.

RULES FOR INEQUALITIES

Rule	Description
1. $A \leq B \iff A + C \leq B + C$	**Adding** the same quantity to each side of an inequality gives an equivalent inequality.
2. $A \leq B \iff A - C \leq B - C$	**Subtracting** the same quantity from each side of an inequality gives an equivalent inequality.
3. If $C > 0$, then $A \leq B \iff CA \leq CB$	**Multiplying** each side of an inequality by the same *positive* quantity gives an equivalent inequality.
4. If $C < 0$, then $A \leq B \iff CA \geq CB$	**Multiplying** each side of an inequality by the same *negative* quantity *reverses the direction* of the inequality.
5. If $A > 0$ and $B > 0$, then $A \leq B \iff \dfrac{1}{A} \geq \dfrac{1}{B}$	**Taking reciprocals** of each side of an inequality involving *positive* quantities *reverses the direction* of the inequality.
6. If $A \leq B$ and $C \leq D$, then $A + C \leq B + D$	Inequalities can be added.

 Pay special attention to Rules 3 and 4. Rule 3 says that we can multiply (or divide) each side of an inequality by a *positive* number, but Rule 4 says that if we multiply each side of an inequality by a *negative* number, then we reverse the direction of the inequality. For example, if we start with the inequality

$$3 < 5$$

and multiply by 2, we get

$$6 < 10$$

but if we multiply by −2, we get

$$-6 > -10$$

▼ Linear Inequalities

An inequality is **linear** if each term is constant or a multiple of the variable.

EXAMPLE 1 | Solving a Linear Inequality

Solve the inequality $3x < 9x + 4$ and sketch the solution set.

SOLUTION

$$3x < 9x + 4$$
$$3x - 9x < 9x + 4 - 9x \qquad \text{Subtract } 9x$$
$$-6x < 4 \qquad \text{Simplify}$$
$$\left(-\tfrac{1}{6}\right)(-6x) > -\tfrac{1}{6}(4) \qquad \text{Multiply by } -\tfrac{1}{6} \text{ (or divide by } -6)$$
$$x > -\tfrac{2}{3} \qquad \text{Simplify}$$

The solution set consists of all numbers greater than $-\tfrac{2}{3}$. In other words the solution of the inequality is the interval $\left(-\tfrac{2}{3}, \infty\right)$. It is graphed in Figure 1.

FIGURE 1

EXAMPLE 2 | Solving a Pair of Simultaneous Inequalities

Solve the inequalities $4 \le 3x - 2 < 13$.

SOLUTION

The solution set consists of all values of x that satisfy both inequalities. Using Rules 1 and 3, we see that the following inequalities are equivalent:

$$4 \le 3x - 2 < 13$$
$$6 \le 3x < 15 \qquad \text{Add 2}$$
$$2 \le x < 5 \qquad \text{Divide by 3}$$

Therefore, the solution set is $[2, 5)$, as shown in Figure 2.

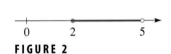

FIGURE 2

EXAMPLE 3 | Relationship between Fahrenheit and Celsius Scales

The instructions on a box of film indicate that the box should be stored at a temperature between 5°C and 30°C. What range of temperatures does this correspond to on the Fahrenheit scale?

SOLUTION

The relationship between degrees Celsius (C) and degrees Fahrenheit (F) is given by the equation $C = \tfrac{5}{9}(F - 32)$. Expressing the statement on the box in terms of inequalities, we have

$$5 < C < 30$$

so the corresponding Fahrenheit temperatures satisfy the inequalities

$$5 < \tfrac{5}{9}(F - 32) < 30$$
$$\tfrac{9}{5} \cdot 5 < F - 32 < \tfrac{9}{5} \cdot 30 \qquad \text{Multiply by } \tfrac{9}{5}$$
$$9 < F - 32 < 54 \qquad \text{Simplify}$$
$$9 + 32 < F < 54 + 32 \qquad \text{Add 32}$$
$$41 < F < 86 \qquad \text{Simplify}$$

The film should be stored at a temperature between 41°F and 86°F.

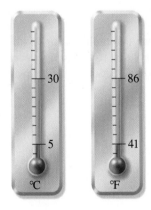

▼ Absolute Value Inequalites

We use the following properties to solve inequalities that involve absolute value.

PROPERTIES OF ABSOLUTE VALUE INEQUALITIES		
Inequality	**Equivalent form**	**Graph**
1. $\lvert x \rvert < c$	$-c < x < c$	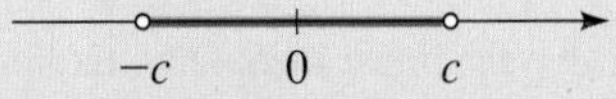
2. $\lvert x \rvert \le c$	$-c \le x \le c$	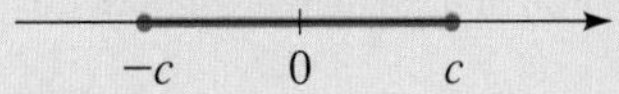
3. $\lvert x \rvert > c$	$x < -c$ or $c < x$	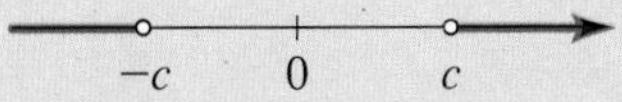
4. $\lvert x \rvert \ge c$	$x \le -c$ or $c \le x$	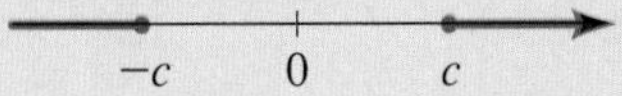

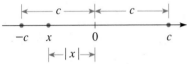

FIGURE 3

These properties can be proved using the definition of absolute value. To prove Property 1, for example, we see that the inequality $\lvert x \rvert < c$ says that the distance from x to 0 is less than c, and from Figure 3 you can see that this is true if and only if x is between c and $-c$.

EXAMPLE 4 | Solving an Absolute Value Inequality

Solve the inequality $\lvert x - 5 \rvert < 2$.

SOLUTION 1

The inequality $\lvert x - 5 \rvert < 2$ is equivalent to

$$-2 < x - 5 < 2 \qquad \text{Property 1}$$
$$3 < x < 7 \qquad \text{Add 5}$$

The solution set is the open interval $(3, 7)$.

SOLUTION 2

Geometrically, the solution set consists of all numbers x whose distance from 5 is less than 2. From Figure 4 we see that this is the interval $(3, 7)$.

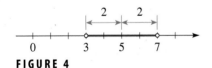

FIGURE 4

EXAMPLE 5 | Solving an Absolute Value Inequality

Solve the inequality $\lvert 3x + 2 \rvert \ge 4$.

SOLUTION

By Property 4 the inequality $\lvert 3x + 2 \rvert \ge 4$ is equivalent to

$$3x + 2 \ge 4 \qquad \text{or} \qquad 3x + 2 \le -4$$
$$3x \ge 2 \qquad\qquad 3x \le -6 \qquad \text{Subtract 2}$$
$$x \ge \tfrac{2}{3} \qquad\qquad x \le -2 \qquad \text{Divide by 3}$$

So the solution set is

$$\left\{ x \mid x \le -2 \ \text{ or } \ x \ge \tfrac{2}{3} \right\} = (-\infty, -2] \cup \left[\tfrac{2}{3}, \infty\right)$$

FIGURE 5

The set is graphed in Figure 5.

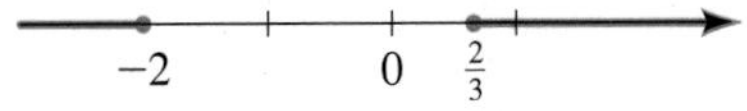

A.5 EXERCISES

1–4 ■ Let $S = \left\{-1, 0, \frac{1}{2}, \sqrt{2}, 2\right\}$. Use substitution to determine which elements of S satisfy the inequality.

1. $2x + 10 > 8$

2. $4x + 1 \le 2x$

3. $\dfrac{1}{x} \le \dfrac{1}{2}$

4. $x^2 + 2 < 4$

5–24 ■ Solve the linear inequality. Express the solution using interval notation and graph the solution set.

5. $2x \le 7$

6. $-4x \ge 10$

7. $2x - 5 > 3$

8. $3x + 11 < 5$

9. $7 - x \ge 5$

10. $5 - 3x \le -16$

11. $2x + 1 < 0$

12. $0 < 5 - 2x$

13. $3x + 11 \le 6x + 8$

14. $6 - x \ge 2x + 9$

15. $\frac{1}{2}x - \frac{2}{3} > 2$

16. $\frac{2}{5}x + 1 < \frac{1}{5} - 2x$

17. $4 - 3x \le -(1 + 8x)$

18. $2(7x - 3) \le 12x + 16$

19. $2 \le x + 5 < 4$

20. $5 \le 3x - 4 \le 14$

21. $-1 < 2x - 5 < 7$

22. $1 < 3x + 4 \le 16$

23. $-2 < 8 - 2x \le -1$

24. $-3 \le 3x + 7 \le \frac{1}{2}$

25–36 ■ Solve the inequality. Express the solution using interval notation and graph the solution set.

25. $|x| < 7$

26. $|x| \ge 3$

27. $|x - 5| \le 3$

28. $|x - 9| > 9$

29. $|x + 5| < 2$

30. $|x + 1| \ge 3$

31. $|2x - 3| \le 0.4$

32. $|5x - 2| < 6$

33. $\left|\dfrac{x - 2}{3}\right| < 2$

34. $\left|\dfrac{x + 1}{2}\right| \ge 4$

35. $4|x + 2| - 3 < 13$

36. $3 - |2x + 4| \le 1$

37. Use the relationship between C and F given in Example 3 to find the interval on the Fahrenheit scale corresponding to the temperature range $20 \le C \le 30$.

38. What interval on the Celsius scale corresponds to the temperature range $50 \le F \le 95$?

39. As dry air moves upward, it expands and in so doing cools at a rate of about $1°C$ for each 100 m rise, up to about 12 km.
   (a) If the ground temperature is $20°C$, write a formula for the temperature at height h.
   (b) What range of temperatures can be expected if a plane takes off and reaches a maximum height of 5 km?

40. A coffee merchant sells a customer 3 lb of Hawaiian Kona at $6.50 per pound. His scale is accurate to within ±0.03 lb. By how much could the customer have been overcharged or undercharged because of possible inaccuracy in the scale?

B.1 CONGRUENCE AND SIMILARITY OF TRIANGLES

| Congruent Triangles ▶ Similar Triangles

In this appendix we review the concepts of congruence and similarity, which are essential in the study of trigonometry.

▼ Congruent Triangles

In general, two geometric figures are congruent if they have the same shape and size. In particular, two line segments are congruent if they have the same length, and two angles are congruent if they have the same measure. For triangles, we have the following definition.

CONGRUENT TRIANGLES

Two triangles are **congruent** if their vertices can be matched up so that corresponding sides and angles are congruent.

We write $\triangle ABC \cong \triangle PQR$ to mean that triangle ABC is congruent to triangle PQR and that the sides and angles correspond as follows:

$AB = PQ \quad \angle A = \angle P$

$BC = QR \quad \angle B = \angle Q$

$AC = PR \quad \angle C = \angle R$

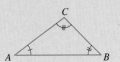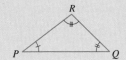

(a) SSS

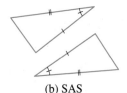

(b) SAS

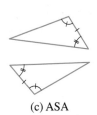

(c) ASA

FIGURE 1

To prove that two triangles are congruent, we don't need to show that all six corresponding parts (side and angles) are congruent. For instance, if all three sides are congruent, then all three angles must also be congruent. You can easily see why the following properties lead to congruent triangles.

- **Side-Side-Side (SSS).** If each side of one triangle is congruent to the corresponding side of another triangle, then the two triangles are congruent. See Figure 1(a).

- **Side-Angle-Side (SAS).** If two sides and the included angle in one triangle are congruent to the corresponding sides and angle in another triangle, then the two triangles are congruent. See Figure 1(b).

- **Angle-Side-Angle (ASA).** If two angles and the included side in one triangle are congruent to the corresponding angles and side in another triangle, then the triangles are congruent. See Figure 1(c).

EXAMPLE 1 | Congruent Triangles

(a) $\triangle ADB \cong \triangle CBD$ by SSS.

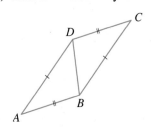

(b) $\triangle ABE \cong \triangle CBD$ by SAS.

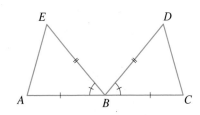

525

(c) $\triangle ABD \cong \triangle CBD$ by ASA.

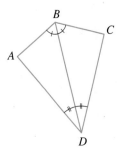

(d) These triangles are not necessarily congruent. "Side-side-angle" does *not* determine congruence.

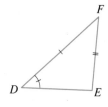

▼ Similar Triangles

Two geometric figures are similar if they have the same shape, but not necessarily the same size. (See the Discovery Project *Similarity* referenced on page 194.) In the case of triangles, we can define similarity as follows.

SIMILAR TRIANGLES

Two triangles are **similar** if their vertices can be matched up so that corresponding angles are congruent. In this case, corresponding sides are proportional.

We write $\triangle ABC \sim \triangle PQR$ to mean that triangle *ABC* is similar to triangle *PQR* and that the following conditions hold.

The angles correspond as follows:

$$\angle A = \angle P, \ \angle B = \angle Q, \ \angle C = \angle R$$

The sides are proportional as follows:

$$\frac{AB}{PQ} = \frac{BC}{QR} = \frac{AC}{PR}$$

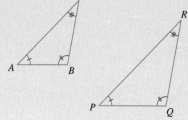

The sum of the angles in any triangle is 180°. So, if we know two angles in a triangle, the third is determined. Thus, to prove that two triangles are similar, we need only show that two angles in one are congruent to two angles in the other.

EXAMPLE 2 | Similar Triangles

Find all pairs of similar triangles in the figures.

(a)

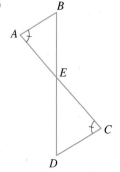

(b)

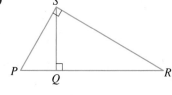

SOLUTION

(a) Since $\angle AEB$ and $\angle CED$ are opposite angles, they are equal. Thus, $\triangle AEB \sim \triangle CED$.

(b) Since all triangles in the figure are right triangles, we have

$$\angle QSR + \angle QRS = 90°$$
$$\angle QSR + \angle QSP = 90°$$

Subtracting these equations we find that $\angle QSP = \angle QRS$. Thus

$$\triangle PQS \sim \triangle SQR \sim \triangle PSR$$

EXAMPLE 3 | Proportional Sides in Similar Triangles

Given that the triangles in the figure are similar, find the lengths x and y.

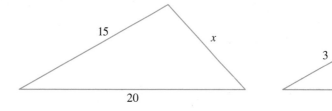

SOLUTION

By similarity, we know that the lengths of corresponding sides in the triangles are proportional. First we find x:

$$\frac{x}{2} = \frac{15}{3} \qquad \text{Corresponding sides are proportional}$$

$$x = \frac{2 \cdot 15}{3} = 10 \qquad \text{Solve for } x$$

Now we find y:

$$\frac{15}{3} = \frac{20}{y} \qquad \text{Corresponding sides are proportional}$$

$$y = \frac{20 \cdot 3}{15} = 4 \qquad \text{Solve for } y$$

B.1 EXERCISES

1–4 ■ Determine whether the pair of triangles is congruent. If so, state the congruence principle you are using.

1.

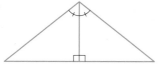

2.

3.

4.

5–8 ■ Determine whether the pair of triangles is similar.

5.

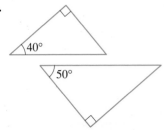

6. **7.**

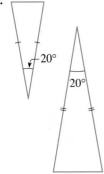

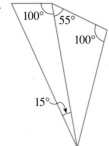

8.

11. **12.**

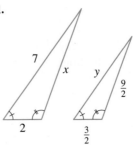

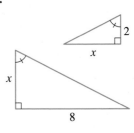

13–14 ■ Express x in terms of a, b, and c.

13.

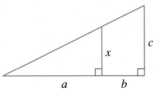

14.

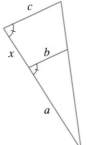

15. In the figure *CDEF* is a rectangle. Prove that $\triangle ABC \sim \triangle AED \sim \triangle EBF$.

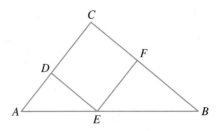

16. In the figure *DEFG* is a square. Prove the following:
   (a) $\triangle ADG \sim \triangle GCF$
   (b) $\triangle ADG \sim \triangle FEB$
   (c) $AD \cdot EB = DG \cdot FE$
   (d) $DE = \sqrt{AD \cdot EB}$

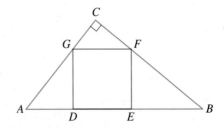

9–12 ■ Given that the pair of triangles is similar, find the length(s) x and/or y.

9.

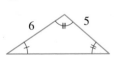

10.

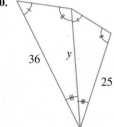

17. Two vertical poles, one 8 ft tall and the other 24 ft tall, have ropes stretched from the top of each to the base of the other (see the figure). How high above the ground is the point where the ropes cross?

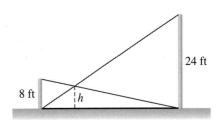

B.2 THE PYTHAGOREAN THEOREM

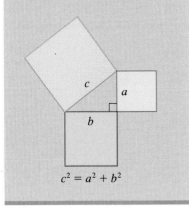

$$c^2 = a^2 + b^2$$

In a right triangle, the side opposite the right angle is called the **hypotenuse** and the other two sides are called the **legs**.

THE PYTHAGOREAN THEOREM

In a right triangle the square of the hypotenuse is equal to the sum of the squares of the legs. That is, in triangle ABC in the figure

$$a^2 + b^2 = c^2$$

EXAMPLE 1 | Using the Pythagorean Theorem

Find the lengths x and y in the right triangles shown.

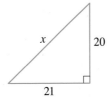

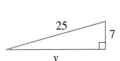

SOLUTION

(a) We use the Pythagorean theorem with $a = 20$, and $b = 21$, and $c = x$. Then $x^2 = 20^2 + 21^2 = 841$. So $x = \sqrt{841} = 29$.

(b) We use the Pythagorean theorem with $c = 25$, $a = 7$, and $b = y$. Then $25^2 = 7^2 + y^2$, so $y^2 = 25^2 - 7^2 = 576$. Thus, $y = \sqrt{576} = 24$.

The converse of the Pythagorean theorem is also true.

CONVERSE OF THE PYTHAGOREAN THEOREM

If the square of one side of a triangle is equal to the sum of the squares of the other two sides, then the triangle is a right triangle.

EXAMPLE 2 | Proving That a Triangle Is a Right Triangle

Prove that the triangle with sides of length 8, 15, and 17 is a right triangle.

SOLUTION

You can check that $8^2 + 15^2 = 17^2$. So the triangle must be a right triangle by the converse of the Pythagorean theorem.

B.2 EXERCISES

1–6 ■ In the given right triangle, find the side labeled x.

1.
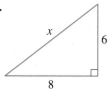

2.
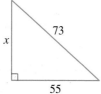

3.

4.
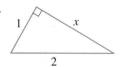

5.
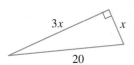

6.
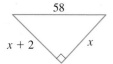

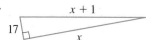

7–12 ■ The lengths of the sides of a triangle are given. Determine whether the triangle is a right triangle.

7. 5, 12, 13   **8.** 15, 20, 25

9. 8, 10, 12   **10.** 6, 17, 18

11. 48, 55, 73   **12.** 13, 84, 85

13. One leg of a right triangle measures 11 cm. The hypotenuse is 1 cm longer than the other leg. Find the length of the hypotenuse.

14. The length of a rectangle is 1 ft greater than its width. Each diagonal is 169 ft long. Find the dimensions of the rectangle.

15. Each of the diagonals of a quadrilateral is 27 cm long. Two adjacent sides measure 17 cm and 21 cm. Is the quadrilateral a rectangle?

16. Find the height h of the right triangle ABC shown in the figure. [*Hint*: Find the area of triangle ABC in two different ways.]

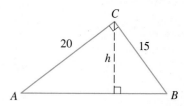

17. Find the length of the diagonal of the rectangular box shown in the figure.

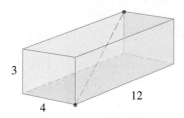

18. If a, b, c are positive integers such that $a^2 + b^2 = c^2$, then (a, b, c) is called a **Pythagorean triple**.

(a) Let m and n be positive integers with $m > n$. Let $a = m^2 - n^2$, $b = 2mn$, and $c = m^2 + n^2$. Show that (a, b, c) is a Pythagorean triple.

(b) Use part (a) to find the rest of the Pythagorean triples in the table.

m	n	(a, b, c)
2	1	(3, 4, 5)
3	1	(8, 6, 10)
3	2	
4	1	
4	2	
4	3	
5	1	
5	2	
5	3	
5	4	

C.1 USING A GRAPHING CALCULATOR

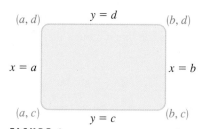

FIGURE 1
The viewing rectangle $[a, b]$ by $[c, d]$

A graphing calculator or computer displays a rectangular portion of the graph of an equation in a display window or viewing screen, which we call a **viewing rectangle**. The default screen often gives an incomplete or misleading picture, so it is important to choose the viewing rectangle with care. If we choose the x-values to range from a minimum value of **Xmin** $= a$ to a maximum value of **Xmax** $= b$ and the y-values to range from a minimum value of **Ymin** $= c$ to a maximum value of **Ymax** $= d$, then the displayed portion of the graph lies in the rectangle

$$[a, b] \times [c, d] = \{(x, y) \mid a \leqslant x \leqslant b, c \leqslant x \leqslant d\}$$

as shown in Figure 1. We refer to this as the $[a, b]$ by $[c, d]$ viewing rectangle.

The graphing device draws the graph of an equation much as you would. It plots points of the form (x, y) for a certain number of values of x, equally spaced between a and b. If the equation is not defined for an x-value, or if the corresponding y-value lies outside the viewing rectangle, the device ignores this value and moves on to the next x-value. The machine connects each point to the preceding plotted point to form a representation of the graph of the equation.

EXAMPLE 1 | Choosing an Appropriate Viewing Rectangle

Graph the equation $y = x^2 + 3$ in an appropriate viewing rectangle.

SOLUTION

Let's experiment with different viewing rectangles. We'll start with the viewing rectangle $[-2, 2]$ by $[-2, 2]$, so we set

$$\mathbf{Xmin} = -2 \qquad \mathbf{Ymin} = -2$$
$$\mathbf{Xmax} = 2 \qquad \mathbf{Ymax} = 2$$

The resulting graph in Figure 2(a) is blank! This is because $x^2 \geqslant 0$, so $x^2 + 3 \geqslant 3$ for all x. Thus, the graph lies entirely above the viewing rectangle, so this viewing rectangle is not appropriate. If we enlarge the viewing rectangle to $[-4, 4]$ by $[-4, 4]$, as in Figure 2(b), we begin to see a portion of the graph.

Now let's try the viewing rectangle $[-10, 10]$ by $[-5, 30]$. The graph in Figure 2(c) seems to give a more complete view of the graph. If we enlarge the viewing rectangle even further, as in Figure 2(d), the graph doesn't show clearly that the y-intercept is 3.

So, the viewing rectangle $[-10, 10]$ by $[-5, 30]$ gives an appropriate representation of the graph.

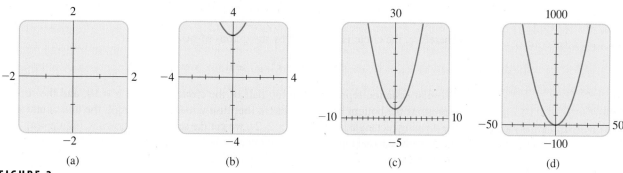

(a)   (b)   (c)   (d)

FIGURE 2
Graphs of $y = x^2 + 3$

Alan Turing (1912–1954) was at the center of two pivotal events of the 20th century—World War II and the invention of computers. At the age of 23 Turing made his mark on mathematics by solving an important problem in the foundations of mathematics that was posed by David Hilbert at the 1928 International Congress of Mathematicians. In this research he invented a theoretical machine, now called a Turing machine, which was the inspiration for modern digital computers. During World War II Turing was in charge of the British effort to decipher secret German codes. His complete success in this endeavor played a decisive role in the Allies' victory. To carry out the numerous logical steps required to break a coded message, Turing developed decision procedures similar to modern computer programs. After the war he helped develop the first electronic computers in Britain. He also did pioneering work on artificial intelligence and computer models of biological processes. At the age of 42 Turing died of cyanide poisoning under mysterious circumstances.

EXAMPLE 2 | Two Graphs on the Same Screen

Graph the equations $y = 3x^2 - 6x + 1$ and $y = 0.23x - 2.25$ together in the viewing rectangle $[-1, 3]$ by $[-2.5, 1.5]$. Do the graphs intersect in this viewing rectangle?

SOLUTION

Figure 3(a) shows the essential features of both graphs. One is a parabola and the other is a line. It looks as if the graphs intersect near the point $(1, -2)$. However, if we zoom in on the area around this point as shown in Figure 3(b), we see that although the graphs almost touch, they don't actually intersect.

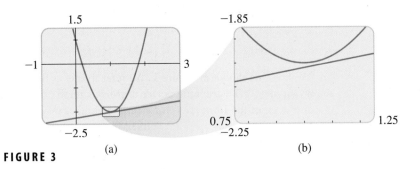

FIGURE 3   (a)   (b)

You can see from Examples 1 and 2 that the choice of a viewing rectangle makes a big difference in the appearance of a graph. If you want an overview of the essential features of a graph, you must choose a relatively large viewing rectangle to obtain a global view of the graph. If you want to investigate the details of a graph, you must zoom in to a small viewing rectangle that shows just the feature of interest.

Most graphing calculators can only graph equations in which y is isolated on one side of the equal sign. The next example shows how to graph equations that don't have this property.

EXAMPLE 3 | Graphing a Circle

Graph the circle $x^2 + y^2 = 1$.

SOLUTION

We first solve for y, to isolate it on one side of the equal sign.

$$y^2 = 1 - x^2$$
$$y = \pm\sqrt{1 - x^2}$$

Therefore, the circle is described by the graphs of *two* equations:

$$y = \sqrt{1 - x^2} \qquad \text{and} \qquad y = -\sqrt{1 - x^2}$$

The first equation represents the top half of the circle (because $y \geq 0$), and the second represents the bottom half of the circle (because $y \leq 0$). If we graph the first equation in the viewing rectangle $[-2, 2]$ by $[-2, 2]$, we get the semicircle shown in Figure 4(a). The graph of the second equation is the semicircle in Figure 4(b). Graphing these semicircles together on the same viewing screen, we get the full circle in Figure 4(c).

The graph in Figure 4(c) looks somewhat flattened. Most graphing calculators allow you to set the scales on the axes so that circles really look like circles. On the TI-82 and TI-83, from the ZOOM menu, choose ZSquare to set the scales appropriately. (On the TI-85 the command is Zsq.)

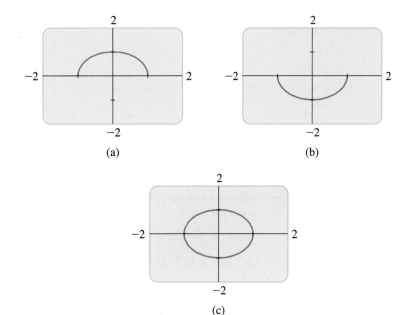

FIGURE 4
Graphing the equation $x^2 + y^2 + 1$

(a)   (b)   (c)

Other uses and capabilities of the graphing calculator are discussed in Appendix C.2 and throughout this textbook.

C.1 EXERCISES

1–6 ■ Use a graphing calculator or computer to decide which viewing rectangle (a)–(d) produces the most appropriate graph of the equation.

1. $y = x^4 + 2$
  (a) $[-2, 2]$ by $[-2, 2]$
  (b) $[0, 4]$ by $[0, 4]$
  (c) $[-8, 8]$ by $[-4, 40]$
  (d) $[-40, 40]$ by $[-80, 800]$

2. $y = x^2 + 7x + 6$
  (a) $[-5, 5]$ by $[-5, 5]$
  (b) $[0, 10]$ by $[-20, 100]$
  (c) $[-15, 8]$ by $[-20, 100]$
  (d) $[-10, 3]$ by $[-100, 20]$

3. $y = 100 - x^2$
  (a) $[-4, 4]$ by $[-4, 4]$
  (b) $[-10, 10]$ by $[-10, 10]$
  (c) $[-15, 15]$ by $[-30, 110]$
  (d) $[-4, 4]$ by $[-30, 110]$

4. $y = 2x^2 - 1000$
  (a) $[-10, 10]$ by $[-10, 10]$
  (b) $[-10, 10]$ by $[-100, 100]$
  (c) $[-10, 10]$ by $[-1000, 1000]$
  (d) $[-25, 25]$ by $[-1200, 200]$

5. $y = 10 + 25x - x^3$
  (a) $[-4, 4]$ by $[-4, 4]$
  (b) $[-10, 10]$ by $[-10, 10]$
  (c) $[-20, 20]$ by $[-100, 100]$
  (d) $[-100, 100]$ by $[-200, 200]$

6. $y = \sqrt{8x - x^2}$
  (a) $[-4, 4]$ by $[-4, 4]$
  (b) $[-5, 5]$ by $[0, 100]$
  (c) $[-10, 10]$ by $[-10, 40]$
  (d) $[-2, 10]$ by $[-2, 6]$

7–18 ■ Determine an appropriate viewing rectangle for the equation and use it to draw the graph.

7. $y = 100x^2$ **8.** $y = -100x^2$

9. $y = 4 + 6x - x^2$ **10.** $y = 0.3x^2 + 1.7x - 3$

11. $y = \sqrt[4]{256 - x^2}$ **12.** $y = \sqrt{12x - 17}$

13. $y = 0.01x^3 - x^2 + 5$ **14.** $y = x(x + 6)(x - 9)$

15. $y = x^4 - 4x^3$ **16.** $y = \dfrac{x}{x^2 + 25}$

17. $y = 1 + |x - 1|$ **18.** $y = 2x - |x^2 - 5|$

19. Graph the circle $x^2 + y^2 = 9$ by solving for y and graphing two equations as in Example 3.

20. Graph the circle $(y - 1)^2 + x^2 = 1$ by solving for y and graphing two equations as in Example 3.

21. Graph the equation $4x^2 + 2y^2 = 1$ by solving for y and graphing two equations corresponding to the negative and positive square roots. (This graph is called an *ellipse*.)

22. Graph the equation $y^2 - 9x^2 = 1$ by solving for y and graphing the two equations corresponding to the positive and negative square roots. (This graph is called a *hyperbola*.)

23–26 ■ Do the graphs intersect in the given viewing rectangle? If they do, how many points of intersection are there?

23. $y = -3x^2 + 6x - \frac{1}{2}$, $y = \sqrt{7 - \frac{7}{12}x^2}$; $[-4, 4]$ by $[-1, 3]$

24. $y = \sqrt{49 - x^2}$, $y = \frac{1}{5}(41 - 3x)$; $[-8, 8]$ by $[-1, 8]$

25. $y = 6 - 4x - x^2$, $y = 3x + 18$; $[-6, 2]$ by $[-5, 20]$

26. $y = x^3 - 4x$, $y = x + 5$; $[-4, 4]$ by $[-15, 15]$

27. When you enter the following equations into your calculator, how does what you see on the screen differ from the usual way of writing the equations? (Check your user's manual if you're not sure.)

(a) $y = |x|$ (b) $y = \sqrt[5]{x}$

(c) $y = \dfrac{x}{x - 1}$ (d) $y = x^3 + \sqrt[3]{x + 2}$

28. A student wishes to graph the equations

$$y = x^{1/3} \quad \text{and} \quad y = \frac{x}{x + 4}$$

on the same screen, so he enters the following information into his calculator:

$$Y_1 = X^\wedge 1/3 \qquad Y_2 = X/X + 4$$

The calculator graphs two lines instead of the equations he wanted. What went wrong? Why are the graphs lines?

C.2 SOLVING EQUATIONS AND INEQUALITIES GRAPHICALLY

| Solving Equations Graphically ▶ Solving Inequalities Graphically

In Appendix A.4 and A.5 we solved equations and inequalities algebraically. In this appendix we use graphs to solve equations and inequalities.

▼ Solving Equations Graphically

One way to solve an equation like

$$3x - 5 = 0$$

is to use the **algebraic method**. This means we use the rules of algebra to isolate x on one side of the equation. We view x as an *unknown* and we use the rules of algebra to hunt it down. Here are the steps in the solution:

$$3x - 5 = 0$$
$$3x = 5 \qquad \text{Add 5}$$
$$x = \tfrac{5}{3} \qquad \text{Divide by 3}$$

So the solution is $x = \frac{5}{3}$.

We can also solve this equation by the **graphical method**. In this method we view x as a *variable* and sketch the graph of the equation

$$y = 3x - 5$$

Different values for x give different values for y. Our goal is to find the value of x for which $y = 0$. From the graph in Figure 1 we see that $y = 0$ when $x \approx 1.7$. Thus, the solution is $x \approx 1.7$. Note that from the graph we obtain an approximate solution.

We summarize these methods in the following box.

"Algebra is a merry science," Uncle Jakob would say. "We go hunting for a little animal whose name we don't know, so we call it x. When we bag our game we pounce on it and give it its right name."

ALBERT EINSTEIN

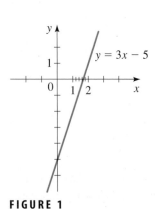

FIGURE 1

SOLVING AN EQUATION

Algebraic Method	**Graphical Method**

Use the rules of algebra to isolate the unknown x on one side of the equation.

Move all terms to one side and set equal to y. Sketch the graph to find the value of x where $y = 0$.

Example: $2x = 6 - x$

$$3x = 6 \qquad \text{Add } x$$
$$x = 2 \qquad \text{Divide by 3}$$

The solution is $x = 2$.

Example: $2x = 6 - x$

$$0 = 6 - 3x.$$

Set $y = 6 - 3x$ and graph.

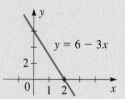

From the graph the solution is $x \approx 2$.

The advantage of the algebraic method is that it gives exact answers. Also, the process of unraveling the equation to arrive at the answer helps us understand the algebraic structure of the equation. On the other hand, for many equations it is difficult or impossible to isolate x.

The graphical method gives a numerical approximation to the answer. This is an advantage when a numerical answer is desired. (For example, an engineer might find an answer expressed as $x \approx 2.6$ more immediately useful than $x = \sqrt{7}$.) Also, graphing an equation helps us visualize how the solution is related to other values of the variable.

EXAMPLE 1 | Solving an Equation Algebraically and Graphically

Solve the equations algebraically and graphically.

(a) $x^2 - 4x + 2 = 0$ **(b)** $x^2 - 4x + 4 = 0$ **(c)** $x^2 - 4x + 6 = 0$

SOLUTION 1: Algebraic

We use the quadratic formula to solve each equation.

The quadratic formula is discussed on page 515.

(a) $x = \dfrac{-(-4) \pm \sqrt{(-4)^2 - 4 \cdot 1 \cdot 2}}{2} = \dfrac{4 \pm \sqrt{8}}{2} = 2 \pm \sqrt{2}$

There are two solutions, $x = 2 + \sqrt{2}$ and $x = 2 - \sqrt{2}$.

(b) $x = \dfrac{-(-4) \pm \sqrt{(-4)^2 - 4 \cdot 1 \cdot 4}}{2} = \dfrac{4 \pm \sqrt{0}}{2} = 2$

There is just one solution, $x = 2$.

(c) $x = \dfrac{-(-4) \pm \sqrt{(-4)^2 - 4 \cdot 1 \cdot 6}}{2} = \dfrac{4 \pm \sqrt{-8}}{2}$

There is no real solution.

SOLUTION 2: Graphical

We graph the equations $y = x^2 - 4x + 2$, $y = x^2 - 4x + 4$, and $y = x^2 - 4x + 6$ in Figure 2. By determining the x-intercepts of the graphs, we find the following solutions.

(a) $x \approx 0.6$ and $x \approx 3.4$ **(b)** $x = 2$

(c) There is no x-intercept, so the equation has no solution.

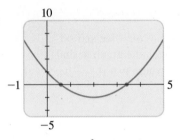

(a) $y = x^2 - 4x + 2$

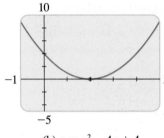

(b) $y = x^2 - 4x + 4$

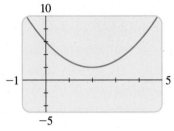

(c) $y = x^2 - 4x + 6$

FIGURE 2

The graphs in Figure 2 show visually why a quadratic equation may have two solutions, one solution, or no real solution. We proved this fact algebraically in Appendix A.4 when we studied the discriminant.

EXAMPLE 2 │ Another Graphical Method

Solve the equation algebraically and graphically: $5 - 3x = 8x - 20$

SOLUTION 1: Algebraic

$$5 - 3x = 8x - 20$$

$$-3x = 8x - 25 \qquad \text{Subtract 5}$$

$$-11x = -25 \qquad \text{Subtract } 8x$$

$$x = \frac{-25}{-11} = 2\tfrac{3}{11} \qquad \text{Divide by } -11 \text{ and simplify}$$

SOLUTION 2: Graphical

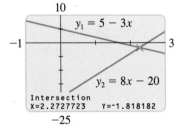

FIGURE 3

We could move all terms to one side of the equal sign, set the result equal to y, and graph the resulting equation. But to avoid all this algebra, we graph two equations instead:

$$y_1 = 5 - 3x \qquad \text{and} \qquad y_2 = 8x - 20$$

The solution of the original equation will be the value of x that makes y_1 equal to y_2; that is, the solution is the x-coordinate of the intersection point of the two graphs. Using the ⎡TRACE⎤ feature or the **intersect** command on a graphing calculator, we see from Figure 3 that the solution is $x \approx 2.27$.

In the next example we use the graphical method to solve an equation that is extremely difficult to solve algebraically.

EXAMPLE 3 │ Solving an Equation in an Interval

Solve the equation

$$x^3 - 6x^2 + 9x = \sqrt{x}$$

in the interval $[1, 6]$.

SOLUTION

We are asked to find all solutions x that satisfy $1 \le x \le 6$, so we will graph the equation in a viewing rectangle for which the x-values are restricted to this interval.

$$x^3 - 6x^2 + 9x = \sqrt{x}$$

$$x^3 - 6x^2 + 9x - \sqrt{x} = 0 \qquad \text{Subtract } \sqrt{x}$$

We can also use the zero command to find the solutions, as shown in Figures 4(a) and 4(b).

Figure 4 shows the graph of the equation $y = x^3 - 6x^2 + 9x - \sqrt{x}$ in the viewing rectangle $[1, 6]$ by $[-5, 5]$. There are two x-intercepts in this viewing rectangle; zooming in we see that the solutions are $x \approx 2.18$ and $x \approx 3.72$.

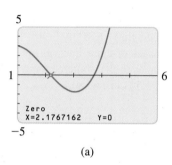

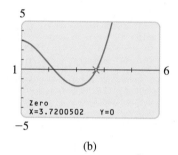

FIGURE 4

(a)

(b)

The equation in Example 3 actually has four solutions. You are asked to find the other two in Exercise 31.

▼ Solving Inequalities Graphically

Inequalities can be solved graphically. To describe the method we solve

$$x^2 - 5x + 6 \le 0$$

To solve this inequality graphically, we draw the graph of

$$y = x^2 - 5x + 6$$

Our goal is to find those values of x for which $y \le 0$. These are simply the x-values for which the graph lies below the x-axis. From Figure 5 we see that the solution of the inequality is the interval $[2, 3]$.

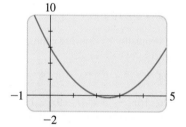

FIGURE 5
$x^2 - 5x + 6 \le 0$

EXAMPLE 4 | Solving an Inequality Graphically

Solve the inequality $3.7x^2 + 1.3x - 1.9 \le 2.0 - 1.4x$.

SOLUTION

We graph the equations

$$y_1 = 3.7x^2 + 1.3x - 1.9$$
$$y_2 = 2.0 - 1.4x$$

in the same viewing rectangle in Figure 6. We are interested in those values of x for which $y_1 \le y_2$; these are points for which the graph of y_2 lies on or above the graph of y_1. To determine the appropriate interval, we look for the x-coordinates of points where the graphs intersect. We conclude that the solution is (approximately) the interval $[-1.45, 0.72]$.

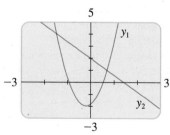

FIGURE 6
$y_1 = 3.7x^2 + 1.3x - 1.9$
$y_2 = 2.0 - 1.4x$

EXAMPLE 5 | Solving an Inequality Graphically

Solve the inequality $x^3 - 5x^2 \geq -8$.

SOLUTION

We write the inequality as

$$x^3 - 5x^2 + 8 \geq 0$$

and then graph the equation

$$y = x^3 - 5x^2 + 8$$

in the viewing rectangle $[-6, 6]$ by $[-15, 15]$, as shown in Figure 7. The solution of the inequality consists of those intervals on which the graph lies on or above the x-axis. By moving the cursor to the x-intercepts we find that, correct to one decimal place, the solution is $[-1.1, 1.5] \cup [4.6, \infty)$.

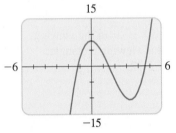

FIGURE 7
$x^3 - 5x^2 + 8 \geq 0$

C.2 EXERCISES

1–10 ■ Solve the equation both algebraically and graphically.

1. $x - 4 = 5x + 12$

2. $\frac{1}{2}x - 3 = 6 + 2x$

3. $\frac{2}{x} + \frac{1}{2x} = 7$

4. $\frac{4}{x+2} - \frac{6}{2x} = \frac{5}{2x+4}$

5. $x^2 - 32 = 0$

6. $x^3 + 16 = 0$

7. $16x^4 = 625$

8. $2x^5 - 243 = 0$

9. $(x - 5)^4 - 80 = 0$

10. $6(x + 2)^5 = 64$

11–18 ■ Solve the equation graphically in the given interval. State each answer correct to two decimals.

11. $x^2 - 7x + 12 = 0$; $[0, 6]$

12. $x^2 - 0.75x + 0.125 = 0$; $[-2, 2]$

13. $x^3 - 6x^2 + 11x - 6 = 0$; $[-1, 4]$

14. $16x^3 + 16x^2 = x + 1$; $[-2, 2]$

15. $x - \sqrt{x + 1} = 0$; $[-1, 5]$

16. $1 + \sqrt{x} = \sqrt{1 + x^2}$; $[-1, 5]$

17. $x^{1/3} - x = 0$; $[-3, 3]$

18. $x^{1/2} + x^{1/3} - x = 0$; $[-1, 5]$

19–22 ■ Find all real solutions of the equation, correct to two decimals.

19. $x^3 - 2x^2 - x - 1 = 0$

20. $x^4 - 8x^2 + 2 = 0$

21. $x(x - 1)(x + 2) = \frac{1}{6}x$

22. $x^4 = 16 - x^3$

23–30 ■ Find the solutions of the inequality by drawing appropriate graphs. State each answer correct to two decimals.

23. $x^2 - 3x - 10 \leq 0$

24. $0.5x^2 + 0.875x \leq 0.25$

25. $x^3 + 11x \leq 6x^2 + 6$

26. $16x^3 + 24x^2 > -9x - 1$

27. $x^{1/3} < x$

28. $\sqrt{0.5x^2 + 1} \leq 2|x|$

29. $(x + 1)^2 < (x - 1)^2$

30. $(x + 1)^2 \leq x^3$

31. In Example 3 we found two solutions of the equation $x^3 - 6x^2 + 9x = \sqrt{x}$, the solutions that lie between 1 and 6. Find two more solutions, correct to two decimals.

32. Consider the family of equations $x^3 - 3x = k$.
   (a) Draw the graphs of $y_1 = x^3 - 3x$ and $y_2 = k$ in the same viewing rectangle, in the cases $k = -4, -2,$ 0, 2, and 4. How many solutions of the equation $x^3 - 3x = k$ are there in each case? Find the solutions correct to two decimals.
   (b) For what ranges of values of k does the equation have one solution? two solutions? three solutions?

APPENDIX D Complex Numbers

D.1 WORKING WITH COMPLEX NUMBERS

> Arithmetic Operations on Complex Numbers ▶ Square Roots of Negative Numbers ▶ Complex Solutions of Quadratic Equations

In Section 1.5 we saw that if the discriminant of a quadratic equation is negative, the equation has no real solution. For example, the equation

$$x^2 + 4 = 0$$

has no real solution. If we try to solve this equation, we get $x^2 = -4$, so

$$x = \pm\sqrt{-4}$$

But this is impossible, since the square of any real number is positive. [For example, $(-2)^2 = 4$, a positive number.] Thus, negative numbers don't have real square roots.

See the note on Cardano (page 542) for an example of how complex numbers are used to find real solutions of polynomial equations.

To make it possible to solve *all* quadratic equations, mathematicians invented an expanded number system, called the *complex number system*. First they defined the new number

$$i = \sqrt{-1}$$

This means that $i^2 = -1$. A complex number is then a number of the form $a + bi$, where a and b are real numbers.

DEFINITION OF COMPLEX NUMBERS

A **complex number** is an expression of the form

$$a + bi$$

where a and b are real numbers and $i^2 = -1$. The **real part** of this complex number is a and the **imaginary part** is b. Two complex numbers are **equal** if and only if their real parts are equal and their imaginary parts are equal.

Note that both the real and imaginary parts of a complex number are real numbers.

EXAMPLE 1 | Complex Numbers

The following are examples of complex numbers.

$3 + 4i$	Real part 3, imaginary part 4
$\frac{1}{2} - \frac{2}{3}i$	Real part $\frac{1}{2}$, imaginary part $-\frac{2}{3}$
$6i$	Real part 0, imaginary part 6
-7	Real part -7, imaginary part 0

✎ NOW TRY EXERCISES **1** AND **5** ∎

A number such as $6i$, which has real part 0, is called a **pure imaginary number**. A real number such as -7 can be thought of as a complex number with imaginary part 0.

In the complex number system every quadratic equation has solutions. The numbers $2i$ and $-2i$ are solutions of $x^2 = -4$ because

$$(2i)^2 = 2^2 i^2 = 4(-1) = -4 \qquad \text{and} \qquad (-2i)^2 = (-2)^2 i^2 = 4(-1) = -4$$

Although we use the term *imaginary* in this context, imaginary numbers should not be thought of as any less "real" (in the ordinary rather than the mathematical sense of that word) than negative numbers or irrational numbers. All numbers (except possibly the positive integers) are creations of the human mind—the numbers -1 and $\sqrt{2}$ as well as the number i. We study complex numbers because they complete, in a useful and elegant fashion, our study of the solutions of equations. In fact, imaginary numbers are useful not only in algebra and mathematics, but in the other sciences as well. To give just one example, in electrical theory the *reactance* of a circuit is a quantity whose measure is an imaginary number.

▼ Arithmetic Operations on Complex Numbers

Complex numbers are added, subtracted, multiplied, and divided just as we would any number of the form $a + b\sqrt{c}$. The only difference that we need to keep in mind is that $i^2 = -1$. Thus, the following calculations are valid.

$$(a + bi)(c + di) = ac + (ad + bc)i + bdi^2 \qquad \text{Multiply and collect like terms}$$
$$= ac + (ad + bc)i + bd(-1) \qquad i^2 = -1$$
$$= (ac - bd) + (ad + bc)i \qquad \text{Combine real and imaginary parts}$$

We therefore define the sum, difference, and product of complex numbers as follows.

ADDING, SUBTRACTING, AND MULTIPLYING COMPLEX NUMBERS

Definition	**Description**
Addition	
$(a + bi) + (c + di) = (a + c) + (b + d)i$	To add complex numbers, add the real parts and the imaginary parts.
Subtraction	
$(a + bi) - (c + di) = (a - c) + (b - d)i$	To subtract complex numbers, subtract the real parts and the imaginary parts.
Multiplication	
$(a + bi) \cdot (c + di) = (ac - bd) + (ad + bc)i$	Multiply complex numbers like binomials, using $i^2 = -1$.

EXAMPLE 2 | Adding, Subtracting, and Multiplying Complex Numbers

Express the following in the form $a + bi$.

(a) $(3 + 5i) + (4 - 2i)$ **(b)** $(3 + 5i) - (4 - 2i)$

(c) $(3 + 5i)(4 - 2i)$ **(d)** i^{23}

Graphing calculators can perform arithmetic operations on complex numbers.

```
(3+5i)+(4-2i)
                7+3i
(3+5i)*(4-2i)
               22+14i
```

SOLUTION

(a) According to the definition, we add the real parts and we add the imaginary parts.

$$(3 + 5i) + (4 - 2i) = (3 + 4) + (5 - 2)i = 7 + 3i$$

(b) $(3 + 5i) - (4 - 2i) = (3 - 4) + [5 - (-2)]i = -1 + 7i$

(c) $(3 + 5i)(4 - 2i) = [3 \cdot 4 - 5(-2)] + [3(-2) + 5 \cdot 4]i = 22 + 14i$

(d) $i^{23} = i^{22+1} = (i^2)^{11}i = (-1)^{11}i = (-1)i = -i$

✎ NOW TRY EXERCISES **11**, **15**, **21**, AND **29** ∎

Division of complex numbers is much like rationalizing the denominator of a radical expression, which we considered in Section 1.4. For the complex number $z = a + bi$ we define its **complex conjugate** to be $\bar{z} = a - bi$. Note that

$$z \cdot \bar{z} = (a + bi)(a - bi) = a^2 + b^2$$

Complex Conjugates

Number	Conjugate
$3 + 2i$	$3 - 2i$
$1 - i$	$1 + i$
$4i$	$-4i$
5	5

Library of Congress

LEONHARD EULER (1707–1783) was born in Basel, Switzerland, the son of a pastor. When Euler was 13, his father sent him to the University at Basel to study theology, but Euler soon decided to devote himself to the sciences. Besides theology he studied mathematics, medicine, astronomy, physics, and Asian languages. It is said that Euler could calculate as effortlessly as "men breathe or as eagles fly." One hundred years before Euler, Fermat (see page 241) had conjectured that $2^{2^n} + 1$ is a prime number for all n. The first five of these numbers are 5, 17, 257, 65537, and 4,294,967,297. It is easy to show that the first four are prime. The fifth was also thought to be prime until Euler, with his phenomenal calculating ability, showed that it is the product $641 \times 6{,}700{,}417$ and so is not prime. Euler published more than any other mathematician in history. His collected works comprise 75 large volumes. Although he was blind for the last 17 years of his life, he continued to work and publish. In his writings he popularized the use of the symbols π, e, and i, which you will find in this textbook. One of Euler's most lasting contributions is his development of complex numbers.

So the product of a complex number and its conjugate is always a nonnegative real number. We use this property to divide complex numbers.

DIVIDING COMPLEX NUMBERS

To simplify the quotient $\dfrac{a + bi}{c + di}$, multiply the numerator and the denominator by the complex conjugate of the denominator:

$$\frac{a + bi}{c + di} = \left(\frac{a + bi}{c + di}\right)\left(\frac{c - di}{c - di}\right) = \frac{(ac + bd) + (bc - ad)i}{c^2 + d^2}$$

Rather than memorizing this entire formula, it is easier to just remember the first step and then multiply out the numerator and the denominator as usual.

EXAMPLE 3 | Dividing Complex Numbers

Express the following in the form $a + bi$.

(a) $\dfrac{3 + 5i}{1 - 2i}$ **(b)** $\dfrac{7 + 3i}{4i}$

SOLUTION We multiply both the numerator and denominator by the complex conjugate of the denominator to make the new denominator a real number.

(a) The complex conjugate of $1 - 2i$ is $\overline{1 - 2i} = 1 + 2i$.

$$\frac{3 + 5i}{1 - 2i} = \left(\frac{3 + 5i}{1 - 2i}\right)\left(\frac{1 + 2i}{1 + 2i}\right) = \frac{-7 + 11i}{5} = -\frac{7}{5} + \frac{11}{5}i$$

(b) The complex conjugate of $4i$ is $-4i$. Therefore,

$$\frac{7 + 3i}{4i} = \left(\frac{7 + 3i}{4i}\right)\left(\frac{-4i}{-4i}\right) = \frac{12 - 28i}{16} = \frac{3}{4} - \frac{7}{4}i$$

NOW TRY EXERCISES 33 AND 39

▼ Square Roots of Negative Numbers

Just as every positive real number r has two square roots ($\sqrt{r}$ and $-\sqrt{r}$), every negative number has two square roots as well. If $-r$ is a negative number, then its square roots are $\pm i\sqrt{r}$, because $(i\sqrt{r})^2 = i^2 r = -r$ and $(-i\sqrt{r})^2 = (-1)^2 i^2 r = -r$.

SQUARE ROOTS OF NEGATIVE NUMBERS

If $-r$ is negative, then the **principal square root** of $-r$ is

$$\sqrt{-r} = i\sqrt{r}$$

The two square roots of $-r$ are $i\sqrt{r}$ and $-i\sqrt{r}$.

We usually write $i\sqrt{b}$ instead of $\sqrt{b}i$ to avoid confusion with $\sqrt{bi}$.

EXAMPLE 4 | Square Roots of Negative Numbers

(a) $\sqrt{-1} = i\sqrt{1} = i$ **(b)** $\sqrt{-16} = i\sqrt{16} = 4i$ **(c)** $\sqrt{-3} = i\sqrt{3}$

NOW TRY EXERCISES 43 AND 45

Special care must be taken in performing calculations that involve square roots of negative numbers. Although $\sqrt{a} \cdot \sqrt{b} = \sqrt{ab}$ when a and b are positive, this is *not* true when both are negative. For example,

$$\sqrt{-2} \cdot \sqrt{-3} = i\sqrt{2} \cdot i\sqrt{3} = i^2\sqrt{6} = -\sqrt{6}$$

but

$$\sqrt{(-2)(-3)} = \sqrt{6}$$

so

$$\sqrt{-2} \cdot \sqrt{-3} \ne \sqrt{(-2)(-3)}$$

 When multiplying radicals of negative numbers, express them first in the form $i\sqrt{r}$ (where $r > 0$) to avoid possible errors of this type.

EXAMPLE 5 | Using Square Roots of Negative Numbers

Evaluate $(\sqrt{12} - \sqrt{-3})(3 + \sqrt{-4})$ and express in the form $a + bi$.

SOLUTION

$$
\begin{aligned}
(\sqrt{12} - \sqrt{-3})(3 + \sqrt{-4}) &= (\sqrt{12} - i\sqrt{3})(3 + i\sqrt{4}) \\
&= (2\sqrt{3} - i\sqrt{3})(3 + 2i) \\
&= (6\sqrt{3} + 2\sqrt{3}) + i(2 \cdot 2\sqrt{3} - 3\sqrt{3}) \\
&= 8\sqrt{3} + i\sqrt{3}
\end{aligned}
$$

✎ NOW TRY EXERCISE 47 ■

▼ Complex Solutions of Quadratic Equations

We have already seen that if $a \ne 0$, then the solutions of the quadratic equation $ax^2 + bx + c = 0$ are

$$x = \frac{-b \pm \sqrt{b^2 - 4ac}}{2a}$$

If $b^2 - 4ac < 0$, then the equation has no real solution. But in the complex number system, this equation will always have solutions, because negative numbers have square roots in this expanded setting.

EXAMPLE 6 | Quadratic Equations with Complex Solutions

Solve each equation.

(a) $x^2 + 9 = 0$ **(b)** $x^2 + 4x + 5 = 0$

SOLUTION

(a) The equation $x^2 + 9 = 0$ means $x^2 = -9$, so

$$x = \pm\sqrt{-9} = \pm i\sqrt{9} = \pm 3i$$

The solutions are therefore $3i$ and $-3i$.

(b) By the Quadratic Formula we have

$$
\begin{aligned}
x &= \frac{-4 \pm \sqrt{4^2 - 4 \cdot 5}}{2} \\
&= \frac{-4 \pm \sqrt{-4}}{2} \\
&= \frac{-4 \pm 2i}{2} = \frac{2(-2 \pm i)}{2} = -2 \pm i
\end{aligned}
$$

So the solutions are $-2 + i$ and $-2 - i$.

✎ NOW TRY EXERCISES **53** AND **55** ■

GEROLAMO CARDANO
(1501–1576) is certainly one of the most colorful figures in the history of mathematics. He was the best-known physician in Europe in his day, yet throughout his life he was plagued by numerous maladies, including ruptures, hemorrhoids, and an irrational fear of encountering rabid dogs. He was a doting father, but his beloved sons broke his heart—his favorite was eventually beheaded for murdering his own wife. Cardano was also a compulsive gambler; indeed, this vice might have driven him to write the *Book on Games of Chance*, the first study of probability from a mathematical point of view.

In Cardano's major mathematical work, the *Ars Magna*, he detailed the solution of the general third- and fourth-degree polynomial equations. At the time of its publication, mathematicians were uncomfortable even with negative numbers, but Cardano's formulas paved the way for the acceptance not just of negative numbers, but also of imaginary numbers, because they occurred naturally in solving polynomial equations. For example, for the cubic equation

$$x^3 - 15x - 4 = 0$$

one of his formulas gives the solution

$$x = \sqrt[3]{2 + \sqrt{-121}} + \sqrt[3]{2 - \sqrt{-121}}$$

This value for x actually turns out to be the *integer* 4, yet to find it, Cardano had to use the imaginary number $\sqrt{-121} = 11i$.

We see from Example 6 that if a quadratic equation with real coefficients has complex solutions, then these solutions are complex conjugates of each other. So if $a + bi$ is a solution, then $a - bi$ is also a solution.

EXAMPLE 7 | Complex Conjugates as Solutions of a Quadratic

Show that the solutions of the equation

$$4x^2 - 24x + 37 = 0$$

are complex conjugates of each other.

SOLUTION   We use the Quadratic Formula to get

$$x = \frac{24 \pm \sqrt{(24)^2 - 4(4)(37)}}{2(4)}$$

$$= \frac{24 \pm \sqrt{-16}}{8} = \frac{24 \pm 4i}{8} = 3 \pm \frac{1}{2}i$$

So the solutions are $3 + \frac{1}{2}i$ and $3 - \frac{1}{2}i$, and these are complex conjugates.

✎. NOW TRY EXERCISE **61**

D.1 EXERCISES

1–10 ■ Find the real and imaginary parts of the complex number.

1. $5 - 7i$

2. $-6 + 4i$

3. $\dfrac{-2 - 5i}{3}$

4. $\dfrac{4 + 7i}{2}$

5. 3

6. $-\frac{1}{2}$

7. $-\frac{2}{3}i$

8. $i\sqrt{3}$

9. $\sqrt{3} + \sqrt{-4}$

10. $2 - \sqrt{-5}$

11–42 ■ Evaluate the expression and write the result in the form $a + bi$.

11. $(2 - 5i) + (3 + 4i)$

12. $(2 + 5i) + (4 - 6i)$

13. $(-6 + 6i) + (9 - i)$

14. $(3 - 2i) + \left(-5 - \frac{1}{3}i\right)$

15. $\left(7 - \frac{1}{2}i\right) - \left(5 + \frac{3}{2}i\right)$

16. $(-4 + i) - (2 - 5i)$

17. $(-12 + 8i) - (7 + 4i)$

18. $6i - (4 - i)$

19. $4(-1 + 2i)$

20. $2i\left(\frac{1}{2} - i\right)$

21. $(7 - i)(4 + 2i)$

22. $(5 - 3i)(1 + i)$

23. $(3 - 4i)(5 - 12i)$

24. $\left(\frac{2}{3} + 12i\right)\left(\frac{1}{6} + 24i\right)$

25. $(6 + 5i)(2 - 3i)$

26. $(-2 + i)(3 - 7i)$

27. i^3

28. $(2i)^4$

29. i^{100}

30. i^{1002}

31. $\dfrac{1}{i}$

32. $\dfrac{1}{1 + i}$

33. $\dfrac{2 - 3i}{1 - 2i}$

34. $\dfrac{5 - i}{3 + 4i}$

35. $\dfrac{26 + 39i}{2 - 3i}$

36. $\dfrac{25}{4 - 3i}$

37. $\dfrac{10i}{1 - 2i}$

38. $(2 - 3i)^{-1}$

39. $\dfrac{4 + 6i}{3i}$

40. $\dfrac{-3 + 5i}{15i}$

41. $\dfrac{1}{1 + i} - \dfrac{1}{1 - i}$

42. $\dfrac{(1 + 2i)(3 - i)}{2 + i}$

43–52 ■ Evaluate the radical expression and express the result in the form $a + bi$.

43. $\sqrt{-25}$

44. $\sqrt{\dfrac{-9}{4}}$

45. $\sqrt{-3}\sqrt{-12}$

46. $\sqrt{\frac{1}{3}}\sqrt{-27}$

47. $(3 - \sqrt{-5})(1 + \sqrt{-1})$

48. $(\sqrt{3} - \sqrt{-4})(\sqrt{6} - \sqrt{-8})$

49. $\dfrac{2 + \sqrt{-8}}{1 + \sqrt{-2}}$

50. $\dfrac{1 - \sqrt{-1}}{1 + \sqrt{-1}}$

51. $\dfrac{\sqrt{-36}}{\sqrt{-2}\sqrt{-9}}$

52. $\dfrac{\sqrt{-7}\sqrt{-49}}{\sqrt{28}}$

53–68 ■ Find all solutions of the equation and express them in the form $a + bi$.

53. $x^2 + 49 = 0$

54. $9x^2 + 4 = 0$

55. $x^2 - 4x + 5 = 0$

56. $x^2 + 2x + 2 = 0$

57. $x^2 + 2x + 5 = 0$

58. $x^2 - 6x + 10 = 0$

59. $x^2 + x + 1 = 0$

60. $x^2 - 3x + 3 = 0$

61. $2x^2 - 2x + 1 = 0$

62. $2x^2 + 3 = 2x$

63. $t + 3 + \dfrac{3}{t} = 0$

64. $z + 4 + \dfrac{12}{z} = 0$

65. $6x^2 + 12x + 7 = 0$

66. $4x^2 - 16x + 19 = 0$

67. $\frac{1}{2}x^2 - x + 5 = 0$

68. $x^2 + \frac{1}{2}x + 1 = 0$

69–76 ■ Recall that the symbol $\bar{z}$ represents the complex conjugate of z. If $z = a + bi$ and $w = c + di$, prove each statement.

69. $\bar{z} + \bar{w} = \overline{z + w}$

70. $\overline{zw} = \bar{z} \cdot \bar{w}$

71. $(\bar{z})^2 = \overline{z^2}$

72. $\bar{\bar{z}} = z$

73. $z + \bar{z}$ is a real number.

74. $z - \bar{z}$ is a pure imaginary number.

75. $z \cdot \bar{z}$ is a real number.

76. $z = \bar{z}$ if and only if z is real.

PROLOGUE ■ PAGE P4

1. It can't go fast enough. **2.** 40% discount **3.** 427, $3n + 1$
4. 57 min **5.** No, not necessarily **6.** The same amount **7.** 2π
8. The North Pole is one such point; there are infinitely many
others near the South Pole.

CHAPTER 1

SECTION 1.1 ■ PAGE 11

1. $(3, -5)$ **2.** $\sqrt{(c - a)^2 + (d - b)^2}$; 10

3. $\left(\dfrac{a + c}{2}, \dfrac{b + d}{2}\right)$; $(4, 6)$ **4.** 2; 3; No **5. (a)** y; x; -1

(b) x; y; $\frac{1}{2}$ **6.** $(1, 2)$; 3

7.

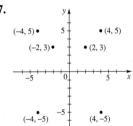

9. (a) $\sqrt{13}$ **(b)** $\left(\frac{3}{2}, 1\right)$ **11. (a)** 10 **(b)** $(1, 0)$
13. (a) **15. (a)**

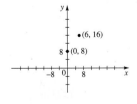

(b) 10 **(c)** $(3, 12)$ **(b)** 25 **(c)** $\left(\frac{1}{2}, 6\right)$
17. (a) **19.** 24

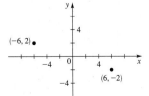

(b) $4\sqrt{10}$ **(c)** $(0, 0)$

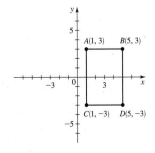

21. Trapezoid, area $= 9$ **23.**

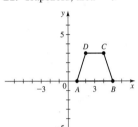
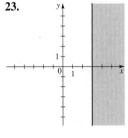

25. **27.**

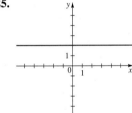
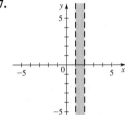

29. **31.**

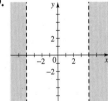
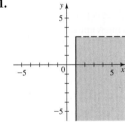

33. $A(6, 7)$ **35.** $Q(-1, 3)$ **39. (b)** 10 **43.** $(0, -4)$
45. $(2, -3)$

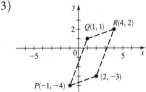

47. (a) **(b)** $\left(\frac{5}{2}, 3\right)$, $\left(\frac{5}{2}, 3\right)$

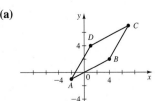

49. No, yes, yes   **51.** Yes, no, yes
53. x-intercepts 0, 4; y-intercept 0
55. x-intercepts -2, 2; y-intercepts -4, 4
57. x-intercept 4,   **59.** x-intercept 3,
  y-intercept 4, y-intercept -6,
  no symmetry   no symmetry

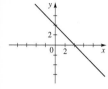   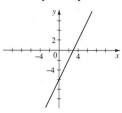

61. x-intercepts ± 1,   **63.** x-intercept 0,
  y-intercept 1, y-intercept 0,
  symmetry about y-axis   symmetry about y-axis

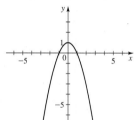   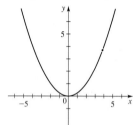

65. x-intercepts ± 3,   **67.** No intercepts,
  y-intercept -9,   symmetry about origin
  symmetry about y-axis

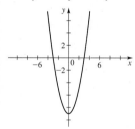   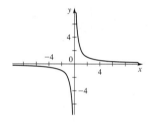

69. x-intercepts ± 2,   **71.** x-intercept 4,
  y-intercept 2, y-intercepts -2, 2,
  symmetry about y-axis   symmetry about x-axis

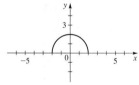   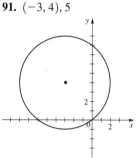

73. x-intercepts ± 2,   **75.** x-intercepts ± 4,
  y-intercept 16, y-intercept 4,
  symmetry about y-axis   symmetry about y-axis

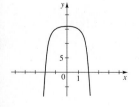   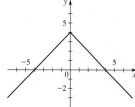

77. Symmetry about y-axis
79. Symmetry about origin
81. Symmetry about origin
83.   **85.**

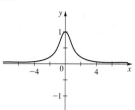   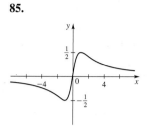

87. $(0, 0)$, 3   **89.** $(3, 0)$, 4

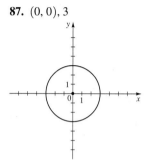   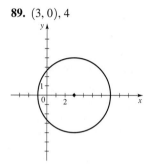

91. $(-3, 4)$, 5

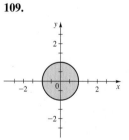

93. $(x - 2)^2 + (y + 1)^2 = 9$ **95.** $x^2 + y^2 = 65$
97. $(x - 2)^2 + (y - 5)^2 = 25$ **99.** $(x - 7)^2 + (y + 3)^2 = 9$
101. $(x + 2)^2 + (y - 2)^2 = 4$ **103.** $(2, -5), 4$ **105.** $\left(\frac{1}{4}, -\frac{1}{4}\right), \frac{1}{2}$
107. $\left(\frac{3}{4}, 0\right), \frac{3}{4}$
109.

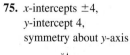

111. 12π **113.** (a) 5   (b) 31; 25   (c) Points P and Q must either be on the same street or the same avenue.
115. (a) 2 Mm, 8 Mm   (b) $-1.33, 7.33$; 2.40 Mm, 7.60 Mm

SECTION 1.2 ■ PAGE 24

1. y; x; 2   **2.** (a) 3   (b) 3   (c) $-\frac{1}{3}$ **3.** $y - 2 = 3(x - 1)$
4. (a) 0; $y = 3$ (b) Undefined; $x = 2$ **5.** $\frac{1}{2}$ **7.** $\frac{1}{6}$
9. $-\frac{1}{2}$ **11.** $-\frac{9}{2}$ **13.** $-2, \frac{1}{2}, 3, -\frac{1}{4}$ **15.** $x + y - 4 = 0$
17. $3x - 2y - 6 = 0$ **19.** $5x - y - 7 = 0$
21. $2x - 3y + 19 = 0$ **23.** $5x + y - 11 = 0$
25. $3x - y - 2 = 0$ **27.** $3x - y - 3 = 0$ **29.** $y = 5$
31. $x + 2y + 11 = 0$ **33.** $x = -1$ **35.** $5x - 2y + 1 = 0$

37. $x - y + 6 = 0$
39. (a)   **(b)** $3x - 2y + 8 = 0$

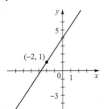

41. They all have the same slope.

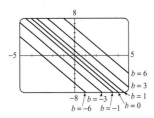

43. They all have the same x-intercept.   **45.** $-1, 3$

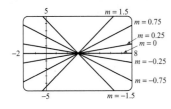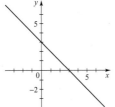

47. $-\frac{1}{3}, 0$ **49.** $\frac{3}{2}, 3$

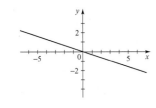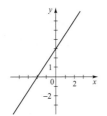

51. $0, 4$ **53.** $\frac{3}{4}, -3$

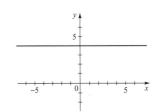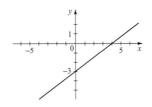

55. $-\frac{3}{4}, \frac{1}{4}$

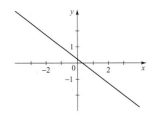

61. $x - y - 3 = 0$ **63. (b)** $4x - 3y - 24 = 0$ **65.** 16,667 ft
67. (a) 8.34; the slope represents the increase in dosage for a one-year increase in age.   **(b)** 8.34 mg
69. (a)

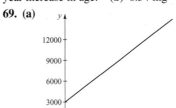

(b) The slope represents production cost per toaster; the y-intercept represents monthly fixed cost.
71. (a) $t = \frac{5}{24}n + 45$ **(b)** $76°F$
73. (a) $P = 0.434d + 15$, where P is pressure in lb/in^2 and d is depth in feet
(b)

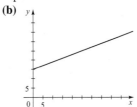

(c) The slope is the rate of increase in water pressure, and the y-intercept is the air pressure at the surface.   **(d)** 196 ft
75. (a) $C = \frac{1}{4}d + 260$
(b) $635
(c) The slope represents cost per mile.
(d) The y-intercept represents monthly fixed cost.

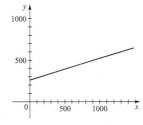

SECTION 1.3 ■ PAGE 35

1. value   **2.** domain, range   **3. (a)** f and g
(b) $f(5) = 10, g(5) = 0$ **4. (a)** square, add 3

(b)

x	0	2	4	6
$f(x)$	19	7	3	7

5. $f(x) = 2(x + 3)$ **7.** $f(x) = (x - 5)^2$ **9.** Square, then add 2
11. Subtract 4, then divide by 3

13.

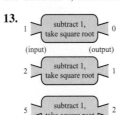

15.

x	$f(x)$
-1	8
0	2
1	0
2	2
3	8

17. $3, 3, -6, -\frac{23}{4}, 94$ **19.** $3, -3, 2, 2a + 1, -2a + 1, 2a + 2b + 1$

21. $-\frac{1}{3}, -3, \frac{1}{3}, \frac{1-a}{1+a}, \frac{2-a}{a}$, undefined

23. $-4, 10, -2, 3\sqrt{2}, 2x^2 + 7x + 1, 2x^2 - 3x - 4$

25. $6, 2, 1, 2, 2|x|, 2(x^2 + 1)$ **27.** $4, 1, 1, 2, 3$

29. $8, -\frac{3}{4}, -1, 0, -1$ **31.** $x^2 + 4x + 5, x^2 + 6$

33. $x^2 + 4, x^2 + 8x + 16$ **35.** $3a + 2, 3(a + h) + 2, 3$

37. $5, 5, 0$ **39.** $\dfrac{a}{a + 1}, \dfrac{a + h}{a + h + 1}, \dfrac{1}{(a + h + 1)(a + 1)}$

41. $3 - 5a + 4a^2, 3 - 5a - 5h + 4a^2 + 8ah + 4h^2, -5 + 8a + 4h$

43. $(-\infty, \infty)$ **45.** $[-1, 5]$ **47.** $\{x \mid x \neq 3\}$ **49.** $\{x \mid x \neq \pm 1\}$

51. $[5, \infty)$ **53.** $(-\infty, \infty)$ **55.** $[\frac{5}{2}, \infty)$ **57.** $[-2, 3) \cup (3, \infty)$

59. $(-\infty, 0] \cup [6, \infty)$ **61.** $(4, \infty)$ **63.** $(\frac{1}{2}, \infty)$

65. (a) $f(x) = \dfrac{x}{3} + \dfrac{2}{3}$

(b)

x	$f(x)$
2	$\frac{4}{3}$
4	2
6	$\frac{8}{3}$
8	$\frac{10}{3}$

(c)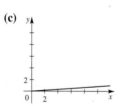

67. (a) $T(x) = 0.08x$

(b)

x	$T(x)$
2	0.16
4	0.32
6	0.48
8	0.64

(c)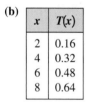

69. (a) $C(10) = 1532.1, C(100) = 2100$ **(b)** The cost of producing 10 yd and 100 yd   **(c)** $C(0) = 1500$ **71. (a)** $50, 0$
(b) $V(0)$ is the volume of the full tank, and $V(20)$ is the volume of the empty tank, 20 minutes later.

(c)

x	$V(x)$
0	50
5	28.125
10	12.5
15	3.125
20	0

73. (a) $v(0.1) = 4440, v(0.4) = 1665$
(b) Flow is faster near central axis.

(c)

r	$v(r)$
0	4625
0.1	4440
0.2	3885
0.3	2960
0.4	1665
0.5	0

75. (a) 8.66 m, 6.61 m, 4.36 m
(b) It will appear to get shorter.
77. (a) $90, $105, $100, $105
(b) Total cost of an order, including shipping

79. (a) $F(x) = \begin{cases} 15(40 - x) & \text{if } 0 < x < 40 \\ 0 & \text{if } 40 \le x \le 65 \\ 15(x - 65) & \text{if } x > 65 \end{cases}$

(b) $150, $0, $150   **(c)** Fines for violating the speed limits

81.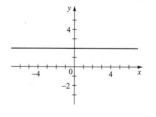

83.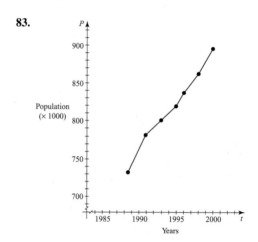

SECTION 1.4 ■ PAGE 45

1. $f(x), x^3 + 2, 10, 10$ **2.** 3   **3.** 3
4. (a) IV   **(b)** II   **(c)** I   **(d)** III

5.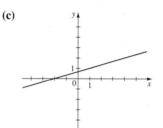

7.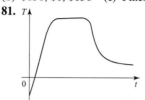

9.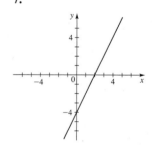

11.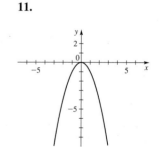

13.

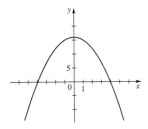

15.

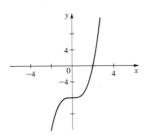

17.

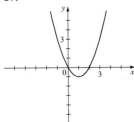

19.

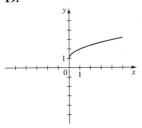

21.

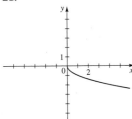

23.

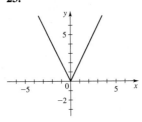

25.

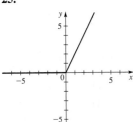

27.

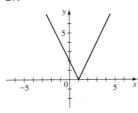

29. (a)

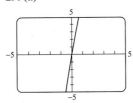

(b)

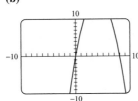

(c)

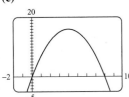

(d)

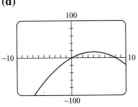

Graph (c) is the most appropriate.

31. (a)

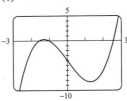

(b)

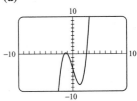

(c)

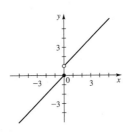

(d)

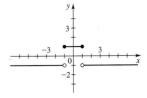

Graph (c) is the most appropriate.

33.

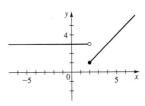

35.

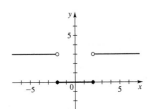

37.

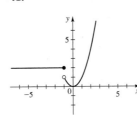

39.

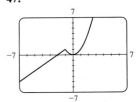

41.

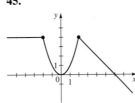

43.

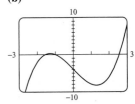

45.

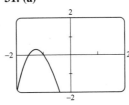

47.

49. $f(x) = \begin{cases} -2 & \text{if } x < -2 \\ x & \text{if } -2 \le x \le 2 \\ 2 & \text{if } x > 2 \end{cases}$

51. (a) Yes **(b)** No **(c)** Yes **(d)** No
53. Function, domain $[-3, 2]$, range $[-2, 2]$ **55.** Not a function
57. Yes **59.** No **61.** No **63.** Yes **65.** Yes **67.** Yes
69. (a) **(b)**

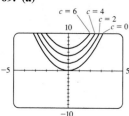

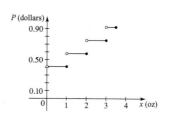

(c) If $c > 0$, then the graph of $f(x) = x^2 + c$ is the same as the
graph of $y = x^2$ shifted upward c units. If $c < 0$, then the graph of
$f(x) = x^2 + c$ is the same as the graph of $y = x^2$ shifted downward
c units.
71. (a) **(b)**

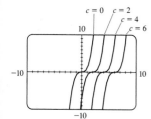

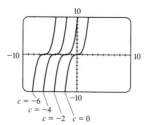

(c) If $c > 0$, then the graph of $f(x) = (x - c)^3$ is the same as the
graph of $y = x^3$ shifted to the right c units. If $c < 0$, then the graph
of $f(x) = (x - c)^3$ is the same as the graph of $y = x^3$ shifted to the
left c units.
73. (a) **(b)**

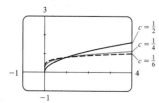

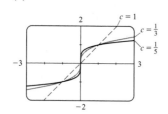

(c) Graphs of even roots are similar to $\sqrt{x}$; graphs of odd roots are
similar to $\sqrt[3]{x}$. As c increases, the graph of $y = \sqrt[c]{x}$ becomes
steeper near 0 and flatter when $x > 1$.
75. $f(x) = -\frac{7}{6}x - \frac{4}{3}$, $-2 \le x \le 4$
77. $f(x) = \sqrt{9 - x^2}$, $-3 \le x \le 3$
79.

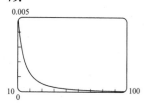

81. (a) $E(x) = \begin{cases} 6 + 0.10x & 0 \le x \le 300 \\ 36 + 0.06(x - 300), & x > 300 \end{cases}$

(b)

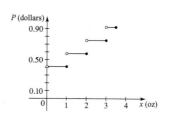

83. $P(x) = \begin{cases} 0.44 & \text{if } 0 < x \le 1 \\ 0.61 & \text{if } 1 < x \le 2 \\ 0.78 & \text{if } 2 < x \le 3 \\ 0.95 & \text{if } 3 < x \le 3.5 \end{cases}$

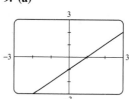

SECTION 1.5 ■ PAGE 54

1. a, 4 **2.** x, y, $[1, 6]$, $[1, 7]$
3. (a) increase, $[1, 2]$, $[4, 5]$ **(b)** decrease, $[2, 4]$, $[5, 6]$
4. (a) largest, 7, 2 **(b)** smallest, 2, 4
5. (a) 1, -1, 3, 4 **(b)** Domain $[-3, 4]$, range $[-1, 4]$
(c) $-3, 2, 4$ **(d)** $-3 \le x \le 2$ and $x = 4$
7. (a) 3, 2, -2, 1, 0 **(b)** Domain $[-4, 4]$, range $[-2, 3]$
9. (a) **11. (a)**

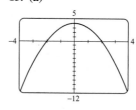

(b) Domain $(-\infty, \infty)$, **(b)** Domain $[1, 3]$,
range $(-\infty, \infty)$ range $\{4\}$
13. (a) **15. (a)**

(b) Domain $(-\infty, \infty)$, **(b)** Domain $[-4, 4]$,
range $(-\infty, 4]$ range $[0, 4]$
17. (a)

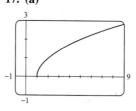

(b) Domain $[1, \infty)$,
range $[0, \infty)$

19. (a) $[-1, 1], [2, 4]$ **(b)** $[1, 2]$
21. (a) $[-2, -1], [1, 2]$ **(b)** $[-3, -2], [-1, 1], [2, 3]$
23. (a)   **25. (a)**

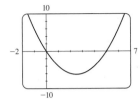   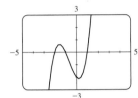

(b) Increasing on $[2.5, \infty)$;   **(b)** Increasing on $(-\infty, -1]$,
decreasing on $(-\infty, 2.5]$ $[2, \infty)$; decreasing on $[-1, 2]$
27. (a)   **29. (a)**

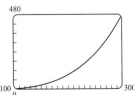   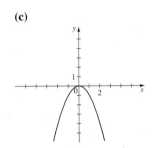

(b) Increasing on   **(b)** Increasing on $[0, \infty)$;
$(-\infty, -1.55], [0.22, \infty)$;   decreasing on $(-\infty, 0]$
decreasing on $[-1.55, 0.22]$
31. (a) Local maximum 2 when $x = 0$; local minimum -1 when
$x = -2$, local minimum 0 when $x = 2$ **(b)** Increasing on
$[-2, 0] \cup [2, \infty)$; decreasing on $(-\infty, -2] \cup [0, 2]$
33. (a) Local maximum 0 when $x = 0$; local maximum 1 when
$x = 3$, local minimum -2 when $x = -2$, local minimum -1 when
$x = 1$ **(b)** Increasing on $[-2, 0] \cup [1, 3]$; decreasing on
$(-\infty, -2] \cup [0, 1] \cup [3, \infty)$ **35. (a)** Local maximum ≈ 0.38
when $x \approx -0.58$; local minimum ≈ -0.38 when $x \approx 0.58$
(b) Increasing on $(-\infty, -0.58] \cup [0.58, \infty)$; decreasing on
$[-0.58, 0.58]$ **37. (a)** Local maximum ≈ 0 when $x = 0$; local
minimum ≈ -13.61 when $x \approx -1.71$, local minimum ≈ -73.32
when $x \approx 3.21$ **(b)** Increasing on $[-1.71, 0] \cup [3.21, \infty)$; decreasing on $(-\infty, -1.71] \cup [0, 3.21]$ **39. (a)** Local maximum
≈ 5.66 when $x \approx 4.00$ **(b)** Increasing on $(-\infty, 4.00]$; decreasing
on $[4.00, 6.00]$ **41. (a)** Local maximum ≈ 0.38 when
$x \approx -1.73$; local minimum ≈ -0.38 when $x \approx 1.73$
(b) Increasing on $(-\infty, -1.73] \cup [1.73, \infty]$; decreasing on
$[-1.73, 0) \cup (0, 1.73]$ **43. (a)** 500 MW, 725 MW
(b) Between 3:00 A.M. and 4:00 A.M.   **(c)** Just before noon
45. (a) Increasing on $[0, 30] \cup [32, 68]$; decreasing on $[30, 32]$
(b) He went on a crash diet and lost weight, only to regain
it again later.   **47. (a)** Increasing on $[0, 150] \cup [300, \infty)$;
decreasing on $[150, 300]$ **(b)** Local maximum when $x = 150$;
local minimum when $x = 300$ **49.** Runner A won the race. All
runners finished. Runner B fell but got up again to finish second.
51. (a)

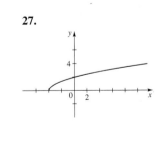

(b) Increases   **53.** 20 mi/h   **55.** $r \approx 0.67$ cm

SECTION 1.6 ▪ PAGE 65
1. (a) up   **(b)** left   **2. (a)** down   **(b)** right   **3. (a)** x-axis
(b) y-axis   **4. (a)** II   **(b)** IV   **(c)** I   **(d)** III   **5. (a)** Shift
downward 5 units   **(b)** Shift to the right 5 units   **7. (a)** Reflect
in the x-axis   **(b)** Reflect in the y-axis   **9. (a)** Reflect in the x-
axis, then shift upward 5 units   **(b)** Stretch vertically by a factor
of 3, then shift downward 5 units   **11. (a)** Shift to the left 1 unit,
stretch vertically by a factor of 2, then shift downward 3 units
(b) Shift to the right 1 unit, stretch vertically by a factor of 2, then
shift upward 3 units   **13. (a)** Shrink horizontally by a factor of $\frac{1}{4}$
(b) Stretch horizontally by a factor of 4   **15. (a)** Shift to the left
2 units   **(b)** Shift upward 2 units   **17. (a)** Shift to the left
2 units, then shift downward 2 units   **(b)** Shift to the right
2 units, then shift upward 2 units
19. (a)   **(b)**

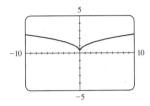   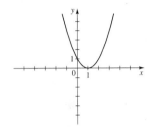

(c)   **(d)**

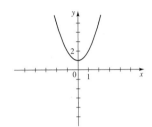   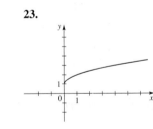

21.   **23.**

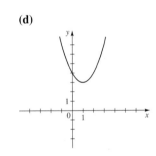

25.   **27.**

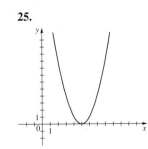   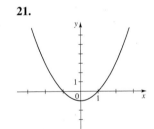

29.

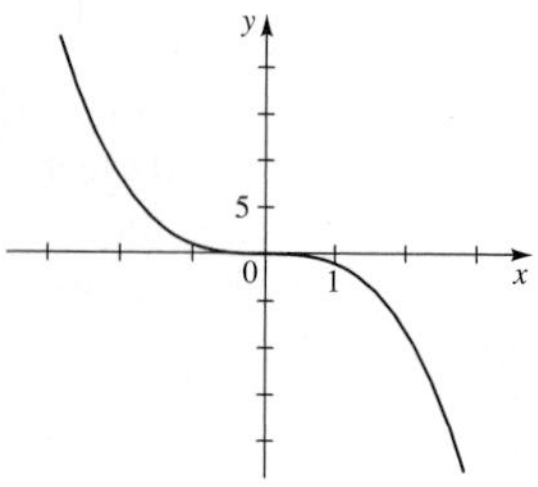

31.

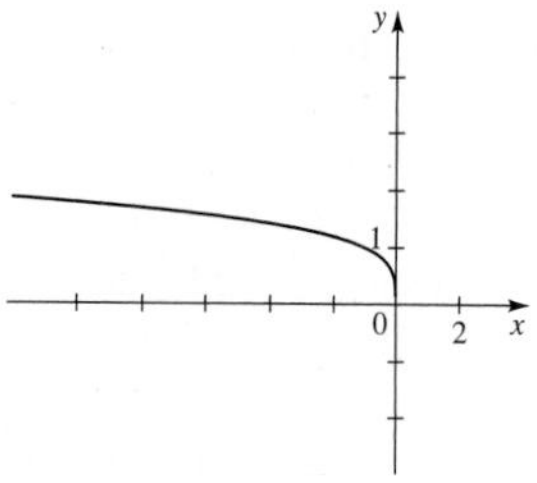

33.

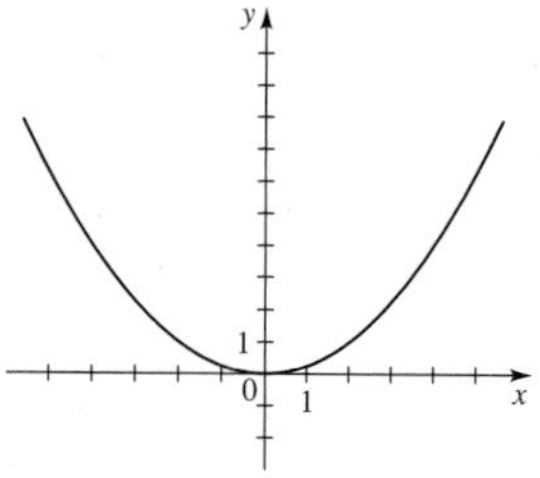

35.

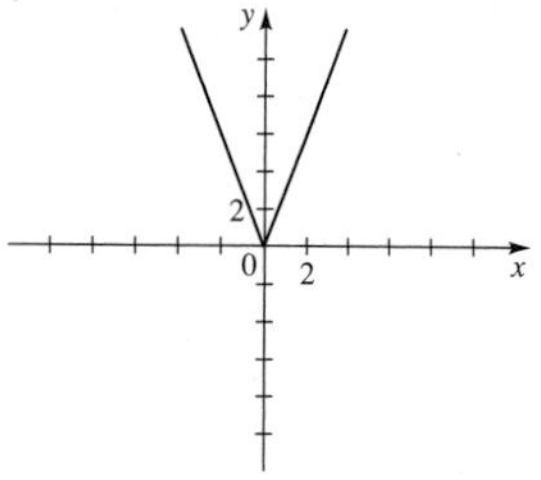

37.

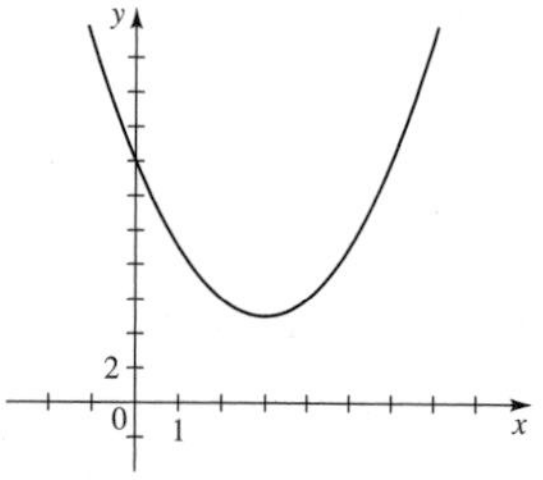

39.

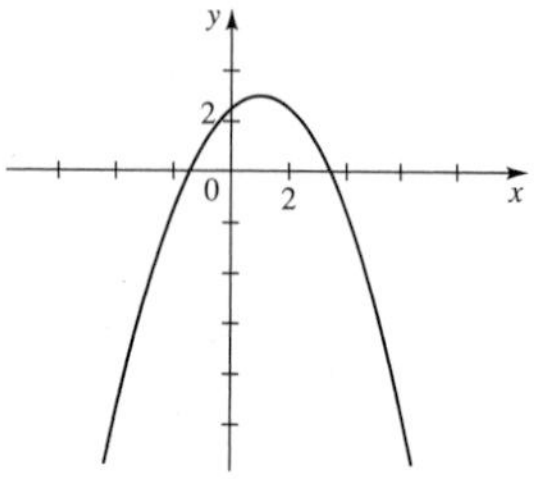

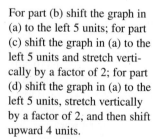

41.

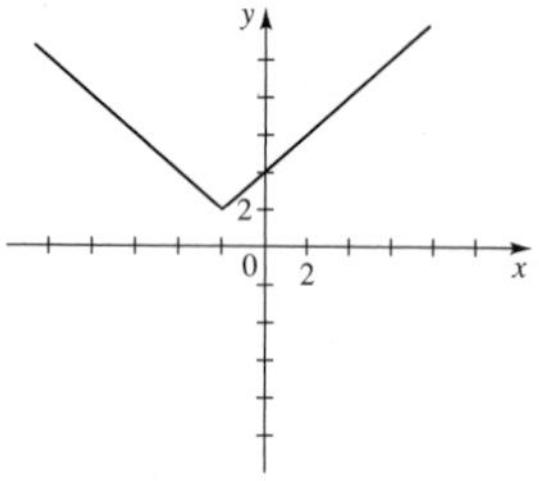

43.

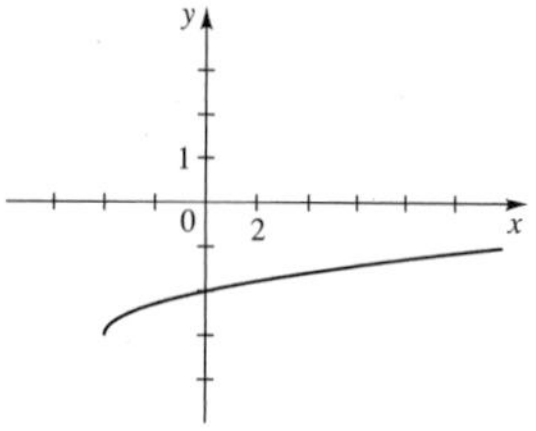

45. $f(x) = x^2 + 3$ **47.** $f(x) = \sqrt{x + 2}$
49. $f(x) = |x - 3| + 1$ **51.** $f(x) = \sqrt[4]{-x} + 1$
53. $f(x) = 2(x - 3)^2 - 2$ **55.** $g(x) = (x - 2)^2$
57. $g(x) = |x + 1| + 2$ **59.** $g(x) = -\sqrt{x + 2}$
61. **(a)** 3 **(b)** 1 **(c)** 2 **(d)** 4

63. **(a)**

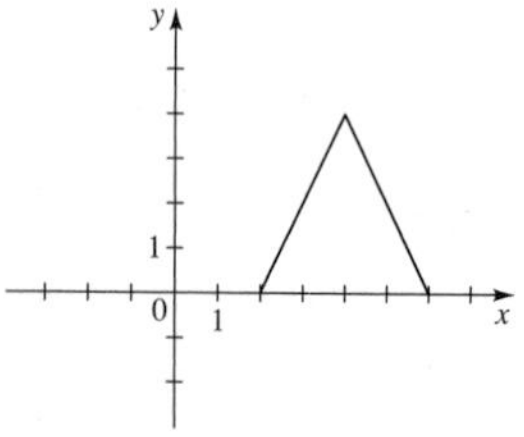

(b)

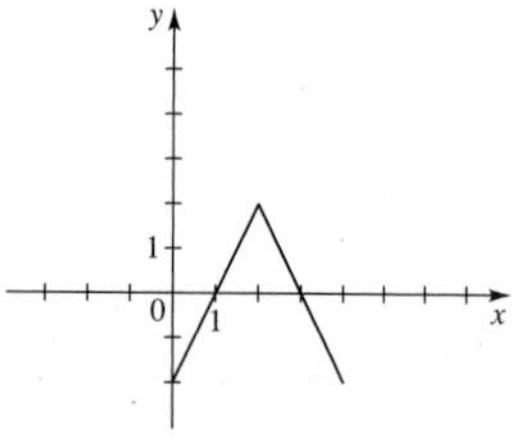

(c)

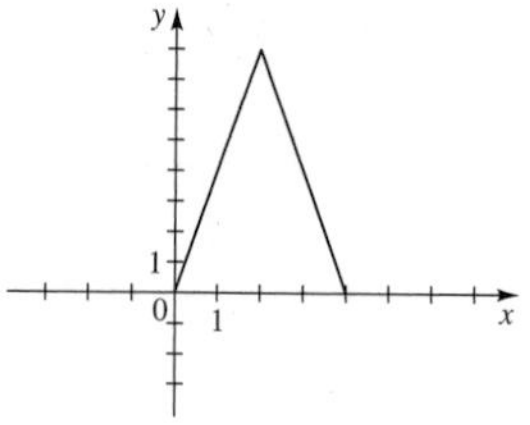

(d)

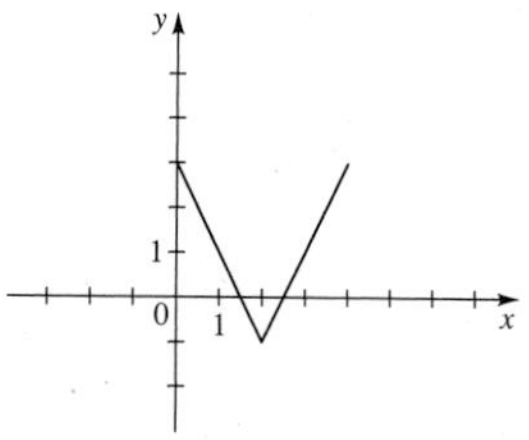

(e)

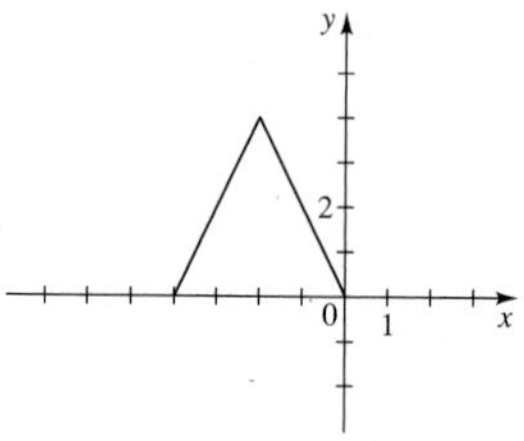

(f)

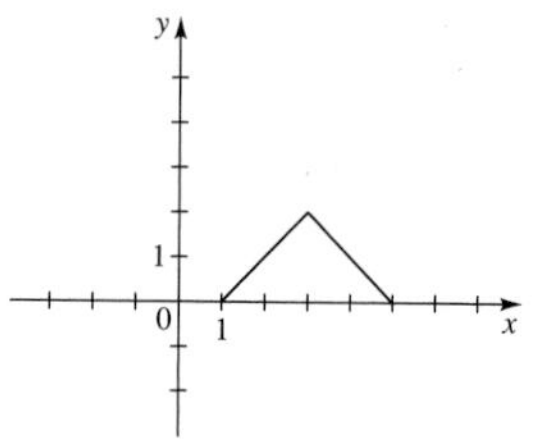

65. **(a)**

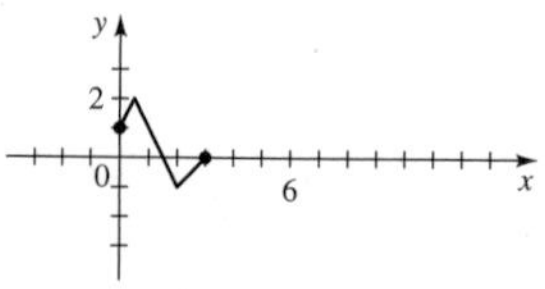

(b)

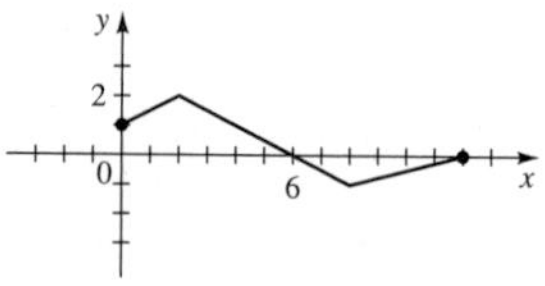

67.

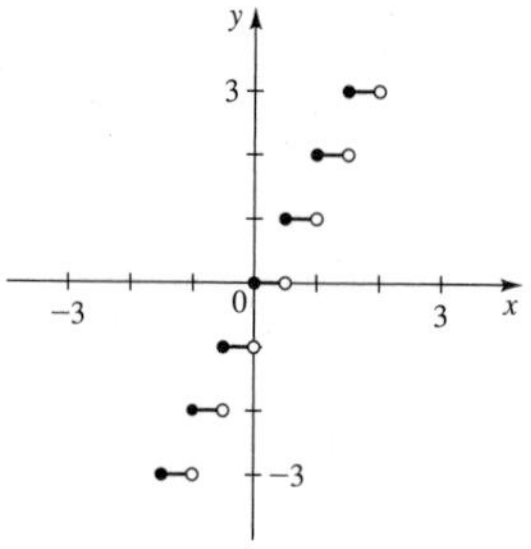

69.

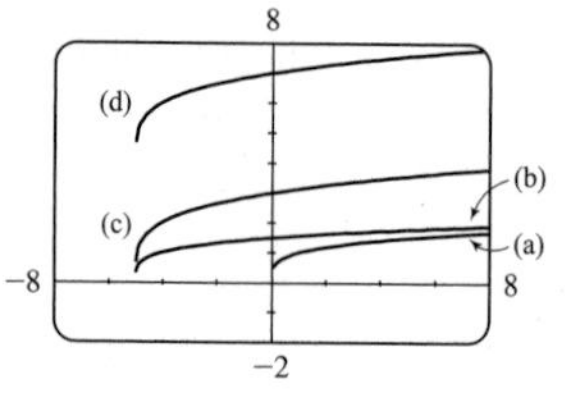

For part (b) shift the graph in (a) to the left 5 units; for part (c) shift the graph in (a) to the left 5 units and stretch vertically by a factor of 2; for part (d) shift the graph in (a) to the left 5 units, stretch vertically by a factor of 2, and then shift upward 4 units.

71.

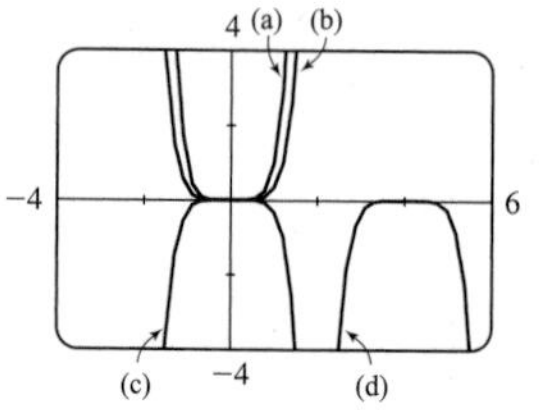

For part (b) shrink the graph in (a) vertically by a factor of $\frac{1}{3}$; for part (c) shrink the graph in (a) vertically by a factor of $\frac{1}{3}$ and reflect in the x-axis; for part (d) shift the graph in (a) to the right 4 units, shrink vertically by a factor of $\frac{1}{3}$, and then reflect in the x-axis.

73.

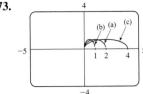

The graph in part (b) is shrunk horizontally by a factor of $\frac{1}{2}$ and the graph in part (c) is stretched by a factor of 2.

75. Even

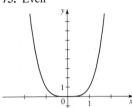

77. Neither

79. Odd

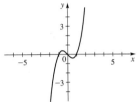

81. Neither

83. (a)

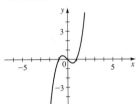

(b)

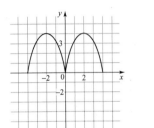

85. To obtain the graph of g, reflect in the x-axis the part of the graph of f that is below the x-axis.

87. (a)

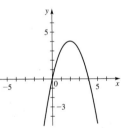

(b)

89. (a) Shift upward 4 units, shrink vertically by a factor of 0.01
(b) Shift to the left 10 units; $g(t) = 4 + 0.01(t + 10)^2$

SECTION 1.7 ■ PAGE 74

1. $8, -2, 15, \dfrac{3}{5}$ **2.** $f(g(x)), 12$ **3.** Multiply by 2, then add 1;
Add 1, then multiply by 2 **4.** $x + 1, 2x, 2x + 1, 2(x + 1)$

5. $(f + g)(x) = x^2 + x - 3, (-\infty, \infty)$;
$(f - g)(x) = -x^2 + x - 3, (-\infty, \infty)$;

$(fg)(x) = x^3 - 3x^2, (-\infty, \infty)$;
$\left(\dfrac{f}{g}\right)(x) = \dfrac{x - 3}{x^2}, (-\infty, 0) \cup (0, \infty)$

7. $(f + g)(x) = \sqrt{4 - x^2} + \sqrt{1 + x}, [-1, 2]$;
$(f - g)(x) = \sqrt{4 - x^2} - \sqrt{1 + x}, [-1, 2]$;
$(fg)(x) = \sqrt{-x^3 - x^2 + 4x + 4}, [-1, 2]$;
$\left(\dfrac{f}{g}\right)(x) = \sqrt{\dfrac{4 - x^2}{1 + x}}, (-1, 2]$

9. $(f + g)(x) = \dfrac{6x + 8}{x^2 + 4x}, x \neq -4, x \neq 0$;

$(f - g)(x) = \dfrac{-2x + 8}{x^2 + 4x}, x \neq -4, x \neq 0$;

$(fg)(x) = \dfrac{8}{x^2 + 4x}, x \neq -4, x \neq 0$;

$\left(\dfrac{f}{g}\right)(x) = \dfrac{x + 4}{2x}, x \neq -4, x \neq 0$

11. $[0, 1]$ **13.** $(3, \infty)$

15.

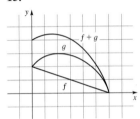

17.

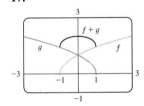

19.

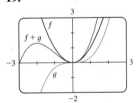

21. (a) 1 **(b)** -23 **23. (a)** -11 **(b)** -119
25. (a) $-3x^2 + 1$ **(b)** $-9x^2 + 30x - 23$
27. 4 **29.** 5 **31.** 4
33. $(f \circ g)(x) = 8x + 1, (-\infty, \infty)$;
$(g \circ f)(x) = 8x + 11, (-\infty, \infty)$; $(f \circ f)(x) = 4x + 9, (-\infty, \infty)$;
$(g \circ g)(x) = 16x - 5, (-\infty, \infty)$
35. $(f \circ g)(x) = (x + 1)^2, (-\infty, \infty)$;
$(g \circ f)(x) = x^2 + 1, (-\infty, \infty)$; $(f \circ f)(x) = x^4, (-\infty, \infty)$;
$(g \circ g)(x) = x + 2, (-\infty, \infty)$
37. $(f \circ g)(x) = \dfrac{1}{2x + 4}, x \neq -2; (g \circ f)(x) = \dfrac{2}{x} + 4, x \neq 0$;

$(f \circ f)(x) = x, x \neq 0, (g \circ g)(x) = 4x + 12, (-\infty, \infty)$
39. $(f \circ g)(x) = |2x + 3|, (-\infty, \infty)$;
$(g \circ f)(x) = 2|x| + 3, (-\infty, \infty)$; $(f \circ f)(x) = |x|, (-\infty, \infty)$;
$(g \circ g)(x) = 4x + 9, (-\infty, \infty)$

41. $(f \circ g)(x) = \dfrac{2x - 1}{2x}, x \neq 0; (g \circ f)(x) = \dfrac{2x}{x + 1} - 1, x \neq -1;$

$(f \circ f)(x) = \dfrac{x}{2x + 1}, x \neq -1, x \neq -\frac{1}{2};$

$(g \circ g)(x) = 4x - 3, (-\infty, \infty)$

43. $(f \circ g)(x) = \dfrac{1}{x + 1}, x \neq -1, x \neq 0; (g \circ f)(x) = \dfrac{x + 1}{x},$

$x \neq -1, x \neq 0; (f \circ f)(x) = \dfrac{x}{2x + 1}, x \neq -1, x \neq -\frac{1}{2};$

$(g \circ g)(x) = x, x \neq 0$

45. $(f \circ g \circ h)(x) = \sqrt{x - 1} - 1$

47. $(f \circ g \circ h)(x) = (\sqrt{x} - 5)^4 + 1$

49. $g(x) = x - 9, f(x) = x^5$

51. $g(x) = x^2, f(x) = x/(x + 4)$

53. $g(x) = 1 - x^3, f(x) = |x|$

55. $h(x) = x^2, g(x) = x + 1, f(x) = 1/x$

57. $h(x) = \sqrt[3]{x}, g(x) = 4 + x, f(x) = x^9$

59. $R(x) = 0.15x - 0.000002x^2$

61. (a) $g(t) = 60t$ (b) $f(r) = \pi r^2$ (c) $(f \circ g)(t) = 3600\pi t^2$

63. $A(t) = 16\pi t^2$ **65.** (a) $f(x) = 0.9x$ (b) $g(x) = x - 100$

(c) $(f \circ g)(x) = 0.9x - 90, (g \circ f)(x) = 0.9x - 100, f \circ g$:

first rebate, then discount, $g \circ f$:

first discount, then rebate, $g \circ f$ is the better deal

SECTION 1.8 ■ PAGE 82

1. different, Horizontal Line **2.** (a) one-to-one, $g(x) = x^3$

(b) $g^{-1}(x) = x^{1/3}$ **3.** (a) Take the cube root, subtract 5, then

divide the result by 3. (b) $f(x) = (3x + 5)^3, f^{-1}(x) = \dfrac{x^{1/3} - 5}{3}$

4. (a) False (b) True **5.** No **7.** Yes **9.** No **11.** Yes

13. Yes **15.** No **17.** No **19.** No **21.** (a) 2 (b) 3 **23.** 1

37. $f^{-1}(x) = \frac{1}{2}(x - 1)$ **39.** $f^{-1}(x) = \frac{1}{4}(x - 7)$

41. $f^{-1}(x) = \sqrt[3]{\frac{1}{4}(5 - x)}$ **43.** $f^{-1}(x) = (1/x) - 2$

45. $f^{-1}(x) = \dfrac{4x}{1 - x}$ **47.** $f^{-1}(x) = \dfrac{7x + 5}{x - 2}$

49. $f^{-1}(x) = (5x - 1)/(2x + 3)$

51. $f^{-1}(x) = \frac{1}{5}(x^2 - 2), x \geq 0$

53. $f^{-1}(x) = \sqrt{4 - x}, x \leq 4$

55. $f^{-1}(x) = (x - 4)^3$

57. $f^{-1}(x) = x^2 - 2x, x \geq 1$

59. $f^{-1}(x) = \sqrt[4]{x}$

61. (a)

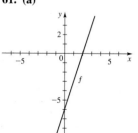

(b)

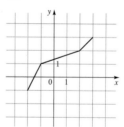

(c) $f^{-1}(x) = \frac{1}{3}(x + 6)$

63. (a)

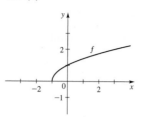

(b)

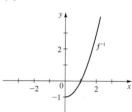

(c) $f^{-1}(x) = x^2 - 1, x \geq 0$

65. Not one-to-one

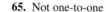
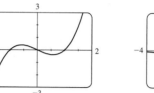

67. One-to-one

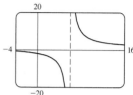

69. Not one-to-one

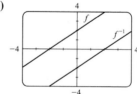

71. (a) $f^{-1}(x) = x - 2$

(b)

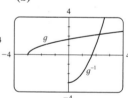

73. (a) $g^{-1}(x) = x^2 - 3, x \geq 0$

(b)

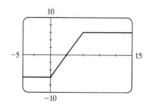

75. $x \geq 0, f^{-1}(x) = \sqrt{4 - x}$ **77.** $x \geq -2, h^{-1}(x) = \sqrt{x} - 2$

79.

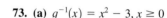

81. (a) $f(x) = 500 + 80x$ (b) $f^{-1}(x) = \frac{1}{80}(x - 500)$, the

number of hours worked as a function of the fee

(c) 9; if he charges \$1220, he worked 9 h

83. (a) $v^{-1}(t) = \sqrt{0.25 - \dfrac{t}{18,500}}$ (b) 0.498; at a distance 0.498

from the central axis the velocity is 30

85. (a) $F^{-1}(x) = \frac{5}{9}(x - 32)$; the Celsius temperature when the

Fahrenheit temperature is x (b) $F^{-1}(86) = 30$; when the

temperature is 86°F, it is 30°C

87. (a) $f(x) = \begin{cases} 0.1x & \text{if } 0 \le x \le 20{,}000 \\ 2000 + 0.2(x - 20{,}000) & \text{if } x > 20{,}000 \end{cases}$

(b) $f^{-1}(x) = \begin{cases} 10x & \text{if } 0 \le x \le 2000 \\ 10{,}000 + 5x & \text{if } x > 2000 \end{cases}$

If you pay x euros (€) in taxes, your income is $f^{-1}(x)$.
(c) $f^{-1}(10{,}000) = €\,60{,}000$
89. $f^{-1}(x) = \frac{1}{2}(x - 7)$. A pizza costing x dollars has $f^{-1}(x)$ toppings.

CHAPTER 1 REVIEW ▪ PAGE 86

1. (a)

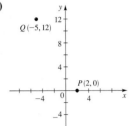

(b) $\sqrt{193}$
(d) $y = -\frac{12}{7}x + \frac{24}{7}$

(c) $\left(-\frac{3}{2}, 6\right)$
(e) $(x - 2)^2 + y^2 = 193$

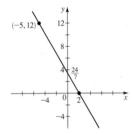   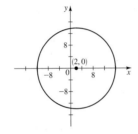

3.

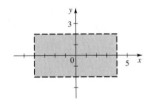

5. B **7.** $(x + 5)^2 + (y + 1)^2 = 26$
9. Circle, center $(-1, 3)$, radius 1   **11.** No graph
13. No symmetry   **15.** No symmetry

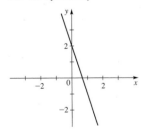   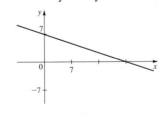

17. Symmetry about y-axis   **19.** No symmetry

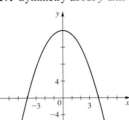   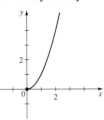

21. $2x - 3y - 16 = 0$ **23.** $3x + y - 12 = 0$
25. $x + 5y = 0$ **27.** $x^2 + y^2 = 169, 5x - 12y + 169 = 0$
29. (a) The slope represents the amount the spring lengthens for a one-pound increase in weight. The S-intercept represents the unstretched length of the spring.   **(b)** 4 in.
31. $f(x) = x^2 - 5$

33. Add 10, then multiply the result by 3.

35.

x	$g(x)$
-1	5
0	0
1	-3
2	-4
3	-3

37. (a) $C(1000) = 34{,}000, C(10{,}000) = 205{,}000$ **(b)** The costs of printing 1000 and 10,000 copies of the book   **(c)** $C(0) = 5000$; fixed costs   **39.** $6, 2, 18, a^2 - 4a + 6, a^2 + 4a + 6, x^2 - 2x + 3, 4x^2 - 8x + 6, 2x^2 - 8x + 10$ **41. (a)** Not a function
(b) Function   **(c)** Function, one-to-one   **(d)** Not a function
43. Domain $[-3, \infty)$, range $[0, \infty)$ **45.** $(-\infty, \infty)$
47. $[-4, \infty)$ **49.** $\{x \mid x \ne -2, -1, 0\}$ **51.** $(-\infty, -1] \cup [1, 4]$
53.   **55.**

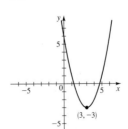   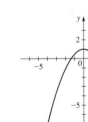

57.   **59.**

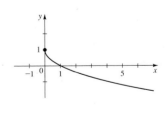

61.

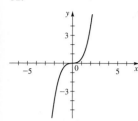

63.

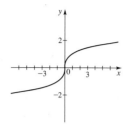

65.

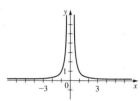

67.

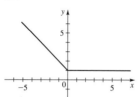

69.

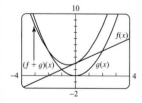

71. No **73.** Yes **75.** $[-2.1, 0.2] \cup [1.9, \infty)$

77.

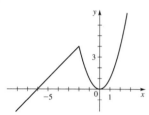

Increasing on $(-\infty, 0]$, $[2.67, \infty)$; decreasing on $[0, 2.67]$

79. (a) Shift upward 8 units **(b)** Shift to the left 8 units
(c) Stretch vertically by a factor of 2, then shift upward 1 unit
(d) Shift to the right 2 units and downward 2 units **(e)** Reflect in
y-axis **(f)** Reflect in y-axis, then in x-axis **(g)** Reflect in x-axis
(h) Reflect in line $y = x$
81. (a) Neither **(b)** Odd **(c)** Even **(d)** Neither
83. $g(-1) = -7$ **85.** 68 ft **87.** Local maximum ≈ 3.79
when $x \approx 0.46$; local minimum ≈ 2.81 when $x \approx -0.46$
89.

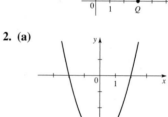

91. (a) $(f + g)(x) = x^2 - 6x + 6$ **(b)** $(f - g)(x) = x^2 - 2$
(c) $(fg)(x) = -3x^3 + 13x^2 - 18x + 8$
(d) $(f/g)(x) = (x^2 - 3x + 2)/(4 - 3x)$
(e) $(f \circ g)(x) = 9x^2 - 15x + 6$ **(f)** $(g \circ f)(x) = -3x^2 + 9x - 2$

93. $(f \circ g)(x) = -3x^2 + 6x - 1, (-\infty, \infty);$
$(g \circ f)(x) = -9x^2 + 12x - 3, (-\infty, \infty); (f \circ f)(x) = 9x - 4,$
$(-\infty, \infty); (g \circ g)(x) = -x^4 + 4x^3 - 6x^2 + 4x, (-\infty, \infty)$
95. $(f \circ g \circ h)(x) = 1 + \sqrt{x}$ **97.** Yes **99.** No **101.** No
103. $f^{-1}(x) = \dfrac{x + 2}{3}$ **105.** $f^{-1}(x) = \sqrt[3]{x} - 1$
107. (a), (b)

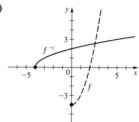

(c) $f^{-1}(x) = \sqrt{x + 4}$

CHAPTER 1 TEST ■ PAGE 90

1. (a) $S(3, 6)$ **(b)** 18

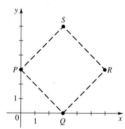

2. (a) **(b)** x-intercepts $-2, 2$
y-intercept -4
(c) Symmetric about
y-axis

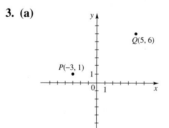

3. (a)

(b) $\sqrt{89}$ **(c)** $\left(1, \frac{7}{2}\right)$ **(d)** $\frac{5}{8}$ **(e)** $y = -\frac{8}{5}x + \frac{51}{10}$
(f) $(x - 1)^2 + \left(y - \frac{7}{2}\right)^2 = \frac{89}{4}$

4. (a) $(0,0), 5$

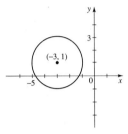

(b) $(2, -1), 3$

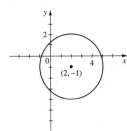

(c) $(-3, 1), 2$

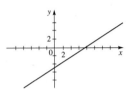

5. $y = \frac{2}{3}x - 5$

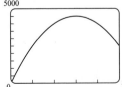

slope $\frac{2}{3}$; y-intercept -5
6. (a) $3x + y - 3 = 0$ **(b)** $2x + 3y - 12 = 0$
7. (a) and **(b)** are graphs of functions, (a) is one-to-one
8. (a) $2/3$, $\sqrt{6}/5$, $\sqrt{a}/(a-1)$ **(b)** $[-1, 0) \cup (0, \infty)$
9. (a) $f(x) = (x - 2)^3$

(b)

x	$f(x)$
-1	-27
0	-8
1	-1
2	0
3	1
4	8

(c)

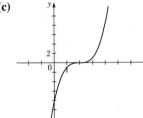

(d) By the Horizontal Line Test; take the cube root, then add 2
(e) $f^{-1}(x) = x^{1/3} + 2$ **10. (a)** $R(2) = 4000, R(4) = 4000$; total sales revenue with prices of $2 and $4
(b)

Revenue increases until price reaches $3, then decreases

(c) $4500; $3

11. (a)

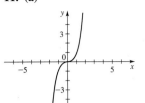

(b)

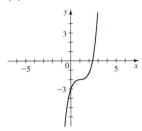

12. (a) Shift to the right 3 units, then shift upward 2 units
(b) Reflect in y-axis **13. (a)** 3, 0
(b)

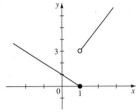

14. (a) $(f \circ g)(x) = (x - 3)^2 + 1$ **(b)** $(g \circ f)(x) = x^2 - 2$
(c) 2 **(d)** 2 **(e)** $(g \circ g \circ g)(x) = x - 9$
15. (a) $f^{-1}(x) = 3 - x^2, x \geq 0$

(b)

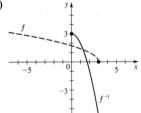

16. (a) Domain $[0, 6]$, range $[1, 7]$
(b)

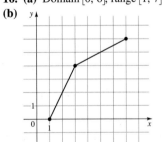

17. (a)

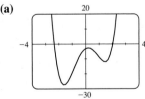

(b) No

(c) Local minimum ≈ -27.18 when $x \approx -1.61$;
local maximum ≈ -2.55 when $x \approx 0.18$;
local minimum ≈ -11.93 when $x \approx 1.43$
(d) $[-27.18, \infty)$ **(e)** Increasing on $[-1.61, 0.18] \cup [1.43, \infty)$;
decreasing on $(-\infty, -1.61] \cup [0.18, 1.43]$

FOCUS ON MODELING ▪ PAGE 97

1. (a)

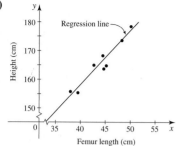

(b) $y = 1.8807x + 82.65$
(c) 191.7 cm

3. (a)

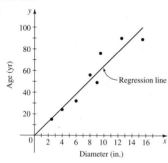

(b) $y = 6.451x - 0.1523$
(c) 116 years

5. (a)

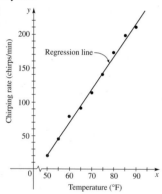

(b) $y = 4.857x - 220.97$
(c) 265 chirps/min

7. (a)

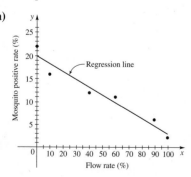

(b) $y = -0.168x + 19.89$ **(c)** 8.13%

9. (a)

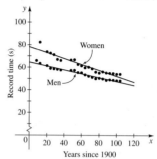

(b) $y = 0.2708x - 462.9$ **(c)** 80.3 years
11. (a) Men: $y = -0.1703x + 64.61$,
women $y = -0.2603x + 78.27$; x represents years since 1900
    (b) 2052

CHAPTER 2

SECTION 2.1 ▪ PAGE 109

1. (a) $(0, 0)$, 1   **(b)** $x^2 + y^2 = 1$ **(c)** (i) 0   (ii) 0   (iii) 0
(iv) 0   **2. (a)** terminal   **(b)** $(0, 1), (-1, 0), (0, -1), (1, 0)$
9. $-\frac{4}{5}$ **11.** $-2\sqrt{2}/3$ **13.** $3\sqrt{5}/7$ **15.** $P\left(\frac{4}{5}, \frac{3}{5}\right)$
17. $P\left(-\sqrt{5}/3, \frac{2}{3}\right)$ **19.** $P\left(-\sqrt{2}/3, -\sqrt{7}/3\right)$
21. $t = \pi/4, \left(\sqrt{2}/2, \sqrt{2}/2\right); t = \pi/2, (0, 1);$
$t = 3\pi/4, \left(-\sqrt{2}/2, \sqrt{2}/2\right); t = \pi, (-1, 0);$
$t = 5\pi/4, \left(-\sqrt{2}/2, -\sqrt{2}/2\right); t = 3\pi/2, (0, -1);$
$t = 7\pi/4, \left(\sqrt{2}/2, -\sqrt{2}/2\right); t = 2\pi, (1, 0)$
23. $(0, 1)$ **25.** $\left(-\sqrt{3}/2, \frac{1}{2}\right)$ **27.** $\left(\frac{1}{2}, -\sqrt{3}/2\right)$
29. $\left(-\frac{1}{2}, \sqrt{3}/2\right)$ **31.** $\left(-\sqrt{2}/2, -\sqrt{2}/2\right)$
33. (a) $\left(-\frac{3}{5}, \frac{4}{5}\right)$ **(b)** $\left(\frac{3}{5}, -\frac{4}{5}\right)$ **(c)** $\left(-\frac{3}{5}, -\frac{4}{5}\right)$ **(d)** $\left(\frac{3}{5}, \frac{4}{5}\right)$
35. (a) $\pi/4$ **(b)** $\pi/3$ **(c)** $\pi/3$ **(d)** $\pi/6$
37. (a) $2\pi/7$ **(b)** $2\pi/9$ **(c)** $\pi - 3 \approx 0.14$ **(d)** $2\pi - 5 \approx 1.28$
39. (a) $\pi/3$ **(b)** $\left(-\frac{1}{2}, \sqrt{3}/2\right)$
41. (a) $\pi/4$ **(b)** $\left(-\sqrt{2}/2, \sqrt{2}/2\right)$
43. (a) $\pi/3$ **(b)** $\left(-\frac{1}{2}, -\sqrt{3}/2\right)$
45. (a) $\pi/4$ **(b)** $\left(-\sqrt{2}/2, -\sqrt{2}/2\right)$
47. (a) $\pi/6$ **(b)** $\left(-\sqrt{3}/2, -\frac{1}{2}\right)$
49. (a) $\pi/3$ **(b)** $\left(\frac{1}{2}, \sqrt{3}/2\right)$ **51. (a)** $\pi/3$ **(b)** $\left(-\frac{1}{2}, -\sqrt{3}/2\right)$
53. $(0.5, 0.8)$ **55.** $(0.5, -0.9)$

SECTION 2.2 ■ PAGE 118

1. $y, x, y/x$ **2.** 1, 1   **3.** $t = \pi/4$, $\sin t = \sqrt{2}/2$, $\cos t = \sqrt{2}/2$;
$t = \pi/2$, $\sin t = 1$, $\cos t = 0$; $t = 3\pi/4$,
$\sin t = \sqrt{2}/2$, $\cos t = -\sqrt{2}/2$;
$t = \pi$, $\sin t = 0$, $\cos t = -1$; $t = 5\pi/4$,
$\sin t = -\sqrt{2}/2$, $\cos t = -\sqrt{2}/2$; $t = 3\pi/2$, $\sin t = -1$,
$\cos t = 0$; $t = 7\pi/4$, $\sin t = -\sqrt{2}/2$, $\cos t = \sqrt{2}/2$;
$t = 2\pi$, $\sin t = 0$, $\cos t = 1$
5. (a) $\sqrt{3}/2$ (b) $-1/2$ (c) $-\sqrt{3}$
7. (a) $-1/2$ (b) $-1/2$ (c) $-1/2$
9. (a) $-\sqrt{2}/2$ (b) $-\sqrt{2}/2$ (c) $\sqrt{2}/2$
11. (a) $\sqrt{3}/2$ (b) $2\sqrt{3}/3$ (c) $\sqrt{3}/3$
13. (a) -1 (b) 0 (c) 0
15. (a) 2 (b) $-2\sqrt{3}/3$ (c) 2
17. (a) $-\sqrt{3}/3$ (b) $\sqrt{3}/3$ (c) $-\sqrt{3}/3$
19. (a) $\sqrt{2}/2$ (b) $-\sqrt{2}$ (c) -1
21. (a) -1 (b) 1 (c) -1 **23.** (a) 0 (b) 1 (c) 0
25. $\sin 0 = 0$, $\cos 0 = 1$, $\tan 0 = 0$, $\sec 0 = 1$,
others undefined
27. $\sin \pi = 0$, $\cos \pi = -1$, $\tan \pi = 0$, $\sec \pi = -1$,
others undefined
29. $\frac{4}{5}, \frac{3}{5}, \frac{4}{3}$ **31.** $-\sqrt{11}/4$, $\sqrt{5}/4$, $-\sqrt{55}/5$
33. $\sqrt{13}/7$, $-6/7$, $-\sqrt{13}/6$ **35.** $-\frac{12}{13}, -\frac{5}{13}, \frac{12}{5}$ **37.** $\frac{21}{29}, -\frac{20}{29}, -\frac{21}{20}$
39. (a) 0.8 (b) 0.84147   **41.** (a) 0.9 (b) 0.93204
43. (a) 1 (b) 1.02964   **45.** (a) -0.6 (b) -0.57482
47. Negative   **49.** Negative   **51.** II   **53.** II
55. $\sin t = \sqrt{1 - \cos^2 t}$
57. $\tan t = (\sin t)/\sqrt{1 - \sin^2 t}$
59. $\sec t = -\sqrt{1 + \tan^2 t}$
61. $\tan t = \sqrt{\sec^2 t - 1}$
63. $\tan^2 t = (\sin^2 t)/(1 - \sin^2 t)$
65. $\cos t = -\frac{4}{5}$, $\tan t = -\frac{3}{4}$, $\csc t = \frac{5}{3}$, $\sec t = -\frac{5}{4}$, $\cot t = -\frac{4}{3}$
67. $\sin t = -2\sqrt{2}/3$, $\cos t = \frac{1}{3}$, $\tan t = -2\sqrt{2}$,
$\csc t = -\frac{3}{4}\sqrt{2}$, $\cot t = -\sqrt{2}/4$
69. $\sin t = -\frac{3}{5}$, $\cos t = \frac{4}{5}$, $\csc t = -\frac{5}{3}$, $\sec t = \frac{5}{4}$, $\cot t = -\frac{4}{3}$
71. $\cos t = -\sqrt{15}/4$, $\tan t = \sqrt{15}/15$, $\csc t = -4$,
$\sec t = -4\sqrt{15}/15$, $\cot t = \sqrt{15}$
73. Odd   **75.** Odd   **77.** Even   **79.** Neither
81. $y(0) = 4$, $y(0.25) = -2.828$, $y(0.50) = 0$,
$y(0.75) = 2.828$, $y(1.00) = -4$, $y(1.25) = 2.828$
83. (a) 0.49870 amp (b) -0.17117 amp

SECTION 2.3 ■ PAGE 130

1. 1, 2π

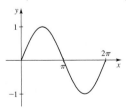

2. 3, π

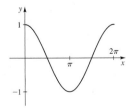

3.

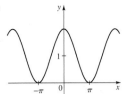

5.

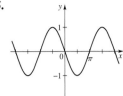

7.

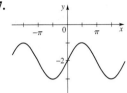

9.

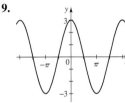

11.

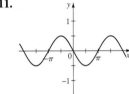

13.

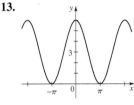

15.

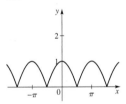

17. 1, π

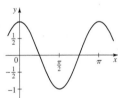

19. 3, $2\pi/3$

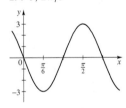

21. 10, 4π

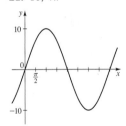

23. $\frac{1}{3}$, 6π

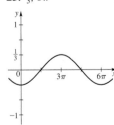

25. 2, 1

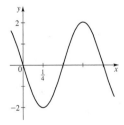

27. $\frac{1}{2}$, 2

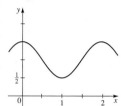

29. 1, 2π, $\pi/2$

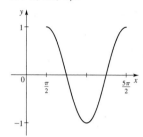

31. 2, 2π, $\pi/6$

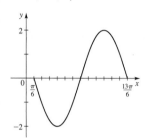

33. 4, π, $-\pi/2$

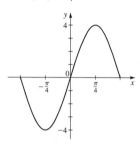

35. 5, $2\pi/3$, $\pi/12$

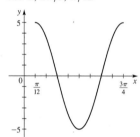

37. $\frac{1}{2}$, π, $\pi/6$

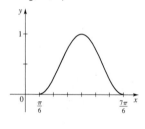

39. 3, 2, $-\frac{1}{2}$

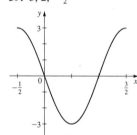

41. 1, $2\pi/3$, $-\pi/3$

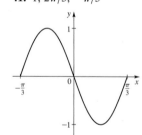

43. (a) 4, 2π, 0   (b) $y = 4 \sin x$
45. (a) $\frac{3}{2}$, $\frac{2\pi}{3}$, 0   (b) $y = \frac{3}{2} \cos 3x$
47. (a) $\frac{1}{2}$, π, $-\frac{\pi}{3}$ (b) $y = -\frac{1}{2} \cos 2(x + \pi/3)$
49. (a) 4, $\frac{3}{2}$, $-\frac{1}{2}$ (b) $y = 4 \sin \frac{4\pi}{3}(x + \frac{1}{2})$

51.

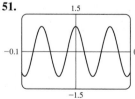

53.

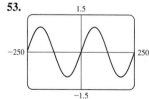

55.

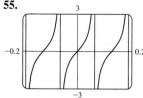

57.

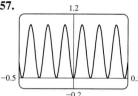

59.

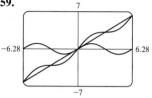

61.

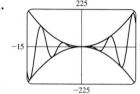

$y = x^2 \sin x$ is a sine curve that lies between the graphs of $y = x^2$ and $y = -x^2$

63.

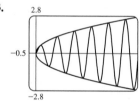

$y = \sqrt{x} \sin 5\pi x$ is a sine curve that lies between the graphs of $y = \sqrt{x}$ and $y = -\sqrt{x}$

65.

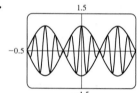

$y = \cos 3\pi x \cos 21\pi x$ is a cosine curve that lies between the graphs of $y = \cos 3\pi x$ and $y = -\cos 3\pi x$

67. Maximum value 1.76 when $x \approx 0.94$, minimum value -1.76 when $x \approx -0.94$ (The same maximum and minimum values occur at infinitely many other values of x.)
69. Maximum value 3.00 when $x \approx 1.57$, minimum value -1.00 when $x \approx -1.57$ (The same maximum and minimum values occur at infinitely many other values of x.)
71. 1.16   **73.** 0.34, 2.80
75. (a) Odd   (b) 0, $\pm 2\pi$, $\pm 4\pi$, $\pm 6\pi$, ...
(c)

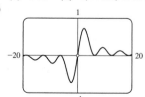

(d) $f(x)$ approaches 0
(e) $f(x)$ approaches 0

77. (a) 20 s **(b)** 6 ft
79. (a) $\frac{1}{80}$ min **(b)** 80
(c)

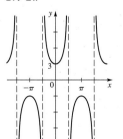

(d) $\frac{140}{90}$; it is higher than normal

SECTION 2.4 ■ PAGE 139

1. π; $\frac{\pi}{2} + n\pi$, n an integer   **2.** 2π; $n\pi$, n an integer

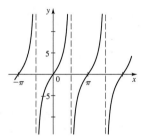   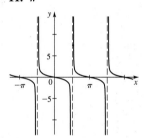

3. II   **5.** VI   **7.** IV
9. π **11.** π

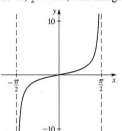   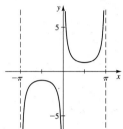

13. π **15.** 2π

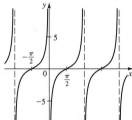   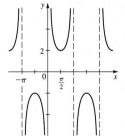

17. 2π **19.** π

   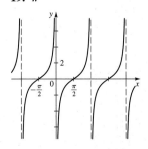

21. 2π **23.** π

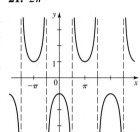   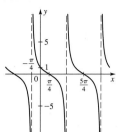

25. 2π **27.** $\pi/4$

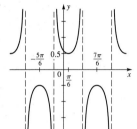   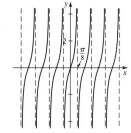

29. 4   **31.** π

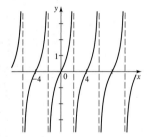   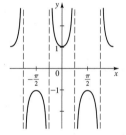

33. $\pi/2$ **35.** $\frac{1}{3}$

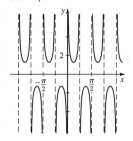   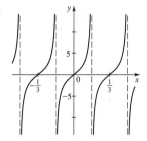

37. $\frac{4}{3}$ **39.** $\pi/2$

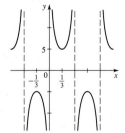   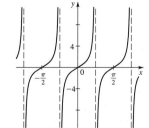

41. $\pi/2$

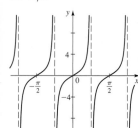

43. $\pi/2$

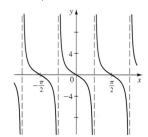

45. 2

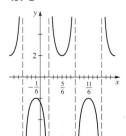

47. $2\pi/3$

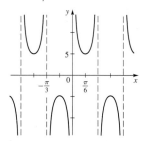

49. $3\pi/2$

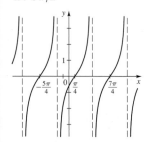

51. 2

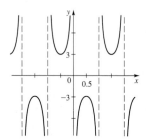

53. $\pi/2$

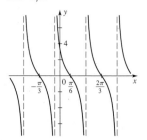

57. (a) 1.53 mi, 3.00 mi, 18.94 mi
(b)

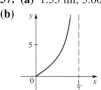

(c) $d(t)$ approaches ∞

SECTION 2.5 ■ PAGE 145

1. (a) $[-\pi/2, \pi/2], y, x, \pi/6, \pi/6, \frac{1}{2}$
(b) $[0, \pi]; y, x, \pi/3, \pi/3, \frac{1}{2}$ **2.** $[-1, 1]$; (b)
3. (a) $\pi/2$ **(b)** $\pi/3$ **(c)** Undefined
5. (a) π **(b)** $\pi/3$ **(c)** $5\pi/6$
7. (a) $-\pi/4$ **(b)** $\pi/3$ **(c)** $\pi/6$

9. (a) $2\pi/3$ **(b)** $-\pi/4$ **(c)** $\pi/4$ **11.** 0.72973
13. 2.01371   **15.** 2.75876   **17.** 1.47113   **19.** 0.88998
21. -0.26005 **23.** $\frac{1}{4}$ **25.** 5
27. Undefined   **29.** $5\pi/6$ **31.** $-\pi/6$ **33.** $\pi/6$ **35.** $\pi/6$
37. $-\pi/3$ **39.** $\sqrt{3}/3$ **41.** $\frac{1}{2}$ **43.** $-\sqrt{2}/2$

SECTION 2.6 ■ PAGE 152

1. $a \sin \omega t$ **2.** $a \cos \omega t$
3. (a) $2, 2\pi/3, 3/(2\pi)$ **5. (a)** $1, 20\pi/3, 3/(20\pi)$
(b)

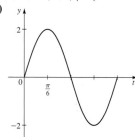

(b)

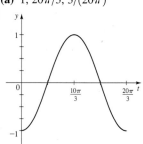

7. (a) $\frac{1}{4}, 4\pi/3, 3/(4\pi)$
(b)

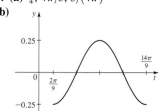

9. (a) $5, 3\pi, 1/(3\pi)$
(b)

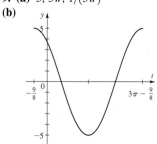

11. $y = 10 \sin\left(\dfrac{2\pi}{3}t\right)$ **13.** $y = 6 \sin(10t)$
15. $y = 60 \cos(4\pi t)$ **17.** $y = 2.4 \cos(1500\pi t)$

19. (a) 10 cycles per minute
(b)   **(c)** 8.2 m

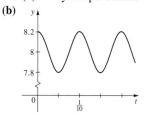

21. (a) 25, 0.0125, 80 **(b)**

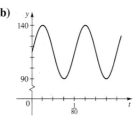

(c) The period decreases and the frequency increases.
23. $d(t) = 5 \sin(5\pi t)$
25. $y = 21 \sin\left(\dfrac{\pi}{6}t\right)$

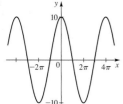

27. $y = 5 \cos(2\pi t)$ **29.** $y = 11 + 10 \sin\left(\dfrac{\pi t}{10}\right)$

31. $y = 3.8 + 0.2 \sin\left(\dfrac{\pi}{5}t\right)$

33. $f(t) = 10 \sin\left(\dfrac{\pi}{12}(t - 8)\right) + 90$

35. (a) 45 V **(b)** 40 **(c)** 40 **(d)** $E(t) = 45 \cos(80\pi t)$

CHAPTER 2 REVIEW ▪ PAGE 155

1. (b) $\frac{1}{2}, -\sqrt{3}/2, -\sqrt{3}/3$ **3. (a)** $\pi/3$ **(b)** $\left(-\frac{1}{2}, \sqrt{3}/2\right)$
(c) $\sin t = \sqrt{3}/2, \cos t = -\frac{1}{2}, \tan t = -\sqrt{3}, \csc t = 2\sqrt{3}/3,$
$\sec t = -2, \cot t = -\sqrt{3}/3$
5. (a) $\pi/4$ **(b)** $\left(-\sqrt{2}/2, -\sqrt{2}/2\right)$
(c) $\sin t = -\sqrt{2}/2, \cos t = -\sqrt{2}/2,$
$\tan t = 1, \csc t = -\sqrt{2}, \sec t = -\sqrt{2}, \cot t = 1$
7. (a) $\sqrt{2}/2$ **(b)** $-\sqrt{2}/2$ **9. (a)** 0.89121 **(b)** 0.45360
11. (a) 0 **(b)** Undefined **13. (a)** Undefined **(b)** 0
15. (a) $-\sqrt{3}/3$ **(b)** $-\sqrt{3}$ **17.** $(\sin t)/(1 - \sin^2 t)$
19. $(\sin t)/\sqrt{1 - \sin^2 t}$
21. $\tan t = -\frac{5}{12}, \csc t = \frac{13}{5}, \sec t = -\frac{13}{12}, \cot t = -\frac{12}{5}$
23. $\sin t = 2\sqrt{5}/5, \cos t = -\sqrt{5}/5,$
$\tan t = -2, \sec t = -\sqrt{5}$
25. $(16 - \sqrt{17})/4$ **27.** 3

29. (a) $10, 4\pi, 0$ **(b)**

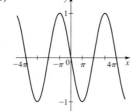

31. (a) $1, 4\pi, 0$ **(b)**

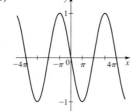

33. (a) $3, \pi, 1$
(b)

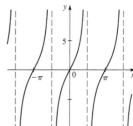
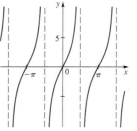

35. (a) $1, 4, -\frac{1}{3}$
(b)

37. $y = 5 \sin 4x$
41. π

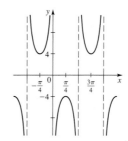

39. $y = \frac{1}{2} \sin 2\pi\left(x + \frac{1}{3}\right)$
43. π

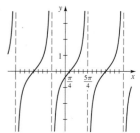

45. π

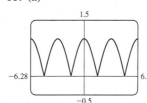

47. 2π

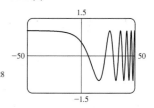

49. $\frac{\pi}{2}$ **51.** $\frac{\pi}{6}$
53. (a)

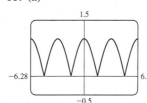

55. (a)

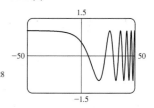

(b) Period π
(c) Even
57. (a)

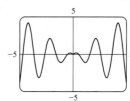

(b) Not periodic
(c) Neither
(b) Not periodic
(c) Even

59.

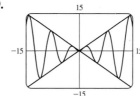

$y = x \sin x$ is a sine function whose graph lies between those of $y = x$ and $y = -x$

61.

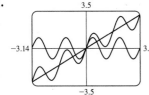

The graphs are related by graphical addition.

63. 1.76, −1.76   **65.** 0.30, 2.84
67. (a) Odd   **(b)** 0, ±π, ±2π, . . .
(c)

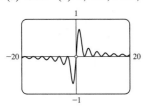

(d) $f(x)$ approaches 0   **(e)** $f(x)$ approaches 0
69. $y = 50 \cos(16\pi t)$ **71.** $y = 4 \cos\left(\frac{\pi}{6}t\right)$

CHAPTER 2 TEST ▪ PAGE 158

1. $y = -\frac{5}{6}$ **2. (a)** $\frac{4}{5}$ **(b)** $-\frac{3}{5}$ **(c)** $-\frac{4}{3}$ **(d)** $-\frac{5}{3}$
3. (a) $-\frac{1}{2}$ **(b)** $-\sqrt{2}/2$ **(c)** $\sqrt{3}$ **(d)** -1
4. $\tan t = -(\sin t)/\sqrt{1 - \sin^2 t}$ **5.** $-\frac{2}{15}$
6. (a) 5, $\pi/2$, 0   **7. (a)** 2, 4π, $\pi/3$
(b)   **(b)**

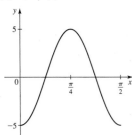   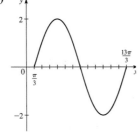

8. π **9.** $\pi/2$

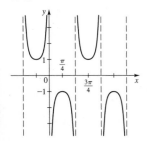   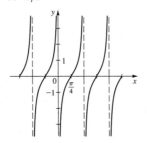

10. (a) $\pi/4$ **(b)** $5\pi/6$ **(c)** 0   **(d)** 1/2
11. $y = 2 \sin 2(x + \pi/3)$
12. (a)

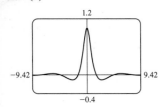

(b) Even
(c) Minimum value -0.11 when $x \approx \pm 2.54$, maximum value 1 when $x = 0$

13. $y = 5 \sin(4\pi t)$

FOCUS ON MODELING ▪ PAGE 162

1. (a) and **(c)**

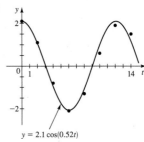

(b) $y = 2.1 \cos(0.52t)$
(d) $y = 2.05 \sin(0.50t + 1.55) - 0.01$ **(e)** The formula of (d) reduces to $y = 2.05 \cos(0.50t - 0.02) - 0.01$. Same as (b), rounded to one decimal.
3. (a) and **(c)**

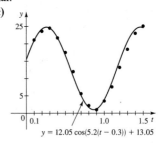

(b) $y = 12.05 \cos(5.2(t - 0.3)) + 13.05$
(d) $y = 11.72 \sin(5.05t + 0.24) + 12.96$ **(e)** The formula of (d) reduces to $y = 11.72 \cos(5.05(t - 0.26)) + 12.96$. Close, but not identical, to (b).
5. (a) and **(c)**

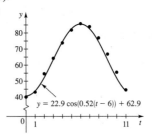

(b) $y = 22.9 \cos(0.52(t - 6)) + 62.9$, where y is temperature (°F) and t is months (January = 0)
(d) $y = 23.4 \sin(0.48t - 1.36) + 62.2$
7. (a) and **(c)**

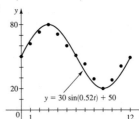

(b) $y = 30 \sin(0.52t) + 50$ where y is the owl population in year t **(d)** $y = 25.8 \sin(0.52t - 0.02) + 50.6$

9. (a) and **(c)**

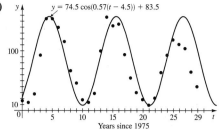

$y = 74.5 \cos(0.57(t - 4.5)) + 83.5$

Years since 1975

(b) $y = 74.5 \cos(0.57(t - 4.5)) + 83.5$, where y is the average daily sunspot count, and t is the years since 1975
(d) $y = 67.65 \sin(0.62t - 1.65) + 74.5$

CHAPTER 3

SECTION 3.1 ■ PAGE 172

1. (a) arc, 1   **(b)** $\pi/180$ **(c)** $180/\pi$ **2. (a)** $r\theta$ **(b)** $\frac{1}{2}r^2\theta$
3. $2\pi/5 \approx 1.257$ rad   **5.** $-\pi/4 \approx -0.785$ rad
7. $-5\pi/12 \approx -1.309$ rad   **9.** $6\pi \approx 18.850$ rad
11. $8\pi/15 \approx 1.676$ rad   **13.** $\pi/24 \approx 0.131$ rad   **15.** $210°$
17. $-225°$ **19.** $540/\pi \approx 171.9°$ **21.** $-216/\pi \approx -68.8°$
23. $18°$ **25.** $-24°$ **27.** $410°, 770°, -310°, -670°$
29. $11\pi/4, 19\pi/4, -5\pi/4, -13\pi/4$
31. $7\pi/4, 15\pi/4, -9\pi/4, -17\pi/4$ **33.** Yes   **35.** Yes   **37.** Yes
39. $13°$ **41.** $30°$ **43.** $280°$ **45.** $5\pi/6$ **47.** π **49.** $\pi/4$
51. $55\pi/9 \approx 19.2$ **53.** 4   **55.** 4 mi   **57.** 2 rad $\approx 114.6°$
59. $36/\pi \approx 11.459$ m   **61. (a)** 44.68   **(b)** 25   **63.** 50 m^2
65. 4 m   **67.** 6 cm^2 **69.** 13.9 mi   **71.** 330π mi ≈ 1037 mi
73. 1.6 million mi   **75.** 1.15 mi   **77.** 360π in$^2 \approx 1130.97$ in^2
79. (a) 90π rad/min   **(b)** 1440π in./min ≈ 4523.9 in./min
81. $32\pi/15$ ft/s ≈ 6.7 ft/s   **83.** 1039.6 mi/h   **85.** 2.1 m/s
87. (a) 10π cm ≈ 31.4 cm   **(b)** 5 cm   **(c)** 3.32 cm
(d) 86.8 cm^3

SECTION 3.2 ■ PAGE 180

1. (a)

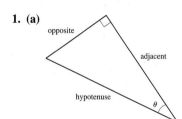

(b) $\dfrac{\text{opposite}}{\text{hypotenuse}}, \dfrac{\text{adjacent}}{\text{hypotenuse}}, \dfrac{\text{opposite}}{\text{adjacent}}$ **(c)** similar

2. $\sin\theta, \cos\theta, \tan\theta$
3. $\sin\theta = \frac{4}{5}, \cos\theta = \frac{3}{5}, \tan\theta = \frac{4}{3}, \csc\theta = \frac{5}{4}, \sec\theta = \frac{5}{3}, \cot\theta = \frac{3}{4}$
5. $\sin\theta = \frac{40}{41}, \cos\theta = \frac{9}{41}, \tan\theta = \frac{40}{9}, \csc\theta = \frac{41}{40}, \sec\theta = \frac{41}{9}, \cot\theta = \frac{9}{40}$
7. $\sin\theta = 2\sqrt{13}/13, \cos\theta = 3\sqrt{13}/13, \tan\theta = \frac{2}{3}$,
$\csc\theta = \sqrt{13}/2, \sec\theta = \sqrt{13}/3, \cot\theta = \frac{3}{2}$
9. (a) $3\sqrt{34}/34, 3\sqrt{34}/34$ **(b)** $\frac{3}{5}, \frac{3}{5}$ **(c)** $\sqrt{34}/5, \sqrt{34}/5$
11. $\frac{25}{2}$ **13.** $13\sqrt{3}/2$ **15.** 16.51658
17. $x = 28\cos\theta, y = 28\sin\theta$

19. $\cos\theta = \frac{4}{5}, \tan\theta = \frac{3}{4}, \csc\theta = \frac{5}{3}, \sec\theta = \frac{5}{4}, \cot\theta = \frac{4}{3}$

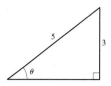

21. $\sin\theta = \sqrt{2}/2, \cos\theta = \sqrt{2}/2, \tan\theta = 1$,
$\csc\theta = \sqrt{2}, \sec\theta = \sqrt{2}$

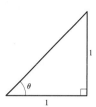

23. $\sin\theta = 3\sqrt{5}/7, \cos\theta = \frac{2}{7}, \tan\theta = 3\sqrt{5}/2$,
$\csc\theta = 7\sqrt{5}/15, \cot\theta = 2\sqrt{5}/15$

25. $(1 + \sqrt{3})/2$ **27.** 1   **29.** $\frac{1}{2}$
31.   **33.**

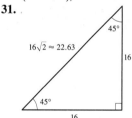

35.   **37.**

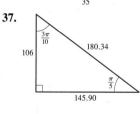

39. $\sin\theta \approx 0.45, \cos\theta \approx 0.89, \tan\theta = 0.50, \csc\theta \approx 2.24$,
$\sec\theta \approx 1.12, \cot\theta = 2.00$ **41.** 230.9   **43.** 63.7
45. $x = 10\tan\theta\sin\theta$ **47.** 1026 ft   **49. (a)** 2100 mi   **(b)** No
51. 19 ft   **53.** 345 ft   **55.** 415 ft, 152 ft   **57.** 2570 ft
59. 5808 ft   **61.** 91.7 million mi   **63.** 3960 mi   **65.** 0.723 AU

SECTION 3.3 ■ PAGE 191

1. $y/r, x/r, y/x$ **2.** quadrant, positive, negative, negative
3. (a) $30°$ **(b)** $30°$ **(c)** $30°$ **5. (a)** $45°$ **(b)** $90°$ **(c)** $75°$
7. (a) $\pi/4$ **(b)** $\pi/6$ **(c)** $\pi/3$ **9. (a)** $2\pi/7$ **(b)** 0.4π **(c)** 1.4
11. $\frac{1}{2}$ **13.** $-\sqrt{3}/2$ **15.** $-\sqrt{3}$ **17.** 1   **19.** $-\sqrt{3}/2$
21. $\sqrt{3}/3$ **23.** $\sqrt{3}/2$ **25.** -1 **27.** $\frac{1}{2}$ **29.** 2   **31.** -1
33. Undefined   **35.** III   **37.** IV
39. $\tan\theta = -\sqrt{1 - \cos^2\theta}/\cos\theta$
41. $\cos\theta = \sqrt{1 - \sin^2\theta}$

43. $\sec\theta = -\sqrt{1 + \tan^2\theta}$

45. $\cos\theta = -\frac{4}{5}$, $\tan\theta = -\frac{3}{4}$, $\csc\theta = \frac{5}{3}$, $\sec\theta = -\frac{5}{4}$, $\cot\theta = -\frac{4}{3}$

47. $\sin\theta = -\frac{3}{5}$, $\cos\theta = \frac{4}{5}$, $\csc\theta = -\frac{5}{3}$, $\sec\theta = \frac{5}{4}$, $\cot\theta = -\frac{4}{3}$

49. $\sin\theta = \frac{1}{2}$, $\cos\theta = \sqrt{3}/2$, $\tan\theta = \sqrt{3}/3$,
$\sec\theta = 2\sqrt{3}/3$, $\cot\theta = \sqrt{3}$

51. $\sin\theta = 3\sqrt{5}/7$, $\tan\theta = -3\sqrt{5}/2$, $\csc\theta = 7\sqrt{5}/15$,
$\sec\theta = -\frac{7}{2}$, $\cot\theta = -2\sqrt{5}/15$

53. (a) $\sqrt{3}/2$, $\sqrt{3}$ **(b)** $\frac{1}{2}$, $\sqrt{3}/4$ **(c)** $\frac{3}{4}$, 0.88967 **55.** 19.1

57. 66.1° **59.** $(4\pi/3) - \sqrt{3} \approx 2.46$

63. (b)

θ	20°	60°	80°	85°
h	1922	9145	29,944	60,351

65. (a) $A(\theta) = 400\sin\theta\cos\theta$

(b)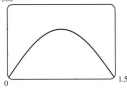

(c) width = depth $\approx$ 14.14 in.

67. (a) $9\sqrt{3}/4 \approx 3.897$ ft, $\frac{9}{16}$ ft = 0.5625 ft

(b) 23.982 ft, 3.462 ft

69. (a) 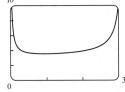 **(b)** 0.946 rad or 54°

SECTION 3.4 ■ PAGE 199

1. (a) $[-1, 1]$, $[-\pi/2, \pi/2]$ **(b)** $[-1, 1]$, $[0, \pi]$
(c) $\mathbb{R}$, $(-\pi/2, \pi/2)$ **2. (a)** $\frac{8}{10}$ **(b)** $\frac{6}{10}$ **(c)** $\frac{8}{6}$ **3. (a)** $\pi/6$
(b) $5\pi/6$ **(c)** $-\pi/4$ **5. (a)** $-\pi/6$ **(b)** $\pi/3$ **(c)** $\pi/6$
7. 0.46677 **9.** 1.82348 **11.** 1.24905 **13.** Undefined
15. 36.9° **17.** 34.7° **19.** 34.9° **21.** 30°, 150°
23. 44.4°, 135.6° **25.** 45.6° **27.** $\frac{4}{5}$ **29.** $\frac{13}{5}$ **31.** $\frac{12}{5}$
33. $\sqrt{1 - x^2}$ **35.** $x/\sqrt{1 - x^2}$ **37.** 72.5°, 19 ft
39. (a) $h = 2\tan\theta$ **(b)** $\theta = \tan^{-1}(h/2)$
41. (a) $\theta = \sin^{-1}(h/680)$ **(b)** $\theta = 0.826$ rad
43. (a) 54.1° **(b)** 48.3°, 32.2°, 24.5°. The function $\sin^{-1}$ is
undefined for values outside the interval $[-1, 1]$.

SECTION 3.5 ■ PAGE 205

1. $\dfrac{\sin A}{a} = \dfrac{\sin B}{b} = \dfrac{\sin C}{c}$ **2.** ASA, SSA **3.** 318.8 **5.** 24.8

7. 44° **9.** $\angle C = 114°$, $a \approx 51$, $b \approx 24$ **11.** $\angle A = 44°$,
$\angle B = 68°$, $a \approx 8.99$ **13.** $\angle C = 62°$, $a \approx 200$, $b \approx 242$

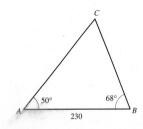

15. $\angle B = 85°$, $a \approx 5$, $c \approx 9$

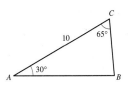

17. $\angle A = 100°$, $a \approx 89$, $c \approx 71$

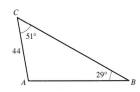

19. $\angle B \approx 30°$, $\angle C \approx 40°$, $c \approx 19$ **21.** No solution
23. $\angle A_1 \approx 125°$, $\angle C_1 \approx 30°$, $a_1 \approx 49$;
$\angle A_2 \approx 5°$, $\angle C_2 \approx 150°$, $a_2 \approx 5.6$ **25.** No solution
27. $\angle A_1 \approx 57.2°$, $\angle B_1 \approx 93.8°$, $b_1 \approx 30.9$;
$\angle A_2 \approx 122.8°$, $\angle B_2 \approx 28.2°$, $b_2 \approx 14.6$
29. (a) 91.146° **(b)** 14.427° **33. (a)** 1018 mi **(b)** 1017 mi
35. 219 ft **37.** 55.9 m **39.** 175 ft **41.** 192 m
43. 0.427 AU, 1.119 AU

SECTION 3.6 ■ PAGE 212

1. $a^2 + b^2 - 2ab\cos C$ **2.** SSS, SAS **3.** 28.9 **5.** 47
7. 29.89° **9.** 15 **11.** $\angle A \approx 39.4°$, $\angle B \approx 20.6°$, $c \approx 24.6$
13. $\angle A \approx 48°$, $\angle B \approx 79°$, $c \approx 3.2$
15. $\angle A \approx 50°$, $\angle B \approx 73°$, $\angle C \approx 57°$
17. $\angle A_1 \approx 83.6°$, $\angle C_1 \approx 56.4°$, $a_1 \approx 193$;
$\angle A_2 \approx 16.4°$, $\angle C_2 \approx 123.6$, $a_2 \approx 54.9$ **19.** No such triangle
21. 2 **23.** 25.4 **25.** 89.2° **27.** 24.3 **29.** 54 **31.** 26.83
33. 5.33 **35.** 40.77 **37.** 3.85 cm^2 **39.** 2.30 mi **41.** 23.1 mi
43. 2179 mi **45. (a)** 62.6 mi **(b)** S 18.2° E **47.** 96°
49. 211 ft **51.** 3835 ft **53.** $165,554

CHAPTER 3 REVIEW ■ PAGE 215

1. (a) $\pi/3$ **(b)** $11\pi/6$ **(c)** $-3\pi/4$ **(d)** $-\pi/2$
3. (a) 450° **(b)** $-30°$ **(c)** 405° **(d)** $(558/\pi)° \approx 177.6°$
5. 8 m **7.** 82 ft **9.** 0.619 rad $\approx$ 35.4° **11.** 18,151 ft^2
13. 300π rad/min $\approx$ 942.5 rad/min,
7539.8 in./min = 628.3 ft/min
15. $\sin\theta = 5/\sqrt{74}$, $\cos\theta = 7/\sqrt{74}$, $\tan\theta = \frac{5}{7}$,
$\csc\theta = \sqrt{74}/5$, $\sec\theta = \sqrt{74}/7$, $\cot\theta = \frac{7}{5}$
17. $x \approx 3.83$, $y \approx 3.21$ **19.** $x \approx 2.92$, $y \approx 3.11$
21. $A = 70°$, $a \approx 2.819$, $b \approx 1.026$
23. $A \approx 16.3°$, $C \approx 73.7°$, $c = 24$
25. $a = \cot\theta$, $b = \csc\theta$ **27.** 48 m **29.** 1076 mi **31.** $-\sqrt{2}/2$
33. 1 **35.** $-\sqrt{3}/3$ **37.** $-\sqrt{2}/2$ **39.** $2\sqrt{3}/3$ **41.** $-\sqrt{3}$
43. $\sin\theta = \frac{12}{13}$, $\cos\theta = -\frac{5}{13}$, $\tan\theta = -\frac{12}{5}$,
$\csc\theta = \frac{13}{12}$, $\sec\theta = -\frac{13}{5}$, $\cot\theta = -\frac{5}{12}$ **45.** 60°
47. $\tan\theta = -\sqrt{1 - \cos^2\theta}/\cos\theta$
49. $\tan^2\theta = \sin^2\theta/(1 - \sin^2\theta)$
51. $\sin\theta = \sqrt{7}/4$, $\cos\theta = \frac{3}{4}$, $\csc\theta = 4\sqrt{7}/7$, $\cot\theta = 3\sqrt{7}/7$
53. $\cos\theta = -\frac{4}{5}$, $\tan\theta = -\frac{3}{4}$, $\csc\theta = \frac{5}{3}$, $\sec\theta = -\frac{5}{4}$, $\cot\theta = -\frac{4}{3}$
55. $-\sqrt{5}/5$ **57.** 1 **59.** $\pi/3$ **61.** $2/\sqrt{21}$ **63.** $x/\sqrt{1 + x^2}$
65. $\theta = \cos^{-1}(x/3)$ **67.** 5.32 **69.** 148.07 **71.** 9.17
73. 54.1° or 125.9° **75.** 80.4° **77.** 77.3 mi **79.** 3.9 mi
81. 32.12

CHAPTER 3 TEST ▪ PAGE 219

1. $11\pi/6, -3\pi/4$ **2.** $240°, -74.5°$
3. (a) 240π rad/min ≈ 753.98 rad/min
(b) 12,063.7 ft/min = 137 mi/h   **4. (a)** $\sqrt{2}/2$
(b) $\sqrt{3}/3$ **(c)** 2   **(d)** 1   **5.** $(26 + 6\sqrt{13})/39$
6. $a = 24 \sin \theta, b = 24 \cos \theta$ **7.** $(4 - 3\sqrt{2})/4$
8. $-\frac{13}{12}$ **9.** $\tan \theta = -\sqrt{\sec^2\theta - 1}$ **10.** 19.6 ft
11. (a) $\theta = \tan^{-1}(x/4)$ **(b)** $\theta = \cos^{-1}(3/x)$ **12.** $\frac{40}{41}$
13. 9.1   **14.** 250.5   **15.** 8.4   **16.** 19.5   **17.** 78.6°   **18.** 40.2°
19. (a) 15.3 m^2 **(b)** 24.3 m   **20. (a)** 129.9°   **(b)** 44.9
21. 554 ft

FOCUS ON MODELING ▪ PAGE 222

1. 1.41 mi   **3.** 14.3 m   **5. (c)** 2349.8 ft
7.

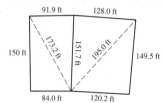

CHAPTER 4

SECTION 4.1 ▪ PAGE 230

1. all; 1   **2.** $\cos(-x) = \cos x$ **3.** $\sin t$ **5.** $\tan \theta$ **7.** -1
9. $\csc u$ **11.** $\tan \theta$ **13.** 1   **15.** $\cos y$ **17.** $\sin^2 x$ **19.** $\sec x$
21. $2 \sec u$ **23.** $\cos^2 x$ **25.** $\cos \theta$

27. (a) LHS $= \dfrac{1 - \sin^2 x}{\sin x} =$ RHS

29. LHS $= \sin \theta \dfrac{\cos \theta}{\sin \theta} =$ RHS

31. LHS $= \cos u \dfrac{1}{\cos u} \cot u =$ RHS

33. LHS $= \sin B + \cos B \dfrac{\cos B}{\sin B}$
$= \dfrac{\sin^2 B + \cos^2 B}{\sin B} = \dfrac{1}{\sin B} =$ RHS

35. LHS $= -\dfrac{\cos \alpha}{\sin \alpha} \cos \alpha - \sin \alpha = \dfrac{-\cos^2\alpha - \sin^2\alpha}{\sin \alpha}$
$= \dfrac{-1}{\sin \alpha} =$ RHS

37. LHS $= \dfrac{\sin \theta}{\cos \theta} + \dfrac{\cos \theta}{\sin \theta} = \dfrac{\sin^2\theta + \cos^2\theta}{\cos \theta \sin \theta}$
$= \dfrac{1}{\cos \theta \sin \theta} =$ RHS

39. LHS $= 1 - \cos^2\beta = \sin^2\beta =$ RHS
41. LHS $= \dfrac{(\sin x + \cos x)^2}{(\sin x + \cos x)(\sin x - \cos x)} = \dfrac{\sin x + \cos x}{\sin x - \cos x}$
$= \dfrac{(\sin x + \cos x)(\sin x - \cos x)}{(\sin x - \cos x)(\sin x - \cos x)} =$ RHS

43. LHS $= \dfrac{\frac{1}{\cos t} - \cos t}{\frac{1}{\cos t}} \cdot \dfrac{\cos t}{\cos t} = \dfrac{1 - \cos^2 t}{1} =$ RHS

45. LHS $= \dfrac{1}{\cos^2 y} = \sec^2 y =$ RHS

47. LHS $= \cot x \cos x + \cot x - \csc x \cos x - \csc x$
$= \dfrac{\cos^2 x}{\sin x} + \dfrac{\cos x}{\sin x} - \dfrac{\cos x}{\sin x} - \dfrac{1}{\sin x} = \dfrac{\cos^2 x - 1}{\sin x}$
$= \dfrac{-\sin^2 x}{\sin x} =$ RHS

49. LHS $= \sin^2 x\left(1 + \dfrac{\cos^2 x}{\sin^2 x}\right) = \sin^2 x + \cos^2 x =$ RHS

51. LHS $= 2(1 - \sin^2 x) - 1 = 2 - 2\sin^2 x - 1 =$ RHS

53. LHS $= \dfrac{1 - \cos \alpha}{\sin \alpha} \cdot \dfrac{1 + \cos \alpha}{1 + \cos \alpha}$
$= \dfrac{1 - \cos^2\alpha}{\sin \alpha(1 + \cos \alpha)} = \dfrac{\sin^2\alpha}{\sin \alpha(1 + \cos \alpha)} =$ RHS

55. LHS $= \dfrac{\sin^2\theta}{\cos^2\theta} - \dfrac{\sin^2\theta \cos^2\theta}{\cos^2\theta}$
$= \dfrac{\sin^2\theta(1 - \cos^2\theta)}{\cos^2\theta} = \dfrac{\sin^2\theta \sin^2\theta}{\cos^2\theta} =$ RHS

57. LHS $= \dfrac{\sin x - 1}{\sin x + 1} \cdot \dfrac{\sin x + 1}{\sin x + 1} = \dfrac{\sin^2 x - 1}{(\sin x + 1)^2} =$ RHS

59. LHS $= \dfrac{\sin^2 t + 2 \sin t \cos t + \cos^2 t}{\sin t \cos t}$
$= \dfrac{\sin^2 t + \cos^2 t}{\sin t \cos t} + \dfrac{2 \sin t \cos t}{\sin t \cos t} = \dfrac{1}{\sin t \cos t} + 2$
$=$ RHS

61. LHS $= \dfrac{1 + \frac{\sin^2 u}{\cos^2 u}}{1 - \frac{\sin^2 u}{\cos^2 u}} \cdot \dfrac{\cos^2 u}{\cos^2 u} = \dfrac{\cos^2 u + \sin^2 u}{\cos^2 u - \sin^2 u} =$ RHS

63. LHS $= \dfrac{\sec x}{\sec x - \tan x} \cdot \dfrac{\sec x + \tan x}{\sec x + \tan x}$
$= \dfrac{\sec x(\sec x + \tan x)}{\sec^2 x - \tan^2 x} =$ RHS

65. LHS $= (\sec v - \tan v) \cdot \dfrac{\sec v + \tan v}{\sec v + \tan v}$
$= \dfrac{\sec^2 v - \tan^2 v}{\sec v + \tan v} =$ RHS

67. LHS $= \dfrac{\sin x + \cos x}{\frac{1}{\cos x} + \frac{1}{\sin x}} = \dfrac{\sin x + \cos x}{\frac{\sin x + \cos x}{\cos x \sin x}}$
$= (\sin x + \cos x)\dfrac{\cos x \sin x}{\sin x + \cos x} =$ RHS

69. LHS $= \dfrac{\frac{1}{\sin x} - \frac{\cos x}{\sin x}}{\frac{1}{\cos x} - 1} \cdot \dfrac{\sin x \cos x}{\sin x \cos x} = \dfrac{\cos x(1 - \cos x)}{\sin x(1 - \cos x)}$
$= \dfrac{\cos x}{\sin x} =$ RHS

71. LHS $= \dfrac{\sin^2 u}{\cos^2 u} - \dfrac{\sin^2 u \cos^2 u}{\cos^2 u} = \dfrac{\sin^2 u}{\cos^2 u}(1 - \cos^2 u) =$ RHS

73. LHS $= (\sec^2 x - \tan^2 x)(\sec^2 x + \tan^2 x) =$ RHS

75. RHS $= \dfrac{\sin \theta - \frac{1}{\sin \theta}}{\cos \theta - \frac{\cos \theta}{\sin \theta}} = \dfrac{\frac{\sin^2\theta - 1}{\sin \theta}}{\frac{\cos \theta \sin \theta - \cos \theta}{\sin \theta}}$
$= \dfrac{\cos^2\theta}{\cos \theta(\sin \theta - 1)} =$ LHS

77. LHS $= \dfrac{-\sin^2 t + \tan^2 t}{\sin^2 t} = -1 + \dfrac{\sin^2 t}{\cos^2 t} \cdot \dfrac{1}{\sin^2 t}$

$= -1 + \sec^2 t = $ RHS

79. LHS $= \dfrac{\sec x - \tan x + \sec x + \tan x}{(\sec x + \tan x)(\sec x - \tan x)}$

$= \dfrac{2 \sec x}{\sec^2 x - \tan^2 x} = $ RHS

81. LHS $= \tan^2 x + 2 \tan x \cot x + \cot^2 x = \tan^2 x + 2 + \cot^2 x$

$= (\tan^2 x + 1) + (\cot^2 x + 1) = $ RHS

83. LHS $= \dfrac{\frac{1}{\cos u} - 1}{\frac{1}{\cos u} + 1} \cdot \dfrac{\cos u}{\cos u} = $ RHS

85. LHS $= \dfrac{(\sin x + \cos x)(\sin^2 x - \sin x \cos x + \cos^2 x)}{\sin x + \cos x}$

$= \sin^2 x - \sin x \cos x + \cos^2 x = $ RHS

87. LHS $= \dfrac{1 + \sin x}{1 - \sin x} \cdot \dfrac{1 + \sin x}{1 + \sin x} = \dfrac{(1 + \sin x)^2}{1 - \sin^2 x}$

$= \dfrac{(1 + \sin x)^2}{\cos^2 x} = \left(\dfrac{1 + \sin x}{\cos x} \right)^2 = $ RHS

89. LHS $= \left(\dfrac{\sin x}{\cos x} + \dfrac{\cos x}{\sin x} \right)^4 = \left(\dfrac{\sin^2 x + \cos^2 x}{\sin x \cos x} \right)^4$

$= \left(\dfrac{1}{\sin x \cos x} \right)^4 = $ RHS

91. $\tan \theta$ **93.** $\tan \theta$ **95.** $3 \cos \theta$

97. 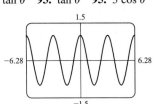 Yes

99. 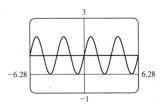 No

SECTION 4.2 ▪ PAGE 237

1. addition; $\sin x \cos y + \cos x \sin y$

2. subtraction; $\cos x \cos y + \sin x \sin y$

3. $\dfrac{\sqrt{6} + \sqrt{2}}{4}$ **5.** $\dfrac{\sqrt{2} - \sqrt{6}}{4}$ **7.** $2 - \sqrt{3}$ **9.** $-\dfrac{\sqrt{6} + \sqrt{2}}{4}$

11. $\sqrt{3} - 2$ **13.** $-\dfrac{\sqrt{6} + \sqrt{2}}{4}$ **15.** $\sqrt{2}/2$ **17.** $\frac{1}{2}$ **19.** $\sqrt{3}$

21. LHS $= \dfrac{\sin\left(\frac{\pi}{2} - u\right)}{\cos\left(\frac{\pi}{2} - u\right)} = \dfrac{\sin \frac{\pi}{2} \cos u - \cos \frac{\pi}{2} \sin u}{\cos \frac{\pi}{2} \cos u + \sin \frac{\pi}{2} \sin u}$

$= \dfrac{\cos u}{\sin u} = $ RHS

23. LHS $= \dfrac{1}{\cos\left(\frac{\pi}{2} - u\right)} = \dfrac{1}{\cos \frac{\pi}{2} \cos u + \sin \frac{\pi}{2} \sin u}$

$= \dfrac{1}{\sin u} = $ RHS

25. LHS $= \sin x \cos \frac{\pi}{2} - \cos x \sin \frac{\pi}{2} = $ RHS

27. LHS $= \sin x \cos \pi - \cos x \sin \pi = $ RHS

29. LHS $= \dfrac{\tan x - \tan \pi}{1 + \tan x \tan \pi} = $ RHS

31. LHS $= \cos x \cos \frac{\pi}{6} - \sin x \sin \frac{\pi}{6} + \sin x \cos \frac{\pi}{3} - \cos x \sin \frac{\pi}{3}$

$= \dfrac{\sqrt{3}}{2} \cos x - \frac{1}{2} \sin x + \frac{1}{2} \sin x - \dfrac{\sqrt{3}}{2} \cos x = $ RHS

33. LHS $= \sin x \cos y + \cos x \sin y$

$- (\sin x \cos y - \cos x \sin y) = $ RHS

35. LHS $= \dfrac{1}{\tan(x - y)} = \dfrac{1 + \tan x \tan y}{\tan x - \tan y}$

$= \dfrac{1 + \frac{1}{\cot x} \frac{1}{\cot y}}{\frac{1}{\cot x} - \frac{1}{\cot y}} \cdot \dfrac{\cot x \cot y}{\cot x \cot y} = $ RHS

37. LHS $= \dfrac{\sin x}{\cos x} - \dfrac{\sin y}{\cos y} = \dfrac{\sin x \cos y - \cos x \sin y}{\cos x \cos y} = $ RHS

39. LHS $= \dfrac{\sin x \cos y + \cos x \sin y - (\sin x \cos y - \cos x \sin y)}{\cos x \cos y - \sin x \sin y + \cos x \cos y + \sin x \sin y}$

$= \dfrac{2 \cos x \sin y}{2 \cos x \cos y} = $ RHS

41. LHS $= \sin((x + y) + z)$

$= \sin(x + y) \cos z + \cos(x + y) \sin z$

$= \cos z [\sin x \cos y + \cos x \sin y]$

$+ \sin z [\cos x \cos y - \sin x \sin y] = $ RHS

43. $\dfrac{\sqrt{1 - x^2} + xy}{\sqrt{1 + y^2}}$ **45.** $\dfrac{x - y}{\sqrt{1 + x^2} \sqrt{1 + y^2}}$

47. $\frac{1}{4} (\sqrt{6} + \sqrt{2})$ **49.** $\dfrac{3 - 2\sqrt{14}}{\sqrt{7} + 6\sqrt{2}}$ **51.** $-\frac{1}{10}(3 + 4\sqrt{3})$

53. $2\sqrt{5}/65$ **55.** $2 \sin\left(x + \dfrac{5\pi}{6} \right)$ **57.** $5\sqrt{2} \sin\left(2x + \dfrac{7\pi}{4} \right)$

59. (a) $g(x) = 2 \sin 2\left(x + \dfrac{\pi}{12} \right)$

(b)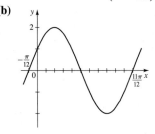

63. $\tan \gamma = \frac{17}{6}$

65. (a)

$\sin^2\left(x + \dfrac{\pi}{4} \right) + \sin^2\left(x - \dfrac{\pi}{4} \right) = 1$

67. $\pi/2$ **69.** (b) $k = 10\sqrt{3}$, $\phi = \pi/6$

SECTION 4.3 ■ PAGE 246

1. Double-Angle; $2 \sin x \cos x$

2. Half-Angle; $\pm \sqrt{(1 - \cos x)/2}$

3. $\frac{120}{169}, \frac{119}{169}, \frac{120}{119}$ **5.** $-\frac{24}{25}, \frac{7}{25}, -\frac{24}{7}$ **7.** $\frac{24}{25}, \frac{7}{25}, \frac{24}{7}$ **9.** $-\frac{3}{5}, \frac{4}{5}, -\frac{3}{4}$

11. $\frac{1}{2}\left(\frac{3}{4} - \cos 2x + \frac{1}{4} \cos 4x\right)$

13. $\frac{1}{16}(1 - \cos 2x - \cos 4x + \cos 2x \cos 4x)$

15. $\frac{1}{32}\left(\frac{3}{4} - \cos 4x + \frac{1}{4} \cos 8x\right)$

17. $\frac{1}{2}\sqrt{2 - \sqrt{3}}$ **19.** $\sqrt{2} - 1$ **21.** $-\frac{1}{2}\sqrt{2 + \sqrt{3}}$

23. $\sqrt{2} - 1$ **25.** $\frac{1}{2}\sqrt{2 + \sqrt{3}}$ **27.** $-\frac{1}{2}\sqrt{2 - \sqrt{2}}$

29. (a) $\sin 36°$ (b) $\sin 6\theta$ **31.** (a) $\cos 68°$ (b) $\cos 10\theta$

33. (a) $\tan 4°$ (b) $\tan 2\theta$ **37.** $\sqrt{10}/10, 3\sqrt{10}/10, \frac{1}{3}$

39. $\sqrt{(3 + 2\sqrt{2})/6}, \sqrt{(3 - 2\sqrt{2})/6}, 3 + 2\sqrt{2}$

41. $\sqrt{6}/6, -\sqrt{30}/6, -\sqrt{5}/5$

43. $\dfrac{2x}{1 + x^2}$ **45.** $\sqrt{\dfrac{1 - x}{2}}$ **47.** $\frac{336}{625}$ **49.** $\frac{8}{7}$ **51.** $\frac{7}{25}$

53. $-8\sqrt{3}/49$ **55.** $\frac{1}{2}(\sin 5x - \sin x)$ **57.** $\frac{1}{2}(\sin 5x + \sin 3x)$

59. $\frac{3}{2}(\cos 11x + \cos 3x)$ **61.** $2 \sin 4x \cos x$

63. $2 \sin 5x \sin x$ **65.** $-2 \cos \frac{9}{2}x \sin \frac{5}{2}x$ **67.** $(\sqrt{2} + \sqrt{3})/2$

69. $\frac{1}{4}(\sqrt{2} - 1)$ **71.** $\sqrt{2}/2$ **73.** LHS $= \cos(2 \cdot 5x) =$ RHS

75. LHS $= \sin^2 x + 2 \sin x \cos x + \cos^2 x$
$= 1 + 2 \sin x \cos x =$ RHS

77. LHS $= \dfrac{2 \sin 2x \cos 2x}{\sin x} = \dfrac{2(2 \sin x \cos x)(\cos 2x)}{\sin x} =$ RHS

79. LHS $= \dfrac{2(\tan x - \cot x)}{(\tan x + \cot x)(\tan x - \cot x)} = \dfrac{2}{\tan x + \cot x}$

$= \dfrac{2}{\frac{\sin x}{\cos x} + \frac{\cos x}{\sin x}} \cdot \dfrac{\sin x \cos x}{\sin x \cos x} = \dfrac{2 \sin x \cos x}{\sin^2 x + \cos^2 x}$

$= 2 \sin x \cos x =$ RHS

81. LHS $= \tan(2x + x) = \dfrac{\tan 2x + \tan x}{1 - \tan 2x \tan x}$

$= \dfrac{\frac{2 \tan x}{1 - \tan^2 x} + \tan x}{1 - \frac{2 \tan x}{1 - \tan^2 x} \tan x}$

$= \dfrac{2 \tan x + \tan x(1 - \tan^2 x)}{1 - \tan^2 x - 2 \tan x \tan x} =$ RHS

83. LHS $= (\cos^2 x + \sin^2 x)(\cos^2 x - \sin^2 x)$
$= \cos^2 x - \sin^2 x =$ RHS

85. LHS $= \dfrac{2 \sin 3x \cos 2x}{2 \cos 3x \cos 2x} = \dfrac{\sin 3x}{\cos 3x} =$ RHS

87. LHS $= \dfrac{2 \sin 5x \cos 5x}{2 \sin 5x \cos 4x} =$ RHS

89. LHS $= \dfrac{2 \sin\left(\frac{x + y}{2}\right) \cos\left(\frac{x - y}{2}\right)}{2 \cos\left(\frac{x + y}{2}\right) \cos\left(\frac{x - y}{2}\right)}$

$= \dfrac{\sin\left(\frac{x + y}{2}\right)}{\cos\left(\frac{x + y}{2}\right)} =$ RHS

95. LHS $= \dfrac{(\sin x + \sin 5x) + (\sin 2x + \sin 4x) + \sin 3x}{(\cos x + \cos 5x) + (\cos 2x + \cos 4x) + \cos 3x}$

$= \dfrac{2 \sin 3x \cos 2x + 2 \sin 3x \cos x + \sin 3x}{2 \cos 3x \cos 2x + 2 \cos 3x \cos x + \cos 3x}$

$= \dfrac{\sin 3x(2 \cos 2x + 2 \cos x + 1)}{\cos 3x(2 \cos 2x + 2 \cos x + 1)} =$ RHS

97. (a)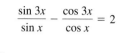

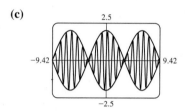

$\dfrac{\sin 3x}{\sin x} - \dfrac{\cos 3x}{\cos x} = 2$

99. (a)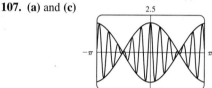

(c) The graph of $y = f(x)$ lies between the two other graphs.

101. (a) $P(t) = 8t^4 - 8t^2 + 1$ (b) $Q(t) = 16t^5 - 20t^3 + 5t$

107. (a) and (c)

The graph of f lies between the graphs of $y = 2 \cos t$ and $y = -2 \cos t$. Thus, the loudness of the sound varies between $y = \pm 2 \cos t$.

SECTION 4.4 ■ PAGE 254

1. infinitely many **2.** no, infinitely many

3. $0.3; x \approx -9.7, -6.0, -3.4, 0.3, 2.8, 6.6, 9.1$

4. (a) $0.30, 2.84$ (b) $2\pi, 0.30 + 2k\pi, 2.84 + 2k\pi$

5. $\dfrac{\pi}{3} + 2k\pi, \dfrac{2\pi}{3} + 2k\pi$

7. $(2k + 1)\pi$ **9.** $1.32 + 2k\pi, 4.97 + 2k\pi$

11. $-0.47 + 2k\pi, 3.61 + 2k\pi$

13. $-\dfrac{\pi}{3} + k\pi$ **15.** $1.37 + k\pi$

17. $\dfrac{5\pi}{6} + 2k\pi, \dfrac{7\pi}{6} + 2k\pi$;

$-7\pi/6, -5\pi/6, 5\pi/6, 7\pi/6, 17\pi/6, 19\pi/6$

19. $\dfrac{\pi}{4} + 2k\pi, \dfrac{3\pi}{4} + 2k\pi$; $-7\pi/4, -5\pi/4, \pi/4, 3\pi/4, 9\pi/4, 11\pi/4$

21. $1.29 + 2k\pi, 5.00 + 2k\pi$; $-5.00, -1.29, 1.29, 5.00, 7.57, 11.28$

23. $-1.47 + k\pi$; $-7.75, -4.61, -1.47, 1.67, 4.81, 7.95$

25. $(2k + 1)\pi$ **27.** $-\dfrac{\pi}{4} + 2k\pi, \dfrac{5\pi}{4} + 2k\pi$

29. $0.20 + 2k\pi, 2.94 + 2k\pi$ **31.** $-\dfrac{\pi}{6} + k\pi, \dfrac{\pi}{6} + k\pi$

33. $\dfrac{\pi}{4} + k\pi, \dfrac{3\pi}{4} + k\pi$ **35.** $-1.11 + k\pi, 1.11 + k\pi$

37. $\dfrac{\pi}{4} + k\pi, \dfrac{3\pi}{4} + k\pi$

39. $-1.11 + k\pi, 1.11 + k\pi, \dfrac{2\pi}{3} + 2k\pi, \dfrac{4\pi}{3} + 2k\pi$

41. $\dfrac{\pi}{3} + 2k\pi, \dfrac{5\pi}{3} + 2k\pi$ **43.** $0.34 + 2k\pi, 2.80 + 2k\pi$

45. $\dfrac{\pi}{3} + 2k\pi, \dfrac{5\pi}{3} + 2k\pi$ **47.** No solution   **49.** $\dfrac{3\pi}{2} + 2k\pi$

51. $\dfrac{\pi}{2} + k\pi, \dfrac{7\pi}{6} + 2k\pi, \dfrac{11\pi}{6} + 2k\pi$

53. $\dfrac{\pi}{2} + k\pi$ **55.** $k\pi, 0.73 + 2k\pi, 2.41 + 2k\pi$ **57.** $44.95°$

59. (a) $0°$ **(b)** $60°, 120°$ **(c)** $90°, 270°$ **(d)** $180°$

SECTION 4.5 ▪ PAGE 260

1. $\sin x = 0, k\pi$ **2.** $\sin x + 2 \sin x \cos x = 0$,

$\sin x = 0, 1 + 2 \cos x = 0$ **3.** $-\dfrac{\pi}{6} + 2k\pi, \dfrac{7\pi}{6} + 2k\pi, \dfrac{\pi}{2} + 2k\pi$

5. $(2k + 1)\pi, 1.23 + 2k\pi, 5.05 + 2k\pi$

7. $k\pi, 0.72 + 2k\pi, 5.56 + 2k\pi$ **9.** $\dfrac{\pi}{6} + 2k\pi, \dfrac{5\pi}{6} + 2k\pi$

11. $\dfrac{\pi}{3} + 2k\pi, \dfrac{5\pi}{3} + 2k\pi, (2k + 1)\pi$ **13.** $(2k + 1)\pi, \dfrac{\pi}{2} + 2k\pi$

15. $2k\pi$ **17. (a)** $\dfrac{\pi}{9} + \dfrac{2k\pi}{3}, \dfrac{5\pi}{9} + \dfrac{2k\pi}{3}$

(b) $\pi/9, 5\pi/9, 7\pi/9, 11\pi/9, 13\pi/9, 17\pi/9$

19. (a) $\dfrac{\pi}{3} + k\pi, \dfrac{2\pi}{3} + k\pi$ **(b)** $\pi/3, 2\pi/3, 4\pi/3, 5\pi/3$

21. (a) $\dfrac{5\pi}{18} + \dfrac{k\pi}{3}$ **(b)** $5\pi/18, 11\pi/18, 17\pi/18, 23\pi/18$,

$29\pi/18, 35\pi/18$ **23. (a)** $4k\pi$ **(b)** 0

25. (a) $4\pi + 6k\pi, 5\pi + 6k\pi$ **(b)** None

27. (a) $0.62 + \dfrac{k\pi}{2}$ **(b)** $0.62, 2.19, 3.76, 5.33$

29. (a) $k\pi$ **(b)** $0, \pi$ **31. (a)** $\dfrac{\pi}{6} + k\pi, \dfrac{\pi}{4} + k\pi, \dfrac{5\pi}{6} + k\pi$

(b) $\pi/6, \pi/4, 5\pi/6, 7\pi/6, 5\pi/4, 11\pi/6$

33. (a) $\dfrac{\pi}{6} + 2k\pi, \dfrac{5\pi}{6} + 2k\pi, \dfrac{3\pi}{4} + k\pi$

(b) $\pi/6, 3\pi/4, 5\pi/6, 7\pi/4$

35. (a)   **37. (a)**

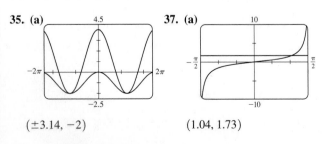

$(\pm 3.14, -2)$ $(1.04, 1.73)$

(b) $((2k + 1)\pi, -2)$ **(b)** $\left(\dfrac{\pi}{3} + k\pi, \sqrt{3}\right)$

39. $\pi/8, 3\pi/8, 5\pi/8, 7\pi/8, 9\pi/8, 11\pi/8, 13\pi/8, 15\pi/8$
41. $\pi/3, 2\pi/3$ **43.** $\pi/2, 7\pi/6, 3\pi/2, 11\pi/6$ **45.** 0
47. $0, \pi$ **49.** $0, \pi/3, 2\pi/3, \pi, 4\pi/3, 5\pi/3$ **51.** $\pi/6, 3\pi/2$

53. $k\pi/2$ **55.** $\dfrac{\pi}{2} + k\pi, \dfrac{\pi}{9} + \dfrac{2k\pi}{3}, \dfrac{5\pi}{9} + \dfrac{2k\pi}{3}$

57. $0, \pm 0.95$ **59.** 1.92 **61.** ± 0.71
63. $0.94721°$ or $89.05279°$ **65. (a)** 34th day (February 3), 308th
day (November 4)   **(b)** 275 days

CHAPTER 4 REVIEW ▪ PAGE 262

1. LHS $= \sin\theta \left(\dfrac{\cos\theta}{\sin\theta} + \dfrac{\sin\theta}{\cos\theta} \right) = \cos\theta + \dfrac{\sin^2\theta}{\cos\theta}$

$= \dfrac{\cos^2\theta + \sin^2\theta}{\cos\theta} = $ RHS

3. LHS $= (1 - \sin^2 x) \csc x - \csc x$

$= \csc x - \sin^2 x \csc x - \csc x$

$= -\sin^2 x \dfrac{1}{\sin x} = $ RHS

5. LHS $= \dfrac{\cos^2 x}{\sin^2 x} - \dfrac{\tan^2 x}{\sin^2 x} = \cot^2 x - \dfrac{1}{\cos^2 x} = $ RHS

7. LHS $= \dfrac{\cos x}{\frac{1}{\cos x}(1 - \sin x)} = \dfrac{\cos x}{\frac{1}{\cos x} - \frac{\sin x}{\cos x}} = $ RHS

9. LHS $= \sin^2 x \dfrac{\cos^2 x}{\sin^2 x} + \cos^2 x \dfrac{\sin^2 x}{\cos^2 x} = \cos^2 x + \sin^2 x = $ RHS

11. LHS $= \dfrac{2 \sin x \cos x}{1 + 2\cos^2 x - 1} = \dfrac{2 \sin x \cos x}{2\cos^2 x} = \dfrac{2 \sin x}{2 \cos x} = $ RHS

13. LHS $= \dfrac{1 - \cos x}{\sin x} = \dfrac{1}{\sin x} - \dfrac{\cos x}{\sin x} = $ RHS

15. LHS $= \frac{1}{2}[\cos((x + y) - (x - y))$

$\quad - \cos((x + y) + (x - y))]$

$= \frac{1}{2}(\cos 2y - \cos 2x)$

$= \frac{1}{2}[1 - 2\sin^2 y - (1 - 2\sin^2 x)]$

$= \frac{1}{2}(2\sin^2 x - 2\sin^2 y) = $ RHS

17. LHS $= 1 + \dfrac{\sin x}{\cos x} \cdot \dfrac{1 - \cos x}{\sin x} = 1 + \dfrac{1 - \cos x}{\cos x}$

$= 1 + \dfrac{1}{\cos x} - 1 = $ RHS

19. LHS $= \cos^2 \frac{x}{2} - 2 \sin \frac{x}{2} \cos \frac{x}{2} + \sin^2 \frac{x}{2}$

$= 1 - \sin\left(2 \cdot \frac{x}{2}\right) = $ RHS

21. LHS $= \dfrac{2 \sin x \cos x}{\sin x} - \dfrac{2 \cos^2 x - 1}{\cos x}$

$= 2 \cos x - 2 \cos x + \dfrac{1}{\cos x} = $ RHS

23. LHS $= \dfrac{\tan x + \tan \frac{\pi}{4}}{1 - \tan x \tan \frac{\pi}{4}} = $ RHS

25. (a)   **(b)** Yes

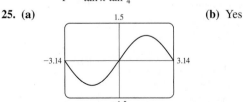

27. (a)

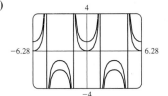

(b) No

29. (a)

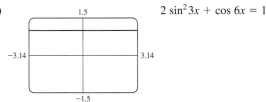

$2 \sin^2 3x + \cos 6x = 1$

31. 0.85, 2.29   **33.** 0, π **35.** $\pi/6, 5\pi/6$ **37.** $\pi/3, 5\pi/3$
39. $2\pi/3, 4\pi/3$ **41.** $\pi/3, 2\pi/3, 3\pi/4, 4\pi/3, 5\pi/3, 7\pi/4$
43. $\pi/6, \pi/2, 5\pi/6, 7\pi/6, 3\pi/2, 11\pi/6$ **45.** $\pi/6$
47. 1.18   **49. (a)** 63.4°   **(b)** No   **(c)** 90°   **51.** $\frac{1}{2}\sqrt{2 + \sqrt{3}}$
53. $\sqrt{2} - 1$ **55.** $\sqrt{2}/2$ **57.** $\sqrt{2}/2$ **59.** $\dfrac{\sqrt{2} + \sqrt{3}}{4}$
61. $2\dfrac{\sqrt{10} + 1}{9}$ **63.** $\frac{2}{3}(\sqrt{2} + \sqrt{5})$ **65.** $\sqrt{(3 + 2\sqrt{2})/6}$
67. $-\dfrac{12\sqrt{10}}{31}$ **69.** $\dfrac{2x}{1 - x^2}$ **71. (a)** $\theta = \tan^{-1}\left(\dfrac{10}{x}\right)$

(b) 286.4 ft

CHAPTER 4 TEST ▪ PAGE 264

1. (a) LHS $= \dfrac{\sin \theta}{\cos \theta} \sin \theta + \cos \theta = \dfrac{\sin^2\theta + \cos^2\theta}{\cos \theta} =$ RHS

(b) LHS $= \dfrac{\tan x}{1 - \cos x} \cdot \dfrac{1 + \cos x}{1 + \cos x} = \dfrac{\tan x(1 + \cos x)}{1 - \cos^2x}$

$= \dfrac{\frac{\sin x}{\cos x}(1 + \cos x)}{\sin^2x} = \dfrac{1}{\sin x} \cdot \dfrac{1 + \cos x}{\cos x} =$ RHS

(c) LHS $= \dfrac{2 \tan x}{\sec^2x} = \dfrac{2 \sin x}{\cos x} \cdot \cos^2x = 2 \sin x \cos x =$ RHS

2. $\tan \theta$ **3. (a)** $\frac{1}{2}$ **(b)** $\dfrac{\sqrt{2} + \sqrt{6}}{4}$ **(c)** $\frac{1}{2}\sqrt{2 - \sqrt{3}}$
4. $(10 - 2\sqrt{5})/15$ **5. (a)** $\frac{1}{2}(\sin 8x - \sin 2x)$
(b) $-2 \cos \frac{7}{2}x \sin \frac{3}{2}x$ **6.** -2 **7. (a)** 0.34, 2.80
(b) $\pi/3, \pi/2, 5\pi/3$ **(c)** $2\pi/3, 4\pi/3$
(d) $\pi/6, \pi/2, 5\pi/6, 3\pi/2$ **8.** 0.58, 2.56, 3.72, 5.70   **9.** $\frac{1519}{1681}$
10. $\dfrac{\sqrt{1 - x^2} - xy}{\sqrt{1 + y^2}}$

FOCUS ON MODELING ▪ PAGE 268

1. (a) $y = -5 \sin\left(\dfrac{\pi}{2} t\right)$

(b)

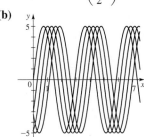

Yes, it is a traveling wave.

(c) $v = \pi/4$

3. $y(x, t) = 2.7 \sin(0.68x - 4.10t)$
5. $y(x, t) = 0.6 \sin(\pi x) \cos(40\pi t)$ **7. (a)** 1, 2, 3, 4
(b) 5:

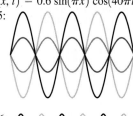

6:

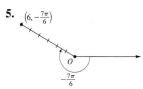

(c) 880π **(d)** $y(x, t) = \sin x \cos(880\pi t)$;
$y(x, t) = \sin(2x) \cos(880\pi t)$; $y(x, t) = \sin(3x) \cos(880\pi t)$;
$y(x, t) = \sin(4x) \cos(880\pi t)$

CHAPTER 5

SECTION 5.1 ▪ PAGE 276

1. coordinate; $(1, 1), (\sqrt{2}, \pi/4)$ **2. (a)** $r \cos \theta, r \sin \theta$
(b) $x^2 + y^2, y/x$
3.

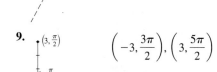

5. $\left(6, -\frac{7\pi}{6}\right)$

7. $\left(-2, \frac{4\pi}{3}\right)$

9. $\left(3, \frac{\pi}{2}\right)$ $\left(-3, \dfrac{3\pi}{2}\right), \left(3, \dfrac{5\pi}{2}\right)$

11. $\left(-1, \frac{7\pi}{6}\right)$ $\left(-1, -\dfrac{5\pi}{6}\right), \left(1, \dfrac{\pi}{6}\right)$

13. $(-5, 0)$ $(-5, 2\pi), (5, \pi)$

15. Q **17.** Q **19.** P **21.** P **23.** $(3\sqrt{2}, 3\pi/4)$
25. $\left(-\dfrac{5}{2}, -\dfrac{5\sqrt{3}}{2}\right)$ **27.** $(2\sqrt{3}, 2)$ **29.** $(1, -1)$ **31.** $(-5, 0)$
33. $(3\sqrt{6}, -3\sqrt{2})$ **35.** $(\sqrt{2}, 3\pi/4)$ **37.** $(4, \pi/4)$
39. $(5, \tan^{-1}\frac{4}{3})$ **41.** $(6, \pi)$ **43.** $\theta = \pi/4$ **45.** $r = \tan \theta \sec \theta$
47. $r = 4 \sec \theta$ **49.** $x^2 + y^2 = 49$ **51.** $x = 0$ **53.** $x = 6$
55. $x^2 + (y - 2)^2 = 4$ **57.** $x^2 + y^2 = (x^2 + y^2 - x)^2$
59. $(x^2 + y^2 - 2y)^2 = x^2 + y^2$ **61.** $y - x = 1$
63. $x^2 - 3y^2 + 16y - 16 = 0$ **65.** $x^2 + y^2 = \dfrac{y}{x}$
67. $y = \pm\sqrt{3}x$

SECTION 5.2 ■ PAGE 283

1. circles, rays   **2. (a)** satisfy   **(b)** circle, 3, pole; line, pole, 1
3. VI   **5.** II   **7.** I   **9.** Symmetric about $\theta = \pi/2$
11. Symmetric about the polar axis
13. Symmetric about $\theta = \pi/2$
15. All three types of symmetry

17.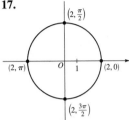

$x^2 + y^2 = 4$

19.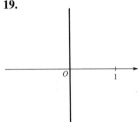

$x = 0$

21.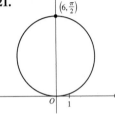

$x^2 + (y - 3)^2 = 9$

23.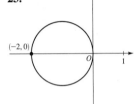

25.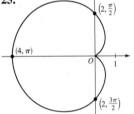

27.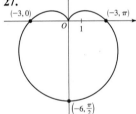

29.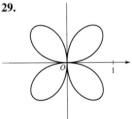

31.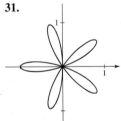

33.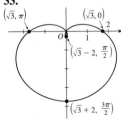

35.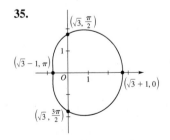

37.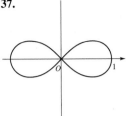

39.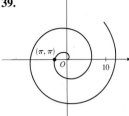

41.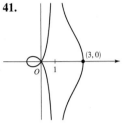

43. $0 \le \theta \le 4\pi$

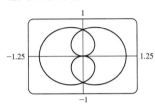

45. $0 \le \theta \le 4\pi$

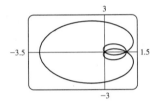

47. The graph of $r = 1 + \sin n\theta$ has n loops.   **49.** IV   **51.** III

53.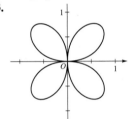

55.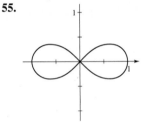

57. $\left(\dfrac{a}{2}, \dfrac{b}{2} \right), \dfrac{\sqrt{a^2 + b^2}}{2}$

59. (a) Elliptical

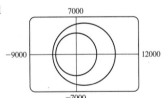

(b) π; 540 mi

SECTION 5.3 ■ PAGE 292

1. real, imaginary, (a, b) **2. (a)** $\sqrt{a^2 + b^2}$, b/a
(b) $r(\cos \theta + i \sin \theta)$
3. (a) $\sqrt{2}\left(\cos \dfrac{3\pi}{4} + i \sin \dfrac{3\pi}{4} \right); \sqrt{3} + i$
(b) $1 + i, \sqrt{2}\left(\cos \dfrac{\pi}{4} + i \sin \dfrac{\pi}{4} \right)$

4. n; four; 2, 2i, -2, $-2i$; 2

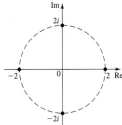

5. 4

7. 2

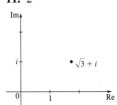

9. $\sqrt{29}$

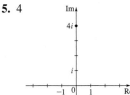

11. 2

13. 1

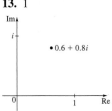

15.

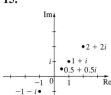

7.

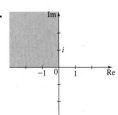

19.

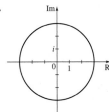

21.

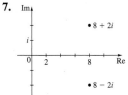

23.

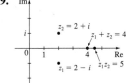

25.

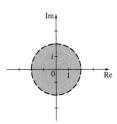

27.

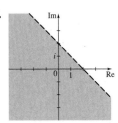

29. $\sqrt{2}\left(\cos\dfrac{\pi}{4} + i\sin\dfrac{\pi}{4}\right)$ **31.** $2\left(\cos\dfrac{7\pi}{4} + i\sin\dfrac{7\pi}{4}\right)$

33. $4\left(\cos\dfrac{11\pi}{6} + i\sin\dfrac{11\pi}{6}\right)$ **35.** $3\left(\cos\dfrac{3\pi}{2} + i\sin\dfrac{3\pi}{2}\right)$

37. $5\sqrt{2}\left(\cos\dfrac{\pi}{4} + i\sin\dfrac{\pi}{4}\right)$ **39.** $8\left(\cos\dfrac{11\pi}{6} + i\sin\dfrac{11\pi}{6}\right)$

41. $20(\cos\pi + i\sin\pi)$ **43.** $5\left[\cos\left(\tan^{-1}\tfrac{4}{3}\right) + i\sin\left(\tan^{-1}\tfrac{4}{3}\right)\right]$

45. $3\sqrt{2}\left(\cos\dfrac{3\pi}{4} + i\sin\dfrac{3\pi}{4}\right)$ **47.** $8\left(\cos\dfrac{\pi}{6} + i\sin\dfrac{\pi}{6}\right)$

49. $\sqrt{5}\left[\cos\left(\tan^{-1}\tfrac{1}{2}\right) + i\sin\left(\tan^{-1}\tfrac{1}{2}\right)\right]$

51. $2\left(\cos\dfrac{\pi}{4} + i\sin\dfrac{\pi}{4}\right)$

53. $z_1z_2 = \cos\dfrac{4\pi}{3} + i\sin\dfrac{4\pi}{3}$

$\dfrac{z_1}{z_2} = \cos\dfrac{2\pi}{3} + i\sin\dfrac{2\pi}{3}$

55. $z_1z_2 = 15\left(\cos\dfrac{3\pi}{2} + i\sin\dfrac{3\pi}{2}\right)$

$\dfrac{z_1}{z_2} = \dfrac{3}{5}\left(\cos\dfrac{7\pi}{6} - i\sin\dfrac{7\pi}{6}\right)$

57. $z_1z_2 = 8(\cos 150° + i\sin 150°)$

$z_1/z_2 = 2(\cos 90° + i\sin 90°)$

59. $z_1z_2 = 100(\cos 350° + i\sin 350°)$

$z_1/z_2 = \tfrac{4}{25}(\cos 50° + i\sin 50°)$

61. $z_1 = 2\left(\cos\dfrac{\pi}{6} + i\sin\dfrac{\pi}{6}\right)$

$z_2 = 2\left(\cos\dfrac{\pi}{3} + i\sin\dfrac{\pi}{3}\right)$

$z_1z_2 = 4\left(\cos\dfrac{\pi}{2} + i\sin\dfrac{\pi}{2}\right)$

$\dfrac{z_1}{z_2} = \cos\dfrac{\pi}{6} - i\sin\dfrac{\pi}{6}$

$\dfrac{1}{z_1} = \dfrac{1}{2}\left(\cos\dfrac{\pi}{6} - i\sin\dfrac{\pi}{6}\right)$

63. $z_1 = 4\left(\cos\dfrac{11\pi}{6} + i\sin\dfrac{11\pi}{6}\right)$

$z_2 = \sqrt{2}\left(\cos\dfrac{3\pi}{4} + i\sin\dfrac{3\pi}{4}\right)$

$z_1z_2 = 4\sqrt{2}\left(\cos\dfrac{7\pi}{12} + i\sin\dfrac{7\pi}{12}\right)$

$\dfrac{z_1}{z_2} = 2\sqrt{2}\left(\cos\dfrac{13\pi}{12} + i\sin\dfrac{13\pi}{12}\right)$

$\dfrac{1}{z_1} = \dfrac{1}{4}\left(\cos\dfrac{11\pi}{6} - i\sin\dfrac{11\pi}{6}\right)$

65. $z_1 = 5\sqrt{2}\left(\cos\dfrac{\pi}{4} + i\sin\dfrac{\pi}{4}\right)$

$z_2 = 4(\cos 0 + i\sin 0)$

$z_1z_2 = 20\sqrt{2}\left(\cos\dfrac{\pi}{4} + i\sin\dfrac{\pi}{4}\right)$

$\dfrac{z_1}{z_2} = \dfrac{5\sqrt{2}}{4}\left(\cos\dfrac{\pi}{4} + i\sin\dfrac{\pi}{4}\right)$

$\dfrac{1}{z_1} = \dfrac{\sqrt{2}}{10}\left(\cos\dfrac{\pi}{4} - i\sin\dfrac{\pi}{4}\right)$

67. $z_1 = 20(\cos \pi + i \sin \pi)$

$z_2 = 2\left(\cos \dfrac{\pi}{6} + i \sin \dfrac{\pi}{6}\right)$

$z_1 z_2 = 40\left(\cos \dfrac{7\pi}{6} + i \sin \dfrac{7\pi}{6}\right)$

$\dfrac{z_1}{z_2} = 10\left(\cos \dfrac{5\pi}{6} + i \sin \dfrac{5\pi}{6}\right)$

$\dfrac{1}{z_1} = \frac{1}{20}(\cos \pi - i \sin \pi)$

69. -1024 **71.** $512(-\sqrt{3} + i)$ **73.** -1 **75.** 4096

77. $8(-1 + i)$ **79.** $\frac{1}{2048}(-\sqrt{3} - i)$

81. $2\sqrt{2}\left(\cos \dfrac{\pi}{12} + i \sin \dfrac{\pi}{12}\right)$,

$2\sqrt{2}\left(\cos \dfrac{13\pi}{12} + i \sin \dfrac{13\pi}{12}\right)$

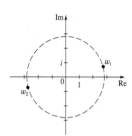

83. $3\left(\cos \dfrac{3\pi}{8} + i \sin \dfrac{3\pi}{8}\right)$,

$3\left(\cos \dfrac{7\pi}{8} + i \sin \dfrac{7\pi}{8}\right)$,

$3\left(\cos \dfrac{11\pi}{8} + i \sin \dfrac{11\pi}{8}\right)$,

$3\left(\cos \dfrac{15\pi}{8} + i \sin \dfrac{15\pi}{8}\right)$

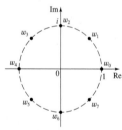

85. $\pm 1, \pm i, \pm \dfrac{\sqrt{2}}{2} \pm \dfrac{\sqrt{2}}{2}i$

87. $\dfrac{\sqrt{3}}{2} + \dfrac{1}{2}i, -\dfrac{\sqrt{3}}{2} + \dfrac{1}{2}i, -i$

89. $\pm \dfrac{\sqrt{2}}{2} \pm \dfrac{\sqrt{2}}{2}i$

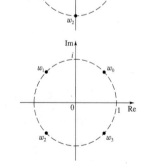

91. $\pm \dfrac{\sqrt{2}}{2} \pm \dfrac{\sqrt{2}}{2}i$

93. $2\left(\cos \dfrac{\pi}{18} + i \sin \dfrac{\pi}{18}\right), 2\left(\cos \dfrac{13\pi}{18} + i \sin \dfrac{13\pi}{18}\right),$

$2\left(\cos \dfrac{25\pi}{18} + i \sin \dfrac{25\pi}{18}\right)$

95. $2^{1/6}\left(\cos \dfrac{5\pi}{12} + i \sin \dfrac{5\pi}{12}\right), 2^{1/6}\left(\cos \dfrac{13\pi}{12} + i \sin \dfrac{13\pi}{12}\right),$

$2^{1/6}\left(\cos \dfrac{21\pi}{12} + i \sin \dfrac{21\pi}{12}\right)$

SECTION 5.4 ■ PAGE 299

1. (a) parameter   **(b)** $(0, 0), (1, 1)$ **(c)** x^2; parabola
2. (a) True   **(b)** $(0, 0), (2, 4)$ **(c)** x^2; path

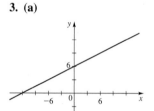

3. (a)

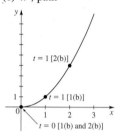

5. (a)

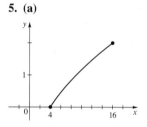

(b) $x - 2y + 12 = 0$
7. (a)

(b) $x = (y + 2)^2$
9. (a)

(b) $x = \sqrt{1 - y}$

(b) $y = \dfrac{1}{x} + 1$

11. (a)

13. (a)

(b) $x^3 = y^2$

(b) $x^2 + y^2 = 4, x \geq 0$

15. (a)

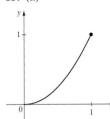

17. (a)

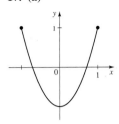

47.

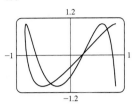

(b) $y = x^2, 0 \le x \le 1$

(b) $y = 2x^2 - 1, -1 \le x \le 1$

49. (a) $x = 2^{t/12} \cos t, y = 2^{t/12} \sin t$
(b)

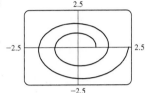

19. (a)

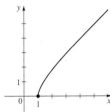

21. (a)

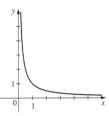

(b) $x^2 - y^2 = 1, x \ge 1, y \ge 0$ **(b)** $xy = 1, x \ge 0$

51. (a) $x = \dfrac{4 \cos t}{2 - \cos t}, y = \dfrac{4 \sin t}{2 - \cos t}$

(b)

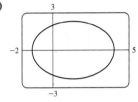

23. (a)

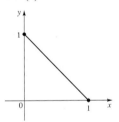

53. III   55. II

57.

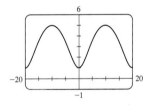

59. (b) $x^{2/3} + y^{2/3} = a^{2/3}$

(b) $x + y = 1, 0 \le x \le 1$
25. 3, (3, 0), counterclockwise, 2π
27. 1, (0, 1), clockwise, π
29. $x = 4 + t, y = -1 + \frac{1}{2}t$
31. $x = 6 + t, y = 7 + t$
33. $x = a \cos t, y = a \sin t$
37.

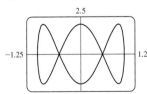

39.

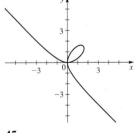

61. $x = a(\sin \theta \cos \theta + \cot \theta), y = a(1 + \sin^2\theta)$
63. (a) $x = a \sec \theta, y = b \sin \theta$
(b)

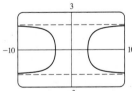

43.

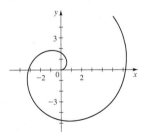

45.

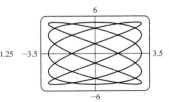

65. $y = a - a \cos\left(\dfrac{x + \sqrt{2ay - y^2}}{a}\right)$

67. (b)

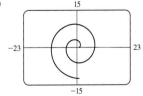

CHAPTER 5 REVIEW ■ PAGE 302

1. (a)

(b) $\left(6\sqrt{3}, 6\right)$

3. (a)

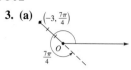

(b) $\left(\dfrac{-3\sqrt{2}}{2}, \dfrac{3\sqrt{2}}{2}\right)$

5. (a)

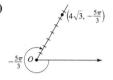

(b) $\left(2\sqrt{3}, 6\right)$

7. (a)

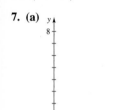

(b) $\left(8\sqrt{2}, \dfrac{\pi}{4}\right)$

(c) $\left(-8\sqrt{2}, \dfrac{5\pi}{4}\right)$

9. (a)

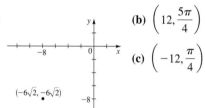

(b) $\left(12, \dfrac{5\pi}{4}\right)$

(c) $\left(-12, \dfrac{\pi}{4}\right)$

11. (a)

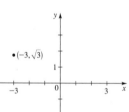

(b) $\left(2\sqrt{3}, \dfrac{5\pi}{6}\right)$

(c) $\left(-2\sqrt{3}, -\dfrac{\pi}{6}\right)$

13. (a) $r = \dfrac{4}{\cos\theta + \sin\theta}$

(b)

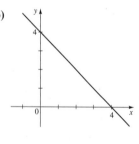

15. (a) $r = 4(\cos\theta + \sin\theta)$

(b)

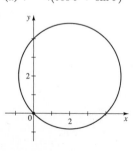

17. (a)

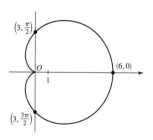

(b) $(x^2 + y^2 - 3x)^2 = 9(x^2 + y^2)$

19. (a)

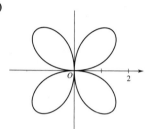

(b) $(x^2 + y^2)^3 = 16x^2y^2$

21. (a)

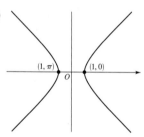

(b) $x^2 - y^2 = 1$

23. (a)

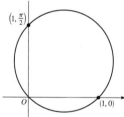

(b) $x^2 + y^2 = x + y$

25. $0 \le \theta \le 6\pi$

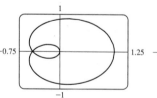

27. $0 \le \theta \le 6\pi$

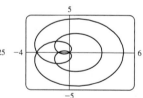

29. (a)

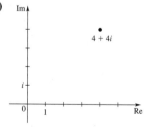

(b) $4\sqrt{2}, \dfrac{\pi}{4}$ **(c)** $4\sqrt{2}\left(\cos\dfrac{\pi}{4} + i\sin\dfrac{\pi}{4}\right)$

31. (a)

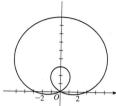

(b) $\sqrt{34}$, $\tan^{-1}\left(\frac{3}{5}\right)$ **(c)** $\sqrt{34}\left[\cos\left(\tan^{-1}\frac{3}{5}\right) + i\sin\left(\tan^{-1}\frac{3}{5}\right)\right]$

33. (a)

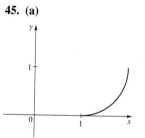

(b) $\sqrt{2}, \dfrac{3\pi}{4}$ **(c)** $\sqrt{2}\left(\cos\dfrac{3\pi}{4} + i\sin\dfrac{3\pi}{4}\right)$

35. $8\left(-1 + i\sqrt{3}\right)$ **37.** $-\frac{1}{32}\left(1 + i\sqrt{3}\right)$ **39.** $\pm 2\sqrt{2}(1 - i)$

41. $\pm 1, \pm\frac{1}{2} \pm \dfrac{\sqrt{3}}{2}i$

43. (a)

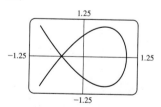

(b) $x = 2y - y^2$

45. (a)

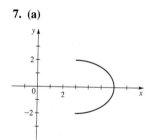

(b) $(x - 1)^2 + (y - 1)^2 = 1$, $1 \le x \le 2, 0 \le y \le 1$

47.

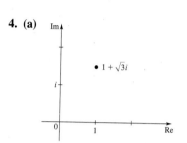

49. $x = \frac{1}{2}(1 + \cos\theta), y = \frac{1}{2}(\sin\theta + \tan\theta)$

CHAPTER 5 TEST ■ PAGE 304

1. (a) $\left(-4\sqrt{2}, -4\sqrt{2}\right)$ **(b)** $\left(4\sqrt{3}, 5\pi/6\right), \left(-4\sqrt{3}, 11\pi/6\right)$

2. (a)   circle

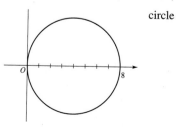

(b) $(x - 4)^2 + y^2 = 16$

3.   limaçon

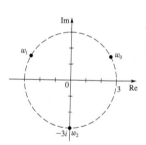

4. (a)

(b) $2\left(\cos\dfrac{\pi}{3} + i\sin\dfrac{\pi}{3}\right)$ **(c)** -512

5. $-8, \sqrt{3} + i$

6. $-3i, 3\left(\pm\dfrac{\sqrt{3}}{2} + \dfrac{1}{2}i\right)$

7. (a)

(b) $\dfrac{(x - 3)^2}{9} + \dfrac{y^2}{4} = 1, x \ge 3$

8. $x = 3 + t, y = 5 + 2t$

FOCUS ON MODELING ■ PAGE 307

1. $y = -\left(\dfrac{g}{2v_0^2\cos^2\theta}\right)x^2 + (\tan\theta)x$

3. (a) 5.45 s   **(b)** 118.7 ft   **(c)** 5426.5 ft

(d)

5. $\dfrac{v_0^2 \sin^2\theta}{2g}$ **7.** No, $\theta \approx 23°$

CHAPTER 6

SECTION 6.1 ▪ PAGE 317

1. (a) A, B

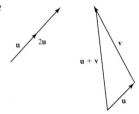

(b) $(2, 1), (4, 3), \langle 2, 2\rangle, \langle -3, 6\rangle, \langle 4, 4\rangle, \langle -1, 8\rangle$
2. (a) $\sqrt{a^2 + b^2}, 2\sqrt{2}$ **(b)** $(|\mathbf{w}|\cos\theta, |\mathbf{w}|\sin\theta)$
3.   **5.**

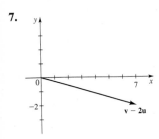

7.

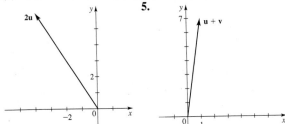

9. $\langle 3, 3\rangle$ **11.** $\langle 3, -1\rangle$ **13.** $\langle 5, 7\rangle$ **15.** $\langle -4, -3\rangle$ **17.** $\langle 0, 2\rangle$
19.   **21.**

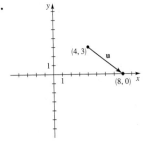

23.   **25.**

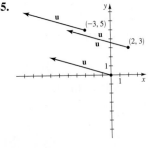

27. $\mathbf{i} + 4\mathbf{j}$ **29.** $3\mathbf{i}$ **31.** $\langle 4, 14\rangle, \langle -9, -3\rangle, \langle 5, 8\rangle, \langle -6, 17\rangle$
33. $\langle 0, -2\rangle, \langle 6, 0\rangle, \langle -2, -1\rangle, \langle 8, -3\rangle$

35. $4\mathbf{i}, -9\mathbf{i} + 6\mathbf{j}, 5\mathbf{i} - 2\mathbf{j}, -6\mathbf{i} + 8\mathbf{j}$
37. $\sqrt{5}, \sqrt{13}, 2\sqrt{5}, \frac{1}{2}\sqrt{13}, \sqrt{26}, \sqrt{10}, \sqrt{5} - \sqrt{13}$
39. $\sqrt{101}, 2\sqrt{2}, 2\sqrt{101}, \sqrt{2}, \sqrt{73}, \sqrt{145}, \sqrt{101} - 2\sqrt{2}$
41. $20\sqrt{3}\,\mathbf{i} + 20\mathbf{j}$ **43.** $-\dfrac{\sqrt{2}}{2}\mathbf{i} - \dfrac{\sqrt{2}}{2}\mathbf{j}$
45. $4\cos 10°\mathbf{i} + 4\sin 10°\mathbf{j} \approx 3.94\mathbf{i} + 0.69\mathbf{j}$
47. $5, 53.13°$ **49.** $13, 157.38°$ **51.** $2, 60°$ **53.** $15\sqrt{3}, -15$
55. $2\mathbf{i} - 3\mathbf{j}$ **57.** S $84.26°$ W   **59. (a)** $40\mathbf{j}$ **(b)** $425\mathbf{i}$
(c) $425\mathbf{i} + 40\mathbf{j}$ **(d)** 427 mi/h, N $84.6°$ E
61. 794 mi/h, N $26.6°$ W   **63. (a)** $10\mathbf{i}$ **(b)** $10\mathbf{i} + 17.32\mathbf{j}$
(c) $20\mathbf{i} + 17.32\mathbf{j}$ **(d)** 26.5 mi/h, N $49.1°$ E
65. (a) $22.8\mathbf{i} + 7.4\mathbf{j}$ **(b)** 7.4 mi/h, 22.8 mi/h
67. (a) $\langle 5, -3\rangle$ **(b)** $\langle -5, 3\rangle$ **69. (a)** $-4\mathbf{j}$ **(b)** $4\mathbf{j}$
71. (a) $\langle -7.57, 10.61\rangle$ **(b)** $\langle 7.57, -10.61\rangle$
73. $\mathbf{T}_1 \approx -56.5\mathbf{i} + 67.4\mathbf{j}, \mathbf{T}_2 \approx 56.5\mathbf{i} + 32.6\mathbf{j}$

SECTION 6.2 ▪ PAGE 325

1. $a_1a_2 + b_1b_2$; real number or scalar   **2.** $\dfrac{\mathbf{a}\cdot\mathbf{b}}{|\mathbf{a}||\mathbf{b}|}$; perpendicular

3. (a) $\dfrac{\mathbf{a}\cdot\mathbf{b}}{|\mathbf{b}|}$ **(b)** $\left(\dfrac{\mathbf{a}\cdot\mathbf{b}}{|\mathbf{b}|^2}\right)\mathbf{b}$

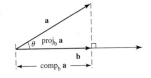

4. $\mathbf{F}\cdot\mathbf{D}$ **5. (a)** 2 **(b)** $45°$ **7. (a)** 13 **(b)** $56°$ **9. (a)** -1
(b) $97°$ **11. (a)** $5\sqrt{3}$ **(b)** $30°$ **13. (a)** 1 **(b)** $86°$ **15.** Yes
17. No   **19.** Yes   **21.** 9 **23.** -5 **25.** $-\frac{12}{5}$ **27.** -24
29. (a) $\langle 1, 1\rangle$ **(b)** $\mathbf{u}_1 = \langle 1, 1\rangle, \mathbf{u}_2 = \langle -3, 3\rangle$
31. (a) $\langle -\frac{1}{2}, \frac{3}{2}\rangle$ **(b)** $\mathbf{u}_1 = \langle -\frac{1}{2}, \frac{3}{2}\rangle, \mathbf{u}_2 = \langle \frac{3}{2}, \frac{1}{2}\rangle$
33. (a) $\langle -\frac{18}{5}, \frac{24}{5}\rangle$ **(b)** $\mathbf{u}_1 = \langle -\frac{18}{5}, \frac{24}{5}\rangle, \mathbf{u}_2 = \langle \frac{28}{5}, \frac{21}{5}\rangle$
35. -28 **37.** 25 **45.** 16 ft-lb   **47.** 8660 ft-lb   **49.** 1164 lb
51. $23.6°$

SECTION 6.3 ▪ PAGE 332

1. x, y, z; $(5, 2, 3)$; $y = 2$

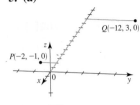

2. $\sqrt{(x_2 - x_1)^2 + (y_2 - y_1)^2 + (z_2 - z_1)^2}$;
$\sqrt{38}$; $(x - 5)^2 + (y - 2)^2 + (z - 3)^2 = 9$
3. (a)   **5. (a)**

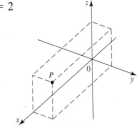

(b) $\sqrt{42}$ **(b)** $2\sqrt{29}$

7. Plane parallel to the
yz-plane

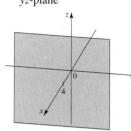

9. Plane parallel to the
xy-plane

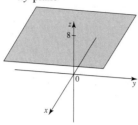

15. (a) $x + y - z = 5$ **(b)** x-intercept 5, y-intercept 5,
z-intercept -5

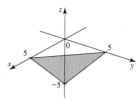

11. $(x - 2)^2 + (y + 5)^2 + (z - 3)^2 = 25$
13. $(x - 3)^2 + (y + 1)^2 + z^2 = 6$
15. Center: $(5, -1, -4)$, radius: $\sqrt{51}$
17. Center: $(6, 1, 0)$, radius: $\sqrt{37}$
19. (a) Circle, center: $(0, 2, -10)$, radius: $3\sqrt{11}$
(b) Circle, center: $(4, 2, -10)$, radius: $5\sqrt{3}$ **21. (a)** 3

17. (a) $6x - z = 4$ **(b)** x-intercept $\frac{2}{3}$, no y-intercept, z-intercept -4

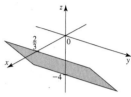

SECTION 6.4 ▪ PAGE 338

1. unit, $a_1\mathbf{i} + a_2\mathbf{j} + a_3\mathbf{k}$; $\sqrt{a_1^2 + a_2^2 + a_3^2}$; 4, (-2), 4, $\langle 0, 7, -24 \rangle$

2. $\dfrac{\mathbf{u} \cdot \mathbf{v}}{|\mathbf{u}||\mathbf{v}|}$; 0; 0, perpendicular   **3.** $\langle -1, -1, 5 \rangle$ **5.** $\langle -6, -2, 0 \rangle$

7. $(5, 4, -1)$ **9.** $(1, 0, -1)$ **11.** 3   **13.** $5\sqrt{2}$

15. $\langle 2, -3, 2 \rangle$, $\langle 2, -11, 4 \rangle$, $\langle 6, -23, \frac{19}{2} \rangle$
17. $\mathbf{i} - 2\mathbf{k}$, $\mathbf{i} + 2\mathbf{j} + 2\mathbf{k}$, $3\mathbf{i} + \frac{7}{2}\mathbf{j} + \mathbf{k}$ **19.** $12\mathbf{i} + 2\mathbf{k}$
21. $3\mathbf{i} - 3\mathbf{j}$ **23. (a)** $\langle 3, 1, -2 \rangle$ **(b)** $3\mathbf{i} + \mathbf{j} - 2\mathbf{k}$ **25.** -4
27. 1   **29.** Yes   **31.** No   **33.** $116.4°$ **35.** $100.9°$
37. $\alpha \approx 65°$, $\beta \approx 56°$, $\gamma = 45°$ **39.** $\alpha \approx 73°$, $\beta \approx 65°$, $\gamma \approx 149°$
41. $\pi/4$ **43.** $125°$ **47. (a)** $-7\mathbf{i} - 24\mathbf{j} + 25\mathbf{k}$ **(b)** $25\sqrt{2}$

19. (a) $3x - y + 2z = -8$ **(b)** x-intercept $-\frac{8}{3}$, y-intercept 8,
z-intercept -4

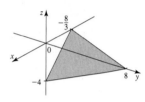

SECTION 6.5 ▪ PAGE 345

1. $\begin{vmatrix} \mathbf{i} & \mathbf{j} & \mathbf{k} \\ a_1 & a_2 & a_3 \\ b_1 & b_2 & b_3 \end{vmatrix} = (a_2b_3 - a_3b_2)\mathbf{i} + (a_3b_1 - a_1b_3)\mathbf{j} + (a_1b_2 - a_2b_1)\mathbf{k}$, $-3\mathbf{i} + 2\mathbf{j} + 3\mathbf{k}$

2. perpendicular; perpendicular   **3.** $9\mathbf{i} - 6\mathbf{j} + 3\mathbf{k}$

5. 0   **7.** $-4\mathbf{i} + 7\mathbf{j} - 3\mathbf{k}$ **9. (a)** $\langle 0, 2, 2 \rangle$ **(b)** $\left\langle 0, \dfrac{\sqrt{2}}{2}, \dfrac{\sqrt{2}}{2} \right\rangle$

11. (a) $14\mathbf{i} + 7\mathbf{j}$ **(b)** $\dfrac{2\sqrt{5}}{5}\mathbf{i} + \dfrac{\sqrt{5}}{5}\mathbf{j}$

13. $\dfrac{3\sqrt{3}}{2}$ **15.** 100   **17.** $\langle 0, 2, 2 \rangle$ **19.** $\langle 10, -10, 0 \rangle$ **21.** $4\sqrt{6}$

23. $\dfrac{5\sqrt{14}}{2}$ **25.** $\sqrt{14}$ **27.** $18\sqrt{3}$ **29. (a)** 0   **(b)** Yes

31. (a) 55   **(b)** No, 55   **33. (a)** -2 **(b)** No, 2
35. (a) $2{,}700{,}000\sqrt{3}$ **(b)** 4677 liters

21. $5x - 3y - z = 35$ **23.** $x - 3y = 2$ **25.** $2x - 3y - 9z = 0$
27. $x = 2t$, $y = 5t$, $z = 4 - 4t$ **29.** $x = 2$, $y = -1 + t$, $z = 5$
31. $12x + 4y + 3z = 12$ **33.** $x - 2y + 4z = 0$

CHAPTER 6 REVIEW ▪ PAGE 351

1. $\sqrt{13}$, $\langle 6, 4 \rangle$, $\langle -10, 2 \rangle$, $\langle -4, 6 \rangle$, $\langle -22, 7 \rangle$
3. $\sqrt{5}$, $3\mathbf{i} - \mathbf{j}$, $\mathbf{i} + 3\mathbf{j}$, $4\mathbf{i} + 2\mathbf{j}$, $4\mathbf{i} + 7\mathbf{j}$
5. $\langle 3, -4 \rangle$ **7.** 4, $120°$ **9.** $\langle 10, 10\sqrt{3} \rangle$
11. (a) $(4.8\mathbf{i} + 0.4\mathbf{j}) \times 10^4$ **(b)** 4.8×10^4 lb, N $85.2°$ E
13. 5, 25, 60   **15.** $2\sqrt{2}$, 8, 0   **17.** Yes   **19.** No, $45°$

21. (a) $\dfrac{17\sqrt{37}}{37}$ **(b)** $\langle \frac{102}{37}, -\frac{17}{37} \rangle$

(c) $\mathbf{u}_1 = \langle \frac{102}{37}, -\frac{17}{37} \rangle$, $\mathbf{u}_2 = \langle \frac{9}{37}, \frac{54}{37} \rangle$

23. (a) $-\dfrac{14\sqrt{97}}{97}$ **(b)** $-\frac{56}{97}\mathbf{i} + \frac{126}{97}\mathbf{j}$

(c) $\mathbf{u}_1 = -\frac{56}{97}\mathbf{i} + \frac{126}{97}\mathbf{j}$, $\mathbf{u}_2 = \frac{153}{97}\mathbf{i} + \frac{68}{97}\mathbf{j}$
25. 3

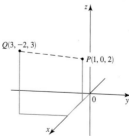

SECTION 6.6 ▪ PAGE 349

1. parametric; $x = x_0 + at$, $y = y_0 + bt$, $z = z_0 + ct$
2. $a(x - x_0) + b(y - y_0) + c(z - z_0) = 0$
3. $x = 1 + 3t$, $y = 2t$, $z = -2 - 3t$
5. $x = 3$, $y = 2 - 4t$, $z = 1 + 2t$
7. $x = 1 + 2t$, $y = 0$, $z = -2 - 5t$
9. $x = 1 + t$, $y = -3 + 4t$, $z = 2 - 3t$
11. $x = 1 - t$, $y = 1 + t$, $z = 2t$
13. $x = 3 + 4t$, $y = 7 - 4t$, $z = -5$

27. $x^2 + y^2 + z^2 = 36$

29. Center: $(1, 3, -2)$, radius: 4

31. $6, \langle 6, 1, 3 \rangle, \langle 2, -5, 5 \rangle, \langle -1, -\frac{15}{2}, 5 \rangle$

33. (a) -1 (b) No, $92.8°$ **35.** (a) 0 (b) Yes

37. (a) $\langle -2, 17, -5 \rangle$ (b) $\left\langle -\dfrac{\sqrt{318}}{159}, \dfrac{17\sqrt{318}}{318}, -\dfrac{5\sqrt{318}}{318} \right\rangle$

39. (a) $\mathbf{i} + \mathbf{j} + 2\mathbf{k}$ (b) $\dfrac{\sqrt{6}}{6}\mathbf{i} + \dfrac{\sqrt{6}}{6}\mathbf{j} + \dfrac{\sqrt{6}}{3}\mathbf{k}$

41. $\frac{15}{2}$ **43.** 9 **45.** $x = 2 + 3t, y = t, z = -6$

47. $x = 6 - 2t, y = -2 + 3t, z = -3 + t$

49. $2x + 3y - 5z = 2$ **51.** $7x + 7y + 6z = 20$

53. $x = 2 - 2t, y = 0, z = -4t$

CHAPTER 6 TEST ▪ PAGE 353

1. (a) 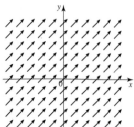   (b) $-6\mathbf{i} + 10\mathbf{j}$ (c) $2\sqrt{34}$

2. (a) $\langle 19, -3 \rangle$ (b) $5\sqrt{2}$ (c) 0 (d) Yes

3. (a) 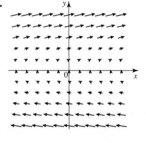   (b) $8, 150°$

4. (a) $14\mathbf{i} + 6\sqrt{3}\mathbf{j}$ (b) 17.4 mi/h, N $53.4°$ E **5.** (a) $45.0°$

(b) $\dfrac{\sqrt{26}}{2}$ (c) $\frac{5}{2}\mathbf{i} - \frac{1}{2}\mathbf{j}$ **6.** 90 **7.** (a) 6

(b) $(x - 4)^2 + (y - 3)^2 + (z + 1)^2 = 36$

(c) $\langle 2, -4, 4 \rangle = 2\mathbf{i} - 4\mathbf{j} + 4\mathbf{k}$ **8.** (a) $11\mathbf{i} - 4\mathbf{j} - \mathbf{k}$ (b) $\sqrt{6}$

(c) -1 (d) $-3\mathbf{i} - 7\mathbf{j} - 5\mathbf{k}$ (e) $3\sqrt{35}$ (f) 18 (g) $96.3°$

9. $\left\langle \dfrac{7\sqrt{6}}{18}, \dfrac{\sqrt{6}}{9}, -\dfrac{\sqrt{6}}{18} \right\rangle, \left\langle -\dfrac{7\sqrt{6}}{18}, -\dfrac{\sqrt{6}}{9}, \dfrac{\sqrt{6}}{18} \right\rangle$

10. (a) $\langle 4, -3, 4 \rangle$ (b) $4x - 3y + 4z = 4$ (c) $\dfrac{\sqrt{41}}{2}$

11. $x = 2 - 2t, y = -4 + t, z = 7 - 2t$

FOCUS ON MODELING ▪ PAGE 356

1. 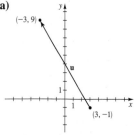   **3.**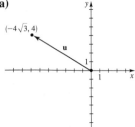

5.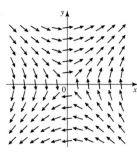

7.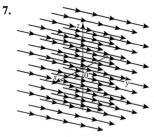

9.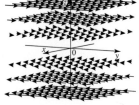

11. II **13.** I **15.** IV **17.** III

19.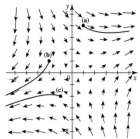

CHAPTER 7

SECTION 7.1 ▪ PAGE 366

1. focus, directrix **2.** $F(0, p), y = -p, F(0, 3), y = -3$

3. $F(p, 0), x = -p, F(3, 0), x = -3$

4. (a)   (b)

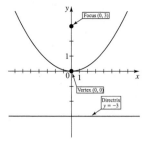

5. III **7.** II **9.** VI

Order of answers: focus; directrix; focal diameter

11. $F\left(0, \frac{9}{4}\right); y = -\frac{9}{4}; 9$ **13.** $F(1, 0); x = -1; 4$

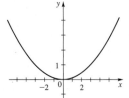   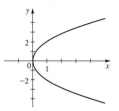

15. $F\left(0, \frac{1}{20}\right)$; $y = -\frac{1}{20}$; $\frac{1}{5}$

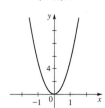

17. $F\left(-\frac{1}{32}, 0\right)$; $x = \frac{1}{32}$; $\frac{1}{8}$

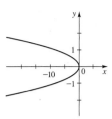

4. (a)

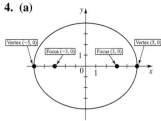

(b)

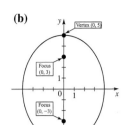

5. II **7.** I

*Order of answers: vertices; foci; eccentricity; major axis
and minor axis*

19. $F\left(0, -\frac{3}{2}\right)$; $y = \frac{3}{2}$; 6

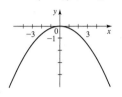

21. $F\left(-\frac{5}{12}, 0\right)$; $x = \frac{5}{12}$; $\frac{5}{3}$

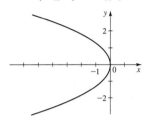

9. $V(\pm 5, 0)$; $F(\pm 4, 0)$;
$\frac{4}{5}$; 10, 6

11. $V(0, \pm 3)$; $F(0, \pm\sqrt{5})$;
$\sqrt{5}/3$; 6, 4

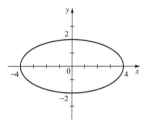

23.

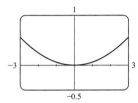

25.

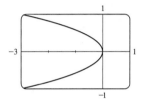

13. $V(\pm 4, 0)$; $F(\pm 2\sqrt{3}, 0)$;
$\sqrt{3}/2$; 8, 4

15. $V(0, \pm\sqrt{3})$; $F(0, \pm\sqrt{3/2})$;
$1/\sqrt{2}$; $2\sqrt{3}$, $\sqrt{6}$

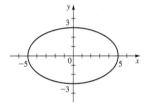

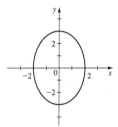

27.

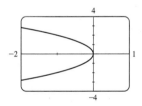

17. $V(\pm 1, 0)$; $F(\pm\sqrt{3}/2, 0)$;
$\sqrt{3}/2$; 2, 1

19. $V(0, \pm\sqrt{2})$; $F(0, \pm\sqrt{3/2})$;
$\sqrt{3}/2$; $2\sqrt{2}$, $\sqrt{2}$

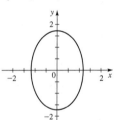
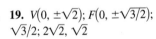

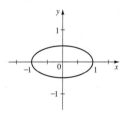

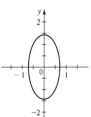

29. $x^2 = 8y$ **31.** $y^2 = -32x$ **33.** $y^2 = -8x$ **35.** $x^2 = 40y$
37. $y^2 = 4x$ **39.** $x^2 = 20y$ **41.** $x^2 = 8y$ **43.** $y^2 = -16x$
45. $y^2 = -3x$ **47.** $x = y^2$ **49.** $x^2 = -4\sqrt{2}y$
51. (a) $x^2 = -4py$, $p = \frac{1}{2}$, 1, 4, and 8

(b) The closer the directrix to the
vertex, the steeper the parabola.

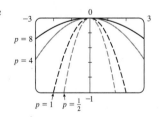

21. $V(0, \pm 1)$; $F(0, \pm 1/\sqrt{2})$;
$1/\sqrt{2}$; 2, $\sqrt{2}$

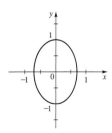

53. (a) $y^2 = 12x$ **(b)** $8\sqrt{15} \approx 31$ cm **55.** $x^2 = 600y$

SECTION 7.2 ■ PAGE 374

1. sum; foci
2. $(a, 0), (-a, 0)$; $c = \sqrt{a^2 - b^2}$; $(5, 0), (-5, 0), (3, 0), (-3, 0)$
3. $(0, a), (0, -a)$; $c = \sqrt{a^2 - b^2}$; $(0, 5), (0, -5), (0, 3), (0, -3)$

23. $\dfrac{x^2}{25} + \dfrac{y^2}{16} = 1$ **25.** $\dfrac{x^2}{4} + \dfrac{y^2}{8} = 1$ **27.** $\dfrac{x^2}{256} + \dfrac{y^2}{48} = 1$

29.

31.

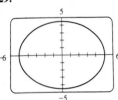

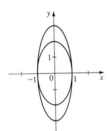

33. $\dfrac{x^2}{25} + \dfrac{y^2}{9} = 1$ **35.** $x^2 + \dfrac{y^2}{4} = 1$ **37.** $\dfrac{x^2}{9} + \dfrac{y^2}{13} = 1$

39. $\dfrac{x^2}{100} + \dfrac{y^2}{91} = 1$ **41.** $\dfrac{x^2}{25} + \dfrac{y^2}{5} = 1$ **43.** $\dfrac{64x^2}{225} + \dfrac{64y^2}{81} = 1$

45. $(0, \pm 2)$ **47.** $(\pm 1, 0)$

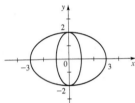

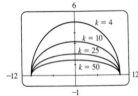

49. (a)

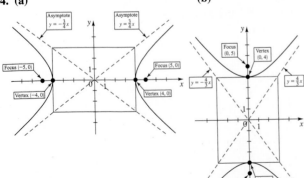

(b) Common major axes and vertices; eccentricity increases as k increases.

51. $\dfrac{x^2}{2.2500 \times 10^{16}} + \dfrac{y^2}{2.2491 \times 10^{16}} = 1$

53. $\dfrac{x^2}{1{,}455{,}642} + \dfrac{y^2}{1{,}451{,}610} = 1$

55. $5\sqrt{39}/2 \approx 15.6$ in.

SECTION 7.3 ■ PAGE 383

1. difference; foci
2. $(-a, 0), (a, 0)$; $\sqrt{a^2 + b^2}$; $(-4, 0), (4, 0), (-5, 0), (5, 0)$
3. $(0, -a), (0, a)$; $\sqrt{a^2 + b^2}$; $(0, -4), (0, 4), (0, -5), (0, 5)$
4. (a)   **(b)**

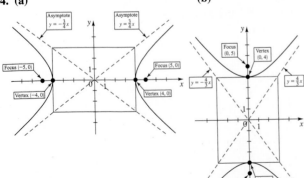

5. III   **7.** II

Order of answers: vertices; foci; asymptotes
9. $V(\pm 2, 0)$; $F(\pm 2\sqrt{5}, 0)$; $y = \pm 2x$

11. $V(0, \pm 1)$; $F(0, \pm\sqrt{26})$; $y = \pm\frac{1}{5}x$

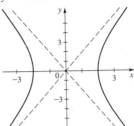

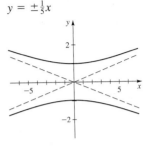

13. $V(\pm 1, 0)$; $F(\pm\sqrt{2}, 0)$; $y = \pm x$

15. $V(0, \pm 3)$; $F(0, \pm\sqrt{34})$; $y = \pm\frac{3}{5}x$

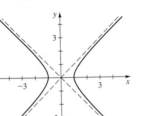

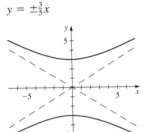

17. $V(\pm 2\sqrt{2}, 0)$; $F(\pm\sqrt{10}, 0)$; $y = \pm\frac{1}{2}x$

19. $V(0, \pm\frac{1}{2})$; $F(0, \pm\sqrt{5}/2)$; $y = \pm\frac{1}{2}x$

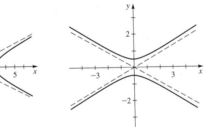

21. $\dfrac{x^2}{4} - \dfrac{y^2}{12} = 1$ **23.** $\dfrac{y^2}{16} - \dfrac{x^2}{16} = 1$ **25.** $\dfrac{x^2}{9} - \dfrac{4y^2}{9} = 1$

27.   **29.**

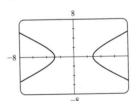

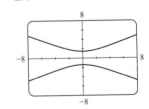

31. $\dfrac{x^2}{9} - \dfrac{y^2}{16} = 1$ **33.** $y^2 - \dfrac{x^2}{3} = 1$ **35.** $x^2 - \dfrac{y^2}{25} = 1$

37. $\dfrac{5y^2}{64} - \dfrac{5x^2}{256} = 1$ **39.** $\dfrac{x^2}{16} - \dfrac{y^2}{16} = 1$ **41.** $\dfrac{x^2}{9} - \dfrac{y^2}{16} = 1$

43. (b) $x^2 - y^2 = c^2/2$

47. (b)

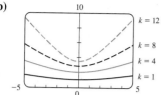

$k = 12$

$k = 8$

$k = 4$

$k = 1$

As k increases, the asymptotes get steeper.

49. $x^2 - y^2 = 2.3 \times 10^{19}$

SECTION 7.4 ■ PAGE 391

1. (a) right; left **(b)** upward; downward

2.

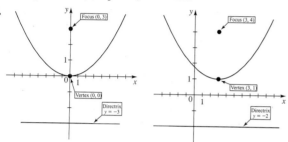

3.

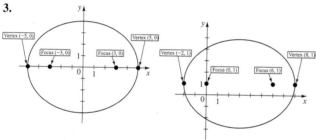

4.

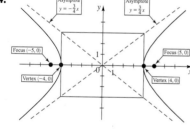

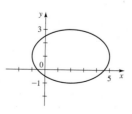

5. Center $C(2, 1)$;
foci $F(2 \pm \sqrt{5}, 1)$;
vertices $V_1(-1, 1)$, $V_2(5, 1)$;
major axis 6, minor axis 4

7. Center $C(0, -5)$;
foci $F_1(0, -1)$, $F_2(0, -9)$;
vertices $V_1(0, 0)$, $V_2(0, -10)$;
major axis 10, minor axis 6

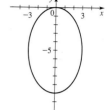

9. Vertex $V(3, -1)$;
focus $F(3, 1)$;
directrix $y = -3$

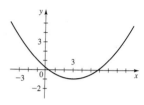

11. Vertex $V\left(-\frac{1}{2}, 0\right)$;
focus $F\left(-\frac{1}{2}, -\frac{1}{16}\right)$;
directrix $y = \frac{1}{16}$

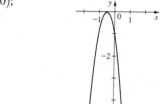

13. Center $C(-1, 3)$;
foci $F_1(-6, 3)$, $F_2(4, 3)$;
vertices $V_1(-4, 3)$, $V_2(2, 3)$;
asymptotes
$y = \pm \frac{4}{3}(x + 1) + 3$

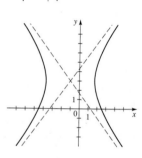

15. Center $C(-1, 0)$;
foci $F(-1, \pm\sqrt{5})$;
vertices $V(-1, \pm 1)$;
asymptotes $y = \pm\frac{1}{2}(x + 1)$

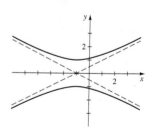

17. $x^2 = -\frac{1}{4}(y - 4)$ **19.** $\dfrac{(x - 5)^2}{25} + \dfrac{y^2}{16} = 1$

21. $(y - 1)^2 - x^2 = 1$

23. Parabola;
$V(-4, 4)$;
$F(-3, 4)$;
$x = -5$

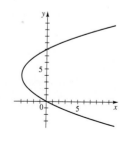

25. Hyperbola;
$C(1, 2)$; $F_1\left(-\frac{3}{2}, 2\right)$, $F_2\left(\frac{7}{2}, 2\right)$;
$V(1 \pm \sqrt{5}, 2)$; asymptotes
$y = \pm\frac{1}{2}(x - 1) + 2$

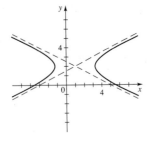

27. Ellipse; $C(3, -5)$;
$F(3 \pm \sqrt{21}, -5)$;
$V_1(-2, -5)$, $V_1(8, -5)$;
major axis 10,
minor axis 4

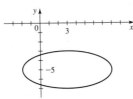

29. Hyperbola; $C(3, 0)$;
$F(3, \pm 5)$; $V(3, \pm 4)$;
asymptotes $y = \pm\frac{4}{3}(x - 3)$

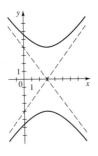

31. Degenerate conic
(pair of lines),
$y = \pm\frac{1}{2}(x - 4)$

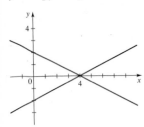

33. Point $(1, 3)$

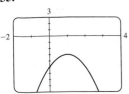

35.

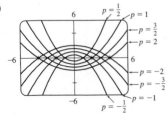

37.

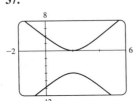

39. (a) $F < 17$ **(b)** $F = 17$ **(c)** $F > 17$
41. (a)

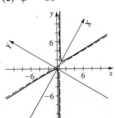

(c) The parabolas become narrower.

43. $\dfrac{(x + 150)^2}{18{,}062{,}500} + \dfrac{y^2}{18{,}040{,}000} = 1$

SECTION 7.5 ■ PAGE 400

1. $x = X \cos \phi - Y \sin \phi$, $y = X \sin \phi + Y \cos \phi$,
$X = x \cos \phi + y \sin \phi$, $Y = -x \sin \phi + y \cos \phi$
2. (a) conic section **(b)** $(A - C)/B$
(c) $B^2 - 4AC$, parabola, ellipse, hyperbola **3.** $\left(\sqrt{2}, 0\right)$
5. $\left(0, -2\sqrt{3}\right)$ **7.** $(1.6383, 1.1472)$
9. $X^2 + \sqrt{3}XY + 2 = 0$
11. $7Y^2 - 48XY - 7X^2 - 40X - 30Y = 0$
13. $X^2 - Y^2 = 2$
15. (a) Hyberbola **(b)** $X^2 - Y^2 = 16$
(c) $\phi = 45°$

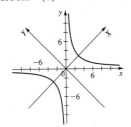

17. (a) Hyberbola
(b) $Y^2 - X^2 = 1$
(c) $\phi = 30°$

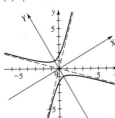

19. (a) Hyberbola
(b) $\dfrac{X^2}{4} - Y^2 = 1$
(c) $\phi \approx 53°$

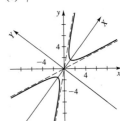

21. (a) Hyberbola
(b) $3X^2 - Y^2 = 2\sqrt{3}$
(c) $\phi = 30°$

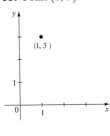

23. (a) Parabola
(b) $Y = \sqrt{2}X^2$
(c) $\phi = 45°$

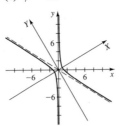

25. (a) Hyberbola
(b) $(X - 1)^2 - 3Y^2 = 1$
(c) $\phi = 60°$

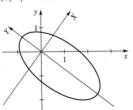

27. (a) Ellipse
(b) $X^2 + \dfrac{(Y + 1)^2}{4} = 1$
(c) $\phi \approx 53°$

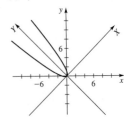

29. (a) Parabola
(b)

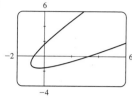

31. (a) Hyperbola
(b)

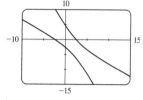

33. (a) $(X - 5)^2 - Y^2 = 1$
(b) XY-coordinates:
$C(5,0)$; $V_1(6,0)$, $V_2(4,0)$; $F(5 \pm \sqrt{2}, 0)$;
xy-coordinates:
$C(4,3)$; $V_1\left(\frac{24}{5}, \frac{18}{5}\right)$, $V_2\left(\frac{16}{5}, \frac{12}{5}\right)$; $F_1\left(4 + \frac{4}{5}\sqrt{2}, 3 + \frac{3}{5}\sqrt{2}\right)$,
$F_2\left(4 - \frac{4}{5}\sqrt{2}, 3 - \frac{3}{5}\sqrt{2}\right)$
(c) $Y = \pm(X - 5)$; $7x - y - 25 = 0$, $x + 7y - 25 = 0$
35. $X = x \cos \phi + y \sin \phi$; $Y = -x \sin \phi + y \cos \phi$

SECTION 7.6 ▪ PAGE 406

1. focus, directrix; $\dfrac{\text{distance from } P \text{ to } F}{\text{distance from } P \text{ to } \ell}$, conic section; parabola, ellipse, hyperbola, eccentricity

2. $\dfrac{ed}{1 \pm e \cos \theta}$, $\dfrac{ed}{1 \pm e \sin \theta}$
3. $r = 6/(3 + 2 \cos \theta)$
5. $r = 2/(1 + \sin \theta)$
7. $r = 20/(1 + 4 \cos \theta)$
9. $r = 10/(1 + \sin \theta)$
11. II   **13.** VI   **15.** IV

17.

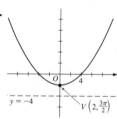

19.

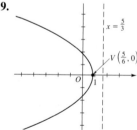

21. (a), (b)

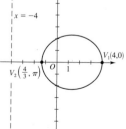

23. (a), (b)

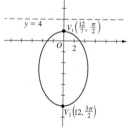

(c) $C\left(\frac{4}{3}, 0\right)$, major axis: $\frac{16}{3}$, minor axis: $\frac{8\sqrt{3}}{3}$

(c) $C\left(\frac{36}{7}, \frac{3\pi}{2}\right)$, major axis: $\frac{96}{7}$, minor axis: $\frac{24\sqrt{7}}{7}$

25. (a), (b)

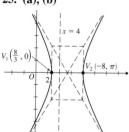

(c) $\left(\frac{16}{3}, 0\right)$

27. (a), (b)

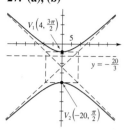

(c) $\left(12, \frac{3\pi}{2}\right)$

29. (a) 3, hyperbola
(b)

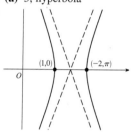

31. (a) 1, parabola
(b)

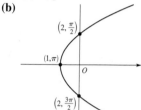

33. (a) $\frac{1}{2}$, ellipse
(b)

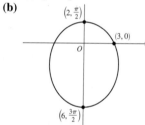

35. (a) $\frac{5}{2}$, hyperbola
(b)

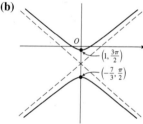

37. (a) eccentricity $\frac{3}{4}$, directrix $x = -\frac{1}{3}$
(b) $r = \dfrac{1}{4 - 3\cos\left(\theta - \frac{\pi}{3}\right)}$
(c)

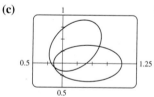

39. (a) eccentricity 1, directrix $y = 2$
(b) $r = \dfrac{2}{1 + \sin\left(\theta + \frac{\pi}{4}\right)}$
(c)

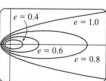

41. The ellipse is nearly circular when e is close to 0 and becomes more elongated as $e \to 1^-$. At $e = 1$, the curve becomes a parabola.

43. (b) $r = (1.49 \times 10^8)/(1 - 0.017 \cos \theta)$
45. 0.25

CHAPTER 7 REVIEW ■ PAGE 409

1. $V(0,0)$; $F(1,0)$; $x = -1$

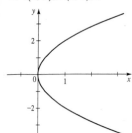

3. $V(0,0)$; $F(0,-2)$; $y = 2$

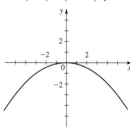

5. $V(-2,2)$; $F\left(-\frac{7}{4},2\right)$; $x = -\frac{9}{4}$

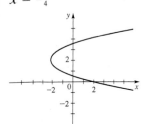

7. $V(-2,-3)$; $F(-2,-2)$; $y = -4$

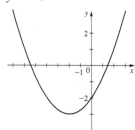

9. $C(0,0)$; $V(0,\pm5)$; $F(0,\pm4)$; axes 10, 6

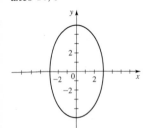

11. $C(0,0)$; $V(\pm4,0)$; $F(\pm2\sqrt{3},0)$; axes 8, 4

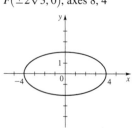

13. $C(3,0)$; $V(3,\pm4)$; $F(3,\pm\sqrt{7})$; axes 8, 6

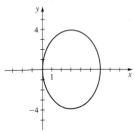

15. $C(0,2)$; $V(\pm3,2)$; $F(\pm\sqrt{5},2)$; axes 6, 4

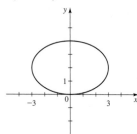

17. $C(0,0)$; $V(0,\pm4)$; $F(0,\pm5)$; asymptotes $y = \pm\frac{4}{3}x$

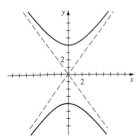

19. $C(0,0)$; $V(\pm4,0)$; $F(\pm2\sqrt{6},0)$; asymptotes $y = \pm\dfrac{1}{\sqrt{2}}x$

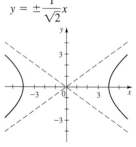

21. $C(-4,0)$; $V_1(-8,0)$, $V_2(0,0)$; $F(-4 \pm 4\sqrt{2},0)$; asymptotes $y = \pm(x+4)$

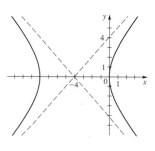

23. $C(-3,-1)$; $V(-3,-1 \pm \sqrt{2})$; $F(-3,-1 \pm 2\sqrt{5})$; asymptotes $y = \frac{1}{3}x$, $y = -\frac{1}{3}x - 2$

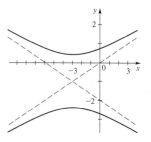

25. $y^2 = 8x$ **27.** $\dfrac{y^2}{16} - \dfrac{x^2}{9} = 1$ **29.** $\dfrac{(x-4)^2}{16} + \dfrac{(y-2)^2}{4} = 1$

31. Parabola; $F(0,-2)$; $V(0,1)$

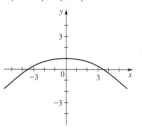

33. Hyperbola; $F(0, \pm12\sqrt{2})$; $V(0, \pm12)$

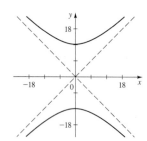

35. Ellipse; $F(1, 4 \pm \sqrt{15})$; $V(1, 4 \pm 2\sqrt{5})$

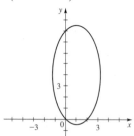

37. Parabola; $F\left(-\frac{255}{4}, 8\right)$; $V(-64, 8)$

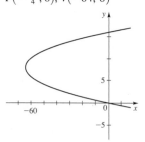

39. Ellipse;
$F(3, -3 \pm 1/\sqrt{2})$;
$V_1(3, -4)$, $V_2(3, -2)$

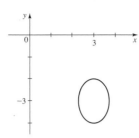

41. Has no graph

43. $x^2 = 4y$ **45.** $\dfrac{y^2}{4} - \dfrac{x^2}{16} = 1$

47. $\dfrac{(x-1)^2}{3} + \dfrac{(y-2)^2}{4} = 1$

49. $\dfrac{4(x-7)^2}{225} + \dfrac{(y-2)^2}{100} = 1$

51. (a) 91,419,000 mi **(b)** 94,581,000 mi

53. (a)

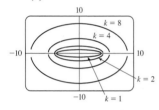

55. (a) Hyperbola **(b)** $3X^2 - Y^2 = 1$
(c) $\phi = 45°$

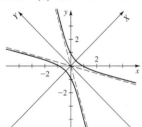

57. (a) Ellipse
(b) $(X - 1)^2 + 4Y^2 = 1$
(c) $\phi = 30°$

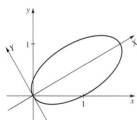

59. Ellipse

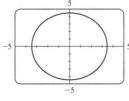

61. Parabola

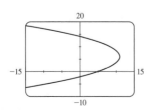

63. (a) $e = 1$, parabola
(b)

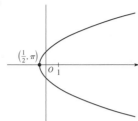

65. (a) $e = 2$, hyperbola **(b)**

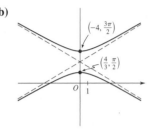

CHAPTER 7 TEST ■ PAGE 411

1. $F(0, -3)$, $y = 3$

2. $V(\pm 4, 0)$; $F(\pm 2\sqrt{3}, 0)$; 8, 4

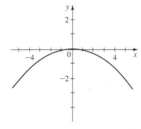

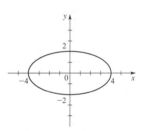

3. $V(0, \pm 3)$; $F(0, \pm 5)$; $y = \pm\frac{3}{4}x$

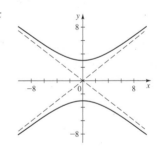

4. $y^2 = -x$ **5.** $\dfrac{x^2}{16} + \dfrac{(y-3)^2}{9} = 1$ **6.** $(x-2)^2 - \dfrac{y^2}{3} = 1$

7. $\dfrac{(x-3)^2}{9} + \dfrac{(y+\frac{1}{2})^2}{4} = 1$ **8.** $\dfrac{(x+2)^2}{8} - \dfrac{(y-4)^2}{9} = 1$

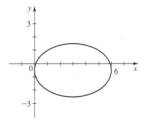

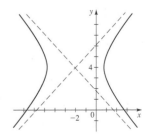

9. $(y + 4)^2 = -2(x - 4)$

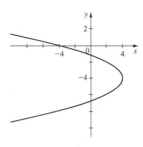

9.

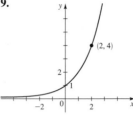

11.

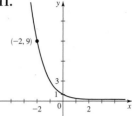

10. $\dfrac{y^2}{9} - \dfrac{x^2}{16} = 1$ **11.** $x^2 - 4x - 8y + 20 = 0$ **12.** $\frac{3}{4}$ in.

13. (a) Ellipse   **(b)** $\dfrac{X^2}{3} + \dfrac{Y^2}{18} = 1$

(c) $\phi \approx 27°$

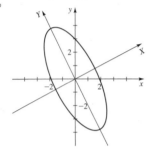

13.

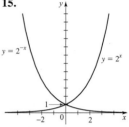

15.

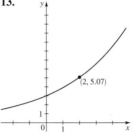

(d) $\left(-3\sqrt{2/5}, 6\sqrt{2/5}\right), \left(3\sqrt{2/5}, -6\sqrt{2/5}\right)$

14. (a) $r = \dfrac{1}{1 + 0.5\cos\theta}$

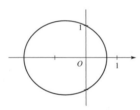

17.

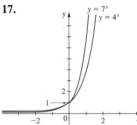

19. $f(x) = 3^x$ **21.** $f(x) = \left(\frac{1}{4}\right)^x$ **23.** II

25. $\mathbb{R}, (-\infty, 0), y = 0$ **27.** $\mathbb{R}, (-3, \infty), y = -3$

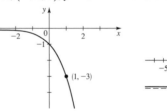

(b) Ellipse

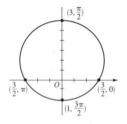

29. $\mathbb{R}, (4, \infty), y = 4$ **31.** $\mathbb{R}, (0, \infty), y = 0$

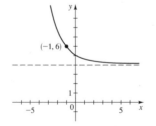

FOCUS ON MODELING ▪ PAGE 414

5. (c) $x^2 - mx + (ma - a^2) = 0$,
discriminant $m^2 - 4ma + 4a^2 = (m - 2a)^2$, $m = 2a$

CHAPTER 8

SECTION 8.1 ▪ PAGE 423

1. $5; \frac{1}{25}, 1, 25, 15{,}625$ **2. (a)** III   **(b)** I   **(c)** II   **(d)** IV
3. (a) downward   **(b)** right   **4.** principal, interest rate per year,
number of times interest is compounded per year, number of years,
amount after t years; \$112.65   **5.** 2.000, 7.103, 77.880, 1.587
7. 0.885, 0.606, 0.117, 1.837

33. $\mathbb{R}, (1, \infty), y = 1$

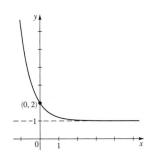

35. $\mathbb{R}, (-\infty, 3), y = 3$

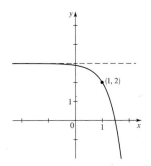

37. (a)

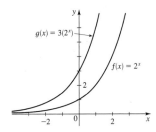

(b) The graph of g is steeper than that of f.

39.

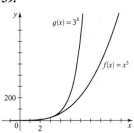

41. (a)

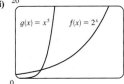

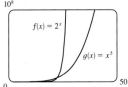

The graph of f ultimately increases much more quickly than that of g.

(b) 1.2, 22.4

43.

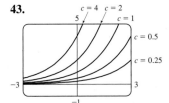

The larger the value of c, the more rapidly the graph increases.

45. (a) Increasing on $(-\infty, 0.50]$; decreasing on $[0.50, \infty)$
(b) $(0, 1.78]$ **47. (a)** $1500 \cdot 2^t$ **(b)** 25,165,824,000
49. \$5203.71, \$5415.71, \$5636.36, \$5865.99, \$6104.98, \$6353.71
51. (a) \$11,605.41   **(b)** \$13,468.55   **(c)** \$15,630.80
53. (a) \$519.02   **(b)** \$538.75   **(c)** \$726.23   **55.** \$7678.96
57. 8.30%

SECTION 8.2 ▪ PAGE 428

1. natural; 2.71828   **2.** principal, interest rate per year, number of years; amount after t years; \$112.75
3. 20.085, 1.259, 2.718, 0.135

5.

x	y
-2	0.41
-1	1.10
-0.5	1.82
0	3
0.5	4.95
1	8.15
2	22.17

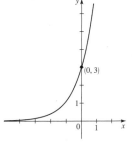

7. $\mathbb{R}, (-\infty, 0), y = 0$

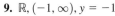
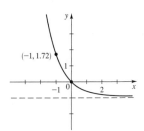

9. $\mathbb{R}, (-1, \infty), y = -1$

11. $\mathbb{R}, (0, \infty), y = 0$

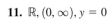
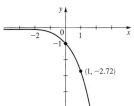

13. $\mathbb{R}, (-3, \infty), y = -3$

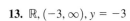
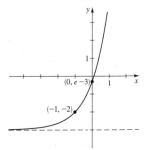

15. (a)

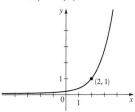

17. (a)

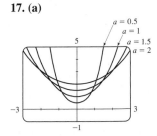

(b) The larger the value of a, the wider the graph.

19. Local minimum $\approx (0.27, 1.75)$
21. (a) 13 kg   **(b)** 6.6 kg

23. (a) 0 **(b)** 50.6 ft/s, 69.2 ft/s
(c)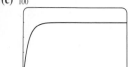
(d) 80 ft/s

25. (a) 100 **(b)** 482, 999, 1168 **(c)** 1200
27. (a) 11.79 billion, 11.97 billion
(b) **(c)** 12 billion
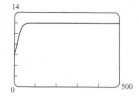

29. $7213.18, $7432.86, $7659.22, $7892.48, $8132.84, $8380.52
31. (a) $2145.02 **(b)** $2300.55 **(c)** $3043.92 **33. (a)** $768.05
(b) $769.22 **(c)** $769.82 **(d)** $770.42 **35. (a)** is best.
37. (a) $A(t) = 5000e^{0.09t}$ **(b)**

(c) After 17.88 yr

SECTION 8.3 ■ PAGE 438

1. x

x	10^3	10^2	10^1	10^0	10^{-1}	10^{-2}	10^{-3}	$10^{1/2}$
$\log x$	3	2	1	0	-1	-2	-3	$\frac{1}{2}$

2. $9; 1, 0, -1, 2, \frac{1}{2}$
3. (a) $\log_5 125 = 3$ **(b)** $5^2 = 25$ **4. (a)** III **(b)** II **(c)** I
(d) IV
5.

Logarithmic form	Exponential form
$\log_8 8 = 1$	$8^1 = 8$
$\log_8 64 = 2$	$8^2 = 64$
$\log_8 4 = \frac{2}{3}$	$8^{2/3} = 4$
$\log_8 512 = 3$	$8^3 = 512$
$\log_8 \frac{1}{8} = -1$	$8^{-1} = \frac{1}{8}$
$\log_8 \frac{1}{64} = -2$	$8^{-2} = \frac{1}{64}$

7. (a) $5^2 = 25$ **(b)** $5^0 = 1$ **9. (a)** $8^{1/3} = 2$ **(b)** $2^{-3} = \frac{1}{8}$
11. (a) $e^x = 5$ **(b)** $e^5 = y$ **13. (a)** $\log_5 125 = 3$
(b) $\log_{10} 0.0001 = -4$ **15. (a)** $\log_8 \frac{1}{8} = -1$ **(b)** $\log_2 \frac{1}{8} = -3$
17. (a) $\ln 2 = x$ **(b)** $\ln y = 3$ **19. (a)** 1 **(b)** 0 **(c)** 2
21. (a) 2 **(b)** 2 **(c)** 10 **23. (a)** -3 **(b)** $\frac{1}{2}$ **(c)** -1
25. (a) 37 **(b)** 8 **(c)** $\sqrt{5}$ **27. (a)** $-\frac{2}{3}$ **(b)** 4 **(c)** -1
29. (a) 32 **(b)** 4 **31. (a)** 5 **(b)** 27 **33. (a)** 100 **(b)** 25
35. (a) 2 **(b)** 4 **37. (a)** 0.3010 **(b)** 1.5465 **(c)** -0.1761
39. (a) 1.6094 **(b)** 3.2308 **(c)** 1.0051

41.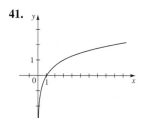

43.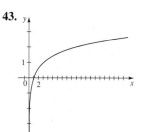

45. $y = \log_5 x$ **47.** $y = \log_9 x$ **49.** I
51.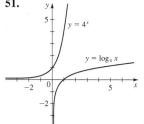
53. $(4, \infty), \mathbb{R}, x = 4$
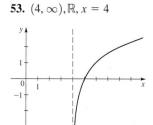

55. $(-\infty, 0), \mathbb{R}, x = 0$ **57.** $(0, \infty), \mathbb{R}, x = 0$
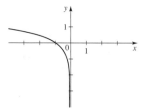 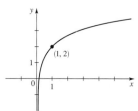

59. $(0, \infty), \mathbb{R}, x = 0$ **61.** $(0, \infty), [0, \infty), x = 0$
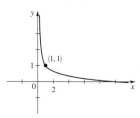 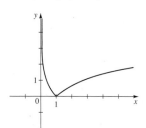

63. $(-3, \infty)$ **65.** $(-\infty, -1) \cup (1, \infty)$ **67.** $(0, 2)$
69.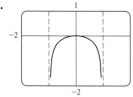
domain $(-1, 1)$
vertical asymptotes $x = 1$,
$x = -1$
local maximum $(0, 0)$

71.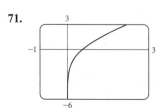
domain $(0, \infty)$
vertical asymptote $x = 0$
no maximum or minimum

73.

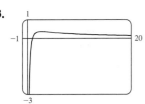

domain $(0, \infty)$
vertical asymptote $x = 0$
horizontal asymptote $y = 0$
local maximum
$\approx (2.72, 0.37)$

75. $(f \circ g)(x) = 2^{x+1}, (-\infty, \infty); (g \circ f)(x) = 2^x + 1, (-\infty, \infty)$
77. $(f \circ g)(x) = \log_2 (x - 2), (2, \infty);$
$(g \circ f)(x) = \log_2 x - 2, (0, \infty)$
79. The graph of f grows more slowly than g.
81. (a)

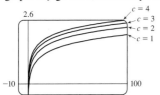

(b) The graph of
$f(x) = \log(cx)$
is the graph of
$f(x) = \log(x)$
shifted upward
$\log c$ units.

83. (a) $(1, \infty)$ **(b)** $f^{-1}(x) = 10^{2^x}$

85. (a) $f^{-1}(x) = \log_2\left(\dfrac{x}{1-x}\right)$ **(b)** $(0, 1)$ **87.** 2602 yr

89. 11.5 yr, 9.9 yr, 8.7 yr   **91.** 5.32, 4.32

SECTION 8.4 ■ PAGE 445

1. sum; $\log_5 25 + \log_5 125 = 2 + 3$
2. difference; $\log_5 25 - \log_5 125 = 2 - 3$
3. times; $10 \cdot \log_5 25$
4. (a) $2 \log x + \log y - \log z$

(b) $\log\left(\dfrac{x^2 y}{z}\right)$

5. 10, e; Change of Base; $\log_7 12 = \dfrac{\log 12}{\log 7} = 1.277$

6. True   **7.** $\frac{3}{2}$ **9.** 2   **11.** 3   **13.** 3   **15.** 200   **17.** 4
19. $1 + \log_2 x$ **21.** $\log_2 x + \log_2 (x - 1)$
23. $10 \log 6$ **25.** $\log_2 A + 2 \log_2 B$ **27.** $\log_3 x + \frac{1}{2} \log_3 y$
29. $\frac{1}{3} \log_5 (x^2 + 1)$ **31.** $\frac{1}{2}(\ln a + \ln b)$
33. $3 \log x + 4 \log y - 6 \log z$
35. $\log_2 x + \log_2 (x^2 + 1) - \frac{1}{2} \log_2 (x^2 - 1)$
37. $\ln x + \frac{1}{2}(\ln y - \ln z)$ **39.** $\frac{1}{4} \log(x^2 + y^2)$
41. $\frac{1}{2}[\log(x^2 + 4) - \log(x^2 + 1) - 2 \log(x^3 - 7)]$
43. $3 \ln x + \frac{1}{2} \ln(x - 1) - \ln(3x + 4)$ **45.** $\log_3 160$

47. $\log_2(AB/C^2)$ **49.** $\log\left(\dfrac{x^4(x-1)^2}{\sqrt[3]{x^2+1}}\right)$

51. $\ln(5x^2(x^2 + 5)^3)$

53. $\log\left(\dfrac{x^2}{x-3}\right)$ **55.** 2.321928   **57.** 2.523719

59. 0.493008   **61.** 3.482892
63.

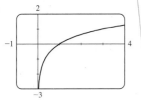

69. (a) $P = c/W^k$ **(b)** 1866, 64
71. (a) $M = -2.5 \log B + 2.5 \log B_0$

SECTION 8.5 ■ PAGE 454

1. (a) $e^x = 25$ **(b)** $x = \ln 25$ **(c)** 3.219
2. (a) $\log 3(x - 2) = \log x$ **(b)** $3(x - 2) = x$ **(c)** 3
3. 1.3979   **5.** −0.9730   **7.** −0.5850   **9.** 1.2040   **11.** 0.0767
13. 0.2524   **15.** 1.9349   **17.** −43.0677   **19.** 2.1492
21. 6.2126   **23.** −2.9469   **25.** −2.4423   **27.** 14.0055
29. $\ln 2 \approx 0.6931, 0$ **31.** $\frac{1}{2} \ln 3 \approx 0.5493$ **33.** ± 1 **35.** $0, \frac{4}{3}$
37. $e^{10} \approx 22026$ **39.** 0.01   **41.** $\frac{95}{3}$ **43.** −7   **45.** 5   **47.** 5
49. $\frac{13}{12}$ **51.** 4   **53.** 6   **55.** $\frac{3}{2}$ **57.** $1/\sqrt{5} \approx 0.4472$ **59.** 2.21
61. 0.00, 1.14   **63.** −0.57   **65.** 0.36
67. $2 < x < 4$ or $7 < x < 9$ **69.** $\log 2 < x < \log 5$

71. $f^{-1}(x) = \dfrac{\ln x}{2 \ln 2}$ **73.** $f^{-1}(x) = 2^x + 1$

75. (a) \$6435.09   **(b)** 8.24 yr   **77.** 6.33 yr   **79.** 8.15 yr
81. 13 days   **83. (a)** 7337   **(b)** 1.73 yr   **85. (a)** $P = P_0 e^{-h/k}$
(b) 56.47 kPa   **87. (a)** $t = -\frac{5}{13} \ln(1 - \frac{13}{60} I)$ **(b)** 0.218 s

SECTION 8.6 ■ PAGE 466

1. (a) $n(t) = 10 \cdot 2^{2t/3}$ **(b)** 1.06×10^8 **(c)** After 14.9 h
3. (a) 3125   **(b)** 317,480
(c) n (millions)

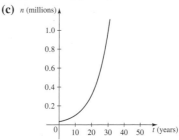

5. (a) $n(t) = 18,000 e^{0.08t}$ **(b)** 34,137

(c) $n(t)$

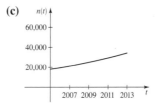

7. (a) 233 million   **(b)** 181 million

9. (a) $n(t) = 112,000 \cdot 2^{t/18}$ **(b)** $n(t) = 112,000 e^{0.0385t}$
(c) n (millions)   **(d)** In the year 2045

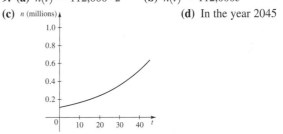

11. (a) 20,000   **(b)** $n(t) = 20,000 e^{0.1096t}$ **(c)** About 48,000
(d) 2017   **13. (a)** $n(t) = 8600 e^{0.1508t}$ **(b)** About 11,600
(c) 4.6 h   **15. (a)** $n(t) = 29.76 e^{0.012936t}$ million
(b) 53.5 yr   **(c)** 38.55 million   **17. (a)** $m(t) = 22 \cdot 2^{-t/1600}$
(b) $m(t) = 22 e^{-0.000433t}$ **(c)** 3.9 mg   **(d)** 463.4 yr
19. 18 yr   **21.** 149 h   **23.** 3560 yr
25. (a) $210°F$ **(b)** $153°F$ **(c)** 28 min

7. (a) 137°F (b) 116 min
29. (a) 2.3 (b) 3.5 (c) 8.3
31. (a) 10^{-3} M (b) 3.2×10^{-7} M
33. $4.8 \leq \text{pH} \leq 6.4$ **35.** $\log 20 \approx 1.3$ **37.** Twice as intense
39. 8.2 **41.** 73 dB **43.** (b) 106 dB

SECTION 8.7 ■ PAGE 472

1. $y = ke^{-ct} \sin wt$ **2.** $y = ke^{-ct} \cos wt$
3. (a) $y = 2e^{-1.5t} \cos 6\pi t$ (b)

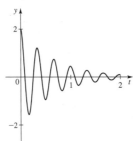

5. (a) $y = 100e^{-0.05t} \cos \dfrac{\pi}{2} t$
(b)

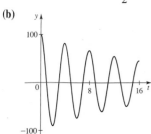

7. (a) $y = 7e^{-10t} \sin 12t$ (b)

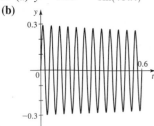

9. (a) $y = 0.3e^{-0.2t} \sin(40\pi t)$
(b)

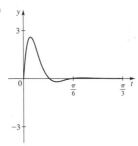

11. $y = 6.114e^{-0.128t} \sin t$ **13.** $y = 100e^{-0.115t} \cos \frac{\pi}{4} t$

15. $f(t) = e^{-0.9t} \sin \pi t$ **17.** $c = \dfrac{1}{3} \ln 4 \approx 0.46$

CHAPTER 8 REVIEW ■ PAGE 474

1. 0.089, 9.739, 55.902 **3.** 11.954, 2.989, 2.518
5. $\mathbb{R}, (0, \infty), y = 0$ **7.** $\mathbb{R}, (3, \infty), y = 3$

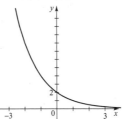
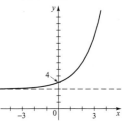

9. $(1, \infty), \mathbb{R}, x = 1$ **11.** $(0, \infty), \mathbb{R}, x = 0$

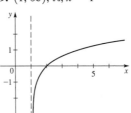
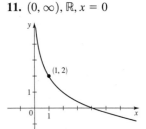

13. $\mathbb{R}, (-1, \infty), y = -1$ **15.** $(0, \infty), \mathbb{R}, x = 0$

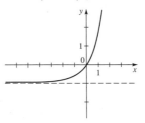
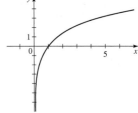

17. $\left(-\infty, \frac{1}{2}\right)$ **19.** $(-\infty, -2) \cup (2, \infty)$ **21.** $2^{10} = 1024$
23. $10^y = x$ **25.** $\log_2 64 = 6$ **27.** $\log 74 = x$ **29.** 7 **31.** 45
33. 6 **35.** -3 **37.** $\frac{1}{2}$ **39.** 2 **41.** 92 **43.** $\frac{2}{3}$
45. $\log A + 2 \log B + 3 \log C$ **47.** $\frac{1}{2}[\ln(x^2 - 1) - \ln(x^2 + 1)]$
49. $2 \log_5 x + \frac{3}{2} \log_5(1 - 5x) - \frac{1}{2} \log_5(x^3 - x)$
51. $\log 96$ **53.** $\log_2\left(\dfrac{(x - y)^{3/2}}{(x^2 + y^2)^2}\right)$ **55.** $\log\left(\dfrac{x^2 - 4}{\sqrt{x^2 + 4}}\right)$
57. 5 **59.** 2.60 **61.** -1.15 **63.** $-4, 2$
65. -15 **67.** 3 **69.** 0.430618
71. 2.303600
73.

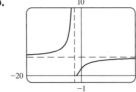

vertical asymptote
$x = -2$
horizontal asymptote
$y = 2.72$
no maximum or minimum

75.

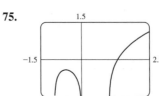

vertical asymptotes
$x = -1, x = 0, x = 1$
local maximum
$\approx (-0.58, -0.41)$

77. 2.42 **79.** $0.16 < x < 3.15$
81. Increasing on $(-\infty, 0]$ and $[1.10, \infty)$, decreasing on $[0, 1.10]$
83. 1.953445 **85.** -0.579352 **87.** $\log_4 258$
89. (a) \$16,081.15 (b) \$16,178.18 (c) \$16,197.64
(d) \$16,198.31 **91.** 1.83 yr **93.** 4.341%
95. (a) $n(t) = 30e^{0.15t}$ (b) 55 (c) 19 yr
97. (a) 9.97 mg (b) 1.39×10^5 yr
99. (a) $n(t) = 150e^{-0.0004359t}$ (b) 97.0 mg (c) 2520 yr
101. (a) $n(t) = 1500e^{0.1515t}$ (b) 7940
103. 7.9, basic **105.** 8.0
107. (a) $y = 16e^{-0.72t} \cos 2.8\pi t$

(b)

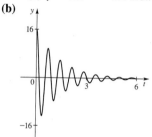

(c) 0.012 cm

CHAPTER 8 TEST ▪ PAGE 477

1. (a) $\mathbb{R}, (4, \infty), y = 4$ (b) $(-3, \infty), \mathbb{R}, x = -3$

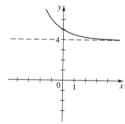 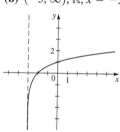

2. (a) $\log_6 25 = 2x$ (b) $e^3 = A$
3. (a) 36 (b) 3 (c) $\frac{3}{2}$ (d) 3 (e) $\frac{2}{3}$ (f) 2
4. $\frac{1}{3}[\log(x + 2) - 4 \log x - \log(x^2 + 4)]$
5. $\ln\left(\dfrac{x\sqrt{3 - x^4}}{(x^2 + 1)^2}\right)$ **6.** (a) 4.32 (b) 0.77 (c) 5.39 (d) 2

7. (a) $n(t) = 1000e^{2.07944t}$
(b) 22,627 (c) 1.3 h
(d)

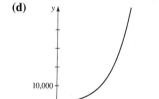

8. (a) $A(t) = 12,000\left(1 + \dfrac{0.056}{12}\right)^{12t}$

(b) \$14,195.06 (c) 9.249 yr

9. (a) $A(t) = 3e^{-0.069t}$ (b) 0.048 g (c) after 3.6 min
10. 1995 times more intense
11. $y = 16e^{-0.1t} \cos 24\pi t$

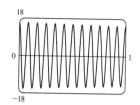

FOCUS ON MODELING ▪ PAGE 484

1. (a)

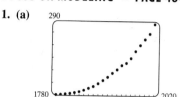

(b) $y = ab^t$, where $a = 1.180609 \times 10^{-15}$, $b = 1.0204139$, and y is the population in millions in the year t (c) 515.9 million
(d) 207.8 million (e) No
3. (a) Yes (b) Yes, the scatter plot appears linear.

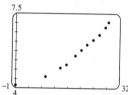

(c) $\ln E = 4.551436 + 0.092383t$, where t is years since 1970 and E is expenditure in billions of dollars
(d) $E = 94.76838139e^{at}$, where $a = 0.0923827621$
(e) 3478.5 billion dollars

5. (a) $I_0 = 22.7586444$, $k = 0.1062398$
(b) (c) 47.3 ft

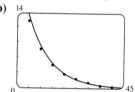

7. (a) $S = 0.14A^{0.64}$
(b) (c) 4 species

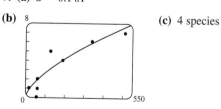

. (a)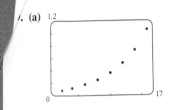

(b)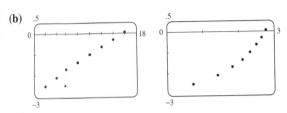

(c) Exponential function
(d) $y = ab^x$ where $a = 0.057697$ and $b = 1.200236$

11. (a) $y = \dfrac{c}{1 + ae^{-bx}}$, where $a = 49.10976596$,

$b = 0.4981144989$, and $c = 500.855793$ **(b)** 10.58 days

APPENDIX

APPENDIX A.1 ■ PAGE 494

1. Commutative Property for addition
3. Associative Property for addition   **5.** Distributive Property
7. $3x + 3y$ **9.** $8m$ **11.** $-5x + 10y$
13. (a) False   **(b)** True   **15. (a)** False   **(b)** True
17. (a) $x > 0$ **(b)** $t < 4$ **(c)** $a \geq \pi$ **(d)** $-5 < x < \frac{1}{3}$
(e) $|p - 3| \leq 5$
19. (a) $\mathbb{R}$ **(b)** $\{x \mid -2 \leq x < 4\}$
21. (a) $\{x \mid x \leq 5\}$ **(b)** $\{x \mid -1 < x < 4\}$
23. $-3 < x < 0$ **25.** $2 \leq x < 8$
27. $x \geq 2$ **29.** $(-\infty, 1]$
31. $(-2, 1]$ **33.** $(-1, \infty)$
35.   **37.**
39.
41. (a) 100   **(b)** 73   **43. (a)** 2   **(b)** -1
45. (a) 12   **(b)** 5   **47. (a)** 15   **(b)** 24   **(c)** $\frac{67}{40}$

APPENDIX A.2 ■ PAGE 500

1. $17^{1/2}$ **3.** $\sqrt[3]{4^2}$ **5.** $\sqrt[5]{a^3}$ **7. (a)** 16   **(b)** -16 **(c)** 1
9. (a) $\frac{16}{25}$ **(b)** 1000   **(c)** 1024   **11. (a)** $\frac{2}{3}$ **(b)** 4   **(c)** $\frac{1}{2}$
13. (a) $\frac{3}{2}$ **(b)** 4   **(c)** $-\frac{1}{5}$ **15.** 5   **17.** 14   **19.** $\sqrt[3]{4}$
21. $2\sqrt{5}$ **23.** a^4 **25.** $6x^7y^5$ **27.** $16x^{10}$ **29.** $4/b^2$
31. $64r^7s$ **33.** $648y^7$ **35.** $\dfrac{x^3}{y}$ **37.** $\dfrac{y^2z^9}{x^5}$ **39.** $\dfrac{s^3}{q^7r^6}$
41. $x^{13/15}$ **43.** $16b^{9/10}$ **45.** $\dfrac{1}{c^{2/3}d}$ **47.** $y^{1/2}$ **49.** $\dfrac{32x^{12}}{y^{16/15}}$

51. $\dfrac{x^{15}}{y^{15/2}}$ **53.** $\dfrac{4a^2}{3b^{1/3}}$ **55.** $\dfrac{3t^{25/6}}{s^{1/2}}$ **57.** $|x|$ **59.** $x\sqrt[3]{y}$

61. $ab\sqrt[5]{ab^2}$ **63.** $2|x|$ **65. (a)** $\dfrac{\sqrt{6}}{6}$ **(b)** $\dfrac{\sqrt{3xy}}{3y}$ **(c)** $\dfrac{\sqrt{15}}{10}$

67. (a) $\dfrac{\sqrt[3]{x^2}}{x}$ **(b)** $\dfrac{\sqrt[5]{x^3}}{x}$ **(c)** $\dfrac{\sqrt[7]{x^4}}{x}$

APPENDIX A.3 ■ PAGE 509

1. $5x^2 - 2x - 4$ **3.** $9x + 103$
5. $-t^4 + t^3 - t^2 - 10t + 5$ **7.** $x^{3/2} - x$
9. $3x^2 + 5xy - 2y^2$ **11.** $1 - 4y + 4y^2$
13. $4x^4 + 12x^2y^2 + 9y^4$ **15.** $2x^3 - 7x^2 + 7x - 5$
17. $x^4 - a^4$ **19.** $a - \dfrac{1}{b^2}$ **21.** $x^5 + x^4 - 3x^3 + 3x - 2$
23. $1 - x^{2/3} + x^{4/3} - x^2$
25. $3x^4y^4 + 7x^3y^5 - 6x^2y^3 - 14xy^4$
27. $6x(2x^2 + 3)$ **29.** $3y^3(2y - 5)$ **31.** $(x - 4)(x + 2)$
33. $(y - 3)(y - 5)$ **35.** $(2x + 3)(x + 1)$ **37.** $9(x - 5)(x + 1)$
39. $(3x + 2)(2x - 3)$ **41.** $(2t - 3)^2$ **43.** $(r - 3s)^2$
45. $(x + 6)(x - 6)$ **47.** $(7 + 2y)(7 - 2y)$
49. $4ab$ **51.** $(x + 3)(x - 3)(x + 1)(x - 1)$
53. $(t + 1)(t^2 - t + 1)$ **55.** $x(x + 1)^2$
57. $3(x - 1)(x + 2)$ **59.** $y^4(y + 2)^3(y + 1)^2$

61. $\dfrac{1}{x + 2}$ **63.** $\dfrac{x + 2}{x + 1}$ **65.** $\dfrac{y}{y - 1}$ **67.** $\dfrac{x(2x + 3)}{2x - 3}$

69. $\dfrac{1}{t^2 + 9}$ **71.** $\dfrac{x + 4}{x + 1}$ **73.** $\dfrac{(2x + 1)(2x - 1)}{(x + 5)^2}$

75. $\dfrac{x}{yz}$ **77.** $\dfrac{3x + 7}{(x - 3)(x + 5)}$ **79.** $\dfrac{1}{(x + 1)(x + 2)}$

81. $\dfrac{3x + 2}{(x + 1)^2}$ **83.** $\dfrac{u^2 + 3u + 1}{u + 1}$ **85.** $\dfrac{2x + 1}{x^2(x + 1)}$

87. $\dfrac{2x + 7}{(x + 3)(x + 4)}$ **89.** $\dfrac{x - 2}{(x + 3)(x - 3)}$

91. $\dfrac{5x - 6}{x(x - 1)}$ **93.** $\dfrac{-5}{(x + 1)(x + 2)(x - 3)}$ **95.** $-xy$

97. $\dfrac{c}{c - 2}$ **99.** $\dfrac{y - x}{xy}$ **101.** $\dfrac{3 - \sqrt{5}}{2}$ **103.** $\dfrac{2(\sqrt{7} - \sqrt{2})}{5}$

105. $\dfrac{-4}{3(1 + \sqrt{5})}$ **107.** $\dfrac{r - 2}{5(\sqrt{r} - \sqrt{2})}$ **109.** $\dfrac{1}{\sqrt{x^2 + 1} + x}$

APPENDIX A.4 ■ PAGE 519

1. (a) Yes   **(b)** No   **3. (a)** Yes   **(b)** No   **5.** 4   **7.** -9
9. -3 **11.** 12   **13.** $-\frac{3}{4}$ **15.** $\frac{32}{9}$ **17.** $-4, 2$
19. $-3, -\dfrac{1}{2}$ **21.** $-\dfrac{4}{3}, \dfrac{1}{2}$ **23.** $-1 \pm \sqrt{3}$ **25.** $-\dfrac{3}{2}, \dfrac{1}{2}$
27. $-2 \pm \dfrac{\sqrt{14}}{2}$ **29.** $-6 \pm 3\sqrt{7}$ **31.** $\dfrac{-3 \pm 2\sqrt{6}}{3}$
33. $\dfrac{1 \pm \sqrt{5}}{4}$ **35.** $\dfrac{8 \pm \sqrt{14}}{10}$ **37.** No real solution   **39.** $-\dfrac{7}{5}, 2$
41. $-50, 100$ **43.** -4 **45.** 4   **47.** 3   **49.** $\pm 2\sqrt{2}, \pm\sqrt{5}$
51. No real solution   **53.** $\pm 3\sqrt{3}, \pm 2\sqrt{2}$ **55.** $-1, 0, 3$
57. 27, 729   **59.** $-\dfrac{3}{2}, \dfrac{3}{2}$ **61.** 3.99, 4.01   **63.** $R = \dfrac{PV}{nT}$
65. $R_1 = \dfrac{RR_2}{R_2 - R}$ **67.** $x = \dfrac{2d - b}{a - 2c}$ **69.** $x = \dfrac{1 - a}{a^2 - a - 1}$
71. $r = \pm\sqrt{\dfrac{3V}{\pi h}}$ **73.** $b = \pm\sqrt{c^2 - a^2}$ **75.** 2   **77.** 1

79. 4.24 s **81. (a)** After 17 yr, on Jan. 1, 2009
(b) After 18.612 yr, on Aug. 12, 2010

APPENDIX A.5 ■ PAGE 524

1. $\{0, \frac{1}{2}, \sqrt{2}, 2\}$ **3.** $\{-1, 2\}$ **5.** $\left(-\infty, \frac{7}{2}\right]$

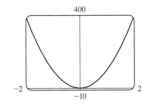

7. $(4, \infty)$

9. $(-\infty, 2]$

11. $\left(-\infty, -\frac{1}{2}\right)$

13. $[1, \infty)$

15. $\left(\frac{16}{3}, \infty\right)$

17. $(-\infty, -1]$

19. $[-3, -1)$

21. $(2, 6)$

23. $\left[\frac{9}{2}, 5\right)$

25. $(-7, 7)$

27. $[2, 8]$

29. $(-7, -3)$

31. $[1.3, 1.7]$

33. $(-4, 8)$

35. $(-6, 2)$

37. $68 \leq F \leq 86$
39. (a) $T = -0.01h + 20$ **(b)** $-30 \leq T \leq 20$

APPENDIX B.1 ■ PAGE 527

1. Congruent, ASA **3.** Not necessarily congruent

5. Similar **7.** Similar **9.** $x = 125$ **11.** $x = 6, y = \dfrac{21}{4}$

13. $x = \dfrac{ac}{a + b}$ **17.** $h = 6$

APPENDIX B.2 ■ PAGE 530

1. $x = 10$ **3.** $x = \sqrt{3}$ **5.** $x = 40$ **7.** Yes **9.** No
11. Yes **13.** 61 cm **15.** No **17.** 13

APPENDIX C.1 ■ PAGE 533

1. (c) **3.** (c) **5.** (c)
7.

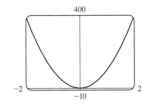

9.

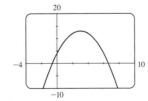

11.

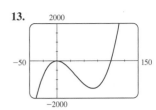

13.

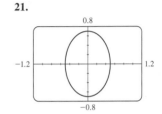

15.

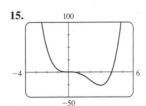

17.

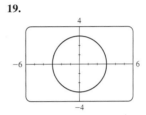

19.

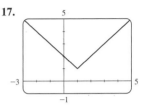

21.

23. No **25.** Yes, 2

APPENDIX C.2 ■ PAGE 538

1. -4 **3.** $\dfrac{5}{14}$ **5.** $\pm 4\sqrt{2} \approx \pm 5.7$ **7.** $2.5, -2.5$
9. $5 + 2\sqrt[4]{5} \approx 7.99, 5 - 2\sqrt[4]{5} \approx 2.01$ **11.** $3.00, 4.00$
13. $1.00, 2.00, 3.00$ **15.** 1.62 **17.** $-1.00, 0.00, 1.00$
19. 2.55 **21.** $-2.05, 0, 1.05$ **23.** $[-2.00, 5.00]$
25. $(-\infty, 1.00] \cup [2.00, 3.00]$ **27.** $(-1.00, 0) \cup (1.00, \infty)$
29. $(-\infty, 0)$ **31.** $0, 0.01$

APPENDIX D.1 ■ PAGE 543

1. Real part 5, imaginary part -7 **3.** Real part $-\frac{2}{3}$, imaginary part $-\frac{5}{3}$ **5.** Real part 3, imaginary part 0 **7.** Real part 0, imaginary part $-\frac{2}{3}$ **9.** Real part $\sqrt{3}$, imaginary part 2 **11.** $5 - i$
13. $3 + 5i$ **15.** $2 - 2i$ **17.** $-19 + 4i$ **19.** $-4 + 8i$
21. $30 + 10i$ **23.** $-33 - 56i$ **25.** $27 - 8i$ **27.** $-i$ **29.** 1
31. $-i$ **33.** $\frac{8}{5} + \frac{1}{5}i$ **35.** $-5 + 12i$ **37.** $-4 + 2i$ **39.** $2 - \frac{4}{3}i$
41. $-i$ **43.** $5i$ **45.** -6 **47.** $(3 + \sqrt{5}) + (3 - \sqrt{5})i$
49. 2 **51.** $-i\sqrt{2}$ **53.** $\pm 7i$ **55.** $2 \pm i$ **57.** $-1 \pm 2i$
59. $-\dfrac{1}{2} \pm \dfrac{\sqrt{3}}{2}i$ **61.** $\frac{1}{2} \pm \frac{1}{2}i$ **63.** $-\dfrac{3}{2} \pm \dfrac{\sqrt{3}}{2}i$ **65.** $\dfrac{-6 \pm \sqrt{6}i}{6}$
67. $1 \pm 3i$

EXPONENTS AND RADICALS

$x^m x^n = x^{m+n}$

$\dfrac{x^m}{x^n} = x^{m-n}$

$(x^m)^n = x^{mn}$

$x^{-n} = \dfrac{1}{x^n}$

$(xy)^n = x^n y^n$

$\left(\dfrac{x}{y}\right)^n = \dfrac{x^n}{y^n}$

$x^{1/n} = \sqrt[n]{x}$

$x^{m/n} = \sqrt[n]{x^m} = \left(\sqrt[n]{x}\right)^m$

$\sqrt[n]{xy} = \sqrt[n]{x}\,\sqrt[n]{y}$

$\sqrt[n]{\dfrac{x}{y}} = \dfrac{\sqrt[n]{x}}{\sqrt[n]{y}}$

$\sqrt[m]{\sqrt[n]{x}} = \sqrt[n]{\sqrt[m]{x}} = \sqrt[mn]{x}$

SPECIAL PRODUCTS

$(x + y)^2 = x^2 + 2xy + y^2$

$(x - y)^2 = x^2 - 2xy + y^2$

$(x + y)^3 = x^3 + 3x^2y + 3xy^2 + y^3$

$(x - y)^3 = x^3 - 3x^2y + 3xy^2 - y^3$

FACTORING FORMULAS

$x^2 - y^2 = (x + y)(x - y)$

$x^2 + 2xy + y^2 = (x + y)^2$

$x^2 - 2xy + y^2 = (x - y)^2$

$x^3 + y^3 = (x + y)(x^2 - xy + y^2)$

$x^3 - y^3 = (x - y)(x^2 + xy + y^2)$

QUADRATIC FORMULA

If $ax^2 + bx + c = 0$, then

$$x = \dfrac{-b \pm \sqrt{b^2 - 4ac}}{2a}$$

INEQUALITIES AND ABSOLUTE VALUE

If $a < b$ and $b < c$, then $a < c$.

If $a < b$, then $a + c < b + c$.

If $a < b$ and $c > 0$, then $ca < cb$.

If $a < b$ and $c < 0$, then $ca > cb$.

If $a > 0$, then

$|x| = a$ means $x = a$ or $x = -a$.

$|x| < a$ means $-a < x < a$.

$|x| > a$ means $x > a$ or $x < -a$.

DISTANCE AND MIDPOINT FORMULAS

Distance between $P_1(x_1, y_1)$ and $P_2(x_2, y_2)$:

$$d = \sqrt{(x_2 - x_1)^2 + (y_2 - y_1)^2}$$

Midpoint of P_1P_2: $\left(\dfrac{x_1 + x_2}{2}, \dfrac{y_1 + y_2}{2}\right)$

LINES

Slope of line through $P_1(x_1, y_1)$ and $P_2(x_2, y_2)$

$m = \dfrac{y_2 - y_1}{x_2 - x_1}$

Point-slope equation of line through $P_1(x_1, y_1)$ with slope m

$y - y_1 = m(x - x_1)$

Slope-intercept equation of line with slope m and y-intercept b

$y = mx + b$

Two-intercept equation of line with x-intercept a and y-intercept b

$\dfrac{x}{a} + \dfrac{y}{b} = 1$

LOGARITHMS

$y = \log_a x$ means $a^y = x$

$\log_a a^x = x$

$a^{\log_a x} = x$

$\log_a 1 = 0$

$\log_a a = 1$

$\log x = \log_{10} x$

$\ln x = \log_e x$

$\log_a xy = \log_a x + \log_a y$

$\log_a\left(\dfrac{x}{y}\right) = \log_a x - \log_a y$

$\log_a x^b = b \log_a x$

$\log_b x = \dfrac{\log_a x}{\log_a b}$

EXPONENTIAL AND LOGARITHMIC FUNCTIONS

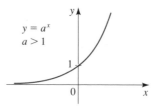

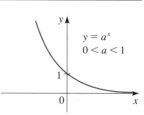

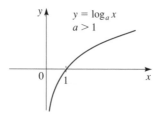

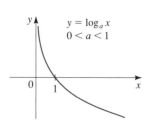